Handbook of Sustainable Biodiversity Management

Mounting perils posed by climate change, habitat degradation, pollution, and overexploitation jeopardize numerous species, including microbes. This book strives to shed light on the pressing necessity for sustainable biodiversity practices that safeguard our planet's natural legacy. By delving into diverse strategies, scientific paradigms, and case studies, the authors reveal how holistic approaches to biodiversity can cultivate flourishing ecosystems, wherein wildlife and human societies coexist harmoniously. This approach sets this book apart from others and presents the tools and knowledge needed to ensure the resilience and thriving of ecosystems for future generations.

Key Features:

- This book provides a holistic view of how biodiversity impacts ecosystem functions and human well-being.
- 'Diverse expertise of editors ensures a well-rounded perspective on sustainable biodiversity management. They present a variety of sustainable management principles, conservation strategies, and policy frameworks'.
- This book emphasizes the often-overlooked microbial communities, known as microbial dark matter, and their linkage to larger organisms.
- It delves into the ecological significance of microbes, their interactions, and their crucial roles in nutrient cycling and soil vitality, highlighting their importance for ecosystem sustainability.
- It includes real-world case studies and examples of successful biodiversity practices, providing readers with practical tools and actionable insights for effective biodiversity management.
- It emphasizes the importance of understanding basic principles, fostering global collaboration, embracing technological innovations, and encouraging community involvement to ensure the resilience and sustainability of ecosystems for future generations.

Handbook of Sustainable Biodiversity Management

Strategies for a Thriving Planet

Edited by
Ming Hung Wong, Xun Wen Chen, and Masayuki Ushio

CRC Press
Taylor & Francis Group
Boca Raton London New York

CRC Press is an imprint of the
Taylor & Francis Group, an **informa** business

Designed cover image: Cheung Yau Chun, Cyrus (張優駿), Flora Conservation Department, Kadoorie Farm and Botanic Garden

First edition published 2027
by CRC Press
2385 NW Executive Center Drive, Suite 320, Boca Raton FL 33431

and by CRC Press
4 Park Square, Milton Park, Abingdon, Oxon, OX14 4RN

CRC Press is an imprint of Taylor & Francis Group, LLC

ISBN: 9781041075141 (hbk)
ISBN: 9781041075158 (pbk)
ISBN: 9781003640899 (ebk)

DOI: 10.1201/9781003640899

Typeset in Times
by codeMantra

Contents

PART III Microbial and Aquatic Ecosystems

PART IV Terrestrial Biodiversity Conservation

Preface

The defining challenge of our age lies in reconciling human advancement with the integrity of the natural world upon which all progress ultimately depends. *Handbook of Sustainable Management of Biodiversity: Strategies for a Thriving Planet* confronts this challenge head-on, offering a vital synthesis of the multifaceted strategies required to nurture life in all its forms. Moving beyond catalogues of loss, this volume presents a forward-looking tapestry of insights, weaving together global policy, microbial ecology, forest restoration, and urban planning to chart pathways toward resilience. Our objective is to both clarify the complex interdependencies that bind human societies to ecological systems and provide actionable frameworks for their preservation.

Part I: Global Strategies and Governance opens with a critical analysis of the political landscapes shaping our collective response. While global biodiversity governance is often fragmented across numerous institutions, it has undeniably been mainstreamed into the agenda of international cooperation. One chapter traces the evolution from isolated environmental initiatives to integrated paradigms, such as the UNESCO Man and the Biosphere Programme, which seeks to bridge conservation and development through science. This journey, however, remains incomplete without the inclusion of diverse ways of knowing. Another contribution examines the pioneering efforts of IPBES to incorporate Indigenous knowledge systems, arguing that such inclusion—though fraught with challenges—is essential for crafting effective and equitable global policies.

Translating these global ambitions into action is the focus of Part II: Regional Policy and Management. Here, readers encounter the persistent gap between protective designations and meaningful conservation outcomes, a discrepancy explored through proposals for more connected and effectively managed protected area networks. The dynamic norm-building process between China and ASEAN illustrates how regional cooperation can evolve from dialogue to shared identity, offering a model of 'South–South' collaboration. As the world accelerates its shift to renewable energy, a comparative policy review sounds a cautionary note, revealing significant variation in how biodiversity safeguards are integrated into this transition. The section concludes by advocating for a strategic fusion of economic instruments and ecological restoration, a combination poised to create systems that are not only viable but equitable.

Part III: Microbial and Aquatic Ecosystems delves into the foundational, yet often invisible, layers of biodiversity. Aquatic microbes, it reveals, are far more than passive inhabitants; they are keystone engineers of biogeochemical cycles, agents of bioremediation, and sentinels of ecosystem stress. Understanding their complex interactions—from symbiosis to pathogenesis—is crucial for managing ecosystem health. A paradigm-shifting discussion introduces microbial rhodopsins, a vast and overlooked solar energy pathway in the oceans that operates independently of traditional carbon fixation. This hidden world is, however, increasingly threatened by pollutants like microplastics, whose pervasive infiltration into freshwater systems and impacts on species are systematically detailed. Against these threats, the section also highlights proactive, innovative strategies for restoring visible marine habitats—from coral gardening to regenerative seaweed farming—demonstrating that recovery is possible when science allies with community engagement.

The restoration and stewardship of land-based ecosystems form the core of Part IV: Terrestrial Biodiversity Conservation. It begins by making a compelling case for uniting two often-separate scientific pursuits: the study of how biodiversity originates and the research on how it underpins ecosystem functions and human well-being. This integrated perspective is powerfully embodied in a detailed account of subtropical forest restoration in Hong Kong, where decades of botanical expertise fuel the painstaking reassembly of a high-diversity, fully functional ecosystem. Such efforts must constantly contend with the threat of invasive alien species, whose management now employs drones, machine learning, and climate models for early detection and control. Below the surface, the critical roles of soil biota come to the fore; one chapter traces how bacterial communities and their nutrient-cycling functions evolve during forest recovery, while another champions the conservation of arbuscular mycorrhizal fungi, proposing specific, scale-dependent strategies to protect these essential plant symbionts from global change.

Finally, Part V: Biodiversity in Urban Environments redefines the conservation frontier, arguing that cities must be transformed from ecological casualties to rich, hybrid habitats. Conserving urban wildlife demands novel approaches—from leveraging remote sensing and trait-based ecology to designing policies that value genetic diversity. A case study of the Dasha River in Shenzhen exemplifies a tailored, diagnostic method for restoring ecological functions in heavily engineered urban waterways, even proposing circular economies to repurpose invasive plant biomass. The urban dimension also encompasses the demands societies place on wildlife beyond their borders, as underscored by a longitudinal study of Hong Kong's songbird trade. This research, identifying species of persistent commercial concern, illuminates the intricate links between local markets, regional poaching pressures, and the enforcement of conservation laws.

In assembling this volume, we have sought to reflect the complexity and interconnectivity of the biodiversity crisis itself. The chapters collectively demonstrate that sustainable management is an interdisciplinary endeavour, demanding not only scientific innovation but also political will, cultural sensitivity, and economic ingenuity. We are profoundly grateful to our contributors for their scholarly rigour and to CRC Press for their partnership. May this work serve as both a cornerstone reference and a catalyst for the integrated, determined action our thriving planet requires.

Ming Hung Wong, Xun Wen Chen, and Masayuki Ushio
The Editors
30 January 2026

About the Editors

Ming Hung Wong is Foreign Member of the Russian Academy of Sciences, Member of the European Academy of Sciences and Arts, and Changjiang Chair Professor of the Ministry of Education, China. He served as the Editor-in-Chief of *Environmental Geochemistry and Health* for 20 years (2002–2023). He was the Coordinator of Central and North-East Asia of 'Regionally Based Assessment of Persistent Toxic Substances' and Panel Member (of three experts) of 'Chemicals Management Issues of Developing Countries and Countries with Economies in Transition', sponsored by UNEP/GEF, during 2001–2003 and 2010–2012, respectively. His research areas included 'Environmental toxicology', 'Ecological restoration', and 'Resource reuse'. In addition to his PhD (Durham), he was awarded two higher doctoral degrees based on published papers in 1992 and 2004: DSc (Durham) and DSc (Strathclyde). Recently, he was awarded an Honorary Doctor of Science (DSc) from the Southern Federal University, Russia (2024), in recognition of his involvement as the Lead Scientist in the Mega Project on Bioremediation of Polluted Ecosystems, sponsored by the Russian Government. He is ranked sixth for 3 years, seventh for 1 year, and eighth for 2 years (career-long ranking) under Environmental Science, according to the World's Top 2% Scientists (Stanford University, 2020–2025).

Xun Wen Chen is an associate professor in the Department of Ecology, College of Life Science and Technology at Jinan University, China. He obtained his PhD at the Hong Kong University of Science and Technology (HKUST). His interdisciplinary research focuses on the complex interactions between plants and microbes under stress, including exposure to chemical and physical stressors. The goal of his research is to gain a deeper understanding of the complex plant–microbe interactions and to extract general principles that can be applied to advance ecological restoration and sustainable development.

Masayuki Ushio is an assistant professor in the Department of Ocean Science at the Hong Kong University of Science and Technology (HKUST). He began his academic career as a forest ecologist in 2005, earned his PhD in 2010 from Kyoto University, and has been actively involved in research in ecology, covering a wide range of topics, including tropical forest ecology, nutrient cycling, community assembly, environmental DNA analysis, and statistical ecology. He moved from Kyoto University to Hong Kong in 2022 and began ecological research in local coastal regions. In Hong Kong, his research team is currently developing environmental DNA-based methods to monitor cetacean and fish communities in coastal ecosystems.

Contributors

Noor Adelyna Mohammed Akib
School of Biological Sciences
Universiti Sains Malaysia
Penang, Malaysia

Basit Akolade Adigun
Department of Science and Environmental Studies
The Education University of Hong Kong
Hong Kong SAR, China

Md. Iqram Uddin Al Amran
Department of Science and Environmental Studies
The Education University of Hong Kong
Hong Kong SAR, China

Cornelia Amon
Department of Forest Biodiversity & Nature Conservation
Austrian Research Centre for Forests
Vienna, Austria

Su Tina Bai
Department of Politics and International Relations
University of Oxford
Oxford, United Kingdom

Wenzhi Cao
Key Laboratory of the Ministry of Education for Coastal and Wetland Ecosystems
College of Environment and Ecology, Xiamen University
Xiamen, China

Fernanda C. G. Cardoso
Flora Conservation Department
Kadoorie Farm and Botanic Garden
Tai Po, New Territories, Hong Kong SAR, China

Grace Y. L. Chan
Department of Construction, Environment and Engineering
Technological and Higher Education Institute of Hong Kong (THEi)
Hong Kong SAR, China

Su Yin Chee
School of Biological Sciences
Universiti Sains Malaysia
Penang, Malaysia

Xun Wen Chen
Department of Ecology, College of Life Science and Technology
Jinan University
Guangzhou, China

Yan Laam Cheng
Department of Science and Environmental Studies
The Education University of Hong Kong
Hong Kong SAR, China

Mang Lung Cheuk
Flora Conservation Department
Kadoorie Farm and Botanic Garden
Tai Po, New Territories, Hong Kong SAR, China

Lingyun Ding
College of Health Science and Environmental Engineering
Shenzhen Technology University
Shenzhen, China

Yuehan Dou
Department of Health and Environmental Sciences
School of Science
Xi'an-Jiaotong Liverpool University
Suzhou, China

Marcin K. Dyderski
Institute of Dendrology
Polish Academy of Sciences
Kórnik, Poland

Yanqiu Feng
College of Environmental Science and Engineering
Southern University of Science and Technology
Shenzhen, China

Guiping Fu
College of Life Sciences and Oceanography
Shenzhen University
Shenzhen, China

Stephan W. Gale
Flora Conservation Department
Kadoorie Farm and Botanic Garden
Tai Po, New Territories, Hong Kong SAR, China

Zhipeng Guo
College of Life Sciences and Oceanography
Shenzhen University
Shenzhen, China

Md. Kamrul Haque
School of Ecology and Applied Meteorology
Nanjing University of Information Science & Technology
Nanjing, China

Ruijia Hu
Department of Geography and Geographic Information Science
University of Cincinnati
Cincinnati, Ohio, United States

Zhishan Huang
School of Environment and Energy
Peking University Shenzhen Graduate School
Shenzhen, China

Guanhui Jin
School of International Studies
Zhengzhou University
Zhengzhou, China

Ki-Hyun Kim
Department of Civil and Environmental Engineering
Hanyang University
Seoul, Republic of Korea

Tapos Kormoker
Department of Science and Environmental Studies
The Education University of Hong Kong
Hong Kong SAR, China

Katharina Lapin
Department of Forest Biodiversity & Nature Conservation
Austrian Research Centre for Forests
Vienna, Austria

Louis S. H. Lee
Department of Construction, Environment and Engineering
Technological and Higher Education Institute of Hong Kong (THEi)
Hong Kong SAR, China

Hui Li
Department of Ecology, College of Life Science and Technology
Jinan University
Guangzhou, China

Pengsong Li
College of Environmental Science and Engineering
Beijing Forestry University
Beijing, China

Wai Chin Li
Department of Science and Environmental Studies
The Education University of Hong Kong
Hong Kong SAR, China

Fanjing Liu
Department of Health and Environmental Sciences
School of Science, Xi'an-Jiaotong Liverpool University
Suzhou, China

Keyan Miao
Department of Bioinformatics
School of Science, Xi'an-Jiaotong Liverpool University
Suzhou, China

Akira S. Mori
Research Center for Advanced Science and Technology
The University of Tokyo
Tokyo, Japan

Keita Nishizawa
Research Center for Advanced Science and Technology
The University of Tokyo
Tokyo, Japan

Xuejiao Qiao
School of Environment and Energy
Peking University Shenzhen Graduate School
Shenzhen, China

Jinglin Qiu
Paris School of International Affairs
Sciences Po
Paris, France

Wenhui Qiu
College of Environmental Science and Engineering
Southern University of Science and Technology
Shenzhen, China

Md. Rokonuzzaman
Department of Agricultural Extension Education
Sylhet Agricultural University
Sylhet-3100, Bangladesh

Dai Saito
Research Center for Advanced Science and Technology
The University of Tokyo
Tokyo, Japan

Kejia Shen
Department of Humanities and Social Sciences
University of Exeter
Exeter, United Kingdom

Yee Ming Siu
Department of Science and Environmental Studies
The Education University of Hong Kong
Hong Kong SAR, China

Susanna T. Y. Tong
Department of Geography and Geographic Information Science
University of Cincinnati
Cincinnati, Ohio, United States

Yiu Fai Tsang
Department of Science and Environmental Studies
The Education University of Hong Kong
Hong Kong SAR, China

Minhaz Uddin
Department of Science and Environmental Studies
The Education University of Hong Kong
Hong Kong SAR, China

Masayuki Ushio
Department of Ocean Science
The Hong Kong University of Science and Technology
Hong Kong SAR, China

Feifei Wang
Key Laboratory of the Ministry of Education for Coastal and Wetland Ecosystems
College of Environment and Ecology, Xiamen University
Xiamen, China

Sidan Wang
Institute of International Relations
China Foreign Affairs University
Beijing, China

Ming Hung Wong
Consortium on Health, Environment, Education, and Research (CHEER)
The Education University of Hong Kong
Hong Kong SAR, China

Zhen Wu
Institute of International Relations
China Foreign Affairs University
Beijing, China

Lingyun Xiao
Department of Health and Environmental Sciences
School of Science, Xi'an-Jiaotong Liverpool University
Suzhou, China

Nan Xu
School of Environment and Energy
Peking University Shenzhen Graduate School
Shenzhen, China

Xinyu Xu
Research Center for Advanced Science and Technology
The University of Tokyo
Tokyo, Japan

Xiaohan Yi
College of Life Sciences and Oceanography
Shenzhen University
Shenzhen, China

Susumu Yoshizawa
Graduate School of Frontier Sciences
The University of Tokyo
Chiba, Japan
and
Atmosphere and Ocean Research Institute
The University of Tokyo
Chiba, Japan

Jinlong Zhang
Flora Conservation Department
Kadoorie Farm and Botanic Garden
Tai Po, New Territories, Hong Kong SAR, China

Lijuan Zhang
School of Environment and Energy
Peking University Shenzhen Graduate School
Shenzhen, China
and
Advanced Institute for Ocean Research
Shenzhen Ocean University Preparatory Office
Southern University of Science and Technology
Shenzhen, China

Miao Miao Zhang
Department of Ecology
College of Life Science and Technology
Jinan University
Guangzhou, China

Chunmiao Zheng
College of Environmental Science and Engineering
Southern University of Science and Technology
Shenzhen, China

Tang Rong Zhou
Department of Ecology
College of Life Science and Technology, Jinan University
Guangzhou, China

Huiling Zhu
Flora Conservation Department
Kadoorie Farm and Botanic Garden
Tai Po, New Territories, Hong Kong SAR, China

Yue Yu Zou
Ecological and Earth Sciences Division
Natural Sciences Sector
The United Nations Educational Scientific and Cultural Organization (UNESCO)
Paris, France

Part I

Global Strategies and Governance

1 Global Biodiversity Politics and Governance

Sidan Wang and Zhen Wu

1.1 INTRODUCTION

Biodiversity loss has been seen as a significant global challenge, and it requires a range of assessments for addressing issues of biodiversity and ecosystems (IPBES et al. 2019). Human activities have various impacts on biodiversity (Keck et al. 2025). In order to address the biodiversity challenges, global biodiversity governance has been constructed around different international institutions. The Convention on Biological Diversity (CBD) has functioned as a core framework for guiding the global biodiversity governance. This is mainly because the CBD reflects the international institutional arrangements under the UN framework, and it thus determines its significant international legitimacy. The CBD was established in 1992 when other significant international environmental treaties, particularly the United Nations Framework Convention on Climate Change (UNFCCC) and the UN Convention to Combat Desertification (UNCCD), had been designed to address global environmental issues (Convention on Biological Diversity 2025). This moment of witnessing the establishment of the three conventions had been recorded during the 1992 Rio Earth Summit. However, in a sharp contrast to the high politicisation of climate change issues, the global biodiversity governance has seen its gradual growth in the international political agenda. The UN biodiversity conferences and their rounds of negotiations have received limited attention from political stakeholders.

This distinctive nature of global biodiversity governance is substantially determined by its institutional complexity. This work finds that global biodiversity governance has been constructed across a range of international institutions and the involvement of different stakeholders. It mainly argues that while the global biodiversity governance has been fragmented across existing international institutions, it has been identified within a wide range of various international environmental institutions, and it has even been mainstreamed into the agenda of global governance. On the one hand, this work summarises the main features of global biodiversity governance and particularly identifies a range of international institutions addressing biodiversity issues. This raises the institutional complexity of the global biodiversity governance. On this basis, the global biodiversity actions have been fragmented across the different international institutions. On the other hand, the global biodiversity governance has been institutionally evolving beyond the CBD into various international institutions. This shows a positive trend towards mainstreaming biodiversity into different sectors and global governance agenda. However, there are a few challenges for enhancing the global biodiversity governance. Strengthening the coordination across the different international institutions is a hard work. And, sustaining biodiversity conservation on the global governance agenda seems to be challenged by global political crises, particularly conflicts and traditional security issues.

On the basis of the main work of observing the global biodiversity governance, this work shows the biodiversity-related international institutions. And, following this section, it observes how the biodiversity governance has been mainstreamed into other international environmental institutions and global governance agenda. Finally, the main features of the global biodiversity are summarised and discussed.

DOI: 10.1201/9781003640899-2

1.2 GLOBAL BIODIVERSITY GOVERNANCE AND ITS INTERNATIONAL INSTITUTIONS

Global biodiversity has been targeted by a number of international institutions (see Table 1.1). Due to the complexity of biodiversity issues on our planet, there are both comprehensive conventions, like the CBD, and international institutional agreements addressing specific fields. The Agreement under the United Nations Convention on the Law of the Sea (UNCLOS) on the Conservation and Sustainable Use of Marine Biological Diversity of Areas beyond National Jurisdiction (BBNJ Agreement), well known as the High Seas Treaty, is addressing marine biological diversity through the law of the sea (United Nations 2023). This marine conservation and the protection of marine life can be traced back to a much early stage, around the establishment of the International Whaling Commission, to specifically target the control of whaling operations (Iliff 2008; Fitzmaurice 1946). These cases of the international institutional arrangements demonstrate that the functions and tasks of addressing global biodiversity issues have been redistributed across a range of different international organisations and agreements.

It is vital to discuss the role of the CBD simply because it has been a core framework of guiding the global biodiversity governance, reflecting the international arrangements under the UN framework, and thus determining its significant international legitimacy. As the climate change issue has been highly politicised, the global biodiversity governance has also been gradually engaged in the international political agenda. In 1992, the CBD was established as a historic commitment to biodiversity conservation and environmental development, when other significant international environmental treaties, particularly the UNFCCC and the UNCCD, were created by a large number of countries. As the first comprehensive treaty focusing on a comprehensive approach to the conservation of biological diversity and sustainable development, the CBD goes beyond the traditional conversation about biodiversity conservation, and covers a wide range of biodiversity-related issues, including biological technology and genetic resources (Glowka et al. 1994). In practice, compared with the other two conventions related to environmental protection, developing countries have relatively bargaining power in the biodiversity negotiations because of their abundant

TABLE 1.1
Biodiversity-Related International Agreements and Institutions

Year	Full Name of Treaty/Organisation (Abbreviation)	Functions of Addressing Global Biodiversity
1946	International Convention for the Regulation of Whaling	Controlling whaling operations
1951	International Plant Protection Convention (IPPC)	Protecting plant diversity, especially crops
1971	Convention on Wetlands of International Importance especially as Waterfowl Habitat (Ramsar Convention on Wetlands)	Providing a framework for the conservation of wetlands' biodiversity
1973	Convention on International Trade in Endangered Species of Wild Fauna and Flora (CITES)	Focusing on endangered species of wild animals and plants
1979	Convention on the Conservation of Migratory Species of Wild Animals (CMS)	Addressing the conservation of migratory species and habitats
1992	Convention on Biological Diversity (CBD)	Functioning as a core framework guiding the global biodiversity governance for all major issues
2001	International Treaty on Plant Genetic Resources for Food and Agriculture (ITPGRFA)	Addressing plant genetic resources of food and agriculture
2023	Agreement under the United Nations Convention on the Law of the Sea on the Conservation and Sustainable Use of Marine Biological Diversity of Areas beyond National Jurisdiction (BBNJ Agreement)	Addressing marine biological diversity

natural and biological resources (Park 2024). However, this shows a two-edged sword for the negotiations because a wide range of issues and themes can be debated across developed and developing countries. Following the contradiction across a large number of countries and stakeholders, a range of international environmental negotiations can be witnessing more difficulties and divergences. Besides, not surprisingly, there is a gap between the agreed goals of the international environmental agreements and their actual targets and actions at a national level, although the agreements have clear goals (Bull et al. 2019). While the 2015 Paris Conference witnessed the establishment of the Paris Climate Change Agreement, which has been recognised as an important milestone in global environment governance, the UN biodiversity conferences and their rounds of negotiations have not received enough and equal attention from political stakeholders. Even though biodiversity governance has been growing on the agenda of global governance, its global governance remains weak and fragmented. In 2002, global leaders had promised to achieve a significant reduction in the rate of biodiversity loss, but this commitment failed to be delivered and achieved successfully (Butchart et al. 2010). And, in 2010, the global biodiversity governance witnessed the coming of the Aichi Targets to set the objectives of addressing biodiversity issues during the period between 2011 and 2020. While this institutional establishment shows the political willingness of global biodiversity actions, a majority of the targets had not been achieved finally (Maney et al. 2024).

It is also very important to raise global efforts to conserve marine biodiversity. There is a common understanding of recognising the marine biodiversity conservation as an integral part of the global biodiversity governance. But the negotiations on the marine biodiversity conservation have been substantially institutionalised beyond the CBD framework. Under the UN framework, the BBNJ Agreement was adopted in 2023 to be an important legislative arrangement in the law of the sea, having filled the space of the law in terms of marine biodiversity. Following the procedure of establishing and signing the BBNJ agreement, it has already been a landmark for marine biodiversity conservation. The BBNJ Agreement addresses a range of core biodiversity issues and affairs, including marine genetic resources, area-based management tools, environmental impact assessments and capacity building and the transfer of marine technology (Shi 2020). While the BBNJ Agreement marks a progressive achievement in global environmental governance, its path to implementation has complex challenges. The agreement contains provisions which are primarily principled and lack operational details. These uncertain issues and gaps can be significant topics to be debated in the coming negotiations. Besides, given the complexity of marine conservation and ocean affairs, it may be tough for the BBNJ Agreement to coordinate with other international institutions (Friedman 2024).

There are other international institutions related to biodiversity, focusing on smaller areas and specific biological resources. The Convention on Wetlands of International Importance Especially as Waterfowl Habitat, widely known as the Ramsar Convention on Wetlands, provides a framework for the conservation of biodiversity on wetlands (UNESCO 2024). Being signed in 1971 and officially entered into force in 1975, the Ramsar Convention on Wetlands has been an intergovernmental treaty dedicated to the specific ecosystem, undertaking responsibility for supporting conservation of wetlands-related ecosystem. Due to the features impact over enhancing of the Ramsar Convention, its disadvantages have been obviously observed in terms of its implementation. In the Ramsar Sites under the implementation of the convention, while the wetland-related species seem to be conserved properly, their populations have declined in tropical areas (Gardner et al. 2015). Although the Convention has been established, the institutional construction of conservation towards biodiversity of wetlands requires more works. Particularly, various international institutions and a wide range of stakeholders such as environmental non-governmental organisations (NGOs) and the academic community have an important impact on enhancing local implementation of the Ramsar Convention (Mauerhofer et al. 2015).

It is worth noting that a range of international institutions, including international organisations and international treaties, share their overlapping concerns about biodiversity. However, they often show significant differences in the details of their focus on conservation frameworks and policy implementation. For instance, while wildlife conservation and food security have their institutional roots in ecological sector and agricultural sector, respectively, they are both closely linked to biodiversity governance. The International Plant Protection Convention (IPPC) makes an effort to protect plants, especially crop diversity (IPPC Secretariat 2024). Comparing with the IPPC, the International Treaty on Plant Genetic Resources for Food and Agriculture (ITPGRAF) is addressing the genetic resources of food and agriculture (FAO 2025). While they are both held by the Food and Agriculture Organization of the United Nations (FAO) and share their focus on the biodiversity and agricultural plants, the IPPC and the ITPGRAF have slight differences in their functions. The IPPC and the ITPGRAF sometimes may be confusing over their shared functions of addressing plant diversity, including crop issues. While they both have been supervised by the FAO, they have different perspectives of conserving agricultural plants. The IPPC pays attention to protecting plants' health. For example, it issued *Strategic framework for the International Plant Protection Convention (IPPC) 2020–2030* to show its targets of controlling the spread of pests and adopting international plant health standards (IPPC Secretariat 2021). The ITPGRAF regulates the use of genetic resources and raises a focus on the conservation and sustainable utilisation of crop diversity (FAO 2012).

This kind of institutional overlap can be identified in other cases, such as the conservation of wildlife species and endangered species. The Convention on the Conservation of Migratory Species of Wild Animals (CMS) accurately addresses the conservation of migratory species and habitats of animals (UNEP 2024a). The CMS has its attention on biodiversity conservation by controlling species extinction, conserving genetic diversity, and managing trade and sustainability of wild species (Convention on Biological Diversity 2024a). For the endangered species of wild animals and plants, the Convention on International Trade in Endangered Species of Wild Fauna and Flora (CITES) undertakes supervision and management of threatened species (IUCN 2025). The CMS and the CITES are the international environmental agreements concerning and conserving wildlife species, and they have been widely recognised and implemented at a global level (Hensz and Soberón 2018). Compared with the CMS, the CITES mechanism focuses on controlling the trading of the endangered species and establishing licensing systems, having a direct restriction on regulations of the commercial trade (IUCN 2025; Ugochukwu et al. 2018). In this sense, the CMS and the CITES have their different functions of improving the conservation of wildlife species. The CITES focuses on managing the international trade of the endangered species and imposing the restrictions over the global commercial activities, while the CMS focuses on the international cooperation for controlling and managing transboundary migratory species. Furthermore, the coordination between the CMS and the CITES has been growing. For instance, the Joint CITES-CMS African Carnivores Initiative seeks to improve collaboration and efficiency (Hellinx 2020). Generally speaking, these international environmental institutions have supported and expanded biodiversity conservation in different ways, while having been coordinated to make a great contribution to global biodiversity governance.

On the international stage, the positions and actions of various stakeholders exert a considerable influence on global biodiversity governance. While sovereign states remain the key actors in multilateral negotiations, a number of different actors, including the NGOs, business groups, and local pressure groups, exert a substantial influence on policy-making and implementation of the biodiversity governance. Compared to member states of the international agreements, various non-state actors have no such power and status. However, they have a substantial potential for making a great contribution to the global biodiversity governance in various ways. It is important to note that they supplement and evaluate the existing intergovernmental cooperation and promote the adoption and implementation of global actions on biodiversity conservation (Krug et al. 2020). Given their distinctive role in global environmental governance and activities, NGOs have been involved in scientific research and evaluation, have raised public awareness and encouraged people and society to pay attention to the importance of biodiversity conservation and thus play an important role in local and national implementation of the international agreements (Ulloa 2023; Buijs et al. 2024). In terms of the negotiations, funding is often one of the key interests for all actors, and it thus becomes an obstacle to reaching a consensus. In such circumstances, the business groups and actors have their important roles in enhancing global biodiversity governance. For instance, private management has a role in positively influencing the outcomes of wildlife conservation (Denny et al. 2024). Being different to other global environmental governance, the international biodiversity negotiations raise indigenous people and local communities (IPLCs) as an important group in shaping the global biodiversity governance. The IPLCs are critical to local environmental protection and usually possess long-established traditional and sustainable ways of living with local natural environments in harmony (IUCN 2022). Moreover, as their livelihoods are intrinsically and tightly linked to the local ecosystem, the IPLCs demonstrably possess an urgent motivation for biodiversity conservation. However, in the multilateral negotiations and governance practices, the IPLCs have failed to represent their voices and interests to meet their expectations, particularly in terms of receiving funding and mobilising financial resources (International Institute for Environment and Development 2022). In practice, the requirements and interests of different stakeholders may be conflicting with each other, particularly like the tension between indigenous people and governmental management. However, overall, involving various stakeholders in the initiatives and programmes of biodiversity conservation proves that the global biodiversity governance has increasingly been linked to a wide range of actors.

1.3 LINKING BIODIVERSITY WITHIN INTERNATIONAL ENVIRONMENTAL INSTITUTIONS

While biodiversity conservation has been complex across the different international institutions, it seems to be one of the most important issues in global environmental politics. As Table 1.2 shows, it has been repeatedly referred to in various international environmental institutions and negotiations. Biodiversity loss has been highly associated with other environmental issues, including climate issues and plastic pollution. These linkages require international environmental actions and their reflection in the existing international institutions. Meanwhile, it is quite positive to see that the public and international attention to biodiversity conservation has been continuously increasing (De Oliveira Caetano et al. 2023). Similarly, there are discussions on biodiversity conservation around the existing international environmental institutions. By identifying and using terms particularly like "biodiversity", this section searches the keywords in a number of reports of recording and reflecting significant international environmental negotiations to observe the issues of biodiversity

TABLE 1.2
International Environmental Negotiations and Their Linkages to Biodiversity Issues

International Negotiations	Linkages between Biodiversity and Other International Environmental Negotiations
United Nations Framework Convention on Climate Change (UNFCCC)	Enhancing cooperation among the three Rio Conventions
	Close correlation between biodiversity and climate change
United Nations Convention to Combat Desertification (UNCCD)	Compound effect of biodiversity loss
	Addressing desertification, land degradation, and drought in achieving the objectives of addressing biodiversity issues
Minamata Convention on Mercury	Profound linkage between biodiversity loss and mercury pollution
	Calling for addressing hazardous chemicals and mercury in the global biodiversity framework
Plastics negotiations	Plastic pollution harms biodiversity
	Biodiversity is linked to health and ecosystems
UN Water Conference	Interlinkages between water management and biodiversity loss

conservation. As a result, it shows that the international negotiations related to environmental issues have focused on solving the consequences of biodiversity loss and enhancing collaborative efforts on biodiversity conservation. It is worth noting that the biodiversity issues and actions have been mentioned widely in the existing international environmental institutions. Establishing a co-governance system with the global biodiversity governance seems to be a potential solution to addressing biodiversity and other global environmental issues. It can be seen from this that global biodiversity governance is a crucial part of environmental protection and governance, involving multiple dimensions such as protection measures and the consequences of biodiversity loss.

1.3.1 The UNFCCC and Its Links to Biodiversity Governance

The UNFCCC holds the Conference of Parties (COP) once a year to discuss and amend the convention to achieve climate objectives. During the Baku Climate Change Conference (COP-29) held in 2024, while the topic related to mountain ecosystem did show a call for advancing the agenda to protect the vulnerability of mountain ecosystem to climate change, biodiversity issues were not explicitly identified in the summary document (International Institute for Sustainable Development 2024a). This lack of association between biodiversity and climate change is mainly because the COP-29 focuses on climate finance. However, this situation does not mean that the UNFCCC cares little about biodiversity. According to the 2024 summary report of recording the negotiations on the Dubai Climate Change Conference (COP-28), the search result shows that biodiversity issues were mentioned explicitly and particularly by developing countries (International Institute for Sustainable Development 2023a). Besides, it is worth noticing that climate mitigation and adaptation can be related to biodiversity issues. Addressing biodiversity issues can be coordinated with climate actions and governance. Besides, during the 2025 Bonn Climate Conference, being the pioneer for the COP30, the Subsidiary Body for Scientific and Technological Advice recalled the synergies and cooperation among the three Rio Conventions, referring to climate change, biodiversity, and desertification (International Institute for Sustainable Development 2025). It is also worth noticing that there is so-called Bern Process, being recognised as an informal and party-led approach to enhancing cooperation among the Secretariats of various conventions and other relevant environmental agreements and biodiversity-related stakeholders, with the objective of facilitating the adoption and effective implementation of the Kunming-Montreal Global Biodiversity Framework (UNEP 2025). Generally speaking, biodiversity conservation has a close correlation with climate change. Therefore, the issue of biodiversity seems to be one topic mentioned in the international climate negotiations.

1.3.2 The UNCCD and Its Links to Biodiversity Governance

The United Nations UNCCD is a legally binding international agreement linking the environment and development to sustainable land management. The primary objective of the UNCCD and its conferences is the mitigation of desertification and land degradation in arid regions (UNCCD 2024). With the ultimate aim of the protection of livelihoods and ecosystems, the UNCCD has addressed biodiversity as one of its key issues in different perspectives. During the Sixteenth Session of the Conference of the Parties to the UNCCD held at the end of 2024, the keywords related to biodiversity were mentioned very explicitly to show profound interlinkage between biodiversity and desertification. Both as the part of the Rio Conventions which is the foundation of international environmental governance, biodiversity conservation, and combating desertification share common objectives of addressing global environmental challenges. In this context, the negotiation explicitly raised the significant role of addressing desertification, land degradation, and drought in achieving the objectives of addressing biodiversity issues (International Institute for Sustainable Development 2024c). Specifically, according to the 2024 summary report of recording the conference, global collaboration and synergies across the Rio Conventions were discussed by many stakeholders, presenting the importance of interconnections and compounded effects between land and ecosystem. These negotiating positions align with academic discussions on enhancing the governing system of climate change, desertification, and biodiversity (Bouramdane 2025). Furthermore, combating desertification and biodiversity conservation both share a common ground in governance systems, particularly raising a concern about the negative impacts of land degradation on the loss of various ecosystems. In conclusion, the UNCCD has relatively tight interlinkages with biodiversity issues, requiring a holistic approach to ecological conservation.

1.3.3 Other International Environmental Institutional Links to Biodiversity Governance

The Minamata Convention on Mercury is an international agreement aimed at controlling and reducing mercury emissions on a global scale. Biodiversity issues had been obviously observed and identified in the mercury negotiations. The UNEP emphasised that a majority of the targets of the Kunming-Montreal Global Biodiversity Framework reflected the close links between mercury pollution and biodiversity loss during the mercury negotiations (COP-5) in 2023 (International Institute for Sustainable Development 2023b). According to the 2023 summary report of recording the COP-5, the profound linkage between biodiversity and mercury can be observed very clearly because of the straightforward impact on biodiversity caused by mercury pollution. In order to address this issue, the mercury conference even advanced to call for the CBD negotiations to set indicators to address highly hazardous chemicals and mercury (International Institute for Sustainable Development 2023b). Following this finding, it is quite surprising to state that biodiversity issues have been explicitly identified in the mercury negotiations.

The Intergovernmental Negotiating Committee (INC) to Develop an International Legally Binding Instrument on Plastic Pollution is dedicated to solve the problem of addressing the full life cycle of plastic pollution. During the conference of the INC-5 at the end of 2024, biodiversity was identified in the summary report (International Institute for Sustainable Development 2024b). This reveals the fact that the biodiversity loss may be a ripple effect of the expansion of plastic pollution. Besides, according to the 2024 summary report of the INC-5, biodiversity was linked to health and ecosystems by a number of state actors. However, according to the result of searching the keywords in the document, biodiversity has been mentioned in a very general way.

The UN Water Conference in 2023 paid attention to the Kunming-Montreal Global Biodiversity Framework as well. As a global public product, water is doubtlessly important to human survival. In terms of policy dimension, the Water Conference highlighted the interlinkages between water management, land degradation, biodiversity loss, and climate change, raising a serious concern about the loss of biodiversity affected by water damage, especially in the agricultural sector (International Institute for Sustainable Development 2023c).

1.4 MAINSTREAMING BIODIVERSITY INTO GLOBAL GOVERNANCE

Biodiversity conservation has been observed to be adopted into the agenda of global governance beyond environmental and ecological sectors. The global biodiversity governance and its institutional arrangements have been enhanced, particularly following the establishment of the 2022 Kunming-Montreal Global Biodiversity Framework (Rabitz et al. 2025). This positive change has been observed across many international stakeholders, witnessing a process of mainstreaming biodiversity into the agenda of global governance.

International institutional changes and those non-state actors play an important role in enhancing global biodiversity governance (Petersson and Stoett 2022). The World Bank has been committed to support the various programmes on protecting nature and, particularly Kunming-Montreal Global Biodiversity Framework (World Bank Group 2025). Similarly, the International Monetary Fund raised its attention to the nature-related economic risks and linkages between global banks and the biodiversity-related protected areas (Gardes-Landolfini 2024). These international financial institutions have set their functional arrangements to adopt biodiversity-related affairs.

Biodiversity loss can be caused by a range of sectors such as trade. This association triggers involvement of specific international institutions in biodiversity actions. International trade has been found to have a negative impact on local diversity of species, witnessing 1% increase in global export and 2.28% decrease in species diversity (Tian and Cheng 2024). It is estimated that 30% of threats to global species diversity are closely linked to international trade particularly in developing countries (Lenzen et al. 2012). On scientific basis, the World Trade Organization (WTO) has cooperated with and offered observer status for the Secretariat of the Convention on Biological Diversity in the WTO Committee on Trade and Environment, focusing on intellectual property rights, access to genetic resources and the sharing of benefits from the use of those resources, agricultural trade, and technical barriers to trade (Convention on Biological Diversity 2023).

Those biodiversity-related environmental challenges, particularly like biodiversity loss, lead to pressures over health sector and require a holistic approach to the health-biodiversity governance (Willetts et al. 2023). The role of biodiversity conservation has been raised as an important dimension of global environment-health nexus governance (Willetts et al. 2022; Willetts and Grant 2022). The World Health Organization (WHO) has identified global biodiversity challenges as various threats to human health because they are closely related to food systems, health and traditional medicine, infectious diseases, and climate change (World Health Organization 2025). The mechanism of the CBD has established *the Global Action Plan on Biodiversity and Health* to enhance the international institutional interlinkages between health and biodiversity (Convention on Biological Diversity 2024b).

Regional institutional arrangements have reflected biodiversity governance. The EU has established a core framework of guiding and constructing the institutional arrangements of conserving biodiversity. In 2020, the EU issued the EU biodiversity strategy for 2030 on a previous basis of the EU 2020 Biodiversity Strategy, setting a range of objectives to address climate impacts, forest fires, food security, and diseases related to wildlife (European Commission 2025). This strategy has been widely linked to forests, raising attention towards wood-based bioeconomy and a cost-benefit approach to its implementation (Fulvio et al. 2025; Kangas and Ollikainen 2025). And, the policies of nature restoration have been seen as an important means of conserving biodiversity in the forest sector of the EU (Aggestam 2024). However, while the 2030 strategy focuses on a range of issues, particularly forest conservation, it has not yet paid much attention to invasive species and their threats to marine biodiversity (Ojaveer et al. 2025). This contrast demonstrates that the EU has its focus on land-based biodiversity challenges. Other regional intergovernmental institutions have their distinctive attention to the global biodiversity governance. For example, the Association of Southeast Asian Nations (ASEAN) plays a proactive role in raising global marine biodiversity governance and establishing relevant international agreements (Gunasekara and Karim 2021).

And, biodiversity conservation has been explicitly on the global governance agenda. The UN Sustainable Development Goals (SDGs) have been closely and widely linked to biodiversity conservation. The actions on conserving biodiversity have been mainstreamed into a wide range of the SDGs including poverty (SDG-1), food security (SDG-2), health (SDG-3), gender equality (SDG-5), water (SDG-6), economic growth (SDG-8), infrastructure (SDG-9), cities and human settlements (SDG-11), consumption and production (SDG-12), climate change (SDG-13), marine conservation (SDG-14), terrestrial ecosystems (SDG-15), institutions (SDG-16) and global partnerships (SDG-17) (Diz et al. 2018; Adhikari et al. 2023; Convention on Biological Diversity 2018). Besides, the World Economic Forum, being an important platform of attracting global political and economic elites, has recognised the serious challenges to biodiversity and called for shaping a nature-positive economy to achieve economic transitions and thus decrease economic impacts on biodiversity loss (World Economic Forum 2023).

1.5 DISCUSSIONS AND CONCLUSION

This work finds that global biodiversity governance has been constructed across a range of international institutions and involvement of different stakeholders. It supports the main argument that while the global biodiversity governance has been fragmented across existing international institutions, it has been identified within a wide range of various international environmental institutions, and it has even been mainstreamed into the agenda of global governance.

The global biodiversity governance has witnessed its growth in the existing international institutions (Petersson and Stoett 2022; Rabitz et al. 2025). Particularly, this institutional progress can be demonstrated by the establishment of the Kunming-Montreal Global Biodiversity Framework and the BBNJ. By involving a wide range of stakeholders, the global biodiversity governance has been evolving horizontally into different international institutions beyond the CBD and the UNCLOS. However, it is quite important to demonstrate that the global biodiversity governance has been fragmented across different international institutions, including the CMS, the CITES, and the Ramsar Convention on Wetlands. This institutional fragmentation requires a nexus approach to enhancing global biodiversity governance by establishing coordinating mechanisms across the different international institutions. For instance, the implementation of the BBNJ confronts an institutional challenge of integrating the objectives of environmental and ecological conservation into marine issues and ocean governance (Barirani 2022).

Mainstreaming biodiversity into global governance agenda matters. This can be seen that a range of significant international platforms offer their space for talking about biodiversity governance. It demonstrates that the biodiversity issues have been identified and adopted in the global governance agenda. Beyond this discursive adoption of addressing the biodiversity challenges, the existing international environmental institutions have witnessed the biodiversity governance on their agenda and institutional arrangements. Particularly, the various platforms of the international environmental negotiations have more or less raised their concerns about the biodiversity issues and called for strengthening their links to the biodiversity governance. It is really important to note that the UNEP (2024b) has been substantially supporting the institutional coordination across the UNFCCC, the CBD, and the UNCCD.

The biodiversity governance has been vertically mainstreamed into actions at subnational levels. A wide range of stakeholders, particularly the non-state actors, play an important role in translating the biodiversity governance into local and subnational actions (Kok and Ludwig 2022). Integrating local governance into the biodiversity policies seems to be an important factor in determining the effectiveness of the policy implementation (Li et al. 2023). And, local environmental groups and ideas even have the power of influencing national actions on participating in the global biodiversity governance (Shim and Shin 2020). These involvements of the various stakeholders have delivered their support for improving the global biodiversity governance.

However, while the global biodiversity governance has been growing institutionally, it confronts global political and institutional challenges. First, while the CBD has set objectives of addressing biodiversity issues, global extinction of species remains severe (Kim and Bridgewater 2025). This shows a clear gap between institutional design and its implementation and effectiveness in terms of global environmental governance. Second, climate change issues remain around the central point of global environmental governance and politics. This so-called climatisation refers to a situation in which climate change issues have become a dominant topic in other policy fields (Aykut and Maertens 2021). It is not surprising to see that climate issues have been frequently mentioned in a range of international environmental negotiations, including the biodiversity institutions. Third, traditional security issues have been coming back to the central position of global politics. Since the end of the Cold War, the non-traditional security issues have been substantially emerging to be important themes on the global governance agenda. Clearly, terrorism, climate change, and financial crises, beyond traditional military conflicts, have been defined as the non-traditional security issues and they have been seen as important topics of global governance. The biodiversity governance can be linked to the security issues (Wang 2022). But, geopolitical tensions and traditional conflicts have been astonishingly shocking the existing global politics and thus have the potential to influence the global governance agenda. In addition to the global security context, local resistance to and populist approaches to the existing environmental governance pose a threat to the long-term development of global biodiversity governance.

It is very important to conclude that the global biodiversity governance has been rooted in a very complex system across different international institutions. Global political changes should not be ignored within the interdisciplinary discussions on the biodiversity governance. The global biodiversity governance has been clearly influenced by the coordination across the various international institutions, a wide range of vertical and horizontal stakeholders, and global political turbulence. On the basis of this work, it is important to observe how the global biodiversity governance has been evolving to address the urgent challenges of the biodiversity loss. Particularly, more works can be developed around political explanations of the institutional complexity of and mainstreaming of the global biodiversity governance. Meanwhile, attention should be paid to the development of the global biodiversity governance in the context of the global political changes.

ACKNOWLEDGEMENTS

This research gratefully acknowledges the support from the Beijing Philosophy and Social Science Foundation (Grant numbers: 24DTR043)

REFERENCES

Adhikari, Biraj, Davnah Urbach, Nakul Chettri, et al. 2023. 'A Multi-Methods Approach for Assessing How Conserving Biodiversity Interacts with Other Sustainable Development Goals in Nepal'. *Sustainable Development* 31(5): 3239–53. WOS:000975780400001. https://doi.org/10.1002/sd.2582.

Aggestam, Filip. 2024. 'Crosswalking the EU Nature Restoration Regulation and the Kunming-Montreal Global Biodiversity Framework: A Forest-Centred Outlook'. *Sustainability* 16(11). https://doi.org/10.3390/su16114863.

Aykut, Stefan C., and Lucile Maertens. 2021. 'The Climatization of Global Politics: Introduction to the Special Issue'. *International Politics* 58(4): 501–18. https://doi.org/10.1057/s41311-021-00325-0.

Barirani, Dona. 2022. 'A UN Treaty for Marine Biodiversity: Establishing Environmental Policy Integration in Global Governance'. *Global Policy* (London) 13(3): 390–400. https://doi.org/10.1111/1758-5899.13088.

Bouramdane, Ayat-Allah. 2025. 'Optimal Strategies to Combat Land Degradation amid Climate Change: Aligned with UNCCD and Kunming-Montreal Commitments'. *Land Use Policy* 157(October): 107625. https://doi.org/10.1016/j.landusepol.2025.107625.

Buijs, Arjen E., Susan De Koning, Thomas J. M. Mattijssen, Ingeborg W. Smeding, Marie-José Smits, and Nathalie A. Steins. 2024. 'Civil Society for Sustainable Change: Strategies of NGOs and Active Citizens to Contribute to Sustainability Transitions'. *Journal of Environmental Planning and Management* 67(12): 2863–84. https://doi.org/10.1080/09640568.2023.2205571.

Bull, Joseph W., Eleanor J. Milner-Gulland, Prue F. E. Addison, et al. 2019. 'Net Positive Outcomes for Nature'. *Nature Ecology & Evolution* 4(1): 4–7. https://doi.org/10.1038/s41559-019-1022-z.

Butchart, Stuart H. M., Matt Walpole, Ben Collen, et al. 2010. 'Global Biodiversity: Indicators of Recent Declines'. *Science* 328(5982): 1164–68. https://doi.org/10.1126/science.1187512.

Convention on Biological Diversity. 2018. *Biodiversity and the Sustainable Development Goals*. Secretariat of the Convention on Biological Diversity.

Convention on Biological Diversity. 2023. 'Cooperation with WTO'. Secretariat of the Convention on Biological Diversity, March 29. https://www.cbd.int/incentives/coop-wto.shtml.

Convention on Biological Diversity. 2024a. 'Convention on the Conservation of Migratory Species of Wild Animals COP 14--12 to 17 February 2024'. Convention on Biological Diversity, February 12. https://www.cbd.int/article/cms-cop-14-2024.

Convention on Biological Diversity. 2024b. 'Global Action Plan on Biodiversity and Health'. Biodiversity and Health, Secretariat of the Convention on Biological Diversity, December 6. https://www.cbd.int/health/GAP.shtml.

Convention on Biological Diversity. 2025. 'The Rio Conventions'. September 13. https://www.cbd.int/rio.

De Oliveira Caetano, Gabriel Henrique, Reut Vardi, Ivan Jarić, Ricardo A. Correia, Uri Roll, and Diogo Veríssimo. 2023. 'Evaluating Global Interest in Biodiversity and Conservation'. *Conservation Biology* 37(5): e14100. https://doi.org/10.1111/cobi.14100.

Denny, Sean, Gabriel Englander, and Patrick Hunnicutt. 2024. 'Private Management of African Protected Areas Improves Wildlife and Tourism Outcomes but with Security Concerns in Conflict Regions'. *Proceedings of the National Academy of Sciences* 121(29): e2401814121. https://doi.org/10.1073/pnas.2401814121.

Diz, Daniela, David Johnson, Michael Riddell, et al. 2018. 'Mainstreaming Marine Biodiversity into the SDGs: The Role of Other Effective Area-Based Conservation Measures (SDG 14.5)'. *Marine Policy* 93(July 2018): 251–61. https://doi.org/10.1016/j.marpol.2017.08.019.

European Commission. 2025. 'Biodiversity Strategy for 2030'. https://environment.ec.europa.eu/strategy/biodiversity-strategy-2030_en.

FAO. 2012. *Conservation and Sustainable Use under the International Treaty*. Food and Agriculture Organization of the United Nations.

FAO. 2025. *International Treaty on Plant Genetic Resources for Food and Agriculture*. Food and Agriculture Organization of the United Nations. https://www.fao.org/plant-treaty/en.

Fitzmaurice, Malgosia. 1946. *International Convention for the Regulation of Whaling*. United Nations.

Friedman, Shani. 2024. 'The Interaction of the BBNJ Agreement and the Legal Regime of the Area, and Its Influence on the Implementation of the BBNJ Agreement'. *Marine Policy* 167(September): 106235. https://doi.org/10.1016/j.marpol.2024.106235.

Fulvio, Fulvio Di, Tord Snäll, Pekka Lauri, et al. 2025. 'Impact of the EU Biodiversity Strategy for 2030 on the EU Wood-Based Bioeconomy'. *Global Environmental Change* 92(July 2025): 102986. https://doi.org/10.1016/j.gloenvcha.2025.102986.

Gardes-Landolfini, Charlotte. 2024. 'Embedded in Nature'. *Staff Climate Notes* 2024(2): 1. https://doi.org/10.5089/9798400288548.066.

Gardner, Royal C., Stefano Barchiesi, Coralie Beltrame, et al. 2015. *State of the World's Wetlands and Their Services to People: A Compilation of Recent Analyses*. Ramsar Convention Secretariat. https://www.ssrn.com/abstract=2589447.

Glowka, Lyle, Françoise Burhenne-Guilmin, Hugh Synge, Jeffrey A. McNeely, and Lothar Gündling. 1994. *A Guide to the Convention on Biological Diversity*. Vol. 30. Environmental Policy and Law Paper. IUCN the World Conservation Union.

Gunasekara, Sandya Nishanthi, and Md Saiful Karim. 2021. 'The Role of ASEAN and Its Members in Promoting the Norm of Responsible Governance of Marine Biodiversity of Areas beyond National Jurisdiction'. *Review of European Comparative and International Environmental Law* 30(1): 128–37. WOS:000591412200001. https://doi.org/10.1111/reel.12372.

Hellinx, Elke. 2020. 'The CMS-CITES African Carnivore Initiative as an Illustration of Synergies Between MEAs'. *Frontiers in Ecology and Evolution* 8(February): 10. https://doi.org/10.3389/fevo.2020.00010.

Hensz, Christopher Michael, and Jorge Soberón. 2018. 'Participation in the Convention on Migratory Species: A Biogeographic Assessment'. *Ambio* 47(7): 739–46. https://doi.org/10.1007/s13280-018-1024-0.

Iliff, Mike. 2008. 'Modernisation of the International Convention for the Regulation of Whaling'. *Marine Policy* 32(3): 402–7. https://doi.org/10.1016/j.marpol.2007.08.003.

International Institute for Environment and Development. 2022. 'Indigenous Peoples, Local Communities Underrepresented in Governance of Nature Funding'. Biodiversity, December 13. https://www.iied.org/indigenous-peoples-local-communities-underrepresented-governance-nature-funding.

International Institute for Sustainable Development. 2023a. 'Summary of the 2023 Dubai Climate Change Conference: 30 November – 13 December 2023'. *Earth Negotiations Bulletin*, COP 28 Final, 12(842): 1–30.

International Institute for Sustainable Development. 2023b. 'Summary of the Fifth Meeting of the Conference of the Parties to the Minamata Convention on Mercury: 30 October – 3 November 2023'. *Earth Negotiations Bulletin* 28(70): 1–14.

International Institute for Sustainable Development. 2023c. 'Summary of the UN 2023 Water Conference: 22–24 March 2023'. *Earth Negotiations Bulletin* 3(17): 1–9.

International Institute for Sustainable Development. 2024a. 'Summary of the 2024 Baku Climate Change Conference: 11–22 November 2024'. *Earth Negotiations Bulletin*, COP 29 Final, 12(865): 1–25.

International Institute for Sustainable Development. 2024b. 'Summary of the First Session of the Fifth Session of the Intergovernmental Negotiating Committee to Develop an International Legally Binding Instrument on Plastic Pollution: 25 November – 2 December 2024'. *Earth Negotiations Bulletin* 36(34): 1–12.

International Institute for Sustainable Development. 2024c. 'Summary of the Sixteenth Session of the Conference of the Parties to the UN Convention to Combat Desertification: 2–13 December 2024'. *Earth Negotiations Bulletin*, COP 16 Final, 4(313): 1–22.

International Institute for Sustainable Development. 2025. 'Bonn Highlights: Wednesday, 25 June 2025'. *Earth Negotiations Bulletin* 12(875): 1–4.

IPBES, Sandra. Díaz, Josef. Settele, Eduardo S. Brondízio, et al. 2019. *Summary for Policymakers of the Global Assessment Report on Biodiversity and Ecosystem Services*. Version summary for policy makers. Zenodo. https://doi.org/10.5281/ZENODO.3553579.

IPPC Secretariat. 2021. *Strategic Framework for the International Plant Protection Convention (IPPC) 2020–2030. Protecting Global Plant Resources and Facilitating Safe Trade*. FAO on behalf of the Secretariat of the International Plant Protection Convention.

IPPC Secretariat. 2024. *International Plant Protection Convention (1997)*. Food and Agriculture Organization of the United Nations.

IUCN. 2022. 'Indigenous Peoples and Biodiversity'. August 9. https://www.iucn.nl/en/blog/indigenous-peoples-and-biodiversity/.

IUCN. 2025. 'Convention on International Trade in Endangered Species of Wild Fauna and Flora | IUCN'. September 1. https://iucn.org/our-work/informing-policy/international-policy/convention-international-trade-endangered-species.

Kangas, Johanna, and Markku Ollikainen. 2025. 'Towards an Efficient Implementation of the EU Biodiversity Strategy in Forests: An Analysis of Alternative Voluntary Conservation Mechanisms and Selection Criteria'. *Forest Policy and Economics* 172(March 2025): 103448. https://doi.org/10.1016/j.forpol.2025.103448.

Keck, François, Tianna Peller, Roman Alther, et al. 2025. 'The Global Human Impact on Biodiversity'. *Nature* 641(8062): 395–400. https://doi.org/10.1038/s41586-025-08752-2.

Kim, Rakhyun E., and Peter Bridgewater. 2025. 'Environmental Treaties Are Paralysed: Here's How We Can Do Better'. *Nature* 646(8087): 1054–56. https://doi.org/10.1038/d41586-025-03477-8.

Kok, Marcel, and Kathrin Ludwig. 2022. 'Understanding International Non-State and Subnational Actors for Biodiversity and Their Possible Contributions to the Post-2020 CBD Global Biodiversity Framework: Insights from Six International Cooperative Initiatives'. *International Environmental Agreements: Politics, Law and Economics* 22(March): 1–25. https://doi.org/10.1007/s10784-021-09547-2.

Krug, Cornelia B., Eleanor Sterling, Timothy Cadman, et al. 2020. 'Stakeholder Participation in IPBES: Connecting Local Environmental Work with Global Decision Making'. *Ecosystems and People* 16(1): 197–211. https://doi.org/10.1080/26395916.2020.1788643.

Lenzen, Manfred, Daniel D. Moran, Keiichiro Kanemoto, Barney Foran, Leonarda Lobefaro, and Arne Geschke. 2012. 'International Trade Drives Biodiversity Threats in Developing Nations'. *Nature* 486(7401): 109–12. https://doi.org/10.1038/nature11145.

Li, Qingyang, Yingxue Ge, and Jeffrey A. Sayer. 2023. 'Challenges to Implementing the Kunming-Montreal Global Biodiversity Framework'. *Land* 12(12): 2166. https://doi.org/10.3390/land12122166.

Maney, Calum, Daniela Guaras, Jerry Harrison, et al. 2024. 'National Commitments to Aichi Targets and Their Implications for Monitoring the Kunming-Montreal Global Biodiversity Framework'. *NPJ Biodiversity* 3(1): 6. https://doi.org/10.1038/s44185-024-00039-5.

Mauerhofer, Volker, Rakhyun E. Kim, and Casey Stevens. 2015. 'When Implementation Works: A Comparison of Ramsar Convention Implementation in Different Continents'. *Environmental Science & Policy* 51(August): 95–105. https://doi.org/10.1016/j.envsci.2015.03.016.

Ojaveer, Henn, Bella Galil, and Hanno Seebens. 2025. 'The Commitment of the EU Biodiversity Strategy for 2030 Is Unfeasible for Marine Threatened Species Affected by Invasive Alien Species'. *Marine Policy* 173(March 2025): 106582. https://doi.org/10.1016/j.marpol.2024.106582.

Park, HyeMin. 2024. 'Historical and Relative Dominance of the Convention on Biological Diversity Agenda: A Case Analysis of Decisions and Recommendations'. *Environmental Research Communications* 6(8): 085018. https://doi.org/10.1088/2515-7620/ad6d3b.

Petersson, Matilda, and Peter Stoett. 2022. 'Lessons Learnt in Global Biodiversity Governance'. *International Environmental Agreements: Politics, Law and Economics* (Dordrecht) 22(2): 333–52. https://doi.org/10.1007/s10784-022-09565-8.

Rabitz, Florian, Simon Beaudoin, Theresa Jedd, et al. 2025. 'A Turning Point in Global Biodiversity Governance?' *Environmental Politics*, no. Oct 2025: 1–10. https://doi.org/10.1080/09644016.2025.2565867.

Shi, Yubing. 2020. 'Settlement of Disputes in a BBNJ Agreement: Options and Analysis'. *Marine Policy* 122(December): 104156. https://doi.org/10.1016/j.marpol.2020.104156.

Shim, Jae-Mahn, and Eunjung Shin. 2020. 'Drivers of Ratification Rates in Global Biodiversity Governance: Local Environmentalism, Orientation toward Global Governance, and Peer Pressure'. *Environmental Politics* 29(5): 845–65. https://doi.org/10.1080/09644016.2019.1630070.

Tian, Hui, and Chaoyin Cheng. 2024. 'International Trade and Biodiversity: Is Export a Species Killer?' *Ecological Economics* 225(November): 108313. WOS:001291512000001. https://doi.org/10.1016/j.ecolecon.2024.108313.

Ugochukwu, Albert I., Jill E. Hobbs, Peter W. B. Phillips, and William A. Kerr. 2018. 'Technological Solutions to Authenticity Issues in International Trade: The Case of CITES Listed Endangered Species'. *Ecological Economics* 146(April): 730–39. https://doi.org/10.1016/j.ecolecon.2017.12.021.

Ulloa, Ana Maria. 2023. 'Accountability as Constructive Dialogue: Can NGOs Persuade States to Conserve Biodiversity?' *Global Environmental Politics* 23(1): 42–67. https://doi.org/10.1162/glep_a_00673.

UNCCD. 2024. 'Landmark UN Conference to Accelerate Global Efforts for Land Restoration and Drought Resilience'. United Nations Convention to Combat Desertification, November 25. https://www.unccd.int/news-stories/press-releases/landmark-un-conference-accelerate-global-efforts-land-restoration-and.

UNEP. 2024a. 'Landmark UN Report: The World's Migratory Species of Animals Are in Decline, and the Global Extinction Risk Is Increasing'. December 2. https://www.unep.org/news-and-stories/press-release/landmark-un-report-worlds-migratory-species-animals-are-decline-and.

UNEP. 2024b. 'MEA Synergies'. April 11. https://www.unep.org/topics/nature-action/global-biodiversity-framework/mea-synergies.

UNEP. 2025. 'Bern Process | UNEP - UN Environment Programme'. June 2. https://www.unep.org/topics/environmental-law-and-governance/strengthening-institutions/bern-process.

UNESCO. 2024. 'Convention on Wetlands of International Importance Especially as Waterfowl Habitat'. September 12. https://www.unesco.org/en/biodiversity/wetlands.

United Nations. 2023. *Agreement under the United Nations Convention on the Law of the Sea on the Conservation and Sustainable Use of Marine Biological Diversity of Areas beyond National Jurisdiction.* United Nations Headquarters.

Wang, Sidan. 2022. 'Convergence and Divergence of Biosecurity and Biodiversity in China'. *China International Strategy Review* 4(1): 144–55. https://doi.org/10.1007/s42533-022-00096-0.

Willetts, Elizabeth, Liz Grant, Jennifer Bansard, et al. 2022. *Health in the Global Environmental Agenda: A Policy Guide.* IISD Earth Negotiations Bulletin. International Institute for Sustainable Development.

Willetts, Liz, Liane Comeau, Neil Vora, et al. 2023. 'Health in Global Biodiversity Governance: What Is Next?' *The Lancet (British Edition)* (England) 401(10376): 533–36. https://doi.org/10.1016/S0140-6736(23)00130-7.

Willetts, Liz, and Liz Grant. 2022. 'The Health–Environment Nexus: Global Negotiations at a Crossroads'. *The Lancet* 399(10336): 1677–78. https://doi.org/10.1016/S0140-6736(22)00561-X.

World Bank Group. 2025. 'Biodiversity'. World Bank Group, April 12. https://www.worldbank.org/en/topic/biodiversity#2.

World Economic Forum. 2023. '50% of the Global Economy Is under Threat from Biodiversity Loss'. World Economic Forum, February 7. https://www.weforum.org/stories/2023/02/biodiversity-nature-loss-cop15/.

World Health Organization. 2025. 'Biodiversity'. WHO Newsroom, February 18. https://www.who.int/news-room/fact-sheets/detail/biodiversity.

2 Bridging Conservation and Development

Toward a Renewed Framework for Evaluating Human–Earth Relationship

Yue Yu Zou

2.1 INTRODUCTION

2.1.1 Conceptual Frameworks

The human–Earth relationship, as a complex system, is changing more intensely than at any other period in history. The observed phenomena—including extreme climate events, accelerated species extinction, and land degradation—are not merely natural processes, but reflect profound interactions between human production and lifestyle and the Earth's systems (Bliege Bird & Nimmo, 2018). Recognizing and quantifying these interactions and addressing the challenges they pose require both close international collaboration and robust scientific frameworks to inform decision-making. Yet the scientific frameworks are not neutral technical tools; they continuously shape key narratives about "what to conserve" and "by whom": what is considered important, whose interests are taken into account, and how governance objectives are quantified and allocated (Biermann, 2022). In short, the concept and disciplinary system of human–Earth relations have remained one of the major dimensions influencing global environmental governance.

Over the past half-century, scientific research on human–Earth relations has made significant advances. Until the latter half of the 20th century, environmental research primarily focused on the unidirectional impacts of human activities on natural systems, such as pollution and deforestation, as well as on components of the biosphere, including forest ecosystems, marine and coastal ecosystems, wetlands, carbon cycles, and water cycles. Over the past three decades, interdisciplinary research integrating ecology, Earth system science, and social sciences has gradually revealed that human societies and natural ecosystems form an interdependent, feedback-driven coupled social–ecological system (Holling, 2001). This system involves not only exchanges of matter and energy but also the co-evolution of knowledge, culture, and institutions.

However, when it comes to governance practice, the complexity of natural systems and the lack of integration among related disciplines have hindered the establishment of a scientifically robust, objective, and multidimensional assessment system for human–Earth relations. Existing indicator systems primarily focus on human pressures and impacts on the environment, with insufficient quantification of the contributions and feedbacks of the Earth system to human well-being (Ellis et al., 2025). Numerous specialized intergovernmental environmental conventions and platforms have been developing, testing and advancing these theoretical frameworks, such as the United Nations Sustainable Development Goals (SDGs), the United Nations Framework Convention on Climate Change (UNFCCC), the Convention on Biological Diversity (CBD) and its Kunming–Montreal Global Biodiversity Framework (KMGBF) (CBD, 2022), the Man and the Biosphere

DOI: 10.1201/9781003640899-3

(MAB) Programme of the United Nations Educational, Scientific and Cultural Organization (UNESCO) as well as the Intergovernmental Science–Policy Platform on Biodiversity and Ecosystem Services (IPBES) (Escobar-Pemberthy & Ivanova, 2020). Despite the proliferation of international mechanisms, evaluations of human–Earth relations remain fragmented, unidimensional, and anthropocentric, often failing to reflect the bidirectional interactions and co-evolution of humans and the Earth system (Juffe-Bignoli et al., 2014). This undermines both the scientific integrity of assessments and the fairness and effectiveness of governance decisions, including setting policy priorities, allocating ecological responsibilities and benefits among different groups, and designing incentives for biodiversity conservation. Furthermore, methodological differences among frameworks and inconsistent data collection standards have led to fragmented evaluations, thereby limiting the comparability and coherence of global environmental data.

The question is: how can we develop a more scientific and comprehensive approach to measure human–Earth interactions, thereby enhancing the effectiveness of global biodiversity and sustainability governance? To answer this question, this chapter will explore: How has the scientific understanding of human–earth relations evolved over the past five decades, and how has it influenced the construction of global governance systems; through a case study of UNESCO's MAB Programme and other relevant global mechanisms, understand what role did these mechanisms played and can they play in shaping the theoretical and institutional frameworks that bridges nature and development; finally, facing the deficiencies exist in current evaluation systems, will the innovations on theoretical and technical aspects transform the human-Earth relationship evaluation framework that guides humanity to harmonious existence with nature?

2.1.2 Methodology and Data Sources

The methodology combines historical-institutional analysis and comparative research: Section 2.2 will analyze the theoretical evolution of human–earth relationship concepts and their co-development with global biodiversity governance, highlighting the contributions of the MAB Programme, whose evolution mirrors broader transformations in global environmental governance and existing assessment frameworks (e.g., SDG assessments and national/regional biodiversity indicators) in terms of indicator orientation and methodology. Section 2.3 will center on identifying the structural problems and new opportunities in current evaluation systems, particularly in light of technological advancements such as artificial intelligence and big data; Section 2.4 will try to propose a future-oriented integrative framework for human–earth evaluation, emphasizing interdisciplinary collaboration and institutional synergy, while highlighting the potential of biosphere reserve networks as practical testing grounds.

2.2 THE CO-EVOLUTION OF THE CONCEPTUAL FRAMEWORK OF HUMAN–EARTH RELATIONS AND GLOBAL NATURE GOVERNANCE

2.2.1 The Evolution of Human–Earth Relationship Studies

The scientific study of the human–earth relationship has a long history, but its paradigmatic transformation mainly occurred in the mid- to late 20th century. In the 1960s and 1970s, the publication of Silent Spring triggered a global environmental awakening. At that time, research on the human–earth relationship was grounded primarily in the natural sciences, emphasizing the impact of human activities on the environment. The underlying epistemology was dualistic—humans and nature were perceived as separate entities, with the focus placed on pollution control, resource depletion, and the establishment of nature reserves (Steffen et al., 2020). Evaluation criteria have primarily focused on species richness, habitat integrity, and the number of threatened species. While these indicators provide clear guidance for conservation actions, they often overlook the role of social factors and their connections to human well-being.

The 1972 United Nations Conference on the Human Environment (the "Stockholm Conference") catalyzed the emergence of sustainable-development discourse and encouraged researchers to integrate economic, social, and environmental dimensions. Correspondingly, studies of human–environment relationships began to expand toward interdisciplinary approaches, emphasizing the roles of institutions, governance, and value systems. Holling introduced the concept of ecosystem resilience, underscoring the existence of multiple stable states and nonlinear feedback within ecological systems (Holling, 1973).

During the 1980s and 1990s, the rise of systems ecology and complex systems theory marked a turning point. Scholars began to recognize ecosystems as open, non-linear systems characterized by dynamic exchanges of energy and materials, in which human society acts as both a component and a driver (Odum, 1983). The theory of adaptive management subsequently emerged, advocating "learning-based" management approaches to facilitate the co-adaptation of ecological and social systems under conditions of uncertainty (Walters, 1986). Within this framework, ecosystems are conceptualized as open, non-linear systems of energy flow, and human societies are understood both as integral components of these systems and as key drivers influencing their dynamics (Schellnhuber, 1998). The 1987 *Our Common Future* (WCED, 1987) report formally introduced the concept of sustainable development, emphasizing the need to balance ecological, economic, and social dimensions. This marked the integration of social and ecological thinking and laid the groundwork for the global environmental governance agenda.

With the launch of the Millennium Ecosystem Assessment in 2001, research attention shifted toward understanding nature's contributions to society—namely, ecosystem services—and toward expressing ecological value through economic and social indicators (Millennium Ecosystem Assessment, 2005).

The ecosystem-services framework strengthened communication between science and policy, and highlighted potential synergies between conservation and development. Gunderson, Holling, and others (2002) further argued for integrating ecological resilience with adaptive management of human systems within a social–ecological systems framework, emphasizing cross-scale co-adaptation and dynamic interactions.

Entering the 21st century, the integration of Earth System Science and Socio-Ecological Systems (SES) theory further deepened this interdisciplinary synthesis. Elinor Ostrom's Institutional Analysis and Development (IAD) framework (Ostrom, 2005) and her Social–Ecological Systems Framework (SESF) (Ostrom, 2007) have become foundational theoretical approaches in this field. The planetary boundary framework (Richardson et al., 2023) describes the "safe operating space" for humanity, highlighting that in assessing Earth system and biosphere functions (Rockström et al., 2009), human appropriation of net primary production is suggested as a key indicator of functional biosphere integrity (Steffen et al., 2015).

These analytical frameworks conceptualize humans and nature as components of a coupled system in which governance outcomes depend on interactions among institutions, social capital, and ecological resilience. Research in this tradition highlights the roles of community-based governance, property-rights arrangements, and local knowledge in resource management, marking a paradigm shift from "ecological management" to "social–ecological co-governance." With the expansion of global supply chains, trade flows, and remote-sensing observations, scholarship on globalization and cross-scale couplings has continued to develop. Liu and colleagues (2013) proposed the telecoupling framework to analyze environmental and social feedback among geographically distant regions.

The IPBES articulated this approach in its Conceptual Framework (Díaz et al., 2019), which outlined the causal relationship between "nature," "nature's contributions to people," and "human well-being." In terms of assessment methods, a range of approaches has emerged, including system dynamics modeling (Liu et al., 2007) and multi-criteria evaluation and participatory assessment (de Groot et al., 2010). These methods seek to combine quantitative and qualitative approaches to integrate ecological integrity, ecosystem services, governance effectiveness, and human well-being within a single analytical framework (Scholes & Biggs, 2005).

In short, as sustainable development research has deepened, the study of human–earth relations has expanded from ecology into economics, geography, sociology, and political science. This expansion has given rise to several major research pathways:

Ecological and Earth-system pathways: the studies focus on ecosystem processes and planetary boundaries. Representative contributions include Rockström et al. (2009), who proposed the concept of a "safe operating space for humanity," as well as subsequent refinements by Steffen et al. (2015).

Economic and development pathways: this line of research examines human–environment interactions through the valuation of natural capital and ecosystem services, exemplified by Costanza et al. (1997) global assessment of ecosystem service values (Pascual et al., 2023). Socio-ecological resilience analysis has also been used to assess the adaptive capacity of coupled social–ecological systems (Folke, 2006). While widely applied at regional or landscape scales, this approach has yet to produce a globally comparable index.

Social and governance pathways: this line of inquiry highlights local knowledge (Bjork-James et al., 2022), institutional arrangements, and community participation (Berkes et al., 1998; Brondízio et al., 2021). Research on the human–nature connection has also emerged at the intersection of psychology and sociology. For instance, Mayer and Frantz developed the "Connectedness to Nature Scale," which captures individuals' emotional attachment to nature and is widely used in behavioral studies, though it does not function as a global statistical indicator (Mayer & Frantz, 2004).

Technology and data pathways: advances in remote sensing, geographic information systems (GIS), and artificial intelligence have enabled dynamic, cross-scale monitoring of socio-ecological processes (Reyers et al., 2018). The convergence of these trajectories led to the formation of Human–Earth System Science, an approach that enabled a transition from single-dimensional ecological indicators to composite frameworks that capture the complexity of socio-ecological interactions.

2.2.2 The Evolution of Global Biodiversity Governance

Since the 1972 Stockholm Conference on the Human Environment, global environmental governance has gradually transitioned from issue-specific interventions to institutionalized multilateral cooperation. The evolution of biodiversity governance can be roughly divided into three main stages:

The first stage (1970s–1980s) was characterized by conservation-oriented governance. During this period, international instruments such as the Convention on Wetlands of International Importance (Ramsar Convention) and UNESCO's World Heritage Convention established a normative foundation for global conservation efforts, focusing on the intrinsic value of nature. The launch of UNESCO's MAB Programme in 1971 represented a groundbreaking shift. Unlike traditional conservation mechanisms, the MAB Programme incorporated human activity into the ecological framework, emphasizing coexistence rather than exclusion.

The second stage (1990s–2000s) witnessed the rise of sustainable development and comprehensive assessment. The 1992 Rio Earth Summit produced landmark agreements such as the CBD and the UNFCCC, and proposed the Agenda 21 Action Plan. Global governance thus evolved from the protection of ecosystems to the integrated management of socio-ecological systems. During this period, the MAB Programme shifted from being a scientific research project to becoming a policy-support platform. It institutionalized 10-year periodic reviews for biosphere reserves, covering not only ecological monitoring but also socio-economic performance and governance effectiveness.

The third stage (2010s–present) features multi-goal integration and cross-mechanism coordination. The adoption of the SDGs in 2015 and the KMGBF in 2022 marked the move toward comprehensive sustainability governance. The MAB Programme has been explicitly positioned by UNESCO as "a local-level laboratory for the implementation of the SDGs." Through its World

Network of Biosphere Reserves, MAB now serves as a platform for data integration, multi-stakeholder collaboration, and community-based experimentation within global governance.

The evolution of biodiversity governance reflects a mutual reinforcement between scientific paradigms and governance needs: the transformation of ecological science has stimulated institutional innovation, while global policy agendas have, in turn, expanded the frontiers of scientific inquiry.

2.2.3 The Evaluation of the MAB Programme

Established in 1971, UNESCO's MAB Programme marked a milestone in global environmental cooperation. As the first intergovernmental scientific initiative to focus explicitly on human–environment interactions, MAB aimed to reconcile biodiversity conservation with socio-economic development. Practically, the Programme was implemented through globally designated biosphere reserves, spatial units delineated based on ecological and socio-economic characteristics, integrating research, education, international collaboration, and local practice (Ishwaran et al., 2008). As an intergovernmental scientific cooperation platform, MAB has not only established a cross-disciplinary research network but also provided valuable experiences and lessons for subsequent international environmental conventions and assessment methods (Reed & Price, 2020).

At the core of the MAB Programme lies its three-zone model, consisting of the core area, buffer zone, and transition area. This spatial structure embodies the Programme's understanding of human–earth relationships: the core area preserves ecosystem integrity and biodiversity representativeness; the buffer zone accommodates research, education, and limited human activity; and the transition area promotes sustainable socio-economic practices under participatory management. The model thus provides not only a physical demarcation but also a theoretical reflection of how humans and nature can coexist within a structured governance framework.

Since the 1980s, MAB has gradually developed a formal system for biosphere reserve designation and periodic evaluation. The criteria for assessment have expanded from ecological parameters and scientific outputs to include governance structures, social benefits, community participation, and education and cultural functions. The Seville Strategy and the Statutory Framework of the World Network of Biosphere Reserves further institutionalized the requirement for 10-year comprehensive assessments, with an emphasis on policy feedback and information sharing. This has made the MAB network one of the few global mechanisms combining long-term ecological monitoring with social and governance evaluation.

Moreover, the MAB evaluation framework has established close linkages with other international mechanisms. It complements the CBD in biodiversity monitoring and habitat conservation (CBD, 2024), contributes to several SDG targets such as poverty alleviation (Goal 1), quality education (Goal 4), and sustainable cities (Goal 11), and aligns with the UNFCCC in areas such as carbon sequestration and climate adaptation. Through these connections, the MAB Programme serves as both a vertical linkage (connecting global, national, and local levels) and a horizontal bridge (integrating science, policy, and society) (Figure 2.1).

2.2.4 A Comparative Analysis

Although diverse in disciplinary origins, most share the goal of balancing environmental and human dimensions. However, their conceptual emphases and methodological designs vary significantly.

For instance, the Sustainable Development model emphasizes the coordination of economic, social, and environmental aspects, but lacks explicit mechanisms to capture systemic feedback. The Socio-Ecological Systems framework focuses on interactions between resource systems, governance units, and users, highlighting institutional design but facing challenges in quantification. The IPBES Conceptual Framework (2019) provides a multi-value perspective that integrates nature, nature's contributions, and human well-being, yet data consistency remains problematic. The MAB zoning model embodies spatial hierarchy and human participation, but it lacks a comprehensive quantitative system for evaluation.

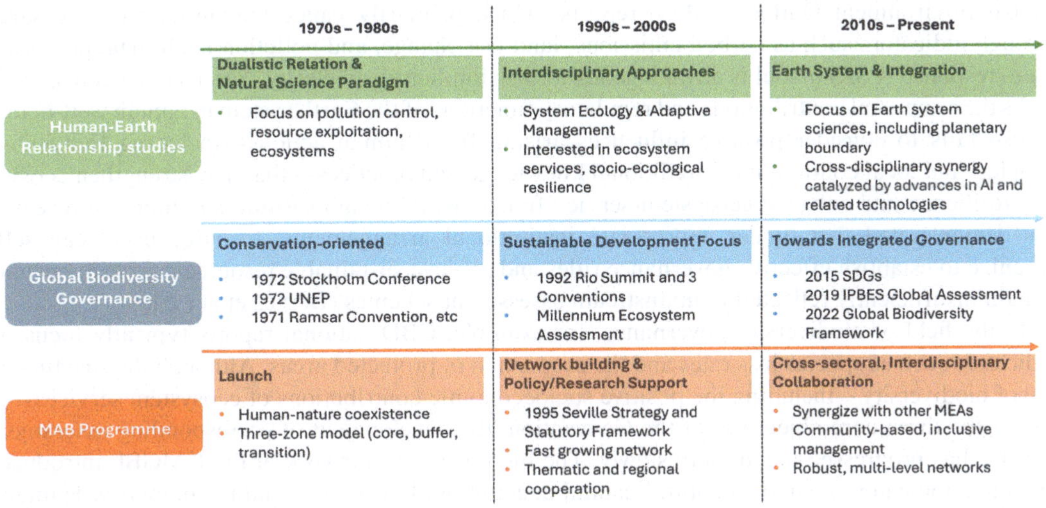

FIGURE 2.1 Co-evolutionary timeline of MAB Programme, Human-Earth Studies and Global Biodiversity Governance.

These comparative insights underscore a common challenge: the lack of a globally recognized, integrative framework that can quantitatively reflect the complex feedbacks within human–earth systems (Mouillot et al., 2024). Future research must therefore focus on constructing models that integrate ecological, social, and institutional dimensions, enabling dynamic evaluation across multiple scales.

Methodologically, the evaluation of human–earth relations should aim for comprehensive coverage that encompasses ecological integrity, socioeconomic welfare, and governance processes, while enabling data comparability and policy applicability. The MAB network, with its interdisciplinary foundation and long-term continuity, provides an unparalleled platform for testing such integrated approaches.

Through collaboration with advances in remote sensing, big data analytics, and machine learning, the MAB Programme can help develop a global-scale, data-driven system for evaluating the coupled dynamics between human societies and the biosphere (Vaz et al., 2020). This will strengthen the scientific underpinnings of global biodiversity governance and advance the transition toward truly integrated sustainability assessment.

2.3 DEFICIENCIES AND NEW OPPORTUNITIES

2.3.1 Critiques on the Current Human–Earth Relationship Evaluation System

2.3.1.1 The One-Dimensionality of Existing Theoretical Frameworks

Overall, current assessment indicators can be broadly categorized into the following types:

- Pressure/driver indicators, such as greenhouse gas emissions, rates of land-use change, and per capita resource consumption;
- State indicators, including ecosystem extent, species richness, and habitat integrity;
- Impact/service indicators, such as quantified ecosystem services (provisioning, regulating, and cultural services) and their contributions to food security, water, and health; and
- Response/governance indicators, including protected-area coverage, governance participation, financial inflows, and policy effectiveness.

These categories reveal that contemporary global environmental assessments predominantly adopt a human-impact-centric logic. A typical example is the United Nations Environment Programme's

Global Environment Outlook (GEO) reports, which primarily gauge human-induced pressures through indicators such as carbon emissions, land-use change, and pollution loads. The pressure–state–response (PSR) model is advantageous for its simplicity and clarity. Yet, its perspective overlooks the reciprocal contributions and regulatory functions of the Earth system to human well-being. It also fails to consider positive influences arising from human actions—including Indigenous Peoples' and local communities' traditional knowledge and practices—that may strengthen ecosystem resilience and enhance ecosystem-service functions. Although substantial empirical research has demonstrated that, under appropriate institutional arrangements, resource users can self-organize to establish effective governance rules and achieve sustainable resource use; such insights remain insufficiently reflected in mainstream assessment schemes (Berkes et al., 2003).

In the field of biodiversity governance, for example, CBD national reports typically focus on indicators such as species loss rates and the proportion of protected areas. Although the sustainable use of biodiversity—including the positive socioeconomic contributions of ecosystem services—is one of the three core objectives of the Convention, the development of corresponding assessment systems has progressed much more slowly. The monitoring framework of the KMGBF introduces reference indicators related to natural capital accounting, but these remain exploratory. Human–nature relations are thus often simplified into a unidirectional structure of "negative intervention → passive protection," rather than being conceptualized as a complex system of bidirectional interactions and co-evolution.

Similarly, although UNESCO's MAB Programme and its "three-zone" model emphasize human–nature coexistence, its assessment practices continue to prioritize ecological monitoring and scientific outputs. Quantitative analysis of "positive indicators"—such as social well-being, cultural values, and governance innovation—remains limited (Ellis et al., 2025). This imbalance hampers the ability of evaluation systems to capture the positive potential and cooperative value of human–nature relationships (Pritchard et al., 2020), and risks steering policy toward crisis response, restoration, and compensation rather than prevention, incentives, and co-benefit management.

2.3.1.2 Comparability across Evaluation Frameworks

By comparing existing frameworks, it is noted that no integrated assessment system has yet been established that is both capable of capturing the complexity of coupled human–environment systems and suitable for quantitative comparison.

Evaluation Frameworks	Key Components	Characterization of Human–Nature Relations	Limitations
SDG Indicator Framework (United Nations, 2017)	Seventeen goals and 169 indicators covering ecological, economic, and social dimensions	Primarily *outcome-based* indicators, e.g., forest cover, air quality index	Lacks process-based and system-level indicators.
Environmental Performance Index (EPI) (Block et al., 2024)	Countries' air and water quality, biodiversity, and climate policy	Emphasizes environmental health and ecosystem vitality (Wendling et al., 2022)	Uneven data quality, limited social context integration.
Environmental-Economic Accounting—Ecosystem Accounting (SEEA-EA) (United Nations, 2021)	Ecosystem extent, condition, and the flow of ecosystem services to the economy	Highlights the economy–ecological interactions	Limited global coverage due to high data and technical requirements. Not capturing non-market and cultural values.
Inclusive Wealth Index (IWI) (UNU-IHDP; UNEP, 2014)	Produced, human, and natural capital	Aggregating produced, human, and natural capital to assess sustainability (Arrow et al., 2012)	Data gaps, assumptions in valuing natural and human capital, and challenges in capturing non-market and ecosystem service values.

(Continued)

(Continued)

Evaluation Frameworks	Key Components	Characterization of Human–Nature Relations	Limitations
IPBES Conceptual Framework (IPBES, 2020)	Nature—Nature's Contributions to People—Human well-being	Emphasizes value pluralism and co-production	Only evaluates global and regional scope, long timeframe.
The Living Planet Index (LPI) (WWF, 2024)	Long-term population trends of species, impact of human activities	Highlights the human pressure on biodiversity	Does not reflect ecological health and the complexity of human–environment relationships (Puurtinen et al., 2021; Toszogyova et al., 2024).
MAB Three-Zone Model	Core, buffer, and transition zones	Emphasizes spatial hierarchy and human participation (Bridgewater, 2016)	Lack of integrated indicators and a reporting mechanism

Despite the recent proliferation of indicators and evaluation frameworks related to biodiversity and human–nature relations, the coexistence of multiple, largely uncoordinated frameworks—each relying on different data definitions, temporal scales, and assessment models—has resulted in overlapping indicators and non-comparable datasets, as well as limited explanatory power. Coordination mechanisms among global environmental conventions have long faced challenges associated with fragmented governance. Biodiversity-related conventions are typically administered by different secretariats, data systems, and reporting cycles, resulting in policy duplication and inefficient resource utilisation. Although the MAB Programme has practical advantages as a cross-sectoral initiative, it has not yet established formal coordination mechanisms with these conventions, and its accumulated experience is therefore difficult to translate into global policy inputs.

Moreover, even within a single platform or within assessments conducted at the same spatial scale, issues of data integration persist. Within MAB itself, for instance, although the Programme encompasses a network of biosphere reserves across more than 140 countries, monitoring indicators, technical standards, and reporting formats vary significantly among participating states, with no unified data architecture in place. This heterogeneity constrains the potential for global-scale comparative analysis and prevents the substantial data resources within MAB from being fully transformed into scientific knowledge or policy influence.

2.3.1.3 Deficiency on Balance and Inclusiveness

Global scientific assessments of human–environment relationships require continuous, multi-temporal, and cross-scale datasets. In practice, however, data coverage across regions is highly uneven. Many developing countries, constrained by limited capacity for observation, data collection, and data application, have long remained in a state of "data marginalization" (Turnhout et al., 2016) within global databases and related assessment processes.

Furthermore, within existing assessment frameworks, human participation, cultural perspectives, and social knowledge systems are often reduced to mere "governance tools" rather than treated as core system variables. This limitation contributes to a tendency toward technocratic governance in global environmental decision-making, as scientific models and quantitative indicators dominate. At the same time, cultural ecosystem services and local knowledge are difficult to quantify or integrate into mainstream indicator systems. As a result, the knowledge systems and value judgments of local communities are frequently marginalized. This not only weakens the social legitimacy of policies but also undermines the cultural diversity that is essential for sustainable governance.

2.3.1.4 Insufficient Integration of System Feedback and Nonlinear Dynamics

Most existing assessment systems rely on linear causal logics—such as input–output models—while real human–environment systems exhibit nonlinear dynamics, threshold effects, and multi-level feedbacks. Traditional indicators often fail to capture sudden shifts in ecology or institutions. For example, long-term incremental changes in forest cover may eventually trigger abrupt failures in carbon-cycle regulation, a nonlinear process that requires dynamic modeling and continuous monitoring rather than static indicators (Scheffer et al., 2001).

2.3.2 Advances in Research and Practice

In recent years, academic research and local initiatives have undertaken valuable explorations into ecosystem services evaluation and accounting (Ouyang & Zheng, 2020), institutional resilience, and civic participation, generating innovative indicator sets and methodological tools. However, most of these advances remain at an experimental or pilot stage and have yet to be systematically integrated into international governance mechanisms.

Emerging technologies—including artificial intelligence (AI) (Pollock et al., 2025), big data analytics (Trantas et al., 2025), remote sensing, and the Internet of Things—offer new means for filling data gaps and enabling continuous, cross-scale monitoring. For example, high-resolution satellite imagery combined with machine-learning techniques can support large-scale monitoring of land cover, forest dynamics (Ma, 2024), and marine biomass. Social media data and mobile phone records can be used to detect patterns of human activity and track the use of cultural ecosystem services. Sensor networks such as bioacoustic monitors, along with long-term ecological research stations, can generate time-series data that captures ecological processes more comprehensively.

Particularly noteworthy is the development of social–ecological digital twin models (van den Homberg et al., 2024), which integrate multi-source remote-sensing data, ground-based sensors, economic activity datasets, and social information to construct multi-scale dynamic representations of human–environment systems (Li et al., 2023). Such models can be used for policy scenario simulation and real-time early warning—for instance, to conduct synchronized monitoring and decision experiments on carbon, water, and livelihoods in watersheds or biosphere reserves, or to evaluate the actual effectiveness of nature-based solutions (NbS) in enhancing carbon sequestration, livelihoods, and resilience. These approaches are opening up new frontiers for research on human–nature interactions.

There is little doubt that AI possesses strong potential for data processing, trend prediction, and scenario modeling, and can support the development of dynamic and coupled evaluation systems for human–environment relations. Yet technology alone is not a panacea: data quality, political and privacy concerns, algorithmic transparency and fairness, and limited capacity in many developing countries all constitute significant challenges (Tao et al., 2024). Effective technological solutions must therefore be advanced in parallel with governance arrangements, capacity-building efforts, and mechanisms for knowledge sharing, in order to achieve meaningful application on a global scale.

2.4 FUTURE RECOMMENDATIONS: BUILDING AN INTEGRATED ASSESSMENT SYSTEM FOR HUMAN–EARTH RELATIONS FOR GLOBAL GOVERNANCE

Over the past 50 years, scientific understanding and governance practices of human–Earth relations have made significant progress, evolving from single-focus ecological conservation to integrated social–ecological system governance, and from local actions to global governance frameworks. UNESCO's MAB Programme, as an interdisciplinary and cross-scale practical network, provides an institutionalized platform and long-term monitoring resources that have supported this evolution. However, current assessment systems still face issues of fragmentation, unidimensionality, and data gaps, which hinder the translation of global objectives into coordinated policy and transformative societal action. Recent technological revolutions offer new possibilities for integrating assessment systems of human–Earth relations, and it is recommended to actively explore the following directions using emerging tools such as artificial intelligence:

- **Develop a "positive indicators" system via the MAB Programme**: Future human–Earth relation assessment systems should not only measure "degrees of degradation" but also the "degree of co-benefit," shifting focus from "damage to planetary health" toward "enhancement of both the Earth and human well-being." Indicators could include community engagement in ecological stewardship, perceptions of ecosystem services, and biocultural diversity indices, providing a comprehensive reflection of the transition from passive governance to active co-existence, and giving practical force to the concept of "harmony between humans and nature."
- **Construct a "Human–Earth Harmony Index" based on ecosystem units (e.g., biosphere reserves) rather than administrative units**: This approach fully captures the contribution of intact ecosystems to local climate resilience and ecological security, as well as changes driven by social behaviors, thereby motivating local communities to participate effectively in integrated conservation and development governance. Respecting natural systems and emphasizing regional and landscape scales, rather than mechanically nested administrative scales, improves scientific rigor and comparability of indicators, and enables the creation of long-term, multi-site comparative networks to standardize metrics across diverse social–ecological contexts and evaluate governance innovations.
- **Strengthen data development and global coordination**: Since 2020, there has been a trend toward cross-mechanism data collaboration within the UN system, including UNDP's SDG Integration Hub, UNEP's World Environment Situation Room, and UNESCO's Sites Navigator for World Heritage Sites, Biosphere Reserves, and Geoparks. Future efforts should aim to develop a Global Human–Earth Relations Integrated Assessment Platform, enabling interactive indicators and cross-sectoral data sharing.
- **Pilot intelligent governance in biosphere reserves**: Transitioning from "monitoring" to "governance intelligence" and from "ex-post assessment" to "real-time warning and adaptive management" can be facilitated through the integration of AI and data science. Predictive management could proactively identify ecological risk areas; scenario simulation and policy testing could evaluate systemic feedback of different policy options; and participatory intelligent platforms could allow local communities to contribute data and feedback to governance processes. Intelligent transformation can improve decision-making efficiency, enhance inclusivity and sustainability of governance, and provide scientific and data-based support for green finance and nature-positive investment.

DECLARATION OF GENERATIVE AI USE

During the preparation of this manuscript, the author used AI tools (ChatGPT and Deepseek) for the English grammar check and reference formality checks. The author subsequently reviewed and edited the content and takes full responsibility for the final version of the published chapter.

REFERENCES

Arrow, K., Dasgupta, P., Goulder, L., Mumford, K. J., Oleson, K. Sustainability and the Measurement of Wealth. *Environment and Development Economics*, 2012, 17(3), 317–353. https://doi.org/10.1017/S1355770X12000113

Berkes, F., Colding, J., Folke, C., Eds. *Linking Social and Ecological Systems: Management Practices and Social Mechanisms for Building Resilience*. Cambridge University Press: Cambridge, 1998.

Berkes, F., Colding, J., Folke, C. *Navigating Social–Ecological Systems*. Cambridge University Press: Cambridge, 2003.

Biermann, F. The future of 'environmental' policy in the Anthropocene. In *Trajectories in Environmental Politics*; Hayes, G., Jinnah, S., Kashwan, P., Eds.; Routledge, 2022. https://doi.org/10.4324/9781003213321-4

Bjork-James, C., Checker, M., Edelman, M. Transnational Social Movements: Environmentalist, Indigenous, and Agrarian Visions for Planetary Futures. *Annual Review of Environment and Resources*, 2022, 47, 583–608. https://doi.org/10.1146/annurev-environ-112320-084822

Bliege Bird, R., Nimmo, D. Restore the Lost Ecological Functions of People. *Nature Ecology & Evolution*, 2018, 2, 1050–1052. https://doi.org/10.1038/s41559-018-0585-4

Block, S., Emerson, J. W., Esty, D. C., de Sherbinin, A., Wendling, Z. A. *2024 Environmental Performance Index*. Yale Center for Environmental Law & Policy: New Haven, CT, 2024.

Bridgewater, P. The Man and Biosphere Programme of UNESCO: Rambunctious Child of the Sixties, but Was the Promise Fulfilled? *Current Opinion in Environmental Sustainability*, 2016, 19, 1–6. https://doi.org/10.1016/j.cosust.2016.01.001

Brondízio, E. S., Aumeeruddy-Thomas, Y., Bates, P., Carino, J., Fernández-Llamazares, Á., Ferrari, M. F., et al. Locally Based, Regionally Manifested, and Globally Relevant: Indigenous and Local Knowledge, Values, and Practices for Nature. *Annual Review of Environment and Resources*, 2021, 46, 481–509. https://doi.org/10.1146/annurev-environ-012220-012127

CBD. *Monitoring Framework for the Kunming-Montreal Global Biodiversity Framework*. CBD Secretariat: Montreal, 2022.

CBD. *Mechanisms for Planning, Monitoring, Reporting and Review, Including the Global Review of Collective Progress in the Implementation of the Kunming-Montreal Global Biodiversity Framework to be Conducted at the Seventeenth and Nineteenth Meetings of the Conference of the Parties*. CBD Secretariat: Montreal, 2024.

Costanza, R., d'Arge, R., de Groot, R., Farber, S., Grasso, M., Hannon, B., et al. The Value of the World's Ecosystem Services and Natural Capital. *Nature*, 1997, 387, 253–260. https://doi.org/10.1038/387253a0

de Groot, R. S., Alkemade, R., Braat, L., Hein, L., Willemen, L. Challenges in Integrating the Concept of Ecosystem Services and Values in Landscape Planning, Management and Decision Making. *Ecological Complexity*, 2010, 7(3), 260–272. https://doi.org/10.1016/j.ecocom.2009.10.006

Díaz, S., Settele, J., Brondízio, E. S., Ngo, H. T., Guèze, M., Agard, J., et al. Pervasive Human-Driven Decline of Life on Earth Points to the Need for Transformative Change. *Science*, 2019, 366, eaax3100. https://doi.org/10.1126/science.aax3100

Ellis, E. C., Malhi, Y., Ritchie, H., Montana, J., Díaz, S., Obura, D., et al. An Aspirational Approach to Planetary Futures. *Nature*, 2025, 642, 889–899. https://doi.org/10.1038/s41586-025-09080-1

Escobar-Pemberthy, N., Ivanova, M. Implementation of Multilateral Environmental Agreements: Rationale and Design of the Environmental Conventions Index. *Sustainability. Global Environmental Policy and Governance in Sustainability*, 2020, 12(17), 7098. https://doi.org/10.3390/su12177098

Folke, C. Resilience: The Emergence of a Perspective for Social–Ecological Systems Analyses. *Global Environmental Change*, 2006, 16(3), 253–267. https://doi.org/10.1016/j.gloenvcha.2006.04.002

Gunderson, L.H., Holling, C.S. (Eds.) *Panarchy: Understanding Transformations in Human and Natural Systems* (pp. 25–62). Island Press: Washington, DC, 2002.

Holling, C. S. Resilience and Stability of Ecological Systems. *Annual Review of Ecology and Systematics*, 1973, 4, 1–23. https://doi.org/10.1146/annurev.es.04.110173.000245

Holling, C. S. Understanding the Complexity of Economic, Ecological, and Social Systems. *Ecosystems*, 2001, 4, 390–405. https://doi.org/10.1007/s10021-001-0101-5

IPBES. *Global Assessment Report on Biodiversity and Ecosystem Services*. IPBES: Bonn, 2019. https://doi.org/10.5281/zenodo.3831673

IPBES Multidisciplinary Expert Panel. *Information note on applying "nature's contributions to people"*. 2020. https://files.ipbes.net/ipbes-web-prod-public-files/inline-files/ipbes_mep_note%20on%20NCP%20by%20MEP.pdf

Ishwaran, N., Persic, A., Tri, N. H. Concept and Practice: The Case of UNESCO Biosphere Reserves. *International Journal of Environment and Sustainable Development*, 2008, 7(2), 118–131. https://doi.org/10.1504/IJESD.2008.018358

Juffe-Bignoli, D., Burgess, N. D., Bingham, H., Belle, E. M. S., de Lima, M. G., Deguignet, M., et al. *Protected Planet Report 2014: Tracking Progress Towards Global Targets for Protected Areas*. UNEP-WCMC: Cambridge, 2014.

Li, X., Feng, M., Ran, Y., Su, J., Feng, L., Huang, C., et al. Big Data in Earth System Science and Progress towards a Digital Twin. *Nature Reviews Earth & Environment*, 2023, 4, 319–336. https://doi.org/10.1038/s43017-023-00409-w

Liu, J., Dietz, T., Carpenter, S. R., Alberti, M., Folke, C., Moran, E., et al. Complexity of Coupled Human and Natural Systems. *Science*, 2007, 317(5844), 1513–1516. https://doi.org/10.1126/science.1144004

Liu, J., Hull, V., Batistella, M., DeFries, R., Dietz, T., Fu, F., et al. Framing Sustainability in a Telecoupled World. *Ecology and Society*, 2013, 18(2), 26. https://doi.org/10.5751/ES-05873-180226

Ma, K. The Chinese Forest Biodiversity Monitoring Network (CForBio) Plays a Significant Role in Advancing Chinese Ecology onto the Global Stage. *Biodiversity Science*, 2024, 32(12), 25039. https://doi.org/10.17520/biods.2025039

Mayer, F. S., Frantz, C. M. The Connectedness to Nature Scale. *Journal of Environmental Psychology*, 2004, 24(4), 503–515. https://doi.org/10.1016/j.jenvp.2004.10.001

Millennium Ecosystem Assessment (MA). *Ecosystems and Human Well-being: Synthesis*. Island Press: Washington, DC, 2005.

Mouillot, D., Velez, L., Albouy, C., Casajus, N., Claudet, J., Delbar, V., et al. The Socioeconomic and Environmental Niche of Protected Areas Reveals Global Conservation Gaps and Opportunities. *Nature Communications*, 2024, 15(1), 9007.

Odum, E. P. *Systems Ecology: An Introduction*. Wiley: New York, 1983.

Ostrom, E. *Understanding Institutional Diversity*. Princeton University Press: Princeton, 2005.

Ostrom, E. A Diagnostic Approach for Going Beyond Panaceas. *Proceedings of the National Academy of Sciences*, 2007, 104(39), 15181–15187. https://doi.org/10.1073/pnas.0702288104

Ouyang, Z., Zheng, H. Research Progress and Perspectives of Ecosystem Services Assessment. *Research of Environmental Sciences*, 2020, 33(5), 1197–1209.

Pascual, U., Balvanera, P., Díaz, S., Pataki, G., Roth, E., Stenseke, M., et al. Diverse Values of Nature for Sustainability. *Nature*, 2023, 620, 813–823. https://doi.org/10.1038/s41586-023-06406-9

Pollock, L. J., Kitzes, J., Beery, S., Gaynor, K. M., Jarzyna, M. A., Aodha, O. M., et al. Harnessing Artificial Intelligence to Fill Global Shortfalls in Biodiversity Knowledge. *Nature Reviews Biodiversity*, 2025, 1(3), 166–182. https://doi.org/10.1038/s44358-025-00022-3

Pritchard, A., Richardson, M., Sheffield, D., McEwan, K. The Relationship between Nature Connectedness and Eudaimonic Well-Being: A Meta-Analysis. *Journal of Happiness Studies*, 2020, 21, 1145–1167. https://doi.org/10.1007/s10902-019-00118-6

Puurtinen, M., Elo, M., Kotiaho, J.S., The Living Planet Index Does Not Measure Abundance. *Nature*, 2021, 598, 313–316. https://doi.org/10.1038/s41586-021-03708-8

Reed, M. G., Price, M. F. *UNESCO Biosphere Reserves: Supporting Biocultural Diversity, Sustainability and Society*. Routledge: Abingdon, 2020.

Reyers, B., Folke, C., Moore, M.-L., Biggs, R., Galaz, V. Social-Ecological Systems Insights for Navigating the Dynamics of the Anthropocene. *Annual Review of Environment and Resources*, 2018, 43, 267–289. https://doi.org/10.1146/annurev-environ-110615-085349

Richardson, K., Steffen, W., Lucht, W., Bendtsen, J., Cornell, S. E., Donges, J. F., et al. Earth Beyond Six of Nine Planetary Boundaries. *Science Advances*, 2023, 9, eadh2458. https://doi.org/10.1126/sciadv.adh2458

Rockström, J., Steffen, W., Noone, K., Persson, A., Stuart Chapin III, F., Lambin, E, F., et al. A Safe Operating Space for Humanity. *Nature*, 2009, 461, 472–475. https://doi.org/10.1038/461472a

Scheffer, M., Carpenter, S., Foley, J. A., Folke, C., Walker, B. Catastrophic Shifts in Ecosystems. *Nature*, 2001, 413(6856), 591–596. https://doi.org/10.1038/35098000

Schellnhuber, H.-J. Discourse: Earth System Analysis: The Scope of the Challenge. In *Earth System Analysis: Integrating Science for Sustainability* (pp. 3–195); Schellnhuber, H.-J., Wenzel, V., Eds.; Springer: Heidelberg, 1998.

Scholes, R. J., Biggs, R. A Biodiversity Intactness Index. *Nature*, 2005, 434, 45–49. https://doi.org/10.1038/nature03289

Steffen, W., Richardson, K., Rockström, J., Cornell, S. E., Fetzer, I., Bennett, E. M., et al. Planetary Boundaries: Guiding Human Development on a Changing Planet. *Science*, 2015, 347, 1259855. https://doi.org/10.1126/science.1259855

Steffen, W., Richardson, K., Rockström, J., Schellnhuber, H. J., Dube, O. P., Dutreuil, S., et al. The Emergence and Evolution of Earth System Science. *Nature Reviews Earth & Environment*, 2020, 1, 54–63. https://doi.org/10.1038/s43017-019-0005-6

Tao, F., Qi, Q., Liu, A., Kusiak, A. Responsible Artificial Intelligence Governance: A Review and Future Directions. *Journal of Cleaner Production*, 2024, 436, 143220. https://doi.org/10.1016/j.jclepro.2024.143220

Toszogyova, A., Smyčka, J., Storch, D. Mathematical Biases in the Calculation of the Living Planet Index Lead to Overestimation of Vertebrate Population Decline. *Nature Communications*, 2024, 15, 5295. https://doi.org/10.1038/s41467-024-49070-x

Trantas, A., Mensio, M., Stasinos, S., Gribincea, S., Khan, T., Podareanu, D., et al. *BioAnalyst: A Foundation Model for Biodiversity*. arXiv 2507.09080, 2025.

Turnhout, E., Dewulf, A., Hulme, M. What Does Policy-Relevant Global Environmental Knowledge Do? The Cases of Climate and Biodiversity. *Current Opinion in Environmental Sustainability*, 2016, 18, 65–72. https://doi.org/10.1016/j.cosust.2015.09.004

United Nations. *Resolution Adopted by the General Assembly on Work of the Statistical Commission Pertaining to the 2030 Agenda for Sustainable Development (A/RES/71/313), Annex*, 2017. https://docs.un.org/en/A/RES/71/313

United Nations. *System of Environmental-Economic Accounting: Ecosystem Accounting (SEEA EA), Experimental Ecosystem Accounting*. United Nations: New York, 2021.

UNU-IHDP; UNEP. *Inclusive Wealth Report 2014: Measuring Progress toward Sustainability.* Cambridge University Press: Cambridge, 2014.

vanden Homberg, W., Aerts, J. P. M., Bauer, P., Bierkens, M. F. P., Camps-Valls, G., Dekkar, M., et al. Digital Twins of the Earth with and for Humans. *Nature,* 2024. https://doi.org/10.1038/s43247-024-01626-x

Vaz, A. S., Moreno-Llorca, R. A., Gonçalves, J. F., Vicente, J. R., Méndez, P. F., Revilla, E., et al. Digital Conservation in Biosphere Reserves: Earth Observations, Social Media, and Nature's Cultural Contributions to People. *Conservation Letters*, 2020, 13(3), e12704. https://doi.org/10.1111/conl.12704

Walters, C. J. *Adaptive Management of Renewable Resources*. Macmillan: New York, 1986.

Wendling, Z. A., Emerson, J. W., de Sherbinin, A., Esty, D. C., Srebotnjak, T. *2022 Environmental Performance Index*. Yale Center for Environmental Law & Policy: New Haven, 2022.

World Commission on Environment and Development (WCED). *Our Common Future.* Oxford University Press: Oxford, 1987.

WWF. *Living Planet Report 2024: A System in Peril*. WWF: Gland, 2024.

3 Empowering Indigenous Peoples and Local Communities in Environmental Management

Toward a Framework of Knowledge Co-Production

Su Tina Bai

3.1 INTRODUCTION

Residing in some of the Earth's most biologically diverse and vulnerable regions, Indigenous Peoples and local communities (IPLCs) play crucial roles in environmental management. Rather than viewing IPLCs as mere victims of environmental degradation, recent academic literature and policy reports increasingly acknowledge that IPLCs possess leadership capabilities to mitigate and adapt to environmental change. This chapter analyzes how IPLCs can effectively contribute to sustainable environmental management and assesses how international environmental institutions, such as the United Nations Framework Convention on Climate Change (UNFCCC) and the Intergovernmental Science-Policy Platform on Biodiversity and Ecosystem Services (IPBES), have meaningfully engaged with IPLCs. The chapter argues that Co-production of Knowledge (CPK) between Indigenous and scientific knowledge systems offers demonstrable benefits for local environmental management; however, meaningful IPLC engagement within international environmental regimes through the CPK framework remains constrained by institutional, procedural, and linguistic barriers.

The chapter proceeds as follows. Section 3.1 outlines the theoretical framework of the CPK to illustrate how Indigenous knowledge and governing systems can complement the scientific community in informing environmental management. Section 3.2 draws on case studies from the Arctic and forest management to demonstrate how IPLCs enrich ecological data, promote sustainable practices, and contribute to a collective management system that enhances environmental outcomes. Section 3.3 examines publications from the UNFCCC and the IPBES, as well as secondary literature and my fieldwork experience in the 2024 COP29 Baku, to illustrate the progress and shortcomings of international environmental institutions in engaging with IPLCs. It discusses potential ways forward to overcome institutional, procedural, and language barriers in applying the CPK to encourage more meaningful IPLC engagement in international environmental regimes.

3.2 KNOWLEDGE CO-PRODUCTION BETWEEN SCIENTISTS AND IPLCs

Indigenous and scientific knowledge appear to represent fundamentally different epistemologies; nevertheless, closer scrutiny reveals that they share similarities in how they generate and validate knowledge and can complement each other to enhance knowledge production. The United Nations

DOI: 10.1201/9781003640899-4

defines Indigenous Peoples as those who have a historical and communal attachment to land, or who self-identify as Indigenous (Ijatuyi et al., 2025). Indigenous and local knowledge (ILK) is acquired through observation, experimentation, and adaptation over centuries of lived experience. These communities orally pass down knowledge from generation to generation through songs, stories, and rituals. ILK is holistic, which means that environmental practices are inseparable from cultural, ethical, and spiritual values that define human-nature interactions. For example, IPLCs often maintain sacred sites, where human activity is restricted, for spiritual rather than environmental conservation purposes (Artelle et al., 2019). Considering that Indigenous knowledge and practices are intergenerational, Indigenous practices are often inherently "sustainable", even before this term entered global environmental discourse.

In contrast, Western scientific knowledge adopts a positivist epistemology that emphasizes objectivity, quantification, and causal explanation (Parks and Tsioumani, 2023). Science is conventionally perceived as holding universal truth because it claims to produce empirically verifiable and replicable knowledge. The four Scientific Ethos—universalism, communism, disinterestedness, and skepticism—Indicate how science is supposed to reflect factual reality rather than normative beliefs (Grundmann, 2013). Scientific knowledge relies on observation, experimentation, and interactive learning, which is indeed similar to the IKL. Nevertheless, scientific knowledge is spatially broad but temporally narrow, whereas the ILK—given its intergenerational characteristics—is spatially shallow but temporally broad. Despite rapid advances in modern science technologies, many aspects of the environment remain unknown to scientists, such as the consequences of global warming on ice thickness and biodiversity in the Himalayan and Arctic regions (Krishnan et al., 2019; Moore and Hauser, 2019). From this perspective, Indigenous and scientific knowledge have the potential to complement each other to generate environmental knowledge.

Contemporary society is embedded with a knowledge hierarchy that privileges the Western scientific way of knowing (Wiegleb and Bruns, 2025). In environmental policy-making, the orthodox view suggests that Western science alone provides legitimate knowledge because of its supposed capacity to generate "objective and universal truths". Indeed, scientific knowledge has profoundly shaped the trajectory of global environmental governance since the 1970s. International regimes such as the Intergovernmental Panel on Climate Change (IPCC) and the IPBES rely heavily on scientific expertise to define problems, set agendas, and legitimize policy action (Ryghaug and Skjølsvold, 2010). For example, the Hockey Stick Graph and the *Fourth IPCC Assessment Report* frame climate change as anthropogenic, resulting from the Industrial Revolution, and thereby call for urgent global collective action. Climate targets, such as limiting global warming to well below 2°C, are directly informed by climate models in IPCC assessments (Grundmann, 2013).

The field of International Relations recognizes that scientific knowledge and institutions exercise certain political power in shaping the global environmental agenda. Haas (1992) suggests that scientists form an epistemic community—a network of professionals with recognized expertise and authority to influence policy—in climate politics. Scientific epistemic communities can exercise what Barnett and Duvall (2005) conceptualize as discursive and institutional power: Discursively, scientific communities shape how environmental issues are framed and problematized; institutionally, environmental regimes like the UNFCCC have designated gateways for scientific bodies like the IPCC to inform policy agendas. Furthermore, Falkner (2021) suggest that environmental stewardship emerged in Anglo-American societies and then spread to the Global South. This process can foster the spread of Western environmental norms, while sometimes leaving alternative ways of knowing—especially Indigenous Knowledge—silenced, marginalized, and neglected.

Social constructivists challenge scientific knowledge's exclusive expertise in informing environmental knowledge by arguing that science represents only one among multiple ways of knowing (Latour, 2009; Jones, 2002). They suggest that knowledge is socially situated and mediated by institutional and cultural contexts. From this perspective, ILK offers an alternative and complementary way of knowing human-nature relations. Unfortunately, despite its sophistication, ILK was long dismissed as "unscientific"; in fact, IPLCs were often blamed for causing environmental degradation. For instance, the Indonesian and Malaysian governments used to scapegoat Indigenous swidden agriculture practices for deforestation, while ignoring illegal logging and migration patterns (Dauvergne, 1993). In the Himalayas, the Sherpa people's grazing practices were previously misrepresented as unsustainable land use (Krishnan et al., 2019).

Nonetheless, recent studies have demonstrated the environmental effectiveness of Indigenous management. For instance, Vergara-Asenjo and Potvin (2014) find that Indigenous land tenure is associated with lower deforestation rates than in non-Indigenous areas. Schuster et al. (2019) show that Indigenous-managed lands in Australia, Brazil, and Canada protect as many or more endangered species as formal protected areas. Likewise, Zhang et al.'s study (2023) finds that cooperation with Indigenous Peoples tends to improve both environmental and social outcomes. These findings challenge Western science's monopoly over informing environmental practices by showing that Indigenous management practices are at least equally suited.

Acknowledging the value of IPLCs' knowledge and governing systems, academics and practitioners have sought to bridge scientific and Indigenous knowledge in environmental management. Knowledge co-production is essential to address sustainability challenges where issues are often interconnected and involve multiple stakeholders (Norström et al., 2020). The CPK framework developed by Yua et al. (2022) aims to bring multiple knowledge systems into environmental management on equal grounds. The concept of "co-production" stems from Jasanoff's (2004) idea that nature and society are co-produced. Similarly, Ostrom uses the idea of "co-production" to illustrate that authorities and citizens are both involved in the maintenance of public goods (Nakashima et al., 2022). In essence, knowledge co-production proposes a participatory approach from multiple stakeholders to engage with the knowledge-policy interface. Because IPLCs reside on biologically diverse and vulnerable lands, their expertise about local environmental management practices makes valuable contributions to the knowledge co-production framework.

The CPK emphasizes the importance of equal power-sharing between scientists and IPLCs in generating knowledge. The focus on equity distinguishes CPK from other collaborative frameworks because previous attempts to collaborate with IPLCs often resulted in their exploitation in the name of scientific research (Yua et al., 2022; Wylie and Murillo, 2023). Nakashima et al. (2022) emphasize that knowledge co-production requires a decolonial perspective to view IPLCs as equal knowledge holders rather than research subjects. For this reason, the CPK requires the engagement of IPLC at all stages of the research—from research question identification, method development, data gathering, result analysis, and policy development (Yua et al., 2022, p. 34; Figure 3.1). The CPK framework should follow an iterative decision-making process to enhance mutual, inclusive participation and to build long-term partnerships (Norström et al., 2020; Moallemi et al., 2023). Additionally, it is important to seek free, prior, and informed consent (FRIC) when using Indigenous knowledge (Norström et al., 2020). Overall, CPK seeks to equally integrate diverse knowledge systems by valuing Indigenous and scientific knowledge.

3.3 CASE STUDIES OF IPLC ENVIRONMENTAL CONTRIBUTIONS

This section explores how IPLCs contribute to improved environmental management through knowledge co-production frameworks. The Arctic case illustrates how ILK improves environmental monitoring in remote landscapes. The forest-fire management case shows how co-management can produce better environmental outcomes.

The Arctic provides an informative case about knowledge co-production between scientists and IPLCs. Rapid warming and pronounced seasonal variability make the region difficult to monitor using conventional scientific methods. Nevertheless, Arctic Indigenous communities, such as the Inuit and Saami, have maintained intricate understandings of ice formation, animal migrations, and weather patterns (Moore and Hauser, 2019). These observations, transmitted through oral traditions and community practices, constitute a long-term environmental archive that extends beyond the temporal scope of modern scientific data. Here, IKL crucially complements modern science to

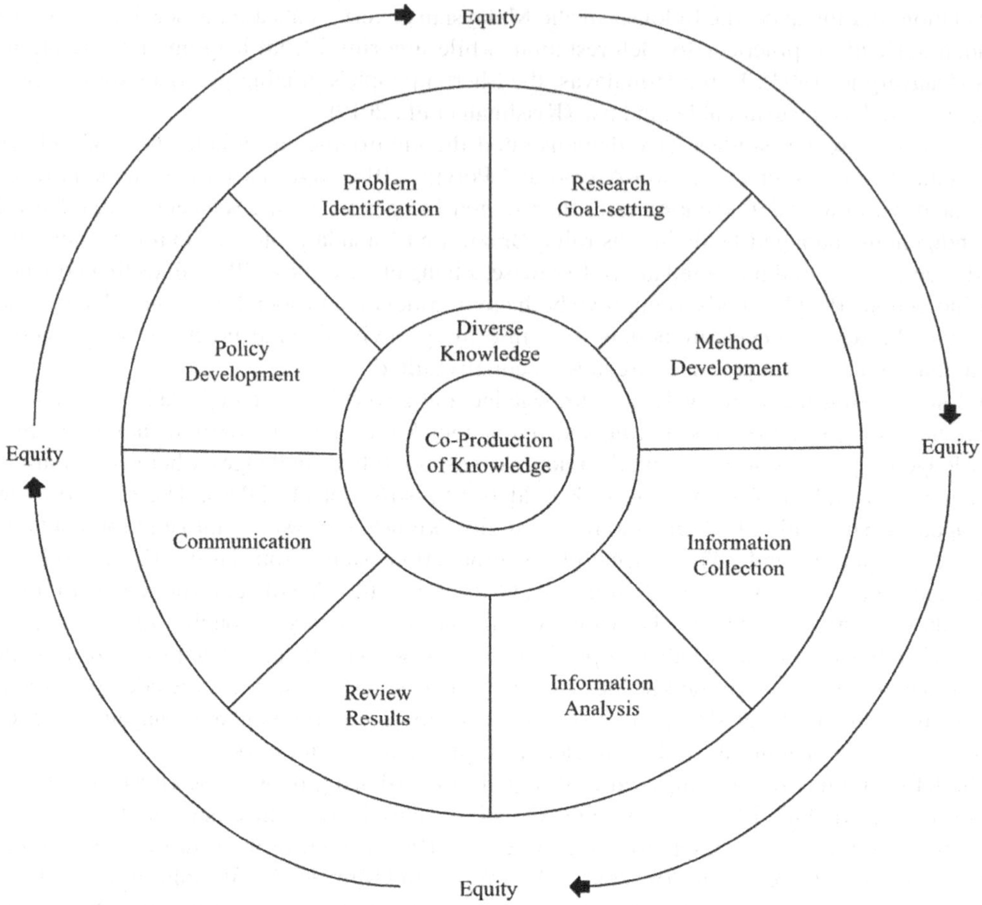

FIGURE 3.1 Ideal process of the co-production of knowledge. (Diagram adapted from Yua et al., 2022, p. 33.)

monitor environmental change. For example, the Siku-Inuit-Hila Project and Exchange for Local Observations and Knowledge of the Arctic encourage Indigenous hunters and elders to collaborate with scientists to document and interpret changes in sea ice. The Skolt Sámi community in Finland contributes to biodiversity data on Atlantic salmon spawning cycles (Mustonen and Feodoroff, 2018, p. 114). These partnerships enrich data on climate change and biodiversity by integrating continuous local observations with satellite and remote-sensing data.

The Climate Impact on Terrestrial Environments (CITE) Initiative demonstrates deeper collaboration between the Arctic Monitoring and Assessment Programme and the Saami Council in assessing climate impacts across Sápmi (Saami Council, 2025). CITE places the CPK framework at its core by enabling Indigenous Peoples to participate directly in setting research agendas. CITE also relies on multilingual communication so that Indigenous Peoples can describe conditions of "snow" more precisely in their native language. By incorporating Indigenous knowledge that documents human–environment interactions, the project enriches GIS datasets that often treat landscapes as separate from human activity. The project has developed a digital tool, the CITE Maptamaire, to map reindeer herders' observations of landscape and seasonal change. The integration of these distinct knowledge systems supports more informed decision-making on biodiversity conservation and local adaptation to environmental change. This project not merely engaged with IPLCs but, more importantly, positioned Indigenous experts as co-authors, decision-makers, and communicators of environmental change.

Beyond the Arctic, Indigenous forest management practices offer further evidence of the value of engaging with IPLCs' knowledge and governing systems. Western scientific paradigms often dismiss the use of fire in forests; however, growing evidence demonstrates that traditional burning practices indeed play vital roles in ecosystem maintenance. Fire suppression has contributed to ecological disturbances such as late-summer wildfires and pest outbreaks (Russell-Smith et al., 2013). Australia provides a notable example of how revitalizing traditional practices through co-management among Indigenous communities, scientists, and government agencies has improved forest management.

In northern Australia, forced migration of Indigenous Peoples following colonization and urbanization led to the decline of Indigenous land management practices, disrupting local ecological balance. Suppressing traditional fire use, such as early-season burning, led to the accumulation of dry fuel and the emergence of highly destructive late-season wildfires that generate substantial greenhouse gas emissions and habitat loss (Russell-Smith et al., 2013). The West Arnhem Land Fire Abatement (WALFA) project represents a successful model of Indigenous–scientific co-governance. WALFA reintroduces early-season burning practices informed by traditional knowledge, and brings together Indigenous landowners, scientists, energy companies, and government agencies to co-manage the fire application. Scientists helped design geospatial mapping and emissions modeling, while Indigenous fire practitioners used traditional techniques to conduct controlled burning practices (Heckbert et al., 2011). Between 2005 and 2011, the proportion of land burned in the early dry season rose from 7.6% to 20.9%, while late-season fires declined from 32% to 10.9% (Russell-Smith et al., 2013). This shift reduced carbon emissions by approximately 147,000 metric tons annually and supported ecological recovery through vegetation regrowth and species preservation (Robinson et al., 2016). Beyond environmental benefits, WALFA revitalized cultural practices and generated new employment opportunities for IPLCs.

WALFA illustrates how environmental co-management with Indigenous Peoples enhances environmental outcomes. Rather than limiting Indigenous actors to mere consultative roles, the project established governance mechanisms and long-term partnerships between scientists and traditional knowledge holders. The success of this governing structure exemplifies Elinor Ostrom's (2015) theory of common-pool resource management, which suggests that environmental goods are better governed through collective community management rather than top-down approaches, as the latter often leads to sanctioning errors when policymakers lack local knowledge. Engagement with IPLCs creates bottom-up governing systems that enhance environmental management outcomes in local contexts.

The Arctic and Australian case studies underscore how knowledge co-production frameworks between IPLCs and the scientific communities enrich environmental monitoring and management practices. However, these local successes contrast with the challenges of implementing the CPK within the rigid architectures of international environmental regimes.

3.4 ASSESSING IPLC ENGAGEMENT IN ENVIRONMENTAL REGIMES

This section examines how influential international climate and biodiversity regimes have engaged with IPLCs. It focuses first on the UNFCCC and its Local Communities and Indigenous Peoples Platform (LCIPP), followed by a comparative reflection on the IPBES. The analysis reveals both progress and institutional shortcomings in encouraging IPLCs' participation in global environmental governance.

The UNFCCC, established in 1992, was relatively slow to recognize the contributions of IPLCs. For many years, the IPCC—a scientific body under the United Nations—dominated the Convention's agenda-setting and policy discussions. While the 2015 Paris Agreement began to acknowledge the importance of ILK systems for climate adaptation and mitigation, formal institutional mechanisms remained limited until the establishment of the Local Communities and Indigenous Peoples Platform (LCIPP) in the 2018 Katowice Climate Agreement.

The LCIPP (2025) was created to "strengthen knowledge, technologies, practices, and efforts of local communities and Indigenous Peoples related to addressing and responding to climate change". It operates under a 3-year work plan organized around three functions: knowledge, capacity for engagement, and policies and actions. The Platform is divided into seven UN sociocultural regions (Africa, Asia, Central and Eastern Europe, Central and South America, the Arctic, North America, and the Pacific) and governed by a Facilitative Working Group (FWG) composed of 14 members, with regional groups nominating their own representatives. In practice, the LCIPP convenes annual gatherings, knowledge-holder dialogues, youth roundtables, and FWG meetings alongside the annual Conferences of the Parties. It also manages a public knowledge platform that disseminates case studies and good practices in IPLC's engagement in climate adaptation and mitigation. Through these initiatives, the LCIPP has advanced IPLC visibility within the UNFCCC to increase dialogue and engagement among IPLCs, party members, and other UNFCCC constitutive bodies.

Despite these achievements, both insider reflections and my fieldwork observations at the 2024 Baku COP29 reveal persistent institutional barriers that constrain IPLC participation. As the former head of the Indigenous Peoples' Forum on Climate Change (2014–2016), Hindou Oumarou Ibrahim (2022) reflected that coordinating IPLCs' engagement with UNFCCC processes was challenging. She reflected on the experience of operationalizing a platform to foster the exchange of traditional knowledge during the 2016 COP22 in Marrakech. Ibrahim noted that the IPLC representatives were informed only 1 day in advance of the formal negotiations, so the representatives were not prepared for the meetings. Additionally, IPLCs could only "observe" and were not allowed to "take the floor" until all Party members had spoken. Ibrahim critiqued such practices as "speaking about us, for us, but without us" (p. 171).

Ibrahim's reflections echoed my fieldwork observations at COP29 in Baku. Some IPLC participants only held "Invitee" badges, which restricts their access to negotiation rooms where decisions are made. IPLCs with "Observer" badges can speak in negotiations but still do not hold decision-making power. In comparison, "Party member" badges allow one to enter the key negotiation rooms and influence decision-making. This badge system reflects how the UNFCCC remains, by design, a Party-driven process in which only state delegates hold formal negotiating authority. IPLCs' engagement thus depends heavily on national governments' willingness to include them in the national delegation team. In countries where IPLCs' participation is politically sensitive, IPLCs' voices are limited to informal spaces such as side events or LCIPP-hosted dialogues. Even then, some LCIPP invitees requested not to be photographed as they feared political repercussions at home for attending international meetings without formal government endorsement. Although the LCIPP provides a venue for IPLC participation and engagement, the country-centric institutional design continues to determine who can meaningfully shape negotiation outcomes.

Moreover, language and communication processes restrict IPLCs' participation. The UNFCCC operates in six official UN languages, with English serving as the default working language for the LCIPP. However, Indigenous knowledge is often transmitted orally—through songs, storytelling, or ritual—rather than in written technical reports. This mode of knowledge transmission sharply contrasts with the textual and bureaucratic communication of the UN processes. Moreover, translation services are limited to UN languages, so IPLCs who cannot comfortably speak these languages cannot meaningfully participate. The language barrier prevents many true knowledge holders, especially the elderly, from participating in the UN processes. Indeed, many IPLC representatives come from more privileged backgrounds—educated, multilingual, and often already employed in global institutions.

The LCIPP has attempted to bridge this gap using creative initiatives. For instance, at COP29, the Platform used photo exhibitions in the Conference Hall to convey IPLC relationships with land and nature. The Annual Gathering of Knowledge Holders began with hymns and songs that symbolize IPLC's spiritual ties with nature. While these efforts increase cultural visibility, they remain largely symbolic rather than transformative within formal policymaking.

There also exist issues of regional variation in representation. Regions such as the Arctic and the Americas—where Indigenous groups are more institutionally recognized—are more active in LCIPP processes. In contrast, IPLCs in parts of Asia, Africa, and Eastern Europe often lack formal organizations and therefore lack effective participation channels. Some Asian governments resist recognition of Indigenous groups due to national sovereignty concerns. As a result, these IPLC representatives often depend on NGO or academic intermediaries to participate in the UNFCCC processes. This reproduces a pattern in which countries with established Indigenous organizations—such as Australia, Canada, the Americas, and Nordic countries—dominate LCIPP and broader UNFCCC engagements more than other regional IPLC groups.

In sum, the LCIPP provides a platform to incorporate and transmit IPLC's knowledge into international climate negotiation systems. However, the UNFCCC's state-centric design restricts IPLCs' ability to influence negotiations or shape policy outcomes equally with Party members and scientific bodies such as the IPCC. The LCIPP's institutional design still falls short in realizing the CPK Framework that enables IPLCs to participate equally in international environmental regimes, falling short on what the CPK framework promotes.

My fieldwork observations at COP29 align with existing studies of IPLC engagement under the IPBES, which is often regarded as performing better than the UNFCCC in integrating IPLCs. Established in 2014, the IPBES explicitly called for multidisciplinary and multi-knowledge approaches to strengthen the knowledge–policy interface. It developed a multiple evidence-based approach to facilitate CPK across stakeholders (Díaz et al., 2015). The IPBES assesses and evaluates the inclusion of IPLC using the Global Assessment. The reports found that ILK helped advance the concept of "nature's contributions to people", which fosters understanding of nature-human interactions beyond the utilitarian concept of "ecosystem services" (McElwee et al., 2020). The Global Assessment documented concrete outcomes: ILK-informed management practices helped improve pollination and revitalize beekeeping, predict and adapt to extreme events, and identify new plant species that could be used as medicines (McElwee et al., 2020, p. 1670).

Nevertheless, the IPBES faces similar limitations. Primarily, language barriers persist, as most official documentation and communication are in English (Athayde et al., 2023). Despite calls for including more diverse knowledge systems, the selection criteria for authorship based on proven academic performance still systematically bias experts from the English-speaking sphere (Wiegleb and Bruns, 2025). Under this circumstance, ILK is often interpreted through a scientific lens by non-IPLC members, while time constraints and procedural norms limit meaningful engagement with IPLC representatives. For instance, one IPBES pollination assessment gave IPLCs only 5 days to contribute; this short notice does not provide sufficient time to obtain informed consent or to conduct collective knowledge-sharing processes (Krug et al., 2020, p. 204). These cases reveal that while IPLCs' engagement and visibility have improved, more transformative changes, such as realizing the CPK and the establishment of equal, long-term partnerships, remain to be achieved.

Both my fieldwork on the UNFCCC and the secondary literature on the IPBES reveal persistent shortcomings in current environmental regimes' engagement with IPLCs. On the positive side, these institutions increasingly recognize the importance of IPLC participation and have established formal procedures to facilitate their involvement. However, a fundamental paradox remains: the UNFCCC and the IPBES are, by design, intergovernmental organizations. This means that despite growing acknowledgment of non-state actors, states continue to hold exclusive authority over formal negotiations and decision-making. National differences in how states define and relate to IPLCs result in regional variations in IPLC engagement. Procedural constraints, such as the reliance on English and written reports, further limit inclusivity and effective communication between IPLCs and other bodies. As a result, deeper collaborations that reflect the essence of knowledge co-production are far more challenging to realize at the global level, given the sheer diversity of countries and Indigenous groups involved. Nevertheless, the increasing international recognition of IPLCs has strengthened the normative expectation of their participation. For many representatives, especially those from countries where IPLC voices are politically constrained, platforms like the LCIPP at least provide opportunities to articulate their perspectives and gain visibility on the international stage. Table 3.1 presents potential ways forward to enhance IPLC engagement in international environmental regimes with respect to the CPK framework.

TABLE 3.1
Potential Ways to Enhance IPLC Engagement in International Environmental Regimes

Current Barriers	Ways Forward
Language	• Increase funding to support translation from Indigenous and local languages into the official UN languages • Using more photos and other visual representations in meetings, plenaries, and conferences
Written reports	• Consider accepting IPLC oral histories and place-based knowledge in technical reports • Provide technical training for IKL knowledge-holders to engage in the report-writing process • Recognize IPLC knowledge holders as co-authors • Credit Indigenous intellectual property
Project procedures	• Increase the project time frame to ensure extensive consultation with IPLC partners and align with IPLCs' decision-making processes • Ensure enough time to acquire FRIC consent
Institutional designs	• Encourage having Indigenous co-chairs in working groups • Increase Indigenous representation in expert panels
Unequal representation of certain IPLC groups	• Environmental institutions can actively reach out and encourage participation from currently underrepresented IPLC groups • Increase funding for travel expenses and associated costs for underrepresented IPLC groups to participate in international environmental institutions

3.5 CONCLUSION

This chapter examined how IPLCs can co-produce knowledge with scientists to improve environmental management practices and evaluated the approaches of key international environmental regimes toward IPLC engagement. Adopting a social constructivist perspective challenges the presumed universalism and objectivity of scientific knowledge, as it only constitutes one among many ways of knowing. The CPK framework illustrates how Indigenous and scientific knowledge systems can complement one another to achieve better environmental outcomes grounded in equity and long-term partnerships.

Empirical cases from the Arctic and Australia's forest management demonstrate that adopting a knowledge co-production approach enhances environmental management by enriching ecological data collection and fostering collective governance approaches. However, evidence from fieldwork at UNFCCC COP29 and analyses of IPBES processes indicate that, despite increased rhetorical commitments to IPLC inclusion, their influence within international environmental institutions remains constrained by institutional, procedural, and linguistic barriers. Addressing these limitations requires more than symbolic participation; it demands reforms that recognize IPLCs as co-producers of knowledge and governance actors. Without such reforms, international environmental regimes risk reproducing knowledge hierarchies that undermine both equity and environmental effectiveness.

DECLARATION OF GENERATIVE AI USE

During the preparation of this manuscript, the author(s) used ChatGPT to improve English grammar. The author(s) subsequently reviewed and edited the content and take(s) full responsibility for the final version of the published chapter.

REFERENCES

Artelle, K. A., Zurba, M., Bhattacharyya, J., Chan, D. E., Brown, K., Housty, J., & Moola, F. (2019). Supporting resurgent Indigenous-led governance: A nascent mechanism for just and effective conservation. *Biological Conservation*, *240*, 108284. https://doi.org/10.1016/j.biocon.2019.108284

Athayde, S., Cariño, J., Howard, P., Flores, A., Bates, P., & François, F. (2023). *Review of the inclusion of Indigenous and local knowledge in IPBES functions and deliverables*. IPBES task force. https://files.ipbes.net/ipbes-web-prod-public-files/2023-06/Review%20of%20Indigenous%20and%20local%20knowledge%20in%20IPBES.pdf

Barnett, M., & Duvall, R. (2005). Power in international politics. *International Organization*, *59*(1), 39–75.

Dauvergne, P. (1993). The politics of deforestation in Indonesia. *Pacific Affairs*, *66*(4), 497–518. https://doi.org/10.2307/2760676

Díaz, S., Demissew, S., Carabias, J., Joly, C., Lonsdale, M., Ash, N., Larigauderie, A., Adhikari, J. R., Arico, S., Báldi, A., Bartuska, A., Baste, I. A., Bilgin, A., Brondizio, E., Chan, K. M., Figueroa, V. E., Duraiappah, A., Fischer, M., Hill, R., … Zlatanova, D. (2015). The IPBES conceptual framework: Connecting nature and people. *Current Opinion in Environmental Sustainability*, *14*, 1–16. https://doi.org/10.1016/j.cosust.2014.11.002

Falkner, R. (2021). *Environmentalism and Global International Society*. Cambridge University Press.

Grundmann, R. (2013). "Climategate" and the scientific ethos. *Science, Technology, & Human Values*, *38*(1), 67–93.

Haas, P. M. (1992). Introduction: Epistemic communities and international policy coordination. *International Organization*, *46*(1), 1–35.

Heckbert, S., Russell-Smith, J., Reeson, A., & James, G. (2011). Indigenous Australians Fight Climate Change with Fire. *Portland State University Institute for Sustainable Solutions Publications and Presentations*.

Ibrahim, H. O. (2022). The climate agreements: What we have achieved and the gaps that remain. In D. Nakashima, I. Krupnik, & M. Roué (Eds.), *Resilience through Knowledge Co-Production: Indigenous Knowledge, Science, and Global Environmental Change* (pp. 165–174). Cambridge University Press. https://doi.org/10.1017/9781108974349.011

Ijatuyi, E. J., Lamm, A., Yessoufou, K., Suinyuy, T., & Patrick, H. O. (2025). Integration of indigenous knowledge with scientific knowledge: A systematic review. *Environmental Science & Policy*, *170*, 104119. https://doi.org/10.1016/j.envsci.2025.104119

Jasanoff, S. (Ed.). (2004). *States of Knowledge: The Co-Production of Science and the Social Order*. Routledge. https://doi.org/10.4324/9780203413845

Jones, S. (2002). Social constructionism and the environment: Through the quagmire. *Global Environmental Change*, *12*(4), 247–251. https://doi.org/10.1016/S0959-3780(02)00062-6

Krishnan, R., Shrestha, A. B., Ren, G., Rajbhandari, R., Saeed, S., Sanjay, J., Syed, Md. A., Vellore, R., Xu, Y., You, Q., & Ren, Y. (2019). Unravelling climate change in the Hindu Kush Himalaya: Rapid warming in the mountains and increasing extremes. In P. Wester, A. Mishra, A. Mukherji, & A. B. Shrestha (Eds.), *The Hindu Kush Himalaya Assessment* (pp. 57–97). Springer International Publishing. https://doi.org/10.1007/978-3-319-92288-1_3

Krug, C. B., Sterling, E., Cadman, T., Geschke, J., Drummond de Castro, P. F., Schliep, R., Osemwegie, I., Muller-Karger, F. E., & Maraseni, T. (2020). Stakeholder participation in IPBES: Connecting local environmental work with global decision making. *Ecosystems and People*, *16*(1), 197–211. https://doi.org/10.1080/26395916.2020.1788643

Latour, B. (2009). 1. Why political ecology has to let go of nature. In *Politics of Nature* (pp. 9–52). Harvard University Press. https://www.degruyterbrill.com/document/doi/10.4159/9780674039964-003/html?lang=en&srsltid=AfmBOop9fbjiYGbXWxcFgOd-MK4ytZemCp521KD6tBj2L9fQXTf3AlW3

LCIPP. (2025). *Local Communities and Indigenous Peoples Platform*. United Nations Climate Change. https://lcipp.unfccc.int/homepage

McElwee, P., Fernández-Llamazares, Á., Aumeeruddy-Thomas, Y., Babai, D., Bates, P., Galvin, K., Guèze, M., Liu, J., Molnár, Z., Ngo, H. T., Reyes-García, V., Roy Chowdhury, R., Samakov, A., Shrestha, U. B., Díaz, S., & Brondízio, E. S. (2020). Working with Indigenous and local knowledge (ILK) in large-scale ecological assessments: Reviewing the experience of the IPBES Global Assessment. *Journal of Applied Ecology*, *57*(9), 1666–1676. https://doi.org/10.1111/1365-2664.13705

Moallemi, E. A., Zare, F., Hebinck, A., Szetey, K., Molina-Perez, E., Zyngier, R. L., Hadjikakou, M., Kwakkel, J., Haasnoot, M., Miller, K. K., Groves, D. G., Leith, P., & Bryan, B. A. (2023). Knowledge co-production for decision-making in human-natural systems under uncertainty. *Global Environmental Change*, *82*, 102727. https://doi.org/10.1016/j.gloenvcha.2023.102727

Moore, S. E., & Hauser, D. D. W. (2019). Marine mammal ecology and health: Finding common ground between conventional science and indigenous knowledge to track arctic ecosystem variability. *Environmental Research Letters*, *14*(7), 075001. https://doi.org/10.1088/1748-9326/ab20d8

Mustonen, T., & Feodoroff, P. (with the Skolt Sámi Fishermen of Sevettijärvi). (2018). Skolt Sámi and Atlantic Salmon Collaborative Management of Näätämö Watershed, Finland as a case of indigenous evaluation and knowledge in the Eurasian Arctic. *New Directions for Evaluation*, *2018*(159), 107–119. https://doi.org/10.1002/ev.20334

Nakashima, D., Krupnik, I., & Roué, M. (Eds.). (2022). Introduction. In *Resilience through Knowledge Co-Production: Indigenous Knowledge, Science, and Global Environmental Change* (pp. 1–24). Cambridge University Press. https://doi.org/10.1017/9781108974349.001

Norström, A. V., Cvitanovic, C., Löf, M. F., West, S., Wyborn, C., Balvanera, P., Bednarek, A. T., Bennett, E. M., Biggs, R., de Bremond, A., Campbell, B. M., Canadell, J. G., Carpenter, S. R., Folke, C., Fulton, E. A., Gaffney, O., Gelcich, S., Jouffray, J.-B., Leach, M., … Österblom, H. (2020). Principles for knowledge co-production in sustainability research. *Nature Sustainability*, *3*(3), 182–190. https://doi.org/10.1038/s41893-019-0448-2

Ostrom, E. (2015). *Governing the Commons: The Evolution of Institutions for Collective Action*. Cambridge University Press. https://doi.org/10.1017/CBO9781316423936

Parks, L., & Tsioumani, E. (2023). Transforming biodiversity governance? Indigenous peoples' contributions to the convention on biological diversity. *Biological Conservation*, *280*, 109933. https://doi.org/10.1016/j.biocon.2023.109933

Robinson, C. J., Renwick, A. R., May, T., Gerrard, E., Foley, R., Battaglia, M., Possingham, H., Griggs, D., & Walker, D. (2016). Indigenous benefits and carbon offset schemes: An Australian case study. *Environmental Science & Policy*, *56*, 129–134. https://doi.org/10.1016/j.envsci.2015.11.007

Russell-Smith, J., Cook, G. D., Cooke, P. M., Edwards, A. C., Lendrum, M., Meyer, C. (Mick), & Whitehead, P. J. (2013). Managing fire regimes in north Australian savannas: Applying Aboriginal approaches to contemporary global problems. *Frontiers in Ecology and the Environment*, *11*(s1), e55–e63. https://doi.org/10.1890/120251

Ryghaug, M., & Skjølsvold, T. M. (2010). The global warming of climate science: Climategate and the construction of scientific facts. *International Studies in the Philosophy of Science*, *24*(3), 287–307. https://doi.org/10.1080/02698595.2010.522411

Saami Council. (2025, May 12). *Co-Production of Knowledge in the Arctic: Bridging Indigenous and Scientific Perspectives*. Arctic Council. https://arctic-council.org/news/co-production-of-knowledge-in-the-arctic-bridging-indigenous-and-scientific-perspectives/

Schuster, R., Germain, R. R., Bennett, J. R., Reo, N. J., & Arcese, P. (2019). Vertebrate biodiversity on indigenous-managed lands in Australia, Brazil, and Canada equals that in protected areas. *Environmental Science & Policy*, *101*, 1–6. https://doi.org/10.1016/j.envsci.2019.07.002

Vergara-Asenjo, G., & Potvin, C. (2014). Forest protection and tenure status: The key role of indigenous peoples and protected areas in Panama. *Global Environmental Change*, *28*, 205–215. https://doi.org/10.1016/j.gloenvcha.2014.07.002

Wiegleb, V., & Bruns, A. (2025). Whose knowledge counts? Unpacking the uneven geographies and politics of knowledge co-production in IPBES. *Human Ecology*, *53*(1), 73–86. https://doi.org/10.1007/s10745-025-00578-w

Wylie, C., & Murillo, L. F. R. (2023). Care-fully?: The question of "knowledge co-production" in Arctic science. *Catalyst: Feminism, Theory, Technoscience*, *9*(2). https://doi.org/10.28968/cftt.v9i2.39359

Yua, E., Raymond-Yakoubian, J., Daniel, R. A., & Behe, C. (2022). A framework for co-production of knowledge in the context of Arctic research. *Ecology and Society*, *27*(1), art34. https://doi.org/10.5751/ES-12960-270134

Zhang, Y., West, P., Thakholi, L., Suryawanshi, K., Supuma, M., Straub, D., Sithole, S. S., Sharma, R., Schleicher, J., Ruli, B., Rodríguez-Rodríguez, D., Rasmussen, M. B., Ramenzoni, V. C., Qin, S., Pugley, D. D., Palfrey, R., Oldekop, J., Nuesiri, E. O., Nguyen, V. H. T., … Agyei, F. K. (2023). Governance and conservation effectiveness in protected areas and Indigenous and locally managed areas. *Annual Review of Environment and Resources*, *48*, 559–588. https://doi.org/10.1146/annurev-environ-112321-081348

Part II

Regional Policy and Management

4 Effective Conservation
Strategies for Ecosystems and Protected Areas

Wenhui Qiu, Yanqiu Feng, and Chunmiao Zheng

4.1 INTRODUCTION

Global climate change is intensifying, and the impact of human activities on the natural environment continues to increase in intensity. Under the combined impact of these two factors, ecosystem degradation and biodiversity loss have become core environmental issues threatening human survival and development (Díaz et al., 2019). As the core carrier of the Earth's life support system, the structural integrity and functional stability of ecosystems are directly related to the supply level of key ecological services such as carbon cycle regulation, water resource supply, and disaster prevention and control (Díaz et al., 2019). Protected areas, as the core practical carriers of ecological protection, can effectively curb this degradation trend and maintain biodiversity (Langhammer et al., 2024). Therefore, "effective protection of ecosystems and protected areas" has not only become a core research field in ecology and environmental management but also been integrated into the important agenda of global sustainable development. Its research covers various ecosystem types such as forests, wetlands, and oceans, and runs through the entire chain of practical links from basic theoretical construction and protected area planning and design to cross-regional collaborative governance.

Current academic circles have formed a systematic cognitive framework for the protection of ecosystems and protected areas, gradually breaking away from the limitations of early single-species protection or isolated area guardianship, and shifting to the core cognition of "holistic view of ecosystems" and "synergetic development between humans and nature" (Sharma et al., 2025). This means that the academic community generally recognises that ecological protection is not simply "banning development", but achieving the dialectical unity of protection and utilisation based on respecting the operating laws of ecosystems (Sharma et al., 2025). Specifically, due to differences in structure and function, different ecosystems face different core threats and protection needs. Forest ecosystems need to focus on addressing habitat fragmentation, marine ecosystems need to prioritise the prevention and control of overfishing and pollution governance, and wetland ecosystems must maintain hydrological connectivity as their core task (Lacher et al., 2025) (Figure 4.1). The deepening of this cognitive level has not only promoted the transformation of protection practices from local intervention to systematic governance but also laid a solid foundation for the development of interdisciplinary research.

The mainstream theories supporting research and practice in this field have formed a diversified and complementary system. Among them, the Ecosystem-Based Management (EBM) theory is the core dominant paradigm. This theory is guided by the overall function of ecosystems, integrates the interactive relationships between biological components, physical environments, and human activities, and realises the synergetic advancement of protection and utilisation. The marine carrying capacity theory provides quantitative support for the sustainable utilisation of resources (Zheng & Wang, 2025), valuating the carrying capacity of ecosystems. It provides a scientific basis for the formulation of control standards, such as fishing quotas and development intensity. The cross-border governance theory specifically addresses the issues of responsibility division and collaboration in the protection

DOI: 10.1201/9781003640899-6

FIGURE 4.1 Human-induced marine pollution poses a continuous threat to marine ecosystems, especially tropical mangrove ecosystems, which are highly vulnerable to pollution from both inland and marine sources due to the lack of effective waste management.

of transboundary ecosystems, providing theoretical guidance for synergistic protection at the global or regional scale. At the same time, the three principles of integrity, dynamic balance, and sustainable utilisation constitute the basic theoretical core of ecological protection, which are deeply integrated into the entire process of designing and implementing various protection strategies (Díaz et al., 2019).

In terms of research methods, this field has developed a multidimensional, integrated system of "field investigation-remote sensing monitoring-model simulation-interdisciplinary integration" (Yan et al., 2024). As a basic link, field investigation obtains first-hand data on the status of ecosystems through underwater observation, vegetation quadrats, species tracking, and other means, providing core support for accurately identifying protection needs. Remote sensing technologies, including satellite remote sensing and UAV inspections, enable dynamic monitoring due to their advantages of wide coverage and timeliness, and have become efficient tools for early warning of ecological degradation and supervision of illegal activities. The application of ecosystem dynamics models, food web models, and other models can simulate ecological responses under disturbance scenarios, providing scientific predictions for the optimisation of protection strategies (IPBES, 2019). At the same time, research methods from disciplines such as sociology and economics are constantly integrated, including community interviews, stakeholder analysis, and cost-benefit assessment, which promote more precise, synergistic adaptation between protection practices, community development, and economic benefits (Montgomery, 2025).

The academic and practical fields have established several core consensus on ecological protection. The most fundamental one is that the superimposed effect of overdevelopment, habitat destruction, and climate change is the main driving factor for global ecosystem degradation. A fundamental one is that the combined effect of overdevelopment, habitat destruction, and climate change is the primary driving factor for global ecosystem degradation (Maron et al., 2017). However, although the degradation mechanisms of different ecosystems are different, the prevention and control of core threats still need to take both pertinence and systematisms into account. From the planning level, scientific planning and design, and connectivity construction are the keys to ensuring the effective operation of protected areas. Through zonal management and ecological corridor construction, the phenomenon of "ecological islands" can be avoided, and species gene exchange can be guaranteed (Brennan et al., 2022). From the perspective of practical guarantee, community participation and benefit sharing are the core support for the sustainability of protection work. Only when local communities truly benefit from protection actions can their enthusiasm for active participation be fully stimulated, thereby resolving the contradiction between "protection and development" (He & Min, 2020). For the protection of transboundary ecosystems, it is more necessary to rely on the international treaty framework and regional synergistic mechanisms, and to realise the overall protection of the entire river basin or region through unified standards, shared data, and joint law enforcement (Sun et al., 2024). Based on the above understanding and consensus, this paper conducts a systematic discussion on the core issues of ecosystem and protected area protection, providing scientific guidance and strategic reference for effective protection at the practical level.

Centring on the core theme of "Effective Protection: Strategies for Ecosystems and Protected Areas", this paper constructs a complete ecological protection system from basic theory to practical application, and from local measures to global collaboration. Specifically, the paper first elaborates on the basic principles and three-level goal system of ecosystem protection, including the three principles of integrity, dynamic balance, and sustainable utilisation. At the same time, it analyses the unique protection needs of different ecosystems, such as forests, wetlands, and oceans, as well as practical obstacles, including funding shortages, policy gaps, and illegal activities. On this basis, it systematically explains the scientific design of protected areas, covering key contents such as core planning elements and connectivity construction. Among them, connectivity construction focuses on constructing ecological corridors and buffer zones and draws on the best practice experiences from successful cases around the world (Figure 4.2). Furthermore, it explores the sustainable management model of multi-functional protected areas, clarifies the integration path of ecological protection with human activities such as agriculture, fishery, and tourism, and emphasises the importance of community participation and cross-departmental collaboration. Subsequently, it introduces the integration innovation and application effects of traditional ecological knowledge and modern technologies, where contemporary technologies include cutting-edge means such as remote sensing, genomics, and UAVs. Finally, it elaborates on the necessity of cross-border ecological protection and provides in-depth analyses of relevant international treaties and transnational cooperation cases. Supported by numerous examples and data, the chapter offers scientific and practical advice for ecological protection.

4.2 FOUNDATIONS OF ECOSYSTEM CONSERVATION: STRATEGIES, GOALS, AND CHALLENGES

An ecosystem is an organism that results from the interactions between biological communities and the abiotic environment through processes such as material cycling, energy flow, and information transfer. Its health directly sustains the survival foundation and sustainable development prospects of all life on Earth. It is essential to emphasise that ecological protection extends beyond merely

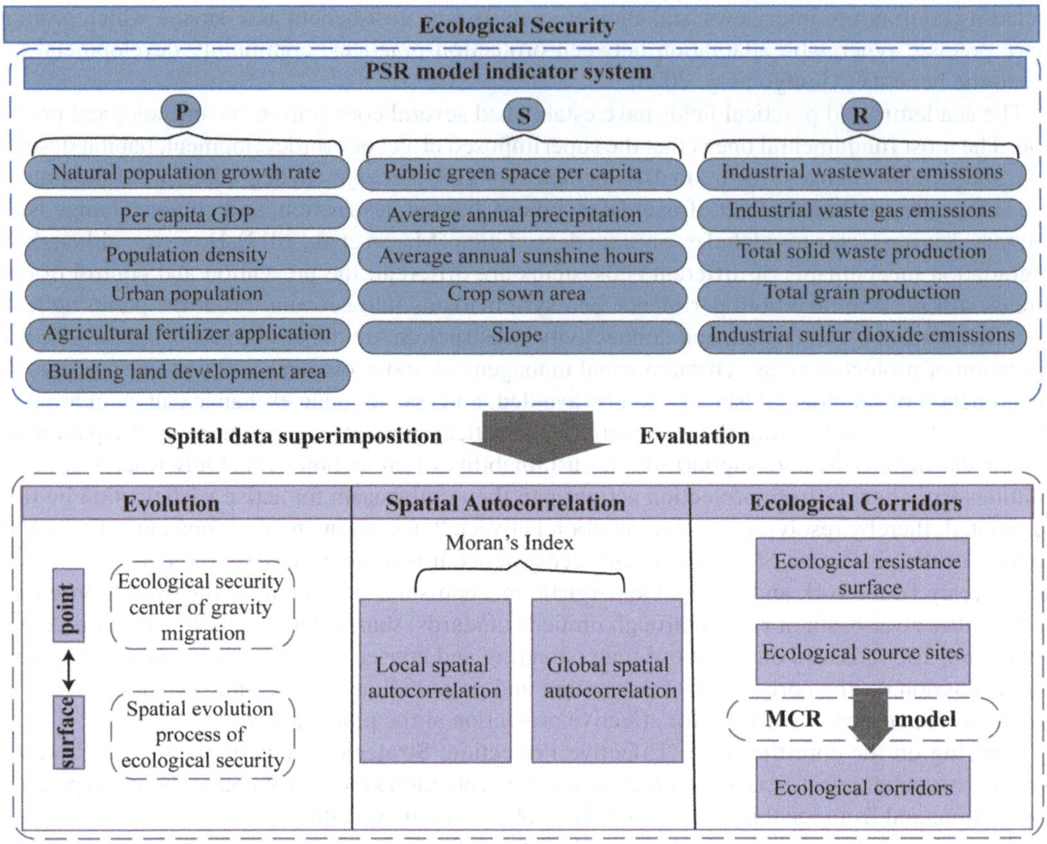

FIGURE 4.2 How to construct ecological corridors (ECs), as well as the process and effects of constructing ecological corridors, include: protecting ecological diversity to maintain the balance of the ecosystem and promote sustainable development.

preserving specific species or locations—it aims to sustain and restore the broader functions of entire ecosystems. Therefore, only by accurately grasping the core principles and objectives of ecological protection and thoroughly clarifying the diverse challenges faced in practice can effective conservation strategies be developed and implemented.

4.2.1 The Basic Principles and Goals of Ecosystem Conservation, Emphasising the Maintenance and Restoration of Ecosystem Functions

The fundamental principles of ecosystem protection are based on in-depth research and a summary of the operational laws of ecosystems, serving as the fundamental guidelines for all protection actions. The principle of integrity is the primary principle of ecosystem protection (Hughes, 2020). All components of an ecosystem, whether biological components such as plants, animals, and microorganisms, or non-biological components such as soil, water, air, and sunlight, are interdependent and interact with each other. For example, in the Serengeti grassland ecosystem in Africa, lions, as top predators, control the population of zebras and wildebeests; the grazing activities of zebras and wildebeests affect the distribution and growth of grassland vegetation; the condition of grassland vegetation determines the number of insects, which provide food for birds; the faeces of birds and the carcasses of animals, after decomposition by decomposers, can provide nutrients for the soil and nourish the grassland vegetation (Sinclair et al., 2007). If humans overhunt lions, it will lead to a sharp increase in the population of zebras and wildebeests, resulting in overgrazing of grassland vegetation and a series of chain reactions, such as a decrease in insect populations and the migration of birds, ultimately destroying the balance of the entire grassland ecosystem. Therefore, ecological protection must start from the whole, pay attention to the connections between various elements in the system, and not just target a single species or a single link.

The principle of dynamic balance emphasises that ecosystems have self-regulating capabilities (Tuljapurkar & Semura, 1977). A healthy ecosystem can withstand a certain degree of natural or human interference and restore itself to a relatively stable state through its own regulatory mechanisms. Take the forest ecosystem in Yellowstone National Park in the United States, for example (Turner et al., 2016). Lodgepole pine is one of the primary tree species in the park, and its cones require high temperatures to crack and release their seeds. Natural forest fires, although they burn some trees, also provide the high temperature needed for the cones to crack, and the ashes after the fire can provide rich nutrients for the germination of new seeds (Perkins, 2020). If humans overly intervene in natural fires to protect trees, it will lead to a large accumulation of aging lodgepole pines, an increase in flammable materials under the forest, and once a fire occurs, it will evolve into a highly destructive, large-scale fire, which will instead cause more severe damage to the forest ecosystem. Therefore, the goal of ecological protection is not to prevent all changes, but to maintain the dynamic balance and self-regulating ability of the ecosystem, ensuring that it can restore stability when faced with disturbances.

The principle of sustainable utilisation reflects the dialectical relationship between ecological protection and human development (Nicholson et al., 2021). Ecosystems provide humans with a diverse range of natural resources, including food, timber, water, and medicinal materials. Therefore, it is vital that resource development and utilisation strictly adhere to the rigid limits of ecosystem carrying capacity, to maintain the long-term sustainability of resource supply and ecosystem services. Such as the cod fisheries in northern Norway (Brattland et al., 2019). Local authorities accurately estimate the natural reproductive potential of the cod population, based on which they establish a catch quota system, specifying that the annual catch should not exceed 40% of the population's reproductive capacity. This regulatory measure has proven effective, ensuring stable economic returns from the fishery (with an average annual export of about $8 billion) while maintaining the dynamic balance of cod populations, achieving a coordinated development of resource use and ecological protection (Utne, 2007).

Based on the above basic principles, the core goals of ecological protection can be divided into a three-level progressive system. The primary goal is to preserve the integrity of the ecosystem's structure. This requires ensuring the completeness of the ecosystem's components, sufficient and connected habitat areas, and stable population levels of key species at a reasonable level (Nicholson et al., 2021). For example, the Wolong Nature Reserve in China has set "maintaining the connected area of the giant panda habitat at over 1,500 square kilometers" as the basic goal, because only a large and connected habitat can meet the foraging, breeding, and activity needs of giant pandas and ensure the continuation of their population (Liu et al., 2001). If the habitat area is too small or fragmented, the activity range of giant pandas will be restricted, and gene exchange between populations will be impossible, putting them at risk of extinction. The primary objective is to ensure the normal functioning of ecosystem processes. Ecosystem functions include material cycling (such as carbon, nitrogen, and water cycles), energy flow, information transmission, and self-regulation of the ecosystem (Barabás et al., 2017). The core goals of the Amazon rainforest reserve include "annual carbon sequestration of no less than 200 million tons" and "stable regulation of runoff during the rainy season". The Amazon rainforest plays a crucial role in the global carbon cycle, absorbing and fixing a large amount of carbon dioxide, which is of great significance for mitigating global climate change. At the same time, the rainforest can store a large amount of water during the rainy season and release it slowly during the dry season, playing a key role in regulating the regional climate and maintaining river flow (Ahlstrom et al., 2017). If these ecological functions are disrupted, it will not only affect the ecological balance of the rainforest itself but also have a profound impact on the global ecological environment. The ultimate goal is to achieve a sustainable supply of ecosystem services. Ecosystem services refer to various benefits provided by ecosystems to humans, including provisioning services (such as providing food, fresh water, wood, medicinal materials, etc.), regulating services (such as climate regulation, water purification, flood control and disaster reduction, etc.), supporting services (such as soil formation, nutrient cycling, pollination, etc.) and cultural services (such as tourism and recreation, scientific research and education, cultural inheritance, etc.) (Daily & Matson, 2008). The Swiss Alpine Nature Reserve has promoted the diversified transformation of ecosystem services through systematic protection of alpine meadows and forest ecosystems. On the one hand, the area serves as a stable water source barrier for downstream cities, supplying approximately 1.2 billion tons of water annually; on the other hand, the ecotourism industry, developed based on its unique natural landscapes, generates approximately 3 billion euros in annual revenue. Meanwhile, the Swiss Alpine Nature Reserve's pristine ecological foundation makes it a natural scientific observation site, providing a valuable in situ research platform for related fields and ultimately achieving the sustainable and diversified provision of ecosystem services (Pichler-Koban & Jungmeier, 2017).

4.2.2 The Unique Needs and Challenges of Protecting Different Ecosystems, Such as Forests, Wetlands, and Marine Environments

The surface of the Earth is dotted with numerous ecosystems, including forests, wetlands, and oceans. Among them, forests are important carbon sinks, providing habitats for many species. Wetlands play a prominent role in the field of carbon sequestration and water purification. The ocean is the primary source of oxygen in the world and plays a crucial role in ensuring food security. However, various ecosystems are facing differentiated threats: tropical rainforests in the forest ecosystem continue to degrade due to commercial logging and agricultural expansion, and the problems of deforestation and habitat fragmentation are becoming increasingly severe; Temperate forests encounter the dual compression of urban expansion and the replacement of natural forests by plantations, which significantly increases the complexity of conservation work. Boreal forests are affected by the combined impact of climate change and pest outbreaks, with the melting of permafrost and frequent wildfires further exacerbating ecological damage. In the wetland ecosystem, inland wetlands continue to shrink due to excessive water withdrawal and damming in the upstream areas, causing a hydrological imbalance. Coastal wetlands are gradually degraded under the pressure of land development and pollution. In terms of marine ecosystems, coral reefs and seagrass beds in shallow waters have been impacted by ocean warming and destructive fishing. Meanwhile, the open sea area faces threats such as fisheries management disorder, plastic pollution, and emerging deep-sea mining activities.

The Earth's surface ecosystems are diverse, which directly determines their unique conservation needs. Among various ecosystems, forest ecosystems have become the core area of global ecological protection due to their high biodiversity and key ecological functions - not only serving as important carbon sinks but also playing a crucial role in oxygen production. Studies have shown that global forests store approximately 45% of terrestrial carbon and provide support for the survival and reproduction of 80% of terrestrial species (Dixon et al., 1994). It is worth noting that there are significant differences in the conservation needs and challenges faced by different types of forests. Tropical rainforests are primarily distributed in the Amazon Basin, Congo Basin, and Southeast Asia, and their core characteristics include extremely high biodiversity and irreplaceable ecological functions. Therefore, curbing large-scale deforestation and habitat fragmentation is the primary task of protecting such ecosystems. However, commercial logging and agricultural expansion have become significant threats to the survival of tropical rainforests. Approximately 10 million ha of rainforest are cut down for timber harvesting globally each year, and 2 million ha of forests are converted into soybean plantations in Brazil alone (Maeda et al., 2021). At the same time, habitat fragmentation is further aggravated by road construction and land reclamation, resulting in large areas of continuous forest being divided into isolated patches. Research by Lawrence confirmed that when rainforest patches shrink to less than 100 km^2, bird species diversity decreases by 30%. Species that rely on large areas of continuous habitat are at direct risk of extinction. Additionally, the destruction of tropical rainforests can lead to soil erosion and climate anomalies (Laurance et al., 2000). Temperate forests are mainly distributed in mid-latitude regions. The core of protecting temperate forests lies in balancing the conflict between reforestation of artificial forests and urban expansion (Geedicke et al., 2018). Although fast-growing poplar and fir plantations can quickly restore vegetation cover in the short term and help mitigate soil erosion, their single-species composition can easily lead to a decline in regional biodiversity. However, their monoculture composition, typically consisting of only one or two tree species, fails to provide the rich ecological functions of natural forests (Caron et al., 2023). For instance, red pine forests produce pine nuts that are essential food sources for squirrels and black bears, which plantations cannot adequately support. It also reduced temperate forest areas, with Beijing's surrounding forests shrinking by approximately 50 km^2 annually, significantly impairing their ecological functions (UNEP-FI, 2025).

Boreal forests, located in high-latitude regions of the Northern Hemisphere such as the Siberian Taiga and Canadian Boreal Forest, face threats from climate change and pest infestations (Venäläinen et al., 2020). From 1938 to 2009, the average summer temperature in north-eastern Siberia increased by 1.0°C (UNEP-FI, 2025), expanding the range of pests, such as the pine wood nematode, which has increased the affected area tenfold. First, nematode infestations can directly cause pine trees to wither and die, severely damaging the integrity of coniferous forest ecosystems. Second, rising temperatures drive permafrost thawing, which not only leads to waterlogging in forest lands but also significantly affects the normal growth of dominant species, such as larch. More importantly, northern forests, as a major global carbon sink, may release large amounts of carbon dioxide due to the increased frequency of wildfires caused by climate warming, forming a negative feedback loop of "climate warming—more frequent wildfires—increased carbon release", which further aggravates global climate change.

Although wetland ecosystems cover only 1% of the Earth's surface, they play an indispensable role in the global ecosystem, with core ecological functions reflected in three aspects: first, they contribute 20% of terrestrial carbon storage, serving as an important "carbon reservoir"; second, they provide breeding and habitat spaces for 25% of species, supporting the maintenance of biodiversity; third, they provide crucial services in water purification and flood regulation. Existing studies clearly indicate that maintaining hydrological connectivity is a core requirement for wetland protection and a prerequisite for ensuring the normal functioning of their ecological services (Temmink et al., 2022).

From a specific type of perspective, the core threat to inland wetlands is concentrated on water loss. China's Poyang Lake and Africa's Okavango Delta are typical examples of this issue. Among them, Poyang Lake, as a key freshwater wetland in China, is a critical winter habitat for a vast number of migratory birds. However, due to the continuous increase in upstream agricultural water use, the dry season has shortened by 2 months compared to the past. The drop in water levels has directly led to a 40% reduction in wintering habitat for rare birds such as the Siberian Crane, while severely disrupting their food resource supply system (Li et al., 2023). The Okavango Delta, one of Africa's largest inland wetlands, relies heavily on rainfall from the Angolan highlands for water supply. The construction of upstream dams has significantly reduced the inflow to the delta, causing continuous wetland shrinkage and posing a serious threat to many plant and animal species that depend on the wetland for survival (Bartlam-Brooks et al., 2013).

Coastal wetlands (including mangroves and salt marshes) are mainly pressured by development encroachment and environmental pollution (Lovelock et al., 2022). In mangrove ecosystems, globally, mangroves are disappearing at a rate of 0.62% per year (about 130,000 ha), with the main drivers being aquaculture conversion (such as shrimp pond development in Southeast Asia) and urban expansion (UNEP-FI, 2025). Nutrient pollution from agricultural fertiliser runoff and industrial wastewater discharge leads to algal blooms, which both inhibit the growth of mangrove seedlings and disrupt the original ecological balance of wetlands. Besides, salt marsh wetlands are also faced with the survival pressures of the combined effects of sea level rise and land reclamation. Rising sea levels inundate marsh areas, while land reclamation directly reduces marsh extent; the effect of these factors further accelerates the degradation of salt marsh ecosystems (Valiela et al., 2018).

Marine ecosystems occupy approximately 71% of the Earth's surface and represent a vital component of the global life support system. Their ecological functions are irreplaceable, as they generate roughly 50% of the world's oxygen and supply 20% of the animal protein consumed by humans. Studies have highlighted that the core conservation demand of this ecosystem lies in addressing two primary problems: overfishing and habitat degradation (Agardy, 2000).

From the perspective of ecosystem types, coral reefs and seagrass beds in offshore ecosystems are often referred to as "tropical rainforests in the ocean" due to their characteristic enrichment of biodiversity. Among them, coral reefs account for only 0.1% of the total ocean area, but they host 25% of the survival and reproduction of marine species (Hinrichsen, 1997). However, the global situation regarding coral reef degradation is extremely severe, and 50% of coral reefs have degraded to varying degrees. The main driving factors include coral bleaching induced by sea warming, as well as direct damage to the physical structure of coral reefs by destructive fishing practices such as fish frying. As the world's largest coral reef system, Australia's Great Barrier Reef experienced five large-scale bleaching events between 2016 and 2022, primarily due to rising sea temperatures, resulting in a decline in its coral cover to 13% (Hughes et al., 2017). Similarly, seagrass beds, which are key habitats for fish spawning and reproduction, also have efficient carbon sink functions; however, their area is shrinking due to multiple pressures, including coastal development, environmental pollution, and overfishing (Rosalina et al., 2023).

Compared to offshore ecosystems, pelagic ecosystems (including high-sea fisheries and deep-sea areas) face the dual stress of a lack of regulation and plastic pollution. Due to unclear jurisdiction and the lack of regulatory mechanisms, fisheries management in the high seas is chaotic, with approximately 30% of fish stocks being overexploited, resulting in a continuous decline in fishery resources (Jackson et al., 2001). Besides the disorder in fisheries, plastic pollution is also a serious ecological threat in the high seas that urgently needs to be addressed. It was found that approximately 8 million tonnes of plastic waste enter the ocean each year, gradually breaking down into microplastics. When microplastics are ingested by marine organisms, such as fish and sea turtles, they directly threaten the survival of these organisms and accumulate through the food chain, ultimately posing risks to human health (Jambeck et al., 2015).

4.2.3 Major Obstacles in Conservation Efforts, Including Insufficient Funding, Policy Gaps, and Illegal Activities

With the advancement of ecological protection efforts, various obstacles are gradually emerging. Among these obstacles, the primary issue is the lack of funding in the current global ecological protection development process.

According to statistics, the annual global funding required for ecological protection is about \$700 billion, while the actual investment is only \$300 billion. Lack of funding directly leads to many planned protection actions being difficult to initiate, and on-going projects often come to a halt (UNEP-FI, 2025).

From a regional perspective, the lack of funding is particularly pronounced in developing countries, where many protected areas struggle with inadequate infrastructure and insufficient personnel. In Malawi's Liwonde National Park in Africa, the lack of monitoring equipment, such as infrared surveillance, makes it difficult to stop poaching activities in a timely manner, resulting in a decline in the black rhinoceros' population from 100 in 2000 to 30 in 2023. In addition to causing a lack of equipment, the shortage of funds also seriously affects the continuity and integrity of ecological restoration projects. For instance, the Ganges Wetland Restoration Project in India was halted due to insufficient funds, completing only 40% of the vegetation reconstruction, with the wetland's purification function falling short and resulting in minimal improvement in water quality (CAG, 2025). Please cite this properly.

Apart from the insufficient funding, the lack of policy frameworks and weak law enforcement pose challenges to ecological protection efforts. At the policy-making level, some countries have yet to establish complete frameworks for ecological protection. For example, between 2019 and 2022, Brazil relaxed regulations related to rainforest protection and reduced penalties for illegal logging, which directly led to a 60% surge in illegal logging (Brancalion et al., 2018). Even in countries where relevant laws and regulations have been enacted, weak policy enforcement is a universal issue. The shortcomings of policy and enforcement further fuel illegal activities and the proliferation of organised crime. Currently, illegal poaching, illegal logging, and wildlife trafficking have formed large chains, with annual trade exceeding $200 billion; the huge economic gains serve as the driving force behind these illegal activities. Specifically, African elephant poaching is rampant, with about 110,000 elephants illegally killed between 2010 and 2020, and ivory entering Asia and other regions through smuggling networks for profit (Schlossberg et al., 2018). The endangered rosewood species in Madagascar are also illegally logged and trafficked to European markets, with prices reaching up to $10,000 m^2. Attracted by such profits, over-exploitation of rosewood has become severe (Rakotonirina et al., 2023). It should be noted that these illegal activities often coexist with corruption, violence, and organised crime, which greatly increase the difficulty of law enforcement and regulation.

In addition, climate change further exacerbates the practical challenges of ecological protection (Luque et al., 2013). Climate change leads to global temperature increases and alters precipitation patterns, as well as increases the frequency of extreme weather events, including heavy rainfall, droughts, and hurricanes (Weiskopf et al., 2020). In the Arctic tundra, rising temperatures promote the spread of shrubs, which both compress the living space of local tundra vegetation and directly reduce the food resources for reindeer (John Bryant et al., 2014). In addition, climate change-induced sea level rise directly threatens coastal wetlands and low-lying island ecosystems. At the same time, climate change aggravates the frequency of forest fires and the severity of pests and diseases, severely damaging the structural integrity and functional stability of forest ecosystems (Schuerch et al., 2018).

4.3 PROTECTED AREA DESIGN AND OPTIMISATION: EFFECTIVE PLANNING AND IMPLEMENTATION

As a core means of ecological protection, nature reserves are a key way to maintaining biodiversity and the integrity of ecosystems. Scientifically planned and reasonably arranged reserves can maximise conservation effectiveness.

4.3.1 The Planning, Design, and Implementation of Protected Areas, Covering Species Protection, Habitat Restoration, and Ecosystem Services Maintenance

It is worth noting that the core focus of nature reserve planning will vary significantly depending on the conservation objectives. Areas with species protection as the primary goal should focus on managing key ecological indicators. All human activities must be strictly prohibited within the core zone. All human activities must be strictly prohibited within the core zone. Efforts should be made to expand suitable habitats for species, and conservation effectiveness should be assessed using quantitative indicators. Areas focusing on habitat restoration require measures such as controlling pollution sources and restoring the substrate environment. Areas aimed at maintaining ecosystem services should prioritise ecosystem protection and strengthen the effectiveness of regional ecological barriers.

Based on these differentiated goal orientations, zoned management becomes a universal core design framework for nature reserve planning. The core zone enforces the strictest protection standards, prohibiting all human interference. In the buffer zone, low-impact activities, such as scientific monitoring, may be permitted to balance conservation objectives with research needs. In the outer zone, limited sustainable use activities are supported under the premise of maintaining ecological balance.

The primary task of protected area planning, with species protection as the core, is to accurately identify the core habitat needs of key species, encompassing essential elements such as food resource supply, water source security, suitability of breeding sites, and sufficiency of activity space. Taking the giant panda protected area planning as an example (Zhu et al., 2024), since giant pandas primarily use bamboo as their food source, bamboo forest density (≥200 plants/ha) is identified as a core planning indicator. At the same time, the suitable altitude range for giant pandas to survive is 2,600–3,000 m, which provides the most suitable climatic conditions and can effectively avoid extreme climate stress, such as severe cold and heat. In addition, the terrain slope (≤25°) is also a key planning factor that cannot be ignored; gentle terrain is more conducive to the movement, foraging, and activity of giant pandas (Kang et al., 2011). To further strengthen the effectiveness of protection, all human activities, such as logging and tourism, are completely prohibited in the core area of the giant panda reserve to minimise the impact of human interference on the survival of giant pandas. In the process of practical promotion, through the implementation of ecological restoration projects such as "returning farmland to forests", the bamboo forest habitat area of China's Giant Panda National Park has been expanded by 120 km^2, providing more abundant living space for the giant panda population. At the same time, to systematically evaluate the effectiveness of conservation and the implementation of planning, relevant studies have set up quantitative monitoring indicators to carry out dynamic tracking, in which the faecal density of pandas can reflect their activity frequency. The survival rate of pups can characterise the reproductive effect of the population (Li et al., 2013).

Different from the species conservation orientation, protected area planning with habitat restoration as the core mainly targets degraded ecosystems and needs to formulate phased and targeted restoration plans based on the degree of degradation and core problems. The restoration plan of Dianchi Lake in Yunnan Province is a typical example. The wetland has faced severe ecological degradation problems, which are manifested in the deterioration of water quality (prominent eutrophication), the extinction of emergent plants, and the sharp decline in fish diversity (Xiang et al., 2013). Based on this, the restoration plan is carried out in an orderly manner in four stages: first, through the construction of sewage treatment plants and sewage interception pipe networks to implement source pollution control, successfully intercept 70% of the surrounding domestic sewage, and reduce the pollution load from the root; Secondly, carry out substrate restoration work to improve the substrate conditions of wetland water quality by adsorbing pollutants in water bodies by injecting activated carbon. In the third stage, promote the vegetation reconstruction project by screening and planting native aquatic plants, such as reeds and calamus. This not only facilitates the absorption and purification of nutrients in the water body but also provides habitat for aquatic organisms. This not only facilitates the absorption and purification of nutrients in the water body but also provides a basic habitat environment for wetland organisms. Finally, filter-feeding fish such as silver carp and bighead carp were released to consume plankton and organic detritus in the water body with the help of their feeding behaviour, further strengthening the water purification function. After 5 years of systematic restoration, the wetland ecosystem of Dianchi Lake has been significantly improved: the water quality level has been upgraded from poor Class V to Class IV, the number of bird species has increased from 50 to 130, and the ecological function of the wetland has been gradually restored, and the restoration effect has been remarkable (Liu et al., 1991).

Ecosystem service-oriented spatial planning focuses on prioritising the protection of ecological functional areas that play a crucial role in human well-being, thereby ensuring the sustainable provision of ecosystem services. The New York City Watershed Protection Area is a typical example of an ecologically service-oriented conservation area, with its core function being to ensure the drinking water supply for 8 million city residents, providing an annual water supply of 1.5 billion cm^2 (Mehaffey et al., 2005). Centred on ensuring the drinking water supply, the conservation area focuses on the upstream forest zones that provide water source conservation functions, and the riparian ecosystems that can filter pollutants. To ensure the achievement of these objectives, New York City has implemented a series of targeted measures, purchasing upstream forest lands and strictly prohibiting development in restricted areas to maintain the hydrological regulation capability of the forest ecosystem, using financial support to guide upstream farmers to transition to organic farming, reducing the use of fertilisers and pesticides, thereby achieving precise control of agricultural non-point source pollution. These measures not only ensure the stability of drinking water quality but also avoid the high construction and operating costs of large-scale water treatment facilities, cumulatively saving up to $6 billion (Schuerch et al., 2018).

It is worth noting that, to balance the dual needs of ecological protection and scientific management, the zoning management model has become a universal framework for planning and operating protected areas worldwide. Among these, the core zone serves as the ecological heart of the protected area, implementing the strictest protection policies, completely prohibiting all types of development, construction, and human interference activities, thus providing an undisturbed environment for the natural evolution of ecosystems and the survival and reproduction of species (Liu et al., 2022). For example, in the Giant Panda Nature Reserve, all human activities, such as logging, tourism, and farming, are strictly prohibited in core habitats to maximise the safeguarding of the natural living conditions of giant pandas. From the perspective of delineation logic, the boundaries of core zones are generally defined based on the principle of maintaining stable populations of the protected species and the integrity of the ecosystem structure, while scientifically delineating according to the specific ecological requirements of the protected objects (such as habitat suitability, activity range, etc.) (Kang et al., 2011).

In the zoning management system of nature reserves, there are some differences between the strict protection modes of buffer zones and core zones. Serving as the transitional area between the core zone and the surrounding regions, the buffer zone is located on the periphery of the core zone and is subject to relatively relaxed management. Specifically, limited human activities such as scientific research, ecological monitoring, and low-intensity ecotourism can be carried out within the buffer zone (Liu et al., 2022, 1991). For example, the bird-watching area of Dianchi Wetland is situated within the buffer zone, where human disturbance is minimised by restricting bird-watching activities and regulating visitor behaviour.

As the outermost level of the zoning management system, the peripheral zone is located outside the buffer zone and serves as a link between the reserve and surrounding communities. The peripheral zone allows activities such as agricultural production, aquaculture, and the traditional collection of wild resources (Gao et al., 2020). However, it must be noted that these activities must be controlled within the ecological carrying capacity and must not damage the ecosystem of the reserve. In some forest reserve peripheral zones, local communities are permitted to forage for wild vegetables and mushrooms or engage in low-intensity grazing, but large-scale logging and reclamation are prohibited (Straede & Treue, 2006).

The implementation of zoned management depends on the planning and design of nature reserves. This process requires consideration of both natural and socio-economic dimensions. Natural elements include topography and landforms, hydrological characteristics, and vegetation types. Socio-economic elements include transportation conditions and population distribution (Karel Mokany, 2013). For example, in areas with complex terrain, natural barriers such as mountains and rivers can be utilised to delineate reserve boundaries, thereby reducing human interference by leveraging geographical features.

Scientific planning and design must be transformed into practical conservation results through effective implementation. Therefore, the implementation phase of nature reserves is crucial (Slocombe, 1992). During implementation, monitoring and evaluation must be strengthened to accurately identify problems and optimise measures. For example, during the construction phase of a reserve, monitoring vegetation recovery and species population changes can guide adjustments to ecological restoration measures and management strategies (Christian Kitungwa Lugoma, 2023).

4.3.2 The Connectivity between and beyond Protected Areas, Such as Ecological Corridors and Buffer Zones, Ensures Species Migration and Gene Flow

Connectivity in nature reserves refers to the degree of connection between different reserves and surrounding ecosystems. Connectivity is crucial for ensuring the smooth passage of wildlife migration corridors, promoting gene flow among populations, and maintaining the integrity of ecosystem functions. Therefore, constructing a well-connected network has become a prerequisite for improving conservation effectiveness, and exploring pathways to achieve connectivity is a key task in implementing this prerequisite (Saura et al., 2017). Among the various pathways to achieve connectivity, constructing ecological corridors is one of the most widely used and effective core approaches. Ecological corridors, as ecological passages connecting different reserves or ecosystems (Beltrão et al., 2024), primarily function to provide safe paths for species migration, promote gene flow among populations, and thus maintain ecosystem connectivity.

In the design of terrestrial ecological corridors, precision planning based on species-specific characteristics is particularly evident. Terrestrial corridor design must align fully with the mobility and behavioural traits of various species (Lin et al., 2022). Large mammals, such as tigers and lions, with wide activity ranges and high migration needs, typically require corridors at least 500 m wide to meet their survival and migration requirements. Smaller birds and rodents have lower corridor width requirements, and simple corridors such as hedgerows or tree lines can achieve connectivity (Merkle et al., 2023). Such as forest corridors in India's Bandhavgarh National Park, by building a 10-km-long forest corridor connecting two isolated reserves, the tiger population's activity range expanded from 300 to 800 km^2, with a 50% increase in population over 10 years, ensuring stable tiger migration, promoting gene flow among the population (Thatte et al., 2018).

In addition to terrestrial ecosystems, ensuring the connectivity of aquatic ecosystems is equally crucial. Unlike terrestrial corridors that focus on species mobility and adaptation, the design of aquatic ecological corridors needs to be kept to ensure the connectivity and continuity of the river and lake system. In natural water systems, hydraulic facilities such as dams and slices often block the migration routes of fish, affecting their reproductive cycles and survival status. For example, in the Columbia River Basin in the United States, many dams in the basin prevent salmon migration, resulting in a decline in salmon populations. To restore migration corridors, local measures removed four blocking dams and added dedicated fishways, which increased the number of migrating salmon from 50,000 to 300,000/year (Dambacher et al., 2001). Notably, the recovery of the salmon population not only restored the population size but also provided a stable food source for predators such as bears and eagles, thereby promoting the balance of the entire river ecosystem.

Ecological corridors serve as a core pathway is insufficient to ensure connectivity. Buffer zones play a crucial role in enhancing the connectivity of nature reserves (Burger et al., 2024). Buffer zones can filter pollutants and intercept external pollutants, such as pesticides, fertilisers, and industrial wastewater, thereby reducing their harmful impact on the core area's ecological environment. Planting mixed tree and shrub buffer zones around agricultural areas near reserves can absorb and filter pollutants from farmland runoff, decrease the pollution load on water and soil within the reserve (Prosser et al., 2020). Previous research has shown that a forested buffer zone with a width of up to 2 km can achieve over 70% filtering efficiency for agricultural pesticides. Besides, buffer zones provide additional habitats and food resources for species, thereby expanding their activity range (Lin et al., 2011). An example is the Serengeti National Park in Africa: this area, with its interwoven grassland and woodland zones and abundant vegetation resources, provides ample food for giraffes during the rainy season. Thanks to the ecological support of the buffer zone, giraffe populations increased by 30% compared to relying solely on the core area (Denninger Snyder et al., 2019). At the same time, buffer zones can serve as transit areas for migratory species, offering essential support for their successful migration and further strengthening the connectivity foundation of the reserve.

4.3.3 Case Studies of Successful Protected Areas, Analysing How They Succeeded in Specific Environments and Discussing Best Practices

Case 1: Yellowstone National Park, USA—A Model of Ecological Self-Regulation

Around the world, numerous successful examples of nature reserve construction and management have emerged. These cases have achieved significant accomplishments in planning and design, management measures, and conservation outcomes, providing valuable experiences that can be referenced in the practice of other reserves globally. Among these successful examples of nature reserve construction, Yellowstone National Park in the United States served as a significant model of ecosystem self-regulation (Stahler et al., 2006).

Yellowstone National Park thrives by adhering to the natural rules of its ecosystem. In the early 20th century, the park culled wolves to protect the elk population, which almost led to the extinction of wolves. Without elk's natural predator, the elk population rapidly surged, grazing on willow and other vegetation, causing a reduction in willow stands. The loss of willow will then affect beaver survival, leading to a decline in beaver numbers and subsequent wetland shrinkage, as well as the disruption of ecosystem balance. It was not until the 1990s that Yellowstone National Park repaired its ecology by reintroducing wolves. The return of the top predator reshaped the regional ecological structure. Wolves controlled the elk population through predation, and they also altered the elk's range of activity, preventing overgrazing in specific areas. In this way, willow forests gradually recovered, providing beavers with the building materials they needed for dams. As the beaver population increased, their dam-building activity created more wetlands, which in turn promoted the breeding of fish and waterfowl, establishing a positive ecological cycle of wolves, and elk vegetation wetlands (Stahler et al., 2006).

Yellowstone adopts a zoned management model. The core is strictly protected, allowing only scientific research activities. The buffer zone conducts low-impact tourism activities, such as designated trail hiking and wildlife observation. The peripheral area enables indigenous people to continue their traditional hunting and gathering activities. This zoned management model not only protects the core ecosystem but also meets the public's tourism needs and the traditional living needs of indigenous people (Jensen, 2000).

Case 2: Qinghai Lake National Nature Reserve, China—A Successful Practice of Plateau Wetland Restoration

Qinghai Lake is the largest inland lake in China and an important plateau wetland ecosystem. Due to climate drought and overgrazing, the Qinghai Lake ecosystem faced severe degradation: the lake area shrank, the naked carp (a second-class protected animal in China) was on the verge of extinction, and the number and variety of birds decreased (Wood et al., 2007; Zhang et al., 2022).

Qinghai Lake Nature Reserve has implemented a series of effective conservation measures to promote ecosystem improvement along two core dimensions: habitat restoration and species protection. In terms of habitat restoration, the reserve has restored 300,000 acres of cultivated land to wetland ecosystems by promoting the "returning farmland to wetland" project, while planting drought-resistant vegetation, such as sea buckthorn, to curb the spread of desertification (Dong et al., 2020). These measures have effectively improved the vegetation coverage of the lake area and promoted the sustainable development of the ecological environment. At the level of species protection, for the protection of Huangyu, the reserve has successfully restored the natural migration path of Huangyu by removing the barrages on its migratory channel and building a special fish-crossing channel (Weng et al., 2023). Huangyu resources have also increased significantly from less than 1,000 tonnes in 2002 to 114,000 tonnes in 2023. At the same time, the reserve has strengthened bird protection by establishing bird monitoring stations to ensure a sufficient food supply and safe habitats. It provides a strong guarantee for the survival and reproduction of birds.

It is worth noting that community participation is a key supporting element for the success of protecting Qinghai Lake Reserve. The reserve innovatively hires local herders as ecological rangers to undertake core responsibilities, such as patrolling and monitoring, and the average annual income of herders increases by approximately ¥30,000. This measure not only significantly enhances the initiative and enthusiasm of people to participate in ecological protection but also strengthens the awareness of ecological protection among residents, promoting herders to realise an identity transformation from "bystanders" to "participants" and "beneficiaries" of ecological protection. The change in resident identity has further effectively reduced illegal fishing, overgrazing, and other ecologically destructive behaviours, forming a benign interaction between protection and development.

After years of systematic protection, the ecological environment of Qinghai Lake has undergone significant improvement: the lake area has recovered to 4,625 km^2, reaching its highest value in the past 40 years; the number of waterbird species has increased from 189 to 225 (Zhang et al., 2024). Based on the successful practice, the best model of protected area design and management can be summarised, including the following four points: first, follow the law of ecosystem evolution, respect nature's self-regulation ability, and avoid excessive human intervention; second, clarify the core protection objectives and formulate targeted protection measures and management strategies accordingly; Third, establish scientific mechanisms such as zoning management and community participation to coordinate and balance the relationship between ecological protection and regional development; Fourth, strengthen dynamic monitoring and adaptive adjustment, regularly evaluate the effectiveness of protection, and optimise protection measures in a timely manner according to dynamic changes in the ecosystem.

4.4 MULTIFUNCTIONAL PROTECTED AREAS AND SUSTAINABLE MANAGEMENT: BALANCING ECOLOGICAL PROTECTION WITH HUMAN NEEDS

Multi-functional protected areas are an innovative model in nature reserve management. The principle of multi-functional protected areas is achieving a balance between conservation and utilisation through scientific planning. Based on this principle, we maintain the integrity of ecosystem functions to strengthen the ecological foundation, utilising ecological resources to reconcile the development of local communities with broader societal needs.

4.4.1 Combine Ecological Protection with Human Activities to Create Multifunctional Protected Areas

Multi-functional protected areas make reasonable use of natural resources, allowing different zones to accommodate various human activities by delineating functional zones, formulating systematic plans, and adhering to ecological protection boundaries. The key to achieving a balance of protection and utilisation depends on establishing community participation and benefit-sharing mechanisms, enabling local communities to benefit from conservation efforts and thus participate in ecological protection actions (Coetzer et al., 2015).

Among the various practical models that aim to balance protection and utilisation, the integrated agriculture and conservation model is a notable example. An integrated agriculture and conservation model can enhance land-use efficiency and increase community economic income while maintaining the integrity of ecosystem structure (Jat et al., 2020). In the agroforestry system implemented in the protected areas of the Brazilian Amazon rainforest, 70% of the original rainforest is preserved to consolidate the ecological baseline, while intercropping Brazil nuts and beans under the canopy (Figure 4.3). The market price of Brazil nuts can reach up to $15/kg, providing farmers with economic income, while beans improve soil fertility through nitrogen fixation, benefiting forest vegetation growth. It is found that farmers' incomes increased by 30% compared to single-crop soybean cultivation, while maintaining rainforest coverage and reducing illegal logging by 60%. During harvest periods, farmers protect Brazil nut host trees, further curbing illegal logging activities (Maezumi et al., 2018). Ultimately, this model enhances farmers' economic benefits while ensuring the functionality of the rainforest ecosystem.

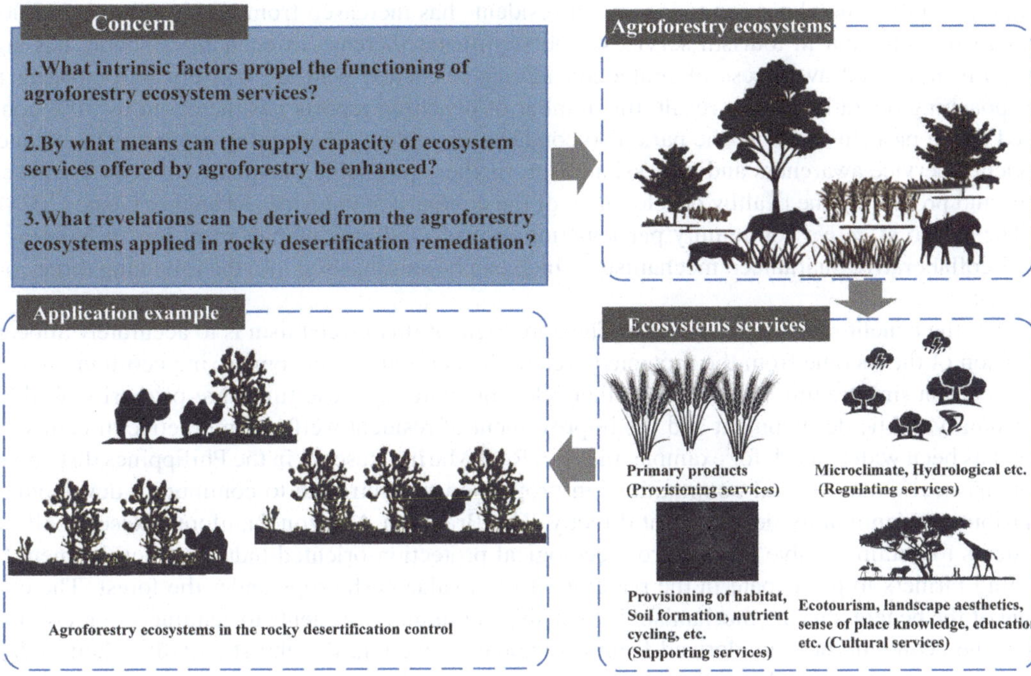

FIGURE 4.3 The structure of traditional agroforestry systems and the services they provide in the process of ecosystem restoration.

The coordinated development of fisheries and conservation requires scientific management measures to regulate fishing behaviour and ensure the sustainable supply of fishery resources. The combined management model of "no-catch zone-fishing area" adopted by the Apo Reef Marine Reserve in the Philippines has shown significant results (Alcala & Russ, 2006). Specifically, the reserve designates 20% of the area as a core no-catch zone, a key area for fish breeding, spawning, and juvenile breeding. Buffer zones while enabling communities to engage in these activities, while destructive operations, such as blast fishing, are strictly prohibited. The fish in the core no-catch area will naturally spread to the buffer fishing area, effectively replenishing fishery resource reserves and increasing the community's catch by 40%. At the same time, the reserve generates economic benefits through the rational development of eco-tourism and allocates 20% of the tourism income for community dividends, particularly funding the construction of local schools and infrastructure improvements, which directly enhances the community's living standards. This model enables community residents to gain a deep understanding of the importance of fishery resource protection and actively participate in the management and maintenance of protected areas, thereby forming a virtuous circle of protection and utilisation (Perkins, 2020).

The sustainable development of ecotourism is an important path for multi-functional protected areas to achieve economic benefits. By guiding tourists to experience natural landscapes and watch wild animals, ecotourism can not only create direct economic benefits but also subtly enhance the public's awareness of ecological protection. Chitwan National Park in Nepal has gained extensive experience in sustainable development of ecotourism (Upadhaya et al., 2022). In order to avoid the interference of tourism activities on the ecosystem, the park implements strict control over tourism behaviour: limiting the daily number of tourists in the core area to less than 200, prohibiting the use of noisy means of transportation such as motor boats to reduce disturbance to wild animals, and formulating clear tourist behaviour norms, strictly prohibiting feeding animals and discarding garbage at will.

Community participation is the cornerstone of sustainable ecotourism development in Chitwan National Park. The local Taru community dominates 70% of the tour guide services and homestay formats, and the annual per capita income of residents has increased from US$2,000 to US$5,000 through participation in tourism services. The significant increase in economic income has significantly improved awareness of protection among residents, who are actively participating in anti-poaching operations. As a result, the number of poaching reports has increased by 70% compared to the past. In addition, the park also conducts targeted skills training programs to enhance residents' service awareness and professionalism, further optimising the quality of ecotourism services and promoting the healthy development of the ecotourism industry (Khanal & Gosai, 2024).

The key to effective community participation in protected area management lies in building a multi-collaborative guaranteed mechanism, which can be summarised into the following three core aspects:

First, the benefit-sharing mechanism. The core logic of this mechanism is to accurately allocate a portion of the income from the protected area to the community, encompassing eco-tourism revenue, carbon sink trading income, and other relevant sources. These funds are primarily used for community public development and the improvement of resident welfare. In practice, this mechanism has been widely used: for example, the Apo Reef Marine Reserve in the Philippines distributes ecotourism revenue to local communities in proportion to contribute to community development contribute to community development directly. The Brazilian Amazon Rainforest Reserve allows residents to obtain a stable income from ecological protection-oriented industrial development by guiding farmers to participate in the planting of high-value cash crops under the forest. The core value of the benefit-sharing mechanism is to enable community residents to genuinely perceive and enjoy the economic benefits derived from ecological protection, thereby stimulating their endogenous motivation to actively participate in conservation work.

Second, the decision-making participation mechanism. By absorbing community representatives in the protected area management committee, it is the core essence of the mechanism to allow the community to participate in the whole process of planning, daily management and key decision-making. Community residents have the best understanding of the local natural environment, resource endowments and their own development needs, and their participation in decision-making can ensure that management measures are more in line with the actual local situation, effectively avoid the problem of disconnection between policy and practice, and then improve the overall management efficiency of the protected area. Taking Chitwan National Park in Nepal as an example, the park management incorporates representatives of the Talu community to participate in management decision-making, and community representatives provide targeted suggestions for core policies such as tourism development planning and sustainable use of resources based on local conditions, which significantly improves the feasibility and implementation effect of the policy.

Third, a capacity-building mechanism. Through systematic training projects, the awareness and professional skills of community residents will be enhanced simultaneously, laying the groundwork for their deeper participation in the management and sustainable use of resources in protected areas.

4.4.2 Integrated Management Models That Emphasise Cross-Sector Collaboration and Long-Term Sustainability

Ecosystems possess both complexity and diversity, a characteristic that necessitates a multi-department collaborative approach in the management of nature reserves. This is because a model where a single department manages independently is often limited by insufficient resources and a narrow scope of responsibilities, making it difficult to address the various challenges faced in ecological protection. Multi-department collaboration can effectively integrate the core resources and expertise of all participating departments, consolidate and unify management efforts, and thereby enhance the overall management effectiveness of nature reserves (Caves et al., 2013).

In the process of cross-departmental collaboration, a multi-departmental coordination mechanism is the core support. A multi-departmental coordination mechanism establishes a unified coordinating body, clarifies the responsibilities and roles of each participating department, and facilitates joint action. The integrated management model of the Great Barrier Reef in Australia serves as a successful example for a multi-departmental mechanism in practice (Olsson et al., 2008). With the Great Barrier Reef Marine Park Authority as the core leadership institution, it coordinates the federal government, state governments, research institutions, tourism enterprises, and Indigenous communities to create a multi-faceted collaborative management framework. In this multi-faceted collaborative management framework, the federal government is responsible for formulating laws, regulations, and top-level policies related to the protection of the Great Barrier Reef, state governments undertake specific management duties such as fisheries management and pollution control, research institutions conduct continuous monitoring of key issues and submitting quarterly reports to provide scientific support for decision-making, tourism enterprises implement an eco-certified vessel operation model to reduce marine pollutants. Indigenous communities participate in protection efforts based on traditional ecological knowledge, such as assessing water quality by observing fish behavior to supplement ecological monitoring data (Qu Yichen, 2022). To further ensure the effectiveness of collaboration, all parties continue to communicate and coordinate through regular consultation mechanisms and information sharing platforms, creating a closed-loop management system of monitoring, assessment, decision-making implementation (Bay et al., 2023). Since the implementation of the Great Barrier Reef 2050 Plan, through the coordinated efforts of multiple departments, the annual decline rate of coral cover in the Great Barrier Reef has dropped from 2% to 1%.

Moreover, establishing a long-term and stable funding mechanism is a core prerequisite for achieving sustainable management of protected areas. It is urgent to expand diversified funding sources, beyond government budget allocations, to provide stable financial support for the continued advancement of conservation work. The Watershed Protection Payment Mechanism in New York City offers valuable lessons in this regard (Yu, 2017). The forests of the Catskill Mountains serve as the core water source area for New York City, supplying 90% of the city's drinking water. New York City raises a dedicated fund of $150 million each year by imposing a "water quality protection fee" of $0.05/m^3$ of water. In terms of fund use, its core investments focus on two main areas: first, purchasing upstream forest land and strictly prohibiting development activities through property rights management, thereby strengthening the ecological barrier of the water source area; second, subsidising upstream farmers to transition to organic farming, reducing the use of fertilisers and pesticides, and mitigating agricultural non-point source pollution at its origin (National Academies of Sciences, 2020). The implementation of this funding mechanism has not only effectively maintained the ecological integrity of the Catskill Mountains Forest and ensured the long-term stability of New York City's drinking water quality but has also successfully saved up to US$6 billion in the construction costs of large water treatment plants, achieving synergistic gains in ecological and economic benefits.

Complementing the financial guarantee mechanism, the comprehensive management model is the key path to promoting the long-term development of protected areas. The essence of this model is to "integrate protection into the overall planning of regional development", promote the deep integration of ecological protection and regional social and economic development, rather than promoting protection as an isolated project. Our country's Qiandao Lake Protected Area is a typical practice example of this model. With "water quality protection" as its core goal, the Qiandao Lake Protected Area has specially formulated the "Qiandao Lake Ecological Economic Zone Plan" to fully integrate ecological protection requirements into all aspects and links of regional development.

At the level of industrial development, the Qiandao Lake Protected Area strictly prohibits the settlement of highly polluting industries, such as the chemical industry and printing and dyeing. It focuses on supporting green and low-carbon industries such as organic fisheries and eco-tourism, and builds an industrial system that is compatible with ecological protection (Sun Mengting et al., 2020). Among them, the large-scale development of organic fisheries has enabled the annual output value of dried fish in Qiandao Lake to exceed 1 billion yuan, and eco-tourism has attracted many tourists with its high-quality natural landscape, directly driving regional economic growth.

At the level of management measures, all departments have formed a collaborative linkage pattern: the environmental protection department takes the lead in the dynamic monitoring and evaluation of the water quality of Qiandao Lake to ensure early detection and early treatment of pollution problems; The agricultural department focuses on the breeding link, actively promotes ecological breeding technology, guides fishermen to breed scientifically, and reduces the impact of aquaculture activities on water quality. The tourism department strengthens control over sewage discharge from tourist ships, requiring all ships to be equipped with sewage treatment devices and strictly banning all types of contaminants from entering the lake (Zhang et al., 2023). All departments gather management synergy through information-sharing platforms and joint law enforcement mechanisms to ensure collaboration and achieve management synergy, thereby enhancing the effectiveness of various protection measures.

Practical results show that the above measures not only maintain the water quality of Qiandao Lake at the Class II standard, but also increase the per capita income of surrounding residents to 1.2 times the average level of Zhejiang Province, thereby achieving a truly win-win situation of ecological protection and economic development. It should be emphasised that the long-term sustainability of the integrated governance model also relies on a scientific evaluation and adjustment mechanism (Wang Jiaotuan et al., 2008). Specifically, a comprehensive assessment of the ecological status, economic benefits, and social effects of the protected area should be carried out regularly, and management measures and development plans should be dynamically optimised accordingly. Management measures and development plans should be dynamically optimised in response to the assessment results. At the same time, it is necessary to strengthen public awareness and education, guiding the entire society to foster a positive atmosphere for joint participation in protection, and laying a social foundation for the long-term operation of the comprehensive governance model.

4.5 INNOVATIVE CONSERVATION STRATEGIES: INTEGRATING TRADITIONAL KNOWLEDGE WITH MODERN TECHNOLOGY

Traditional ecological knowledge embodies the practical wisdom and experience humans have accumulated through long-term practice. Modern technology has created tools and methodologies that combine precision and efficiency, thereby overcoming the limitations of traditional conservation approaches. Integrating traditional ecological knowledge with modern technology can improve the overall effectiveness and comprehensive benefits of ecological protection efforts.

This paper focuses on the innovative protection strategies derived from the integration of traditional ecological knowledge and modern technology. Specifically, modern technical tools such as UAV remote sensing, satellite monitoring, genomic analysis, Internet of Things perception, and big data mining have greatly strengthened the level of dynamic fire control, accurate monitoring of species, and refined management of protected areas by improving the accuracy, implementation efficiency, and rapid response ability of ecological protection work. At the same time, these modern technologies have also played a positive role in mitigating climate change and protecting ecological and cultural heritage.

A number of case studies further confirm the synergistic effect of the integration of local traditional knowledge and modern technological innovation: in the bird monitoring work in the Western Ghats of India, the traditional bird identification experience of the local community is combined with drone aerial photography and data modelling technology to improve the comprehensiveness and accuracy of monitoring; The anti-poaching operation in the Maasai Mara Reserve in Kenya has built a three-dimensional anti-poaching network by integrating the traditional patrol experience of the Maasai people with modern infrared monitoring and satellite positioning technology, which has significantly improved the efficiency of law enforcement.

4.5.1 How Traditional Ecological Knowledge Can Be Integrated with Modern Technologies to Drive Innovative Conservation Strategies

The integrated application of modern technologies has enhanced the precision and effectiveness of fire management measures (Edwards et al., 2021). Specifically, drones are widely used to monitor vegetation density and can accurately identify areas that require controlled burning. Satellite remote sensing technology can dynamically assess fire spread trends, providing data support for the timely optimisation of fire prevention strategies (Mukhiddinov et al., 2022). Additionally, infrared thermal imaging can track wildlife activity after fires, offering critical information for a comprehensive assessment of the fire's impact on the ecosystem.

In the field of species monitoring, the organic combination of traditional knowledge and modern technology has significantly enhanced work efficiency and data accuracy. In the case of bird monitoring in the Western Ghats of India, local tribes have accumulated deep traditional knowledge of ornithology and recorded bird migration patterns in the form of traditional songs, including an accurate description of the grey-breasted wren returning to the bamboo forest 3 weeks before the monsoon (Praveen & Nameer, 2009). To verify the scientific validity of this traditional knowledge, the researchers introduced modern monitoring technology to systematically record the migration trajectory of the grey-breasted wren warbler through a combination of bird-ringing logging and satellite tracking. Monitoring results confirm that predictions about migration timing in traditional songs are accurate to within 85%. Relying on this precise traditional knowledge, protected area managers can deploy targeted conservation measures in advance: during the critical season for the migration and breeding of the grey-breasted wren, tourism activities are completely prohibited in core areas, effectively reducing human disturbance and ultimately increasing the breeding success rate of the species by 40% (Fischer et al., 2025).

It is worth emphasising that the core premise of successfully integrating traditional knowledge and modern technology is to respect the centrality of traditional knowledge as a fundamental framework for formulating protection strategies, and that modern technology plays a synergistic role as an auxiliary tool, not a replacement for traditional knowledge. For instance, the anti-poaching practices in Kenya's Maasai Mara National Reserve, Maasai rangers combine traditional tracking techniques with modern tools to accurately locate lion prides, then record and upload location information in real-time to the reserve's management system through a smartphone application (Dolrenry et al., 2016). It has reduced anti-poaching response times from 24 to 4 hours, enhancing law enforcement efficiency. Additionally, it provides Maasai members involved in tracking with a monthly payment of US$300, which greatly stimulates local community participation and enthusiasm in ecological conservation.

4.5.2 The Use of Remote Sensing, Genomics, Drones, and Other Modern Technologies in Protected Area Management

The rapid development of modern technology has provided diversified new tools and innovative methods for nature reserve management. These technologies play a key supporting role in core management areas such as monitoring and early warning, which effectively improve the accuracy and efficiency of reserve management.

Among these new tools from modern technology, remote sensing technology, with its advantage of large-scale dynamic monitoring, can promptly obtain key ecological information such as vegetation cover, land-use changes, and fire hazards (Gilmour et al., 2025). Particularly, satellite remote sensing technology, which combines wide coverage with high data acquisition efficiency. Satellite remote sensing technology, such as the Sentinel-2 satellite, with a resolution of up to 10m, enables weekly routine monitoring of global nature reserves. In the management of Amazon rainforest reserves, authorities utilise the Sentinel-2 satellite for monitoring deforestation, identifying illegal logging hotspots, and transmitting monitoring data to law enforcement agencies, providing location support for enforcement (Lechler et al., 2020).

As a complement to satellite remote sensing for large-scale monitoring, drone remote sensing technology is better suited for small-scale and detailed monitoring scenarios due to its high flexibility and controllable cost advantages. In Kenya's anti-poaching operations, patrol teams utilise drones equipped with thermal imaging cameras for nighttime monitoring, allowing them to detect the movements of poachers in the dark and ensure timely intervention (Liu, 2020). In China's Qilian Mountains Nature Reserve, authorities utilise drones to monitor glacier retreat, capturing glacier dynamics with 1-m precision, thereby providing a basis for informed conservation management decisions regarding the glacier ecosystem (Shang, 2018).

In addition to remote sensing technology, genomics also holds important application value in species conservation management. Gnomics's core advantage lies in analysing species' genetic information to assess population health, thereby providing guidance for formulating targeted conservation strategies. Although the population of crested ibis has recovered from just seven individuals in 1981 to more than 9,000 today, genomic analysis results indicate that the genetic diversity of its population remains at a low level, making it vulnerable to the dual threats of disease attack and environmental change (Feng et al., 2019). To improve the genetic diversity of ibis, the reserve has implemented a special gene exchange program to increase the frequency of gene exchange in the population by pairing crested ibises from different breeding centres across regions, which has been proven to be effective in improving the survival rate of young birds (Feng et al., 2019). In addition, genomic technology can also be applied to the traceability of illicit trade species, and by comparing the genetic information of seized ivory with the genetic database of known elephant herds, the origin of ivory can be accurately determined, providing key clues for combating poaching and smuggling (Wasser et al., 2015). It is reported that a batch of seized ivory was successfully identified as originating from a protected area in Tanzania through genetic analysis, and law enforcement agencies strengthened anti-poaching controls in the area (Wasser et al., 2015).

The Internet of Things and big data technology realise intelligent monitoring and in-depth data analysis of environmental parameters and species activity status in protected areas through the collaborative mode of sensor deployment and data platform construction (Li et al., 2015). Real-time monitoring is the core application scenario of Internet of Things technology, and Yellowstone National Park in the United States is a typical practice case: the management team deploys sensors on a large scale in the reserve to continuously track environmental parameters such as water temperature, humidity, and soil nutrients, while simultaneously monitoring animal activity dynamics (Bruggeman et al., 2007). The massive data collected by these sensors is transmitted to the big data analysis platform through the cloud for integration and analysis: through in-depth mining of wolf pack activity trajectory data, the migration route of the deer pack can be accurately predicted, and then early warning and guidance of tourists to avoid it, effectively reducing human–animal conflicts; By analysing the data collected by the water temperature sensor, the change trend of the hot spring ecosystem can be predicted, which provides support for timely targeted protection measures (Colman et al., 2021).

4.5.3 The Critical Role of Local Communities and Indigenous Knowledge in Conservation Management, Especially in the Context of Cultural Heritage and Ecosystem Preservation

Indigenous communities, through long-term interactions with the natural environment, have accumulated knowledge about the survival habits of ecosystems and species, which informs the survival habits of ecosystems and species, forming a traditional knowledge system adapted to their local region (Lorz & Trueman, 2025).

Resource management wisdom is an essential component of indigenous traditional knowledge systems, which reflect their understanding of and practical experience in the sustainable use of natural resources (Kenny et al., 2018). Such as the Inuit reindeer hunting taboos in Canada, through hunting practices, the Inuit have developed strict hunting regulations, including prohibiting hunting during the reindeer breeding season and forbidding the killing of lactating female reindeer. These strict hunting regulations ensure the normal reproduction of reindeer populations. At the same time, hunting areas are adjusted according to the seasonal distribution of snow, avoiding deep snow regions and providing them as natural refuge areas for reindeer. These traditional regulations were closely aligned with modern ecological principles, and hunting prohibitions during the breeding season maintain population stability (Ljubicic et al., 2018). This hunting wisdom ensures a stable food source for the Inuit and achieves sustainable use of the reindeer resource (Ian Leiper et al., 2018).

In addition to wisdom in resource management, indigenous ecological early warning knowledge also provides support for the management of protected areas by helping them anticipate and respond to risks posed by ecological disasters. Indigenous communities are adept at predicting trends in ecological changes by observing subtle shifts in natural phenomena (Porto, 2024). The Amazonian Yanomami people can predict droughts by monitoring the flowering time of indicator plants, such as warning vines. If the warning vines bloom earlier than usual, it signals a prolonged dry season. Based on indigenous ecological knowledge, protected area managers can optimise fire prevention measures in advance, such as clearing firebreaks early and stockpiling firefighting materials, reducing the probability and impact of forest fires (Tourneau, 2007).

Additionally, integrating Indigenous cultural preservation with ecological protection can enhance local community awareness of conservation (Southgate, 1994). Australian Aboriginal Dreamtime regards elements of nature, such as land and species, as embodiments of ancestors, endowing conservation behaviours with meaning related to reverence for ancestors and cultural transmission. This cultural level guidance and constraint stimulate conservation consciousness among Indigenous communities. Species extinction rates are 50% lower in protected areas managed primarily by Indigenous peoples compared to different management models (Ian Leiper et al., 2018), highlighting the effectiveness of integrating Indigenous cultural preservation with ecological protection.

It is essential to acknowledge that leveraging the conservation value of Indigenous traditional knowledge necessitates respect for Indigenous knowledge. This ensures that Indigenous communities can participate in the use and management of traditional knowledge, thereby preventing misuse and rights violations from the outset (Jennifer Laughlin et al., 2013). For example, the cooperation agreement between the Costa Rican government and the Bribri people stipulates that when companies utilise the traditional herbal knowledge of the Bribri to develop medicinal plants from the rainforest, 10% of the patent revenues must be paid to the tribe. This measure safeguards the knowledge rights and economic benefits of Indigenous peoples, enhancing their initiative and enthusiasm for participating in ecological conservation work and creating a cycle of protection and development (Krol, 2024).

4.6 GLOBAL COLLABORATION AND TRANSBOUNDARY CONSERVATION: ECOSYSTEM PROTECTION BEYOND BORDERS

The distribution of ecosystems naturally has cross-border properties, and many core ecosystems (such as cross-border rivers, contiguous forests, and transnational wetlands) often span the administrative boundaries of multiple countries. At the same time, the phenomenon of cross-border species migration is also widespread. Based on this objective reality, ecological protection work must transcend national boundaries and actively promote the development and implementation of global collaborative cooperation and cross-border protection mechanisms. From a practical perspective, global cooperation can effectively integrate the resource endowments and professional advantages of various countries to form a joint force to address cross-border ecological challenges. Targeted cross-border conservation measures can ensure the structural integrity and functional connectivity of transnational ecosystems, providing a solid guarantee for the survival and reproduction of migratory species (Mason et al., 2020).

At present, through a series of key international treaties and regional agreements, a relatively complete legal coordination framework for cross-border ecological protection has been gradually built, providing institutional support for the coordinated actions of all countries: among them, the Convention on Biological Diversity and the Ramsar Convention provide core guidance for the formulation and implementation of cross-border conservation plans, and promote the development of multinational joint conservation actions; The Convention on International Trade in Endangered Species of Wild Fauna and Flora (CITES) effectively curbs the threat of illegal wildlife trade to cross-border species by regulating the international trade order of endangered species. At the regional level, various regional cooperation frameworks have also played a significant role, such as the African Sahel-Saharan Conservation Initiative and the European Alpine Convention, which have targeted and addressed significant regional cross-border ecological challenges.

At the practical level, successful cases of multinational joint protection continue to emerge: the Mekong River Basin covers six countries, and countries have significantly enhanced the resilience of river ecosystems through coordinated measures such as building a unified ecological monitoring system and carrying out joint anti-poaching operations; Coincidentally, eight countries in the Alps have joined hands to build a network of cross-border protected areas, and have achieved remarkable results in two major areas: biodiversity conservation and regional landscape maintenance by promoting the construction of ecological corridors, establishing standardised monitoring systems, and implementing coordinated tourism management.

4.6.1 The Importance of Global Collaboration in Ecosystem and Species Protection, Particularly in Transboundary Ecosystems

International treaties and agreements are the legal cornerstone for building a global system of collaborative cooperation and cross-border ecological protection. International treaties and agreement instruments effectively coordinate the dispersed conservation actions of various countries by establishing unified protection goals, implementation standards, and operational measures.

Among numerous international treaties, the Convention on Biological Diversity (CBD) is the foundational treaty in the field of global biodiversity conservation. CBD has been ratified by 196 parties, which have collectively ratified the CBD and provide the central framework for global cooperative efforts to protect biodiversity. The objectives of CBD cover conserving biological diversity, sustainably using its components, and fairly and equitably sharing the benefits arising from the utilisation of genetic resources (Nations, 1992). To facilitate the achievement of the objectives of the CBD, the Post-2020 Global Biodiversity Framework has been established, which introduces the 30×30 Target aiming to ensure that at least 30% of terrestrial and marine areas are conserved by 2030 (Grumbine & Xu, 2021). Under the requirements of the CBD Convention, each party is obliged to develop national biodiversity action plans based on this framework, regularly update the progress of implementation, and submit reports to the Convention Secretariat. This mechanism provides clear guidance and target benchmarks for global biodiversity conservation and enhances the coordination and effectiveness of international cooperation (Grumbine & Xu, 2021).

This coordinating role of international cooperation is evident in cross-border conservation practices, while the CBD also plays a leading role (Dunning, 2022). For example, under the guidance of the CBD convention, Costa Rica and Panama negotiated to establish the Cordillera de Talamanca transboundary protected area. This transboundary protected area focused on the joint conservation of migratory species such as green turtles and provided an example of international collaboration in cross-border ecological protection. By implementing unified protection standards and synchronised monitoring measures, Costa Rica and Panama have effectively improved the breeding environment of green sea turtles, significantly increased nesting rates, and gradually stabilised population numbers, providing a successful example for the conservation of cross-border migratory species.

The Ramsar Convention, which focuses on the ecological protection of wetlands, currently has 172 Parties participating (Pittock, 2021). The core mechanism of the convention is to inscribe wetlands with important ecological functions on the list of "wetlands of international importance" and requires Parties to implement special protection measures for wetlands on the list, while actively promoting international cooperation on cross-border wetlands. Among the 2,400 internationally important wetlands identified in the world, 30% are transboundary wetlands, and the Sino-Russian cross-border Xingkai Lake wetland is a typical representative (Fengkun Wang et al., 2007). Relying on the Ramsar Convention, China and Russia have established a cooperation mechanism to carry out annual joint statistical monitoring of waterbirds in Xingkai Lake Wetland, effectively improving the wintering environment of waterbirds by sharing monitoring data and implementing unified protection and control measures, promoting the steady growth of the number of wintering waterbirds, and effectively maintaining the integrity and functional stability of the wetland ecosystem.

Compared with the universal norms of global conventions, regional agreements focus on the ecological problems of specific regions and formulate differentiated protection measures and cooperation mechanisms based on regional natural endowments and development needs, which play an irreplaceable key role in regional ecological protection (Russian Federation and the Government of the People's Republic of China, 1996). For example, the Sahel-Sahara Protected Areas Network in Africa brings together the strengths of 14 countries with the core goal of protecting the unique desert ecosystems and rare species resources of the Sahel and Sahara region (Brito et al., 2016). The network has significantly improved the effectiveness of regional anti-poaching law enforcement by standardising anti-poaching patrol equipment configuration standards and establishing an intelligence-sharing platform. Thanks to this cross-border cooperation, the number of endangered species such as African wild dogs has grown steadily over the past decade (Brito et al., 2016).

The European Alpine Convention, co-signed by eight Alpine countries, focuses on regional ecological challenges, including glacier melting, hindered species migration, and environmental pressures caused by excessive tourism development (Chauvier-Mendes et al., 2024). To address these challenges, a standardised monitoring system has been established under the framework of the Convention, unifying the measurement of glacier retreat rates and species statistics standards, and promoting the comparison and sharing of monitoring data across various countries. At the same time, by coordinating tourism management policies in various countries and implementing control measures such as restricting the disorderly expansion of ski resorts, the disturbance of tourism activities on the ecosystem in the Alps has been effectively reduced (Chauvier-Mendes et al., 2024).

In general, international treaties and agreements have greatly reduced the communication and implementation costs of global cooperation and cross-border protection by establishing a unified standard system and effectively resolving the obstacles to coordination between countries. Taking the CITES as an example, the convention has reduced global illegal wildlife trade by 30% by regulating the international trade order of endangered species and clearly implementing control measures such as the ivory trade ban, providing a strong guarantee for the survival of many endangered species (United Nations Office on Drugs and Crime, 2024).

4.6.2 Examples of Transnational Collaborations, Analysing How These Partnerships Achieve Regional or Global Conservation Goals

Transnational cooperation serves as a pivotal approach to achieving cross-border conservation. It establishes collaborative mechanisms, delineates responsibilities and divisions of labour among nations, and implements joint actions to protect transboundary ecosystems and species collectively.

Successful cases of transnational cooperation can provide valuable references for cross-border conservation efforts in other regions (Mason et al., 2020).

Case 1: Transboundary Conservation in the Mekong River Basin (China, Laos, Myanmar, Thailand, Cambodia, and Vietnam)

The Mekong River, spanning 4,880 km and traversing six countries—China, Laos, Myanmar, Thailand, Cambodia, and Vietnam—is one of the most biodiverse rivers globally (LMEC, 2020). However, the Mekong River Basin faces numerous ecological challenges, including overfishing, habitat degradation, and water resource exploitation (Yu et al., 2024).

To protect the ecosystem of the Mekong River Basin, the 1995 Mekong Agreement, signed by the four signatory countries, decided to cooperate in developing, managing, using, and conserving water resources in areas such as fisheries, flood control, irrigation, hydropower, and navigation. The MRC adopted five procedures on water quality, data sharing, water use monitoring, water flow maintenance, and water use consultation to support implementation of the 1995 Mekong agreement and to help member countries promote fisheries' sustainability, opportunities for agriculture, freedom of navigation, sustainable hydropower, flood management, and preservation and conservation of important ecosystems (MRC, 1995).

In terms of joint actions, the countries standardised fish monitoring methods (e.g., using standardised gillnet surveys), established a data-sharing platform, and shared information on fish population sizes and distributions. Annual cross-border anti-poaching patrols are conducted to dismantle illegal fish cages, confiscate unlawful fishing gear, and combat cross-border illegal fishing activities (Yu et al., 2024). Through these collaborative measures, the declining trend in the capture of endangered species such as the Mekong giant catfish has been curbed, the forest coverage rate in the basin has stabilised at 65%, and the ecosystem of the Mekong River Basin has been protected to a certain extent (Yu et al., 2024).

Case 2: The Transboundary Protected Area Network in the European Alps

Spanning eight countries, including France, Italy, and Switzerland, the Alps are a core ecosystem of great ecological value in Europe, rich not only in biodiversity but also in unique natural landscapes. However, due to the dual impact of global climate change and human activities, the region is facing multiple ecological challenges, including the continued retreat of glaciers (the rate of retreat is about 30 m/decade), obstructed species migration, and ecological pressures caused by excessive tourism development (Chauvier-Mendes et al., 2024; Sommer et al., 2020). In order to systematically address the above ecological challenges, the eight countries have jointly built a network of cross-border protected areas in the Alps and implemented a series of coordinated control measures. In terms of ecological connectivity, the border between France and Italy has been specially preserved to maintain native alpine meadow corridors, providing safe migration paths for region-endemic species, such as the Alpine sheep. As an iconic endemic species in the region, the Alpine sheep relies on this cross-border corridor to facilitate free migration and activities between protected areas in different countries, effectively overcoming the restrictions on population dispersal at administrative boundaries (Kohier et al., 2008).

At the level of ecological monitoring, the eight countries reached a consensus to unify the monitoring standards and technical methods of glacier dynamics and species resources. Taking Alpine sheep monitoring as an example, countries adopt a standardised survey and recording process to systematically track their population size and distribution range and integrate monitoring data through cross-border data sharing platforms to comprehensively and accurately grasp the dynamic changes in populations. In terms of tourism ecological control, countries have coordinated ski resort development plans, strictly limited the scale of new tourism facilities such as cable cars, and actively promoted low-impact tourism activities such as hiking to reduce the disturbance of tourism development to regional ecosystems (Angelini et al., 2023; Haeberli et al., 2007); On the one hand, the frequency of cross-border migration of Alpine sheep has increased significantly, and the genetic exchange between populations has been strengthened, effectively maintaining the genetic diversity of the population. On the other hand, through the implementation of supporting measures such as cross-border water transfer, the water shortage caused by glacier retreat has been successfully alleviated, providing a solid guarantee for regional water security (d'Aosta, 2024).

The region's cross-border cooperation practice has accumulated valuable experience, which can be summarised into three core points: First, adhere to ecological needs-oriented cooperation goals, such as the Mekong River Basin cooperation focuses on fish migration protection and water resource protection, while the Alpine region focuses on core issues such as smooth species migration and glacier retreat; Second, establish a fair and just collaborative governance mechanism, and avoid the risk of strong country-led cooperation through institutional arrangements such as the rotating chairmanship system of the eight countries, so as to ensure that the interests and demands of all parties can be fully respected and reflected. Third, establish a binding benefit-sharing mechanism, enabling participating countries to effectively benefit from cooperation through the cross-border sharing of water resources and the equitable distribution of tourism revenue, thereby significantly enhancing the willingness and initiative of all countries to participate in cross-border protection.

4.7 CONCLUSION

Protection of ecosystems and protected areas is a complex undertaking that requires consideration of ecological, economic, and social factors, along with the simultaneous implementation of diversified strategies and safeguard measures.

We conclude that ecological conservation must adhere to principles such as ecosystem integrity, dynamic equilibrium, and sustainable utilisation. These principles serve as guidelines for various conservation practices, ensuring that protection efforts are aligned with the intrinsic functions of ecosystems. Building upon the adherence to ecosystem integrity, dynamic equilibrium, and sustainable utilisation, it is vital to ensure normal ecosystem functioning and achieve the sustainable provision of ecosystem services. This system will enhance the level and overall effectiveness of conservation efforts.

In ecological conservation efforts, strengthening ecological protection barriers requires scientifically planned and optimised protected areas. The planning of protected areas should be goal-oriented, considering dimensions such as species survival needs, habitat restoration needs, and ecosystem service supply. In protected areas, connectivity-building measures such as ecological corridors and buffer zones can promote cross-border species migration and gene exchange among populations, thereby strengthening the overall network of protected areas.

The process of ecological conservation faces the challenge of balancing environmental protection with human development needs. Multifunctional protected areas and sustainable management models provide approaches to reconcile the conflict between conservation and development. By integrating conservation objectives with the development of industries such as agriculture, fisheries, and eco-tourism during the establishment of protected areas, the scientific and rational utilisation of ecological resources can be achieved. Meanwhile, the establishment of community engagement mechanisms and benefit-sharing systems enhances the initiative and participation of local residents in conservation efforts.

Enhanced effectiveness of conservation efforts represents one of the objectives in contemporary ecological preservation. The integration of traditional knowledge with modern technology has innovative momentum in ecological protection. Traditional knowledge is distinguished by its regional specificity and local adaptability, whereas modern technology offers advantages in precision and efficiency. The synergy between the two can elevate both the efficacy and efficiency of conservation initiatives.

In the context of ecological conservation, cross-border environmental issues have emerged as constraints in contemporary transnational conservation efforts. Global cooperation and transboundary protection represent an approach to addressing these challenges, aligning with the integrated and transboundary nature of ecosystems. International treaties and multilateral cooperation mechanisms provide a legal and institutional foundation for collaborative conservation.

Ecological protection should further enhance collaboration and coordination across various fields and industries, integrate diverse protection strategies and measures, consolidate comprehensive protection efforts, and promote the sustainability and development of ecological protection. On one hand, it is necessary to continuously innovate conservation concepts and technological methods, fully leveraging the advantages of modern technology while respecting and preserving the traditional ecological knowledge of indigenous peoples. On the other hand, global conservation cooperation and exchange need to be deepened to address global ecological challenges. Only in this way can ecological systems achieve sustainable development and lay a solid ecological foundation for the sustainable development of human society.

DECLARATION OF GENERATIVE AI USE

During the preparation of this manuscript, the author(s) utilised DeepSeek and ChatGPT to review and enhance the English grammar. The content was subsequently examined and revised by the author(s), who assume full responsibility for the final version of the published work.

REFERENCES

Agardy, T. (2000). Effects of fisheries on marine ecosystems: a conservationist's perspective. *ICES J Mar Sci, 57*(3), 761–765. https://doi.org/10.1006/jmsc.2000.0721

Ahlstrom, A., Canadell, J. G., Schurgers, G., Wu, M., Berry, J. A., Guan, K., & Jackson, R. B. (2017). Hydrologic resilience and Amazon productivity. *Nat Commun, 8*(1), 387. https://doi.org/10.1038/s41467-017-00306-z

Alcala, A. C., & Russ, G. R. (2006). No-take marine reserves and reef fisheries management in the Philippines: a new people power revolution. *Ambio, 35*(5), 245–254. https://doi.org/10.1579/05-a-054r1.1

Angelini, F., Pollayil, M. J., Valle, B., Borgatti, M. S., Caccianiga, M., & Garabini, M. (2023). Robotic monitoring of Alpine screes: a dataset from the EU Natura2000 habitat 8110 in the Italian Alps. *Sci Data, 10*(1), 855. https://doi.org/10.1038/s41597-023-02764-1

Barabás, G., Michalska-Smith, M. J., & Allesina, S. (2017). Self-regulation and the stability of large ecological networks. *Nat Ecol Evol, 1*(12), 1870–1875. https://doi.org/10.1038/s41559-017-0357-6

Bartlam-Brooks, H. L., Bonyongo, M. C., & Harris, S. (2013). How landscape scale changes affect ecological processes in conservation areas: external factors influence land use by zebra (Equus burchelli) in the Okavango Delta. *Ecol Evol, 3*(9), 2795–2805. https://doi.org/10.1002/ece3.676

Bay, L. K., Gilmour, J., Muir, B., & Hardisty, P. E. (2023). Management approaches to conserve Australia's marine ecosystem under climate change. *Science, 381*(6658), 631–636. https://doi.org/10.1126/science.adi3023

Beltrão, M. G., Gonçalves, C. F., Brancalion, P. H. S., Carmignotto, A. P., Silveira, L. F., Galetti, P. M., Jr., & Galetti, M. (2024). Priority areas and implementation of ecological corridor through forest restoration to safeguard biodiversity. *Sci Rep, 14*(1), 30837. https://doi.org/10.1038/s41598-024-81483-y

Brancalion, P. H. S., de Almeida, D. R. A., Vidal, E., Molin, P. G., Sontag, V. E., Souza, S., & Schulze, M. D. (2018). Fake legal logging in the Brazilian Amazon. *Sci Adv, 4*(8), eaat1192. https://doi.org/10.1126/sciadv.aat1192

Brattland, C., Eythórsson, E., Weines, J., & Sunnanå, K. (2019). Social-ecological timelines to explore human adaptation to coastal change. *Ambio, 48*(12), 1516–1529. https://doi.org/10.1007/s13280-018-1129-5

Brennan, A., Naidoo, R., Greenstreet, L., Mehrabi, Z., Ramankutty, N., & Kremen, C. (2022). Functional connectivity of the world's protected areas. *Science, 376*(6597), 1101–1104. https://doi.org/10.1126/science.abl8974

Brito, J. C., Tarroso, P., Vale, C. G., Martínez-Freiría, F., Boratyński, Z., Campos, J. C., Ferreira, S., Godinho, R., Gonçalves, D. V., Leite, J. V., Lima, V. O., Pereira, P., Santos, X., da Silva, M. J. F., Silva, T. L., Velo-Antón, G., Veríssimo, J., Crochet, P. -A., Pleguezuelos, J. M., & Carvalho, S. B. (2016). Conservation biogeography of the Sahara-Sahel: additional protected areas are needed to secure unique biodiversity. *Divers Distrib, 22*(4), 14. https://doi.org/10.1111/ddi.12416

Bruggeman, J. E., Garrott, R. A., White, P. J., Watson, F. G., & Wallen, R. (2007). Covariates affecting spatial variability in bison travel behavior in Yellowstone National Park. *Ecol Appl, 17*(5), 1411–1423. https://doi.org/10.1890/06-0196.1

Burger, J., Gochfeld, M., Brown, K. G., Ng, K., Cortes, M., & Kosson, D. (2024). The importance of recognizing buffer zones to lands being developed, restored, or remediated: on planning for protection of ecological resources. *J Toxicol Environ Health A*, *87*(4), 133–149. https://doi.org/10.1080/15287394.2023.2285511

CAG. (2025). *CAG Slams Bihar govt for failing to utilise funds under 'Namami Gange' programme.* https://economictimes.indiatimes.com/news/india/cag-slams-bihar-govt-for-failing-to-utilise-funds-under-namami-gange-programme/articleshow/90635730.cms

Caron, S., Garvey, S. M., Gewirtzman, J., Schultz, K., Bhatnagar, J. M., Driscoll, C., Hutyra, L. R., & Templer, P. H. (2023). Urbanization and fragmentation have opposing effects on soil nitrogen availability in temperate forest ecosystems. *Glob Chang Biol*, *29*(8), 2156–2171. https://doi.org/10.1111/gcb.16611

Caves, J. K., Bodner, G. S., Simms, K., Fisher, L. A., & Robertson, T. (2013). Integrating collaboration, adaptive management, and scenario-planning: experiences at las cienegas national conservation area. *Ecol Soc*, *18*(3), Article 43. https://doi.org/10.5751/ES-05749-180343

Chauvier-Mendes, Y., Pollock, L. J., Verburg, P. H., Karger, D. N., Pellissier, L., Lavergne, S., Zimmermann, N. E., & Thuiller, W. (2024). Transnational conservation to anticipate future plant shifts in Europe. *Nat Ecol Evol*, *8*(3), 454–466. https://doi.org/10.1038/s41559-023-02287-3

Christian Kitungwa Lugoma, J. X. (2023). Kundelungu National Park: a case study in environmental education for sustainable management of Congolese National Parks. *Open J Soc Sci*, *11*(7). https://doi.org/10.4236/jss.2023.117028

Coetzer, K., Witkowski, E., & Erasmus, B. (2015). Biosphere reserves as multi-use conservation landscapes: perspectives from the Kruger to Canyons Biosphere Reserve, South Africa. *S Afr J Bot*, *98*, 208. https://doi.org/10.1016/j.sajb.2015.03.147

Colman, D. R., Lindsay, M. R., Harnish, A., Bilbrey, E. M., Amenabar, M. J., Selensky, M. J., Fecteau, K. M., Debes, R. V., 2nd, Stott, M. B., Shock, E. L., & Boyd, E. S. (2021). Seasonal hydrologic and geologic forcing drive hot spring geochemistry and microbial biodiversity. *Environ Microbiol*, *23*(7), 4034–4053. https://doi.org/10.1111/1462-2920.15617

d'Aosta, E. L. L. P. e. V. (2024). *From distressing to addressing: HKA tackles alpine climate crisis.* https://www.innovations-report.com/ecology-the-environment-and-conservation/from-distressing-to-addressing-hka-tackles-alpine-climate-crisis/

Daily, G. C., & Matson, P. A. (2008). Ecosystem services: from theory to implementation. *Proc Natl Acad Sci U S A*, *105*(28), 9455–9456. https://doi.org/10.1073/pnas.0804960105

Dambacher, J. M., Rossignol, P. A., Li, H. W., & Emlen, J. M. (2001). Dam Breaching and Chinook Salmon recovery. *Science*, *291*(5506), 939–939. https://doi.org/10.1126/science.291.5506.939a

Denninger Snyder, K., Mneney, P. B., & Wittemyer, G. (2019). Predicting the risk of illegal activity and evaluating law enforcement interventions in the western Serengeti. *Conserv Sci Pract*, *1*(9), e81. https://doi.org/https://doi.org/10.1111/csp2.81

Díaz, S., Settele, J., Brondízio, E. S., Ngo, H. T., Guèze, M., Agard, J., Arneth, A., Balvanera, P., Brauman, K. A., Butchart, S. H. M., Chan, K. M. A., Garibaldi, L. A., Ichii, K., Liu, J., Subramanian, S. M., Midgley, G. F., Miloslavich, P., Molnár, Z., Obura, D., Pfaff, A., Polasky, S., Purvis, A., Razzaque, J., Reyers, B., Roy Chowdhury, R., Shin, Y. J., Visseren-Hamakers, I. J., Willis, K. J., & Zayas, C. N. (eds.). (2019). *Summary for policymakers of the global assessment report on biodiversity and ecosystem services of the Intergovernmental Science-Policy Platform on Biodiversity and Ecosystem Services.*

Dixon, R. K., Solomon, A. M., Brown, S., Houghton, R. A., Trexier, M. C., & Wisniewski, J. (1994). Carbon pools and flux of global forest ecosystems. *Science*, *263*(5144), 185–190. https://doi.org/10.1126/science.263.5144.185

Dolrenry, S., Hazzah, L., & Frank, L. G. (2016). Conservation and monitoring of a persecuted African lion population by Maasai warriors. *Conserv Biol*, *30*(3), 467–475. https://doi.org/10.1111/cobi.12703

Dong, S., Shang, Z., Gao, J., & Boone, R. B. (2020). Enhancing sustainability of grassland ecosystems through ecological restoration and grazing management in an era of climate change on Qinghai-Tibetan Plateau. *Agric Ecosyst Environ*, *287*, 106684. https://doi.org/10.1016/j.agee.2019.106684

Dunning, K. (2022). Biodiversity conservation policy in megadiverse countries: comparing policy systems for 2020 targets to inform management in the coming decades. *J Environ Manage*, *302*(Pt A), 113815. https://doi.org/10.1016/j.jenvman.2021.113815

Edwards, A., Archer, R., De Bruyn, P., Evans, J., Lewis, B., Vigilante, T., Whyte, S., & Russell-Smith, J. (2021). Transforming fire management in northern Australia through successful implementation of savanna burning emissions reductions projects. *J Environ Manage*, *290*, 112568. https://doi.org/10.1016/j.jenvman.2021.112568

Feng, S., Fang, Q., Barnett, R., Li, C., Han, S., Kuhlwilm, M., Zhou, L., Pan, H., Deng, Y., Chen, G., Gamauf, A., Woog, F., Prys-Jones, R., Marques-Bonet, T., Gilbert, M. T. P., & Zhang, G. (2019). The Genomic footprints of the fall and recovery of the Crested ibis. *Curr Biol*, *29*(2), 340–349.e347. https://doi.org/10.1016/j.cub.2018.12.008

Fischer, S. E., Kramer, G. R., Baumann, M. J., Fahy, E. F., Granillo, K., & Streby, H. M. (2025). Migratory connectivity and potential nonbreeding sexual segregation in Gray Vireos (Vireo vicinior) Conectividad migratoria y potencial segregación sexual no-reproductiva en Vireo vicinior. *Wilson J Ornithol*, *137*(3). https://doi.org/https://doi.org/10.1080/15594491.2025.2497654

Gao, J., Zou, C., Zhang, K., Xu, M., & Wang, Y. (2020). The establishment of Chinese ecological conservation redline and insights into improving international protected areas. *J Environ Manage*, *264*, 110505. https://doi.org/10.1016/j.jenvman.2020.110505

Geedicke, I., Oldeland, J., & Leishman, M. R. (2018). Urban stormwater run-off promotes compression of saltmarshes by freshwater plants and mangrove forests. *Sci Total Environ*, *637–638*, 137–144. https://doi.org/10.1016/j.scitotenv.2018.04.357

Gilmour, M. E., Pollock, K., Adams, J., Block, B. A., Caselle, J. E., Filous, A., Friedlander, A. M., Game, E. T., Hazen, E. L., Hill, M., Holmes, N. D., Lafferty, K. D., Maxwell, S. M., McCauley, D. J., Schallert, R., Shaffer, S. A., Wolff, N. H., & Wegmann, A. (2025). Multi-species telemetry quantifies current and future efficacy of a remote marine protected area. *Glob Chang Biol*, *31*(4), e70138. https://doi.org/10.1111/gcb.70138

Grumbine, R. E., & Xu, J. (2021). Five steps to inject transformative change into the post-2020 global biodiversity framework. *Bioscience*, *71*(6), 637–646. https://doi.org/10.1093/biosci/biab013

Haeberli, W., Hoelzle, M., Paul, F., & Zemp, M. (2007). Integrated monitoring of mountain glaciers as key indicators of global climate change: the European Alps. *Anna Glaciol*, *46*, 150–160. https://doi.org/10.3189/172756407782871512

He, S. Y. L., & Min, Q. (2020). Community participation in nature conservation: the Chinese experience and its implication to National Park Management. *Sustainability*, *12*(11). https://doi.org/10.3390/su12114760

Hinrichsen, D. (1997). Coral reefs in crisis. *People Planet*, *6*(2), 6–11.

Hughes, R. M. (2020). Ecological Integrity: Conceptual Foundations and Application. *Oxford Bibliographies*. https://doi.org/10.1093/obo/9780199363445-0113

Hughes, T. P., Kerry, J. T., Alvarez-Noriega, M., Alvarez-Romero, J. G., Anderson, K. D., Baird, A. H., Babcock, R. C., Beger, M., Bellwood, D. R., Berkelmans, R., Bridge, T. C., Butler, I. R., Byrne, M., Cantin, N. E., Comeau, S., Connolly, S. R., Cumming, G. S., Dalton, S. J., Diaz-Pulido, G., … Wilson, S. K. (2017). Global warming and recurrent mass bleaching of corals. *Nature*, *543*(7645), 373–377. https://doi.org/10.1038/nature21707

Ian Leiper, K. Z., Robinson, C., Carwadine, J., Moggridge, B. J., & Garnett, S. (2018). Quantifying current and potential contributions of Australian Indigenous peoples to threatened species management. *Conserv Biol*, *32*(5), 1038–1047. https://doi.org/10.1111/cobi.13178

IPBES. (2019). *Global Assessment Report on Biodiversity and Ecosystem Services of the Intergovernmental Science-Policy Platform on Biodiversity and Ecosystem Services*. IPBES Secretariat, Bonn, Germany.

Jackson, J. B., Kirby, M. X., Berger, W. H., Bjorndal, K. A., Botsford, L. W., Bourque, B. J., Bradbury, R. H., Cooke, R., Erlandson, J., Estes, J. A., Hughes, T. P., Kidwell, S., Lange, C. B., Lenihan, H. S., Pandolfi, J. M., Peterson, C. H., Steneck, R. S., Tegner, M. J., & Warner, R. R. (2001). Historical overfishing and the recent collapse of coastal ecosystems. *Science*, *293*(5530), 629–637. https://doi.org/10.1126/science.1059199

Jambeck, J. R., Geyer, R., Wilcox, C., Siegler, T. R., Perryman, M., Andrady, A., Narayan, R., & Law, K. L. (2015). Marine pollution. Plastic waste inputs from land into the ocean. *Science*, *347*(6223), 768–771. https://doi.org/10.1126/science.1260352

Jat, H., Datta, A., Choudhary, M., Sharma, P., & Jat, M. (2020). Conservation agriculture: factors and drivers of adoption and scalable innovative practices in Indo-Gangetic plains of India: a review. *Int J Agric Sustain*, *18*. https://doi.org/10.1080/14735903.2020.1817655

Jennifer Laughlin, C. M., Sriskanthan, G., & Kantcheva, N. (2013). *Statement and guidelines on free, prior, informed consent*. UN-REDD Programme Secretariat. https://www.fao.org/family-farming/detail/en/c/1739035/

Jensen, M. O. (2000). Regional management program at Yellowstone National Park. *Environments; Waterloo*, *28*, 37–46.

John Bryant, K. J., Chapin, F., DeAngelis, D. L., & Kielland, K. (2014). Can antibrowsing defense regulate the spread of woody vegetation in arctic tundra? *Ecography*, *37*(3), 204–211. https://doi.org/10.1111/j.1600-0587.2013.00436.x

Kang, D. W., Zhao, Z. J., Guo, W. X., Tan, L. Y., Kang, W., & Li, J. Q. (2011). [Habitat selection attributes of giant panda]. *Ying Yong Sheng Tai Xue Bao*, *22*(2), 519–525.

Karel Mokany, T. D. H., & Ferrier, S. (2013). Comparing habitat configuration strategies for retaining biodiversity under climate change. *J Appl Ecol*, *50*(2), 519–527. https://doi.org/10.1111/1365-2664.12038

Kenny, T. A., Fillion, M., Simpkin, S., Wesche, S. D., & Chan, H. M. (2018). Caribou (Rangifer tarandus) and inuit nutrition security in Canada. *Ecohealth*, *15*(3), 590–607. https://doi.org/10.1007/s10393-018-1348-z

Khanal, B., & Gosai, K. R. (2024). Evaluation of ecotourism practices: a comparative study of lowlands and mountains of Nepal. *J Tourism Himalayan Adventures*, *6*(1), 58–71. https://doi.org/10.3126/jtha.v6i1.67392

Kohier, Y., Plassmann, G., Ullrich, A., Götz, A., Scheurer, T., Hölscher, S., & Savoia, S. (2008). The continuum project: establishing ecological networks throughout the European Alps. *Mt Res Dev*, *28*(2), 168–172. https://www.jstor.org/stable/25164208

Krol, E. (2024). *Natural medicine use in Costa Rica*. Western Michigan University. https://scholarworks.wmich.edu/honors_theses/3938

Lacher, T. E., Butchart, S. H. M., Gumbs, R., Long, B., Lopez-Gallego, C., Raimondo, D., Simkins, A. T., Sunarto, S., & Hoffmann, M. (2025). The status, threats and conservation of critically endangered species. *Nat Rev Biodiver*, *1*(7), 421–438. https://doi.org/10.1038/s44358-025-00059-4

Langhammer, P. F., Bull, J. W., Bicknell, J. E., Oakley, J. L., Brown, M. H., Bruford, M. W., Butchart, S. H. M., Carr, J. A., Church, D., Cooney, R., Cutajar, S., Foden, W., Foster, M. N., Gascon, C., Geldmann, J., Genovesi, P., Hoffmann, M., Howard-McCombe, J., Lewis, T., … Brooks, T. M. (2024). The positive impact of conservation action. *Science*, *384*(6694), 453–458. https://doi.org/10.1126/science.adj6598

Laurance, W. F., Delamônica, P., Laurance, S. G., Vasconcelos, H. L., & Lovejoy, T. E. (2000). Rainforest fragmentation kills big trees. *Nature*, *404*(6780), 836. https://doi.org/10.1038/35009032

Lechler, S., Picoli, M. C. A., Soares, A. R., Sanchez, A., Chaves, M. E. D., & Verstegen, J. (2020). Exploring nasa's harmonized landsat and sentinel-2 (HLS) dataset to monitor deforestation in the amazon rainforest. *Int Arch Photogramm Remote Sens Spatial Inf Sci ISPRS Arch*, *43*(B3), 705–711. https://doi.org/10.5194/isprs-archives-XLIII-B3-2020-705-2020

Li, B., Yang, G., & Wan, R. (2023). Reassessment of the declines in the largest freshwater lake in China (Poyang Lake): uneven trends, risks and underlying causes. *J Environ Manage*, *342*, 118157. https://doi.org/10.1016/j.jenvman.2023.118157

Li, C., Chen, D., Wu, D., & Su, X. (2015). Design of an EIoT system for nature reserves: a case study in Shangri-La County, Yunnan Province, China. *Int J Sustain Develop World Ecol*, *22*(2), 184–188. https://doi.org/10.1080/13504509.2014.943329

Li, Y., Viña, A., Yang, W., Chen, X., Zhang, J., Ouyang, Z., Liang, Z., & Liu, J. (2013). Effects of conservation policies on forest cover change in Giant Panda Habitat Regions, China. *Land Use Policy*, *33*, 42–53. https://doi.org/10.1016/j.landusepol.2012.12.003

Lin, C. H., Lerch, R. N., Goyne, K. W., & Garrett, H. E. (2011). Reducing herbicides and veterinary antibiotics losses from agroecosystems using vegetative buffers. *J Environ Qual*, *40*(3), 791–799. https://doi.org/10.2134/jeq2010.0141

Lin, Q., Song, Y., Zhang, Y., Hao, J. L., & Wu, Z. (2022). Strategies for restoring and managing ecological corridors of freshwater ecosystem. *Int J Environ Res Public Health*, *19*(23). https://doi.org/10.3390/ijerph192315921

Liu, F., Feng, C., Zhou, Y., Zhang, L., Du, J., Huang, W., Luo, J., & Wang, W. (2022). Effectiveness of functional zones in National Nature Reserves for the protection of forest ecosystems in China. *J Environ Manage*, *308*, 114593. https://doi.org/10.1016/j.jenvman.2022.114593

Liu, J. (2020). Providing the "Chinese Solution" for anti-poaching efforts in Africa. *Man and Biosphere* (04), 67. https://kns.cnki.net/kcms2/article/abstract?v=5q1osi_AJOQo0M4rQc2s5h-jkyp7lGoTuXA76QItVIwGlIjFVLoGnOxe0gTD_0C4rYcQBiI50nSAIc5WNm_CTyg6X3E_3GIBKp1kE3WsYHPZN9iuCOLlHE0VFkh0yfs8o4UARKodt6YVybf2620OGQXdG0YIQkFrD9Lt3tUBUijlqFHOBRtcRWD6sGFX_j-e&uniplatform=NZKPT&language=CHS

Liu, J., Linderman, M., Ouyang, Z., An, L., Yang, J., & Zhang, H. (2001). Ecological degradation in protected areas: the case of Wolong Nature Reserve for giant pandas. *Science*, *292*(5514), 98–101. https://doi.org/10.1126/science.1058104

Liu, Y., Taong, Z., Han, M., Zou, L., & Zheng, B. (1991). An ecodynamic model of Lake Dianchi and its application. *Res Environ Sci*, *4*(6), 1–8. https://www.hjkxyj.org.cn/en/article/id/19910601

Ljubicic, G., Okpakok, S., Robertson, S., & Mearns, R. (2018). Uqsuqtuurmiut inuita tuktumi qaujimaningit (Inuit knowledge of caribou from Gjoa Haven, Nunavut): collaborative research contributions to co-management efforts. *Polar Record*, *54*(3), 213–233. https://doi.org/10.1017/S0032247418000372

Lorz, I. T., & Trueman, R. J. (2025). Nature's guardians: harnessing Indigenous wisdom for biodiversity preservation. *Biodiversity*, *26*(3), 197–198. https://doi.org/10.1080/14888386.2025.2517128

Lovelock, C. E., Barbier, E., & Duarte, C. M. (2022). Tackling the mangrove restoration challenge. *PLoS Biol*, *20*(10), e3001836. https://doi.org/10.1371/journal.pbio.3001836

Luque, G. M., Hochberg, M. E., Holyoak, M., Hossaert, M., Gaill, F., & Courchamp, F. (2013). Ecological effects of environmental change. *Ecol Lett*, *16*(Suppl 1), 1–3. https://doi.org/10.1111/ele.12050

Maeda, E. E., Abera, T. A., Siljander, M., Aragao, L., Moura, Y. M., & Heiskanen, J. (2021). Large-scale commodity agriculture exacerbates the climatic impacts of Amazonian deforestation. *Proc Natl Acad Sci U S A*, *118*(7). https://doi.org/10.1073/pnas.2023787118

Maezumi, S. Y., Alves, D., Robinson, M., de Souza, J. G., Levis, C., Barnett, R. L., Almeida de Oliveira, E., Urrego, D., Schaan, D., & Iriarte, J. (2018). The legacy of 4,500 years of polyculture agroforestry in the eastern Amazon. *Nat Plants*, *4*(8), 540–547. https://doi.org/10.1038/s41477-018-0205-y

Maron, M., Mitchell, M. G. E., Runting, R. K., Rhodes, J. R., Mace, G. M., Keith, D. A., & Watson, J. E. M. (2017). Towards a threat assessment framework for ecosystem services. *Trends Ecol Evol*, *32*(4), 240–248. https://doi.org/10.1016/j.tree.2016.12.011

Mason, N., Ward, M., Watson, J. E. M., Venter, O., & Runting, R. K. (2020). Global opportunities and challenges for transboundary conservation. *Nat Ecol Evol*, *4*(5), 694–701. https://doi.org/10.1038/s41559-020-1160-3

Mehaffey, M. H., Nash, M. S., Wade, T. G., Ebert, D. W., Jones, K. B., & Rager, A. (2005). Linking land cover and water quality in New York City's water supply watersheds. *Environ Monit Assess*, *107*(1–3), 29–44. https://doi.org/10.1007/s10661-005-2018-5

Merkle, J. A., Lowrey, B., Wallace, C. F., Hall, L. E., Wilde, L., Kauffman, M. J., & Sawyer, H. (2023). Conserving habitat for migratory ungulates: how wide is a migration corridor? *J Appl Ecol*, *60*(9), 1763–1770. https://doi.org/10.1111/1365-2664.14473

Montgomery, R. (2025). *Cost-benefit analysis of environmental projects: an integrated theoretical framework*. https://doi.org/10.13140/RG.2.2.16962.90562

MRC. (1995). *Agreement on the cooperation for the sustainable development of the Mekong River Basin*. http://www.lmcchina.org

Mukhiddinov, M., Abdusalomov, A. B., & Cho, J. (2022). A wildfire smoke detection system using unmanned aerial vehicle images based on the optimized YOLOv5. *Sensors (Basel)*, *22*(23). https://doi.org/10.3390/s22239384

National Academies of Sciences, Engineering, and Medicine; Division on Earth and Life Studies; Water Science and Technology Board; Committee to Review the New York City Watershed Protection Program. (2020). *Review of the New York City Watershed Protection Program*. Washington (DC): National Academies Press (US). https://www.ncbi.nlm.nih.gov/books/NBK566282/

Nations, U. (1992). Convention on Biological Diversity. https://www.cbd.int/

Nicholson, E., Watermeyer, K. E., Rowland, J. A., Sato, C. F., Stevenson, S. L., Andrade, A., Brooks, T. M., Burgess, N. D., Cheng, S. T., Grantham, H. S., Hill, S. L., Keith, D. A., Maron, M., Metzke, D., Murray, N. J., Nelson, C. R., Obura, D., Plumptre, A., Skowno, A. L., & Watson, J. E. M. (2021). Scientific foundations for an ecosystem goal, milestones and indicators for the post-2020 global biodiversity framework. *Nat Ecol Evol*, *5*(10), 1338–1349. https://doi.org/10.1038/s41559-021-01538-5

Olsson, P., Folke, C., & Hughes, T. P. (2008). Navigating the transition to ecosystem-based management of the Great Barrier Reef, Australia. *Proc Natl Acad Sci U S A*, *105*(28), 9489–9494. https://doi.org/10.1073/pnas.0706905105

Perkins, C. (2020). *Forest fires in yellowstone: the science of burning and regrowth*. https://www.nps.gov/articles/fires-yell.htm

Pichler-Koban, C., & Jungmeier, M. (2017). Alpine parks between yesterday and tomorrow: a conceptual history of Alpine national parks via tourism in charismatic parks in Austria, Germany and Switzerland. *eco.mont*, 9(special issue), 17–28. https://austriaca.at/?arp=0x0034cb8b

Pittock, J. (2021). Use of the Ramsar Convention to protect springs and other wetlands. *Conserv Biol*, *35*(6), 1969–1971. https://doi.org/10.1111/cobi.13799

Porto, R. N. (2024). Ecology under the falling sky: nature, ecology and entropy in Yanomami cosmology. *Anthrop Rev*, *11*(3), 509–524. https://doi.org/10.1177/20530196231211849

Praveen, J., & Nameer, P. O. (2009). Monitoring bird diversity in Western Ghats of Kerala. *Curr Sci*, *96*(10), 1390–1395. https://www.jstor.org/stable/24105380

Prosser, R. S., Hoekstra, P. F., Gene, S., Truman, C., White, M., & Hanson, M. L. (2020). A review of the effectiveness of vegetated buffers to mitigate pesticide and nutrient transport into surface waters from agricultural areas. *J Environ Manage*, *261*, 110210. https://doi.org/10.1016/j.jenvman.2020.110210

QU Yichen, CHEN Chuwen. Research and enlightenment of marine protected area planning based on co-management: a case study of Great Barrier Reef Marine Park[J]. *Forestry Construction*, 2022, *40*(4): 1–6.

Rakotonirina, T. J., Viljoen, E., Rakotonirina, A. H., Leong Pock Tsy, J. M., & Radanielina, T. (2023). A DNA barcode reference library for CITES listed Malagasy Dalbergia species. *Ecol Evol*, *13*(3), e9887. https://doi.org/10.1002/ece3.9887

Russian Federation and the Government of the People's Republic of China. (1996). *Agreement between the Government of the Russian Federation and the Government of the People's Republic of China on Hanka Lake nature reserve.* https://www.informea.org/en/content/legislation/agreement-between-government-russian-federation-and-government-peoples-20

Rosalina, D., Jamil, K., Rombe, K. H., Surachmat, A., Sabilah, A. A., Sari, S. P., & Wahda, A. N. (2023). Estimation of seagrass beds changes in the Barugaiya Village waters over a period of 5 years 2019–2023. *Pak J Biol Sci*, *26*(5), 266–278. https://doi.org/10.3923/pjbs.2023.266.278

Saura, S., Bastin, L., Battistella, L., Mandrici, A., & Dubois, G. (2017). Protected areas in the world's ecoregions: how well connected are they? *Ecol Indic*, *76*, 144–158. https://doi.org/10.1016/j.ecolind.2016.12.047

Schlossberg, S., Chase, M. J., & Griffin, C. R. (2018). Poaching and human encroachment reverse recovery of African savannah elephants in south-east Angola despite 14 years of peace. *PLoS One*, *13*(3), e0193469. https://doi.org/10.1371/journal.pone.0193469

Schuerch, M., Spencer, T., Temmerman, S., Kirwan, M. L., Wolff, C., Lincke, D., McOwen, C. J., Pickering, M. D., Reef, R., Vafeidis, A. T., Hinkel, J., Nicholls, R. J., & Brown, S. (2018). Future response of global coastal wetlands to sea-level rise. *Nature*, *561*(7722), 231–234. https://doi.org/10.1038/s41586-018-0476-5

Shang, J. (2018). *A Study on the Variations of Thermokarst Landslides in the Eboling Permafrost Region of the Qilian Mountains Based on Remote Sensing Technology* [硕士. https://kns.cnki.net/kcms2/article/abstract?v=5q1osi_AJOQLY60JtJ939vyrO1y6PDGfwZ7R99eHCKzhkXXwkCNWvGbCuXULzqXIDiOGXkC1nMKITdFlksolyDPsA72QQoUD19ZyOUin4yKdqTeWC80H-dXTTvatMybEK746o8r21axMWH1IOPfpHhAt8Y7Pl5IA-tix0E-IPZjIVmE427Me17akcMWxGlD&uniplatform=NZKPT&language=CHS

Sharma, S., Singh, D., Gawdiya, S., & Sharma, R. K. (2025). Trade-offs and synergies in biodiversity conservation and food security. In H. S. Jatav, V. D. Raiput, & T. Minkina (Eds.), *Ecologically Mediated Development: Promoting Biodiversity Conservation and Food Security* (pp. 195–218). Springer Nature Singapore. https://doi.org/10.1007/978-981-96-2413-3_9

Sinclair, A. R., Mduma, S. A., Hopcraft, J. G., Fryxell, J. M., Hilborn, R., & Thirgood, S. (2007). Long-term ecosystem dynamics in the Serengeti: lessons for conservation. *Conserv Biol*, *21*(3), 580–590. https://doi.org/10.1111/j.1523-1739.2007.00699.x

Slocombe, D. S. (1992). Environmental monitoring for protected areas: review and prospect. *Environ Monit Assess*, *21*(1), 49–78. https://doi.org/10.1007/bf00400057

Sommer, C., Malz, P., Seehaus, T. C., Lippl, S., Zemp, M., & Braun, M. H. (2020). Rapid glacier retreat and downwasting throughout the European Alps in the early 21(st) century. *Nat Commun*, *11*(1), 3209. https://doi.org/10.1038/s41467-020-16818-0

Southgate, D. A. (1994). Australian dreamtime. *Br J Nutr*, *71*(1), 1–2. https://doi.org/10.1079/bjn19940105

Stahler, D. R., Smith, D. W., & Guernsey, D. S. (2006). Foraging and feeding ecology of the gray wolf (Canis lupus): lessons from Yellowstone National Park, Wyoming, USA. *J Nutr*, *136*(7 Suppl), 1923s–1926s. https://doi.org/10.1093/jn/136.7.1923S

Straede, S., & Treue, T. (2006). Beyond buffer zone protection: a comparative study of park and buffer zone products' importance to villagers living inside Royal Chitwan National Park and to villagers living in its buffer zone. *J Environ Manage*, *78*(3), 251–267. https://doi.org/10.1016/j.jenvman.2005.03.017

Sun, Y., He, J., Xiang, Q., & Zhou, K. (2024). Leveraging intergovernmental data sharing for digital transformation in ecological and environmental protection. *J Clean Product*, *477*, 143780. https://doi.org/10.1016/j.jclepro.2024.143780

Sun Mengting, S. J., Wang, M., Zheng, J., Pan, J., & He, G. (2020). Current status and development prospects of the ecological ranch construction in Qiandao Lake. *Technol Drives Dev*, *2*, 249–253.

Temmink, R. J. M., Lamers, L. P. M., Angelini, C., Bouma, T. J., Fritz, C., van de Koppel, J., Lexmond, R., Rietkerk, M., Silliman, B. R., Joosten, H., & van der Heide, T. (2022). Recovering wetland biogeomorphic feedbacks to restore the world's biotic carbon hotspots. *Science*, *376*(6593), eabn1479. https://doi.org/10.1126/science.abn1479

Thatte, P., Joshi, A., Vaidyanathan, S., Landguth, E., & Ramakrishnan, U. (2018). Maintaining tiger connectivity and minimizing extinction into the next century: insights from landscape genetics and spatially-explicit simulations. *Biol Conserv*, *218*, 181–191. https://doi.org/10.1016/j.biocon.2017.12.022

Tourneau, B. A. F. M. L. (2007). Ethnogeography and resource use among the Yanomami toward a model of "reticular space". *Curr Anthropol*, *48*(4). https://doi.org/10.1086/519914

Tuljapurkar, S. D., & Semura, J. S. (1977). Dynamic equilibrium under periodic perturbations in simple ecosystem models. *J Theor Biol*, *66*(2), 327–343. https://doi.org/10.1016/0022-5193(77)90175-8

Turner, M. G., Whitby, T. G., Tinker, D. B., & Romme, W. H. (2016). Twenty-four years after theYellowstone Fires: are postfire lodgepole pine stands converging in structure and function? *Ecology*, *97*(5), 1260–1273. https://doi.org/10.1890/15-1585.1

UNEP-FI. (2025). *Governments adopt first global strategy to finance biodiversity: implications for financial institutions*. https://www.unepfi.org/themes/ecosystems/governments-adopt-first-global-strategy-to-finance-biodiversity-implications-for-financial-institutions/

United Nations Office on Drugs and Crime. (2024). *World Wildlife Crime Report 2024*. https://www.un-ilibrary.org/content/books/9789211064582

Upadhaya, S., Tiwari, S., Poudyal, B., Godar Chhetri, S., & Dhungana, N. (2022). Local people's perception of the impacts and importance of ecotourism in Central Nepal. *PLoS One*, *17*(5), e0268637. https://doi.org/10.1371/journal.pone.0268637

Utne, I. B. (2007). System evaluation of sustainability in the Norwegian cod-fisheries. *Marine Policy*, *31*(4), 390–401. https://doi.org/10.1016/j.marpol.2006.10.006

Valiela, I., Lloret, J., Bowyer, T., Miner, S., Remsen, D., Elmstrom, E., Cogswell, C., & Robert Thieler, E. (2018). Transient coastal landscapes: rising sea level threatens salt marshes. *Sci Total Environ*, *640–641*, 1148–1156. https://doi.org/10.1016/j.scitotenv.2018.05.235

Venäläinen, A., Lehtonen, I., Laapas, M., Ruosteenoja, K., Tikkanen, O. P., Viiri, H., Ikonen, V. P., & Peltola, H. (2020). Climate change induces multiple risks to boreal forests and forestry in Finland: a literature review. *Glob Chang Biol*, *26*(8), 4178–4196. https://doi.org/10.1111/gcb.15183

Wang, F., Cui, C., & Li, H. (2007). Analysis of the conditions for cross-border conservation of Xingkai Lake between China and Russia. *The Russian, Central Asian, and Eastern European Markets* (09), 35–38. https://kns.cnki.net/kcms2/article/abstract?v=5q1osi_AJOSGK791V1AzNOkBTCnCo7kkjQkC6sw6kumhvSe2mWOcIHViGHHCpMPw1DOR8c6gUSzN5GAWlEJuFAPGkD7f_sXgLJ4qDj64boXYZy-qGDnI-42SttaLn_4ZbYqEEWCPTWkWWP4AUQrAnKEqj0e8nxDo0DdqP2hRoI-11MuAOc8Io7Q==&uniplatform=NZKPT&language=CHS

Wang Jiaotuan, Q. Z., Chaowei, Z., & Jianye, C. (2008). Comprehensive evaluation, prediction, and prevention strategies for the water environment of Qiandao Lake. *Chin Agric Sci Bull*, *1*, 431–435.

Wasser, S. K., Brown, L., Mailand, C., Mondol, S., Clark, W., Laurie, C., & Weir, B. S. (2015). Conservation. Genetic assignment of large seizures of elephant ivory reveals Africa's major poaching hotspots. *Science*, *349*(6243), 84–87. https://doi.org/10.1126/science.aaa2457

Weiskopf, S. R., Rubenstein, M. A., Crozier, L, G., Gaichas, S., Griffis, R., Halofsky, J. E., Hyde, K. J. W., Morelli, T., Morisette, J. T., Munoz, R. C., Pershing, A. J., Peterson, D. L., Poudel, R., Staudinger, M. D., Sutton-Grier, A. E., Thompson, L., Vose, J., Weltzinn, J. F., & Whyte, K. P. (2020). Climate change effects on biodiversity, ecosystems, ecosystem services, and natural resource management in the United States. *Sci Total Environ*, *733*. https://doi.org/10.1016/j.scitotenv.2020.137782

Weng, C., Xu, M., Lei, F., & Rose, K. A. (2023). Management strategy of the naked carp (Gymnocypris przewalskii) in the Qinghai lake using matrix population modeling. *J Environ Manage*, *336*, 117596. https://doi.org/10.1016/j.jenvman.2023.117596

Wood, C. M., Du, J., Rogers, J., Brauner, C. J., Richards, J. G., Semple, J. W., Murray, B. W., Chen, X. Q., & Wang, Y. (2007). Przewalski's naked carp (Gymnocypris przewalskii): an endangered species taking a metabolic holiday in Lake Qinghai, China. *Physiol Biochem Zool*, *80*(1), 59–77. https://doi.org/10.1086/509212

Xiang, X. X., Wu, Z. L., Luo, K., Ding, H. B., & Zhang, H. Y. (2013). [Impacts of human disturbance on the species composition of higher plants in the wetlands around Dianchi Lake, Yunnan Province of Southwest China]. *Ying Yong Sheng Tai Xue Bao*, *24*(9), 2457–2463.

Yan, Z., Jinman, W., Yan, C., Lingxiao, Y., Xu, Z., Yu, F., Jing, & Ming. (2024). Technological path exploration for application of nature-based solutions to integrated ecological conservation and restoration project of mountains, rivers, forests, farmland, lakes, grassland and deserts. *China Land Sci*, *38*(6), 40–49. https://zgtdkxen/article/abstract/20240190

Yu, W. (2017). The "water fund" model for public participation in small watershed conservation. *World Environment* (03), 33–34. https://kns.cnki.net/kcms2/article/abstract?v=5q1osi_AJOSt1Mw88yH78p6vhLDIwWWgFYrztqK6KjWD2L6fRlqb3R4L_5jVQ8LoJP5yHCxqht9tquG-J2b8qQNE6rek7t-nRG9nijs_sZrdYOq3pZ5bSPvgeaQcsM3SDgwV7g4VfqNMvW-epcP-DehK9iH0rFDTfGuw1AFwCEFzLNsq77wMxTJZBSFBDzLbK&uniplatform=NZKPT&language=CHS

Yu, Y., Bo, Y., Castelletti, A., Dumas, P., Gao, J., Cai, X., Liu, J., Kahil, T., Wada, Y., Hu, S., Liu, B., Zhou, F., & Zhao, J. (2024). Transboundary cooperation in infrastructure operation generates economic and environmental co-benefits in the Lancang-Mekong River Basin. *Nat Water*, *2*(6), 589–601. https://doi.org/10.1038/s44221-024-00246-1

Zhang, Z., Liu, Q., Gao, G., Shao, J., Pan, J., He, G., & Hu, Z. (2023). Integrating ecosystem services closely related to human well-being into the restoration and management of deep lakes facing multiple stressors: lessons from long-term practice in Qiandao Lake, China. *Sci Total Environ*, *902*, 166457. https://doi.org/10.1016/j.scitotenv.2023.166457

Zhang, Z., Liu, Y., Li, Y., Wang, X., Li, H., Yang, H., Ding, W., Liao, Y., Tang, N., & He, F. (2022). Lake ecosystem health assessment using a novel hybrid decision-making framework in the Nam Co, Qinghai-Tibet Plateau. *Sci Total Environ*, *808*, 152087. https://doi.org/10.1016/j.scitotenv.2021.152087

Zhang, Z., Wei, Y., Tang, L., Wang, Z., Hu, X., Li, X., Bi, Y., & Huang, B. (2024). Co-management enhances social capital and recognition of protected area: perspectives from indigenous rangers on the Qinghai-Tibet Plateau. *J Environ Manage*, *5*(1). https://doi.org/10.1016/j.jenvman.2024.123346

Zheng, L., & Wang, S. (2025). Assessing marine resource carrying capacity: methods, economic impacts, and management strategies. *Water*, *17*, 691. https://doi.org/10.3390/w17050691

Zhu, J., Li, Z., Yang, J., Yu, K., Zhang, D., & Zhong, J. (2024). Ecological space management and control zoning of Giant Panda National Park from the perspective of ecosystem services and land use. *Sci Rep*, *14*(1), 19951. https://doi.org/10.1038/s41598-024-65344-2

5 Learning and Internalization

Norm-Building in China–ASEAN Biodiversity Governance from a "Global South" Perspective

Guanhui Jin

Biodiversity constitutes a vital foundation for human survival and sustainable development, underpinning both ecosystem stability and long-term human well-being. Yet, the accelerating global loss of biodiversity and the worsening degradation of ecosystems have made biodiversity conservation a major shared challenge for the international community. China and most ASEAN member states, as developing countries, face dual pressures of biodiversity loss and ecosystem degradation amid rapid economic growth, population expansion, and accelerating industrialization and urbanization. Although China and the majority of ASEAN countries became parties to the Convention on Biological Diversity (CBD) relatively early—thereby accepting the international community's basic norms for biodiversity governance—formal cooperation between the two sides in this field was not launched until 2009, when they jointly articulated the governance norm of "Joint Biodiversity Protection." Over the following decade, China–ASEAN biodiversity governance underwent significant transformation: the main governing actors shifted from government dominance to multi-stakeholder participation; governance instruments evolved from administrative control to policy coordination, knowledge sharing, and social mobilization; and governance content expanded from species protection to ecosystem restoration, ecological corridor construction, and sustainable use. These developments indicate that, under the normative framework of "Joint Biodiversity Protection," China and ASEAN have gradually formed a multi-level, polycentric governance structure with distinctive regional characteristics. This process not only reflects institutional innovation in China–ASEAN ecological cooperation but also illustrates the growing capacity of "Global South" countries to shape and diffuse governance norms within the global environmental regime.

In the field of global environmental governance, theories of norm evolution and norm diffusion provide valuable analytical tools for understanding how international rules emerge and spread (Finnemore and Sikkink, 1988; Acharya, 2011). Unlike previous studies that focus on norm transfer in North–South relations, this article introduces the analytical lens of norm evolution into the context of South–South cooperation. It explores how China and ASEAN—through institutional learning, socialization, and policy coordination—have jointly constructed regional environmental governance norms through continuous interaction. Accordingly, this study seeks to address three core questions: How did the governance norm of "Jointly Protecting Biodiversity" emerge and evolve in China–ASEAN cooperation? To answer this question, the paper draws upon the theoretical framework of norm diffusion and evolution in global governance, integrates the analytical perspective of the "Global South," and employs process tracing and discourse analysis to systematically examine the development and underlying mechanisms of China–ASEAN biodiversity governance since 2009. This research carries both theoretical and practical significance. Theoretically, by introducing the theory of norm evolution into the study of environmental governance among "Global South" countries, it seeks to transcend the traditional North-centered paradigm of global governance and reveal the unique mechanisms through which norms are generated, diffused, and internalized

DOI: 10.1201/9781003640899-7

within the South–South cooperation framework. Practically, examining the evolution of biodiversity governance norms between China and ASEAN helps summarize cooperative experience, provides insights for enhancing coordinated biodiversity governance among Global South countries, and offers policy implications for China in advancing green partnerships and implementing the Kunming–Montreal Global Biodiversity Framework (GBF).

5.1 DRIVERS OF CHINA–ASEAN COOPERATION IN BIODIVERSITY CONSERVATION

With the collective rise of the Global South, developing countries have come to play an increasingly important role in global biodiversity governance. As key representatives of the Global South, both China and ASEAN hold significant positions in the global distribution of biodiversity resources while simultaneously confronting mounting ecological and environmental challenges. Against this backdrop, strengthening bilateral cooperation and jointly advancing biodiversity conservation are not only intrinsic requirements for regional ecological security and sustainable development but also expressions of the Global South's growing normative agency and institutional aspirations in international governance.

5.1.1 Strengthening the Role and Normative Agency of Global South Countries in Global Biodiversity Governance

The signing of the CBD in 1992 marked the formal establishment of a global biodiversity governance system. Anchored in the CBD as an institutional foundation and the Conference of the Parties (COP) as its decision-making platform, this global mechanism has become the principal framework for coordinating policies, allocating responsibilities, and promoting collective action among nations. To date, the Convention has 196 parties and has convened 15 COP meetings, forming a relatively comprehensive governance architecture. However, persistent structural disparities between developed and developing countries in terms of governance goals, technological capabilities, and financial resources have long contributed to a pronounced North–South divide—a root cause of the enduring dilemmas in global biodiversity governance (Wang, 2021). Developed countries enjoy clear advantages in species monitoring, protected area management, and the utilization of genetic resources, while developing countries generally suffer from funding shortages and capacity limitations (Liu, 2021). This structural imbalance has led to the relative marginalization of developing countries' voices within the global biodiversity governance system.

As the economic strength and political influence of the Global South have risen, developing countries have increasingly sought institutionalized channels to assert their voices and normative autonomy within global governance. In 2002, 17 developing countries—home to approximately 60%–70% of the world's biodiversity—formed the Like-Minded Megadiverse Countries (LMMCs) group to coordinate their positions and enhance their collective bargaining power in global biodiversity negotiations. In recent years, Global South countries have continued to advance the principle of common but differentiated responsibilities in negotiations under the CBD and its supplementary protocols—such as the Nagoya Protocol and the Cartagena Protocol—calling for fair and inclusive institutional arrangements. Moreover, Global South actors have also played an active role in negotiations on the Agreement on the Conservation and Sustainable Use of Marine Biological Diversity of Areas Beyond National Jurisdiction (BBNJ Agreement), advocating for equitable benefit-sharing of marine genetic resources and collective responsibility in protecting ocean ecosystems (Li, 2023). These efforts collectively reflect a significant transformation—from rule-takers to co-shapers of international norms. At the same time, China's introduction of its research achievements and practical experience in biodiversity conservation to the international community is also aimed at enhancing its global influence and discourse power (Xiangyu et al., 2018). Within this broader movement, China and ASEAN, as pivotal actors of the Global South, serve both as biodiversity-rich regions and as stakeholders deeply invested in ecological protection. Through their joint agenda of "cooperatively protecting biodiversity," the two sides have strengthened the institutional practice of South–South cooperation and provided a normative model for collective action by Global South countries within the realm of global environmental governance.

5.1.2 Responding to Regional Ecological Crises and Advancing Sustainable Development

China and ASEAN are geographically connected and ecologically interlinked, representing two of the most biodiversity-rich regions in Asia. Both possess abundant ecological resources yet face mounting pressures from rapid industrialization and urbanization that have accelerated ecosystem degradation, the ASEAN region, despite covering only about 3% of the world's terrestrial area, harbors roughly 18% of global plant and animal species. However, over the past four decades, the region has lost between 30% and 45% of its forest cover, and numerous key species are now on the brink of extinction (ASEAN Centre for Biodiversity, 2017). In the Mekong River Basin, hydropower development and pollution have led to sharp declines in fish populations, with some migratory species decreasing by as much as 70%. This region is also part of a broader biodiversity hotspot, home to three of the world's 17 recognized megadiverse countries—Indonesia, Malaysia, and the Philippines—which together occupy less than 10% of the Earth's surface but support over 70% of its biodiversity (Uriarte, Simorangkir, and Barrer, 2014). In the marine realm, research shows that 50%–80% of mangroves in this region have been destroyed, with seagrass beds suffering similar losses, while about 48% of coral reefs are under high to very high levels of threat (Kamil et al., 2017). These pressing threats have made biodiversity conservation an urgent prerequisite for regional sustainable development.

As the largest developing country, China's biodiversity is also under significant strain from habitat loss and degradation, overexploitation of natural resources, environmental pollution, invasive alien species, and climate change. These pressures have resulted in varying degrees of degradation and loss across ecosystems, species, and genetic diversity. According to the China Biodiversity Red List, 4,088 higher plant species—classified as critically endangered, endangered, or vulnerable—account for 10.39% of all assessed species; among vertebrates (excluding marine fish), 1,050 species, or 22.02% of those assessed, are considered threatened (Ministry of Ecology and Environment, the People's Republic of China, 2024). Illegal harvesting and exploitation of plant and animal resources, as well as forest degradation, have caused severe "forest hollowing" in many areas of China. Water quality in major rivers continues to deteriorate, groundwater levels are dropping, and pollution and eutrophication are expanding (Yuqin, 2021).

In recent years, China has steadily intensified its international cooperation on biodiversity protection and made substantial financial contributions to global biodiversity governance. Since 2019, China has become the largest contributor among developing countries to the core budget of the CBD and its Protocols, thereby providing strong support for their operations and implementation (The State Council Information Office of the People's Republic of China, 2021). It has also increased its contributions to the Global Environment Facility (GEF), becoming the largest developing-country donor to the GEF and playing a key role in advancing global biodiversity protection. Nevertheless, China continues to face serious ecological challenges and ongoing biodiversity loss. Overall, China and ASEAN share similar ecological pressures but also possess complementary advantages in policy design and technological capacity. China's experience in institutional development and financial support, combined with ASEAN's practical expertise in regional ecosystem management and community-based conservation, provides a realistic foundation for deepening systematic cooperation between the two sides in biodiversity governance.

5.2 THE EMERGENCE AND CONSOLIDATION OF THE NORM OF "JOINT BIODIVERSITY PROTECTION" BETWEEN CHINA AND ASEAN

As environmental cooperation between China and ASEAN has continued to deepen, both sides have gradually recognized the central role of biodiversity conservation in achieving regional sustainable development. Initially introduced as a sub-component within the broader framework of environmental cooperation, biodiversity protection later evolved into a strategic focus and a key area of collaboration. From a historical perspective, the norm of "Joint Biodiversity Protection" between China and ASEAN has evolved through three stages: the emergence of consensus, institutional formation, and progressive deepening. This trajectory reflects not only the institutional innovation capacity of South–South cooperation in global ecological governance but also the learning-oriented and dynamic nature of regional governance norms in practice.

The Evolutionary Process of Norms for "Joint Biodiversity Protection"

Time	Document or Meeting	Content
November 2002	China-ASEAN Comprehensive Economic Cooperation Framework Agreement	Environmental protection was identified as a field of cooperation between the two sides for the first time.
November 2007	11th China-ASEAN Summit	It is proposed to establish the China-ASEAN Environmental Cooperation Center.
October 2009	China-ASEAN Strategy on Environmental Protection Cooperation 2009–2015	Cooperation on biodiversity conservation has been identified as a new area of environmental collaboration between the two sides, and the establishment of the China-ASEAN Environmental Cooperation Center has been confirmed.
May 2016	China-ASEAN Environmental Cooperation Action Plan (2016–2020)	Biodiversity and ecological conservation has been identified as one of the priority areas for cooperation between the two sides.
May 2021	Framework of China-ASEAN Environmental Cooperation Strategy and Action Plan (2021–2025)	Biodiversity conservation and ecosystem management have been identified as one of the Strategic Directions for cooperation between the two sides.
August 2025	Plan of Action to Implement the ASEAN-China Comprehensive Strategic Partnership (2026–2030)	To facilitate the implementation of the Kunming-Montreal Global Biodiversity Framework, improve awareness and capacity in biodiversity conservation policies, and promote protection and sustainable utilization of biological resources.

5.2.1 Stage of Consensus Building: The Emergence of the Issue and Normative Awareness (Before 2009)

In the early phase of China–ASEAN environmental cooperation, joint efforts primarily focused on pollution control, climate change, and sustainable development. As global attention to biodiversity increased, China and ASEAN gradually developed a shared understanding of the importance of biodiversity conservation within both multilateral and regional cooperation frameworks. Although no independent mechanism for cooperation on biodiversity existed during this period, the two sides, through ministerial meetings, policy dialogues, and regional environmental action plans, established an initial consensus that marked the nascent emergence of normative awareness regarding "Joint Biodiversity Protection."

China and ASEAN formally established a dialogue partnership in 1991, marking the beginning of their official engagement. However, given that the cooperation at that time was largely centered on political and economic issues—and that both sides had yet to fully recognize the significance of biodiversity governance—environmental cooperation, including biodiversity conservation, had not yet entered the agenda. It was not until November 2002, with the signing of the Framework Agreement on Comprehensive Economic Cooperation between China and ASEAN, that both sides agreed to extend cooperation into the environmental field (Ministry of Foreign Affairs of the People's Republic of China, 2022). Environmental issues thus began to emerge as an area of China–ASEAN interaction. In August 2004, the First China–ASEAN Environmental Policy Dialogue Seminar was convened, during which both sides exchanged information on their national environmental policies and identified key priorities for future cooperation. Later that year, in December 2004, China and ASEAN jointly issued the Plan of Action to Implement the Joint Declaration on the Strategic Partnership for Peace and Prosperity, designating the environment as a major area of cooperation (China-ASEAN Environmental Cooperation Center, 2011). The plan explicitly called for strengthening subregional cooperation on biodiversity conservation, exploring the establishment of biodiversity corridors, promoting reforestation in the Lancang–Mekong and Red River basins, and addressing soil erosion and land degradation (Association of Southeast Asian Nations, 2020). These developments signified a new stage in China–ASEAN environmental cooperation, as both sides came to recognize the necessity of joint environmental action. Subsequently, a series of activities—such as the China–ASEAN Environmental Management Seminar and the China–ASEAN Seminar on Eco-labeling and Cleaner Production—were organized to enhance policy understanding and practical collaboration. These exchanges deepened mutual consensus, strengthened information sharing, and laid the groundwork for sustained environmental cooperation.

In November 2007, Chinese Premier Wen Jiabao, at the 11th China–ASEAN Leaders' Meeting, proposed the establishment of a China–ASEAN Environmental Protection Cooperation Center to formulate joint strategies and suggested creating a ministerial-level environmental meeting mechanism (Beijing Review, 2008). At this stage, both parties had come to realize that effective environmental cooperation required concrete action plans and institutionalized mechanisms. This awareness culminated in October 2009, when China and ASEAN jointly formulated and released the China–ASEAN Strategy and Action Plan for Environmental Cooperation (2009–2015). The document explicitly identified biodiversity conservation as a new domain of collaboration within their environmental partnership and outlined detailed plans for implementation. Thus, biodiversity protection was, for the first time, formally articulated as an independent issue within the China–ASEAN environmental cooperation framework, laying the groundwork for systematic governance and collaborative mechanisms. This development marked the transition from policy consensus to institutional design, providing the foundational basis for the later formal establishment of the norm of "Joint Biodiversity Protection."

It should be noted that even during this early phase, China–ASEAN environmental cooperation had already touched upon biodiversity-related initiatives and pilot projects. For instance, in 2005, with financial support from the World Bank, China and the Greater Mekong Subregion (GMS) countries jointly launched the GMS Biodiversity Conservation Corridors Initiative. Nevertheless, these dialogues and projects were largely conducted under the general framework of environmental protection rather than as part of a systematic biodiversity governance agenda. The absence of a dedicated mechanism or comprehensive governance plan indicates that, at this stage, biodiversity protection remained an emergent rather than institutionalized area of cooperation.

5.2.2 Stage of Institutional Formation: Building Cooperation Mechanisms and Consolidating the Norm (2010–2020)

After 2009, biodiversity conservation was formally incorporated into the China–ASEAN environmental cooperation agenda. Both sides successively established key institutional platforms—including the China–ASEAN Environmental Cooperation Center (CAEC) and the ASEAN Centre for Biodiversity (ACB)—which laid a solid foundation for the institutionalization of the emerging norm. During this stage, the cooperation framework evolved from conceptual consensus to structured institutional arrangements, demonstrating a clear transition from normative advocacy to institutional embedding.

The 2009 Copenhagen Climate Conference further underscored the importance of biodiversity protection in global climate governance, propelling biodiversity issues to the forefront of the international environmental agenda. In 2010, the CAEC was officially inaugurated, becoming a central platform for advancing joint biodiversity conservation initiatives and implementing cooperative projects. Under its guidance, China and ASEAN jointly organized thematic forums such as "Biodiversity and Regional Green Development" and collaboratively launched several flagship initiatives—including the China–ASEAN Biodiversity and Ecological Conservation Cooperation Program and the China–ASEAN Green Envoys Program—to strengthen institutional linkages and knowledge exchange in biodiversity governance. In terms of concrete action plans, the two sides successively adopted the China–ASEAN Environmental Cooperation Action Plan (2011–2013) and the China–ASEAN Environmental Cooperation Action Plan (2014–2015), which progressively expanded the scope and depth of biodiversity-related collaboration. As cooperation deepened, biodiversity conservation gained increasing prominence within the broader environmental protection agenda.

In May 2016, China's Ministry of Environmental Protection and the environmental authorities of ASEAN member states jointly endorsed the China–ASEAN Environmental Cooperation Strategy (2016–2020), designating biodiversity and ecological protection as one of the priority areas for collaboration. To further institutionalize this agenda, both sides released the China–ASEAN Environmental Cooperation Action Plan (2016–2020) in 2017, which included detailed objectives and implementation pathways for biodiversity conservation. In November 2018, China and ASEAN adopted the China–ASEAN Strategic Partnership Vision 2023, which articulated the goal of building a China–ASEAN Community of Shared Future. The vision called for fostering a Blue Economy Partnership to promote the protection of marine ecosystems and the sustainable use of ocean resources (Association of Southeast Asian Nations, 2018). This initiative elevated China–ASEAN cooperation to a new strategic level and simultaneously expanded the scope of biodiversity governance, reflecting the deepening institutionalization of the "Joint Biodiversity Protection" norm.

As some scholars have noted, China's approach to environmental governance typically follows a top-down model in which government agencies and institutions play a dominant role (Zhu et al., 2024). Similarly, in the construction of the biodiversity governance norm within China–ASEAN cooperation, governmental actors have continued to function as the primary driving force, steering agenda-setting, coordination, and institutional design.

5.2.3 Stage of Deepening and Expansion: Issue Broadening and Norm Consolidation (2021–Present)

The year 2021 marked a watershed moment for both global biodiversity governance and China–ASEAN cooperation in biodiversity protection. As the host and presidency of the Fifteenth Meeting of the COP to the CBD COP15, China played a leading role in steering the global transition toward a new phase of biodiversity governance. Meanwhile, 2021 also marked the 30th anniversary of China–ASEAN dialogue relations and the official establishment of their Comprehensive Strategic Partnership. Within this broader framework, biodiversity conservation was identified as one of the strategic priorities in China–ASEAN environmental cooperation, signifying the deepening institutionalization of the norm of "Joint Biodiversity Protection."

As key members of the Global South, China and ASEAN have made substantial contributions to global biodiversity governance. This shared responsibility has further motivated both sides to strengthen collaboration in implementing the Kunming–Montreal GBF. Since 2021, China and ASEAN have reached a renewed understanding of the strategic importance of biodiversity protection, jointly emphasizing the need to mainstream biodiversity governance, enhance policy awareness and capacity building, and promote the conservation and sustainable use of biological resources (Association of Southeast Asian Nations, 2020). In May 2021, the two sides released the China–ASEAN Environmental Cooperation Strategy and Action Framework (2021–2025), in which biodiversity conservation and ecosystem management were designated as strategic pillars of their environmental partnership. Moreover, other major cooperation themes—such as policy dialogue and capacity building, sustainable cities and marine plastic reduction, climate change mitigation, and air quality improvement—also directly or indirectly intersect with biodiversity protection objectives (China-ASEAN Environmental Cooperation Center, 2020). The same year, in commemorating the 30th anniversary of China–ASEAN dialogue relations, both parties announced the establishment of a meaningful, substantive, and mutually beneficial Comprehensive Strategic Partnership, reaffirming their commitment to building a closer China–ASEAN Community of Shared Future (Association of Southeast Asian Nations, 2021). This milestone not only elevated bilateral relations to a new strategic level but also opened broader horizons for biodiversity cooperation.

To operationalize this partnership, China has encouraged ASEAN countries to participate in the Global Development Initiative and to explore pragmatic cooperation in emerging areas such as the bioeconomy and green economy. These efforts complement and reinforce related ASEAN and global initiatives, including the ASEAN Community Vision 2025, the United Nations 2030 Agenda for Sustainable Development, and the ASEAN Outlook on the Indo-Pacific (Ministry of Foreign Affairs of the People's Republic of China, 2022). Consequently, biodiversity protection has expanded beyond a traditional ecological issue to become an integrated, cross-sectoral agenda closely linked with the bioeconomy, green growth, and clean energy transitions—thus forming a core pillar of regional sustainable development. From the perspective of strategic alignment, China–ASEAN cooperation exhibits a dual orientation. On one hand, it seeks to enhance synergy between biodiversity protection and other multilateral initiatives, ensuring the deep integration of ecological conservation with regional development agendas. On the other hand, guided by the CBD and related international norms, both sides are committed to fulfilling their global obligations through practical cooperation, jointly building mechanisms for the conservation and sustainable use of biological resources. This evolving logic underscores that China and ASEAN are actively transforming their regional practices and governance norms on "Joint Biodiversity Protection" into an integral pathway of global biodiversity governance—contributing a replicable regional model for improving the global ecological governance system.

In summary, the norm of "Joint Biodiversity Protection" between China and ASEAN has undergone a gradual and cumulative evolution—from consensus building, to institutional formation, and finally to practical deepening. Its formation and development rest on two interrelated logics: first, an institutional adaptation process, through which both sides—as parties to the CBD—proactively align with global biodiversity governance frameworks and draw upon advanced international practices; and second, a practice-based learning process, wherein iterative collaboration, evaluation, and adjustment have facilitated the sustained consolidation and internalization of this governance norm.

5.3 THE LEARNING AND INTERNALIZATION OF THE GOVERNANCE NORM OF "JOINT BIODIVERSITY PROTECTION" BETWEEN CHINA AND ASEAN

The emergence and evolution of the governance norm of "Joint Biodiversity Protection" between China and ASEAN represents not only a process of institutional learning but also one driven by social learning. This normative evolution reflects the dynamic process through which Global South countries absorb, reinterpret, and localize global principles of biodiversity governance. Through multi-level dialogue mechanisms, transnational cooperation projects, and participatory social practices, the China–ASEAN norm of "jointly protecting biodiversity" has been continuously deepened, consolidated, and sustained.

5.3.1 Institutional Learning: From Multi-level Dialogue Mechanisms to the Establishment of the CAEC

Institutional learning constitutes a crucial foundation for the generation and internalization of the China–ASEAN "jointly protecting biodiversity" governance norm. Both sides' accession to the CBD marked the starting point of their engagement with global biodiversity governance and the beginning of a learning process through participation. The eventual establishment of the CAEC was the result of accumulated institutional interaction and policy learning, laying a structural foundation for the further consolidation of this governance norm.

Since the establishment of China–ASEAN dialogue relations in 1991, the bilateral partnership has evolved through successive stages—from the Comprehensive Dialogue Partnership, to the Strategic Partnership, and finally to the Comprehensive Strategic Partnership—gradually forming a multi-layered cooperation architecture encompassing politics, security, economy, and the environment. This evolving framework has strengthened the institutional cohesion of the region and created a sustained mechanism for normative learning in ecological and biodiversity governance. In the early stages, cooperation relied primarily on multilateral platforms such as the China–ASEAN Leaders' Meetings, the East Asia Summit, and the ASEAN Plus Three (China, Japan, and South Korea) Leaders' Meetings to discuss global and regional environmental issues. Within the environmental field, ministerial meetings, senior officials' meetings, and environmental cooperation forums became important venues for policy coordination and knowledge sharing. Through these mechanisms, China and ASEAN gradually formed a cooperative principle characterized by joint consultation, mutual respect, and shared benefits. The launch of the China–ASEAN Environmental Cooperation Forum in 2010 marked the institutional embedding of the outcomes of institutional learning. To date, the forum has convened 11 sessions, addressing topics such as biodiversity, green cities, ecological infrastructure, and low-carbon development. It has served as a stable platform for policy dialogue and experience exchange, facilitating the translation and localization of global biodiversity governance ideas at the regional level.

Established in 2011, the CAEC became the institutional vehicle for sustained learning and cooperation. Integrating policy coordination, technical collaboration, capacity building, and knowledge dissemination, CAEC has concretized international norms within regional policy frameworks. For example, initiatives such as the China–ASEAN Eco-City Cooperation Forum, the Symposium on Ecological Civilization, and Marine Ecosystem Protection Training Workshops have transformed abstract international norms into operational regional practices. Beyond technical assistance, CAEC functions as a normative intermediary mechanism, linking the processes of global norm diffusion, regional institutional adaptation, and national policy implementation. This intermediary role ensures a continuous learning cycle of "global norms-regional mechanisms-national policies."

In summary, China and ASEAN's institutional learning has evolved through three progressive stages: from multi-level policy dialogues, to mechanism-based cooperation, and finally to institutional embedding. This process exemplifies how Global South countries, by absorbing and adapting international biodiversity governance experiences, achieve regional norm internalization through sustained institution building and policy integration.

5.3.2 Social Learning through Practice: Diverse Cooperative Activities in Biodiversity Protection

Practical, experience-based learning forms the social foundation for norm diffusion and mutual internalization. Through policy dialogue, joint research, cross-border cooperation, and social mobilization, China and ASEAN have fostered mutual understanding and co-produced knowledge in the field of biodiversity conservation, thereby gradually embedding the norm of "Joint Biodiversity Protection." Local actors are not passive recipients or learners of transnational norms acting under a universal moral framework; instead, they play an active role in shaping and directing norm diffusion in world politics (Acharya, 2004). Rather than merely replicating international practices, both sides have engaged in adaptive processes that reflect their specific developmental priorities and ecological contexts.

First, Policy Dialogue and Capacity Building. Policy dialogue and capacity building have long been central components of China–ASEAN cooperation in biodiversity governance. Since 2011, the China–ASEAN Environmental Cooperation Forum has served as the principal platform for policy communication, facilitating in-depth exchanges on topics such as ecological governance, green city development, and biodiversity conservation. The 2012 forum, themed "Biodiversity and Regional Green Development," marked the first time that biodiversity was incorporated into the regional development agenda. Subsequent workshops and training programs under the forum have strengthened regional policy coordination and improved capacities for biodiversity management. In terms of capacity building, China and ASEAN have jointly organized a series of training sessions, seminars, and technical assistance programs to enhance regional ecological governance. For example, the Biodiversity Mainstreaming and Ecological Security Training Program, jointly implemented by China's Ministry of Ecology and Environment and ASEAN member states, established regional training bases in Yunnan and Guangxi. These initiatives have not only enhanced policy coordination but also facilitated knowledge sharing and mutual learning, forming a vital channel for regional social learning in biodiversity governance.

Secondly, Joint Research and Knowledge Co-creation. Joint research serves as an essential vehicle for social learning. The China-ASEAN Environmental Cooperation Outlook released in 2016 systematically assessed the status and potential of bilateral cooperation in environmental and biodiversity protection. In 2021, in alignment with the implementation of the Kunming–Montreal GBF, China and ASEAN jointly emphasized strengthening scientific collaboration, promoting data sharing, and developing decision-support systems for biodiversity management. The Southeast Asia Biodiversity Research Center under the Chinese Academy of Sciences, in partnership with ASEAN research institutions, has conducted studies on genetic resource conservation, ecosystem monitoring, and the sustainable use of biological resources. At the local level, Yunnan Province has established platforms such as the China–Laos Transboundary Ecological Corridor and the Asian International Rivers Institute to promote cross-border ecosystem monitoring and species protection. These collaborative mechanisms have fostered scientific co-production and localized the learning process of biodiversity governance norms.

In addition, Multi-level Biodiversity Conservation Projects. Beyond policy coordination and joint research, China and ASEAN have implemented a range of practical biodiversity initiatives, including the China–ASEAN Biodiversity and Ecological Conservation Cooperation Program, the China–ASEAN Eco-Friendly City Partnership, and the China–ASEAN Mangrove Conservation Network. These projects have promoted equitable benefit-sharing of genetic resources, coastal wetland conservation, ecosystem valuation, and the development of green infrastructure. Examples such as the GMS Biodiversity Corridor Demonstration Project and the China–Vietnam Cross-Border Joint Protection Mechanism have become model cases of regional ecological cooperation. By facilitating the translation of scientific and policy knowledge into tangible practices, these initiatives have achieved the social embedding of norms—transforming "Joint Biodiversity Protection" into a shared value and collective orientation among regional actors.

Lastly, Marine Biodiversity Protection and the "Blue Partnership." Marine ecosystem governance has recently emerged as a new frontier of China–ASEAN environmental cooperation. In recent years, issues such as plastic pollution, coastal degradation, and declining fishery resources have become pressing common challenges. The China–ASEAN Strategic Partnership Vision 2030 first proposed the creation of a Blue Economy Partnership, highlighting the importance of marine ecosystem protection and sustainable utilization of ocean resources.

In 2019, the China–ASEAN Marine Plastic Reduction Initiative was launched, linking marine biodiversity governance with urban ecological cooperation and marking a concrete step toward building a Blue Community of Shared Future. Building on this initiative, China and ASEAN countries have established marine protected areas to safeguard representative marine ecosystems—such as mangroves, coral reefs, and seagrass beds—in order to maintain their ecological integrity and stability (Chen et al., 2024). Furthermore, the China–ASEAN Environmental Cooperation Strategy and Action Framework (2021–2025) has institutionalized marine ecological protection as a key priority, thereby advancing a more integrated land–sea ecological governance framework.

5.3.3 Norm Internalization: The Formation of a Collective Identity of "Joint Biodiversity Protection"

As cooperation on biodiversity protection has deepened, China and ASEAN have gradually developed a cooperative model centered on government leadership but characterized by the joint participation of multiple stakeholders. This evolving configuration reflects not only the expansion of governance networks but also the social internalization and identity formation surrounding the norm of "Joint Biodiversity Protection" at the regional level. According to norm life-cycle theory, norm internalization occurs at the final stage of the norm cascade, when a norm becomes taken for granted and ceases to be a subject of public debate (Finnemore and Sikkink, 1998). In this context, China and ASEAN have cultivated a shared collective identity around this norm—both sides no longer question its legitimacy or necessity, but instead actively promote biodiversity conservation as a common regional endeavor. This demonstrates that the norm has achieved enduring legitimacy and normative authority within the region.

First, government-led institutional coordination has provided a stable organizational foundation for the internalization of the norm. The China–ASEAN Leaders' Meeting mechanism defines the overall direction for biodiversity protection, while the CAEC translates the principles of international agreements—such as the CBD—into regional policy frameworks through strategic plans and action programs. Beyond the central government level, local governments have played an intermediary role in promoting the internalization of the norm. For example, since 2015, Yunnan Province has organized more than 40 international training programs for South and Southeast Asian participants on topics such as the sustainable use and management of biological resources, biosafety, ecological value assessment, and pollution prevention. Such local initiatives have not only reinforced domestic institutional embedding but also expanded transboundary channels for learning and policy diffusion.

Second, the participation of diverse societal actors has enabled the idea of "Joint Biodiversity Protection" to transcend intergovernmental cooperation and evolve into a shared social value across the region. In recent years, enterprises, foundations, research institutions, and civil society organizations have become increasingly involved in the China–ASEAN ecological cooperation network, forming a co-governance model characterized by knowledge sharing and public participation. In 2021, business leaders, foundations, research institutions, and international organizations from China and ASEAN jointly launched the China–ASEAN Mangrove Conservation Network and issued the China–ASEAN Mangrove Conservation Initiative, promoting cross-sectoral and transnational cooperation at the grassroots level. Non-governmental organizations (NGOs) have also played important roles as knowledge brokers and agenda setters in the process of norm internalization. For instance, the World Wide Fund for Nature (WWF) China Office, through reports such as Supporting Green Development along the Belt and Road: A Handbook on Infrastructure Projects and Biodiversity Conservation, has contributed to building a shared regional understanding of biodiversity-related governance risks. Through information dissemination, public advocacy, and technical collaboration, non-governmental actors have helped diffuse and embed biodiversity conservation values within regional societies.

Finally, the construction of regional identity represents the deeper dimension of norm internalization. As developing countries, China and most ASEAN members share similar developmental challenges and ecological priorities. China's experience as a developing nation enables it to better understand the needs of ASEAN partners, thereby promoting biodiversity governance from a position of equality and empathy. This mutual understanding has fostered emotional resonance and identity convergence between the two sides (Zhu et al., 2024). With increasing interaction on ecological governance issues, "Joint Biodiversity Protection" has gradually evolved into a symbol of collective regional identity.

Through the combined influence of multilateral institutions and civil society networks, China and ASEAN have fostered a shared sense of belonging to an "ecological community," recognizing biodiversity conservation as a core dimension of regional security and shared development. This transformation—from external compliance with international norms to internalized value recognition—demonstrates how the norm has evolved from a set of policy objectives into an intrinsic driver of sustained cooperation in global environmental governance. As constructivist theory suggests, agents may act "appropriately" by learning and performing certain roles—acquiring the knowledge necessary to meet normative expectations—regardless of their personal preferences or agreement with those roles (Checkel, 2005). The internalization of the norm of "Joint Biodiversity Protection" thus reflects a gradual process of mutual learning, accommodation, and coordination, which has endowed the norm with enduring vitality and self-reinforcing momentum.

5.4 LIMITATIONS AND CHALLENGES IN THE NORM-BUILDING OF "JOINTLY PROTECTING BIODIVERSITY" BETWEEN CHINA AND ASEAN

Over the past decade, China and ASEAN have advanced the institutionalization of biodiversity cooperation through policy dialogues, joint research, and practical initiatives, forming a norm-evolution trajectory centered on the logic of "learning–internalization–diffusion." However, in terms of the maturity of the governance system and the sustained influence of the established norm, several structural constraints continue to impede progress. These challenges affect not only the effectiveness of regional cooperation but also the long-term institutional impact of Global South–driven norm diffusion within global environmental governance.

5.4.1 Institutional Coordination Mechanisms Remain Underdeveloped, with Structural Gaps in Policy Alignment

Although China and ASEAN have established relatively comprehensive platforms for policy dialogue—such as the China–ASEAN Environmental Cooperation Forum and the Ministerial Meeting on Environment—the regional biodiversity governance framework remains fragmented. There is insufficient coordination among various departments and institutions in areas such as policy planning, data sharing, and implementation, resulting in an unstable institutional basis for transnational ecological cooperation. For instance, China has enacted more than 20 policy instruments and laws related to biodiversity protection, including the Wildlife Protection Law, the Seed Law, and the Biosafety Law. However, the absence of a unified Biodiversity Conservation Law has led to a lack of coherence and integration among these legal provisions, leaving the overall framework dispersed and loosely connected (Zhu et al., 2024). In practice, several biodiversity conservation projects have faced obstacles such as fragmented financial mechanisms and inconsistent technical standards, which have hindered the translation of project outcomes into durable regional policy tools.

Compared with the European Union's transnational ecological governance model, China–ASEAN cooperation still lags in institutional integration and governance evaluation. Furthermore, there is no unified regional standard for biodiversity data monitoring, information disclosure, or policy feedback mechanisms. The absence of an institutionalized system for policy learning and evaluation has constrained the capacity to assess and refine cooperation outcomes. To achieve a genuine transition from cooperation to co-governance, it is essential to strengthen cross-sectoral coordination and policy integration at the regional level. Institutionalizing mechanisms for data-sharing platforms, policy assessment, and inter-agency coordination will be critical to ensuring the long-term sustainability and effectiveness of the "Joint Biodiversity Protection" governance norm.

5.4.2 Uneven Participation among Multiple Stakeholders: Underdeveloped Networks for Social Learning and Collective Action

Although China–ASEAN biodiversity cooperation has, at the institutional level, taken on the appearance of a multi-stakeholder governance model, in practice, government agencies continue to dominate the cooperation process. The participation of NGOs, academic institutions, and local communities remains relatively limited. This state-centric mode of cooperation has, to some extent, constrained the vitality of social learning and the potential for policy innovation.

On one hand, social organizations face constraints in funding, personnel capacity, and access to international exchange channels, which restrict the depth and breadth of their involvement in cooperative projects. Government-led initiatives, meanwhile, tend to prioritize economic growth objectives and often fail to effectively integrate biodiversity protection into their policy outcomes. Within China–ASEAN relations, numerous national and regional policy agendas explicitly link trade development with environmental considerations; however, the implementation and tangible impacts of such linkages remain weak and inconsistent (Dosch, 2011). On the other hand, the voices of local communities are still insufficiently represented in policy design and project planning. For example, in the establishment of transboundary protected areas or ecological compensation programs, local communities in some ASEAN member states have shown limited enthusiasm for participation due to opaque benefit-sharing mechanisms. This lack of transparency undermines community trust and weakens the social foundation and sustainability of cooperation efforts. Furthermore, the knowledge translation mechanism between academia and policymaking remains underdeveloped. Scientific research outputs are rarely integrated systematically into policy processes, resulting in a persistent gap between evidence-based insights and policy implementation. Consequently, regional cooperation has often remained confined to the policy coordination level rather than evolving into a sustained network of learning, collaboration, and innovation.

Looking ahead, China and ASEAN should enhance the substantive participation of civil society organizations and local communities by building cross-sector collaboration platforms that bridge governments, academia, and civic actors. Such platforms would strengthen the social foundations of cooperation, foster mutual understanding, and cultivate the shared cognitive frameworks necessary for the long-term internalization of the "jointly protecting biodiversity" norm.

5.4.3 Dual Constraints of Geopolitical Rivalry and Asymmetric Power Structures

Within the increasingly complex landscape of global biodiversity governance, China–ASEAN cooperation faces growing tensions shaped by two interrelated constraints: the spillover effects of geopolitical rivalry and the asymmetric power structures that exist between China and ASEAN member states. Together, these dynamics impose both external and internal limitations on the evolution and institutionalization of regional biodiversity governance norms. Geopolitical factors exert tangible constraints on regional cooperation. In the maritime domain, sovereignty and maritime rights disputes between China and certain ASEAN member states—particularly in the South China Sea—have complicated mutual trust and hindered collaboration on marine ecological governance, biodiversity conservation, and blue economy initiatives. For instance, projects related to marine ecosystem protection and transboundary fisheries management have often been delayed or disrupted by unresolved issues concerning maritime delimitation and resource exploitation rights, thereby reducing policy coordination efficiency and project implementation speed.

Moreover, the involvement of external powers in South China Sea affairs has further securitized ecological cooperation. By linking environmental protection issues to broader geopolitical competition narratives, extra-regional actors have introduced political sensitivities that risk undermining the normative consensus and institutional continuity of the "jointly protecting biodiversity" framework. As a result, biodiversity cooperation has at times been politicized, reducing the space for purely ecological or developmental collaboration. At the same time, a persistent structural tension exists between internal capacity-building and external pressures. Significant disparities remain between China and ASEAN member states in terms of ecological governance institutions, technological capabilities, and financial resources. Some ASEAN countries continue to lack stable environmental management systems and effective inter-agency coordination mechanisms. This internal asymmetry constrains the diffusion, localization, and institutionalization of the biodiversity governance norm, limiting its capacity to generate sustained regional momentum. To mitigate these dual constraints, China and ASEAN should prioritize cooperation in low-sensitivity and high-consensus areas, implementing collaborative objectives in a phased and pragmatic manner. Cooperation in such less politically sensitive domains—such as ecological restoration, environmental education, or biodiversity data-sharing—can generate positive spillover effects, gradually expanding trust and facilitating cooperation in more complex or contested areas. This incremental and trust-based approach offers a viable pathway for advancing the institutional consolidation of the "Joint Biodiversity Protection" norm under the evolving geopolitical landscape (Yang, Chandran, and Zhao, 2023).

5.5 CONCLUSION

After more than a decade of exploration and cooperation, China and ASEAN have achieved a significant transition in the field of biodiversity protection—from policy advocacy to institutionalized practice. From the initial stage of consensus building to the establishment of operational mechanisms and finally to the deepening of collaborative initiatives, biodiversity cooperation has evolved into a crucial pillar supporting regional sustainable development. Through both institutional learning and social learning, China and ASEAN have progressively advanced the generation and internalization of the norm of "Joint Biodiversity Protection." This norm has not only guided their environmental cooperation but has also gradually evolved into a form of regional identity that embodies shared values and collective purpose.

As key members of the Global South, the cooperative practices between China and ASEAN provide valuable experience for regional ecological governance and sustainable economic development, while enhancing the discursive power and institutional shaping capacity of Global South countries within global biodiversity governance. This model of norm construction through learning and internalization reflects the broader transformation of Global South actors—from passive rule-takers to proactive co-builders of global norms—and thus carries significant theoretical and practical implications. Nevertheless, the norm-building process of China–ASEAN biodiversity governance continues to face several challenges, including ambiguous conceptual definitions, blurred institutional boundaries, and the absence of systematic evaluation mechanisms. These limitations have hindered the full realization of the norm's institutional effectiveness. Looking ahead, further cooperation between China and ASEAN should be deepened across three interrelated dimensions: At the institutional level, strengthen policy coordination and institutional alignment to improve the regional architecture of biodiversity governance. At the practical level, promote joint research, technology exchange, and data-sharing to enhance the operationality and effectiveness of governance actions. At the cognitive and social level, foster public awareness, policy dialogue, and identity-building to further internalize and diffuse the norm of "Joint Biodiversity Protection."

In sum, China–ASEAN cooperation offers not only a viable model for regional ecological governance and sustainable development but also contributes to empowering Global South countries to play a more active and constructive role in shaping the future of global biodiversity governance.

DECLARATION OF GENERATIVE AI USE

During the preparation of this manuscript, the author used ChatGPT to translate. The author subsequently reviewed and edited the content and takes full responsibility for the final version of the published chapter.

REFERENCES

Acharya, Amitav. How ideas spread: Whose norms matter? Norm localization and institutional change in Asian regionalism. *International Organization*. 2004;58(2):239–275. https://doi.org/10.1017/S0020818304582024

Acharya, Amitav. Norm subsidiarity and regional orders: Sovereignty, regionalism, and rule-making in the third world. *International Studies Quarterly*, 2011;55(1):95–123.

ASEAN Centre for Biodiversity. ASEAN Biodiversity Outlook 2. 2017. https://www.aseanbiodiversity.org/publications/asean-biodiversity-outlook-2/

China-ASEAN Environmental Cooperation Center. Framework of ASEAN-China Environmental Cooperation Strategy and Action Plan 2009–2015. 2011. http://www.lmec.org.cn/chinaaseanenv/zlyxdjh/202108/t20210806_122878.html

Association of Southeast Asian Nations. ASEAN-China Strategic Partnership Vision 2030. 2018. https://asean.org/asean-china-strategic-partnership-vision-2030/

Association of Southeast Asian Nations. Plan of Action to Implement the ASEAN-China Strategic Partnership for Peace and Prosperity (2021–2025). 2020. https://asean.org/plan-of-action-to-implement-the-asean-china-strategic-partnership-for-peace-and-prosperity-2021-2025

Association of Southeast Asian Nations. Joint Statement of the ASEAN-China Special Summit to Commemorate the 30th Anniversary of ASEAN-China Dialogue Relations: Comprehensive Strategic Partnership for Peace, Security, Prosperity and Sustainable Development. November 22, 2021. https://asean.org/wp-content/uploads/2021/11/Joint-Statement-30th-Anniversary-of-ASEAN-China-Dialogue-Relations-Final.pdf

Checkel, Jeffrey T. International institutions and socialization in Europe: Introduction and framework. *International Organization*, 2005;59(4):801–26. https://www.jstor.org/stable/3877829

Chen, Fenggui, et al. Protection pattern and driving mechanism of typical marine ecosystems: A case study of China-ASEAN countries. *Frontiers in Marine Science* 2024;11:1378188. https://doi.org/10.3389/fmars.2024.1378188

China-ASEAN Environmental Cooperation Center. Framework of ASEAN-China Environmental Cooperation Strategy and Action Plan 2021–2025. 2020. http://www.lmec.org.cn/enchinaaseanenv/Strategy_and_Action_Plan/202108/P020210809343365218598.pdf

Dosch, Jorn. Reconciling trade and environmental protection in ASEAN-China relations: More than political window dressing? *Journal of Current Southeast Asian Affairs*, 2011;30(2):9. https://doi.org/10.1177/186810341103000202

Expand Cooperation for Mutual Benefit and Win-Win Progress. Beijing Review. NO.1 JAN.3, 2008. https://www.beijingreview.com.cn/document/txt/2008-01/03/content_94026.htm

Finnemore, Martha, and Kathryn Sikkink. International norm dynamics and political change. *International Organization*, 1998;52(4):887–917. https://www.jstor.org/stable/2601361

Kamil, Khairunnisa Ahmad, et al. An assessment of marine protected areas as a marine management strategy in Southeast Asia: A literature review. *Ocean & Coastal Management*, 2017;145:72–81. https://doi.org/10.1016/j.ocecoaman.2017.05.008

Li, Lingqun. The role of the "global south" in International Ocean Governance-the case of the BBNJ negotiations. *Asia-Pacific Security and Maritime Affairs*, 2023(06):68–89+3–4. https://doi.org/10.19780/j.cnki.ytaq.2023.6.5

Liu, Zhe. Situation and implications of the negotiations on the convention on biological diversity. *International Economic Review*, 2021(03):155–176+8. https://link.cnki.net/urlid/11.3799.F.20210517.1133.016

Ministry of Ecology and Environment the People's Republic of China. The China National Biodiversity Conservation Strategy and Action Plan (2023–2030). 2024. https://www.mee.gov.cn/ywdt/hjywnews/202401/W020240118377427497957.pdf

Ministry of Foreign Affairs of the People's Republic of China. China-ASEAN Plan of Action on Comprehensive Strategic Partnership (2022–2025). 2022. https://www.fmprc.gov.cn/ziliao_674904/1179_674909/202211/t20221111_10972996.shtml

The State Council Information Office of the People's Republic of China. Biodiversity Conservation in China. 2021. http://www.scio.gov.cn/zfbps/ndhf/2021n_2242/202207/t20220704_130703.html

Uriarte, Monina T., Dicky Simorangkir, and Sahlee B. Barrer. Biodiversity conservation beyond borders: The case of ASEAN Heritage Park. *Ecosystems and Development Journal*, 2014;5(1):5.

Wang, Sidan. Dilemma, motivations and prospects of upgrading global biodiversity governance. *Yuejiang Academic Journal*, 2021(5):15–28+120. https://doi.org/10.13878/j.cnki.yjxk.20210914.001

Xiangyu, Jia, et al. The effects of IPBES deliverables on global biodiversity conservation strategy: An analysis based on the U. S. pollinator protection policy. *Biodiversity Science*, 2018;26(5):527–534.

Yang, Zhe, Suseela Devi Chandran, and Yinliang Zhao. ASEAN, China and South China Sea: Alternative approach for security cooperation. *Bussecon Review of Social Sciences*, 2023;4(2):25–31. https://doi.org/10.36096/brss.v4i2.387

Yuqin, Peng. What did China do for biodiversity between COP15? Raoul Wallenberg. 2021. https://rwi.lu.se/what-did-china-do-for-biodiversity-between-cop15/

Zhu, Annah Lake, et al. China's nature-based solutions in the Global South: Evidence from Asia, Africa, and Latin America. *Global Environmental Change*, 2024;86:5. https://doi.org/10.1016/j.gloenvcha.2024.102842

6 Integrating Biodiversity into Renewable Energy Development

A Review of Policies and Laws across Countries

Kejia Shen and Jinglin Qiu

6.1 INTRODUCTION

The urgency of global climate governance has accelerated the transformation of energy systems, with the expansion of renewable energy widely regarded as a key pathway to limiting global warming and achieving the goals of the Paris Agreement (Sonter et al., 2020). Increasing the share of renewables is not only critical for mitigating climate change but is also seen as an important strategy for addressing biodiversity loss. However, growing concerns have emerged regarding the ecological impacts associated with this transition.

The Convention on Biological Diversity (1992) defines biodiversity as the variability among living organisms from all sources, including terrestrial, marine, and other aquatic ecosystems and the ecological complexes of which they are part, encompassing diversity within species, between species, and of ecosystems (Convention on Biological Diversity, 1992). This definition highlights that biodiversity is not only about the number and variability of species, but also about the ecological processes and interrelationships that sustain ecosystem stability and functionality. Renewable energy, by contrast, refers to energy derived from natural resources that are replenished faster than they are consumed, such as solar, wind, hydropower, geothermal, ocean, and bioenergy. Unlike fossil fuels, which are non-renewable, take millions of years to form, and emit large amounts of greenhouse gases when burned, renewable energy offers a low-emission alternative that has become increasingly cost-competitive and job-generating in most countries (United Nations, 2022).

Yet, while the deployment of renewable energy is essential for decarbonization, the rapid expansion of renewable energy infrastructure itself can drive biodiversity loss, thereby undermining ecosystem resilience, carbon sequestration capacity, and adaptive potential in the face of climate change (OECD, 2024).

6.2 CONFLICTS BETWEEN RENEWABLE ENERGY AND BIODIVERSITY PROTECTION: LITERATURE REVIEW

Existing research shows that the primary impacts of renewable energy on biodiversity include habitat loss and fragmentation, direct mortality of species, disruption of ecosystem processes, and the weakening of ecosystem services (Gasparatos et al., 2017; Gibson et al., 2017). Due to the low energy density of renewable technologies, their infrastructure often requires significantly larger land areas, approximately ten times that of conventional fossil fuel facilities, leading to the encroachment on farmland, wilderness, and protected areas, thereby threatening species survival and accelerating

DOI: 10.1201/9781003640899-8

biodiversity loss (Wang et al., 2025). For example, solar facilities occupy and fragment large tracts of land, alter local microclimates, and even affect keystone species in desert ecosystems; wind farms cause hundreds of thousands of bird and bat deaths annually and disrupt migration routes (Loss et al., 2013); hydropower projects inundate vast areas and block riverine migration routes, leading to habitat fragmentation and constrained species distribution, while in tropical regions they also contribute significantly to methane emissions (Froidevaux et al., 2025). Additionally, the rapid growth of renewable energy increases demand for critical mineral resources, driving mining expansion in ecologically sensitive and governance-poor regions, thereby posing new threats to biodiversity (Sonter et al., 2020).

Empirical studies further highlight the spatial overlap between renewable energy infrastructure and conservation areas, with numerous wind, solar, and hydropower projects already located in protected areas, key biodiversity areas, and wilderness zones. Planned projects are projected to increase the number of affected areas by more than 30%, with biodiversity hotspots in Southeast Asia particularly at risk (Rehbein et al., 2020). In Europe, Spain provides a striking example of such conflict: in pursuit of its energy targets, planned wind and solar projects involve hundreds of thousands of hectares, threatening rare species such as steppe birds and raptors, which are underrepresented in existing conservation networks (Serrano et al., 2020). Furthermore, the ecological impacts vary significantly across energy types. Compared with bioenergy and hydropower, wind and solar present lower risks to biodiversity when appropriately sited, but poor planning can lead their impacts to rival those of bioenergy (Santangeli et al., 2016). Bioenergy, in particular, has spurred debates in Europe over its compatibility with biodiversity due to its role in intensifying land use and driving habitat conversion (Pedroli et al., 2013).

Therefore, even low-carbon renewable pathways prioritized under climate targets can still exert extinction pressures on endangered species and cause ecosystem degradation if ecological considerations are absent (Hooper et al., 2012). To address these challenges, scholars have called for systematically integrating biodiversity into energy planning and policy. This includes incorporating ecological sensitivity data into early-stage siting decisions, designating renewable energy zones away from high-risk areas, prioritizing the use of degraded or developed land (such as rooftops and brownfields), and promoting co-location strategies (e.g., agrivoltaics or offshore wind-fisheries integration) to minimize natural habitat encroachment (OECD, 2024; Allison et al., 2014). For wind energy, improved turbine design, optimized siting, and mitigation measures, such as preventing bird and bat collisions, are also considered crucial (Nazir et al., 2020).

Nonetheless, gaps remain in both policy and research: most studies linking renewable energy and biodiversity focus on Europe and North America, while data on offshore wind, migratory species, cumulative impacts, and developing country contexts are limited (OECD, 2024). These deficiencies undermine the comprehensiveness of scientific assessments and hinder the formulation of pathways that integrate energy transition and biodiversity protection (Dhunny et al., 2019).

In sum, while the expansion of renewable energy is indispensable for decarbonization, its ecological costs and conflicts with biodiversity protection cannot be overlooked. Strengthening interdisciplinary scientific assessments, policy coordination, and spatial planning is essential to ensure synergies between renewable energy development and biodiversity conservation.

6.3 SCOPE AND APPROACH

This study aims to systematically examine the integration of biodiversity protection within renewable energy policies and related legal frameworks across different countries and regions, in order to explore the relationship and interactions between energy transition and ecological conservation. The selected cases include the United States, China, the European Union (EU), Norway, and Brazil. These cases encompass developed countries, developing countries, and a supranational governance model represented by the EU, reflecting variations across different stages of economic development and energy structures while offering strong representativeness and comparative value.

As a review-based study, this paper systematically organizes policy texts and legal provisions from these countries and regions to identify the main pathways, legal mechanisms, and implementation features through which biodiversity protection is incorporated into renewable energy policies. Furthermore, through cross-national comparison, it seeks to uncover common challenges and differentiated strategies under varying governance models, providing insights into how to advance renewable energy development while maintaining ecosystem integrity. More specifically, this paper focuses not only on explicit biodiversity protection requirements and implementation tools within national policy frameworks, such as siting restrictions, environmental assessment procedures, and ecological compensation mechanisms, but also on identifying institutional gaps, potential conflicts, and notable best practices through comparative analysis.

Additionally, in terms of case selection, this study focuses specifically on renewable energy policies or broader energy policy and legal documents, rather than standalone biodiversity protection policies or legislation. This scope aligns with the research objective of examining how biodiversity concerns are embedded within the energy policy framework itself. In other words, the study analyzes whether and how biodiversity issues are addressed within the context of energy transition and renewable energy development, rather than providing a comprehensive review of biodiversity-specific legislation. This delineation helps highlight the institutional arrangements and policy logic through which energy policies reconcile climate and ecological objectives, thereby clarifying the synergies and tensions that arise in the context of energy transition.

6.4 CASE STUDIES: REVIEW OF POLICIES AND LAWS ACROSS COUNTRIES

The following section examines the cases of the EU, Norway, the United States, China, and Brazil, analyzing how biodiversity protection is reflected and operationalized within their renewable energy policies and legal frameworks.

6.4.1 European Union

The EU was selected as a case study primarily due to its leading role in the global renewable energy transition and its institutionalized policy framework. In 2024, 46.9% of the EU's net electricity generation came from renewable sources, with member states such as Denmark (88.4%), Portugal (87.5%), and Croatia (73.7%) ranking among the highest globally in wind and hydropower utilization (Eurostat, 2025). This reflects the EU's notable achievements in clean energy deployment. Moreover, encompassing 27 member states, the EU's policy framework not only drives rapid energy structure transitions within its members but also integrates energy and ecological governance at the regional level through cross-border coordination mechanisms. This combination of high renewable energy penetration and institutionalized governance makes the EU an important case for examining the integration of biodiversity considerations into energy policy.

The EU began promoting renewable energy in the 1970s, but it was not until the early 1990s, when climate change became a central policy concern, that its renewable energy policies accelerated. In 1996, the European Commission published the Green Paper on RES, followed by the 1997 White Paper: Energy for the Future: Renewable Sources of Energy, widely regarded as a turning point in the EU's renewable energy policy (Solorio & Bocquillon, 2017). Although the 1997 White Paper (European Commission, 1997) did not establish a systematic framework for biodiversity protection, it addressed the issue indirectly. In its section on the Common Agricultural Policy (CAP) and rural development, it emphasized leveraging agricultural, fiscal, and industrial policies to promote biomass use. The document proposed that future CAP reforms should prioritize alternative uses of agricultural products and encourage member states to support renewable energy within the framework of national aid. It further suggested that programs should align with agri-environmental objectives to mitigate the environmental pressures of biomass production and use, including additional subsidies for energy crops cultivated with reduced irrigation, low-input, or organic methods, or harvested in ways that promote biodiversity (European Commission, 1997). Annex II projected an additional 90 Mtoe of biomass energy by 2010 sourced from agriculture, forestry residues, waste streams, and energy crops, while stressing that "particular attention must be paid to protecting biodiversity in the EU, and strategies should be adopted to minimize impacts on biodiversity" (European Commission, 1997). Notably, these biodiversity references were confined to biomass energy, without specific ecological provisions for other renewable energy forms.

The 2009 Renewable Energy Directive (RED I) established the "20-20-20" targets: a 20% reduction in greenhouse gas emissions from 1990 levels, a 20% share of renewable energy in total consumption, and a 20% improvement in energy efficiency by 2020 (European Parliament & Council of the European Union, 2009). The preamble explicitly stated that the rising demand for biofuels and bioliquids, as well as the associated policy incentives, should not come at the expense of high-biodiversity lands. It mandated sustainability criteria to ensure that raw materials must not originate from areas of high biodiversity value, except where competent authorities confirm that production does not compromise conservation objectives. The directive incorporated definitions from FAO's Global Forest Resources Assessment, designating primary forests and legally protected forests as priority conservation areas while excluding modified, semi-natural, and plantation forests. It also prohibited sourcing from temperate and tropical savannahs, grasslands, shrublands, and pastures of high biodiversity value and tasked the Commission with establishing criteria to scientifically define and identify such lands (European Parliament & Council of the European Union, 2009). Article 17 explicitly barred the use of raw materials from (1) primary forests and other wooded lands of native species with no visible human disturbance, (2) legally designated nature protection areas or sites established to conserve rare or threatened ecosystems or species, and (3) highly biodiverse natural and non-natural grasslands (European Parliament & Council of the European Union, 2009). Article 18 emphasizes strengthening biodiversity protection through international cooperation and certification mechanisms. It requires that bilateral or multilateral agreements with third countries incorporating sustainability criteria consider protecting key ecosystem service areas, restoring degraded land, and avoiding excessive water use in scarce regions. The Commission is also empowered to recognize national or international voluntary certification schemes to verify compliance and provide data on ecosystem protection, land restoration, and resource management, as well as to acknowledge areas identified under international agreements or listed by IUCN for the protection of rare, threatened, or endangered ecosystems and species (European Parliament & Council of the European Union, 2009).

The 2018 amendment, RED II, reinforced these provisions. Article 29 reiterated the prohibition of feedstocks sourced from high-biodiversity lands and authorized the Commission to adopt implementing acts defining criteria for identifying highly biodiverse grasslands based on scientific evidence (European Parliament & Council of the European Union, 2018). The 2023 Renewable Energy Directive (RED III) introduced a more systemic approach to integrating ecological governance. Article 15c established "renewable energy acceleration areas," requiring member states by 2026 to designate such zones primarily on artificial or degraded land—such as rooftops, industrial sites, mines, and brownfields—to avoid new pressures on natural ecosystems. These areas must exclude Natura 2000 sites, nationally protected areas, and major bird and marine mammal migration routes, with delineation based on wildlife sensitivity maps and ecological data. Environmental assessments were mandated prior to approval, and any projects potentially affecting Natura 2000 sites must undergo appropriate assessments to ensure compliance with the Habitats and Birds Directives. The article also required effective mitigation measures during construction in acceleration areas to prevent adverse effects on soil, biodiversity, and aquatic systems, and allowed pilot testing of new mitigation measures to verify effectiveness (European Parliament & Council of the European Union, 2023). Furthermore, Article 29(3) expanded raw material protection to include old-growth forests and heathlands, while refining forestry standards to prohibit stump and root harvesting, restrict operations on fragile soils, enforce sustainable forest management practices, and set thresholds for clearcutting and deadwood retention (European Parliament & Council of the European Union, 2023).

Overall, the EU's renewable energy policy trajectory demonstrates a shift from a biomass-centered focus toward embedding biodiversity safeguards into broader energy planning. Early measures concentrated on sustainability standards for biofuels and bioliquids, primarily regulating feedstock sourcing and protecting high-biodiversity lands. Only with RED III did biodiversity considerations extend to renewable energy siting and infrastructure planning, signaling a proactive response to the ecological risks of energy expansion. This evolution illustrates the EU's transition from sector-specific regulation toward systemic ecological governance in its energy policy, providing a model for reconciling renewable energy development with biodiversity conservation.

6.4.2 Norway

Norway's extensive river networks and complex topography provide natural advantages for hydropower development. Based on these geographical conditions, the construction of dams has long supported industrial activity and ensured a stable supply of hydroelectric power for the national grid (Boasson & Jevnaker, 2022). Norway's electricity mix is dominated by hydropower, which accounts for 89% of its generation, making it the primary source of its clean energy. Currently, 98% of Norway's electricity comes from renewable sources, surpassing the International Energy Agency's "Net Zero" scenario target of 60% renewable electricity globally. Norway is now the largest producer of hydropower in Europe and the sixth largest globally (Ember, 2025).

Given this dominance of hydropower, water resource management laws constitute a critical part of Norway's energy legislation, which is characterized by a well-developed legal framework. Key legal instruments include the Waterfall Rights Act, Watercourse Regulation Act, Water Resources Act, Energy Act, and Offshore Energy Act, which together provide the institutional basis for the country's low-carbon and sustainable energy transition (Energifakta Norge, 2024).

The Energy Act of 1990 serves as the core law regulating electricity generation, transmission, trade, and supply in Norway's energy sector. Chapter 2 of the Act stipulates the licensing process, including mandatory environmental impact assessments (EIAs), which explicitly require ecological evaluation. Chapter 3 further emphasizes the need to avoid harm to natural environments and cultural heritage (Ministry of Energy, 1990).

The Offshore Energy Act of 2010 established a dedicated legal framework for offshore renewable energy, addressing emerging challenges related to offshore wind and marine resource management. Its stated purpose is to ensure that offshore renewable energy production proceeds in an orderly, safe, environmentally responsible, and socially acceptable manner, while balancing energy production with environmental protection, fisheries, shipping, and other maritime uses. Chapter 3 specifies conditions to minimize or prevent harm to the environment, biodiversity, cultural heritage, and competing land uses (Ministry of Energy, 2010). The 2020 implementing regulations provide detailed rules for offshore energy management, including clearer licensing procedures. During the drafting process, various stakeholders, including governmental, scientific, and non-governmental organizations, emphasized biodiversity-related considerations, particularly within project-specific EIAs (Ministry of Energy, 2020b). In the finalized implementing regulations of the Offshore Energy Act, additional provisions were introduced concerning biodiversity protection. Specifically, Section 6 of Chapter 2 of the Forskrift til havenergilova (Regulations to the Offshore Energy Act), which addresses project-specific impact assessments, stipulates that such assessments must provide a detailed description and evaluation of the potential environmental and social impacts of energy facilities. These assessments must include at a minimum: (1) seabed conditions and the aquatic environment, (2) cultural heritage, cultural environments, and landscapes, (3) birds, fish, habitat types, and other biodiversity, and (4) issues related to environmental monitoring (Ministry of Energy, 2020a).

At the policy level, the 2021 White Paper "Putting Energy to Work – Long Term Value Creation from Norwegian Energy Resources" explicitly addressed environmental and biodiversity protection. It emphasized that wind power development must consider impacts on the environment, society, and nearby communities, and it required robust local and regional consultation processes. The paper acknowledged that all energy projects impose some negative effects and stressed that licensing approvals must be based on socio-economic feasibility, granting approval only where benefits outweigh adverse impacts. It provided detailed discussions of specific energy sources: hydropower's ecological effects on river systems, including impacts on birds, fish, and benthic species, were to be mitigated by revising concession conditions; small hydropower projects were identified as affecting habitats, migration routes, and ecological corridors, necessitating review of landscape and cultural factors. Wind power was noted to fragment habitats, disrupt mammals and birds, and pose collision risks, with approval processes required to weigh such effects alongside concerns from communities, recreation, defense, and reindeer herding. Offshore wind permitting would similarly assess impacts on fish, marine mammals, seabirds, migratory birds, and potential conflicts with fisheries and navigation (Ministry of Energy, 2021).

The 2020 White Paper on Wind Power, proposed by the Ministry of Petroleum and Energy, underscored that wind development inevitably entails environmental and social trade-offs. It designated certain areas as unsuitable for wind power due to their environmental importance, halted implementation of the 2019 National Wind Power Framework, and suspended new licensing approvals. Section 4.8 of the 2020 White Paper further underscores the importance of addressing environmental and social impacts in the permitting process. It notes that, with the expansion of wind farm development, policies have increasingly focused on the cumulative effects of projects on landscapes, biodiversity, and reindeer herding. The Ministry of Petroleum and Energy recommends conducting more systematic pre- and post-construction studies and incorporating these findings into the knowledge base. Several wind project permits have already required studies on birds and reindeer herding to assess actual impacts. The document also proposes mitigation measures within and around planned areas, such as placing power lines underground near wind farms to reduce risks to local bird populations. Section 4.9 of the 2020 White Paper specifically addresses reindeer herding in the context of wind power development, framing it as an issue with both cultural and environmental dimensions. Given its strong dependence on land resources, reindeer herding is highly vulnerable to the expansion of energy infrastructure. The document therefore mandates early engagement with reindeer herding communities, the inclusion of county-level herding authorities in the review process, and the involvement of researchers with relevant expertise. While primarily aimed at safeguarding Sámi culture and livelihoods, this provision also plays a role in environmental governance. Because reindeer herding relies on tundra and forest ecosystems, its protection requirements effectively limit energy infrastructure development in ecologically sensitive areas (Ministry of Energy, 2020c).

Overall, Norway integrates biodiversity protection into its energy governance through environmental assessments, stakeholder participation, and balancing of competing interests. Rooted in hydropower regulation and extended through policies on wind and offshore energy, its approach institutionalizes ecological considerations within licensing and decision-making, representing an indirect but systematic embedding of biodiversity protection within energy transition governance.

6.4.3 United States

Since the early 21st century, the U.S. has gradually established a legal and institutional foundation for its energy transition through several major energy and climate acts. The Energy Policy Act of 2005 covered areas including energy efficiency, renewable energy, fossil fuels, nuclear power, and climate technologies, and promoted energy diversification and emissions reduction through loan guarantees, tax incentives, and biofuel blending requirements (U.S. Congress, 2005). Notably, Section 241 introduced new provisions regarding fishways in hydropower licensing, aiming to strengthen fish conservation and biodiversity protection. It required that significant factual disputes concerning fishways be resolved through trial-type hearings within a set timeframe and allowed for the adoption of alternative fishway prescriptions, provided they offered equivalent or better protection at lower cost or improved power operations. This mechanism introduced greater flexibility into decision-making and mandated consideration of energy supply, cost, water resources, and environmental quality, thus fostering alignment between hydropower development and ecological diversity objectives (U.S. Congress, 2005, sec. 241). The Energy Independence and Security Act of 2007 (EISA) advanced energy security and greenhouse gas reduction while strengthening efficiency targets for federal agencies and establishing the Renewable Fuel Standard (U.S. Environmental Protection Agency, 2025). Title II included provisions indirectly related to ecology and biodiversity. Section 232 required the Department of Energy's systems biology program to develop technologies based on genomics, microbiology, and plant functions aimed at cultivating low-resource cellulosic feedstocks, emphasizing sustainable use of soil, water, and forest resources and environmental quality protection. It also amended the Biomass Research and Development Act of 2000 to incorporate life-cycle GHG emissions and land-use change into biofuel assessments and required analysis of expanded biofuel production impacts on forests and human and animal food supplies (U.S. Congress, 2007, Title II, Subtitle B, secs. 232–233). While not biodiversity-focused, these provisions reflected a policy orientation that linked biofuel development with ecosystem and resource sustainability.

Under the Biden administration, multiple policies and laws further promoted clean energy and climate goals. Executive Order 14017, America's Supply Chains, aimed to build resilient, diversified, and secure supply chains (International Energy Agency, 2024). In 2022, the Department of Energy released "America's Strategy to Secure the Supply Chain for a Robust Clean Energy Transition," the first comprehensive federal plan to strengthen the clean energy industrial base (U.S. Department of Energy, 2022). While it addressed hydropower, solar PV, and wind power, its focus remained on developing energy infrastructure rather than biodiversity protection. The Inflation Reduction Act (IRA) of 2022, the largest climate and energy investment in U.S. history, advanced wind, solar, EVs, and carbon capture primarily through tax credits and funding (U.S. Internal Revenue Service, 2025). The Bipartisan Infrastructure Law (BIL) further emphasized energy infrastructure, grid modernization, and technology deployment (U.S. Department of the Interior, 2023). Although these acts included indirect environmental justice and emissions benefits, they lacked explicit provisions on ecosystems or biodiversity.

An examination of key post-2020 renewable energy policies shows that most prioritize energy transition and emissions goals, with few directly addressing biodiversity. However, some legislative efforts, such as the Reinvesting in Shoreline Economies and Ecosystems (RISEE) Act and the Public Lands and Renewable Energy Development Act (PLREDA), have incorporated biodiversity considerations.

The RISEE Act of 2023 aimed to increase revenue sharing from offshore wind, offshore oil and gas leasing, and onshore energy resources with coastal states and mandated use of funds for shoreline restoration, ecological protection, and related infrastructure. Revenues from offshore wind leases signed after 2022 were allocated among the U.S. Treasury, the National Oceans and Coastal Security Fund, and adjacent coastal states to support ecological and resource management (U.S. Congress, 2023). Section 2 of the cited Act specifically included biodiversity and marine conservation measures: (1) the National Oceans and Coastal Security Fund was authorized to support habitat protection and restoration, monitoring and mitigating harm to marine species (birds, mammals, fish), and sustainable fisheries; and (2) states were required to direct portions of offshore wind revenues toward coastal restoration and mitigation of impacts on fish, wildlife, or natural resources, including fisheries science research (U.S. Congress, 2023). These provisions demonstrate how funding mechanisms for coastal ecosystem and biodiversity protection have been directly embedded within the framework of offshore energy development and revenue allocation, making it a representative example in recent U.S. energy legislation of integrating energy development with ecological restoration.

The PLREDA seeks to promote wind and solar development on public lands while safeguarding critical fish and wildlife habitats (National Wildlife Federation, n.d.). As one of the few federal legislative proposals explicitly linking renewable energy expansion and biodiversity protection, it offers a framework for harmonizing energy transition and ecological conservation. Initially introduced in 2021, it focused on designating priority development zones, streamlining permitting, and establishing a conservation fund (Bureau of Land Management, 2021). Its 2024 version clarified revenue-sharing provisions, directing funds to states and counties and earmarking portions for conservation and permitting improvements (U.S. Department of the Interior, 2024). The 2025 version reinforced this approach, with Section 5 of this Act explicitly creating a Renewable Energy Resource Conservation Fund to finance restoration of affected habitats, wildlife corridors, wetlands, streams, rivers, and other natural water bodies in renewable energy development areas (U.S. Congress, 2025). By allocating project revenues for habitat restoration, PLREDA exemplifies a model that integrates renewable energy development with biodiversity protection.

In summary, while certain U.S. energy legislative initiatives, such as RISEE and PLREDA, incorporate biodiversity considerations, such measures remain fragmented and secondary. Recent flagship policies like the IRA and BIL largely prioritize accelerating energy transition and industrial competitiveness, rather than embedding biodiversity into core policy design.

6.4.4 China

The development of China's renewable energy policies and laws dates back to the late 1990s and has gradually evolved in response to energy restructuring, low-carbon transition needs, and technological progress. The promulgation of the Renewable Energy Law in 2005 formally established the strategic status of renewable energy in law, representing a milestone in the development of China's renewable energy legal framework. From 2006 to 2017, technological advancements, efficiency improvements, and cost reductions drove the large-scale deployment of renewable energy, positioning it as a major alternative energy source. Since 2018, renewable energy has achieved rapid growth, becoming China's second-largest energy source (Yang, 2024).

The Renewable Energy Law of 2005, while not directly stipulating provisions on biodiversity protection, explicitly stated in Chapter II, Article 9 that renewable energy development and utilization plans should follow the principle of adapting measures to local conditions. Additionally, Chapter IV, Article 18 emphasized that renewable energy development plans in rural areas should be formulated in conjunction with local economic and social development needs and ecological protection (National Energy Administration, 2017). Although the law did not include specific references to biodiversity, it underscored the need for energy development to comply with ecological protection requirements, laying the foundation for subsequent measures.

The 12th Five-Year Plan for Renewable Energy Development marked the first explicit articulation that renewable energy development should be coordinated with ecological and environmental protection. In its "Guidelines and Objectives" and "Basic Principles," it emphasized integrating energy resource development with ecological protection to establish an energy development model aligned with ecological civilization (National Energy Administration, 2013). The most recent 14th Five-Year Plan for Renewable Energy Development similarly stressed ecological prioritization as a guiding principle, calling for renewable energy development to be coordinated with ecological protection. Regarding hydropower, it explicitly required rigorous EIAs for hydropower planning and projects, as well as enhanced monitoring and evaluation of ecological protection measures to ensure their continuous improvement and effectiveness. The plan also promoted innovative "PV for desertification control" projects, encouraging region-specific approaches and drought-resistant crops to develop photovoltaic bases that combine power generation with ecological restoration, thereby improving vegetation cover and biodiversity while advancing renewable energy deployment. Furthermore, compared with previous plans, the 14th Five-Year Plan strengthened ecological safeguards in Chapter VIII by calling for key research on ecological protection technologies related to renewable energy development, with a focus on assessing and mitigating the ecological impacts of hydropower and wind projects, particularly on aquatic and terrestrial biodiversity (National Development and Reform Commission, 2022). Considering the impacts of hydropower and wind power development on aquatic species and bird migration, this initiative is intended not only to reduce the ecological disturbances caused by energy development but also to help protect biodiversity and ecosystem integrity in affected areas, thereby promoting coordination between energy transition and ecological security.

In addition to these foundational laws and plans, several regulations specifically address biodiversity-related challenges in renewable energy development. The Technical Guidelines for Environmental Impact Assessment of Hydropower Projects (Ministry of Ecology and Environment of the People's Republic of China, 2003) required comprehensive ecological assessments for hydropower projects. It mandated that EIAs include terrestrial and aquatic biodiversity, ecosystem integrity, and sensitive ecological areas. Baseline studies should cover terrestrial vegetation, wildlife distribution, rare species and habitats, and protected areas, while aquatic surveys should include plankton, benthic organisms, aquatic plants, fish communities, spawning grounds, rare aquatic species, and protected water areas. Impact assessments must analyze effects on ecosystem productivity, vegetation succession, rare species, habitats, aquatic organisms, and wetlands. This guideline provided a systematic framework for assessing biodiversity and ecosystem impacts in hydropower projects (Ministry of Ecology and Environment of the People's Republic of China, 2003).

In 2019, the Notice on Standardizing the Use of Forest Land for Wind Power Projects was issued by the National Forestry and Grassland Administration, addressing biodiversity concerns associated with wind power development. It identified prohibited zones for wind power construction, including natural heritage sites, national parks, nature reserves, forest parks, wetland parks, geological parks, scenic areas, major bird migration routes, and coastal shelterbelts (National Forestry and Grassland Administration of the People's Republic of China, 2019). The 2023 Notice on Supporting the Development of the Photovoltaic Power Generation Industry through Standardized Land-Use Management stipulates that project siting should avoid ecological protection redlines, areas of special natural landscape value, and natural forests. For projects involving protected areas, compliance with relevant regulations and policies governing protected areas is also required (General Office of the Ministry of Natural Resources et al., 2023).

In 2024, the Ministry of Ecology and Environment released the Draft Notice on Strengthening Ecological and Environmental Protection in Onshore Wind and Solar Power Projects. Although not yet finalized, it reflects the growing emphasis on ecological protection in renewable energy development. It called for integrating energy security and ecological security in policy and spatial planning, encouraged policy-level ecological impact assessments, and mandated avoidance of ecologically sensitive areas, biodiversity priority zones, ecological redlines, protected areas, primary grasslands, and natural forests. For wind projects, it required avoiding bird migration routes and critical habitats, while PV projects needed to avoid key migration corridors and habitats for rare and endangered species. It also addressed project impacts such as noise, shadow flicker, and vegetation disturbance, especially in arid regions. The draft introduced specific wildlife protection measures, including maintaining a minimum distance of 1.6 km from key bird migration routes, optimizing lighting to reduce bird collisions, installing warning markers and deterrents, and adopting non-contiguous layouts to preserve migration pathways for large mammals. For flora, it mandated avoiding habitats of endemic plants and reducing disturbance to plant communities (Ministry of Ecology and Environment of the People's Republic of China, 2024).

Overall, China's renewable energy policy framework shares similarities with the U.S. in that its overarching laws and plans emphasize harmonizing energy development with ecological protection but offer only general, principle-based language on biodiversity. However, this gap has been partly addressed through targeted regulations and guidance, particularly on wind and solar siting, EIAs, wildlife protection, and land-use management.

6.4.5 Brazil

With its abundant natural resources and unique geographical conditions, Brazil occupies a prominent position in the global renewable energy transition. In 2023, renewable energy accounted for 89% of Brazil's total electricity generation, ranking first among G20 economies and far exceeding the global average of 30% (Ember, 2024). Currently, large hydropower plants supply approximately 80% of Brazil's domestic electricity, while photovoltaic projects are expected to contribute nearly 70% of new installed capacity in the coming years. Meanwhile, natural gas, wind power, and bioenergy are playing increasingly important complementary roles in the energy mix (International Energy Agency, 2025).

In terms of policy and legal frameworks, the Brazilian government has long committed to advancing renewable energy development, supported by a series of plans and legislative measures that underpin its energy transition. Two of its most significant initiatives are the Alternative Energy Sources Incentive Program (PROINFA) and the Ten-Year Energy Expansion Plan (PDE), aimed at fostering domestic industrial development and establishing a sustainable energy supply system in technological, environmental, and economic terms (Pao & Fu, 2013). In 2002, Congress passed Law No. 10,438 (Brazil, 2002), which formally established PROINFA to increase the share of renewables in the power system through wind, biomass, and small hydropower development (International Energy Agency/IRENA Policy and Measures Database, 2024). Notably, this law did not include provisions related to ecological or biodiversity protection.

From 2010 to 2019, the Ministry of Mines and Energy formulated and implemented the Ten-Year Energy Expansion Plan (PDE), aimed at phasing out fossil fuel power plant construction and expanding hydropower and wind power at scale (International Energy Agency/IRENA Policy and Measures Database, 2013). The version, PDE 2031, is particularly notable for its Chapter 10, which introduces a sustainability-based socio-environmental analysis framework to systematically assess major ecological and social issues arising from energy expansion (Ministry of Mines and Energy & Energy Research Company, 2022). This framework explicitly identifies biodiversity as a primary theme, highlighting the direct and indirect pressures of energy development on species, habitats, and ecological processes, including the loss or alteration of terrestrial and aquatic habitats and disruptions to ecosystem functions. The document stresses that concentrated siting of wind and solar projects, alongside associated transmission lines, can lead to cumulative and synergistic effects, which must be mitigated through measures such as avoiding the clearance of native vegetation, implementing forest restoration, and monitoring bird populations. Similarly, hydropower projects in ecologically sensitive or dam-fragmented regions face comparable challenges, requiring integrated environmental assessments and watershed studies, as well as adherence to international standards, including those on ecosystem services and ecological flow maintenance, to strengthen risk management. PDE 2031 also provides region-specific analyses of energy development impacts on ecosystems. In the North, projects are concentrated in highly heterogeneous environments, posing risks to terrestrial and aquatic habitats. In the Northeast, wind projects present collision risks to birds and bats while also threatening high-biodiversity marine ecosystems. The Midwest faces river fragmentation and headwater vegetation loss, creating cumulative pressures on aquatic species. In the Southeast, project proliferation exerts compounding stress on marine ecosystems and disturbs remnants of the Atlantic Forest. Similarly, in the South, cumulative impacts include disturbances to the Atlantic Forest remnants and riverine ecosystems (Ministry of Mines and Energy & Energy Research Company, 2022). Through this integrated approach, PDE 2031 underscores the strategic importance of aligning energy development with biodiversity conservation and reflects a growing policy trend toward embedding ecological considerations within energy planning.

At the legislative level, Bill of Law No. 576/2021 was signed into law, establishing a regulatory framework for offshore energy development across federally governed areas such as territorial seas, exclusive economic zones, and the continental shelf (Presidency of the Republic of Brazil, 2025a). In the specific legal text of Law No. 15,097/2025, Article 4 of this law emphasizes the rational use of natural resources to enhance energy security while safeguarding marine culture, traditional livelihoods, and environmental and scenic values. Furthermore, Article 6 prohibits the designation of energy blocks within environmentally protected areas and landscape preservation zones, embedding ecological protection directly into the governance of offshore energy development (Presidency of the Republic of Brazil, 2025b). However, the recently enacted Energy Transition Acceleration Program (PATEN) in 2025 contains no explicit provisions related to biodiversity protection (Presidency of the Republic of Brazil, 2025c), suggesting that Brazil's latest energy transition policies remain primarily focused on accelerating energy restructuring and meeting climate targets, while the integration of ecological considerations still requires improvement.

Overall, Brazil's renewable energy governance has increasingly incorporated biodiversity considerations, particularly through PDE 2031, which framed biodiversity as a central theme within energy planning. While this reflects progress toward integrating ecological concerns into the policy process, such measures remain largely procedural and lack binding enforcement mechanisms, as evidenced by the absence of biodiversity provisions in the 2025 PATEN law. Consequently, Brazil's approach demonstrates an initial alignment of energy development with biodiversity objectives, but stronger regulatory safeguards are still needed.

6.5 DISCUSSION

The comparative analysis of the five cases: EU, Norway, the United States, China, and Brazil, reveals significant variation in how biodiversity considerations are integrated into renewable energy policy frameworks. These differences are shaped by the legal status of renewable energy instruments, the role of environmental assessment, and the extent to which ecological safeguards are made enforceable.

First, the mode of integration varies across cases. The EU stands out for embedding biodiversity protection directly into binding legislative instruments, most notably through the Renewable Energy Directives (RED I-III), which progressively evolved from regulating biomass feedstock sourcing to addressing broader ecological constraints. Norway, while lacking a standalone biodiversity framework in its energy laws, incorporates ecological safeguards indirectly through mandatory Environmental Impact Assessments (EIA), licensing conditions, and sectoral coordination mechanisms. By contrast, the U.S. primarily relies on targeted legislation initiatives, such as the RISEE Act and the proposed PLREDA, which embed biodiversity funding and habitat restoration measures within broader resource management frameworks. China and Brazil demonstrate a hybrid approach, combining overarching policy statements emphasizing ecological coordination with more specific administrative measures to address localized ecological risks. Second, these cases reveal differences in implementation instruments. The EU's legally binding directives, compliance monitoring, and use of scientific data to define ecological criteria represent the most institutionalized model. Norway and China emphasize procedural tools, such as EIA and permitting, to mediate biodiversity impacts at the project level. The U.S. and Brazil, by contrast, illustrate different non-prohibitive approaches: the U.S. case highlights revenue-sharing and habitat restoration funds, while Brazil relies more on strategic socio-environmental assessments within energy planning.

From a policy perspective, the comparison of these cases highlights several important implications. First, integrating biodiversity considerations early in renewable energy planning, such as the EU's spatial exclusion zones under RED III and China's siting restrictions, can help avoid conflicts and limit ecological trade-offs before they arise. Next, procedural frameworks, such as Norway's EIA system and Brazil's socio-environmental assessments, further illustrate the value of structured evaluation processes in balancing energy development with ecological concerns, though their effectiveness hinges on reliable data and institutional capacity. Meanwhile, the U.S. example shows that linking renewable energy revenues to habitat restoration and ecological funds offers a pragmatic way to align energy expansion with conservation objectives.

The comparison suggests that the key difference across cases lies not only in whether biodiversity is mentioned in renewable energy policy, but also in whether such concerns are translated into binding siting rules, licensing conditions, funding mechanisms, or monitoring requirements. Overall, while the extent to which biodiversity is embedded in renewable energy policies varies significantly across different countries and regions, these cases collectively underscore a growing recognition of the need to integrate renewable energy development with ecological protection. Advancing this alignment will require broader adoption of science-based spatial planning, enforceable ecological safeguards, and stronger cross-sector coordination, so that the global energy transition mitigates rather than exacerbates biodiversity loss.

DECLARATION OF GENERATIVE AI USE

During the preparation of this manuscript, the authors used generative artificial intelligence tools to assist with language editing and stylistic refinement. The authors subsequently reviewed and edited the content and take full responsibility for the final version of the published chapter.

REFERENCES

Allison, T. D., Root, T. L., & Frumhoff, P. C. (2014). Thinking globally and siting locally: renewable energy and biodiversity in a rapidly warming world. *Climatic Change*, 126(1), 1–6. https://doi.org/10.1007/s10584-014-1127-y

Boasson, E. L., & Jevnaker, T. (2022). Energy Governance in Norway. In: Knodt, M., Kemmerzell, J. (eds) *Handbook of Energy Governance in Europe*. Springer, Cham. https://doi.org/10.1007/978-3-030-43250-8_21

Brazil. (2002). Law No. 10,438 of 26 April 2002: Provides for the Emergency Expansion of Electricity Supply, Extraordinary Tariff Recomposition, Creates the Programme of Incentives for Alternative Electricity Sources (PROINFA). Presidency of the Republic – Planalto. Retrieved from https://www.planalto.gov.br/ccivil_03/leis/2002/l10438.htm

Bureau of Land Management. (2021). H.R. 3326: Public Land Renewable Energy Development Act. Retrieved from https://www.blm.gov/sites/default/files/docs/2021-05/H.R.%203326%2C%20Public%20Land%20Renewable%20Energy%20Development%20Act.pdf

Convention on Biological Diversity. (1992). Secretariat of the Convention on Biological Diversity. Retrieved from https://www.cbd.int/doc/legal/cbd-en.pdf

Dhunny, A. Z., Allam, Z., Lobine, D., & Lollchund, M. R. (2019). Sustainable renewable energy planning and wind farming optimization from a biodiversity perspective. *Energy*, 185, 1282–1297. https://doi.org/10.1016/j.energy.2019.07.147

Ember. (2024). Brazil Rises as G20 Renewables Powerhouse. Ember. Retrieved from https://ember-energy.org/latest-insights/brazil-rises-as-g20-renewables-powerhouse/

Ember. (2025). Norway. Ember. Retrieved from https://ember-energy.org/countries-and-regions/norway/

Energifakta Norge. (2024). Regulation of the Energy Sector: The Legal Framework. Norwegian Ministry of Energy. Retrieved from https://energifaktanorge.no/en/regulation-of-the-energy-sector/det-juridiske-rammeverket/

European Commission. (1997). Energy for the future: Renewable sources of energy (White Paper COM 97-599). https://europa.eu/documents/comm/white_papers/pdf/com97_599_en.pdf

European Parliament, & Council of the European Union. (2009). Directive 2009/28/EC of the European Parliament and of the Council of 23 April 2009 on the promotion of the use of energy from renewable sources and amending and subsequently repealing Directives 2001/77/EC and 2003/30/EC. EUR-Lex. https://eur-lex.europa.eu/legal-content/EN/ALL/?uri=celex:32009L0028

European Parliament, & Council of the European Union. (2018). Directive (EU) 2018/2001 of the European Parliament and of the Council of 11 December 2018 on the promotion of the use of energy from renewable sources (RED II). EUR-Lex. https://eur-lex.europa.eu/eli/dir/2018/2001/oj/eng

European Parliament, & Council of the European Union. (2023). Directive (EU) 2023/2413 of the European Parliament and of the Council of 19 December 2023 on the promotion of energy from renewable sources (RED III). EUR-Lex. https://eur-lex.europa.eu/eli/dir/2023/2413/oj/eng

Eurostat. (2025). Electricity from Renewable Sources Reaches 47% in 2024. Eurostat. Retrieved from https://ec.europa.eu/eurostat/web/products-eurostat-news/w/ddn-20250319-1

Froidevaux, J. S. P., Le Viol, I., Barré, K., Bas, Y., & Kerbiriou, C. (2025). A modeling framework for biodiversity assessment in renewable energy development: A case study on European bats and wind turbines. Renewable and Sustainable Energy Reviews, 211, 115323. https://doi.org/10.1016/j.rser.2024.115323

Gasparatos, A., Doll, C. N. H., Esteban, M., Ahmed, A., & Olang, T. A. (2017). Renewable energy and biodiversity: implications for transitioning to a Green Economy. *Renewable and Sustainable Energy Reviews*, 70, 161–184. https://doi.org/10.1016/j.rser.2016.08.030

General Office of the Ministry of Natural Resources, Office of the National Forestry and Grassland Administration, & Department of Renewable Energy of the National Energy Administration of China. (2023). Notice on Supporting the Development of the Photovoltaic Power Generation Industry through Standardized Land Use Management (ZiranZiBanFa [2023] No. 12). Ministry of Natural Resources of the People's Republic of China et al. Retrieved from https://www.gov.cn/zhengce/zhengceku/2023-04/03/content_5749824.htm

Gibson, L., Wilman, E. N., & Laurance, W. F. (2017). How green is 'green' energy? *Trends in Ecology & Evolution*, 32(12), 922–935. https://doi.org/10.1016/j.tree.2017.09.007

Hooper, D. U., Adair, E. C., Cardinale, B. J., Byrnes, J. E. K., Hungate, B. A., Matulich, K. L., ... O'Connor, M. I. (2012). A global synthesis reveals biodiversity loss as a major driver of ecosystem change. *Nature*, 486(7401), 105–108. https://doi.org/10.1038/nature11118

International Energy Agency. (2024). United States 2024: Energy Policy Review. IEA. https://www.iea.org/reports/united-states-2024

International Energy Agency. (2025). Brazil Energy Profile [Webpage]. International Energy Agency. Retrieved from https://www.iea.org/countries/brazil

International Energy Agency/IRENA Policy and Measures Database. (2013). 2010–2019 Plan for Energy Expansion (Brazil). International Energy Agency/IRENA. Retrieved from https://www.iea.org/policies/5070-2010-2019-plan-for-energy-expansion

International Energy Agency/IRENA Policy and Measures Database. (2024). Programme of Incentives for Alternative Electricity Sources (PROINFA). International Energy Agency/IRENA. Retrieved from https://www.iea.org/policies/4019-programme-of-incentives-for-alternative-electricity-sources-programa-de-incentivo-a-fontes-alternativas-de-energia-eletrica-proinfa

Loss, S. R., Will, T., & Marra, P. P. (2013). Estimates of bird collision mortality at wind facilities in the contiguous United States. *Biological Conservation*, 168, 201–209. https://doi.org/10.1016/j.biocon.2013.10.007

Ministry of Ecology and Environment of the People's Republic of China. (2003). Code for Environmental Impact Assessment of Water Conservancy and Hydropower Projects (HJ/T 88–2003). Ministry of Ecology and Environment. Retrieved from https://www.mee.gov.cn/ywgz/fgbz/bz/bzwb/other/pjjsdz/200307/t20030701_63289.htm

Ministry of Ecology and Environment of the People's Republic of China. (2024). Notice on strengthening ecological and environmental protection in onshore wind power and photovoltaic power generation construction projects. Retrieved from https://picpolicy.mofcom.gov.cn/file/20241021/14261729494931573.pdf

Ministry of Energy. (1990). Act No. 50 of 29 June 1990 relating to the generation, conversion, transmission, trading, distribution and use of energy etc. (The Energy Act). https://www.regjeringen.no/globalassets/upload/oed/vedlegg/lover-og-reglement/act_no_50_of_29_june_1990.pdf

Ministry of Energy. (2010). Lov om fornybar energiproduksjon til havs (Offshore Energy Act No. 21 of 2010). Lovdata. Retrieved from https://lovdata.no/dokument/NL/lov/2010-06-04-21

Ministry of Energy. (2020a). Forskrift Til Havenergilova (FOR-2020–06–12–1192). Retrieved from https://lovdata.no/dokument/SF/forskrift/2020-06-12-1192

Ministry of Energy. (2020b). *Royal Decree on Offshore Energy Regulations*, 12 June 2020. https://www.regjeringen.no/contentassets/aaac5c76aec242f09112ffdceabd6c64/royal-decree-offshore-energy-regulation-june-2020.pdf

Ministry of Energy. (2020c). Vindkraft på land: Endringer i konsesjonsbehandlingen (Meld. St. 28 (2019–2020)). Regjeringen.no. https://www.regjeringen.no/no/dokumenter/meld.-st.-28-20192020/id2714775/?ch=1

Ministry of Energy. (2021). Energi til arbeid: Langsiktig verdiskaping fra norske energiressurser (Meld. St. 36 (2020–2021)). Regjeringen.no. https://www.regjeringen.no/no/dokumenter/meld.-st.-36-20202021/id2860081/?ch=1

Ministry of Mines and Energy (MME), & Energy Research Company (EPE). (2022). Socio-environmental analysis (Chapter 10). In 2031 Ten-Year Energy Expansion Plan (PDE 2031). Retrieved from https://www.gov.br/mme/pt-br/assuntos/secretarias/sntep/publicacoes/plano-decenal-de-expansao-de-energia/pde-2031/english-version/relatorio_pde2031_cap10_eus.pdf

National Development and Reform Commission. (2022). 14th Five Year Plan for Renewable Energy Development. National Development and Reform Commission. Retrieved from https://www.ndrc.gov.cn/xxgk/zcfb/ghwb/202206/P020220602315308557623.pdf

National Energy Administration. (2017). Renewable Energy Law of the People's Republic of China. National Energy Administration. Retrieved from https://www.nea.gov.cn/2017-11/02/c_136722869.htm

National Energy Administration. (2013). Notice of the State Council on the Issuance of the "12th Five-Year Plan for Energy Development". National Energy Administration. Retrieved from https://www.nea.gov.cn/2013-01/28/c_132132808.htm

National Forestry and Grassland Administration of the People's Republic of China. (2019). Notice on Standardizing the Use of Forest Land for Wind Power Projects (LinZiFa [2019] No. 17). National Forestry and Grassland Administration. Retrieved from https://www.gov.cn/zhengce/zhengceku/2019-09/30/content_5435366.htm

National Wildlife Federation. (n.d.). Renewable Energy on Public Lands. Retrieved from https://www.nwf.org/Our-Work/Climate/Climate-Change/Renewable-Energy/On-Public-Lands

Nazir, M. S., Bilal, M., Sohail, H. M., Liu, B., Chen, W., & Iqbal, H. M. N. (2020). Impacts of renewable energy atlas: reaping the benefits of renewables and biodiversity threats. *International Journal of Hydrogen Energy*, 45(41), 22113–22124. https://doi.org/10.1016/j.ijhydene.2020.05.195

OECD. (2024). *Mainstreaming Biodiversity into Renewable Power Infrastructure*. OECD Publishing, Paris. https://doi.org/10.1787/357ac474-en

Pao, H.-T., & Fu, H.-C. (2013). Renewable energy, non-renewable energy and economic growth in Brazil. *Renewable and Sustainable Energy Reviews*, 25, 381–392. https://doi.org/10.1016/j.rser.2013.05.004

Pedroli, B., Elbersen, B., Frederiksen, P., Grandin, U., Heikkilä, R., Krogh, P. H., ... & Spijker, J. (2013). Is energy cropping in Europe compatible with biodiversity? Opportunities and threats to biodiversity from land-based production of biomass for bioenergy purposes. *Biomass and Bioenergy*, 55, 73–86. https://doi.org/10.1016/j.biombioe.2012.09.054

Presidency of the Republic of Brazil. (2025a). President Lula Signs Law Creating Renewable Energy Generation from Offshore Wind Turbines. Government of Brazil. Retrieved from https://www.gov.br/planalto/en/latest-news/2025/01/president-lula-signs-law-creating-renewable-energy-generation-from-offshore-wind-turbines

Presidency of the Republic of Brazil. (2025b). Lei nº 15.097, de 10 de janeiro de 2025 [Law No. 15,097 of 10 January 2025]. Planalto. https://www.planalto.gov.br/ccivil_03/_ato2023-2026/2025/lei/l15097.htm

Presidency of the Republic of Brazil. (2025c). President Lula Signs into Law Energy Transition Acceleration Program. Government of Brazil. Retrieved from https://www.gov.br/planalto/en/latest-news/2025/01/president-lula-signs-into-law-energy-transition-acceleration-program

Rehbein, J. A., Watson, J. E. M., Lane, J. L., Sonter, L. J., Venter, O., Atkinson, S. C., & Allan, J. R. (2020). Renewable energy development threatens many globally important biodiversity areas. *Global Change Biology*, 26(5), 3040–3051. https://doi.org/10.1111/gcb.15067

Santangeli, A., Toivonen, T., Pouzols, F. M., Pogson, M., Hastings, A., Smith, P., & Moilanen, A. (2016). Global change synergies and trade-offs between renewable energy and biodiversity. *Global Change Biology Bioenergy*, 8(5), 941–951.

Serrano, D., Margalida, A., Pérez-García, J. M., Juste, J., Traba, J., Valera, F., ... & Donázar, J. A. (2020). Renewables in Spain threaten biodiversity. *Science*, 370, 1282–1283. https://doi.org/10.1126/science.abf6509

Solorio, I., & Bocquillon, P. (2017). Chapter 2 EU renewable energy policy: a brief overview of its history and evolution. In *A Guide to EU Renewable Energy Policy*. Edward Elgar Publishing, Cheltenham. https://doi.org/10.4337/9781783471560.00011

Sonter, L. J., Dade, M. C., Watson, J. E. M., & Valenta, R. K. (2020). Renewable energy production will exacerbate mining threats to biodiversity. *Nature Communications*, 11(1), 4174. https://doi.org/10.1038/s41467-020-17928-5

U.S. Congress. (2005). Energy Policy Act of 2005, Public Law 109–58. Retrieved from https://www.congress.gov/109/plaws/publ58/PLAW-109publ58.pdf

U.S. Congress. (2007). Energy Independence and Security Act of 2007 (EISA), Title II, Subtitle B, Secs. 232–233.

U.S. Congress. (2023). S. 373: Reinvesting in Shoreline Economies and Ecosystems Act of 2023. Congress.gov. Retrieved from https://www.congress.gov/bill/118th-congress/senate-bill/373/text

U.S. Congress. (2025). H.R. 1994: Public Land Renewable Energy Development Act, §5. Retrieved from https://www.congress.gov/bill/119th-congress/house-bill/1994/text

U.S. Department of Energy. (2022). America's Strategy to Secure the Supply Chain for a Robust Clean Energy Transition. Retrieved from https://www.energy.gov/policy/securing-americas-clean-energy-supply-chain

U.S. Department of the Interior. (2023). Meet BIL: How Bipartisan Infrastructure Law supports wildland fire management. Retrieved from https://www.doi.gov/wildlandfire/meet-bil-how-bipartisan-infrastructure-law-supports-wildland-fire-management

U.S. Department of the Interior. (2024). H.R. 8954: Public Land Renewable Energy Development Act. Retrieved from https://www.doi.gov/ocl/hr-8954

U.S. Environmental Protection Agency. (2025). Summary of the Energy Independence and Security Act of 2007. Retrieved from https://www.epa.gov/laws-regulations/summary-energy-independence-and-security-act

U.S. Internal Revenue Service. (2025). Inflation Reduction Act of 2022. Retrieved from https://www.irs.gov/inflation-reduction-act-of-2022

United Nations. (2022). What is renewable energy? United Nations. Retrieved from https://www.un.org/zh/climatechange/what-is-renewable-energy

Wang, J., Li, C., Deng, Z., Carr, J., Stringer, L. C., Li, K., … & Wang, Z. (2025). Biodiversity Impacts of Land Occupation for Renewable Energy Infrastructure in a Globally Connected World. Environmental Science & Technology, 59(19), 9529–9539. https://doi.org/10.1021/acs.est.4c11453

Yang, X. (2024). Current status, challenges and policy suggestions for the development of renewable energy in China. Science Technology and Industry, 24(18), 156–159. https://oversea.cnki.net/kcms2/article/abstract?v=InoXwbPQWprTegRpZexVPUSHS_ZMwPRZqyJP-lC-W7BY6GS-EG0QwMkLg0n3L3P2DLmJtsevWybVJuhxVoo3ewLlNEm2pMKh-A0qyDHmerJdKjglswg0mjwbZzEMUOf9XmZElhYMF9ggloz8Hi1U7q1jVE56BlyKgNVE68og8d2poey6b0Ja-w==&uniplatform=OVERSEA&language=EN

7 Reframing Economic and Ecological Strategies to Sustain Biodiversity and Human Well-Being

Yuehan Dou, Lingyun Xiao, Fanjing Liu, and Keyan Miao

7.1 INTRODUCTION: REFRAMING BIODIVERSITY MANAGEMENT

The decline of biodiversity and the degradation of ecosystems are some of the greatest challenges of our time. Estimates indicate that the rate of species extinction is accelerating globally, at an unprecedented rate in human history, jeopardizing not only the integrity of ecosystems, but also the social systems which derive from them. This crisis of biodiversity is more than ecological; it also underpins livelihoods, cultural practices, and human well-being. Ecosystem degradation resulting from the decline of biodiversity, such as wetlands, forests, and grasslands, impacts not just material services such as food and water security, but also cultural ecosystem services (CES) which are the non-material benefits to people from ecosystem services such as recreation, identity, and spiritual fulfillment (Soga and Gaston, 2020; Dou et al., 2022).

In this regard, the global community, ranging from governments, NGOs, and local communities, has all been involved in different approaches. Among the most well-known approaches are economic incentives, policies, and instruments that inspire biodiversity-friendly behaviors through financial or regulatory motivations, and ecological restoration, which is the physical-direct repair of degraded ecosystems while at the same time altering the livelihoods and social communities to sustain the restoration process. To illustrate, both factors have become the basis of international biodiversity governance and ecosystem management. The Convention on Biological Diversity has long referenced the role of incentives, especially in relation to removing harmful subsidies and increasing positive behaviors (CBD, 2010). In recognition of increased importance, the United Nations has declared 2021–2030 the "decade of ecosystem restoration", an acknowledgement that restoration will be important to making a sustainable future (Mastretta-Yanes et al., 2024).

However, while their importance is increasingly acknowledged, these approaches are often implemented and assessed separately. A typical project would assess incentives based on cost-effectiveness and participation, while restoration would assess ecological indicators, such as species richness or vegetation cover. This separation masks the point that the two approaches are inextricably linked; incentives create the financial scaffolding for restoration while restoration produces ecological and cultural outcomes that create legitimacy and maintain numerous incentive programs. In consideration of orienting readers, we are working towards providing readers a conceptual framework for the ideas we will discuss in this chapter, specifically a socio-ecological feedback loop in which incentives enable restoration and restoration generates biodiversity and CES, which in turn generates human well-being and supports governance structures. This feedback loop will ultimately help readers to understand how together these economic and ecological strategies can foster strong and resilient socio-ecological systems, which we expand upon later in Figure 7.3.

DOI: 10.1201/9781003640899-9

7.1.1 The Promise and Pitfalls of Incentive-Based Approaches

The concept of using economic incentives is powerful: when conservation looks economically viable, people and communities have increased motivation to conserve wildlife and biodiversity. For example, Costa Rica operates a well-known national Payment for Ecosystem Services (PES) program that compensates landowners monetarily for conserving forests (Pagiola, 2008), and China has a Grain for Green Program that subsidizes converting cropland to forest and grassland (Wu et al., 2022). Likewise, much of Europe has implemented agri-environmental schemes (AES) into its Common Agricultural Policy aimed at addressing the biodiversity crisis by changing farming systems to integrate nature-based approaches (Pe'er et al., 2020).

That acknowledged, the historical understanding of incentive schemes is variable. Whether or not they work typically depends on how the designed: payments are at or above local opportunity costs; contracts are long enough to lock in the benefits; and there is monitoring of non-compliance (de Koning et al., 2011). These conditions together can, but do not always, lead to conservation outcomes. When designed poorly, payment may fail or lead to unanticipated outcomes (with respect to livestock-linked subsidies in the Tibetan Plateau, leading to overgrazing (Xiao et al., 2022)). Outside recognitions of stewardship can even undermine intrinsic motivations [what is termed motivation crowding in Bolivia, Peru, Uganda, and Mexico (Narloch et al., 2012; Rico García-Amado et al., 2013)]. These types of lessons serve as a reminder that incentives can be very strong towards conservation, yet they are not silver bullets—to be effective, they depend on the context provided by social, institutional, and ecological factors.

7.1.2 Restoration as Ecological and Cultural Renewal

Meanwhile, ecological restoration has also become a valued method for reversing biodiversity declines and rebuilding the relationship between people and nature. Restoration techniques as diverse as wetland rehabilitation in Europe, and afforestation in Asia are increasingly recognized for their roles in biodiversity recovery and CES (Nowak-Olejnik et al., 2022, Yuan et al. 2022). Restoration provides clean water, climate regulation, and habitats for species (Mason et al., 2023) while creating possibilities for recreation, education, and science, spiritual and cultural identity (Kis et al., 2017).

Despite all the potential benefits of restoration, it can foster conflict and tensions. Certain projects prioritize ecological complexity over the aesthetics or recreational utility typically valued by locals (Soga et al., 2021). Others may not consider cultural heritage, or spiritual values tied to a landscape, especially when local perspectives are excluded from planning (Havinga et al., 2020; Buisson et al., 2022). In addition, monitoring tends to be an ecological endeavor focusing on narrow ecological measures, including species counts, or the cover of vegetation, and assessments typically undervalue cultural and social outcomes (Hirons et al., 2016; Mehta and Chamberlain, 2023). Overall, these complexities suggest the longer term understanding that restoration can best be considered as the business of ecological repair, but also cultural and social renewal.

7.1.3 Why Integration Matters

Though frequently examined separately, economic incentives and ecological restoration are inextricably linked. Incentives create the financial and policy umbrella that helps restoration be bankable at an appropriate geographic scale, whereas restoration provides the ecological and cultural co-benefits that can generate the justification and support for incentive programs (i.e., demonstrate their value beyond just an economic perspective). For instance, forest reforestation funded through a PES program could sequester carbon, enhance biodiversity, and create a recreational landscape that boosts the well-being of nearby communities or even yield tourism profit (de Groot et al., 2002; Wood et al., 2024). Alternatively, poorly conceived incentives can weakly support restoration through superficial outputs like monoculture plantations that provide little ecological or cultural value (Hua et al., 2018a, b).

An understanding of these linkages opens a new perspective on the management of biodiversity. The basic question is no longer whether incentives or restoration "work" in their own right, but how we might find the appropriate ways to make the two, the one in connection with the other, for ecologically sensible, socially just, and culturally relevant results. This will also require a bridge between the disciplines of economy, natural science, and social sciences, and a development of the methodology in terms of evaluation which conveys the full contents of both biodiversity and human well-being results.

7.2 ECONOMIC INCENTIVES: LESSONS FROM GLOBAL EXPERIENCE

Economic incentives have been a prominent feature of biodiversity policy for the past 30 years. The underlying rationale is simple: If landowners, communities, or firms are given direct economic incentives for conservation, activities that are compatible with biodiversity become more appealing than the activities that destroy it. The notion of economic incentives for biodiversity has its roots in economic theory about externalities, but the emergence of economic incentive schemes has been shaped by political will, social and cultural values, as much as ecological necessity.

The notion of compensating landholders for stewardship gained traction in the 1990s, as pilot programs throughout Latin America and Europe illustrated that it was possible to motivate land use change through financial transfers. Costa Rica's PES program is widely accepted as an early example of this type of program, paying farmers to conserve their forests, which was an important contributor to the reversal of deforestation at a national level (Pagiola, 2008). In China, the Grain for Green Program, initiated in 1999, became the world's largest ecological compensation scheme with financial incentives for converting cropland into forests and grasslands (Wu et al., 2022). In Europe, AES within the Common Agricultural Policy emerged and were tied to biodiversity and landscape goals (Pe'er et al., 2020).

These examples demonstrate that incentives were introduced and created from the context of pilot activities to mainstream policy, and, by the early 2000s, had been institutionalized and were present in biodiversity action plans around the world. The political appeal of incentives is in their flexibility, their supposed "win-win" framing, and because they may complement or substitute for stricter regulation. Table 7.1 highlights crucial types of incentives and their strengths and limitations.

Evidence suggests that well-targeted incentives can deliver ecological benefits—including recovering vegetation, improved forest cover, and biodiversity (Cisneros et al., 2022; Wu et al., 2022). Social benefits are also important—if perceived to be fair, incentives can improve rural livelihoods, diversify income, and promote pro-environmental attitude changes (Rico García-Amado et al., 2013). Despite these success stories, evaluating the real proven impact of incentive programs remains a challenge. For many programs, the evaluation is usually input-based, such as the number of contracts signed, number of hectares enrolled in a program, or number of participants, rather than outcome-based measures that evaluate impact on either ecological recovery or social benefits. Because of this reliance on proxies, it is often unclear if the incentive programs provide actual or real "additionality", meaning benefits beyond what would have happened with or without the incentive programs (i.e., "intervention") (Wunder, 2015).

The geographic bias within available studies exacerbates this issue. As illustrated in Figure 7.1, assessments are concentrated in areas such as Latin America, Europe, and East Asia, whereas Africa and a good portion of Southeast Asia are notably absent. This bias limits the generalizability of findings on a global basis, as the programs that are most documented are often situated in areas with a relatively reasonable governance and institutional capacity. Of course, incentives also pose the risk of perverse outcomes; for instance, when compensation payments are too small, too short-term, or not properly monitored, they fail to compensate for opportunity costs

TABLE 7.1
Typology of Economic Incentives for Biodiversity Conservation and Key Insights

Incentive	Examples	Primary Goals	Strengths	Limitations/Risks
Payments for Ecosystem Services (PES)	Costa Rica (Pagiola, 2008); Mexico; China's Grain for Green (Wu et al., 2022)	Compensate landholders for maintaining or restoring ecosystems (forest, water, carbon, biodiversity)	Flexible, direct link between land use and conservation; scalable; politically palatable	Risk of inequity (wealthier landholders benefit most); monitoring challenges; motivation crowding (Narloch et al., 2012)
Agri-Environmental Schemes (AES)	EU Common Agricultural Policy (Pe'er et al., 2020)	Encourage farmers to adopt biodiversity-friendly practices	Integrates biodiversity with agricultural subsidies; large-scale coverage	Payments are often too low; weak ecological outcomes if poorly targeted; and a bureaucratic burden
Ecological Compensation/ Offsets	US Wetland Banking; China's Ecological Redline Policy	Offset biodiversity losses from development with equivalent gains elsewhere	Internalizes biodiversity into development planning; creates compliance mechanisms	Risk of "license to destroy"; ecological equivalence often questionable; long-term monitoring weak (Fisher, 2012)
Conservation Subsidies/ Stewardship Contracts	Australia's BushTender; Uganda's Forest Stewardship	Reward voluntary biodiversity conservation or sustainable use	Encourages participation from private landholders; builds stewardship networks	High transaction costs; short-term contracts limit durability; outcomes are context-dependent

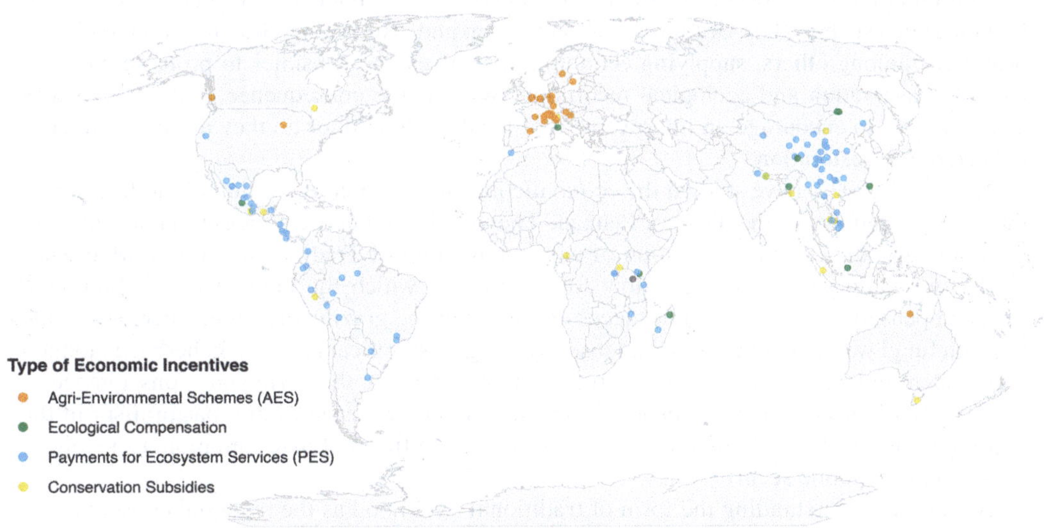

FIGURE 7.1 Geographic distribution of studies evaluating incentive schemes.

or provide a durable conservation (Fisher, 2012; de Koning et al., 2011). They risk producing a perverse incentive. Thus, for instance, subsidy payments based on sheep numbers on the Tibetan Plateau led to overgrazing because of farmers maximizing their number of sheep and undermining desired outcomes for grasslands (Xiao et al., 2022). External payments can also provide an incentive that crowds out intrinsic values of stewardship (motivation crowding) and change the perception of conservation as an act of stewardship to a transactional one. Evidence shows that when external payments cease, the conservation behaviors commonly erode in Bolivia, Peru, Uganda, and Mexico (Narloch, Pascual, and Drucker, 2012; Rico García-Amado et al., 2013). Furthermore, wealth selectivity is a concern as well; generally, resource or scroll-able landholders with secure tenure income capture the benefits of conservation payments, while already marginalized households continue to experience exclusion (Xiao et al., 2022).

Another important problem is the uncertainty of financial sustainability; that is, many incentive programs are based on public budget funding and may be prone to political vagaries and issues of changing political priorities. Barbier et al. (2018) argue that to close the gap in financing biodiversity, it will be necessary to diversify the tools available, such as biodiversity taxes, charges for resource use, and innovative market mechanisms which develop in addition to direct subsidies. Without sustainable financial input, even the best pilot projects will find it difficult to grow or endure. Three decades of international experience show that incentives are not mere financial values but are elements of institutional change that change property rights, clarify responsibilities, and introduce norms for negotiation between communities regarding biodiversity consideration. At their best, incentives promote stewardship and secure biodiversity considerations in governance. At their worst, incentives build iniquity and unproductive, slight ecological outcomes. The major lesson learned is that incentives regarding ecological rehabilitation are most likely to succeed, and supported by governance which embodies equity, accountability, and long-term resilience.

BOX 7.1 GRASSLAND ECO-COMPENSATION ON THE TIBETAN PLATEAU

From around the year 2000 forward. The Chinese state instituted various habitat rehabilitation measures such as the "Grain to Grass" program and "Returning Grazing Land to Grass" programs, among others, supplying economic incentives and subsidies to promote environmental conservation and ecological restitution. What is the consequence of these subsidies and measures to encourage an improvement of local livelihood, or do they directly influence the ecological restoration?

Xiao et al. (2022) have studied the economic incentives for conservation of biodiversity in the alpine meadows in the Tibetan Plateau. They evidenced eco-compensation subsidies in grassland and the commercialization of caterpillar fungi of Traditional Chinese Medicines as the two main sources of cash flow for herders, of which the latter was the dominant. Eco-compensation as such was not successful in curtailing grazing pressure, whereas income through fungi was associated with increased grazing. A significant decline in herbage biomass was evident in heavily grazed, low-altitude grazing, showing the pernicious consequence of grazing. Herders were seen to spend the accumulated cash revenues into pastoralism in the main, because of the traditional role that yaks played in times of asset inflation as a store of capital and economic security.

As well as understanding the form of traditional livelihood as the potential consequences of economic incentives, there comes a need to look at the nature of the social systems and beliefs, conditions of employment, and social service systems, as financial assistance cannot and does not substitute for this (Pollnac and Poggie, 2008). Other studies have also shown the way economic incentives effect the furtherance of traditional activities such as hunting in some areas of Africa (Damania et al., 2005; Buzinde et al., 2014).

7.3 ECOLOGICAL RESTORATION AND HUMAN WELL-BEING

Ecological restoration has recently gained unprecedented visibility as a cornerstone of biodiversity conservation (Wortley et al., 2013). The UN Decade on "Ecosystem Restoration (2021–2030)" positions restoration as essential for halting biodiversity decline and achieving broader Sustainable Development Goals, and the Kunming—Montreal Global Biodiversity Framework (GBF) calls for the restoration of at least 30% of degraded ecosystems by 2030 (Mendes et al., 2023; Mastretta-Yanes et al., 2024). Ecological responses to restoration have a substantial body of literature, and ecological outcomes around restoration across different ecosystems are variable and dependent upon the response group and ecological scale. Some reforestation projects have been demonstrated to recover habitat structure and increase species diversity, while wetland rehabilitation projects are noted for responding with water birds and aquatic biodiversity (Yuan et al., 2022; Mason et al., 2023). Grassland has received attention for returning native vegetation communities and supporting native pollinators (Kis et al., 2017).

The outcomes of ecological restoration depend on the ecological context and the design of the restoration project, both of which have a heavy weighting on the outcomes, for example: species richness in wetlands may recover quickly, while ecological restoration in dryland systems may take decades. Monoculture plantations might provide beneficial carbon for a given landscape but do not deliver benefits for biodiversity nor cultural value (Buisson et al., 2022). Table 7.2 provides a summary of the major restoration activities and outcomes.

A prominent piece of research indicates that restoration not only repairs ecosystems but provides CES, which are the nonmaterial benefits people obtain from ecosystems that connect people to nature (Dou et al., 2022). Restored wetlands and forests often offer opportunities to engage in recreational pursuits and aesthetic experiences, providing venues for walking, seeing birds, and even personal reflection, which can enhance mental health and boost local tourism (Soga and Akasaka, 2019). Rehabilitation of grasslands and restoration in riparian regions reinstates heritage values to some degree and cultural identity—traditional practices, festivals, and ritual—are often tied to landscapes (Martin, 2017). Restoration sites are also being increasingly utilized as a living laboratory in which to educate students, with the added benefit of reinforcing intergenerational knowledge and embracing environmental awareness (Dou et al., 2022,

TABLE 7.2
Restoration Strategies and Their Biodiversity and Cultural Ecosystem Service Outcomes

Restoration Strategy	Biodiversity Outcomes	Cultural Ecosystem Services (CES)	Key Lessons
Wetland restoration	Water quality improvement; recovery of migratory bird populations (Yuan et al., 2022)	Recreation (birdwatching), education, cultural identity	Strong synergies, but outcomes depend on hydrological management
Reforestation/ Afforestation	Forest structure recovery; increased plant and bird diversity (Mason et al., 2023)	Aesthetics, recreation, spiritual values	Risks when monocultures are prioritized; mixed plantations are often better
Grassland rehabilitation	Native vegetation recovery; pollinator habitat (Kis et al. 2017)	Heritage values, pastoral identity, and local festivals	Success depends on integrating cultural practices with ecological design
River and riparian restoration	Fish population recovery; improved ecological connectivity	Recreation (fishing, walking), educational uses	Requires balancing flood control with cultural access to rivers

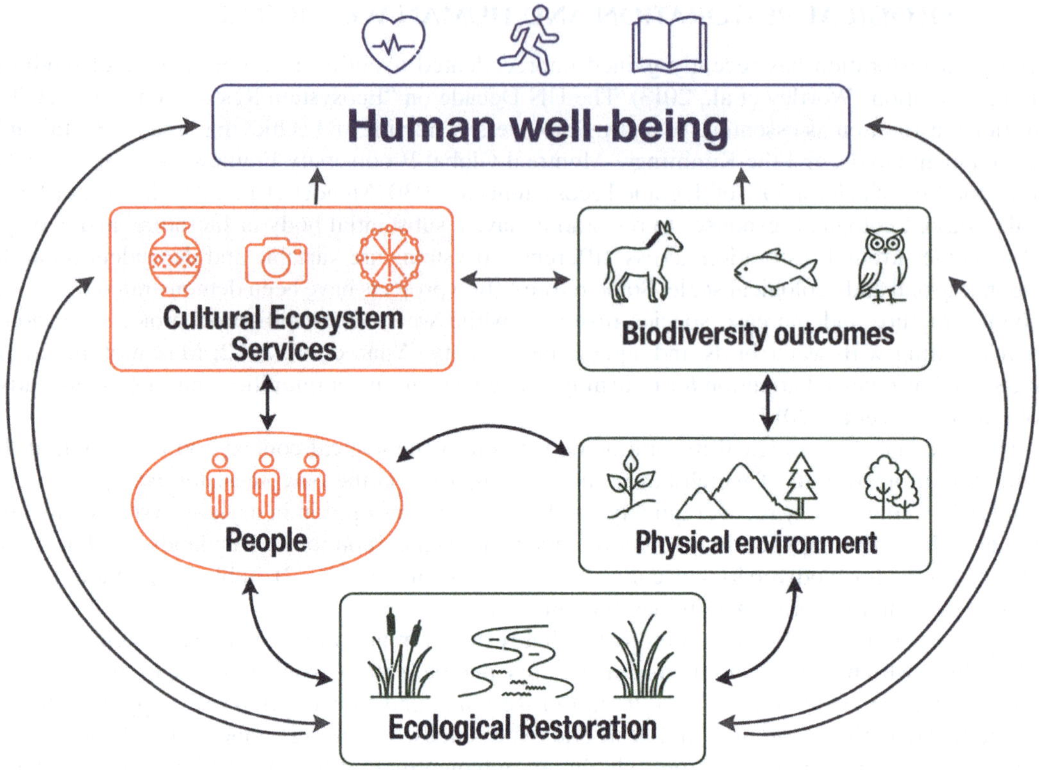

FIGURE 7.2 Conceptual pathways from ecological restoration to human well-being through biodiversity and CES. Directional arrows represent the flow of influence among components, with the grey arrow denoting a one-way relationship.

Havinga et al., 2020; Nowak-Olejnik et al., 2022). These cultural benefits are foundational, not only to engender public support but to begin to link biodiversity conservation with human well-being (Figure 7.2).

Yet restoration has a host of balances to navigate. Projects which maximize ecological complexity often are in tension with local aesthetic or recreational desires (Soga et al., 2021). Large-scale reforestation schemes often favor expeditious carbon sequestration at the expense of monoculture plantations which abode little habitat space and diminish cultural recognizability with native landscapes (Buisson et al., 2022). In certain instances, restoration of habitat's results in a displacement of exploitable means of livelihood. Reflooding agricultural lots for wetland restoration, for example, can inhibit the rural farming community, restricting grazing areas to reduce livestock numbers may challenge herders living safely. These examples above suggest it is fruitless to attempt an examination of restoration strictly based on ecological returns: participatory planning, which integrates cultural and social evaluations, is essential to durable success (Havinga et al., 2020; Mehta and Chamberlain, 2023).

Box 7.2 outlines wetland restoration in Europe, where large-scale restoration has re-instituted adequate aquatic biota and revived bird species while at the same time generating ecotourism advantages and cultural ones as well. However, illicit interrelations with farming communities who are losing arable land illustrate the importance of inclusive governance.

BOX 7.2 WETLAND RESTORATION IN EUROPE

Wetlands are amongst Europe's most degenerated ecosystems and have been largely modified by drainage and land conversion. Large-scale restoration of waterlogged wetlands has been possible through funding from EU LIFE projects since the early 2000s, including in the Danube Delta. Here, the re-flooding of drained polders has created aquatic habitat and brought back avian populations such as pelicans and herons (Nowak-Olejnik et al., 2022).

A restoration of biodiversity has also brought cultural benefits, and the newly restored wetlands are now tourism, outdoor learning, and cultural practice centered at and associated with rivers and fisheries (Martin, 2017). On the other hand, restoration of wetlands has generated tensions. Farmers who have lost arable land have expressed concerns about earning a livelihood. While on the other hand, divergencies between conservation needs and the needs of the community, of biodiversity reclamation versus recreation, have sometimes been in conflict. It was to meet these problems that it has become apparent that planning for successful restoration requires participatory planning, where ecological goals are balanced against cultural and economic requirements (Havinga et al., 2020).

The success of the European wetland experience demonstrates that restoration of both biodiversity and cultural renewal is possible by means of governance structures that include everybody, not just to limit access to the wetlands to a few privileged people. It also demonstrates that restoration is more than ecological repair, but a transformation of society and rebuilding of ecosystems, reconnecting society with its landscape.

Stated simply, ecological restoration is about ecosystem repair and rethinking the human-nature relationship. It produces practical biodiversity benefits and strengthens cultural ties, improves social well-being, and culture. Looking forward, three directions are especially promising.

1. **Integrating CES measurements and biodiversity measures**: so ecological and cultural results can be assessed together.
2. **Associating restoration closely with local authority and cultural factors**: so it is perceived as legitimate, and long-term sustainable.
3. **Using new technologies and citizen science to facilitate participatory assessment and broaden interest.**

In this way, restoration can best be defined as a socio-ecological strategy which restores species and their habitats, whilst facilitating cultural identity and sustainable well-being.

7.4 LINKING INCENTIVES AND RESTORATION TOWARD INTEGRATED STRATEGIES

While both economic incentives and ecological restoration aim to counteract declines in biodiversity, the two sets of strategies have evolved in relative isolation. Incentives are usually assessed with respect to levels of involvement, cost-effectiveness, and institutional arrangements (Fisher, 2012; Wunder, 2015), while restorations are gauged by ecological results, e.g., species richness or vegetation cover (Buisson et al., 2022; Yuan et al., 2022). This compartmentalization has consolidated institutional draws and skyed a way to see how the two approaches can be mutually reinforcing.

However, the experience from various geographic regions suggests that economic incentives provide a basis for restoring degraded habitats on a substantial scale. The Chinese Grain for Green, for example, shows how subsidies and ecological compensation aimed to change croplands into forests

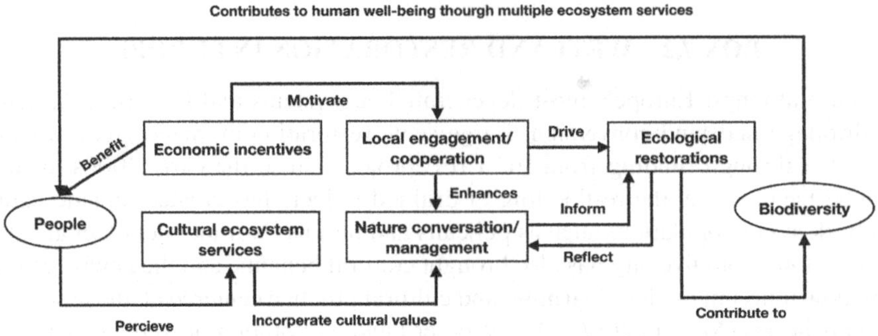

FIGURE 7.3 Integrated pathways linking economic incentives, restoration, and human well-being. (Adapted from Dou et al., 2021.)

and grasslands to a great extent induce the area of vegetation (Wu et al., 2022). In Europe, agri-environment schemes have been launched to help farmers to restore hedgerows, wetlands, and field margins, increasing both biodiversity and cultural landscapes (Pe'er et al., 2020). In Latin America, payments for PET in Costa Rica and Mexico have been applied to fund watershed rehabilitation and jungle recovery, and payments have the only aim of restoring the environment (Pagiola, 2008; Cisneros et al., 2022). These examples prove that incentives decrease opportunity costs, media ecological aims with rural development, and make generally rehabilitation profitable.

The outcomes from restoration hold themselves in place and legitimize the incentive mechanisms used. Given that the results of effective biodiversity and CES restoration can be visually liked, there is a greater likelihood of similar communities being receptive to conservation contracts. For example, wetland restoration in Europe and Asia increased bird and fish populations. Moreover, the possibilities for recreation, education, and cultural identity were restored, giving additional strength to the social legitimation of conservation policies (Nowak-Olejnik et al., 2022). Restoration in this sense is thus a social full stop for the incentives, producing dividends that transcend the financial transfers, and lead to the embedding of conservation into the daily amount of doing business and local identities (Dou et al., 2022).

Where incentives and restoration are designed, therefore, the two together can form a feedback loop of mutually advancing outcomes. Incentives will mobilize financial resources and institutional commitment for restoration. Restoration will bring about biodiversity and CES that will enhance human well-being. These benefits, in turn, will provide political and community support for perpetuating the incentive schemes (Figure 7.3).

Emerging innovations might assist. Results-based incentive schemes where payments are linked to indices of ecological performance encourage greater commitment to restoration (Wunder et al., 2018, Salzman et al. 2018). The models of blended finance, where public funds together with biodiversity credits and private sector investment produce diversified support for long-term projects, are of help (OECD, 2024). Also important, of course, are social innovations – co-designed restoration that incorporates indigenous traditional ecological knowledge and community participation will guarantee that CES are incorporated in the same manner as biodiversity (Havinga et al., 2020; Dou et al., 2022). And technological innovations are also promising for evaluation purposes. Thus, e.g., remote sensing will allow the fine-scale monitoring of land-cover change; AI systems based on acoustic and visual images will allow detection of species more effectively; and citizen science schemes will provide socially relevant data. The ways of combining these resources will lead to strengthened results-based incentive schemes which will link ecological restoration and CES to demonstrate real outcomes from the socio-ecological point of view.

From a standpoint of perspective, the integration of incentives into restoration represents a paradigm shift which is needed. The incentive schemes alone are liable to be an impermanent financial treatment on the problem side, whereas restoration alone may be ecologically sound but socially weak. In combination, they create ecologically common credible, economically sound, and socially legitimated pathways. The lesson from many decades of global experience is a clear one. The management of biodiversity in the Anthropocene must transcend the divisions of disciplines. Three parts of the sciences, i.e., economics, ecology, and culture, must work together in a concerted fashion within a shared framework.

7.5 EMERGING FRONTIERS AND GOVERNANCE FOR INTEGRATED BIODIVERSITY MANAGEMENT

The coordination of governance across different scales, sectors, and social groups that enables effective integration of economic incentives with ecological repair is crucial. Governance in this context involves the formal state regulations but also the networks, norms, and institutions that determine how biodiversity management is done (Buisson et al., 2022; Havinga et al., 2020). Notwithstanding the global trend in the expansion of both incentives and repair, the state of governance is still fragmented. The Ministry of Agriculture led subsidy schemes that are often thought up in isolation from conservation agencies that manage the repair of degraded ecosystems. In such a context, cultural values are inevitably neglected. In this context, the repair is therefore ineffective, and the benefits that might accrue in terms of long-term socio-ecological benefits are missed. To remedy some of the problems there are to be found in the literature, three forms of integrated governance that can be applied (Table 7.3).

Legitimacy is common across the different forms of governance. As Havinga et al. (2020) point out, incentive and repair schemes tend to persist when local wisdom is integrated, and the project aims are congruent with cultural values. Top-down forms of compensation schemes, by way of example, existing at the present time on the Tibetan Plateau are represented as an example of how a mismatch between policy intent and pastoral methods can create unintended ecological outcomes. Thus, overgrazing follows (Hou et al., 2021). Dou et al. (2022) point out, too, that the integrity of CES is also enhanced by adding them to the governance framework as ecological benefits are hence beyond the reach of political control and thereby legitimized, being of tangible human benefit.

The innovations that are coming forward show how adaptive governance can cope with some of these issues. Thus, for example, co-designed repair projects in Europe (Nowak-Olejnik et al., 2022) and China (Yuan et al., 2022) focus on adding psychological values such as recreation and aesthetic appreciation to the aims of the project. Also, in financial investments in biological diversity, there is movement forward. Blended finance schemes that marry together public subsidy, biodiversity credit, and private income. This reduces the reliance on the short-term subsidy. At the same time as this, the systems of digital monitoring and citizen science, both for tracking the flow of ethological benefits, are being focused on with a view to create more transparency and to improve the situation of accountability in these schemes of Payments for Ecosystem Services.

TABLE 7.3
Modes of Integrated Governance for Biodiversity Management

Governance Mode	Key Features	Representative Cases
Multi-level alignment	Links international/national frameworks with local implementation; requires accountability mechanisms.	Brazil's Bolsa Floresta program delivering forest conservation payments to Amazon communities (Börner et al., 2013).
Cross-sector coordination	Coordinates across sectors (water, agriculture, tourism, biodiversity); promotes multifunctional landscapes.	EU LIFE wetland restoration projects in the Danube basin integrating ecological and cultural goals (Tockner et al., 2009; Galewski et al., 2011).
Networked co-governance	Shared authority across multiple actors (communities, NGOs, governments, private actors); balances flexibility and equity.	Ecuador's Socio Bosque program (de Koning et al., 2011); Uganda's PES pilot reducing deforestation under insecure tenure (Jayachandran et al., 2017).

The cases presented here confirm that integrated governance is clearly institutional as well as cultural: institutional integration produces coherence across levels and sectors, and cultural integration bestows legitimacy on these activities by rooting values and practices into decision-making. The frontier in biodiversity governance is not the area of hectares restored, or payments made, but systems that are integrated at multi-levels, cross-sectorally, and networked (Table 7.3).

7.6 SYNTHESIS

The experiences assessed in this chapter demonstrate that no number of economic incentives will counteract the massive scale of the biodiversity crisis currently unfolding, and strictly any number of ecological restorations will in no way address the issue. Economic incentives will enable the means and determine behavior, but will be ephemeral if they are not matched by suitable ecological results on the ground, whereas ecological restoration can take place in all forms of ecosystem rehabilitation, etc., and re-establish an area where cultural links can be connected, they are historically deficient in being able to pay for themselves or being deemed valid. There is no doubt, however, that if the two were to be brought together, they would bolster and help each other; the incentives would make the restorations economic, and as importantly render them visible so that the incentive systems themselves would be reinforced and easily maintained.

The lesson that strikes the most is that financial support must be based around long-term systems and symbiotic systems as opposed to mushroom specialized special projects. Many of the specified pilot studies show how appropriate results can be obtained on paper, but the after-effects are ephemeral when the necessary subsidies are withdrawn or the political motivations changed. The need is for constant facilities financed both from the public and the private sector, governing connections between different aspects of the whole business, and systems to enable not just monitoring of hectares planted but the recovery of biodiversity and cultural support on the ground. In other words, chances for sufficient success will be markedly better where the resulting conservation systems are not reliant on one-off interventions but are built around permanent institutions.

The second is that it is very clear that any aspect of legitimacy in the incentives given must not only be well presented but must not be left to a one-way process where legitimacy is not perceived. Economic systems which can be seen to be inequitable or restorations which overlook the cultural values are unlikely to achieve long-term acceptance. Very much on the other hand, programs which include the various communities in the planning have an appeal to indigenous and local knowledge, and where the benefits are diffused transparently, have long-term chances of success in their processes. It must be always born in mind that legitimacy is not an abstract, but a practical condition, for without it the ecological goals will not remain.

The third lesson is that the conservation of biodiversity as such is one which is developing in an area which will be densely populated with policies. In other words, they impinge on one another, and therefore on diverse concentrations, and departments of agriculture and food, and water, and energy production, and climate adaptation, and development of rural dwellers. Indeed, international policy imperatives such as the Kunming-Montreal Global Biodiversity Framework can be laid down to have initiators, but how feasible they are depends on the efficiency with which local and government authorities are able to integrate them into existing policy systems. Here indeed is where integrated governance, across levels and across cultures, across sections or institutions, is all important.

Further, the real frontier is not the development of constantly more so-called new schemes, but to get there to work in compatible areas with successful ones, and on a broader scale. The incentives ought to be set up so that they are chosen for cultural and ecological outcomes, and not merely compliance. The restorations themselves ought to be seen from not only the angle of recovering nature as such, but also the side of strengthening cultural well-being, identity, and human well-being in the community. The challenge then is to treat these different aspects of ecology and cultural amelioration as possibly two different facets of one single socio-ecology.

7.7 CONCLUSION

The evidence collected here suggests that future biodiversity management will be judged on ecological outcomes and whether it is socially legitimate, and institutionally durable. Incentives and restoration, when woven together under integrated governance, provide a foundation for ecologically credible, socially fair, and resilient over time. The task now is to move from scattered successes to coherent systems that can sustain biodiversity while enriching human well-being in the decades ahead.

DECLARATION OF GENERATIVE AI USE

During the preparation of this manuscript, the author(s) used AI to improve English grammar and language clarity. The author(s) subsequently reviewed and edited the content and take(s) full responsibility for the final version of the published chapter.

REFERENCES

Barbier, E. B., Burgess, J. C., & Dean, T. J. (2018). How to pay for saving biodiversity. *Science, 360*(6388), 486–488. https://doi.org/10.1126/science.aar3454

Börner, J., Wunder, S., Reimer, F., Bakkegaard, R. K., Viana, V., Tezza, J., Pinto, T., Lima, L. S., & Marostica, S. (2013). Promoting forest stewardship in the Bolsa Floresta Programme: Local livelihood strategies and preliminary impacts. CIFOR, Fundação Amazonas Sustentável (FAS), and Zentrum für Entwicklungsforschung (ZEF), Rio de Janeiro, Brazil. https://www.cifor-icraf.org/publications/pdf_files/Books/BBorner1301.pdf

Buisson, E., Archibald, S., Fidelis, A., & Suding, K. N. (2022). Ancient grasslands guide ambitious goals in grassland restoration. *Science, 377*(6606), 594–598. https://doi.org/10.1126/science.abo4605

Buzinde, C. N., Kalavar, J. M., & Melubo, K. (2014). Tourism and community well-being: The case of the Maasai in Tanzania. *Annals of Tourism Research*, 44, 20–35.

Cisneros, E., et al. (2022). Impacts of conservation incentives in protected areas: The case of Bolsa Floresta, Brazul. *Journal of Environmental Economics and Management*, 111, 102572. https://doi.org/10.1016/j.jeem.2021.102572

Convention on Biological Diversity (CBD) (2010). Linking Biodiversity Conservation and Poverty Alleviation: A State of Knowledge Review. CBD Technical Series, No. 55.

Damania, R., Milner-Gulland, E. J., & Crookes, D. J. (2005). A bioeconomic analysis of bushmeat hunting. *Proceedings of the Royal Society B: Biological Sciences*, 272(1560), 259–266.

de Groot, R. S., Wilson, M. A., & Boumans, R. M. J. (2002). A typology for the classification, description and valuation of ecosystem functions, goods and services. *Ecological Economics, 41*(3), 393–408. https://doi.org/10.1016/S0921-8009(02)00089-7

de Koning, F., Aguiñaga, M., Bravo, M., Chiu, M., Lascano, M., Lozada, T., & Suarez, L. (2011). Bridging the gap between forest conservation and poverty alleviation: The Ecuadorian Socio Bosque program. *Environmental Science & Policy, 14*(5), 531–542. https://doi.org/10.1016/j.envsci.2011.04.007

Dou, Y., Yu, X., & Liu, Y. (2021). Rethinking non-material links between people and drylands from a cultural ecosystem services perspective. *Current Opinion in Environmental Sustainability*, *48*, 110–114. https://doi.org/10.1016/j.cosust.2020.12.007

Dou, Y., Zhen, L., Bakker, M., Yu, X., Carsjens, G. J., Liu, J., & De Groot, R. (2022). Investigating the potential impact of ecological restoration strategies on people–landscape interactions through cultural ecosystem services: A case study of Xilin Gol, China. *Journal of Environmental Management, 316*, 115185. https://doi.org/10.1016/j.jenvman.2022.115185

Galewski, T., Collen, B., McRae, L., Loh, J., Grillas, P., Gauthier-Clerc, M., & Devictor, V. (2011). Long-term trends in the abundance of Mediterranean wetland vertebrates: From global recovery to localized declines. Biological Conservation, 144(5), 1392–1399. https://doi.org/10.1016/j.biocon.2010.10.030

Havinga, I., Bogaart, P. W., Hein, L., & Tuia, D. (2020). Defining and spatially modelling cultural ecosystem services using crowdsourced data. *Ecosystem Services, 43*, 101091. https://doi.org/10.1016/j.ecoser.2020.101091

Hirons, M., Comberti, C., & Dunford, R. (2016). Valuing cultural ecosystem services. *Annual Review of Environment and Resources*, 41(1), 54

Hou, L., Xia, F., Chen, Q. et al. Grassland ecological compensation policy in China improves grassland quality and increases herders' income. Nat Commun 12, 4683 (2021). https://doi.org/10.1038/s41467-021-24942-8

Hua, F., Wang, L., Fisher, B., Zheng, X., Wang, X., Yu, D. W., Tang, Y., Zhu, J., & Wilcove, D. S. (2018a). Tree plantations displacing native forests: The nature and drivers of apparent forest recovery on former croplands in Southwestern China from 2000 to 2015. *Biological Conservation, 222*, 113–124. https://doi.org/10.1016/j.biocon.2018.03.034

Hua, F., Xu, J., & Wilcove, D. S. (2018b). A new opportunity to recover native forests in China. *Conservation Letters, 11*, e12396. https://doi.org/10.1111/conl.12396

Jayachandran, S., de Laat, J., Lambin, E. F., Stanton, C. Y., Audy, R., & Thomas, N. E. (2017). Cash for carbon: A randomized trial of payments for ecosystem services to reduce deforestation. *Science, 357*(6348), 267–273. https://doi.org/10.1126/science.aan0568

Kis, J., Barta, S., Elekes, L., Engi, L., Fegyver, T., Kecskeméti, J., ... & Szabó, J. (2017). Traditional herders' knowledge and worldview and their role in managing biodiversity and ecosystem-services of extensive pastures. Knowing our lands and resources: indigenous and local knowledge of biodiversity and ecosystem services in Europe and Central Asia. Paris (France): UNESCO, 56-70.

Mason, V. G., Burden, A., Epstein, G., Jupe, L. L., Wood, K. A., & Skov, M. W. (2023). Blue carbon benefits from global saltmarsh restoration. *Global Change Biology*, 29(23), 6517–6545.

Mastretta-Yanes, A., Da Silva, J. M., Grueber, C. E., Castillo-Reina, L., Köppä, V., Forester, B. R., ... & Hoban, S. (2024). Multinational evaluation of genetic diversity indicators for the Kunming-Montreal Global Biodiversity Framework. Ecology Letters, 27(7), e14461.

Mehta, J. M., & Chamberlain, E. L. (2023). Cultural-ecosystem resilience is vital yet under-considered in coastal restoration. *Humanities and Social Sciences Communications*, 10(1), 265.

Narloch, U., Pascual, U., & Drucker, A. G. (2012). Collective action dynamics under external rewards: Experimental insights from Andean farming communities. World Development, 40(10), 2096–2107. https://doi.org/10.1016/j.worlddev.2012.03.014

Nowak-Olejnik, A., Schirpke, U., & Tappeiner, U. (2022). A systematic review on subjective well-being benefits associated with cultural ecosystem services. *Ecosystem Services, 57*, 101467. https://doi.org/10.1016/j.ecoser.2022.101467

OECD. (2024). *Financing Biodiversity: Policy Guidance and Innovation*. OECD Publishing. https://doi.org/10.1787/d26526ad-en

Pagiola, S. (2008). Payments for environmental services in Costa Rica. *Ecological Economics, 65*(4), 712–724. https://doi.org/10.1016/j.ecolecon.2007.07.033

Pe'er, G., et al. (2020). Action needed for the EU Common Agricultural Policy to address sustainability challenges. *People and Nature,* 2(2), 305–316. https://doi.org/10.1002/pan3.10080

Pollnac, R. B., & Poggie, J. J. (2008). Happiness, well-being and psychocultural adaptation to the stresses associated with marine fishing. *Human Ecology Review*, 194–200.

Rico García-Amado, L., Ruiz Pérez, M., & Barrasa García, S. (2013). Motivation for conservation: Assessing integrated conservation and development projects and payments for environmental services in La Sepultura Biosphere Reserve, Chiapas, Mexico. Ecological Economics, 89, 92–100. https://doi.org/10.1016/j.ecolecon.2013.02.002

Salzman, J., Bennett, G., Carroll, N., Goldstein, A., & Jenkins, M. (2018). The global status and trends of payments for ecosystem services. *Nature Sustainability, 1*(3), 136–144. https://doi.org/10.1038/s41893-018-0033-0

Soga, M., & Akasaka, M. (2019). Multiple landscape-management and social-policy approaches are essential to mitigate the extinction of experience. *Landscape and Urban Planning, 191*, 103634. https://doi.org/10.1016/j.landurbplan.2019.103634

Soga, M., & Gaston, K. J. (2020). The ecology of human–nature interactions. *Proceedings of the Royal Society B: Biological Sciences*, 287(1918).

Soga, M., Fukano, Y., Koyanagi, T. F., & Gaston, K. J. (2021). Species abundance as a determinant of aesthetic values of flowering plant communities. *Urban Forestry & Urban Greening, 63*, 127194. https://doi.org/10.1016/j.ufug.2021.127194

Tockner, K., Robinson, C. T., & Uehlinger, U. (2009). Tockner, K., Tonolla, D., Bremerich, V., Jähnig, S. C., Robinson, C. T., & Zarfl, C. (2022). Introduction to European rivers. In Rivers of Europe (pp. 1-26). Elsevier. . https://doi.org/10.1016/B978-0-08-102612-0.00001-8

Wood, K. A., Jupe, L. L., Aguiar, F. C., Collins, A. M., Davidson, S. J., Freeman, W., ... & Newth, J. L. (2024). A global systematic review of the cultural ecosystem services provided by wetlands. *Ecosystem Services*, 70, 101673.

Wortley, L., Hero, J.-M., & Howes, M. (2013). Evaluating ecological restoration success: A review of the literature. *Restoration Ecology, 21*(5), 537–543. https://doi.org/10.1111/rec.12028

Wu, X., Wang, S., Fu, B., Feng, X., & Chen, Y. (2019). Socio-ecological changes on the Loess Plateau of China after Grain to Green Program. Science of the Total Environment, 678, 565-573. https://doi.org/10.1016/j.scitotenv.2019.05.022

Wunder, S. (2015). Revisiting the concept of payments for environmental services. *Ecological Economics*, 117, 234–243. https://doi.org/10.1016/j.ecolecon.2014.08.016

Wunder, S., Brouwer, R., Engel, S., Ezzine-de-Blas, D., Muradian, R., Pascual, U., & Pinto, R. (2018). From principles to practice in paying for nature's services. *Nature Sustainability, 1*(3), 145–150. https://doi.org/10.1038/s41893-018-0036-x

Xiao, L., Zhao, X., Mei, S., Mishra, C., Alexander, J. S., Weckworth, B., ... & Lu, Z. (2022). When money meets tradition: How new cash incomes could be risky for a vulnerable ecosystem. *Biological Conservation*, 272, 109575. https://doi.org/10.1016/j.biocon.2022.109575

Yuan, L., Liu, D., Tian, B., Yuan, X., Bo, S., Ma, Q., ... & Keesing, J. K. (2022). A solution for restoration of critical wetlands and waterbird habitats in coastal deltaic systems. *Journal of Environmental Management*, 302, 113996. https://doi.org/10.1016/j.jenvman.2021.113996

Part III

Microbial and Aquatic Ecosystems

8 The Hidden World
Aquatic Microbial Diversity, Ecological Functions and Responses to a Changing Planet

Lijuan Zhang, Xuejiao Qiao, Lingyun Ding, Pengsong Li, Zhishan Huang, and Nan Xu

8.1 INTRODUCTION

Aquatic microorganisms represent the largest biome of the global biosphere, inhabiting the freshwater and marine ecosystems, thriving also in extreme ecosystems where environmental conditions deviate from the normal range to which most known life forms are adapted (Figure 8.1).

These microbial communities constitute the core biological capital of aquatic ecosystems, and their diversity and functions directly determine the material cycling, energy flow, and recovery capacity of the ecosystem. The microorganisms in freshwater ecosystems are abundant and present a unique community structure (Sagar et al., 2015). The heterogeneity of microbial communities of freshwater ecosystems is due to differences in their habitat, whereas marine ecosystems are characterized by a vast array of biological species, high biological diversity, and extremely rich biological resources (Alabbosh, 2025). They play a role of vigorous natural purifiers with the power of protecting and recovering ecosystems. In extreme aquatic ecosystems, thermophilic, halophilic, psychrophilic, and other extreme microbial groups manifest the remarkable diversity and robustness of the metabolism, constituting distinctive functional communities (Shu and Huang, 2022). Across different aquatic ecosystems, environmental factors such as temperature, pressure, light, and dissolved oxygen vary widely. In order to cope with fluctuating environments, aquatic bacteria, archaea, and fungi have developed unique and diverse adaptation strategies (Jurdzinski et al., 2023; Zhou et al., 2025). These adaptive mechanisms can improve the metabolic resilience of microorganisms to environmental stress and promote biogenic element cycling and active metabolite production (Liang et al., 2023; Moran et al., 2022).

Global changes, such as climate warming, ocean acidification, and anthropogenic pollution, influence aquatic ecosystems. Microorganisms, serving as the core engines of aquatic biogeochemical cycles, play a critical role in shaping feedback mechanisms that are essential for predicting future climate trajectories (Lakshmi, 2025). However, current research still needs to address two critical issues: (1) clarifying the synergistic response strategies of microbial consortia to multifaceted climate stressors; and (2) establishing a technical framework for translating microbial functions into practical ecological remediation strategies. The recent rapid advancement of molecular biology techniques has provided unprecedented tools for analyzing microbial community structures, functional activities, and adaptation strategies to climate stressors. Technologies such as high-throughput/single-cell sequencing, functional gene analysis, metagenomics, and *in situ* monitoring are elucidating microbiome responses to climate change factors and their feedback effects (Michán et al., 2021). Synthetic biology/ecology has shown great potential in efficient pollutant biodegradation and greenhouse gas mitigation, providing targeted remediation strategies that improve water quality and biodiversity in contaminated ecosystems (Richard, 2024). This enables a more precise and

DOI: 10.1201/9781003640899-11

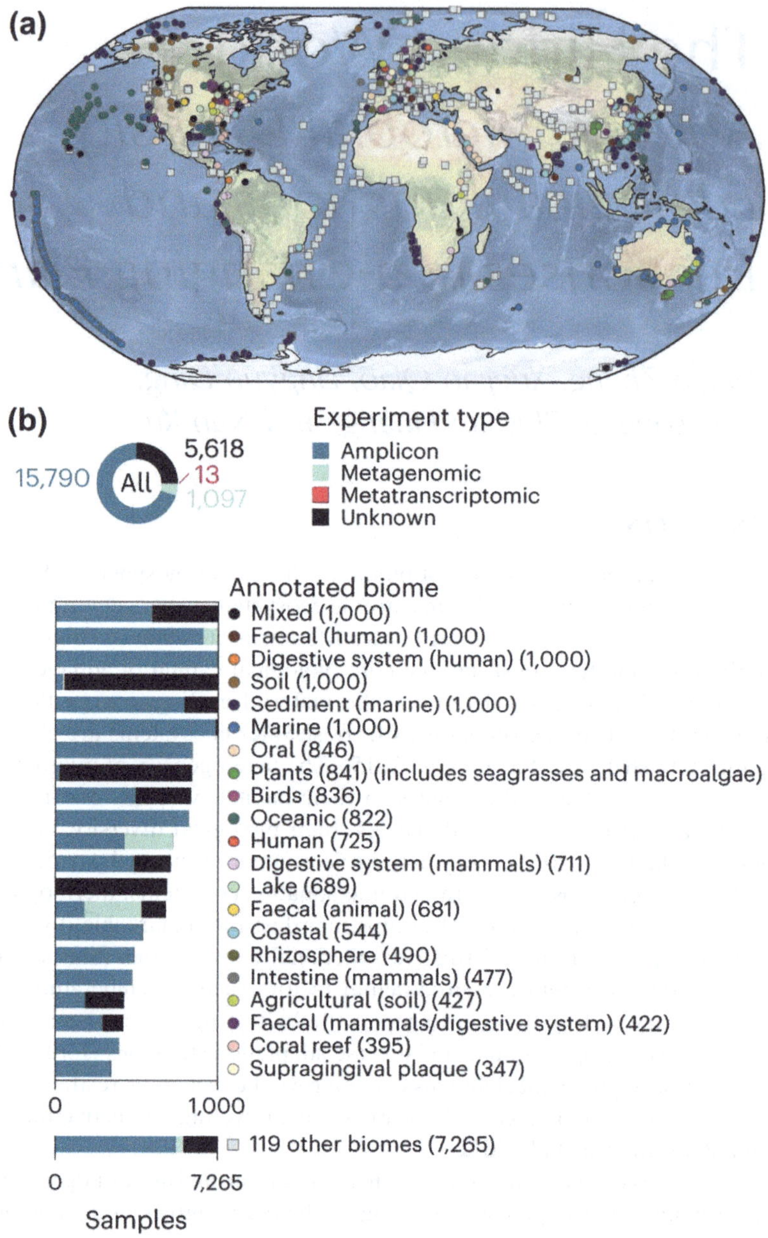

FIGURE 8.1 A diverse and global microbial dataset. (a) Geographical distribution of the samples. (b) The microbial distribution in different habitats. The numbers in parentheses indicate the number of samples within the annotated biome. Annotated biomes with fewer than 347 samples have been grouped as other. (Adapted with permission from von Meijenfeldt et al. (2023). Copyright 2023, the authors. *Nature ecology & evolution*, an open-access article distributed under the terms and conditions of the Creative Commons Attribution (CC BY) license (https://creativecommons.org/licenses/by/4.0/).)

comprehensive dissection of the composition, function, and ecological feedback of microbial communities under global change.

This chapter primarily reviews microbiome diversity in representative aquatic ecosystems and discusses possible mechanisms of stress resistance for extremophilic microorganisms. It then focuses on the ecological functions of aquatic microbes, exploring their roles in biogenic element cycling, valuable metabolite production, as well as environmental bioremediation and ecological stabilization. By illustrating the adaptive responses and functional adjustments of microbes affected by climate change, we propose molecular technologies to address the knowledge gaps in "microbial dark matter". Taking wastewater-based epidemiology (WBE) and synthetic ecology as examples, it thereafter illustrates the feasibility of providing scientific guidance for sustainable ecosystem management.

8.2 MICROBIAL DIVERSITY IN AQUATIC ECOSYSTEMS

Microorganisms in aquatic ecosystems exhibit extremely high diversity, with significant variations in the abundance and distribution patterns of dominant communities across freshwater, marine, and extreme environments.

8.2.1 Microbial Characteristics and Diversity in Aquatic Ecosystems

8.2.1.1 Distribution Patterns of Freshwater and Marine Microorganisms

The freshwater microbial community displays great spatial heterogeneity in distribution. In lake ecosystems, aerobic microbes (such as Actinobacteria) and anaerobic groups (such as Chloroflexi) are enriched in epilimnion and hypolimnion, respectively (Kida et al., 2025; Rodríguez-Gijón et al., 2025). Stratification is also more readily noticeable in deep lakes like Lake Baikal, where the extreme environmental conditions are even sufficient to result in a specific enrichment of the Verrucomicrobia (Cabello-Yeves Pedro et al., 2017). Seasonal dynamics of microbial community structure are notable, in which there are also many environmental factors, including temperature, the intensity of light, nutrient availability, and dissolved oxygen that change seasonally and impact the process of microbial metabolism, microbial reproduction, and community structure (Ji et al., 2024). In the ice, water, and sediments of cold lakes, the bacterial communities are essentially driven by diffusion limitation and drift (Zhang et al., 2024). These studies demonstrate that freshwater microbial diversity and distribution patterns are determined by both environmental selection and random processes.

The distribution of marine bacteria follows both vertical and horizontal patterns. Photoautotrophic or chemolithoheterotrophic bacteria are distributed with large biomass in photic layers, while chemolithoheterotrophic bacteria dominate in aphotic zones. Horizontally, Bacterioplankton abundance is generally higher close to shore compared to that in the open ocean, probably as a result of non-homogeneous organic matter distribution due to direct effects from coastal factors such as land runoff, sewage outfalls, and other anthropic activities (Dang et al., 2019). Similarly, the fungi diversity has been found to be elevated at the shoreline and decreases toward the open ocean (Debeljak and Baltar, 2023; Peng et al., 2024). Proximity and connectivity to estuaries near the shore, along with eutrophication, influence fungal diversity patterns, thereby driving the formation of their unique community structure characteristics (Debeljak and Baltar, 2023). Marine archaea are divided into marine group (MG)I-III, each with unique ecological niche adaptation strategies (Santoro et al., 2019). MGI mainly belongs to the Thaumarchaeota phylum. Both the absolute and relative abundances of Thaumarchaeota generally increase with depth, adapting to deep-sea environments by ammonia oxidation (Bayer et al., 2024; Karner et al., 2001). MGII archaea mainly belong to the Euryarchaeota phylum (Rinke et al., 2018), which are the most abundant in the sunlit waters of the photic zone and decrease with increasing depth (Chen et al., 2024a). Heterotrophic MGIII specializes in the degradation of high-molecular-weight compounds from the photic layer to the mesopelagic zone. Epipelagic MGIII occurs sporadically and in relatively low proportions in marine plankton (Adam et al., 2017).

8.2.1.2 Extremophilic Microorganisms and Their Diversity in Extreme Environments

Extreme environments pose severe challenges to the survival, growth, and reproduction of microorganisms. A wide variety of microorganisms occupy specific ecological niches and exhibit diversity in their preferred environments (Figure 8.2).

Halophiles are the dominant inhabitants of hypersaline environments, with archaea predominating at the highest salinities. Among them, the phylum Euryarchaeota is dominant, with the class Halobacteria being the most representative, including the genera *Halobacterium* and *Haloquadratum* (Oren, 2024). Nanohaloarchaeota are also widely distributed halophilic archaea (Vavourakis et al., 2018), showing extremely small cell sizes (0.2–0.5 μm) and simplified genomes (0.8–1.5 Mb). They rely on hosts (such as halobacteria) for nutrients, with *Candidatus Nanohalarchaeum antarcticus* being a typical representative discovered in the Antarctic salt lakes (Hamm et al., 2019). Deep-sea hydrothermal vents are regions on the seafloor where hot, anoxic, chemical-rich water is released into the cold, oxic, deep ocean. Unlike most ecosystems driven by photosynthesis, hydrothermal vent communities rely on chemosynthesis. Mafic-hosted hydrothermal systems feature sulfur- and hydrogen-respiring *Epsilonproteobacteria*, methanogenic and sulfate-reducing *Euryarchaeota*, and specific *Gammaproteobacteria* clades (Dick, 2019). Microbial communities in ultramafic systems vary internally, with areas like inner chimneys dominated by *Methanosarcinales* archaea and outer parts hosting diverse sulfur and methane-oxidizing bacteria, while hyperthermophilic vent

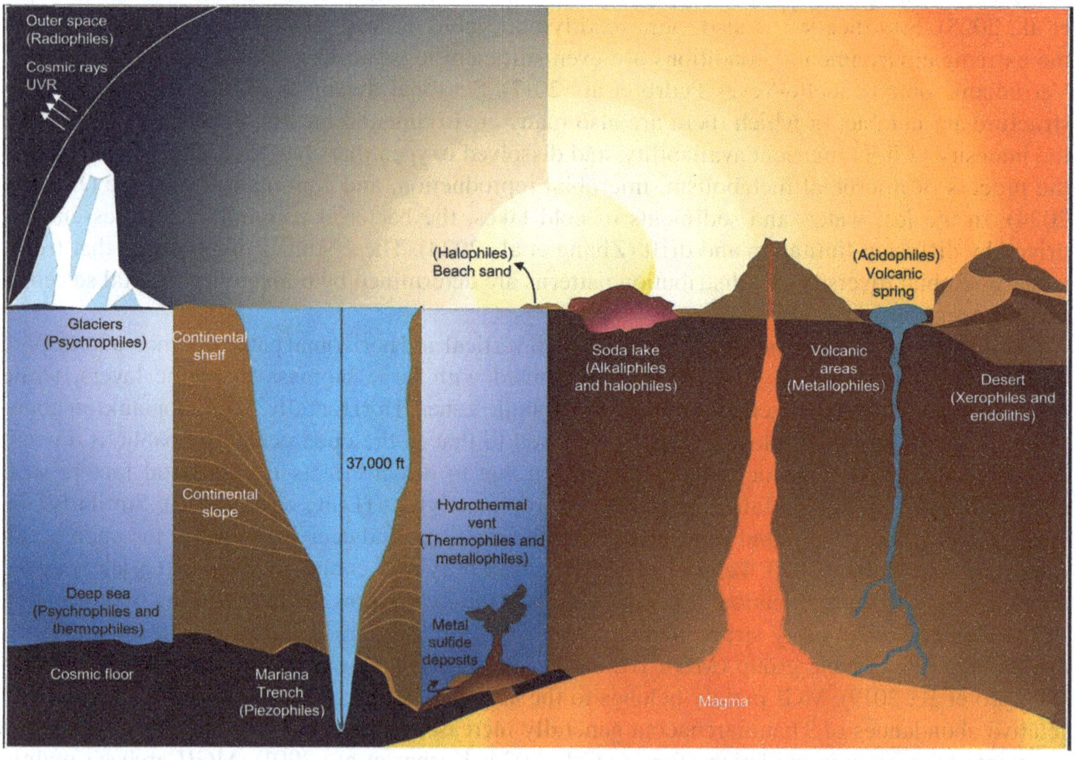

FIGURE 8.2 Different habitats of extremophiles. This cross-sectional diagram illustrates Earth's diverse extreme environments from outer space to the core, along with their corresponding specialized microorganisms (extremophiles). These include the psychrophilic bacteria in glaciers, pressure and thermophilic bacteria in the deep sea and the Mariana Trench, metallophilic bacteria in submarine hydrothermal vents, alkalophilic bacteria in alkaline lakes, acidophilic bacteria in volcanic areas, and xerophilic bacteria in deserts. (Adapted with permission from Chia et al. (2024). Copyright 2024, the authors. *Biointerface Research in Applied Chemistry*, an open-access article distributed under the terms and conditions of the Creative Commons Attribution (CC BY) license (https://creativecommons.org/licenses/by/4.0/).)

fluids host *Thermococcales* and uncultured *Crenarchaeota* (Brazelton et al., 2010). Bacteria are the most abundant microorganisms colonizing ice and snow, with groups such as Cyanobacteria, Proteobacteria, Bacteroidetes, and Actinobacteria having been found on glacier surfaces in various regions (Hu et al., 2022).

8.2.2 Stress Resistance Mechanisms of Extremophilic Microorganisms

8.2.2.1 General Co-adaptation Mechanisms for Environmental Stressors

To survive in the diverse habitats, microorganisms have evolved a fascinating array of biochemical and molecular mechanisms. In studies on adapting to multiple extreme environments, molecular chaperones and compatible solutes are widely reported as key adaptation strategies, representing "co-adaptation" mechanisms for coping with environmental stressors (Wang et al., 2016; Zhao et al., 2020). Molecular chaperones are a class of proteins that bind to biological macromolecules to assist in maintaining their structure and function. These are related to heat shock, cold shock, and acid shock, which facilitate protein refolding under stress conditions (Hartl and Martin, 1995). Compatible solutes, on the other hand, are small-molecule metabolites that increase intracellular water activity to balance osmotic pressure between the cell and its surroundings. These compounds do not interfere with protein structure and function and can mitigate the inhibitory effects of high ion concentrations on enzyme activity, thereby improving tolerance to heat, cold, salt, and drought (Stracke et al., 2025; Yang et al., 2024b). For example, hyperthermophilic archaea in hydrothermal environments can synthesize and accumulate special compatible solutes *de novo*, such as L,L-di-myo-inositol-1,3-phosphate, mannosylglycerate, and cyclic 2,3-bisphospho-D-glycerate, to prevent thermal denaturation (Zhao and Xiao, 2023).

8.2.2.2 Unique Adaptive Strategies in Specific Extreme Environments

Microorganisms in different extreme environments also have their own unique adaptive strategies. Thermophiles maintain the structural integrity of thermophilic enzymes at high temperatures through disulfide bonds, hydrogen bonds, and hydrophobic interactions (Li et al., 2024). Halophiles, particularly extreme halophilic archaea (e.g., *Halobacteriaceae*) and some bacteria, adapt to hypersaline environments via the "salt-in" strategy. This strategy involves intracellular accumulation of KCl at high concentrations to balance external osmotic pressure, and is accompanied by adaptive changes like protein acidification and membrane structure modification (Gutiérrez-Preciado et al., 2024). Acidophiles (e.g., *Acidithiobacillus ferrooxidans*) actively expel intracellular H^+ via K^+/H^+ antiporters while accumulating K^+ to maintain a cytoplasmic pH of approximately 6.5, despite extracellular pH values as low as 1–2 (Zhang and Qiao, 2017). In nutrient-poor glaciers, some microbes synthesize polyhydroxyalkanoate granules as carbon reserves to cope with energy shortages under harsh conditions (Choi et al., 2021). Sulfur-reducing bacteria at hydrothermal vents can simultaneously utilize H_2, CH_4, and organic pollutants for energy, enhancing their survival flexibility and competitiveness through this mixed metabolic strategy (Dick, 2019). These adaptation mechanisms help microorganisms coexist and thrive in extreme environments.

8.3 ECOLOGICAL FUNCTIONS OF AQUATIC MICROBIOME

As the core drivers of the biogeochemical cycle, microorganisms constitute the biological capital of aquatic ecosystems through primary production, biosynthesis of bioactive metabolites, and bioremediation to maintain ecological stability (Figure 8.3).

8.3.1 Biogenic Element Cycling

Aquatic microorganisms play multiple roles in biogeochemical cycling of biogenic elements, among which organic matter decomposition, nitrogen cycling, and photosynthesis are the most critical ecological processes.

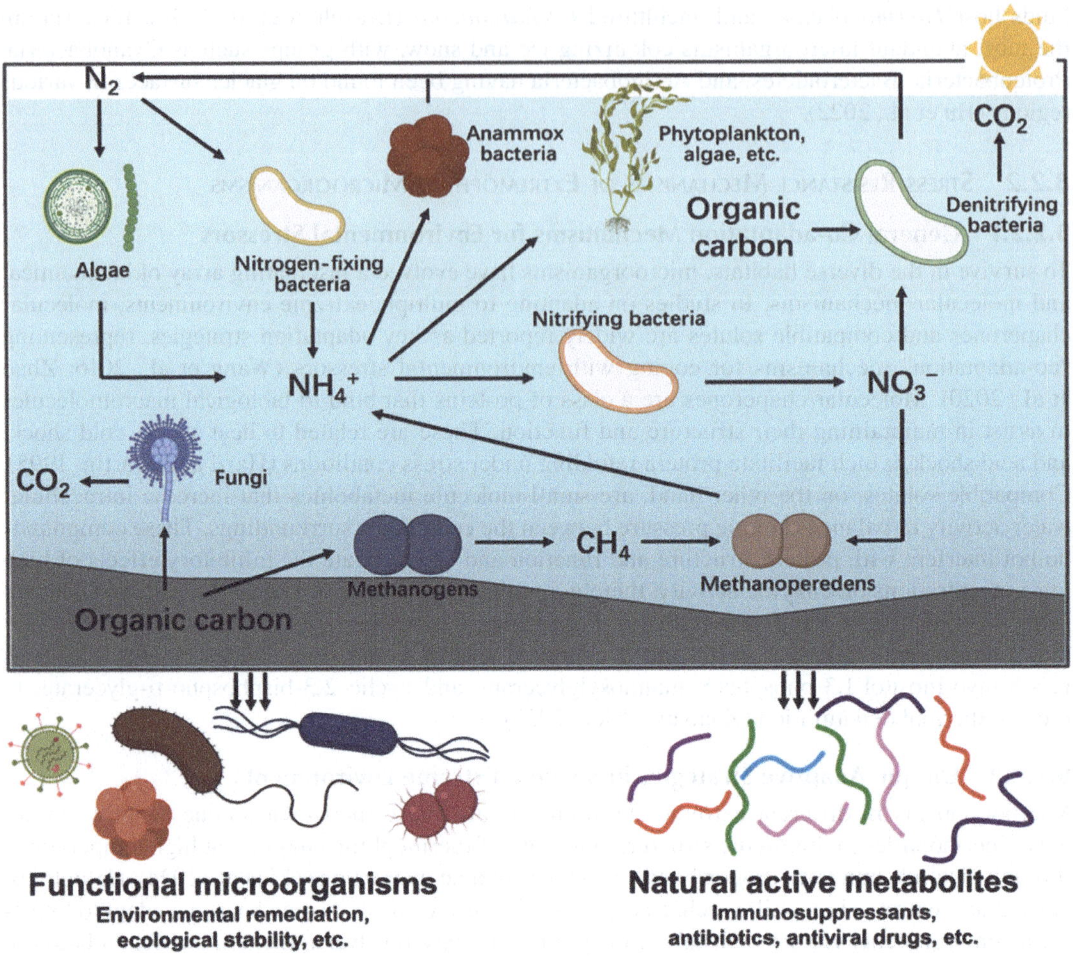

FIGURE 8.3 Biogenic element cycling and ecological functions of aquatic microbiome. Microorganisms, such as algae, nitrogen-fixing bacteria, and nitrifying bacteria, drive the nitrogen and carbon cycles in aquatic ecosystems by transforming key compounds (e.g., N_2, NH_4^+, CO_2, CH_4). Certain functional microorganisms offer significant applications in environmental remediation, maintaining ecological stability, and producing bioactive metabolites.

8.3.1.1 Carbon Cycling

Organic carbon (OC) is one of the largest carbon pools in aquatic ecosystems, and its fate is closely related to complex microbial metabolism (Yang et al., 2024a). Bacteria account for 50% of the oxidation of organic matter by sulfate-reducing species in marine sediments (Liang et al., 2023). They secrete various hydrolytic enzymes to break down complex organic matter into simple compounds such as organic acids, amino acids, and carbonates. These decomposition products not only provide nutrition for bacteria themselves, but also for other aquatic organisms, promoting the step-by-step transfer of carbon and energy in the aquatic food chain (Yang et al., 2024a). Phytoplankton are vital regulators of carbon cycling and climate change. Based on a conceptual biogeochemical mass balance model, phytoplankton photosynthesis and decomposition of sedimentary OC are recognized as the dominant pathways of the carbon cycle, with >70% of fixed carbon emitted in the forms of CO_2 and CH_4, as well as buried plankton biomass in river sediments as a relatively long-term carbon sink (Wang et al., 2025a). Ascomycota, the most species-rich phylum of fungi, can produce asci and ascospores to decompose submerged lignin, hemicellulose, and cellulose, and release extracellular enzymes (amylase, cellulase, laccase, etc.) involved in OC decomposition (Van den Wyngaert et al., 2022). Basidiomycota can produce basidia and basidiospores, while Chytridiomycota is a relatively primitive group of aquatic fungi with unique zoospores for efficient food carbon predation (Qin et al., 2020; Xue et al., 2023). In aquatic microbiome, the carbon flux mediated by bacteria-phytoplankton-fungi interactions, such as bacterial or fungal production of toxins or micronutrients (cobalamin and thiamin) for phytoplankton (Petrou, 2023), relies on microbial diversity and species specificity, emphasizing the importance of seeing broader implications for complex networks of aquatic microbes.

8.3.1.2 Nitrogen Cycling

In various aquatic environments, such as estuaries, lakes, and wastewater systems, novel pathways including anaerobic ammonium oxidation (anammox) and complete ammonia oxidation (comammox) have been confirmed to play critical roles. Among them, anammox bacteria belong to the phylum Planctomycetes and can directly convert nitrite and ammonium into nitrogen gas (Qiao et al., 2025). In oceanic hypoxic zones, freshwater sediments, and wastewater treatment, anammox bacteria account for more than 50% of the total nitrogen loss (Jensen et al., 2007). Comammox bacteria, which belong to the genus *Nitrospira* (van Kessel et al., 2015), accomplish the one-step oxidation of ammonia to nitrate (Costa et al., 2006). Comammox bacteria are widely distributed in freshwater ecosystems and often exhibit a higher abundance in surface water ecosystems compared to the traditional ammonia oxidation process (Palomo et al., 2022; Xu et al., 2020). Previously, it was believed that the main role of Chloroflexi was to decompose macromolecular organic matter (Liu et al., 2021). However, metagenomic analysis has revealed that Chloroflexi possess multiple nitrogen metabolic pathways and actively participate in the nitrogen cycle. Many Chloroflexi contain the gene for nitrous oxide reductase, an enzyme that is key to reducing the potent greenhouse gas N_2O to harmless N_2 (Bovio-Winkler et al., 2024). Therefore, Chloroflexi play an important role in mitigating the greenhouse effect. Additionally, some Chloroflexi have the ability to reduce NO_3^- to NH_4^+. The generated ammonium can be utilized by other microorganisms (Bovio-Winkler et al., 2024), thus retaining nitrogen within the ecosystem.

8.3.1.3 Photosynthesis

Photosynthesis is one of the most important pathways for energy input in aquatic ecosystems. Cyanobacteria are the most important group of photosynthetic bacteria in aquatic ecosystems, contributing to a higher primary productivity and nitrogen fixation efficiency (Sukenik and Kaplan, 2021). Metagenomic studies in extreme environments such as polar regions and hot springs have revealed the diversity of novel photosynthetic microorganisms. For instance, polar cyanobacteria maintain photosynthetic efficiency at low temperatures through unique pigment adaptation mechanisms, such as increasing the proportion of phycobiliproteins and antioxidant systems (Pessi et al., 2023). Anaerobic photosynthetic bacteria can perform photosynthesis without oxygen production, using reducing substances such as H_2S as electron donors to complete energy synthesis (Kushkevych et al., 2024). Due to their multifunctional metabolism, adaptability to extreme conditions, low costs, and high biomass production, these bacteria are promising for wastewater treatment, resource recovery, and biosynthesis of value-added polyhydroxyalkanoates, proteins, and biofertilizers, etc. (George et al., 2020).

8.3.2 Microbial Metabolite Production

8.3.2.1 Bioactive Secondary Metabolites of Aquatic Fungi

Aquatic microorganisms, including phytoplankton, bacteria, grazers, and viruses, can secrete a variety of natural active metabolites (Moran et al., 2022). Particularly, the enormous metabolic diversity of aquatic fungi makes them key microorganisms for producing a variety of bioactive secondary metabolites. A review in 2024 systematically summarized 907 new compounds derived from fungi, including 284 polyketones, 362 terpenoids, 108 alkaloids, and 75 peptide compounds (Bao et al., 2025), which can be used in the production of immunosuppressants, antibiotics, antiviral drugs, etc. (Cong et al., 2025; Yang et al., 2024c). For example, deep-sea sediment fungus *Penicillium commune* SD-118 can produce the key antibacterial compound xanthocillin X, as well as other known steroids, ceramides, and alkaloids (Hasan et al., 2015). Fungi that produce these compounds are widely distributed in marine and freshwater environments, including deep sea, coral reefs, mangroves, etc., as well as organisms such as sponges and macroalgae (Huo et al., 2025), which fully demonstrates that aquatic fungi have extensive ecological distribution and important application potential in the production of bioactive secondary metabolites.

8.3.2.2 The Functional Role of Metabolites in Ecosystems

Metabolite production can link microbial activities, organic matter mineralization, and energy cycling in aquatic ecosystems. Three new compounds, cyclic tripeptide, cyclodidepsipeptide phragamide A, and trioxobutanamide derivative phragamide B, together with 13 known compounds, were extracted from the marine endophytes (Qader et al., 2021). These compounds exhibit marked anti-biofilm effects against both Gram-positive and Gram-negative strains, interfering with bacterial quorum sensing, antibiotic resistance, and colonization of various pathogenic microorganisms. Saprophytic fungi are considered important decomposers of organic matter in all aquatic systems, and some species are able to use the lignin-degrading enzymes to break down the stubborn structure of humus, as important carbon reservoirs. Humus is also active in the redox transformation of complex compounds that affect the distribution and transport of pollutants (Grossart and Rojas-Jimenez, 2016). This not only allocates more energy and metabolic resources stored in humus and releases nutrients for other microorganisms but also maintains the balance and stability of aquatic ecosystems.

8.3.3 Environmental Bioremediation and Ecological Stability

Functional diversity is a key component of biodiversity that quantifies the value, range, and distribution of functional traits in a microbial community. These traits (e.g., physiological, morphological, or behavioral characteristics) define an organism's ecological niche and its influence on ecosystem functioning, such as productivity, nutrient cycling, and stability (Petchey and Gaston, 2006). When multiple microorganisms possess the same or similar functions, even if a certain species disappears due to environmental stress or other factors, the overall function can still be maintained, thereby ensuring the continuous operation of the ecosystem.

8.3.3.1 Environmental Remediation Functions of Aquatic Microbial Communities

Aquatic microorganisms play important roles in mitigating pollution of persistent organic compounds, microplastics, and heavy metals. A variety of fungi (such as *Penicillium*) can also break down a variety of persistent organic pollutants, including pesticides and industrial chemicals (Wolski, 2023). Pseudomonadota, Bacteroidota, Bacillota, and Cyanobacteria were frequently found on plastic biofilms and may degrade plastics. Diverse plastic-degrading bacteria are being isolated from marine environments such as offshore and deep oceanic waters, especially *Pseudomonas* spp., *Bacillus* spp., *Alcanivoras* spp., and Actinomycetes (Lv et al., 2024). Spirulina platensis chelates metals through the carboxyl/phosphate groups of cell wall polysaccharides, and the adsorption capacity of Pb^{2+} reaches 98% (Mahlangu et al., 2024). The functions of these species exhibit considerable redundancy, which can enhance the resilience of the ecosystem. Microplastics, a class of emerging contaminants, can accumulate diverse microbial communities to form biofilms, enhancing the efficiency of hydrocarbon degradation and polymer fragmentation (Chen et al., 2023; Tang and Li, 2025). In anaerobic biofilms, in addition to yeast and filamentous fungi, a large number of uncharacterized Rozellomycota (Rosenqvist et al., 2025) show the potential for biotechnology applications in water treatment.

8.3.3.2 Microbial Interactions and Ecosystem Stability

The interaction among microorganisms further enhances the stability of the marine ecosystem. Fungi often secrete antibiotics or other secondary metabolites to interfere with bacterial cell wall synthesis, protein synthesis, or metabolic pathways, thereby regulating community structure (Chen et al., 2024b). The small molecular products of fungi can be used by bacteria to decompose complex organic matter, and the metabolic products of bacteria also supply the nutrients for fungi, synergistically completing the material cycle and energy conversion (Pawlowska, 2024). Fungi also interact with aquatic plants and animals, with mycorrhizal fungi coexisting with the roots of aquatic plants for nutrient absorption, or fungal endospores serving as a food supplement for planktonic animals (Wang et al., 2025b). Additionally, the coral-associated bacteria, which play the role in preserving coral integrity, imply the necessity of microbial cooperation to keep ecosystem balance (Mohamed et al., 2022).

8.3.3.3 Ecological and Evolutionary Roles of Aquatic Viruses

Aquatic viruses, as crucial components of the microbial food web, play a dual role in ecological regulation and evolutionary driving. At the ecological function level, viruses regulate the structure of the host microbial communities by specific infection, such as cyanobacteria viruses inhibiting excessive cyanobacterial proliferation to prevent water eutrophication (Ke et al., 2025; Martin et al., 2025), while the intracellular organic matter released by lysis can be reused by other microorganisms, promoting the material and energy cycling (Xiao et al., 2021). However, some viruses, e.g., white spot syndrome virus and grass carp reovirus, may cause population collapse in aquaculture organisms due to their pathogenicity (Bai et al., 2025; Yang et al., 2023). At the evolutionary driving level, aquatic viruses are key agents of horizontal gene transfer and auxiliary metabolic gene exchange. They remarkably reprogram host metabolism, affecting photosynthesis, carbon cycling, and aquatic environmental dynamics (Luo et al., 2022; Dominguez-Huerta et al., 2022). The ecological process involving aquatic viruses exhibits spatiotemporal scale effects. Viruses can cause high-speed genetic diversity change on the short time scale and bind host receptors or lyse cells on the nano space scale, respectively (Gazitua et al., 2021). The long-time effect extends to the population and ecological level, ultimately shaping the global biogeochemical cycle and biosphere evolution on a geological time scale.

In addition to the main functions mentioned above, aquatic microorganisms play many other ecological roles, such as producing biological surfactants, extracellular polymers, and other substances that affect the physical and chemical properties of water bodies, thereby maintaining the dynamic stability of aquatic ecosystems.

8.4 AQUATIC MICROBIOME IMPACTED BY ENVIRONMENTAL STRESSORS

Environmental stressors have a profound impact on aquatic ecosystems and their microbial communities, leading to a decline in microbial diversity or a transformation of the community structure, especially under scenarios of climate change. This is essentially a loss of natural capital, which in turn weakens the supply capacity of key ecosystem services, such as water purification and carbon sequestration. Ecological responses of aquatic microorganisms, including stress adaptation, community restructure, and multiple tiers of biological organization during dynamic interactions, depict vulnerable and sensitive ecoregions that need prioritized conservation, thereafter offering targeted guidance for management of large-scale aquatic ecosystems (Figure 8.4).

8.4.1 Aquatic Microbiome Impacted by Global Warming

With increasing annual average surface temperatures, as well as a rising frequency of extreme events such as heatwaves and thawing, the adaptive responses and functional adjustment of aquatic microorganisms are considered crucial indicators to precisely decipher the global warming patterns.

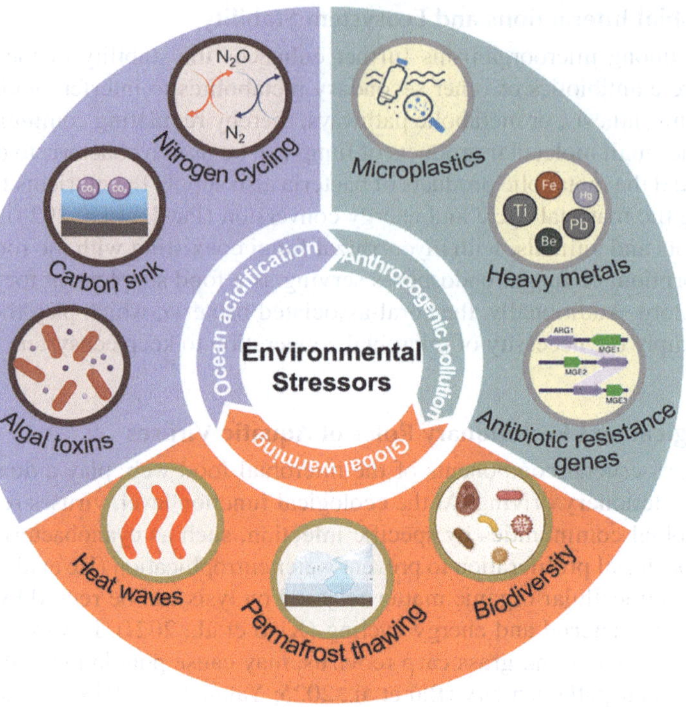

FIGURE 8.4 Impacts of the major environmental stressors on ecosystems, including ocean acidification, anthropogenic pollution, and global warming. Ocean acidification leads to altered nitrogen cycling (e.g., reduced nitrogen fixation and increased N_2O emission), a weakened carbon sink, and algal toxin release; anthropogenic pollution includes microplastics, heavy metals, and ARGs; global warming triggers heatwave stress, permafrost thawing, and biodiversity decline.

Based on ensemble machine learning models, a fast stress maladaptation in microbial population was predicted for almost half of the global river gauge stations during heatwave disturbance from 25°C to 37°C, revealing the loss of the natural capital of microorganisms (Qian and Wang, 2024; Qu et al., 2023). Declining microbiome biodiversity impacts eukaryotic populations, community ecology, and ecosystem nutrient cycling. Elevated phytoplankton biomass and community diversity, in turn, are associated with low microbiome diversity, causing a depletion of total dissolved nitrogen and phosphorus (Dickey et al., 2025). In permafrost thawing zones, thawing creates new water bodies, releases ancient microbes, and drives convergence in microbial functional gene composition and centralization of carbon metabolic pathways, suggesting that under climate disturbance, microbial community structures tend towards simplification, and functional redundancy declines, potentially weakening the buffering capacity of ecosystems (Ernakovich et al., 2022). These indicators of community restructure, nutrient depletion, and genetic function shift point out the hotspots for mitigating microbial collapse, algal bloom hazard, and greenhouse gas emission, especially in deteriorated ecosystems disturbed by climate warming.

8.4.2 Aquatic Microbiome Impacted by Ocean Acidification

The increasing anthropogenic CO_2 emissions result in oceanic pH decline. Ocean acidification, after global warming, is recognized as another significant change that influences microbial propagation, nutritional availability, and toxin production of specific subgroups.

Ocean acidification interferes with microbial functions and element cycling through multiple pathways. First, acidification causes a decrease in cytosolic pH of nitrogen-fixing cyanobacteria such as *Trichodesmium*, forcing proton pumps to consume excessive energy to maintain the transmembrane gradient, directly reducing nitrogen fixation efficiency (Hong et al., 2017). Especially in the iron-limited marine areas, the decline in nitrogen fixation is predicted to be more significant (Luo et al., 2019). Acidification and hypoxia synergistically inhibit nitrification and enhance denitrification, significantly increasing the emission of N_2O (Hutchins and Fu, 2017). Second, the structure of plankton communities tends to shift toward smaller organisms, and calcifying organisms are harmed, weakening the sinking flux of particulate OC and the "biological carbon pump" efficiency, thus reducing the long-term capacity for atmospheric CO_2 sequestration in oceans (Hutchins et al., 2019). In addition, harmful algae are found to produce more neurotoxin domoic acids in increasingly acidic seawater to poison aquatic organisms of higher trophic levels, especially during the nutrient-depleted stationary growth phase (Wingert and Cochlan, 2021; Xu et al., 2023). Therefore, it is urgent to enhance predictive capacity for climate feedback mechanisms through long-term dynamic monitoring.

8.4.3 Aquatic Microbiome Impacted by Anthropogenic Contamination

8.4.3.1 Ecotoxicity and Microbial Responses of Traditional Pollutants

The colonization and succession patterns of aquatic microbiome are shaped by a variety of anthropogenic contaminants and associated chemicals. There has long been interest in the ecological impacts and underlying mechanisms of organic pollutants and heavy metals, targeting the aquatic microbiome. More recent studies have reported the amplified eco-toxicity of organic pollutants and heavy metals by increasing the abundance of antibiotic resistance genes (ARGs) through nonlinear S-curve dynamics in the Yangtze River estuary (Miao et al., 2023). Under the stress of organic pollutants, microbial processes determine the form and bioavailability of heavy metals. Understanding these processes not only advances the fundamentals for assessing the ecological and health risks of organic pollutants and heavy metals, but also elucidates the reciprocal interactions between metal(loid)s and the aquatic microbiome (Lu et al., 2023).

8.4.3.2 Microbial Community Colonization and Potential Risks Driven by Emerging Contaminants

Plastic wastes, especially micro- and nano-plastics, have become a human-made hotspot more recently, owing to their profound influence on community colonization, assembly, survival, and succession (Bos et al., 2023). In the epi-microplastic micro-environments, rare microbial species have been shown to be more common on plastic surfaces than in surrounding water columns. A greater abundance of *Vibrio* species, which are the most common pathogens of aquaculture organisms but rarely thrive in the open seawater, is found inhabiting plastic debris widely from rivers to marine environments (Amaral-Zettler et al., 2020; Delacuvellerie et al., 2022). These pathogenic species may function as reservoirs and vectors of ARGs, and also prolong the persistence of viruses to become more infectious (Liu et al., 2025). Furthermore, the plastisphere exerts a selective pressure on colonizing communities by enriching plastic-degrading microorganisms across a diverse range of aquatic ecosystems (Birnstiel et al., 2022; Chen et al., 2023). There is a significant correlation between the abundance of plastics and their biodegraders, which is in association with geochemical factors (Feng et al., 2023). Therefore, these ecological correlations between microbiome and hydrosphere hold a clue for predicting the microbial adaptability and potential risks under climate change scenarios in the Anthropocene.

In summary, the existing studies extend our knowledge of the microbial ecology and health effects in aquatic ecosystems under environmental stressors. A key goal for future research should be to construct a platform to discover novel microbial taxa and enzymes as well as verify their functions, aiming to tackle the key challenge of aquatic pollution.

8.5 INTERDISCIPLINARY FUSION POWERING SUSTAINABLE MICROBIAL TECHNOLOGIES

Innovations in molecular technologies have profoundly deepened our comprehension of climate feedback from aquatic microorganisms through revealing mechanisms of physiological adjustments, community structural shifts, and evolutionary adaptations (Figure 8.5).

8.5.1 Molecular Ecology: Predicting Microbial Feedbacks to Climate Change

8.5.1.1 High-Throughput Omics Technologies

High-throughput omics technologies have served as indispensable tools in describing the composition of species and functional diversity. The integration of transcriptomics, proteomics, and

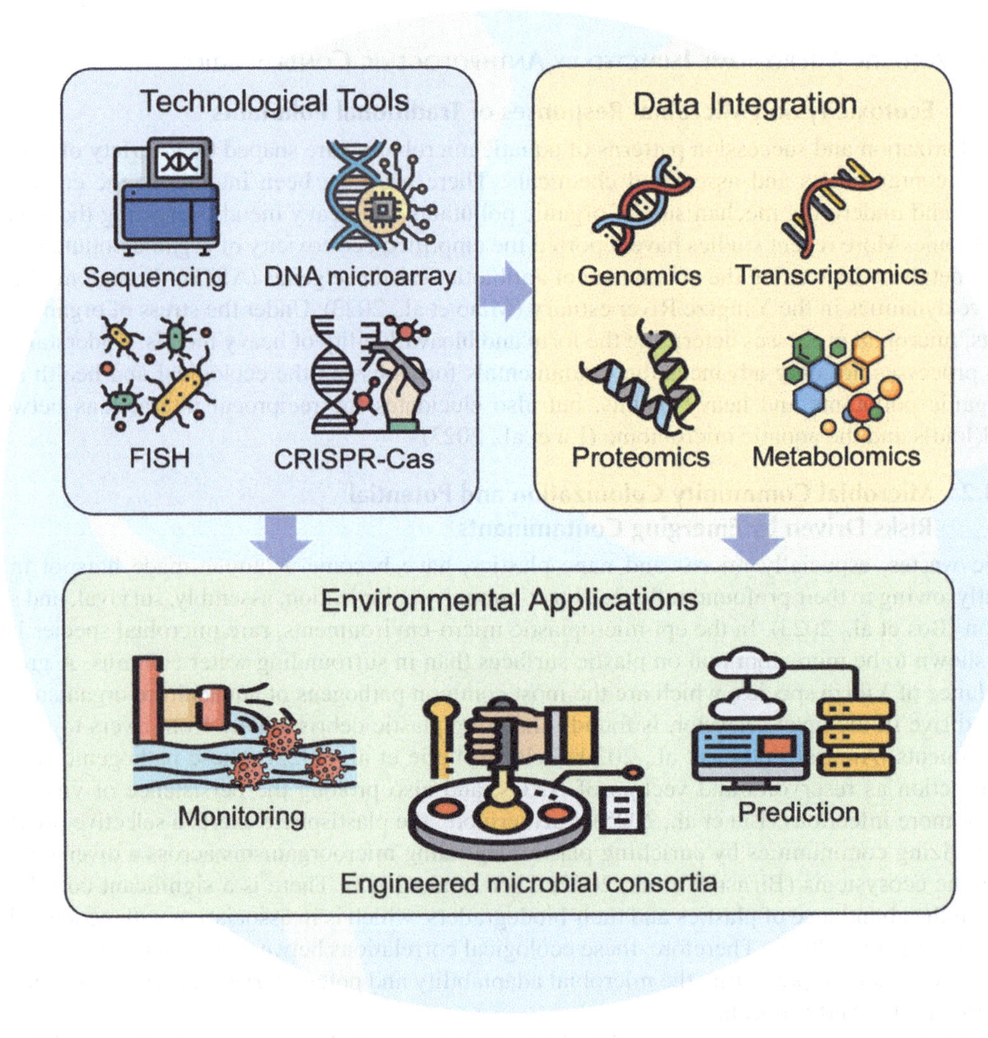

FIGURE 8.5 Interdisciplinary fusion for sustainable environmental applications. The integration of multi-omics technologies (such as high-throughput sequencing, DNA microarrays, and FISH or mass spectrometry analysis) and bioinformatics tools, through the collaborative analysis of genomics, transcriptomics, proteomics and metabolomics data, supports the development and application of environmental pollutant degradation, microbial community function analysis, and ecological restoration technologies.

metabolomics allows us to understand the regulatory mechanisms of microorganisms responding to environmental stress and host interactions (Wu et al., 2024). However, most environmental microorganisms are still uncultured under laboratory conditions. Single-cell sequencing technology is helpful for uncultured species resolution and functional exploration, thereby providing the foundation for understanding the metabolic interactions and stress resistance mechanisms of aquatic microorganisms (Berube et al., 2018). Moreover, cultivation genomics is used to break this bottleneck by simulating *in situ* environmental conditions (e.g., low-nutrient media, extended culture times, and co-cultivation strategies) (Nichols et al., 2010). Not only do these cultivation strategies yield strains for subsequent physiological and biochemical experiments and functional validation, but they may also be cross-verified with single-cell genomics and metagenomic data to enhance the reliability of functional predictions.

8.5.1.2 *In Situ* Monitoring Technologies

To accurately predict microbial feedback to climate change, community structure and genetic potential are far from sufficient. *In situ* monitoring technologies give crucial information on the activity, interactions, and spatiotemporal dynamics of microorganisms under natural conditions. The combination of fluorescence *in situ* hybridization (FISH) and nano-scale secondary ion mass spectrometry (NanoSIMS) enhances tracking metabolic activities in complex microbial communities, successfully revealing mutualistic symbiotic relationships between archaea and sulfate-reducing bacteria in methane anaerobic oxidation, and the rate and pattern of carbon-nitrogen exchange between cyanobacteria and heterotrophic bacteria (Musat et al., 2016). By further integration of stable isotope probing, FISH, Raman spectroscopy, scanning electron microscopy, and NanoSIMS into a powerful multi-technology platform, a comprehensive analysis of functions and structures can be conducted for uncultured multicellular magnetic bacteria (Schaible et al., 2022). In polar oceans, Lin et al. (2021) used 5 years of environmental DNA metabarcoding monitoring and machine learning modeling to show that reduced sea ice led to a decline in plankton eukaryote diversity, and this change in community structure further resulted in reduced carbon export flux, revealing a negative feedback loop where climate change induced shifts in biological community structure that weakened the ocean's carbon sink function.

8.5.1.3 Challenges of Molecular Technology and Its Coping Strategies

The innovation of molecular technologies has greatly deepened our understanding of microbial-climate feedback, but the field still faces several challenges. First, the "microbial dark matter" problem persists. Second, the complexity of natural environments means microbial functions are influenced by both abiotic and biotic factors. Third, most existing studies are short-term or small-scale with limited monitoring data on microbial communities and their functional dynamics, while climate feedbacks are long-term processes. In order to tackle these issues, future research should focus on the following: Microbial ecology should be interdisciplinarily integrated with climate science, computational science, remote sensing technology, and material science. For instance, Kazemi Garajeh et al. (2024) integrated remote sensing information using a Google Earth Engine platform to facilitate large-scale monitoring of water resource dynamics. Long-term and large-scale observation networks should also be established, with standardized sampling and analytical methods to achieve data comparability across time. Moreover, efforts are needed to develop artificial intelligence models that can simulate the functional trait dynamics and ecosystem functions of microbes. As stressed by van Vliet et al. (2023), given the rising incidence of climate change and hydroclimatic extremes, it is necessary to devise new modeling tools that can serve the formulation of sound environmental management strategies.

8.5.2 Environmental Applications: Microbial Bioindicators and Engineered Communities

8.5.2.1 Microbial Indicators in WBE

WBE, which leverages microbial, viral, and chemical markers in municipal wastewater to infer population health status and environmental pollution levels, functions as a "community-level urine test" (Ahmed et al., 2019; Bivins et al., 2020; Zhu et al., 2025). During the COVID-19 outbreak, WBE showed that SARS-Coronavirus-2 RNA abundances in the wastewater of the Netherlands served as a useful early warning (Medema et al., 2020). Computational integration of wastewater data with epidemiological models enhances epidemic trajectory reconstruction (Hart and Halden, 2020), and sustained monitoring protocols improve long-term threat responsiveness (Wade et al., 2022). The spatiotemporal predictive potential of WBE can also be improved by integrating additional socioeconomic, climatologic, and hydraulic parameters (Thompson et al., 2020). For antibiotic resistance, WBE assists in locating controlling nodes such as livestock wastewater streams of high ARG enrichment (Zhu et al., 2013) and municipal wastewater treatment plants where antibiotic-resistant bacteria spread (Rizzo et al., 2013). However, molecular detection faces limitations from hydraulic complexity, PCR inhibitors, and background nucleic acids (Murphy Claire et al., 2024), compounded by region-specific microbiome signatures demanding localized indicator validation. These persisting challenges require interdisciplinary solutions. Efficient risk-screening and source-shifting can be only achieved with rapid-in-the-field, low-cost sensors, and closer interactions between microbiologists, environmental engineers, and public health researchers.

8.5.2.2 Engineering Functional Microbial Consortia through Synthetic Ecology

The emergence of synthetic biology and synthetic ecology provides novel tools and strategies (Jansson et al., 2023), enabling the engineering of microbial systems for targeted environmental outcomes. Advances in CRISPR-Cas gene editing tools facilitate the modification of microbial metabolic pathways for functions like wastewater nitrogen/phosphorus removal, recalcitrant pollutant degradation, and greenhouse gas fixation (Joshi and Sharma, 2025; Yu et al., 2022). However, the application of single microorganisms is difficult in natural environments because of the ecological complexity such as environmental variability, interspecies relationships, and resource competition. To address this issue, synthetic ecology integrates ecology, synthetic biology, and engineering to design and optimize microbial communities (Idowu, 2024; Yan et al., 2025). By engineering synergistic interactions (e.g., metabolic division of labor, signal complementation), synthetic ecology constructs communities with enhanced functionality, stability, and adaptability compared to single-strain systems (Díaz-Zaragoza, 2024). Synthetic ecology relies on genetic engineering as the cornerstone of designing microbial communities. In certain scenarios, improving the performance of specific enzymes through design and optimization can greatly boost the community's efficiency in degrading particular pollutants (Narayanan et al., 2023; Sui et al., 2023). Moreover, by optimizing the interactions among microorganisms, including symbiosis, competition, and cooperation, the resilience of communities to environmental stress can be enhanced (Lechón-Alonso et al., 2021).

8.5.3 Ethical Controversies and Governance Frameworks for Gene-Edited Microbes

Despite advances in pollutant removal with synthetic biology and gene editing, these technologies still pose potential risks if the engineered microbes are released into unrestricted environments (Ou and Guo, 2023). Effective governance of gene-edited microbes requires integrating technical oversight with ethical frameworks. International coordination through agreements like the Cartagena Protocol on Biosafety is essential for standardizing risk assessment procedures and facilitating information sharing on transboundary movement of living modified organisms (Keiper and Atanassova, 2020). Apart from regulation, there are also potential ethical challenges stemming from social and political issues (Børsen, 2025). Technology assessments often adopt exclusionary "expert-led" approaches that marginalize public values, while global technology diffusion risks shifting potential hazards to regions with limited regulatory capacity, such as deploying engineered microbes developed in industrialized nations within developing economies (Covello and Scott Frey, 1990; Parker and Kunjapur, 2020).

Scholars contend that governance frameworks should integrate the precautionary principle and strike a balance among stakeholder interests, thereby addressing both ecological risks and societal equity concerns (Bohua et al., 2023). Concurrently, policy analyses indicate that current biosafety standards and regulatory structures often lag behind scientific innovation, underscoring the need to develop norms for genetically engineered microorganisms and adaptive oversight tools such as real-time biosensors for environmental tracking, and transparent disclosure of engineering objectives and monitoring data (Pei et al., 2022; Shams et al., 2024). Future governance should balance innovation facilitation with precautionary safeguards such as genetic biocontainment strategies to prevent unintended gene flow. Achieving this equilibrium necessitates sustained collaboration across scientific, ethical, and policy domains.

8.6 CONCLUSION

Aquatic microbes, as key process engines of biogeochemical cycles, are the biological foundation of aquatic ecosystems, due to their primary production, bioactive metabolite production, and bioremediation functions. Climate change is largely reshaping microbial functional patterns. On the one hand, it leads to a decline in species and functional diversity, weakening the capacity to sustain ecosystem services and highlighting the urgency of preserving microbial functional diversity; on the other hand, long-term evolutionary adaptations enable microbes to maintain ecological stability and participate in ecosystem restoration even under complex or extreme conditions. Emerging technologies such as metagenomics and single-cell sequencing are rapidly unveiling microbe-driven ecological processes, including nutrient transformation pathways and novel pollutant degradation routes. Accordingly, future research should employ molecular biology techniques to decipher microbial adaptive strategies and biosynthetic pathways in extreme environments, thereby filling the knowledge gap of "microbial dark matter". Meanwhile, it is crucial to quantify microbial functional traits and integrate them into next-generation ecosystem models, which is essential for addressing the pressures of global warming, ocean acidification, and anthropogenic pollution. Successfully addressing these challenges will provide a robust scientific underpinning for ecosystem adaptive management strategies and informed climate policy decisions.

ACKNOWLEDGMENTS

We gratefully acknowledge the support from the National Key Research and Development Program of China (2023YFC3205901), the Natural Science Foundation of Guangdong Province (2025A1515011251 and 2023A1515030263), the Natural Science Foundation of Shenzhen Municipality (JCYJ20250604175815020, JCYJ20241202125905008, and JCYJ20250604175827037), and the Beijing Natural Science Foundation (8252030).

DECLARATION OF GENERATIVE AI USE

During the preparation of this manuscript, the authors utilized ERNIE 4.5 for grammar checking, language polishing, and necessary literature analysis. The authors subsequently reviewed and edited the content and take full responsibility for the final version of the published chapter.

REFERENCES

Adam, P.S., Borrel, G., Brochier-Armanet, C. and Gribaldo, S., 2017. The growing tree of Archaea: New perspectives on their diversity, evolution and ecology. *ISME Journal* 11(11), 2407–2425.

Ahmed, W., Hamilton, K., Toze, S., Cook, S. and Page, D., 2019. A review on microbial contaminants in stormwater runoff and outfalls: Potential health risks and mitigation strategies. *Science of the Total Environment* 692, 1304–1321.

Alabbosh, K.F., 2025. Leveraging the potential of environmental microorganisms: An extensive examination of their capacity in tackling global and local environmental and ecological challenges. *Polish Journal of Environmental Studies* 34(1), 1–11.

Amaral-Zettler, L.A., Zettler, E.R. and Mincer, T.J., 2020. Ecology of the plastisphere. *Nature Reviews Microbiology* 18(3), 139–151.

Bai, Y., Liao, Q., Wang, Z.X., Zhu, C.C., Liu, W., Xiao, W.H. and Liu, X., 2025. *Ctenopharyngodon idella* (grass carp) SIRT7 negatively regulates anti-GCRV innate immunity by promoting IRF3 and IRF7 degradation. *Aquaculture* 606, 742575.

Bao, M., Shi, Y., Gong, X., Guo, Y., Wang, J., Chen, X. and Liu, L., 2025. New bioactive secondary metabolites from fungi: 2024. *Mycology* 16(3), 961–987.

Bayer, B., Kitzinger, K., Paul, N.L., Albers, J.B., Saito, M.A., Wagner, M., Carlson, C.A. and Santoro, A.E., 2024. Contribution of ammonia oxidizers to inorganic carbon fixation in the dark ocean. *bioRxiv*. doi: 10.1101/2024.11.16.623942.

Berube, P.M., Biller, S.J., Hackl, T., Hogle, S.L., Satinsky, B.M., Becker, J.W., Braakman, R., Collins, S.B., Kelly, L., Berta-Thompson, J., Coe, A., Bergauer, K., Bouman, H.A., Browning, T.J., De Corte, D., Hassler, C., Hulata, Y., Jacquot, J.E., Maas, E.W., Reinthaler, T., Sintes, E., Yokokawa, T., Lindell, D., Stepanauskas, R. and Chisholm, S.W., 2018. Single cell genomes of Prochlorococcus, Synechococcus, and sympatric microbes from diverse marine environments. *Scientific Data* 5(1), 180154.

Birnstiel, S., Sebastian, M. and Romera-Castillo, C., 2022. Structure and activity of marine bacterial communities responding to plastic leachates. *Science of the Total Environment* 834, 155264.

Bivins, A., North, D., Ahmad, A., Ahmed, W., Alm, E., Been, F., Bhattacharya, P., Bijlsma, L., Boehm, A.B., Brown, J., Buttiglieri, G., Calabro, V., Carducci, A., Castiglioni, S., Cetecioglu Gurol, Z., Chakraborty, S., Costa, F., Curcio, S., de los Reyes, F.L., III, Delgado Vela, J., Farkas, K., Fernandez-Casi, X., Gerba, C., Gerrity, D., Girones, R., Gonzalez, R., Haramoto, E., Harris, A., Holden, P.A., Islam, M.T., Jones, D.L., Kasprzyk-Hordern, B., Kitajima, M., Kotlarz, N., Kumar, M., Kuroda, K., La Rosa, G., Malpei, F., Mautus, M., McLellan, S.L., Medema, G., Meschke, J.S., Mueller, J., Newton, R.J., Nilsson, D., Noble, R.T., van Nuijs, A., Peccia, J., Perkins, T.A., Pickering, A.J., Rose, J., Sanchez, G., Smith, A., Stadler, L., Stauber, C., Thomas, K., van der Voorn, T., Wigginton, K., Zhu, K. and Bibby, K., 2020. Wastewater-based epidemiology: Global collaborative to maximize contributions in the fight against COVID-19. *Environmental Science & Technology* 54(13), 7754–7757.

Bohua, L., Yuexin, W., Yakun, O., Kunlan, Z., Huan, L. and Ruipeng, L., 2023. Ethical framework on risk governance of synthetic biology. *Journal of Biosafety and Biosecurity* 5(2), 45–56.

Bos, R.P., Kaul, D., Zettler, E.R., Hoffman, J.M., Dupont, C.L., Amaral-Zettler, L.A. and Mincer, T.J., 2023. Plastics select for distinct early colonizing microbial populations with reproducible traits across environmental gradients. *Environmental Microbiology* 25(12), 2761–2775.

Børsen, T., 2025. Technology assessment and postnormal science. *Futures* 166, 103515.

Bovio-Winkler, P., Cabezas, A. and Etchebehere, C., 2024. Unveiling the hidden diversity and functional role of Chloroflexota in full-scale wastewater treatment plants through genome-centric analyses. *ISME Communications* 4(1), ycae050.

Brazelton, W.J., Ludwig, K.A., Sogin, M.L., Andreishcheva, E.N., Kelley, D.S., Shen, C.-C., Edwards, R.L. and Baross, J.A., 2010. Archaea and bacteria with surprising microdiversity show shifts in dominance over 1,000-year time scales in hydrothermal chimneys. *Proceedings of the National Academy of Sciences* 107(4), 1612–1617.

Cabello-Yeves Pedro, J., Zemskaya Tamara, I., Rosselli, R., Coutinho Felipe, H., Zakharenko Alexandra, S., Blinov Vadim, V. and Rodriguez-Valera, F., 2017. Genomes of novel microbial lineages assembled from the sub-ice waters of Lake Baikal. *Applied and Environmental Microbiology* 84(1), e02132–02117.

Chen, B., Zhang, Z., Wang, T., Hu, H., Qin, G., Lu, T., Hong, W., Hu, J., Penuelas, J. and Qian, H., 2023. Global distribution of marine microplastics and potential for biodegradation. *Journal of Hazardous Materials* 451, 131198.

Chen, G., Satake, T., Sugiura, N. and Utsumi, M., 2024a. Vertical distribution and phylogenetic characterization of marine Archaea in the western subarctic Pacific. *Marine Development* 2(1), 5.

Chen, L.J., Dini-Andreote, F., Liu, H.Q., Wang, H.X., Dumbrell, A., Wang, Z.Y., Chen, X.Y., Chen, F.F., Chen, X.L., Wu, L.C. and Jiang, Y.J., 2024b. Integrating variation in bacterial-fungal co-occurrence network with soil carbon dynamics. *Journal of Applied Ecology* 61(1), 36–50.

Chia, X.K., Hadibarata, T., Jusoh, M.N.H., Tan, I.S., Chee, H. and Foo, Y., 2024. Extremophilic microbes for environmental bioremediation: A review. *Biointerface Research in Applied Chemistry* 14(5), 103.

Choi, T.R., Park, Y.L., Song, H.S., Lee, S.M., Park, S.L., Lee, H.S., Kim, H.J., Bhatia, S.K., Gurav, R., Choi, K.Y., Lee, Y.K. and Yang, Y.H., 2021. Fructose-based production of short-chain-length and medium-chain-length polyhydroxyalkanoate copolymer by Arctic *Pseudomonas* sp. B14–6. *Polymers* 13(9), 1398.

Cong, M.J., Li, Y.Y., Li, Y.Q., Chen, X.L., Pang, X.Y., Yan, Y., Tian, X.P., Liu, Y.H., Yang, J. and Wang, J.F., 2025. Spirochrains A-D, four caged 5,6,5 spirocyclic amides from an antarctic fungus Aspergillus *ochraceopetaliformis* SCSIO 05702 with Anti-RSV activities. *Organic Letters* 27(10), 2295–2299.

Costa, E., Pérez, J. and Kreft, J.-U., 2006. Why is metabolic labour divided in nitrification? *Trends in Microbiology* 14(5), 213–219.

Covello, V.T. and Scott Frey, R., 1990. Technology-based environmental health risks in developing nations. *Technological Forecasting and Social Change* 37(2), 159–179.

Dang, H., Klotz, M.G., Lovell, C.R. and Sievert, S.M., 2019. Editorial: The responses of marine microorganisms, communities and ecofunctions to environmental gradients. *Frontiers in Microbiology* 10, 115.

Debeljak, P. and Baltar, F., 2023. Fungal diversity and community composition across ecosystems. *Journal of Fungi* 9(5), 510.

Delacuvellerie, A., Ballerini, T., Frere, L., Matallana-Surget, S., Dumontet, B. and Wattiez, R., 2022. From rivers to marine environments: A constantly evolving microbial community within the plastisphere. *Marine Pollution Bulletin* 179, 113660.

Díaz-Zaragoza, M., 2024. Multifaceted applications of synthetic microbial communities: Advances in biomedicine, bioremediation, and industry. *Microbiology Research* 15(3), 1709–1727.

Dick, G.J., 2019. The microbiomes of deep-sea hydrothermal vents: Distributed globally, shaped locally. *Nature Reviews Microbiology* 17(5), 271–283.

Dickey, J.R., Mercer, N.M., Kuijpers, M.C.M., Props, R. and Jackrel, S.L., 2025. Biodiversity within phytoplankton-associated microbiomes regulates host physiology, host community ecology, and nutrient cycling. *Msystems* 10(2), e01462–01424.

Dominguez-Huerta, G., Zayed, A.A., Wainaina, J.M., Guo, J.R., Tian, F.N., Pratama, A.A., Bolduc, B., Mohssen, M., Zablocki, O., Pelletier, E., Delage, E., Alberti, A., Aury, J.-M., Carradec, Q., da Silva, C., Labadie, K., Poulain, J., Coordinators, T.O., Bowler, C., Eveillard, D., Guidi, L., Karsenti, E., Kuhn, J.H., Ogata, H., Wincker, P., Culley, A., Chaffron, S. and Sullivan, M.B., 2022. Diversity and ecological footprint of Global Ocean RNA viruses. *Science* 376(6598), 1202–1208.

Ernakovich, J.G., Barbato, R.A., Rich, V.I., Schädel, C., Hewitt, R.E., Doherty, S.J., Whalen, E.D., Abbott, B.W., Barta, J., Biasi, C., Chabot, C.L., Hultman, J., Knoblauch, C., Vetter, M.C.Y.L., Leewis, M.-C., Liebner, S., Mackelprang, R., Onstott, T.C., Richter, A., Schütte, U.M.E., Siljanen, H.M.P., Taş, N., Timling, I., Vishnivetskaya, T.A., Waldrop, M.P. and Winkel, M., 2022. Microbiome assembly in thawing permafrost and its feedbacks to climate. *Global Change Biology* 28(17), 5007–5026.

Feng, J., Li, C., Tang, L., Wu, X., Wang, Y., Yang, Z., Yuan, W., Sun, L., Hu, W. and Zhang, S., 2023. Tracing the century-long evolution of microplastics deposition in a cold seep. *Advanced Science* 10(10), 2206120.

Gazitua, M.C., Vik, D.R., Roux, S., Gregory, A.C., Bolduc, B., Widner, B., Mulholland, M.R., Hallam, S.J., Ulloa, O. and Sullivan, M.B., 2021. Potential virus-mediated nitrogen cycling in oxygen-depleted oceanic waters. *ISME Journal* 15(4), 981–998.

George, D.M., Vincent, A.S. and Mackey, H.R., 2020. An overview of anoxygenic phototrophic bacteria and their applications in environmental biotechnology for sustainable resource recovery. *Biotechnology Reports* 28, e00563.

Grossart, H.-P. and Rojas-Jimenez, K., 2016. Aquatic fungi: Targeting the forgotten in microbial ecology. *Current Opinion in Microbiology* 31, 140–145.

Gutiérrez-Preciado, A., Dede, B., Baker, B.A., Eme, L., Moreira, D. and López-García, P., 2024. Extremely acidic proteomes and metabolic flexibility in bacteria and highly diversified archaea thriving in geothermal chaotropic brines. *Nature Ecology & Evolution* 8(10), 1856–1869.

Hamm, J.N., Erdmann, S., Eloe-Fadrosh, E.A., Angeloni, A., Zhong, L., Brownlee, C., Williams, T.J., Barton, K., Carswell, S., Smith, M.A., Brazendale, S., Hancock, A.M., Allen, M.A., Raftery, M.J. and Cavicchioli, R., 2019. Unexpected host dependency of Antarctic Nanohaloarchaeota. *Proceedings of the National Academy of Sciences* 116(29), 14661–14670.

Hart, O.E. and Halden, R.U., 2020. Computational analysis of SARS-CoV-2/COVID-19 surveillance by wastewater-based epidemiology locally and globally: Feasibility, economy, opportunities and challenges. *Science of the Total Environment* 730, 138875.

Hartl, F.U. and Martin, J., 1995. Molecular chaperones in cellular protein folding. *Current Opinion in Structural Biology* 5(1), 92–102.

Hasan, S., Ansari, M.I., Ahmad, A. and Mishra, M., 2015. Major bioactive metabolites from marine fungi: A review. *Bioinformation* 11(4), 176–181.

Hong, H., Shen, R., Zhang, F., Wen, Z., Chang, S., Lin, W., Kranz, S.A., Luo, Y.-W., Kao, S.-J., Morel, F.M.M. and Shi, D., 2017. The complex effects of ocean acidification on the prominent N_2-fixing cyanobacterium *Trichodesmium*. *Science* 356(6337), 527–531.

Hu, Y., Wang, Z., Jiang, H., Chen, Y., Liu, Q., Duan, B. and Lu, X., 2022. Current knowledge and future prospects regarding microorganisms in mountain glacier ecosystems. *Advance in Earth Sciences* 37(9), 899–914.

Huo, R.Y., Tu, Y., Liu, C., Zi, G.R., Shi, Y., Ren, J.W., Cai, L. and Liu, L., 2025. New antifungal chlorinated orsellinic aldehydes from the deep-sea-derived fungus *Acremonium sclerotigenum* LW14. *Mycology* 16, 1–10.

Hutchins, D.A. and Fu, F., 2017. Microorganisms and ocean global change. *Nature Microbiology* 2(6), 17058.

Hutchins, D.A., Jansson, J.K., Remais, J.V., Rich, V.I., Singh, B.K. and Trivedi, P., 2019. Climate change microbiology: Problems and perspectives. *Nature Reviews Microbiology* 17(6), 391–396.

Idowu, E., 2024. Microbial ecology: Understanding the dynamics of microbial communities and their impact on ecosystems. Preprints. doi: 10.20944/preprints202412.0527.v1.

Jansson, J.K., McClure, R. and Egbert, R.G., 2023. Soil microbiome engineering for sustainability in a changing environment. *Nature Biotechnology* 41(12), 1716–1728.

Jensen, M.M., Thamdrup, B. and Dalsgaard, T., 2007. Effects of specific inhibitors on anammox and denitrification in marine sediments. *Applied and Environmental Microbiology* 73(10), 3151–3158.

Ji, L., Zhang, H., Wang, Z., Tian, Y., Tian, W. and Liu, Z., 2024. Temperature alters bacterial community structure in sediment of mountain stream. *Scientific Reports* 14(1), 31159.

Joshi, M. and Sharma, J.G., 2025. The utility of synthetic biology in the treatment of industrial wastewaters. *Nature Environment & Pollution Technology* 24(1), 1–11.

Jurdzinski, K.T., Mehrshad, M., Delgado, L.F., Deng, Z., Bertilsson, S. and Andersson, A.F., 2023. Large-scale phylogenomics of aquatic bacteria reveal molecular mechanisms for adaptation to salinity. *Science Advance* 9(21), eadg2059.

Karner, M.B., DeLong, E.F. and Karl, D.M., 2001. Archaeal dominance in the mesopelagic zone of the Pacific Ocean. *Nature* 409(6819), 507–510.

Kazemi Garajeh, M., Haji, F., Tohidfar, M., Sadeqi, A., Ahmadi, R. and Kariminejad, N., 2024. Spatiotemporal monitoring of climate change impacts on water resources using an integrated approach of remote sensing and Google Earth Engine. *Scientific Reports* 14(1), 5469.

Ke, F., Liu, A.K. and Zhang, Q.Y., 2025. Extra peptidase of a cyanophage confers its stronger lytic effect on bloom-forming *Microcystis aeruginosa*. *International Journal of Biological Macromolecules* 304(2), 140979.

Keiper, F. and Atanassova, A., 2020. Regulation of synthetic biology: Developments under the convention on biological diversity and its protocols. *Frontiers in Bioengineering and Biotechnology* 8, 310.

Kida, M., Ohira, A., Okazaki, Y., Yamaguchi, Y.T., Goto, A.S., Hayakawa, K. and Nishimura, H., 2025. Couples in the deep: Dissolved organic and microbial communities in the oxygenated hypolimnion of a deep freshwater lake. *bioRxiv*. doi: 10.1101/2025.09.23.678040.

Kushkevych, I., Prochazka, V., Vitezova, M., Dordevic, D., Abd El-Salam, M. and Rittmann, S.K.M.R., 2024. Anoxygenic photosynthesis with emphasis on green sulfur bacteria and a perspective for hydrogen sulfide detoxification of anoxic environments. *Frontiers in Microbiology* 15, 1417714.

Lakshmi, P., 2025. Climate change and microbial life: A critical analysis of their reciprocal relationship. *Journal of Advances in Microbiology Research* 6(2), 29–33.

Lechón-Alonso, P., Clegg, T., Cook, J., Smith, T.P. and Pawar, S., 2021. The role of competition versus cooperation in microbial community coalescence. *PLOS Computational Biology* 17(11), e1009584.

Li, Y.J., Li, Y.J., Yang, X.R., Xie, G.Q. and Gao, R.J., 2024. Expanding the boundaries of biocatalysis: The potential and applications of thermophilic enzymes. *The Chemistry of Life* 44(7), 1228–1239.

Liang, Q.Y., Zhang, J.Y., Ning, D., Yu, W.X., Chen, G.J., Tao, X., Zhou, J., Du, Z.J. and Mu, D.S., 2023. Niche modification by sulfate-reducing bacteria drives microbial community assembly in anoxic marine sediments. *mBio* 14(2), e0353522.

Lin, Y., Moreno, C., Marchetti, A., Ducklow, H., Schofield, O., Delage, E., Meredith, M., Li, Z., Eveillard, D., Chaffron, S. and Cassar, N., 2021. Decline in plankton diversity and carbon flux with reduced sea ice extent along the Western Antarctic Peninsula. *Nature Communications* 12(1), 4948.

Liu, R., Wang, L., Wei, X., Song, W. and Fang, J., 2021. Novel Chloroflexi genomes from the challenger deep reveal metabolic strategies for the adaptation to deep-sea habitats. *Microbiome* 10(1), 75.

Liu, Y., Liu, L., Wang, X., Shao, M., Wei, Z., Wang, L., Li, B., Li, C., Luo, X., Li, F. and Zheng, H., 2025. Microplastics enhance the prevalence of antibiotic resistance genes in mariculture sediments by enriching host bacteria and promoting horizontal gene transfer. *Eco-Environment & Health* 4(1), 100136.

Lu, Q.H., Xu, X.P., Fang, W.W., Wang, H.M., Liang, Z.W., Cai, R., Hu, Z.F., Shim, H., Rossetti, S. and Wang, S.Q., 2023. Metal(loid)s in organic-matter-polluted urban rivers in China: Spatial pattern, ecological risk and reciprocal interactions with aquatic microbiome. *Journal of Hazardous Materials* 457, 131781.

Luo, X.Q., Wang, P.D., Li, J.L., Ahmad, M., Duan, L., Yin, L.Z., Deng, Q.Q., Fang, B.Z., Li, S.H. and Li, W.J., 2022. Viral community-wide auxiliary metabolic genes differ by lifestyles, habitats, and hosts. *Microbiome* 10(1), 190.

Luo, Y.W., Shi, D.L., Kranz, S.A., Hopkinson, B.M., Hong, H.Z., Shen, R. and Zhang, F.T., 2019. Reduced nitrogenase efficiency dominates response of the globally important nitrogen fixer Trichodesmium to ocean acidification. *Nature Communications* 10, 1521.

Lv, S., Li, Y., Zhao, S. and Shao, Z., 2024. Biodegradation of typical plastics: From microbial diversity to metabolic mechanisms. *International Journal of Molecular Sciences* 25(1), 593.

Mahlangu, D., Mphahlele, K., De Paola, F. and Mthombeni, N.H., 2024. Microalgae-mediated biosorption for effective heavy metals removal from wastewater: A review. *Water* 16(5), 718.

Martin, R.M., Denison, E.R., Pound, H.L., Barnes, E.A., Chaffin, J.D. and Wilhelm, S.W., 2025. Mitomycin C eliminates cyanobacterial transcription without detectable prophage induction in a *Microcystis*-dominated harmful algal bloom in Lake Erie. *Microbiology Spectrum* 13, e02872–02824.

Medema, G., Heijnen, L., Elsinga, G., Italiaander, R. and Brouwer, A., 2020. Presence of SARS-Coronavirus-2 RNA in sewage and correlation with reported COVID-19 prevalence in the early stage of the epidemic in the Netherlands. *Environmental Science & Technology Letters* 7(7), 511–516.

Miao, J., Ling, Y., Chen, X., Wu, S., Liu, X., Xu, S., Umar, S. and Anderson, B.D., 2023. Assessing the non-linear association of environmental factors with antibiotic resistance genes (ARGs) in the Yangtze River Mouth, China. *Scientific Reports* 13(1), 20367.

Michán, C., Blasco, J. and Alhama, J., 2021. High-throughput molecular analyses of microbiomes as a tool to monitor the wellbeing of aquatic environments. *Microbial Biotechnology* 14(3), 870–885.

Mohamed, H.F., Chen, Y., Abd-Elgawad, A., Cai, R. and Xu, C., 2022. The unseen drivers of coral health; coral microbiome; the hope for effective coral restoration. *Polish Journal of Environmental Studies* 31(2), 989–1006.

Moran, M.A., Kujawinski, E.B., Schroer, W.F., Amin, S.A., Bates, N.R., Bertrand, E.M., Braakman, R., Brown, C.T., Covert, M.W., Doney, S.C., Dyhrman, S.T., Edison, A.S., Eren, A.M., Levine, N.M., Li, L., Ross, A.C., Saito, M.A., Santoro, A.E., Segre, D., Shade, A., Sullivan, M.B. and Vardi, A., 2022. Microbial metabolites in the marine carbon cycle. *Nature Microbiology* 7(4), 508–523.

Murphy Claire, M., Weller Daniel, L., Love Tanzy, M.T., Danyluk Michelle, D. and Strawn Laura, K., 2024. The probability of detecting host-specific microbial source tracking markers in surface waters was strongly associated with method and season. *Microbiology Spectrum* 13(2), e01972–01924.

Musat, N., Musat, F., Weber, P.K. and Pett-Ridge, J., 2016. Tracking microbial interactions with NanoSIMS. *Current Opinion in Biotechnology* 41, 114–121.

Narayanan, M., Ali, S.S. and El-Sheekh, M., 2023. A comprehensive review on the potential of microbial enzymes in multipollutant bioremediation: Mechanisms, challenges, and future prospects. *Journal of Environmental Management* 334, 117532.

Nichols, D., Cahoon, N.C., Trakhtenberg, E.M., Pham, L., Mehta, A., Bélanger, A., Kanigan, T.S., Lewis, K. and Epstein, S.S., 2010. Use of ichip for high-throughput *in situ* cultivation of "uncultivable" microbial species. *Applied and Environmental Microbiology* 76, 2445–2450.

Oren, A., 2024. Novel insights into the diversity of halophilic microorganisms and their functioning in hypersaline ecosystems. *NPJ Biodiversity* 3(1), 18.

Ou, Y. and Guo, S., 2023. Safety risks and ethical governance of biomedical applications of synthetic biology. *Frontiers in Bioengineering and Biotechnology* 11, 1292029.

Palomo, A., Dechesne, A., Pedersen, A.G. and Smets, B.F., 2022. Genomic profiling of Nitrospira species reveals ecological success of comammox Nitrospira. *Microbiome* 10(1), 204.

Parker, M.T. and Kunjapur, A.M., 2020. Deployment of engineered microbes: Contributions to the bioeconomy and considerations for biosecurity. *Health Security* 18(4), 278–296.

Pawlowska, T.E., 2024. Symbioses between fungi and bacteria: From mechanisms to impacts on biodiversity. *Current Opinion in Microbiology* 80, 102496.

Pei, L., Garfinkel, M. and Schmidt, M., 2022. Bottlenecks and opportunities for synthetic biology biosafety standards. *Nature Communications* 13(1), 2175.

Peng, X., Amend, A.S., Baltar, F., Blanco-Bercial, L., Breyer, E., Burgaud, G., Cunliffe, M., Edgcomb, V.P., Grossart, H.-P., Mara, P., Masigol, H., Pang, K.-L., Retter, A., Roberts, C., van Bleijswijk, J., Walker, A.K. and Whitner, S., 2024. Planktonic marine fungi: A review. *Journal of Geophysical Research: Biogeosciences* 129(3), e2023JG007887.

Pessi, I.S., Popin, R.V., Durieu, B., Lara, Y., Tytgat, B., Savaglia, V., Roncero-Ramos, B., Hultman, J., Verleyen, E., Vyverman, W. and Wilmotte, A., 2023. Novel diversity of polar Cyanobacteria revealed by genome-resolved metagenomics. *Microbial Genomics* 9(7), 001056.

Petchey, O.L. and Gaston, K.J., 2006. Functional diversity: Back to basics and looking forward. *Ecology Letters* 9(6), 741–758.

Petrou, K., 2023. Phytoplankton-bacteria interactions 2.0. *Microorganisms* 11(6), 1536.

Qader, M.M., Hamed, A.A., Soldatou, S., Abdelraof, M., Elawady, M.E., Hassane, A.S.I., Belbahri, L., Ebel, R. and Rateb, M.E., 2021. Antimicrobial and antibiofilm activities of the fungal metabolites isolated from the marine endophytes *Epicoccum nigrum* M13 and *Alternaria alternata* 13A. *Marine Drugs* 19(4), 232.

Qian, H. and Wang, Y., 2024. Predicting microbial adaptability to climate change in the Anthropocene. *Global Change Biology* 30(1), e17122.

Qiao, X., Zhang, L., Qiu, Z., Wu, Y., Deng, C., Geng, Y., Zhang, Y., Yan, Y., Li, B., Zhang, L., Zhuang, W.-Q. and Yu, K., 2025. Nitrite impairs bioreactor performance due to decreased replication of Candidatus Brocadia sapporoensis by unbalanced energy allocation. *Water Research* 283, 123806.

Qin, H., Cao, X., Cui, L., Lv, Q. and Chen, T., 2020. The influence of human interference on zooplankton and fungal diversity in Poyang Lake watershed in China. *Diversity* 12(8), 296.

Qu, Q., Xu, J., Kang, W., Feng, R. and Hu, X., 2023. Ensemble learning model identifies adaptation classification and turning points of river microbial communities in response to heatwaves. *Global Change Biology* 29(24), 6988–7000.

Richard, E., 2024. Climate-resilient conservation strategies for environmental microbial ecosystems. *Global Journal of Life Sciences and Biological Research* 10(3), 082.

Rinke, C., Rubino, F., Messer, L.F., Youssef, N., Parks, D.H., Chuvochina, M., Brown, M., Jeffries, T., Tyson, G.W., Seymour, J.R. and Hugenholtz, P., 2018. A phylogenomic and ecological analysis of the globally abundant Marine Group II archaea (*Ca.* Poseidoniales ord. nov.). *ISME Journal* 13(3), 663–675.

Rizzo, L., Manaia, C., Merlin, C., Schwartz, T., Dagot, C., Ploy, M.C., Michael, I. and Fatta-Kassinos, D., 2013. Urban wastewater treatment plants as hotspots for antibiotic resistant bacteria and genes spread into the environment: A review. *Science of the Total Environment* 447, 345–360.

Rodríguez-Gijón, A., Pacheco-Valenciana, A., Milke, F., Dharamshi, J.E., Hampel, J.J., Damashek, J., Wienhausen, G., Rodriguez -R, L.M. and Garcia, S.L., 2025. The ecological success of freshwater microorganisms is mediated by streamlining and biotic interactions. *bioRxiv*. doi: 10.1101/2025.03.24.644981.

Rosenqvist, T., Chan, S. and Paul, C.J., 2025. Uncharacterized members of the phylum Rozellomycota dominate the fungal community of a full-scale slow sand filter for drinking water production. *Water Research* 279, 123447.

Sagar, A., Gaurab, K. and Sunil, P., 2015. Microbial diversity in freshwater and marine environment. *Nepal Journal of Biotechnology* 3(1), 68.

Santoro, A.E., Richter, R.A. and Dupont, C.L., 2019. Planktonic marine archaea. *Annual Review of Marine Science* 11, 131–158.

Schaible, G.A., Kohtz, A.J., Cliff, J. and Hatzenpichler, R., 2022. Correlative SIP-FISH-Raman-SEM-NanoSIMS links identity, morphology, biochemistry, and physiology of environmental microbes. *ISME Communications* 2(1), 52.

Shams, A., Fischer, A., Bodnar, A. and Kliegman, M., 2024. Perspectives on genetically engineered microorganisms and their regulation in the United States. *ACS Synthetic Biology* 13(5), 1412–1423.

Shu, W.-S. and Huang, L.-N., 2022. Microbial diversity in extreme environments. *Nature Reviews Microbiology* 20(4), 219–235.

Stracke, C., Meyer, B.H., De Rose, S.A., Ferrandi, E.E., Kublanov, I.V., Isupov, M.N., Harmer, N.J., Monti, D., Littlechild, J., Müller, F., Snoep, J.L., Bräsen, C. and Siebers, B., 2025. Establishment of an efficient one-step enzymatic synthesis of cyclic-2,3-diphosphoglycerate. *Frontiers in Microbiology* 16, 1601972.

Sui, X., Li, Y., Wang, X., Yu, L., Wang, J. and Ji, H., 2023. Genomics and degradation law of Acinetobacter junii in response to petroleum pollution. *Process Biochemistry* 126, 41–50.

Sukenik, A. and Kaplan, A., 2021. Cyanobacterial harmful algal blooms in aquatic ecosystems: A comprehensive outlook on current and emerging mitigation and control approaches. *Microorganisms* 9(7), 1472.

Tang, K.H.D. and Li, R., 2025. The effects of plastisphere on the physicochemical properties of microplastics. *Bioprocess and Biosystems Engineering* 48(1), 1–15.

Thompson, J.R., Nancharaiah, Y.V., Gu, X., Lee, W.L., Rajal, V.B., Haines, M.B., Girones, R., Ng, L.C., Alm, E.J. and Wuertz, S., 2020. Making waves: Wastewater surveillance of SARS-CoV-2 for population-based health management. *Water Research* 184, 116181.

Van den Wyngaert, S., Ganzert, L., Seto, K., Rojas-Jimenez, K., Agha, R., Berger, S.A., Woodhouse, J., Padisak, J., Wurzbacher, C., Kagami, M. and Grossart, H.-P., 2022. Seasonality of parasitic and saprotrophic zoosporic fungi: Linking sequence data to ecological traits. *ISME Journal* 16(9), 2242–2254.

van Kessel, M.A.H.J., Speth, D.R., Albertsen, M., Nielsen, P.H., Op den Camp, H.J.M., Kartal, B., Jetten, M.S.M. and Lücker, S., 2015. Complete nitrification by a single microorganism. *Nature* 528(7583), 555–559.

van Vliet, M.T.H., Thorslund, J., Strokal, M., Hofstra, N., Flörke, M., Ehalt Macedo, H., Nkwasa, A., Tang, T., Kaushal, S.S., Kumar, R., van Griensven, A., Bouwman, L. and Mosley, L.M., 2023. Global river water quality under climate change and hydroclimatic extremes. *Nature Reviews Earth & Environment* 4(10), 687–702.

Vavourakis, C.D., Andrei, A.-S., Mehrshad, M., Ghai, R., Sorokin, D.Y. and Muyzer, G., 2018. A metagenomics roadmap to the uncultured genome diversity in hypersaline soda lake sediments. *Microbiome* 6(1), 168.

von Meijenfeldt, F.A.B., Hogeweg, P. and Dutilh, B.E., 2023. A social niche breadth score reveals niche range strategies of generalists and specialists. *Nature Ecology & Evolution* 7(5), 768–781.

Wade, M.J., Lo Jacomo, A., Armenise, E., Brown, M.R., Bunce, J.T., Cameron, G.J., Fang, Z., Farkas, K., Gilpin, D.F., Graham, D.W., Grimsley, J.M.S., Hart, A., Hoffmann, T., Jackson, K.J., Jones, D.L., Lilley, C.J., McGrath, J.W., McKinley, J.M., McSparron, C., Nejad, B.F., Morvan, M., Quintela-Baluja, M., Roberts, A.M.I., Singer, A.C., Souque, C., Speight, V.L., Sweetapple, C., Walker, D., Watts, G., Weightman, A. and Kasprzyk-Hordern, B., 2022. Understanding and managing uncertainty and variability for wastewater monitoring beyond the pandemic: Lessons learned from the United Kingdom national COVID-19 surveillance programmes. *Journal of Hazardous Materials* 424, 127456.

Wang, C., Oliver, E.E., Christner, B.C. and Luo, B.H., 2016. Functional analysis of a bacterial antifreeze protein indicates a cooperative effect between its two ice-binding domains. *Biochemistry* 55(28), 3975–3983.

Wang, J., He, B., He, J., Sun, Z., Liu, J., Liu, H., Guo, Y., Zhou, Z. and Wang, X., 2025a. Process-based quantitative description of carbon biogeochemical cycle in a reclaimed water intake area. *Environmental Research* 267, 120716.

Wang, W., Wang, R., Li, Y., Li, Y., Zhang, P., Gao, M., Cao, Y., Fohrer, N., Zhang, Y. and Li, B.L., 2025b. Cross-sectional-dependent microbial assembly and network stability: Bacteria sensitivity response was higher than eukaryotes and fungi in the Danjiangkou Reservoir. *Journal of Environmental Management* 379, 124851.

Wingert, C.J. and Cochlan, W.P., 2021. Effects of ocean acidification on the growth, photosynthetic performance, and domoic acid production of the diatom *Pseudo-nitzschia australis* from the California Current System. *Harmful Algae* 107, 102030.

Wolski, E.A., 2023. The versatility of *Penicillium* species to degrade organic pollutants and its use for wastewater treatment. *Studies in Fungi* 8(1), 1–10.

Wu, X., Yang, X., Dai, Y., Zhao, Z., Zhu, J., Guo, H. and Yang, R., 2024. Single-cell sequencing to multi-omics: Technologies and applications. *Biomarker Research* 12(1), 110.

Xiao, X.L., Guo, W.D., Li, X.L., Wang, C., Chen, X.W., Lin, X.Q., Weinbauer, M.G., Zeng, Q.L., Jiao, N.Z. and Zhang, R., 2021. Viral lysis alters the optical properties and biological availability of dissolved organic matter derived from *Prochlorococcus* Picocyanobacteria. *Applied and Environmental Microbiology* 87(3), e02271–02220.

Xu, D., Zheng, G., Brennan, G., Wang, Z., Jiang, T., Sun, K., Fan, X., Bowler, C., Zhang, X., Zhang, Y., Wang, W., Wang, Y., Li, Y., Wu, H., Li, Y., Fu, F.-X., Hutchins, D.A., Tan, Z. and Ye, N., 2023. Plastic responses lead to increased neurotoxin production in the diatom *Pseudo-nitzschia* under ocean warming and acidification. *ISME Journal* 17(4), 525–536.

Xu, Y., Liu, G., Hua, Y., Wan, X., Hu, J., Zhu, D. and Zhao, J., 2020. The diversity of comammox bacteria and the effect of sewage discharge on their abundance in eutrophic lake sediments. *Journal of Soils and Sediments* 20(5), 2495–2503.

Xue, Z., Wang, L., Geng, X.Y. and Guo, X.F., 2023. Research progress on fungi in lake ecosystems and their roles. *South China Agriculture* 17(19), 70–75.

Yan, Q., Ming, Y., Liu, J., Yin, H., He, Q., Li, J., Huang, M. and He, Z., 2025. Microbial biotechnology: From synthetic biology to synthetic ecology. *Advanced Biotechnology* 3(1), 1.

Yang, L.W., Wang, Z.A., Geng, R., Deng, H.W., Niu, S.W., Zuo, H.L., Weng, S.P., He, J.G. and Xu, X.P., 2023. *White Spot Syndrome Virus* (WSSV) inhibits hippo dignaling and activates Yki to promote its infection in *Penaeus vannamei*. *Microbiology Spectrum* 11(1), e02363–22.

Yang, M.L., Liu, N., Wang, B.L., Li, Y.J., Li, W.Z., Shi, X.J., Yue, X.R. and Liu, C.Q., 2024a. Stepwise degradation of organic matters driven by microbial interactions in China's coastal wetlands: Evidence from carbon isotope analysis. *Water Research* 250, 121062.

Yang, T., Nian, Y., Lin, H., Li, J., Lin, X., Li, T., Wang, R., Wang, L., Beattie, G.A., Zhang, J. and Fan, M., 2024b. Structure and mechanism of the osmoregulated choline transporter BetT. *Science Advances* 10(33), eado6229.

Yang, Y.T., Wang, J.Z., Tian, Y., Li, M., Xu, S.H., Zhang, L.J., Luo, X.W., Tan, Y.H., Liang, H. and Chen, M., 2024c. Equisetin protects from atherosclerosis *in vivo* by binding to STAT3 and inhibiting its activity. *Pharmacological Research* 206, 107289.

Yu, H.Y., Wang, S.G. and Xia, P.F., 2022. Reprogramming microbial CO_2-metabolizing chassis with CRISPR-Cas systems. *Frontiers in Bioengineering and Biotechnology* 10, 897204.

Zhang, Y.M. and Qiao, J.J., 2017. The acid-tolerant pH balance mechanism of acidophilic bacteria and its potential applications. *Chinese Journal of Bioengineering* 37(12), 8.

Zhang, Z., Lu, J., Zhang, S., Tian, Z., Feng, C. and Liu, Y., 2024. Analysis of bacterial community structure, functional variation, and assembly mechanisms in multi-media habitats of lakes during the frozen period. *Ecotoxicology and Environmental Safety* 284, 116903.

Zhao, W., Ma, X., Liu, X., Jian, H., Zhang, Y. and Xiao, X., 2020. Cross-stress adaptation in a piezophilic and hyperthermophilic archaeon from deep sea hydrothermal vent. *Frontiers in Microbiology* 11, 2081.

Zhao, W. and Xiao, X., 2023. Microbiome and environmental adaption mechanisms in deep-sea hydrothermal vents. *Scientia Sinica Vitae* 53(5), 660–671.

Zhou, P., He, X.-Q., Xu, P., Zhang, D.-S. and Wang, C.-S., 2025. Exploring environmental adaptation mechanisms of symbiotic microorganisms in marine reducing ecosystems: Harnessing genomic comparison to unveil the underlying mechanisms. *Frontiers in Marine Science* 12, 1571722.

Zhu, W., Wang, D., Li, P., Deng, H. and Deng, Z., 2025. Advances in wastewater-based epidemiology for pandemic surveillance: Methodological frameworks and future perspectives. *Microorganisms* 13, 1169.

Zhu, Y.-G., Johnson, T.A., Su, J.-Q., Qiao, M., Guo, G.-X., Stedtfeld, R.D., Hashsham, S.A. and Tiedje, J.M., 2013. Diverse and abundant antibiotic resistance genes in Chinese swine farms. *Proceedings of the National Academy of Sciences* 110(9), 3435–3440.

9 Microbial Interactions and Ecosystem Dynamics
From Symbiosis to Pathogenesis

Basit Akolade Adigun, Tapos Kormoker, Yan Laam Cheng, Yee Ming Siu, Ki-Hyun Kim, and Yiu Fai Tsang

9.1 INTRODUCTION

Microbial interactions are not static but are dynamic, and they are context-dependent processes that fundamentally drive ecosystem stability, function, and responses to change. Microbes are a crucial part of every environment; they have been found in nearly all environments, and their influence is of great value to various ecosystems (Mydeen et al., 2023). Microbes in different ecosystems form a microbiome consisting of various species that interact with one another; likewise, they interact with other macroscopic organisms and the abiotic components of their environment. Numerous microbiomes can be found in various habitats, including the soil microbiome, river water microbiome, marine water microbiome, as well as within living organisms, such as the gut microbiome, skin microbiome, oral microbiome, and root or rhizosphere microbiome (Moënne-Loccoz et al., 2014; Hacquard et al., 2015). Microflora in plants and animals perform various functions, including nutrient acquisition, digestion, the secretion of vitamins and hormones, and protection against pathogens (Chen et al., 2024; Prakash et al., 2014). Microbial communities are one of the ecosystem's most valuable and versatile components, playing a crucial role in many processes that sustain ecosystem function. Their role ranges from serving as agents in nutrient cycling via biogeochemical processes, breaking down and extracting organic matter, and causing various infectious diseases (Chen et al., 2024; Singh et al., 2023).

This chapter provides a detailed account of microbial interactions and their role in ecosystem dynamics. It delves into explaining the basic concept of interactions in biomes involving microbial communities and the factors driving these interactions. Drawing cues from published scientific research to elaborate on these interactions and occurrences that reflect how a network of microbial interactions has been implicated on a larger scale, leading to various ecosystem dynamics.

9.2 THE SPECTRUM OF MICROBIAL INTERACTION

Generally, microbial interaction cuts across different associations or relationships, ranging from symbiosis to pathogenesis. Interaction amongst living organisms can be beneficial or positive if it benefits all or some of the participants (e.g., mutualism, commensalism, and cooperation). Or it may be conflictual or negative if it's detrimental to all or some of the participants (e.g., amensalism and competition). Predation and parasitism can be grouped in the intercept of beneficial and conflictual, as it involves organisms that get benefit at the expense of the other. In this chapter, predation and parasitism are grouped under conflictual or negative interaction. Symbiotic interactions can take any of these forms, whereas pathogenesis involves a pathogen that causes diseases in the other organism. These two associations (i.e., symbiosis and pathogenesis) can be dynamic; a transformation or transition may occur between them depending on specific factors and conditions (Figure 9.1).

DOI: 10.1201/9781003640899-12

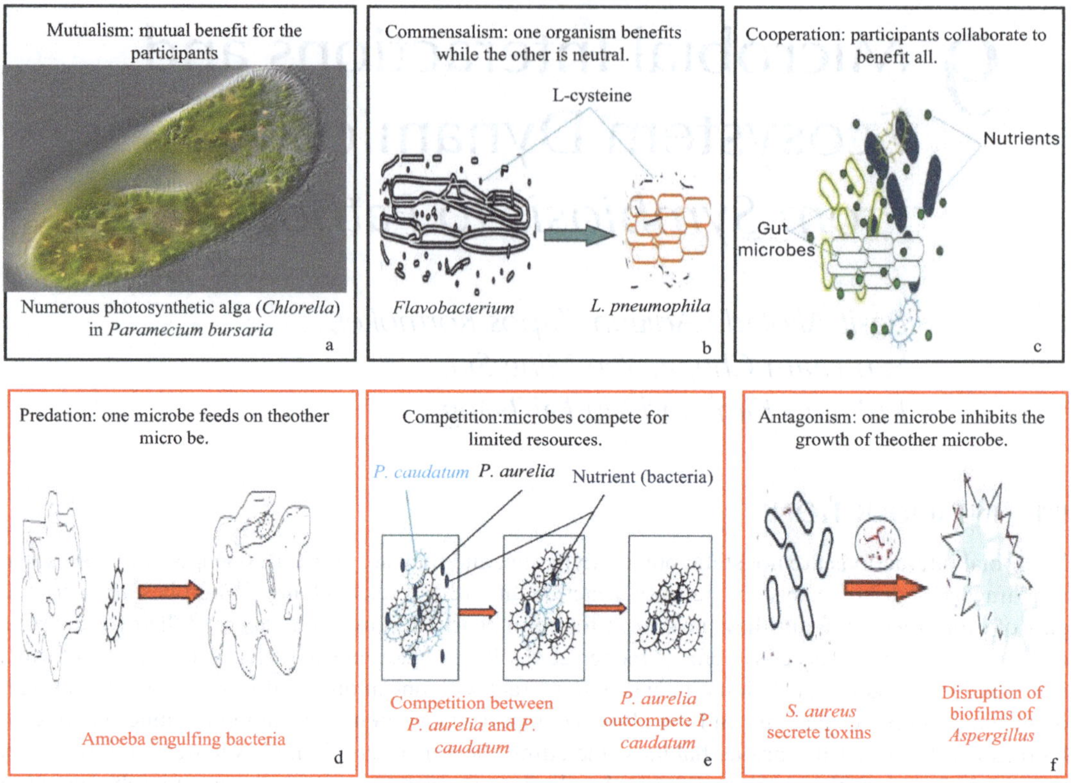

FIGURE 9.1 Examples of microbial interaction (a) mutualism between *Chlorella vulgaris* and *Paramecium bursaria*. (Adapted from Kreutz, 2025) (b) Commensalism between *Flavobacterium* and *Legionella pneumophila* (c) cooperation between the microbiome to secrete essential nutrients, e.g., vitamins, purines, and amino acids (d) predation in *Amoeba* (e) competition between *Paramecium aurelia* and *Paramecium caudatum* (f) antagonism between *Staphylococcus aureus* and *Aspergillus fumigatus*.

9.3 BENEFICIAL OR POSITIVE INTERACTIONS

9.3.1 Mutualistic Interaction

Two or more different organisms living in the same environment and having positive influences on each other. This association is a reciprocal benefit amongst the participants (Marchal et al., 2017). Examples of such interactions include those between animals and their gut microbiome, where the gut microbes perform crucial roles in digestion, nutrient absorption, immunity, metabolism, and pathogen inhibition (Lin et al., 2024). Cooperative interactions have been reported amongst various microbes, where microorganisms collaborate to obtain the necessary materials for their growth and support one another. Cooperation can occur as indirect positive influences among microbes residing in the same habitat, which may take the form of cross-feeding, detoxification, degradative synergy, and the production of public goods (Di Martino et al., 2024; Peterson et al., 2020). Cross-feeding, a form of cooperative interaction in which a microorganism relies on the product of another microorganism (Solowiej-Wedderburn et al., 2025), has been observed between *Acetobacter fabarum* and *Lactobacillus brevis* in the *Drosophila* gut (Sommer & Newell, 2019). Likewise, an interaction that includes concurrent competition for linoleic acid as a carbon source and cooperation through detoxification between *Agrobacterium tumefaciens* and *Comamonas testosteroni* has been reported (Di Martino et al., 2024). In this interaction, *C. testosteroni* plays a crucial role by detoxifying reactive oxygen species (ROS) that accumulate at higher concentrations of linoleic acid, thereby rescuing the ROS-sensitive *A. tumefaciens* (Di Martino et al., 2024).

The interaction between *Rhizobium* and the roots of legumes, in which legumes are provided with utilizable nitrogen forms (nitrate and ammonia), and the *Rhizobium* (bacteria) benefit through photosynthates supplied by the plant, is an example of mutualism. Similarly, in actinorhizal plants, some non-leguminous plants form a symbiotic association with a specific nitrogen-fixing bacterium called *Frankia*. *Frankia* helps transform atmospheric N_2 into usable nitrogen for the plant while getting carbohydrate supply from the plant (Narayanasamy et al., 2020). Grass and endophytic fungi association; the fungi enhance the growth of the host (grass) through tiller formation, conferring insect resistance and drought tolerance, and toxic responses in herbivores, while the fungi receive nutrients and accommodation from the host plant (Adigun et al., 2024; Bacon & Hill, 1996). Fungi and algae in lichen; fungi absorb water and protect the algae while algae photosynthesize and supply carbohydrates to the fungi. Mycorrhiza is an association between plant roots and fungi similar to that of lichens. The plant increases its absorption of water and minerals through the fungus, while the fungus obtains carbon sources from the plant. *Paramecium bursaria* cells harbour hundreds of symbiotic *Chlorella vulgaris* in their perialgal vacuole suspended in their cytoplasm (Kodama & Fujishima, 2010). The alga (*Chlorella vulgaris*) provides carbon and oxygen to the protist (Paramecium), which in turn provides carbon dioxide, protection, mobility, and shelter to the alga cells (Karki, 2018).

9.4 COMMENSALISM

These are interactions in which one organism benefits while the other is neutral, i.e., neither gains nor loses from the association (Ghosh, 2013). During ecological succession, activities of early succession organisms do not benefit the organism themselves, but they prepare the habitat for a succeeding species, which can be commensalism. In the intestinal tract, the facultative anaerobic bacterium *Escherichia coli* utilizes oxygen and creates anaerobic conditions that support the obligatory anaerobe *Bacteroides* (Jones et al., 2007; Moënne-Loccoz et al., 2014). Within the nitrogen cycle, ammonia-oxidizing bacteria (AOB) oxidize ammonia (NH_3) to yield nitrite (NO_2), which serves as a source of nutrients for the nitrite-oxidizing bacteria (NOB) and further oxidize NO_2 to produce nitrate (NO_3) (Daims et al., 2016; Karki, 2018). The AOB supplies the NO_2 to the NOB and does not require the NOB for its metabolism, resulting in a communalistic interaction. The role of NOB helps prevent nitrogen loss in the ecosystem by fixing NO_2 into NO_3, which serves as a usable nitrogen source for living organisms. Its effect can be as significant as fixing over 80% of nitrogen in the marine ecosystem (Daims et al., 2016; Gruber, 2008). In addition, some commensal microflora that inhabit the surface of animal skin can prevent colonization of pathogens in this habitat by creating a simple physical barrier (Moënne-Loccoz et al., 2014). In an aquatic environment, *Flavobacterium* secretes L-cysteine, which is a growth requirement for *Legionella pneumophila*. In nutrient-deficient environments, such as biofilms and hot-water tanks, *Flavobacterium* supports and enhances the survival of *L. pneumophila* by releasing L-cysteine (Ewann & Hoffman, 2006; Karki, 2018; Wadowsky & Yee, 1983).

9.5 CONFLICTUAL OR NEGATIVE INTERACTIONS

9.5.1 Competition

Competition is a type of interaction that stems from a concurrent need by two or more organisms for a limited environmental resource, such as water, nutrients, oxygen, dwelling space, or light (Moënne-Loccoz et al., 2014). Different ecosystems often harbour organisms with similar metabolic demands. For instance, soil bacteria and fungi can metabolize plant-based macromolecules (such as pectin, cellulose, and hemicellulose), and they may compete for these resources (Moënne-Loccoz et al., 2014). *Pichia guilliermondii* engaged in competitive interaction for nitrate and carbohydrates with *Botrytis cinerea* on apples or with *Colletotrichum* spp. on peppers (Fenta et al., 2023; Spadaro & Droby, 2016). As revealed by Gause (1934) *Paramecium caudatum* and *Paramecium aurelia* compete for the same resources (bacteria as food) when found in the same habitat. Typically, *P. aurelia* outcompetes *P. caudatum* and may potentially exterminate the *P. caudatum* population, a phenomenon described as competitive exclusion (Gause, 1934; Karki, 2018; Braude et al., 2010; Müller et al., 2012). However, *Paramecium bursaria* will survive this interaction if found in the same environment as *P. caudatum*, because its niche doesn't overlap with that of *P. caudatum* (i.e., resource partitioning). *P. bursaria* stays at the bottom and therefore feeds on the bacteria in its surroundings together with its endosymbiont algae (*Chlorella vulgaris*) that supply it with oxygen and organic carbon (Braude et al., 2010; Müller et al., 2012).

Competition is highly probable and stronger among individuals within the same population due to niche overlap and similar needs, whereas competition may be less intense among individuals from different populations. Solowiej-Wedderburn et al. (2025) demonstrated that microorganisms formed cooperative interactions in environments with limited resources, but not in environments with abundant resources. The observed trend by Solowiej-Wedderburn et al. (2025) is necessary for the survival of microbes under limited resources, and it is consistent with the stress gradient hypothesis, which implies that species interactions are more beneficial and cooperative under stressful conditions, whereas competition dominates during less stressful conditions (Bertness & Callaway, 1994).

9.6 ANTAGONISM OR AMENSALISM

This is an interaction in which one organism secretes substances that limit the growth or metabolism of the other. The production of antibiotics by microbes to inhibit the growth of the other population is termed antibiosis (Nehra et al., 2022; Hutchings et al., 2019). Antibiotics are compounds produced by microbes to kill other microbes (Hutchings et al., 2019). Ordinarily, microbes produce antibiotics to act as chemical agents that attack competitors, either defensively or offensively, to send signals to close relatives, or to facilitate interaction with eukaryotic hosts (Hutchings et al., 2019). Microbial antagonism has been utilized in the production of antibiotics to control plant and animal diseases. Many antibiotics were discovered during the early 20th century and have been widely accepted and used in medical applications to treat various ailments (Hutchings et al., 2019; Peterson et al., 2020). Amongst the earliest identified antibiotic-producing microbes are *Penicillium*, *Aspergillus*, *Trichoderma* spp., *Streptomyces*, *Pseudomonas*, and *Bacillus* (Nehra et al., 2022; Hutchings et al., 2019). Soil microbes such as *Streptomyces*, *Bacillus*, *Trichoderma*, and *Pseudomonas* have been utilized in biocontrol of phytopathogens as they are capable of synthesizing different antibiotic compounds (Fenta et al., 2023). A broad-spectrum antibiotic compound with antibacterial, antifungal, and nematicidal activity from bacterial species belonging to the genera *Bacillus* and *Streptomyces*, with an antagonistic effect on phytopathogens in the desert ecosystem, was reported by Köberl et al. (2013). *Bacillus* spp. produce antibiotics such as bacillomycin and fengycin, which are antifungal in nature; gramicidin has antibacterial properties, and surfactin has broad-spectrum antimicrobial properties and biosurfactant (Fenta et al., 2023). *Staphylococcus aureus* antagonizes biofilm formation in *Aspergillus fumigatus* through physical interactions and the secretion of specific metabolites (Ramírez Granillo et al., 2015). Other researchers found that *S. aureus* secretion contains substances, such as peptidoglycan fragments and staphylokinase, which can influence biofilm formation and metabolism of *Candida albicans*, supporting the antifungal properties of secretion from *S. aureus* (Eichelberger & Cassat, 2021).

Soil bacteria with antagonistic properties have been studied to understand their interactions within the same population, specifically *Streptomyces sp.*, which secretes small antimicrobial molecules, and *Pseudomonas fluorescens*, which produces bacteriocins (Bruce et al., 2017; Kinkel et al., 2014; Peterson et al., 2020). *Streptomyces sp.* mixture from the same location showed stronger antagonistic interactions than the interaction between geographically isolated strains of *Streptomyces sp.* (Kinkel et al. 2014). Although persistent competition in *P. fluorescens* with an overlapping niche has been demonstrated, increased antagonistic interactions from isolates from different sites compared to those from identical sites, while genetic similarity does not promote inhibitory interactions (Bruce et al., 2017; Peterson et al., 2020). The magnitude of difference among the sites in the *P. fluorescens* experiment was measured in meters, which is very small compared to the nationwide-scale sampling utilized in the Streptomyces experiment (Bruce et al., 2017; Kinkel et al., 2014; Peterson et al., 2020). Notably, it was found in both bacterial species that the capacity to withstand antagonism was far more common than the ability to suppress a different strain (Peterson et al., 2020).

Allelopathic stimulators of ginsenosides on *Pythium irregulare*, *Cylindrocarpon destructans*, *Phytophthora cactorum*, and *Fusarium solani* soilborne phytopathogens of *Panax* were reported (Yang et al., 2015). A similar observation has also been found in medicinal plants, where autotoxicity preferentially stimulates the abundance of pathogenic microbes over beneficial ones (Haq et al., 2025). Autotoxins produced by grapevine roots altered the fungal and bacterial communities in the rhizosphere, which was found to enhance the abundance of potential pathogenic microbes and inhibit the abundance of beneficial strains (Liu et al., 2022). The interaction between these plants and soilborne microbial pathogens, mediated by the discharged autotoxins in the rhizosphere, has a detrimental effect on the growth, development, and health of the plants, while also facilitating pathogenic interactions and infectious diseases.

9.7 PREDATION AND PARASITISM

These types of interactions are only favourable to one group of organisms (the predator and the parasite), while the other participants (the prey or the host) are at a disadvantage. In predation, the predator feeds on the prey, typically resulting in the prey's death. Similarly, in parasitism, the parasite depends on the host, lives in or on the host, and feeds on it while the host is alive. Viruses are one of the major microbial parasites and they are obligate intracellular parasites. Different organisms (bacteria, fungi, algae, protozoans, plants, and animals) have their specific viral inhabitant. In soil ecosystems, protozoans are the major microbial predators (Moënne-Loccoz et al., 2014). Specifically, amoebae have been reported as the most abundant bacterial predator protozoans in soil habitats (Rossine et al., 2022). An amoeba feeds on smaller bacteria or algae in its environment by engulfing the prey and forming a food vacuole for its intracellular digestion. Through feeding, amoebae can serve as biocontrol agents for plant pathogens; however, considering the diverse interactions among different species of amoebae and bacteria, the relationship between amoebae and bacteria can range from mutualistic to parasitic to predatory (Dinda et al., 2024). Also, amoebae have been found to feed on plant pathogenic microbes (fungi and bacteria), thereby suppressing pathogen growth, helping maintain plant health, and reducing the spread and outbreaks of associated infectious diseases (Dinda et al., 2024; Long et al., 2019). The amoeba-bacteria predator-prey interaction influences the bacterial community and nutrient mineralization in soil (Rossine et al., 2022). Mycophagous bacteria and mycoparasitic fungi feed on fungi, and *Bdellovibrio spp.* feeds on root-associated bacteria (Hassani et al., 2018). Some actinobacteria and proteobacteria species were found to feed on other bacteria and fungi (Moënne-Loccoz et al., 2014). Likewise, Paramecium feeds on bacteria while bacteriophage viruses parasitize and feed on bacteria. The bacterium *Pseudomonas tolaasii* parasitizes the fungus *Agaricus bisporus*, which forms browning spots on carpophores, leading to brown blotch disease (Moënne-Loccoz et al., 2014).

Additionally, parasitic interactions can lead to infection or disease, and all pathogens can also be referred to as parasites (Olano et al., 2011). *Plasmodium* and *Trypanosoma* are infectious protozoans that are typically carried by insect vectors, specifically female *Anopheles* mosquitoes and tsetse flies, respectively, and transmitted to their mammalian hosts, where they cause infectious diseases. *Trypanosoma brucei* causes African sleeping sickness in humans, livestock, and wild animals (Hill, 2003; Pollitt et al., 2011), while *Plasmodium falciparum* is responsible for malaria in humans, which was described as the deadliest parasitic disease affecting humans (Kaiser, 2023). Predation and parasitism often lead to undesired consequences for the host and prey. However, they are crucial to a functioning ecosystem, as they are integral to food chains and food webs through which nutrients are transferred from one feeding level to another. Through feeding, predators can help control the population of their prey, thereby maintaining balance in the ecosystem. Additionally, parasitism is linked to various diseases. Understanding the interaction between parasites and their hosts will contribute to offering solutions to the associated diseases.

9.8 PATHOGENESIS: A CASE STUDY IN HOST-MICROBE INTERACTION

Microbes are responsible for a significant number of diseases in living organisms, but not all diseases are associated with microbial influences. An organism's disease and health states are closely linked to its microbiome, interactions, growth, and pathogenicity (Braga et al., 2016; MacAlpine et al., 2023). Pathogenesis is the process by which a disease develops in a living organism. It encompasses all the stages a pathogen undergoes until the disease becomes apparent in the victim. For a pathogen to successfully cause disease, it must go through four key stages: contact or exposure, adhesion or attachment, colonization, invasion, and infection (Liu et al., 2025; Soni et al., 2024). Afterwards, the pathogen escapes from the current host, is transmitted to another host, and initiates another cycle (Liu et al., 2025; Soni et al., 2024).

Exposure or contact is easy for microbes due to their ubiquitous availability in the environment, microscopic size, and ability to transmit through an intermediate host. Adhesion is the ability of a pathogen to attach to the host cell, and this usually involves the use of adhesins (molecular substances that can be proteins or carbohydrates) of the pathogen to bind to specific receptor proteins of the host (Liu et al., 2025). The nature of the interaction between the pathogen and host cells depends largely on the interplay between the adhesin produced by the pathogen and the receptor of the host cells (Nash et al., 2015). Highly humid environments or the presence of moisture and high temperatures enhance pathogen penetration. After penetration, the pathogen interacts with host cells to bypass the host's defence system, relying on nutrients from the host (Cohen et al., 2016). Microbial pathogen interactions for acquiring nutrients from the host include the use of siderophores to sequester iron *by Salmonella* and the secretion of soluble reductants for extracting iron from the host's transferrin, as seen in *Listeria monocytogenes* (Nash et al., 2015). Invasion is the multiplication of the pathogen through reproduction to increase its numbers and to colonize host cells or tissues (Liu et al., 2025). As this interaction continues, the pathogen releases toxins and exoenzymes to damage the host tissue and facilitate the spread of pathogen cells throughout the organism (Liu et al., 2025). The successful invasion and multiplication of the pathogen lead to infection, which can be either local or systemic and symptomatic or asymptomatic, depending on the virulence level of the pathogen (Liu et al., 2025; Soni et al., 2024). Continuous damage to the host caused by the pathogen can lead to diseases and even death in extreme cases (Casadevall & Pirofski, 2000).

9.9 HOST RESPONSE TO PATHOGENIC INTERACTIONS

Throughout the stages of pathogenesis, the host-pathogen interaction is a continuous and dynamic process; as the pathogen releases toxins and enzymes to exert its virulence, the host activates its defence mechanisms and antimicrobial responses against the pathogen (Jo, 2019). One of the early immune system defences in animals is the action of phagocytic cells, which engulf, ingest, and degrade invading pathogens, such as leukocytes (white blood cells). Furthermore, antimicrobial substances and antibody-based responses are triggered in animals.

Phytopathogens release different proteins to weaken the plant's defence system, which are countered by a series of defence responses from the plant (Chisholm et al., 2006). After infection, the plant secretes secondary metabolites that are harmful to pathogens, such as phytoalexins. For example, ferulic acid, sinapic acid, *p*-hydroxybenzoic acid, and *p*-coumaric acid are produced by date palm against *Fusarium oxysporum* infection (Binyamin et al., 2019; El Modafar & El Boustani, 2001), and pisatin in pea against *Ascochyta pisi* (Binyamin et al., 2019; Morkunas et al., 2013). Ethylene, salicylic acid (SA), and jasmonic acid (JA) are plant signalling chemicals used in defence against pathogens, wounding, and herbivory (Chisholm et al., 2006; Kunkel & Brooks, 2002). SA-mediated signalling protects against local and systemic bacterial diseases, while JA-mediated signalling is responsible for wounding and herbivory defence. JA- and SA-mediated signalling interactions are mutually antagonistic (Chisholm et al., 2006; Kunkel & Brooks, 2002). *Pseudomonas* interacts with the plant by exploiting the antagonistic nature of the SA and JA signalling relationship, secreting coronatine that mimics JA to antagonize and prevent the SA-mediated plant defence system (Chisholm et al., 2006). Other phytohormones are also involved in the interaction between the plant and the pathogen. Increased auxin production in the plant promoted some pathogenic disease signs and symptoms, perturbation in host defence, and proliferation of the pathogen (Binyamin et al., 2019; McClerklin et al., 2018). In addition, elevated auxin increased plant susceptibility to pathogens and hampered the plant's SA, JA signalling, and other defensive metabolites (Ye et al., 2019). IAA and benzoxazinoids are associated with plant growth and defence; however, IAA's role in defence may be compromised, whereas benzoxazinoids are likely to enhance defence against infection (Ye et al., 2019). Regulated phytohormone production induced plant disease resistance (Binyamin et al., 2019; Tezuka et al., 2019). SA facilitates disease resistance, while JA, Ethylene, auxin, abscisic acid, Gibberellic acid, and brassinosteroids can have a dual role, either enhancing disease resistance or inducing vulnerability to disease symptoms, depending on the interaction (De Bruyne et al., 2014; Di et al., 2016). Pluriform function of some phytohormones in pathogenic infection (i.e., either enhancing the defence system or promoting disease symptoms) depends on the pathogen's lifestyle, environmental factors, infection methods, stage of interaction, specialized characteristics of interaction, and individual plant involved (De Bruyne et al., 2014; Di et al., 2016).

9.10 THE TRANSITION BETWEEN SYMBIOSIS AND PATHOGENESIS

The transition between symbiosis and pathogenesis is context-dependent, driven by factors ranging from environmental, developmental, and evolutionary processes to host responses and the community composition of the interacting species (Drew et al., 2021). A symbiont could become pathogenic due to factors that compromise the immunity of the other partner, making it vulnerable; likewise, a pathogen can become a symbiont (Overstreet & Lotz, 2016; Taylor et al., 2012). *Candida albican* is a notable commensal and opportunistic asymptomatic colonizer of human gastrointestinal and urogenital mucosa, which switches to an infectious partner due to disturbances in the host's bacterial microbiome, immune suppression, or antibiotic use (Huffnagle & Noverr, 2013; Christian et al., 2015; Lilic, 2012; MacAlpine et al., 2023). Coinfection involving a pathogen and a mutualist partner leads to an outcome that depends on the outcome of the counteraction between the virulent impact of the pathogen and the ameliorative effect of the mutualist, whichever is greater, determining whether the shift is towards pathogenesis, mutualism, or commensalism (Nelson & May, 2017). Mutations leading to disturbances in the carbon catabolite repression pathway transform a free-living *E. coli* into a mutualistic partner in insects (Koga et al., 2022).

9.11 MECHANISMS AND DRIVERS OF MICROBIAL INTERACTIONS

Microbial interactions can result from their structural adaptations (haustoria, conidia, and chlamydospores), the synthesis of certain metabolites (e.g., enzymes and plant hormones) and environmental modifications.

9.12 STRUCTURAL ADAPTATION

In some fungi, growth structures such as mycelia and hyphae are produced, which can extend into their host or nearby organisms, exerting various influences on these organisms. These vegetative growth structures may be highly active in colonizing and establishing fungal species on substrates for some saprophytic fungi and in producing sclerotia for rejuvenation after a dormancy period. In pathogenic fungi, the appressorium and haustorium are specialized structures that enable these fungi to penetrate intercellular spaces, allowing them to absorb nutrients from their hosts for survival and potentially causing diseases (Dean et al., 2012; Mapuranga et al., 2022; Ryder & Talbot, 2015).

Bacteria produce biofilms while interacting with organisms in their environment, which can protect them from unfavourable environmental conditions (Chu et al., 2021; Garnett & Matthews, 2012). Biofilms can help withstand suboptimal conditions, such as extreme dryness, low pH, low nutrients, and predation, by providing a conducive environment for the interaction and exchange of genetic resources and metabolites among individual cells (Dang & Lovell, 2016; Gambino & Cappitelli, 2016). The formation of biofilms enhances growth, survival, resource cycling, dispersal, and resilience to extreme environmental conditions (Yin et al., 2019).

9.13 CHEMICAL SIGNALLING

Interactions may be mediated through the secretion of specific metabolites or chemicals that influence other microbes or organisms in their environment (Cremer et al., 2019). Chemicals secreted by microbes determine the nature of interactions, shape the ecosystem, and fuel ecosystem dynamics. *Pseudomonas aeruginosa* produces quinolones and rhamnolipids through these two groups of chemicals it could drive complex interactions with the host and amongst the microbial community. Quinolones are used for the expression of their virulent factor, while rhamnolipids serve as biosurfactants that help in maintaining communication interactions within the biofilms (Garg et al., 2017). Prodigiosin secreted by *Serratia marcescens* facilitates the invasion of its host (*Mucor irregularis*) by rendering the cell wall of the host permeable (Hazarika et al., 2020; Khalid & Keller, 2021).

The antifungal properties of lipopeptides produced by *Bacillus* have been established in different fungal species (Markelova & Chumak, 2025). However, lipopeptides produced by *Bacillus subtilis* C2 strain stimulate the production of chlamydospores in fungi (*Gliocladium roseum* and *Trichoderma harzianum*) at relatively low concentrations (0.02 mg/mL) while causing disruption of the spores at higher concentrations (0.05 mg/mL) (Khalid & Keller, 2021; Li et al., 2005). Therefore, the concentration of the lipopeptides produced by *Bacillus* determines the direction of its interaction with these fungal species. The addition of acylhomoserine lactone in bacteria has been found to significantly inhibit antagonistic interactions by manipulating the synthesis of antimicrobial metabolites (Armes et al., 2025). Likewise, the addition of lactonase reduces the antagonistic interaction, fitness, and abundance of *Pseudomonas* strains (Armes et al., 2025).

Microbes can compete in their environment through enhanced features towards harvesting of environmental resources or through the secretion of toxins that can impede the development of competing cells (Kern et al., 2021). The secretion of hydrolytic enzymes, such as protease, β-1,3-glucanase, and chitinase, by *Bacillus* sp. and *Trichoderma* species has been reported (Fenta et al., 2023). Biofilm-producing bacteria can inhibit the biofilm production of competing bacteria. For instance, *Pseudomonas aeruginosa*, which adopts surface blanketing to hinder biofilm formation in *Agrobacterium*, while *E. coli*, which secretes extracellular surfactants to retard biofilm formation in both Gram-positive and Gram-negative bacteria (Chu et al., 2021).

In many crop plants, fungi produce mycotoxins as a result of their interaction with the plant during field growth, harvesting, transportation, storage, and processing. As a result, food crops such as grains, fruits, nuts, and vegetables are contaminated (Moënne-Loccoz et al., 2014). Through mycotoxin production, fungal interactions can cause unpalatable effects in plants, and through feeding, it can lead to health challenges in animals. The production of Indole-3-Acetic Acid (IAA) by plant growth-promoting rhizobacteria (PGPR) acts as a molecular signal to fine-tune or circumvent plant defence systems, enabling effective colonization and fostering a beneficial plant-microbe interaction (Moënne-Loccoz et al., 2014). Likewise, PGPR produces biosurfactants and siderophores, which can increase the bioavailability of heavy metals for rhizobacteria and plants through chelation (Yan et al., 2020).

Similarly, the non-microbial partners can also stimulate the microbial responses during interaction. Plants secrete metabolites when they interact with microorganisms in their surroundings, which also affect microbial communities. The effect of Alder (*Alnus nepalensis)* plant growth in soil was found to enhance bacterial and fungal biodiversity and abundance (Krishna et al., 2020; Mortimer et al., 2015). Allelopathic compounds associated with plant autotoxicity have been shown to selectively enhance the abundance of potential phytopathogens (Haq et al., 2025; Liu et al., 2022; Yang et al., 2015). Different plants under the same growth conditions were found to associate with distinct microbial communities, indicating that the plants affect their associated microbial communities (Berg et al., 2016; Braga et al., 2016). Plant exudates (e.g., strigolactones) are strong factors that influence microbial composition and their interactions with the plant, and this may differ within a single plant depending on its growth stage (Braga et al., 2016). Likewise, earthworm is known to play a crucial role in organic matter formation in soil, which benefits the plant, and this is achievable through the interaction between the earthworm and soil microbiome in a process that has been denoted as "Sleeping Beauty Paradox" (Ahmed & Al-Mutairi, 2022; Medina-Sauza et al., 2019). Certain beneficial soil bacteria lie dormant in the soil waiting for favourable environmental conditions, which are triggered by the glycoprotein-rich mucus of earthworms. This stimulates rapid microbial processes for a brief period, "hot moment," and in a considerably small area of soil, "hot spot," at the microscale of biopores or aggregate, which impacts the drilosphere and soil levels (Ahmed & Al-Mutairi, 2022; Medina-Sauza et al., 2019). The interaction between the soil microbiome and the earthworm leads to the enhancement of organic matter formation, mineralization, and nutrient cycling in the soil ecosystem (Ahmed & Al-Mutairi, 2022). Likewise, in aquatic environments, grazers, such as invertebrates and fish, significantly modify the abundance and composition of microbes through selective feeding and physical interactions (Havird et al., 2022). Feeding by atyid shrimps influenced the composition and abundance of microbial communities, specifically, these grazers enhanced algal population in periphyton (Havird et al., 2022; Sterling et al., 2022).

9.14 PUBLIC GOODS

Public goods are extracellular substances produced by certain microbes whose functions are shared amongst microbes in the same environment (Lindsay et al., 2021; Park et al., 2025). Public goods are not limited to extracellular chemical molecules but include behavioural attributes, communication systems, and physical structures that benefit both producers and non-producers in the same vicinity (Smith & Schuster, 2019). Physical structures like motility structures may facilitate collective movement and promote spread across the environment. Various microbial extracellular enzymes are featured as public goods serving the purpose of degrading larger molecules for the consumption of others in their environment. Public goods could be useful in nutrient acquisition and metabolism, mitigating environmental stressors, protection against antibiotics, attack competitors or hosts, formation of fruiting bodies for dispersal, and multicellular macrocysts for sexual reproduction (Lindsay et al., 2021; Park et al., 2025). *Pseudomonas aeruginosa* alone has been reported to exhibit various public good roles, which include production of extracellular enzymes, siderophore for sequestration of iron, biosurfactant for enhanced motility, antibiotics for competition with other microbial populations, toxin to damage host cells for nutrient acquisition, biofilm formation, and quorum sensing (Smith & Schuster, 2019). Microbes in densely populated ecosystems (e.g., wastewater) interact by sharing public goods amongst themselves to survive antibiotics and heavy metal stress (Park et al., 2023; Park et al., 2025). In gut microbiota, a mixed culture of a public good producer and a non-producer enhances the fitness of the producer species compared to a monoculture of the producer, indicating a mutualistic interaction between the bacterial species (Rakoff-Nahoum et al., 2014). Likewise, production of nutrient-rich public goods (e.g., vitamins, purines, and amino acids) was found to promote cooperative microbial interaction and positive ecosystem dynamics (Marinos et al., 2025).

Quorum sensing is a population density-dependent cell-cell signalling in microorganisms, which is induced when the population reaches a critical density threshold and triggers behavioural changes that express their virulent factors (Abisado et al., 2018). This coordinates interactions between the microbes in this environment, as well as the direction of interactions between the microbial community and its host.

9.15 ENVIRONMENTAL MODIFICATIONS

Similar microbial compositions across environments do not necessarily imply the same interactions or functions, as environmental factors influence how microbes interact (Solowiej-Wedderburn et al., 2025; Zhong et al., 2018). Nutrient availability, temperature, moisture content, humidity, pH, and toxins affect microbial interaction (Cheng et al., 2019). Environmental factors affect both the physiology and the interaction of microorganisms. Zhou et al. (2018) found that temperature and salinity levels were the prominent factors determining the dynamics in bacterial and archaeal community structure, respectively, while inorganic nitrogen and inorganic phosphate were responsible for variation in eukaryotic community structures in the salt pond ecosystem. Stable environmental conditions in the amphibian host gut, specifically pH, facilitate beneficial interaction amongst microbes in the gut as microbial diversity increases, as the gut area increases, which implies more resources for microbes, while the skin microbiome was found to engage in negative interaction, such as modifying the pH to hinder the growth of other microbes (Liu et al., 2024). The fragility and exposed nature of the host skin, making this habitat prone to fluctuations in environmental conditions, is suggested to facilitate antagonistic interactions (Liu et al., 2024).

Most environmental factors affect nutrient availability in the environment. Nutrient availability is a critical factor that determines the nature of microbial interaction, whether collaborative or exploitative. While it is not a rule of thumb, as other factors could influence, Qiao et al. (2024) found rhizosphere bacteria with metabolic similarities under insufficient nutrients expressing increased competition, negative interactions, and niche degradation, while sufficient availability of nutrients facilitates collaborative interaction and niche development in about 70% of cases (Qiao et al., 2024). Inhibitive interaction amongst the bacterial species was more dependent on highly competitive interaction rather than secretion of toxins or other inhibitory secondary metabolites (Qiao et al., 2024). However, abundant simple sugar in the human gut was found to induce competitive interactions among the gut microbiota as they do not cooperate or cross-feed but engage in individualism by competing more for the same resources and shifting the dynamics from cross-feeding and metabolic interdependence to exploitative interactions and reduced cooperation (Marinos et al., 2025). The nutrient type is also a factor that can influence dynamics; in the gut, sugar is not considered a limiting nutrient, and the ease of independent metabolism by microorganisms rendered cooperation useless, while vitamins, amino acids, and other nitrogen-containing nutrient molecules foster beneficial interactions such as commensalism, as they are limiting nutrients and considered expensive products (Marinos et al., 2025).

Soil parent materials are also a factor that determines the dynamics in the soil ecosystem because they influence the soil physicochemical properties, which in turn affect the microbial and other communities within the soil. Soil originating from granite was found to have greater microbial population, diversity, and activity than the other parent materials (tertiary red sandstone and quaternary red earth) in subtropical China (Deng et al., 2015).

Microorganisms have also been found to implement environmental factors as a "weapon" while interacting with other microbes. Initial microbial colonizer inhibits the growth of other microbial taxa in amphibian host skin through pH modification (Liu et al., 2024). Bacteria can modify the pH levels in their environment, which may transform their interactions with other organisms, switching from cooperative to antagonistic, or vice versa (Gorter et al., 2020; Ratzke & Gore, 2018). Also, through pH alterations, bacteria were found to affect bistability, successive growth, extended suicide, and stabilization amongst themselves in the same environment (Ratzke & Gore, 2018). Studies have shown that manipulation through the addition or subtraction of one chemical compound in a microbial community can transform the interaction from competition to cooperative or vice versa (Di Martino et al., 2024; Solowiej-Wedderburn et al., 2025). Linoleic acid concentration has been proven to determine the fate of interaction between *A. tumefaciens* and *C. testosteroni*, whether cooperation or competition (Di Martino et al., 2024).

9.16 CONSEQUENCES OF MICROBIAL INTERACTION FOR ECOSYSTEM DYNAMICS

9.16.1 Microbial Interactions as Drivers of Ecosystem Functioning

Biotic interactions are major drivers of functioning ecosystems, significantly impacting their composition and dynamics. Without adequate consideration of these interactions, we risk erroneous interpretations of the causes of various phenomena and inadequate comprehension of ecosystem functioning (Ho et al., 2016). An experiment comparing the effect of propagules and biotic interactions in 12 microcosms under species invasion found that biotic interactions outweigh the propagule pressure in determining the microbial composition and function (Albright et al., 2020). In the real sense, biotic interactions give rise to characteristics of communities and ecosystems that we cannot comprehend by considering each species separately; it distinguishes a community from a random collection of species (Gorter et al., 2020). Monoculture of each species might not be able to forecast the overall productivity, dynamics, or how resilient they are to biotic and abiotic disturbances (Gorter et al., 2020). Bacterial populations have been found to cooperatively secrete metabolites that cannot be produced individually, contributing to the degradation of their environment and leading to their extinction (Ratzke & Gore, 2018). Specific microbial communities exhibit complex actions like collective motion, electrochemical signalling, and chemical reactions that would be impossible for a single species to accomplish (Gorter et al., 2020). Cooperative interactions between various microbial species with different physiological characteristics are necessary for the breakdown of soil organic matter (Guo et al., 2024; Nazaries et al., 2021). Elevated concentration of a specific enzyme (phosphatase) in arid soil was attributed to the deficiency of the associated mineral (phosphorus) facilitating the cooperative efforts of the residing microbes to source this mineral from organic substrates (Lozano et al., 2014).

The consequence of interaction often extends beyond its direct impact on the participants, as it influences the environment and even determines the fate of subsequent interactions among the participants. The existence of some plant populations depends on their interaction with microbes; mycorrhizal plants coexist with their competitors only when their mycorrhizal fungi symbiont is present, offering a competitive advantage to the host (Müller et al., 2012). This mycorrhizal association enhances soil nutrients and enzyme activities during interspecific competition or improves soil carbon and glomalin-related soil protein during intraspecific competition (Wu et al., 2023). Additionally, competition among *P. fluorescens* with four other bacterial species enhanced the morphological diversity of the bacterial species (Chu et al., 2021).

The interaction involving *Trichoderma* demonstrates how bioactive fungi species can influence ecosystem dynamics through their network of interactions in the soil ecosystem. In the rhizosphere under the influence of plant exudate, the *Trichoderma* is attracted to the plant and subsequently interacts with the plant using its hyphae to attach, penetrate, and colonize the plant root (Fazeli-Nasab et al., 2022; Sood et al., 2020). It also engages in cooperative interactions with other beneficial microbes in the soil to support the growth and development of its host plant (Sood et al., 2020).

Trichoderma spp. can modulate the environment through the secretion of organic acids, which lowers the soil pH and enhances the solubility of macronutrients and micronutrients for plant absorption and metabolism (Vinale et al., 2008). Additionally, through the secretion of plant growth-promoting substances, it stimulates growth and enhances tolerance to various abiotic stressors in plants (Fazeli-Nasab et al., 2022; Vinale et al., 2004; Vinale et al., 2008). Simultaneously, it engages in negative interactions (mycoparasitism, competition, and antibiosis) with harmful microbes in its environment (Fazeli-Nasab et al., 2022; Ren et al., 2020; Sood et al., 2020; Vinale et al., 2008).

9.17 MICROBIAL INTERACTIONS MEDIATE ECOSYSTEM BALANCE

Interactions in the ecosystem can help achieve ecological balance, stability, and functions (Mishra, 2024; Verma et al., 2025). Interactions depicting feeding relationships and energy transfer among organisms, such as food chains and food webs, show how herbivory, predation, and carnivory can help regulate the population of preceding-level organisms in the food chain or web (Correia & Lopes, 2023; Mishra, 2024). Saprophytism exhibited by some microbes helps decompose, purify, and even recycle nutrients within the ecosystem (Correia & Lopes, 2023). Soil microbial communities have been reported to mediate various functions in the soil environment, thereby facilitating ecosystem functioning (Guo et al., 2024; Ma et al., 2018; Nazaries et al., 2021). In the biogeochemical cycle of elements (e.g., nitrogen, sulphur, carbon, phosphorus, and iron), some key roles such as production and oxidation of methane, transformation of inorganic nitrogen, metal sulphide oxidation, iron oxidation, and reduction are exclusively performed by microbes and microbial interactions (Gonzalez & Aranda, 2023; Ho et al., 2016). Biofilm structure enhances the evolutionary process by providing protection for newly developed mutant cells or cells with decreased fitness, allowing them to accumulate more mutations that improve their fitness and resilience to environmental disturbances (Penesyan et al., 2021).

Beneficial microbes, such as mycorrhizal fungi and *Frankia* bacteria, have also been introduced into established communities and have been found to disrupt native microbe-plant interactions, aid in the establishment of non-native plants, and enhance nitrogen fixation (Thakur et al., 2019). Inoculation of a CSY-P13 strain of *Acinetobacter calcoaceticus* in cucumber-grown soil alters the whole bacterial community, facilitates the availability of a wide range of soil enzymes and antioxidants, and mitigates ferulic acid and *p*-hydroxybenzoic acid stress (Custer et al., 2022; Wu et al., 2018). These examples illustrate how the interaction between foreign microorganisms and native organisms in ecosystems can influence the abundance and diversity of communities, thereby facilitating ecological dynamics.

9.18 MICROBIAL INTERACTIONS INFLUENCE ON ECOLOGICAL SUCCESSION

During ecological succession, pioneer colonizers, mainly decomposers (fungi and bacteria), act on available substrates by secreting extracellular enzymes to extract nutrients and detoxify them, thereby supporting themselves and succeeding organisms (Ni et al., 2022). The interaction between exterior fungi in association with photosynthetic microorganisms forming lichens is also observed here, as they form a self-sustaining unit and are hardy, capable of thriving on bare surfaces with limited nutrients (Kolita & Gogoi, 2024). Pioneer plant species (*Chamerion angustifolium* and *Dryas drummondii*) was found growing on a thin layer of soil formed by lichens (National Park Service, 2020). Succeeding these plant species are actinorhizal plants, such as Alder, that can form a symbiotic relationship with Actinobacteria (*Frankia*) and thereby nourish the soil, enhance bacterial biodiversity, improve soil formation, and facilitate the biogeochemical cycle (Krishna et al., 2020; National Park Service, 2020). Actinobacteria (*Streptomyces*), which have been proven to be drought-tolerant and capable of utilizing different carbon sources, were prominently found in environments undergoing succession (Köberl et al., 2013; Lozano et al., 2014). It has been demonstrated that the bacterial community in agricultural land exhibits higher biodiversity and supports a functioning ecosystem and plant health compared to the adjacent soil in a desert ecosystem (Köberl et al., 2013). Animal guts from the infant stage are colonized by bacteria as the pioneer colonizer, which then develop with time to accommodate more microbes to form the gut microbiome (Lin et al., 2024; Shao et al., 2024). Dynamics in animal gut microbiota have been observed as these microbiomes change with the interaction and influences of other factors such as age, diet, probiotics, supplements, and genotype (Du et al., 2023).

9.19 ALGA BLOOM: AS A CASE STUDY OF MICROBIAL INTERACTIONS AND ECOSYSTEM DYNAMICS

An algal bloom is a hotspot zone of microbial interactions in an aquatic ecosystem. The cumulative effect of these interactions determines whether the dynamics of the algal bloom can be either sustained or disrupted. A series of biotic interactions occurs within the phycosphere (a biochemically fortified microenvironment), which is highly influenced by the secretion of these microbes (Cirri & Pohnert, 2019) and growth factors such as dissolved oxygen, pH, and light intensity (Zhang et al., 2018a&b). The algae select their interaction and partner through modification of their discharge type and concentration (Abate et al., 2024). The availability of algal-derived organic carbon impacts the metabolic activities of heterotrophic bacteria in the phycosphere (Abate et al., 2024; Hoke et al., 2021; Yarbro et al., 2024). These bacteria, in turn, facilitate nutrient (nitrogen, phosphorus, sulphur, and carbon) cycling, organic matter transformation, production of vitamins and antibiotics, growth hormone biosynthesis, digestion of polysaccharides, and fatty acid metabolism (Abate et al., 2024; Cirri & Pohnert, 2019; Hoke et al., 2021; Krausfeldt et al., 2024; Yarbro et al., 2024; Zhang et al., 2018a). While these dynamics and interactions are ongoing on the surface of the aquatic ecosystem as the niche develops, they have a significant impact on life beneath this layer, as they affect water quality and deplete growth requirements (Zhang et al., 2018a).

At the bloom terminated stage, the interactions within this ecosystem changes, and the roles and interactions of these mutualistic microorganisms become detrimental, leading to the decline in the development of this ecosystem. The presence of *p*-coumaric acid, indicating the senescence stage of the algae, signals this transformation (Abate et al., 2024). Mutualistic bacteria (e.g., *Kordia algicida*) can transition into antagonists by activating algicidal properties, such as inducing oxidative stress and secreting antibiotics and lysis enzymes (Cirri & Pohnert, 2019; Abate et al., 2024). Alternatively, some bacteria (e.g., *Pseudomonas protegens*) hunt, surround, and deflagellate motile algal cells (*Chlamydomonas reinhardtii*) (Aiyar et al., 2017; Cirri & Pohnert, 2019). These antagonistic interactions release different macromolecules and fewer nutrients into the environment, which enhances heterotrophy in the bacterial population (Buchan et al., 2014; Zhou et al., 2018). Actinobacteria nutrient opportunists (Cyanobacteria and Flavobacteriaceae) decompose the dead alga mass and clean off organic matter from the dead algae (Palenik, 2015; Zhou et al., 2018; Zhu et al., 2023). Fungal species from the Ascomycota and Basidiomycota also collaborate to facilitate decomposition through saprotrophic interactions (Zhou et al., 2018). These algal bloom occurrences feature the algae as the key determinant of the interaction at the initial stage. However, at the terminal stage, the bacteria take charge of the nature of the interaction, hijacking the algae and leading to alga death and bloom decline (Zhang et al., 2018a; Zhou et al., 2018). The presence or absence of these interacting species depicting the loss of its ecological role may affect the balance in this ecosystem and shift the dynamics.

9.20 CONCLUSION

This chapter has explored how microbial interactions drive the dynamics and functions of the ecosystem, from beneficial processes, such as mutualism, to detrimental ones, including pathogenesis. While microbial interaction is a crucial factor driving ecological dynamics, it is influenced by environmental gradients and chemical cues, which affect its nature. The case of Trichoderma and the role of earthworms on microbial populations demonstrates how microbes influence the dynamics in the soil ecosystem. At the same time, algal blooms demonstrate how microbial interactions can trigger chain reactions in the aquatic ecosystem, thereby affecting the health and dynamics of this ecosystem.

Future research can further explore the nature of these interactions at the molecular level and characterize some of the chemical compounds and metabolites involved in these interactions, thereby gaining a deeper understanding of this bioactive group of organisms. Knowledge of this interaction serves as a fundamental basis for understanding and applying it in various fields, such as agronomy, food science, environmental conservation and reclamation, pathology and disease control, pharmacy, and biotechnology. By leveraging our insights into microbial interactions and ecosystem dynamics, we can promote healthier ecosystems, improve agricultural practices, and ultimately enhance human health and environmental sustainability.

ACKNOWLEDGEMENT

This study has been financially supported by the Research Grants Council (RGC) under the Hong Kong PhD Fellowship Scheme Scholarship (Reference No.: PF23–88013). This work was also supported by the Multi-disciplinary Research Capacity Building Scheme (No. 04A29) of The Education University of Hong Kong. In addition, this research was supported by the Marine Ecology Enhancement Fund (MEEF) (No. MEEF2023006) and the SKLMP Seed Collaborative Research Fund (No. SKLMP/SCRF/0053).

DECLARATION OF GENERATIVE AI USE

The authors declared that AI tools, including QuillBot and Grammarly, were used for proofreading and paraphrasing the chapter. The authors subsequently reviewed and edited the content and take full responsibility for the final version of the published chapter.

REFERENCES

Abate, R., Oon, Y. L., Oon, Y. S., Bi, Y., Mi, W., Song, G., & Gao, Y. (2024). Diverse interactions between bacteria and microalgae: a review for enhancing harmful algal bloom mitigation and biomass processing efficiency. *Heliyon*, *10*(17).

Abisado, R. G., Benomar, S., Klaus, J. R., Dandekar, A. A., & Chandler, J. R. (2018). Bacterial quorum sensing and microbial community interactions. *MBio*, *9*(3), 10–1128.

Adigun, B. A., Nahar, S. B. M., Oyebamiji, Y. O., Daba, T. M., Sinumvayo, J. P., Adegboyega, T. T., ... & Adebayo, I. A. (2024). Opportunistic fungi, plant, and nematode interactions in agricultural crops. In Akhtar M. S. (ed.), *Opportunistic Fungi, Nematode and Plant Interactions: interplay and Mechanisms* (pp. 11–32). Singapore: Springer Nature Singapore.

Ahmed, N., & Al-Mutairi, K. A. (2022). Earthworms effect on microbial population and soil fertility as well as their interaction with agriculture practices. *Sustainability*, *14*(13), 7803.

Aiyar, P., Schaeme, D., Garcia-Altares, M., Flores, D. C., Dathe, H., Hertweck, C., Sasso, S., & Mittag, M. (2017). Antagonistic bacteria disrupt calcium homeostasis and immobilize algal cells. *Nature Communications 8*, 1756.

Albright, M., Sevanto, S., Gallegos-Graves, L. V., & Dunbar, J. (2020). Biotic interactions are more important than propagule pressure in microbial community invasions. *MBio*, 11(5).

Armes, A., Schaefer, A., Hochanadel, L., Klingeman, D., Carper, D., Abraham, P., … & Pelletier, D. (2025). Quorum sensing modulates microbial community structure through regulation of secondary metabolites. *Msphere*, 10(7), e0105024.

Bacon, C. W., & Hill, N. S. (1996). Symptomless grass endophytes: products of coevolutionary symbioses and their role in the ecological adaptations of grasses. In Redlin S. C., Carris, L. M. (eds) *Entophytic Fungi in Grasses and Woody Plants: Systematics, Ecology, and Evolution* (pp. 155–178). St. Paul, MN: The American Phytopathological Society.

Berg, G., Rybakova, D., Grube, M., & Köberl, M. (2016). The plant microbiome explored: implications for experimental botany. *Journal of Experimental Botany*, 67(4), 995–1002.

Bertness, M. D., & Callaway, R. (1994). Positive interactions in communities. *Trends in Ecology & Evolution*, *9*(5), 191–193.

Binyamin, R., Nadeem, S. M., Akhtar, S., Khan, M. Y., & Anjum, R. (2019). Beneficial and pathogenic plant-microbe interactions: a review. *Soil & Environment*, *38*(2), 127–150.

Braga, R. M., Dourado, M. N., & Araújo, W. L. (2016). Microbial interactions: ecology in a molecular perspective. *Brazilian Journal of Microbiology*, *47*, 86–98.

Braude, S., Scherer, T., & McGaha, R. (2010). 7 Lotka-Volterra competition modeling. In Braude, S. & Low B. S. (eds.), *An Introduction to Methods and Models in Ecology, Evolution, and Conservation Biology* (pp. 69–78). Princeton University Press.

Bruce, J. B., West, S. A., & Griffin, A. S. (2017). Bacteriocins and the assembly of natural Pseudomonas fluorescens populations. *Journal of Evolutionary Biology*, *30*(2), 352–360.

Buchan, A., LeCleir, G. R., Gulvik, C. A., & González, J. M. (2014). Master recyclers: features and functions of bacteria associated with phytoplankton blooms. *Nature Reviews Microbiology*, *12*, 686–698.

Casadevall, A., & Pirofski, L. A. (2000). Host-pathogen interactions: basic concepts of microbial commensalism, colonization, infection, and disease. *Infection and Immunity*, *68*(12), 6511–6518.

Chen, Q., Song, Y., An, Y., Lu, Y., & Zhong, G. (2024). Soil microorganisms: their role in enhancing crop nutrition and health. *Diversity*, *16*(12), 734.

Cheng, Y. T., Zhang, L., & He, S. Y. (2019). Plant-microbe interactions facing environmental challenge. *Cell Host & Microbe*, *26*(2), 183–192.

Chisholm, S. T., Coaker, G., Day, B., & Staskawicz, B. J. (2006). Host-microbe interactions: shaping the evolution of the plant immune response. *Cell*, *124*(4), 803–814.

Christian, N., Whitaker, B. K., & Clay, K. (2015). Microbiomes: unifying animal and plant systems through the lens of community ecology theory. *Frontiers in Microbiology*, *6*, 869.

Chu, X. L., Zhang, Q. G., Buckling, A., & Castledine, M. (2021). Interspecific niche competition increases morphological diversity in multi-species microbial communities. *Frontiers in Microbiology*, *12*, 699190.

Cirri, E., & Pohnert, G. (2019). Algae–bacteria interactions that balance the planktonic microbiome. *New Phytologist*, *223*(1), 100–106.

Cohen, J., Powderly, W. G., & Opal, S. M. (2016). *Infectious Diseases E-Book: Infectious Diseases E-Book.* Elsevier Health Sciences. Elsevier. China.

Correia, A. M., & Lopes, L. F. (2023). Revisiting biodiversity and ecosystem functioning through the lens of complex adaptive systems. *Diversity*, *15*(8), 895.

Cremer, J., Melbinger, A., Wienand, K., Henriquez, T., Jung, H., & Frey, E. (2019). Cooperation in microbial populations: theory and experimental model systems. *Journal of Molecular Biology*, *431*(23), 4599–4644.

Custer, G. F., Bresciani, L., & Dini-Andreote, F. (2022). Ecological and evolutionary implications of microbial dispersal. *Frontiers in Microbiology*, *13*, 855859.

Daims, H., Lücker, S., & Wagner, M. (2016). A new perspective on microbes formerly known as nitrite-oxidizing bacteria. *Trends in Microbiology*, *24*(9), 699–712.

Dang, H., & Lovell, C. R. (2016). Microbial surface colonization and biofilm development in marine environments. *Microbiology and Molecular Biology Reviews*, *80*(1), 91–138.

De Bruyne, L., Höfte, M., & De Vleesschauwer, D. (2014). Connecting growth and defense: the emerging roles of brassinosteroids and gibberellins in plant innate immunity. *Molecular Plant*, 7(6), 943–959.

Dean, R., Van Kan, J. A., Pretorius, Z. A., Hammond-Kosack, K. E., Di Pietro, A., Spanu, P. D., … & Foster, G. D. (2012). The top 10 fungal pathogens in molecular plant pathology. *Molecular Plant Pathology*, *13*(4), 414–430.

Deng, H., Yu, Y. J., Sun, J. E., Zhang, J. B., Cai, Z. C., Guo, G. X., & Zhong, W. H. (2015). Parent materials have stronger effects than land use types on microbial biomass, activity and diversity in red soil in subtropical China. *Pedobiologia*, *58*(2–3), 73–79.

Di, X., Takken, F. L., & Tintor, N. (2016). How phytohormones shape interactions between plants and the soil-borne fungus Fusarium oxysporum. *Frontiers in Plant Science*, *7*, 170.

Di Martino, R., Picot, A., & Mitri, S. (2024). Oxidative stress changes interactions between 2 bacterial species from competitive to facilitative. *PLoS Biology*, *22*(2), e3002482.

Dinda, S. K., Hazra, S., De, A., Datta, A., Das, L., Pattanayak, S., … & Manna, D. (2024). Amoebae: beyond pathogens-exploring their benefits and future potential. *Frontiers in Cellular and Infection Microbiology*, *14*, 1518925.

Drew, G., Stevens, E. J., & King, K. C. (2021). Microbial evolution and transitions along the parasite-mutualist continuum. *Nature Reviews Microbiology*, *19*(10), 623–638.

Du, Y., Gao, Y., Hu, M., Hou, J., Yang, L., Wang, X., ... & Xu, Q. (2023). Colonization and development of the gut microbiome in calves. *Journal of Animal Science and Biotechnology*, *14*(1), 46.

Eichelberger, K. R., & Cassat, J. E. (2021). Metabolic adaptations during *Staphylococcus aureus* and *Candida albicans* co-infection. *Frontiers in Immunology*, *12*, 797550.

El Modafar, C., & El Boustani, E. (2001). Cell wall-bound phenolic acid and lignin contents in date palm as related to its resistance to Fusarium oxysporum. *Biologia Plantarum*, *44*(1), 125–130.

Ewann, F., & Hoffman, P. S. (2006). Cysteine metabolism in Legionella pneumophila: characterization of an L-cystine-utilizing mutant. *Applied and Environmental Microbiology*, *72*(6), 3993–4000.

Fazeli-Nasab, B., Shahraki-Mojahed, L., Piri, R., & Sobhanizadeh, A. (2022). Trichoderma: improving growth and tolerance to biotic and abiotic stresses in plants. In Soni, R., Suyal, D. C., Yadav, A. N., & Goel, R. (eds.), *Trends of Applied Microbiology for Sustainable Economy* (pp. 525–564). Elsevier Academic Press. (London, United Kingdom).

Fenta, L., Mekonnen, H., & Kabtimer, N. (2023). The exploitation of microbial antagonists against postharvest plant pathogens. *Microorganisms*, *11*(4), 1044.

Gambino, M., & Cappitelli, F. (2016). Mini-review: biofilm responses to oxidative stress. *Biofouling*, *32*(2), 167–178.

Garg, N., Wang, M., Hyde, E. R., Silva, R., Melnik, A. V., Protsyuk, I., ... & Dorrestein, P. C. (2017). Three-dimensional microbiome and metabolome cartography of a diseased human lung. *Cell Host & Microbe*, 22(5), 705–716.e4.

Garnett, J. A., & Matthews, S. (2012). Interactions in bacterial biofilm development: a structural perspective. *Current Protein and Peptide Science*, *13*(8), 739–755.

Gause, G. F. (1934). *The Struggle for Existence*. Baltimore: Williams & Wilkins.

Ghosh, A. R. (2013). Appraisal of microbial evolution to commensalism and pathogenicity in humans. *Clinical Medicine Insights: Gastroenterology*, *6*, CGast–S11858.

Gonzalez, J. M., & Aranda, B. (2023). Microbial growth under limiting conditions-future perspectives. *Microorganisms*, *11*(7), 1641.

Gorter, F. A., Manhart, M., & Ackermann, M. (2020). Understanding the evolution of interspecies interactions in microbial communities. *Philosophical Transactions of the Royal Society B*, *375*(1798), 20190256.

Gruber, N. (2008). The marine nitrogen cycle: overview and challenges. *Nitrogen in the Marine Environment*, *2*, 1–50.

Guo, Y., Wang, H., Du, L., Shi, P., Du, S., Xu, Z., ... & Wei, G. (2024). Microbial communities mediate the effect of cover cropping on soil ecosystem functions under precipitation reduction in an agroecosystem. *Science of the Total Environment*, *947*, 174572.

Hacquard, S., Garrido-Oter, R., González, A., Spaepen, S., Ackermann, G., Lebeis, S., ... & Schulze-Lefert, P. (2015). Microbiota and host nutrition across plant and animal kingdoms. *Cell Host & Microbe*, *17*(5), 603–616.

Haq, M. Z. U., Bai, Z., Gu, G., Liu, Y., Yang, D., Yang, H., ... & Wu, Y. (2025). Continuous cropping obstacles in medicinal plants: driven by soil microbial communities and root exudates: a review. *Plant Science*, *359*, 112686.

Hassani, M. A., Durán, P., & Hacquard, S. (2018). Microbial interactions within the plant holobiont. *Microbiome*, *6*(1), 58.

Havird, J. C., Brannock, P. M., Yoshioka, R. M., Vaught, R. C., Carlson, K. A., Edwards, C., ... & Santos, S. R. (2022). Grazing by an endemic atyid shrimp controls microbial communities in the Hawaiian anchialine ecosystem. *Limnology and Oceanography*, *67*(9), 2012–2027.

Hazarika, D. J., Gautom, T., Parveen, A., Goswami, G., Barooah, M., Modi, M. K., & Boro, R. C. (2020). Mechanism of interaction of an endofungal bacterium Serratia marcescens D1 with its host and non-host fungi. *PLoS One*, *15*(4), e0224051.

Hill, K. L. (2003). Biology and mechanism of trypanosome cell motility. *Eukaryotic Cell*, *2*(2), 200–208.

Ho, A., Angel, R., Veraart, A. J., Daebeler, A., Jia, Z., Kim, S. Y., ... & Bodelier, P. L. (2016). Biotic interactions in microbial communities as modulators of biogeochemical processes: methanotrophy as a model system. *Frontiers in Microbiology*, *7*, 1285.

Hoke, A., Reynoso, G., Smith, M., Gardner, M. I., Lockwood, D. J., Gilbert, N. E., ... & Steffen, M. M. (2021). Genomic signatures of lake erie bacteria suggest interaction in the *microcystis phycosphere*. *PLoS One*, *16*(9), e0257017.

Huffnagle, G. B., & Noverr, M. C. (2013). The emerging world of the fungal microbiome. *Trends in Microbiology*, *21*(7), 334–341.

Hutchings, M. I., Truman, A. W., & Wilkinson, B. (2019). Antibiotics: past, present, and future. *Current Opinion in Microbiology*, *51*, 72–80.

Jo, E. K. (2019). Interplay between host and pathogen: immune defense and beyond. *Experimental & Molecular Medicine*, *51*(12), 1–3.

Jones, S. A., Chowdhury, F. Z., Fabich, A. J., Anderson, A., Schreiner, D. M., House, A. L., … & Conway, T. (2007). Respiration of Escherichia coli in the mouse intestine. *Infection and Immunity*, *75*(10), 4891–4899.

Kaiser, A. (2023). The role of spermidine and its key metabolites in important, pathogenic human viruses and in parasitic infections caused by *Plasmodium falciparum* and *Trypanosoma brucei*. *Biomolecules*, *13*(5), 803.

Karki, G. (2018, July 21). Microbial interaction and types (mutualism, syntropism, proto cooperation, commensalism, antagonism, parasitism, predation, competition). Online Biology Notes. Retrieved from https://www.onlinebiologynotes.com/microbial-interaction-and-types-mutualism-syntropism-proto-cooperation-commensalism-antagonism-parasitism-predation-competition/

Kern, L., Abdeen, S. K., Kolodziejczyk, A. A., & Elinav, E. (2021). Commensal inter-bacterial interactions shaping the microbiota. *Current Opinion in Microbiology*, *63*, 158–171.

Khalid, S., & Keller, N. P. (2021). Chemical signals driving bacterial–fungal interactions. *Environmental Microbiology*, *23*(3), 1334–1347.

Kinkel, L. L., Schlatter, D. C., Xiao, K., & Baines, A. D. (2014). Sympatric inhibition and niche differentiation suggest alternative coevolutionary trajectories among Streptomycetes. *The ISME Journal*, *8*(2), 249–256.

Köberl, M., Ramadan, E. M., Adam, M., Cardinale, M., Hallmann, J., Heuer, H., … & Berg, G. (2013). Bacillus and Streptomyces were selected as broad-spectrum antagonists against soilborne pathogens from arid areas in Egypt. *FEMS Microbiology Letters*, *342*(2), 168–178.

Kodama, Y., & Fujishima, M. (2010). Secondary symbiosis between *Paramecium* and *Chlorella* cells. *International Review of Cell and Molecular Biology*, *279*, 33–77.

Koga, R., Moriyama, M., Onodera-Tanifuji, N., Ishii, Y., Takai, H., Mizutani, M., … & Fukatsu, T. (2022). Single mutation makes escherichia coli an insect mutualist. *Nature Microbiology*, 7(8), 1141–1150.

Kolita, B., & Gogoi, R. K. (2024). Lichen as pioneer of natural ecosystem. In Das, A. K., Sharma, A., Kathuria, D., Ansari, M. J. & Bhardwaj G. (eds.), *Chemistry, Biology and Pharmacology of Lichen*. (pp. 81–89). John Wiley & Sons Ltd. Hoboken, New Jersey, United States.

Krausfeldt, L. E., Shmakova, E., Lee, H. W., Mazzei, V., Loftin, K. A., Smith, R. P., … & Lopez, J. V. (2024). Microbial diversity, genomics, and phage–host interactions of cyanobacterial harmful algal blooms. *MSystems*, *9*(7), e00709–23.

Kreutz, M. (2025). Paramecium bursaria. Real Micro Life. Retrieved from https://realmicrolife.com/paramecium-bursaria/

Krishna, M., Gupta, S., Delgado–Baquerizo, M., Morriën, E., Garkoti, S. C., Chaturvedi, R., & Ahmad, S. (2020). Successional trajectory of bacterial communities in soil are shaped by plant-driven changes during secondary succession. *Scientific Reports*, *10*(1), 9864.

Kunkel, B. N., & Brooks, D. M. (2002). Cross-talk between signaling pathways in pathogen defense. *Current Opinion in Plant Biology*, *5*(4), 325–331.

Li, L., Qu, Q., Tian, B., & Zhang, K. Q. (2005). Induction of chlamydospores in Trichoderma harzianum and Gliocladium roseum by antifungal compounds produced by Bacillus subtilis C2. *Journal of Phytopathology*, *153*(11–12), 686–693.

Lilic, D. (2012). Unravelling fungal immunity through primary immune deficiencies. *Current Opinion in Microbiology*, *15*(4), 420–426.

Lin, Q., Lin, S., Fan, Z., Liu, J., Ye, D., & Guo, P. (2024). A review of the mechanisms of bacterial colonization of the mammal gut. *Microorganisms*, *12*(5), 1026.

Lindsay, R. J., Jepson, A., Butt, L., Holder, P. J., Smug, B. J., & Gudelj, I. (2021). Would that it were so simple: interactions between multiple traits undermine classical single-trait based predictions of microbial community function and evolution. *Ecology Letters*, *24*(12), 2775–2795.

Liu, Q., Zhang, L., Wang, L., Wu, Q., Li, K., & Guo, X. (2022). Autotoxin affects the rhizosphere microbial community structure by influencing the secretory characteristics of grapevine roots. *Frontiers in Microbiology*, *13*, 953424.

Liu, Y., Chang, S., Murphy, G., Ajayi-Akinsulire, E., Ardren, I., Guy, I., Johnston, K., Lee, S., & Russell, L. (2025, March 18). 17.1: Stages of pathogenesis. In Introduction to microbiology. Biology LibreTexts. Retrieved from https://bio.libretexts.org/Bookshelves/Microbiology/Introduction_to_Microbiology_(Liu_et_al.)/17%3A_Pathogenicity_and_Virulence_Factors/17.01%3A_Stages_of_Pathogenesis

Liu, Z., Guo, Z., Zhou, J., Guo, X., & Chen, Y. (2024). Biotic interactions and environmental modifications determine symbiotic microbial diversity and stability. *Computational and Structural Biotechnology Journal*, *23*, 2717–2726.

Long, J. J., Luna, E. K., Jackson, M., Wheat, W., Jahn, C. E., & Leach, J. E. (2019). Interactions of free-living amoebae with the rice fungal pathogen, Rhizoctonia solani. *BMC Research Notes*, *12*(1), 746.

Lozano, Y. M., Hortal, S., Armas, C., & Pugnaire, F. I. (2014). Interactions among soil, plants, and microorganisms drive secondary succession in a dry environment. *Soil Biology and Biochemistry*, *78*, 298–306.

Ma, B., Cai, Y., Bork, E. W., & Chang, S. X. (2018). Defoliation intensity and elevated precipitation effects on microbiome and interactome depend on site type in northern mixed-grass prairie. *Soil Biology and Biochemistry*, *122*, 163–172.

MacAlpine, J., Robbins, N., & Cowen, L. E. (2023). Bacterial-fungal interactions and their impact on microbial pathogenesis. *Molecular Ecology*, *32*(10), 2565–2581.

Mapuranga, J., Zhang, L., Zhang, N., & Yang, W. (2022). The haustorium: the root of biotrophic fungal pathogens. *Frontiers in Plant Science*, *13*, 963705.

Marchal, M., Goldschmidt, F., Derksen-Müller, S. N., Panke, S., Ackermann, M., & Johnson, D. R. (2017). A passive mutualistic interaction promotes the evolution of spatial structure within microbial populations. *BMC Evolutionary Biology*, *17*(1), 106.

Marinos, G., Moors, K. A., Rühlemann, M., Waschina, S., Lieb, W., Franke, A., … & Kadibalban, A. S. (2025). Genome-scale metabolic models predict diet-and lifestyle driven shifts of ecological interactions in the gut microbiome. *bioRxiv*, 2025–09.

Markelova, N., & Chumak, A. (2025). Antimicrobial activity of bacillus cyclic lipopeptides and their role in the host adaptive response to changes in environmental conditions. *International Journal of Molecular Sciences*, *26*(1), 336.

McClerklin, S. A., Lee, S. G., Harper, C. P., Nwumeh, R., Jez, J. M., & Kunkel, B. N. (2018). Indole-3-acetaldehyde dehydrogenase-dependent auxin synthesis contributes to virulence of Pseudomona. *PLoS Pathogens*, *4*(1), e1006811.

Medina-Sauza, R. M., Álvarez-Jiménez, M., Delhal, A., Reverchon, F., Blouin, M., Guerrero Analco, J. A., … & Barois, I. (2019). Earthworms building up soil microbiota, a review. *Frontiers in Environmental Science*, *7*, 81.

Mishra, S. (2024). Keystone species and their impact on the ecosystem: current trends. *Journal of Ecological Conservation, 1*(1), 1–10.

Moënne-Loccoz, Y., Mavingui, P., Combes, C., Normand, P., & Steinberg, C. (2014). Microorganisms and biotic interactions. In Bertrand, J. C., Caumette, P., Lebaron, P., Matheron, R., Normand, P., Sime-Ngando, T. (eds), *Environmental Microbiology: Fundamentals and Applications: Microbial Ecology* (pp. 395–444). Dordrecht: Springer Netherlands.

Morkunas, I., Formela, M., Marczak, Ł., Stobiecki, M., & Bednarski, W. (2013). The mobilization of defence mechanisms in the early stages of pea seed germination against Ascochyta pisi. *Protoplasma*, *250*(1), 63–75.

Mortimer, P. E., Gui, H., Xu, J., Zhang, C., Barrios, E., & Hyde, K. D. (2015). Alder trees enhance crop productivity and soil microbial biomass in tea plantations. *Applied Soil Ecology*, *96*, 25–32.

Müller, J. P., Hauzy, C., & Hulot, F. D. (2012). Ingredients for protist coexistence: competition, endosymbiosis and a pinch of biochemical interactions. *Journal of Animal Ecology*, *81*(1), 222–232.

Mydeen, A. K. M., Agnihotri, N., Bahadur, R., Lytand, W., Kumar, N., & Hazarika, S. (2023). Microbial Maestros: unraveling the crucial role of microbes in shaping the environment. *Acta Biology Forum*. *2*(2): 23–28.

Narayanasamy, M., Dhanasekaran, D., & Thajuddin, N. (2020). Frankia. In Amaresan, N., Senthil Kumar, M., Annapurna, K., Krishna Kumar & Sankaranarayanan, A. (eds.), *Beneficial Microbes in Agro-Ecology* (pp. 185–211). Elsevier Academic Press. (London, United State of America).

Nash, A. A., Dalziel, R. G., & Fitzgerald, J. R. (2015). Mims' pathogenesis of infectious disease. Academic Press.

National Park Service. (2020, January 21). *Plant Succession - Kenai Fjords National Park*. US Department of the Interior. Retrieved from nps.gov/kefj/learn/nature/plant succession.htm.

Nazaries, L., Singh, B. P., Sarker, J. R., Fang, Y., Klein, M., & Singh, B. K. (2021). The response of soil multi-functionality to agricultural management practices can be predicted by key soil abiotic and biotic properties. *Agriculture, Ecosystems & Environment*, *307*, 107206.

Nehra, S., Gothwal, R. K., Dhingra, S., Varshney, A. K., Solanki, P. S., Meena, P., … & Ghosh, P. (2022). Mechanism of antagonism: hyperparasitism and antibiosis. In Kumar, A. (eds), *Microbial Biocontrol: Sustainable Agriculture and Phytopathogen Management: Volume 1* (pp. 257–277). Cham: Springer International Publishing.

Nelson, P., & May, G. (2017). Coevolution between mutualists and parasites in symbiotic communities may lead to the evolution of lower virulence. *The American Naturalist*, *190*(6), 803–817.

Ni, G., Lappan, R., Hernández, M., Santini, T., Tomkins, A. G., & Greening, C. (2022). Functional basis of primary succession: traits of the pioneer microbes. *Environmental Microbiology*, *25*(1), 171.

Olano, J. P., Weller, P. F., Guerrant, R. L., & Walker, D. H. (2011). Principles of parasitism: host parasite interactions. In Guerrant, R. L., Walker, D. H. & Weller, P. F. (eds.), *Tropical infectious diseases: Principles, pathogens and practice* (3rd ed., Vol. 1, pp. 1–7). Saunders Elsevier. Edinburgh, Scotland.

Overstreet, R. M., & Lotz, J. M. (2016). Host–symbiont relationships: understanding the change from guest to pest. In Hurst, C. J. (ed.), *The Rasputin Effect: When Commensals and Symbionts Become Parasitic* (pp. 27–64). Cham: Springer International Publishing.

Palenik, B. (2015). Molecular mechanisms by which marine phytoplankton respond to their dynamic chemical environment. *Annual Review of Marine Science*, *7*, 325–340.

Park, Y., Kim, W., Bae, J., & Park, W. (2025). Public goods-mediated bacterial interplay in aquatic ecosystems. *Water Research*, *287*, 124310.

Park, Y., Kim, W., Kim, M., & Park, W. (2023). The β-Lactamase activity at the community level confers β-lactam resistance to bloom-forming *Microcystis aeruginosa* cells. *Journal of Microbiology*, *61*(9), 807–820.

Penesyan, A., Paulsen, I. T., Kjelleberg, S., & Gillings, M. R. (2021). Three faces of biofilms: a microbial lifestyle, a nascent multicellular organism, and an incubator for diversity. *NPJ Biofilms and Microbiomes*, *7*(1), 80.

Peterson, S. B., Bertolli, S. K., & Mougous, J. D. (2020). The central role of interbacterial antagonism in bacterial life. *Current Biology*, *30*(19), R1203–R1214.

Pollitt, L. C., MacGregor, P., Matthews, K., & Reece, S. E. (2011). Malaria and trypanosome transmission: different parasites, same rules? *Trends in Parasitology*, *27*(5), 197–203.

Prakash, O., Sharma, R., Rahi, P., & Karthikeyan, N. (2014). Role of microorganisms in plant nutrition and health. In Rakshit, A., Singh, H. B., Sen, A. (eds), *Nutrient Use Efficiency: From Basics to Advances* (pp. 125–161). New Delhi: Springer India.

Qiao, Y., Huang, Q., Guo, H., Qi, M., Zhang, H., Xu, Q., … & Ling, N. (2024). Nutrient status changes bacterial interactions in a synthetic community. *Applied and Environmental Microbiology*, *90*(1), e01566–23.

Rakoff-Nahoum, S., Coyne, M. J., & Comstock, L. E. (2014). An ecological network of polysaccharide utilization among human intestinal symbionts. *Current Biology*, 24(1), 40–49.

Ramírez Granillo, A., Canales, M. G. M., Espíndola, M. E. S., Martínez Rivera, M. A., de Lucio, V. M. B., & Tovar, A. V. R. (2015). Antibiosis interaction of *Staphylococcus aureus* on Aspergillus fumigatus assessed in vitro by mixed biofilm formation. *BMC Microbiology*, *15*(1), 33.

Ratzke, C., & Gore, J. (2018). Modifying and reacting to the environmental pH can drive bacterial interactions. *PLoS Biology*, *16*(3), e2004248.

Ren, X., Zhang, Q., Zhang, W., Mao, J., & Li, P. (2020). Control of aflatoxigenic molds by antagonistic microorganisms: inhibitory behaviors, bioactive compounds, related mechanisms, and influencing factors. *Toxins*, *12*(1), 24.

Rossine, F. W., Vercelli, G. T., Tarnita, C. E., & Gregor, T. (2022). Structured foraging of soil predators unveils functional responses to bacterial defenses. *Proceedings of the National Academy of Sciences*, *119*(52), e2210995119.

Ryder, L. S., & Talbot, N. J. (2015). Regulation of appressorium development in pathogenic fungi. *Current Opinion in Plant Biology*, *26*, 8–13.

Shao, Y., Garcia-Mauriño, C., Clare, S., Dawson, N. J., Mu, A., Adoum, A., … & Lawley, T. D. (2024). Primary succession of Bifidobacteria drives pathogen resistance in neonatal microbiota assembly. *Nature Microbiology*, *9*(10), 2570–2582.

Singh, N., Srivastava, R., Kanda, T., Yadav, S., Yadav, S., & Atri, N. (2023). The role of microorganisms in ecosystem restoration. In Srivastava, R. & Kumar, P. (eds.), Conserving, Halting and Reinstating Degraded Natural Resources. (pp. 83–104). Nova Science Publishers, Inc. New York, United States of America.

Smith, P., & Schuster, M. (2019). Public goods and cheating in microbes. *Current Biology*, *29*(11), R442–R447.

Solowiej-Wedderburn, J., Pentz, J. T., Lizana, L., Schroeder, B. O., Lind, P. A., & Libby, E. (2025). Competition and cooperation: the plasticity of bacterial interactions across environments. *PLoS Computational Biology*, *21*(7), e1013213.

Sommer, A. J., & Newell, P. D. (2019). Metabolic basis for mutualism between gut bacteria and its impact on the Drosophila melanogaster host. *Applied and Environmental Microbiology*, *85*(2), e01882–18.

Soni, J., Sinha, S., & Pandey, R. (2024). Understanding bacterial pathogenicity: a closer look at the journey of harmful microbes. *Frontiers in Microbiology*, *15*, 1370818.

Sood, M., Kapoor, D., Kumar, V., Sheteiwy, M. S., Ramakrishnan, M., Landi, M., … & Sharma, A. (2020). Trichoderma: the "secrets" of a multitalented biocontrol agent. *Plants*, *9*(6), 762.

Spadaro, D., & Droby, S. (2016). Development of biocontrol products for postharvest diseases of fruit: the importance of elucidating the mechanisms of action of yeast antagonists. *Trends in Food Science & Technology*, *47*, 39–49.

Sterling, J. J., Sakihara, T. S., Brannock, P. M., Pearson, Z. G., Maclaine, K. D., Santos, S. R., & Havird, J. C. (2022). Primary microbial succession in the anchialine ecosystem. *Integrative and Comparative Biology*, *62*(2), 275–287.

Taylor, M., Mediannikov, O., Raoult, D., & Greub, G. (2012). Endosymbiotic bacteria associated with nematodes, ticks and amoebae. *FEMS Immunology & Medical Microbiology*, *64*(1), 21–31.

Thakur, M. P., Van der Putten, W. H., Cobben, M. M., van Kleunen, M., & Geisen, S. (2019). Microbial invasions in terrestrial ecosystems. *Nature Reviews Microbiology*, *17*(10), 621–631.

Verma, K., Dogra, R., & Rathour, N. S. (2025). Ecosystem dynamics: exploring types, components and the forces shaping their transformation. *Journal of Geography, Environment and Earth Science International*, *29*(4), 177–191.

Vinale, F., D'Ambrosio, G., Abadi, K., Scala, F., Marra, R., Woo, S. L., … & Lorito, M. (2004). Application of Trichoderma harzianum (T22) and Trichoderma atroviride (P1) as plant growth promoters, and their compatibility with copper oxychloride. *Journal of Plant Pathology*, *86*, 336–336.

Vinale, F., Sivasithamparam, K., Ghisalberti, E. L., Marra, R., Woo, S. L., & Lorito, M. (2008). Trichoderma–plant–pathogen interactions. *Soil Biology and Biochemistry*, *40*(1), 1–10.

Wadowsky, R. M., & Yee, R. B. (1983). Satellite growth of Legionella pneumophila with an environmental isolate of Flavobacterium breve. *Applied and Environmental Microbiology*, *46*(6), 1447–1449.

Wu, B., Umer, M., Guo, Y., He, M., Han, X., Shen, K., … & He, X. (2023). Positive responses of soil nutrients and enzyme activities to AM fungus under interspecific and intraspecific competitions when associated with litter addition. *Rhizosphere*, *27*, 100728.

Wu, F., An, Y. Q., An, Y., Wang, X. J., Cheng, Z. Y., Zhang, Y., ... & Bai, J. G. (2018). Acinetobacter calcoaceticus CSY-P13 mitigates stress of ferulic and p hydroxybenzoic acids in cucumber by affecting antioxidant enzyme activity and soil bacterial community. *Frontiers in Microbiology*, *9*, 1262.

Yan, A., Wang, Y., Tan, S. N., Mohd Yusof, M. L., Ghosh, S., & Chen, Z. (2020). Phytoremediation: a promising approach for revegetation of heavy metal-polluted land. *Frontiers in Plant Science*, *11*, 359.

Yang, M., Zhang, X., Xu, Y., Mei, X., Jiang, B., Liao, J., … & Zhu, S. (2015). Autotoxic ginsenosides in the rhizosphere contribute to the replant failure of *panax notoginseng*. *PLoS One*, 10(2), e0118555.

Yarbro, J., Khorunzhy, E., & Boyle, N. (2024). The phycosphere and its role in algal biofuel production. *Frontiers in Climate*, *6*, 1277475.

Ye, J., Zhong, T., Zhang, D., Ma, C., Wang, L., Yao, L., … & Xu, M. (2019). The auxin regulated protein ZmAuxRP1 coordinates the balance between root growth and stalk rot disease resistance in maize. *Molecular Plant*, *12*(3), 360–373.

Yin, W., Wang, Y., Liu, L., & He, J. (2019). Biofilms: the microbial "protective clothing" in extreme environments. *International Journal of Molecular Sciences*, *20*(14), 3423.

Zhang, H., Jia, J., Chen, S., Huang, T., Wang, Y., Zhao, Z., … & Ma, X. (2018a). Dynamics of bacterial and fungal communities during the outbreak and decline of an algal bloom in a drinking water reservoir. *International Journal of Environmental Research and Public Health*, *15*(2), 361.

Zhang, H., Wang, K., Shen, L., Chen, H., Hou, F., Zhou, X., … & Zhu, X. (2018b). Microbial community dynamics and assembly follow trajectories of an early-spring diatom bloom in a semienclosed bay. *Applied and Environmental Microbiology*, *84*(18), e01000–18.

Zhong, Y., Yan, W., Wang, R., Wang, W., & Shangguan, Z. (2018). Decreased occurrence of carbon cycle functions in microbial communities along with long-term secondary succession. *Soil Biology and Biochemistry*, *123*, 207–217.

Zhou, J., Richlen, M. L., Sehein, T. R., Kulis, D. M., Anderson, D. M., & Cai, Z. (2018). Microbial community structure and associations during a marine dinoflagellate bloom. *Frontiers in Microbiology*, *9*, 1201.

Zhu, J., Tang, S., Cheng, K., Cai, Z., Chen, G., & Zhou, J. (2023). Microbial community composition and metabolic potential during a succession of algal blooms from *Skeletonema* sp. to *Phaeocystis* sp. *Frontiers in Microbiology*, *14*, 1147187.

10 Preserving Marine Biodiversity
Strategies for a Thriving Ocean Ecosystem

Su Yin Chee and Noor Adelyna Mohammed Akib

10.1 INTRODUCTION

10.1.1 The Crucial Role of Marine Biodiversity in Global Ecosystems

The foundation of life beneath water is formed by marine biodiversity—from minute plankton to the largest whales and the myriad of lifeforms in between. Oceans cover more than 70% of Earth's surface. However, they are not only a vast expanse of water but a living, breathing engine that regulates Earth's climate, supports food security, and drives important biogeochemical cycles (Costello et al., 2010; Mora et al., 2011). The biological diversity that is within marine ecosystems is pivotal to the stability and health of our planet. The ocean's biodiversity guarantees ecosystem resilience, productivity, and the provision of ecosystem services that billions of people depend on for food, livelihoods, and cultural identity. From coral reefs that offer refuge and breeding grounds for innumerable species to deep-sea ecosystems that are still mostly unexplored – they all contribute to the needs of humans and their surrounding environment (Worm et al. 2006; Barbier et al., 2011).

Marine biodiversity directly impacts human well-being. Fisheries and aquaculture supply more than 3 billion people with nearly 20% of their average intake of animal protein (FAO, 2018). Mangroves, coral reefs, and seagrass meadows act as natural coastal erosion control structures and barriers against extreme weather events protecting coastal communities from the increasing threats posed by climate change (UNEP, 2020). Besides that, marine organisms play an important role in regulating atmospheric gases, including carbon dioxide that contributes greatly to climate change, through processes such as photosynthesis, carbon sequestration, and nutrient cycling (Duarte et al., 2005; Falkowski et al., 2008). The ocean's hidden pharmacy—its vast genetic resources—holds promise for new medicines, biotechnologies, and innovations critical to human health and sustainable development (Arrieta et al., 2010; Blasiak et al., 2018).

10.1.2 Alarming Decline: Current Threats to Marine Biodiversity

Despite its value and significance, marine biodiversity is under unprecedented threat due to a combination of human-induced pressures (Bulleri et al., 2020). Leading causes of biodiversity loss in the marine environment include excessive coastal development, unsustainable and unregulated fishing practices, pollution, climate change, and ocean acidification (Halpern et al., 2008; IPBES, 2019). Fishing practices like bottom trawling, fish bombing, and illegal encroachment into marine parks to catch fish have led to the depletion of numerous fish stocks, with cascading effects throughout food webs (Pauly et al., 2002; Pitcher & Cheung, 2013). Bottom trawling not only removes targeted fish but also destroys habitats on the seabed that support underwater communities and complex ecological interactions (Puig et al., 2012).

Coastal development and destructive land-use practices have resulted in the degradation of vital habitats such as coral reefs, estuaries, and mangrove forests (Alongi, 2002; Hughes et al., 2017). Pollution—from plastic waste and agricultural runoff to heavy metals and oil spills—further

DOI: 10.1201/9781003640899-13

compounds stress on marine ecosystems, impairing reproductive and feeding capacities of marine organisms (Rochman et al., 2013; Jambeck et al., 2015). Among these, plastic pollution has emerged as a particularly pervasive threat, with microplastics now found in the deepest ocean trenches and within the tissues of marine species across trophic levels (Woodall et al., 2014; Lusher et al., 2013).

Most critically, the impacts of climate change are becoming increasingly apparent in marine ecosystems over time. Rising sea temperatures have led to more extensive and longer coral bleaching events, while shifting ocean currents are changing species distributions and reproductive timing (Hoegh-Guldberg et al., 2007; Poloczanska et al., 2013; Dong et al., 2025). Ocean acidification, caused by the absorption of excess atmospheric CO_2, is weakening the calcareous structures of shellfish (e.g., oysters, mussels, and cockles), corals, and plankton—organisms that serve as the building blocks of marine food webs (Doney et al., 2009; Kroeker et al., 2010). These pressures act synergistically, creating complex and often irreversible changes that challenge the resilience of marine ecosystems (Crain et al., 2008).

10.1.3 Urgency and Opportunity

The preservation of marine biodiversity is no longer a matter of environmental preference but an urgent global necessity. As the world advances to implementing the post-2020 Global Biodiversity Framework (GBF) and the UN Sustainable Development Goals, particularly Goal 14, "Life Below Water"—there is a growing census that safeguarding marine life is essential for safeguarding planetary health and ensuring socio-economic stability (Friedman et al., 2022; UN, 2023). The future of marine biodiversity depends not only on policy frameworks and scientific innovation but also on local stewardship, international cooperation, and a profound rethinking of humanity's relationship with the ocean.

Using ecological principles, conservation science, and community-based practices, this chapter examines a variety of methods for protecting marine biodiversity. To preserve and restore marine ecosystems and guarantee a healthy ocean for future generations, it offers an integrated perspective on how science, policy, and public involvement must come together.

10.2 THE ROLE OF MARINE ECOSYSTEMS IN BIODIVERSITY

10.2.1 Vital Habitats: The Diversity of Marine Ecosystems

The most biodiverse marine habitats—coral reefs, mangrove forests, seagrass meadows, deep-sea habitats, and other lesser-known habitats—are introduced in this section. Instead of being dispersed equally throughout the ocean, marine biodiversity is concentrated in a number of unique and extremely productive ecosystems. Coral reefs, mangrove forests, seagrass beds, deep-sea habitats, and lesser-known habitats like kelp forests, polar ice ecosystems, sponge grounds, and submarine canyons are some of the most ecologically significant and biodiverse marine habitats. Despite having different structures and purposes, these ecosystems are all essential to preserving ecological balance and biodiversity.

Coral reefs, also known as the "rainforests of the sea", are among the planet's most biologically diverse ecosystems. Despite making up less than 0.1% of the ocean floor, they are home to about 25% of all known marine species (Spalding et al., 2001; Knowlton et al., 2010). For fish, invertebrates, and algae, reefs offer vital habitats as well as a variety of commercially significant species. Additionally, they are structurally complex, providing shelter and spatial partitioning that support ecosystem resilience and species coexistence.

Mangrove forests serve as essential ecological barriers between land and sea and are found in tropical and subtropical intertidal zones. High levels of biodiversity, such as young fish, crustaceans, birds, and marine mammals, are supported by these salt-tolerant trees (Alongi, 2012). In addition to storing large amounts of carbon in their biomass and soils—a crucial role in climate regulation—their dense root systems stabilize coastlines, trap sediments, and filter pollutants (Donato et al., 2011).

Seagrass meadows are underwater flowering plant habitats that form extensive beds in shallow coastal waters. These ecosystems offer food and shelter to a wide variety of organisms, including endangered species such as sea turtles, dugongs, and manatees (Orth et al., 2006). Like mangroves, seagrasses are efficient carbon sinks and contribute to water clarity by trapping sediments and nutrients.

Abyssal plains, hydrothermal vents, and cold seeps are examples of deep-sea habitats that are the world's largest and least studied ecosystems. The deep sea, which was formerly believed to be desolate, is now known to support remarkable biodiversity that is suited to harsh environments like high pressure, low temperatures, and lack of sunlight (Ramirez-Llodra et al., 2010). Hydrothermal vent communities—which support unusual organisms like giant tubeworms and extremophile microbes—are propelled by chemosynthesis rather than photosynthesis (Dick, 2019).

Sponge grounds, polar ice ecosystems, and kelp forests are some of the less well-known habitats that are crucial to the environment. For instance, polar ecosystems support specialized species that are specially adapted to cold temperatures and seasonal productivity, while kelp forests are dynamic systems that offer three-dimensional structure for a variety of marine life (Dayton, 1985; Van de Putte et al., 2021).

At the local, regional, and global levels, these habitats support marine biodiversity through an interconnected network of ecological niches. The loss or deterioration of any one of these ecosystems may have a domino effect on ecosystem services and biodiversity.

10.2.2 Linking Marine Biodiversity and Human Well-Being

Human well-being is inextricably linked to the health of marine ecosystems and the biodiversity they support. A variety of ecosystem services that support economic activity, food security, climate regulation, and cultural identity are provided by marine ecosystems. According to WWF (2015), coral reefs alone bring in an estimated $2.7 trillion a year from tourism, fishing, and coastal protection. In addition to protecting infrastructure from storm surges and rising seas, mangroves and seagrasses support fisheries by acting as nurseries for numerous species that are harvested in offshore waters (Barbier et al., 2011).

The health of marine ecosystems directly affects more than 600 million people who live in coastal regions worldwide (UNEP, 2022). Healthy marine systems are essential to the food, livelihoods, and cultural practices of coastal indigenous communities and small island developing states. A strong bond between humans and the sea is reinforced in many areas where marine biodiversity is essential to traditional ecological knowledge (TEK) and spiritual beliefs (Thurstan & Roberts, 2010).

Additionally, the advancement of biotechnology, medicine, and innovation depends on marine biodiversity. Many biomedical advances have come from marine organisms, such as antiviral and anticancer compounds made from deep-sea bacteria, tunicates, and sponges (Leal et al., 2012). Maintaining biodiversity guarantees both the possibility of future discoveries and ongoing access to these genetic resources.

The advantages of marine ecosystems are not, however, assured. They rely on the diversity and well-being of the ecosystems themselves. The ability of marine ecosystems to provide services decreases as biodiversity decreases as a result of human pressures. For example, degraded coral reefs no longer sustain healthy fisheries or offer effective storm protection. Similarly, mangrove deforestation destroys vital nursery habitats and diminishes shoreline stability (Fillyaw et al., 2021). Vulnerable communities bear a disproportionate share of the social, economic, and environmental costs associated with this loss of function.

Therefore, preserving and replenishing marine biodiversity is both an ecological and a social necessity. It guarantees the continuation of ecosystem services that are essential to human survival and the world economy. Our ability to preserve and restore the very ecosystems that support marine life is essential to the sustainability of ocean resources and the resilience of coastal communities.

10.3 PROTECTING MARINE ECOSYSTEMS: GOVERNANCE, THREATS AND COORDINATED STRATEGIES

10.3.1 Global Frameworks

In marine conservation, international law and policy are crucial in guiding national and regional initiatives. This is achieved through the United Nations Convention on the Law of the Sea (UNCLOS), which establishes the legal framework for maritime zones, environmental conservation, and the management of living resources (Suncls & Cai, 1982; Tanaka, 2023). Part XII of UNCLOS requires all states to protect the marine environment (Article 192) and to control pollution from all sources (Articles 194–196) (United Nations, 1982). These rules form the basis for international marine governance and reinforce the duty of states to protect the health of ecosystems and biodiversity.

The Convention on Biological Diversity (CBD) also promotes the conservation and sustainable use. The Kunming–Montreal GBF was adopted during the fifteenth meeting of the Conference of the Parties (COP15) following a 4-year consultation and negotiation process. This historic framework, which supports the achievement of Sustainable Development Goals and builds on the Convention's previous strategic plans, sets out an ambitious pathway to reach the global vision of a world living in harmony with nature by 2050 (CBD, 2022). The 23 targets for 2030 and the four goals for 2050 are among the framework's key elements, which highlight a commitment from involved countries to protect at least 30% of ocean areas by 2030 through establishing well-managed and ecologically connected regions, known as 30×30. (Stephens, 2023; Streck, 2023).

Supporting Sustainable Development Goal 14 (Life Below Water), which aims to reduce marine pollution, restore ecosystems, and end overfishing and harmful subsidies. It also emphasizes the need to lessen ocean acidification and ensure fair benefits for coastal and island communities, particularly small-scale fishers.

The importance of governance in practice:

- Overfishing remains the most significant challenge when initiating marine management efforts, and as one-third of assessed fish stocks were reported as overexploited despite the implementation of international regulations (Pham et al., 2023; Andersen et al., 2024). For example, the collapse of Northwest Atlantic cod stocks has not been reversed, despite years of restoration efforts, due to the region's weak regulation (Lear, 1998; Pauly et al., 2005; Bavington, 2011). Next, in the decline of Pacific cod in Alaska from 2018 to 2020 also shows that climate events like marine heatwaves can overwhelm fixed rules unless management is flexible (West et al., 2025).
- The persistent plastic pollution, despite global agreements to make plastic pollution the main concern. This can be seen in Southeast Asia, which has weak systems for plastic management and lacks enforcement, leading to slow progress. As a result, plastics remain numerous in the ocean despite commitments from both global and regional bodies (OECD, 2022; World Bank, 2021).

10.3.2 Implementation Pathways

10.3.2.1 Marine Protected Areas (MPAs)

Among the most important tools for meeting these international commitments (mentioned in Section 10.3.1) is the Marine Protected Areas (MPAs). More than 15,000 MPAs have been established worldwide. Together, they cover approximately 7% of the world's marine environment (Lubchenco

& Grorud-Colvert, 2015; UNEP-WCMC & IUCN, 2018). Scientific evidence shows that MPAs influence the social well-being and political power of fishing communities, and impacts (positive and negative) vary within and among social groups (Mascia et al., 2010).

Social impacts are linked to some, but not all, commonly hypothesized explanatory factors that MPAs can provide ecological, economic, and social benefits when properly designed and managed (Mascia et al., 2010; Ban et al., 2019). Therefore, MPAs might be a viable strategy for improving food security and empowering local communities, but current practices negatively impact at least a minority of fishers.

Scientific evidence demonstrates that MPAs influence not only ecosystems but also the social well-being and political power of fishing communities. The impacts, both positive and negative, may vary across and within social groups. However, when properly designed and managed, MPAs can generate ecological, economic, and social benefits by enhancing biodiversity, rebuilding fish biomass, and strengthening ecosystem resilience against climate and human pressures (Sala et al., 2021).

Some examples:

- Well-enforced no-take reserves in the Philippines have enhanced reef fish biomass and provided spillover benefits that sustain small-scale fisheries (Alcala & Russ, 2006).
- A study in the Mediterranean (Di Franco et al., 2016) reported that fish stocks are healthier, and fishermen's incomes are higher. Social acceptance of management practices is fostered if five attributes are present (i.e., high MPA enforcement, presence of a management plan, fishermen's engagement in MPA management, fishermen's representation in the MPA board, and promotion of sustainable fishing). These findings are pivotal to Mediterranean coastal communities as they work to achieve conservation goals while allowing for profitable exploitation of fisheries resources.
- In Fiji, community-managed MPAs under the Locally Managed Marine Area (LMMA) network have increased fish stocks and improved food security and livelihoods (Govan et al., 2024; O'Garra et al., 2023).

Marine protection has been inspired by the Aichi Goals since 2000. As a result, there has been a 650% increase in MPAs, indicating a significant expansion in the marine realm (UNEP-WCMC & IUCN, 2021). Since 2020, about 1.77 million km^2 have been added, meaning 8.4% of the global oceans are protected as of mid-2024 (UNEP-WCMC & IUCN, 2024). However, this differs from terrestrial protection, which saw a moderate increase of 62% during the same period. From the data, we can see that marine protection surpassed 30 million km^2 by 2024, with rapid gains around 2017 exceeding terrestrial coverage (UNEP-WCMC & IUCN, 2024) (Figure 10.1). In addition to formally designated protected areas, progress in conservation is increasingly measured by recognizing Other Effective Area-Based Conservation Measures (OECMs), which include areas not legally protected but delivering sustained biodiversity outcomes, such as community-managed forests, indigenous territories, or sustainably managed military zones (UNEP-WCMC & IUCN, 2024) (Figure 10.1). Together, protected areas and OECMs contribute to the global 30×30 biodiversity target.

Given the global goal to expand MPAs to 30% of the ocean by 2030 (CBD, 2022), identifying governance strategies that deliver ecological and social outcomes is essential. Ecologically effective MPAs restore marine biodiversity and ecosystem health. At the same time, MPAs that support local livelihoods and encourage community participation are more likely to gain broad adoption, advancing effective and equitable marine conservation (Bennet et al., 2019).

However, their success depends on stakeholder involvement, adherence to legal compliance, and the availability of sufficient resources. Ineffective MPAs often face non-compliance and illegal activity, which undermine ecological goals (Giakoumi et al., 2018). Whether through reserves, aquaculture, or MPAs, local communities play a crucial role in marine conservation and resource management (Cinner et al., 2018; Bradford et al., 2020; Johnson et al., 2020; Marriott et al., 2021).

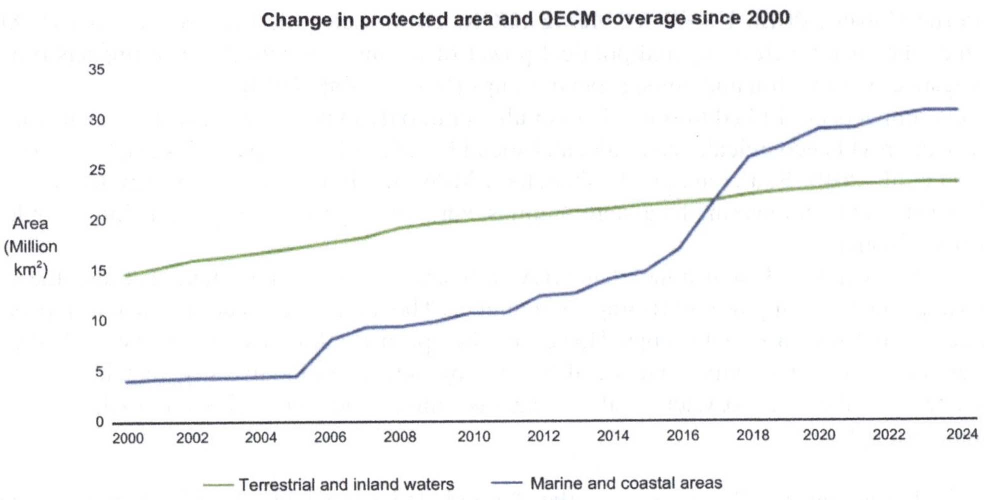

FIGURE 10.1 The growth of protected areas and OECMs (Other Effective Area-Based Conservation Measures) since 2000. OECMs are geographically defined areas that are not formally designated as protected areas but achieve long-term biodiversity conservation. (Source: UNEP-WCMC & IUCN, 2024)

Conservation approaches have been successfully matched with cultural and social contexts through the integration of TEK into Community-Based Resource Management (CBRM) (Gadgil et al.,1993; Berkes, 2017).

Fisheries rules are more successful when communities share decision-making authority and directly benefit from management outcomes. There is evidence that governance co-created with communities improves the performance of MPAs (Cinner et al., 2018). In Southeast Asia, including Malaysia and Indonesia, co-management models where local communities and government agencies share responsibilities have proven especially successful in managing mangroves and coral reefs (Cinner et al., 2018; Johnson et al., 2020). These frameworks support inclusive governance, encourage adaptive responses to environmental changes, and strengthen local capacity (Table 10.1).

10.3.3 Natural and Anthropogenic Threats: Interactions and Comparisons

Tyre-derived microplastics (TDMPs) have become an emerging pollutant because they represent the ongoing influence of human activities concerning global climate challenges. Both anthropogenic and natural factors significantly impact marine ecosystems (Pham et al., 2023). The combined impacts of climate stress, overfishing, pollution, and habitat loss diminish ecosystem resilience and increase risks for coastal communities (IPCC, 2022; Pham et al., 2023).

10.3.3.1 Ocean Acidification and Rising Sea Temperatures

Anthropogenic climate change has further exacerbated natural stressors, such as ocean acidification and rising sea surface temperatures (SST), resulting in weakened marine ecosystems (Doney et al., 2009; IPCC, 2022), where the ocean absorbs roughly 25%–30% of anthropogenic CO_2 emissions, decreasing global pH by about 0.1 since pre-industrial times.

If emissions of CO_2 continue without control, it is projected that an additional 0.3–0.4 decline in pH reading will occur by 2100 (Fabry et al., 2008; Olsen et al., 2018). The decrease in pH will cause ocean acidification, making the ocean more acidic, and this will directly affect calcification processes in corals, molluscs, and other taxa, with tropical species being particularly vulnerable (Figuerola et al., 2021). In coral reefs, declines in aragonite saturation (Ω_arag) below 3.0 by mid-century threaten skeletal formation and long-term reef survival (Klein et al., 2024).

TABLE 10.1
Overview of Community-Based Marine Management Programmes Demonstrating How the Integration of Top-Down Policy Frameworks with Bottom-Up Community Engagement Can Enhance Governance, Strengthen Conservation Outcomes, and Contribute to Regional and Global Biodiversity Targets

No	Focus of Project	Methods	Findings	Source
1	Development of the Pacific Community-based Resource Management (CBRM) toolkit.	Developed participatory tools (e.g., posters, guides, monitoring templates) to support adaptive marine management in Pacific Island nations.	Improved community understanding, encouraged adaptive management, and scaled up CBRM in the Pacific.	Bradford et al. (2020)
2	Co-management models in Southeast Asia.	Studied community-led MPAs and co-managed fisheries in Malaysia and Indonesia.	Inclusive governance enhanced compliance and ecological outcomes, but success relied on addressing socio-economic challenges.	Johnson et al. (2020)
3	Community-based marine management in the Pacific.	Examined customary marine tenure in Melanesia, LMMAs in Fiji, and small-scale fisheries initiatives.	CBRM boosted compliance and ecological effectiveness, though outcomes varied with external pressures.	Marriott et al. (2021)
4	Evaluation of the Fiji Locally Managed Marine Area (FLMMA) network	Conducted a national-scale survey of 350+ communities engaged in the FLMMA network.	Participation increased knowledge and practices, but ecological/social outcomes were uneven.	O'Garra et al. (2023)
5	The Blue Communities programme in Southeast Asia.	Applied marine spatial planning, ecosystem service valuation, and health research across Southeast Asia.	Locally led, interdisciplinary initiatives strengthened conservation and community well-being.	Cheung et al. (2021)

Thermal stress amplifies these effects. Tropical SSTs are rising at 0.2°C–0.3°C per decade, increasingly exceeding thermal thresholds for coral survival (Hopkins et al., 2020). For example, the Great Barrier Reef experienced five mass bleaching events between 1998 and 2020, including the severe 2016 event when 91% of surveyed reefs bleached. These conditions reduce coral fecundity, disrupt larval dispersal, and cause regime shifts towards algae-dominated states (Hughes et al., 2017; Leung et al., 2022). In addition, acidification also alters microbial processes in sediments, weakening nutrient cycling and the ecological functions of benthic habitats (Molari et al., 2018). Such large-scale climate drivers have cascading implications for reef resilience, fish nursery grounds, food security, and livelihoods, and cannot be addressed without global mitigation (Hughes et al., 2017).

10.3.3.2 Anthropogenic Stressors: Overfishing and Plastic Pollution

10.3.3.2.1 Overfishing

Overfishing decreases the number of fish available for harvest and damages seabed (benthic) habitats. It also weakens ecosystems' ability to adapt to climate change (Pham et al., 2023; Issifu et al., 2022). These effects are intensified by destructive practices like bottom trawling, which involves dragging heavy nets across the ocean floor (Pham et al., 2023; Andersen et al., 2024; Issifu et al., 2022). Policy tools such as quotas, gear restrictions, and improved enforcement strategies are crucial to mitigate these impacts. For example, quotas can regulate the quantity of fish that can be harvested, while gear restrictions can minimize habitat destruction. Well-enforced no-take MPAs, where fishing is entirely prohibited, have been shown to restore fish biomass and biodiversity (Edgar et al., 2014), and this was further discussed in the previous section on the MPAs.

10.3.3.2.2 Plastic Pollution

Plastic pollution continues to present a significant anthropogenic challenge. Global plastic production increased by 8.4% annually from 1950 to 2015, reaching approximately 370 million tonnes by 2020, despite major international agreements and initiatives such as the Rio Agreement (1992), Kyoto Protocol (1997), and Paris Agreement (2015) (Geyer et al., 2017; Huerta Lwanga et al., 2017; Borrelle et al., 2017) (Figure 10.2).

Widespread mismanagement of plastic waste persists. In 2022, 6.66 million tonnes of waste were traded internationally, with a substantial portion either incinerated or disposed of in landfills (Asante et al., 2016; Zhao et al., 2025). Southeast Asia is particularly susceptible due to inadequate waste management infrastructure, prevalent open burning, and rapid urbanization (Anuar et al., 2023). Annually, approximately 8 million tonnes of plastic enter the ocean (Jambeck et al., 2015). Microplastics may constitute up to 5% of particulate organic carbon (NOAA Office of Response and Restoration, 2025; Zhao et al., 2025), resulting in disrupted physiological processes, reduced reproductive success, compromised immunity, and trophic-level ingestion, which raises concerns regarding seafood security (Wang et al., 2021). The Great Pacific Garbage Patch, containing an estimated 1.7 trillion particles and weighing 6,400 tonnes, exemplifies the global magnitude of this problem (Lebreton & Andrady, 2019).

Despite the vast amount of data available, there are still many uncertainties about the full extent of plastic pollution. The fact that waste is often underreported and sampling biases are common due to the challenges of tracking it across large areas and in various media can lead to distorted perceptions and analyses. It is crucial to acknowledge these limitations to have a meaningful discussion and create more effective policies.

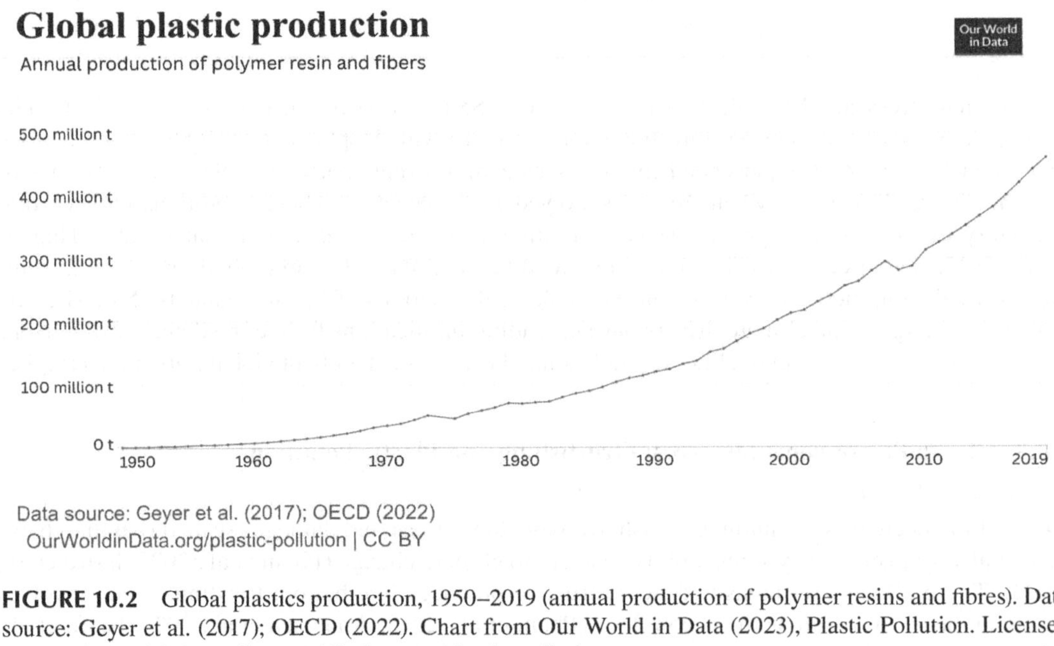

FIGURE 10.2 Global plastics production, 1950–2019 (annual production of polymer resins and fibres). Data source: Geyer et al. (2017); OECD (2022). Chart from Our World in Data (2023), Plastic Pollution. Licensed under CC-BY 4.0. https://ourworldindata.org/plastic-pollution.

10.3.3.3 Targeted Case: TDMPs and Policy Blind Spots

TDMPs account for 5%–10% of global microplastic emissions, approximately 0.81 kg/capita annually, and enter the environment through runoff, road dust, atmospheric deposition, and wastewater (Kole et al., 2017; Wagner et al., 2018). In Peninsular Malaysia, river and estuary surveys attribute TDMP presence primarily to urban runoff and vehicular traffic, with atmospheric deposition also contributing (Sulaiman et al., 2023; Zaki et al., 2024; Abd. Rahim et al., 2025). Despite their prevalence, TDMPs are seldom addressed in policy frameworks, and detection methods such as pyrolysis-GC/MS and FTIR libraries remain limited in application (Ali et al., 2024). The documented ingestion of TDMPs by fish (Noor et al., 2024) highlights the urgent need for source-specific standards and systematic monitoring.

The effectiveness of global and national policies is enhanced when integrated with local source-control measures, such as segregation, collection, and bans. In Southeast Asia, insufficient waste management infrastructure and rapid urbanization continue to drive plastic leakage into marine environments (Anuar et al., 2023). More broadly, effective resource management requires community empowerment and the establishment of connections across multiple governance levels, as illustrated in Table 10.1.

10.3.3.4 Comparative Perspectives and Interactions

Both anthropogenic and natural threats impact marine ecosystem resilience, but they vary in scale and reversibility. Human-related stressors are typically local and can be managed through policies like establishing MPAs, limiting fishing gear, and banning plastics (Raynor et al., 2025; Seguin et al., 2025). Conversely, natural stressors are global and systemic, requiring substantial reductions in greenhouse gas emissions to counteract acidification and warming (Gattuso et al., 2015). Even if the Paris Agreement's 2°C goal is met, considerable ecological disruption is still expected.

These stressors rarely act alone, and their interactions can worsen ecosystem vulnerabilities. For example, coral reefs affected by bleaching are less able to resist overfishing and pollution. Microplastics may also change microbial communities, which can lead to more disease outbreaks and disrupt nutrient cycling, especially at higher temperatures (Zettler et al., 2013; Amaral-Zettler et al., 2020). Therefore, integrated management strategies are needed, combining local efforts to tackle immediate human pressures with global actions to address systemic climate changes.

10.3.3.5 Governance Gaps and Challenges

Although recent advancements are occurring, the expansion of MPAs remains insufficient to achieve effective conservation outcomes. At present, only 5.7% of ocean areas are operationally managed, and a mere 2.8% are fully or highly protected, despite 31 nations reporting that they have achieved 30% protection within their national jurisdictions (UNEP-WCMC & IUCN, 2024). Common shortcomings include unclear objectives, poor enforcement, and limited stakeholder involvement (Giakoumi et al., 2018; Ban et al., 2019).

To prevent habitat destruction and overexploitation of fish populations, measures such as quotas, catch limits, gear restrictions, and no-take MPAs are essential for managing sustainable marine environments (Pham et al., 2023; Andersen et al., 2024; Issifu et al., 2022). Despite these measures, bottom trawling and dredging continue in certain regions, resulting in benthic habitat degradation and the release of carbon from seabeds, which weakens climate regulation (Issifu et al., 2022; Andersen et al., 2024). In addition, exclusion of local communities from management processes can lead to the displacement of fishing effort, increased conflict, and the transfer of environmental and economic costs.

10.4 SCIENTIFIC & TECHNOLOGICAL APPLICATIONS

In the previous sections, issues concerning governance and community engagement for better marine conservation were discussed. Scientific and technological innovation has become one of the key elements in marine conservation because it provides improved monitoring, quicker detection, and more effective ecosystem management.

10.4.1 Genetics Tools

Genetics is an emerging technique used in marine conservation. Genetic tools such as DNA Barcoding and eDNA metabarcoding have become essential practices among scientists and fisheries management.

DNA barcoding has been proposed as a valuable method by Hebert et al. (2003). This technique uses a short, specific segment of the mitochondrial cytochrome c oxidase I (COI) gene, approximately 658 bp, to identify animal species. It is a rapid and reliable method for taxonomy and has been particularly valuable in marine ecosystems for identifying commercially and ecologically important taxa, even from partial specimens or different life stages (Keskİn & Atar, 2013; Sun et al., 2015).

Ward et al. (2005) demonstrated its effectiveness by accurately distinguishing 207 fish species, including teleosts, sharks, and rays, based on COI sequence divergence, with clustering patterns aligning with traditional taxonomy. Building on earlier mitochondrial DNA applications in the 1980s and 1990s, used for species identification, the applications of DNA barcoding as a molecular technique are highly diverse and ultimately important in conservation strategies. The applications of DNA barcoding, among others, include:

- Enables precise identification of cryptic species that are morphologically identical but genetically distinct, common in marine environments (Moura et al., 2008; Brasier et al., 2016). DNA barcoding of 723 individuals from 36 putative Carangidae species in Indo-Malay waters revealed cryptic diversity. This study offers a framework for managing and conserving important fisheries resources in the region (Mat Jaafar et al., 2012).
- Facilitates the identification of different life cycle stages, as most juveniles are very distinct from their adult forms, a task often challenging in marine ecosystems (Chu et al., 2019). Accurate larval identification is vital for fisheries management and habitat protection.
- Assessing seafood authentication and safety (Galimberti et al., 2013). Fish species substitution is common, especially in imported products that are hard to identify after processing and freezing (Filonzi et al., 2010). A study by Lowenstein et al. (2009) found that tuna in sushi served in US restaurants was identified as an endangered species, highlighting cases of food fraud that could pose health risks. Another example of food fraud is a study that discovered 41% of 254 Canadian seafood samples were mislabelled (Hanner et al., 2011).

DNA barcoding has proven to be an effective tool for marine biodiversity assessment and conservation, especially when combined with advanced techniques such as high-throughput sequencing, which can demonstrate multiple applications in marine conservation strategies.

Further advanced techniques such as Environmental DNA (eDNA) and eDNA metabarcoding enhance traditional barcoding by allowing non-invasive biodiversity monitoring through the direct collection of genetic material from seawater (Zhang et al., 2023). Defined as the use of PCR with universal primers followed by high-throughput sequencing, eDNA metabarcoding allows for the detection of multiple species at once (Deiner et al., 2017; Ruppert et al., 2019). Building momentum since the 2000s (Ficetola et al., 2008), eDNA has advanced rapidly between 2020 and 2025, often outperforming or supplementing traditional methods, especially in environments with challenging sampling conditions. Its applications range from fish and invertebrates to elusive marine mammals (Osathanunkul & Suwannapoom, 2023; Morin et al., 2021), with higher species detection efficiency than traditional survey methods in numerous studies (Gold et al., 2021; Roblet et al., 2024).

Examples:

- Spear et al. (2021) reviewed the utility of eDNA for vertebrate biodiversity assessment. They highlighted its usefulness in inaccessible aquatic systems and for species that are nocturnal, elusive, or exist in low densities.

- Bucklin et al. (2020) emphasized its importance in oceanic plankton sampling, particularly when dealing with small, fragile, or morphologically similar taxa that traditional net tows can damage or misidentify.
- Leblanc et al. (2020) developed and validated species-specific qPCR assays for nine invasive marine invertebrates. They observed that eDNA detection was as effective or more effective than traditional biofouling monitoring. This highlights its policy significance, as earlier detection enables quick responses and helps reduce impacts on aquaculture and ecosystems.

10.4.2 Artificial Intelligence (AI)

Advanced analysis of large ocean datasets used in marine monitoring is enabled by revolutionary artificial intelligence (AI) technologies. Research in an advanced interdisciplinary field has facilitated the merging of biology and computer vision (Zhou et al., 2025), a progress that was once impossible prior to AI technology.

Using AI-powered systems, satellite images, and aquatic invasive species (AIS) data can now analyse illegal fishing activities, classify marine species in underwater visuals, and monitor habitat changes in real time (Molina-Molina et al., 2021; MIT Press Reader, 2024).

Recent advances in AI, particularly deep learning methods, have accelerated the development of image-based tools for automatic marine species recognition (Zhou et al., 2025). Deep learning models such as DeepMind's GenCast enhance oceanographic predictions by delivering high accuracy in estimating SSTs and currents (Stowell, 2022; U.S. Naval Institute, 2025). AI-powered acoustic analysis is gaining popularity, utilizing deep learning to automate the identification of marine species from underwater soundscapes (Stowell, 2022).

The technology and advancements in AI have enabled continuous monitoring of vocalizing species, such as cetaceans, and gained insights into their distribution, behaviour, and population dynamics. (Nur Korkmaz et al., 2023; Risch et al., 2019; Sattar, 2023). In addition, AI-driven acoustic identification systems integrated into various autonomous platforms, such as underwater gliders, buoys, and stationary recording units, are able to assist in large-scale and long-term data collection across diverse aquatic environments (Buchan et al., 2023; Han et al., 2021; Li et al., 2020; Yassir et al., 2023).

AI technology also contributed to climate studies by assisting in improving predictions of ocean thermal trends, chlorophyll-a concentrations, and near-surface humidity levels, which are critical parameters for assessing marine ecosystem health and climate vulnerability (Nieves et al., 2023). This is valuable because autonomous drones and smart sensor networks are increasingly used to continuously monitor MPAs and vessel activity (Luo et al., 2022; Molina-Molina et al., 2021).

Enhanced species recognition applications serve as a valuable tool for detecting non-indigenous species before they establish as AIS (Wäldchen & Mäder, 2018). Gaining on this, AI has significantly advanced species identification through the application of machine learning and neural networks, thereby transforming biodiversity monitoring and supporting early detection of non-indigenous species. Examples,

- Platforms like iNaturalist use AI to assign taxonomic names to species based on images, which experts then verify (Van Horn et al., 2018).
- AI is already transforming marine biodiversity monitoring through applications in plankton, fish, and benthic species detection through the RAPID system, which uses Edge AI to classify plankton in real time at sea, reducing reliance on slow manual processing and enabling near-instant ecosystem insights (Pitois et al., 2025).
- Convolutional neural networks trained with synthetic datasets accurately identify fish species, expanding monitoring capacity even when natural training data are limited (Allken et al., 2019)

- For Benthic organisms, platforms like CoralNet, VIAME, and TagLab automate benthic image annotation (Costa et al., 2022), and BenthicNet compiled a large seafloor image dataset that supports deep learning models for species and habitat recognition (Lowe et al., 2025).

Aside from biological monitoring, the use of AI in detecting pollution in the marine environment has also been developed. Examples:

- AI has enhanced Synthetic Aperture Radar (SAR) imagery for semi-automatic oil spill detection. Specifically, methods such as convolutional neural networks (CNNs), Mask R-CNN, and TransUNet have substantially improved segmentation and classification of oil slicks (Solberg et al., 2007; Liu et al., 2019).
- Artificial neural networks (ANNs), Random Forests, and deep learning have been increasingly applied to predict and monitor key water quality parameters, including ammonia (Ghayas et al., 2020), combined biosensors with machine learning models (MLP, Random Forest, XGBoost) to predict mercury (Hg^{2+}) with high precision (Pennacchio et al., 2022).
- Drones, satellites, and shipborne cameras monitor macroplastic pollution. AI algorithms like YOLO and CNNs detect floating plastic debris, including bottles and bags (Kikaki et al., 2020).

10.4.3 Potential Limitations and Challenges in Scientific and Technological Initiatives

Although science and technology have provided practical tools in marine conservation, they have limitations and challenges that should be acknowledged to prevent misinterpretation of results, overreliance on single methods, and gaps in management decisions.

DNA barcoding and eDNA techniques delivered rapid and accurate results, making them a reliable and powerful tool that provides a significant advantage in marine conservation. However, these tools often depend heavily on reference libraries that contain reliable taxonomy and high-quality sequences, such as well-curated reference databases like BOLD and Genbank, and many marine groups remain underrepresented, restricting species-level identification (Bucklin et al., 2011; Weigand et al., 2019). The issue of misidentifications also impacts the database itself, where some species have been incorrectly identified and, as a consequence, wrongly uploaded sequences, thereby challenging the integrity of the reference database (Bemis et al., 2023).

Adding to the database challenges, the use of eDNA also encounters further methodological limitations. eDNA is prone to contamination, which can cause false positives, situations where a species is detected in a sample but is not actually present in the environment (Ficetola et al., 2016). Such errors may arise from DNA contamination during laboratory handling, the transport of aged eDNA by water currents, or the persistence of non-viable DNA from dead organisms (Ficetola et al., 2016). Furthermore, DNA degradation rates vary with environmental factors such as temperature, salinity, and microbial activity, influencing detection accuracy (Goldberg et al., 2015; Collins et al., 2018). These factors can bias detection and complicate interpretation. Moreover, the lack of standardized global protocols for sampling and analysis makes comparisons across studies difficult (Deiner et al., 2017).

Although AI has created new opportunities for species recognition, monitoring, and forecasting, it also encounters significant challenges. To develop reliable models, large, high-quality datasets are essential, but marine data are often limited or biased towards specific regions and taxa (Norouzzadeh et al., 2018; Valavi et al., 2021). The "black box" nature of deep learning models also makes it difficult for managers to interpret results and confidently incorporate them into policy (Carvalho et al., 2019; Alotaibi & Nassif, 2024). Training and deploying AI systems require substantial computational resources and technical infrastructure, which may be unavailable in resource-limited regions (Alotaibi & Nassif, 2024; Crosby et al., 2023) and might not be practical in environments with constrained resources. Moreover, over-reliance on automated systems risk marginalizing TEK and local participation, which are crucial for effective governance and conservation. While Berkes (2017) highlights the importance of TEK in sustainable management, recent research on AI for ocean science emphasizes that equitable, human-centred design is necessary to prevent excluding local stakeholders (Crosby et al., 2023).

These challenges collectively highlight the necessity of integrating genetic and AI methodologies with traditional monitoring and community-based approaches, which have been discussed in previous sections. The integration of advanced technologies with field ecology, governance structures, and stakeholder engagement is expected to produce more robust and equitable outcomes in marine conservation.

10.5 RESTORATION AND CONSERVATION EFFORTS

Coastal biodiversity and resilience depend on ecosystems like mangrove forests and coral reefs. By offering vital ecosystem services like fisheries production, coastal protection, carbon sequestration, and climate regulation, they support both ecological processes and human well-being. However, widespread degradation has been caused by increasing anthropogenic pressures and the effects of climate change, prompting urgent calls for restoration on a global scale.

10.5.1 Coral Reef Restoration

Using methods like coral gardening, larval propagation, micro-fragmentation, and substrate stabilization to rebuild damaged structures and restore ecosystem function, coral restoration has emerged as a key tactic to combat reef decline (Boström-Einarsson et al., 2020; Bayraktarov et al., 2020). In addition to restoring reef biodiversity, these interventions seek to improve resilience and guarantee vital services like coastal defense and fisheries support.

Initiatives for coral restoration face ongoing difficulties despite their potential. Impact is limited by low post-transplant survival, limited scalability, insufficient long-term monitoring, and funding shortages (Shaver et al., 2022). Although genetic diversity is frequently disregarded, it is essential for the ability to adapt to disease and environmental stressors. Furthermore, failure may result from improper site selection, such as planting in regions with poor water quality or inappropriate hydrodynamics (Bayraktarov et al., 2020).

Successful models, however, show what can happen when ecological science, community involvement, and policy support come together. The largest coral restoration nation in the world, Indonesia, has led the way in large-scale, community-led initiatives that have enhanced coastal resilience, biodiversity, and reef structure (Watt-Pringle et al., 2024). Out-planted corals have endured and expanded in the Caribbean and Pacific, adding to the complexity of the habitat and attracting (Boström-Einarsson et al., 2020; Razak et al., 2022). As global climate action advances, restoration is increasingly framed within ecosystem-based management and climate adaptation strategies, acting as a supplement to emissions mitigation rather than as a replacement for it (Lamont et al., 2022).

10.5.2 Mangrove Reforestation

Another essential component of coastal resilience is mangrove forests, which are most prevalent in Southeast Asia. Mangroves are vital for both biodiversity conservation and climate change mitigation because they are highly productive and ecologically significant. They buffer storm surges, support fisheries, prevent shoreline erosion, and act as significant carbon sinks (Friess et al., 2016; Alongi, 2018). In addition to providing water filtration, sediment stabilization, nutrient cycling, and a variety of materials and resources for coastal communities, they act as nurseries for commercially significant species (Bosire et al., 2008; Su et al., 2021).

Despite their significance, the amount of mangrove cover worldwide—currently 147,000 km^2 spread across 123 countries—continues to decrease as a result of pollution, agriculture, coastal development, aquaculture expansion, and sea level rise (Global Mangrove Watch, 2022; Lovelock et al., 2022). Nearly 25% of the world's mangroves are found in Indonesia alone, but degradation has continued and more than one-third of the global extent has been lost in recent decades (Bunting et al., 2022).

With goals ranging from reestablishing forests in deforested areas to restoring degraded stands, restoration has become a priority response. However, species appropriateness, hydrological compatibility, and careful site selection are necessary for success. Monoculture planting, disregard for tidal dynamics, and inadequate local participation are common mistakes (Bosire et al., 2008). The concepts of ecological alignment, long-term monitoring, and community co-management are increasingly serving as the foundation for restoration initiatives.

Well-planned initiatives show positive results. Effective reforestation has been achieved in the Philippines and Malaysia through government-NGO-community partnerships, especially when connected to local livelihoods like ecotourism or sustainable harvesting (Gerona-Daga & Salmo, 2022; Azman et al., 2021). In India, adaptive co-management strategies have also resulted in increased community well-being and ecological benefits (Su et al., 2021). Mangrove restoration is now acknowledged as a key natural solution by international frameworks like the Blue Carbon Initiative (2023) and the UN Decade on Ecosystem Restoration (UNEP & FAO, 2020).

Ultimately, coral and mangrove restoration illustrate how science, community stewardship, and policy innovation can converge to recover ecosystem services and biodiversity. While not substitutes for addressing root drivers of degradation, they are vital components of integrated coastal resilience strategies that support both thriving marine ecosystems and sustainable human development.

10.5.3 Seagrass and Seaweed Restoration

Important coastal ecosystems that support fisheries, biodiversity, and carbon sequestration are seagrass meadows and seaweed beds. Seagrasses stabilize sediments and improve water clarity while providing vital nursery habitats for commercially significant fish, invertebrates, and endangered species like dugongs and green turtles. Seaweeds, including commercially significant species like Eucheuma and Kappaphycus, are essential to food webs, coastal productivity, and nutrient cycling. These habitats are disappearing both internationally and locally, despite their ecological and economic significance. Large-scale seagrass meadows in Johor (Pulau Tinggi, Pulau Sibu, and Tanjung Kupang) in Malaysia have been deteriorated by sedimentation, reclamation, and coastal development, endangering biodiversity and coastal communities' means of subsistence.

Through seeding, transplantation, and integration with aquaculture methods, restoration efforts have started to address these losses. In order to preserve dugong habitats, community and non-governmental organization-led initiatives in Pulau Tinggi, Malaysia, have tested seagrass transplantation in addition to awareness campaigns. Regenerative farming programs for seaweed in the Philippines, like those started by Coast 4C, are fusing livelihood diversification and conservation. These projects improve ecosystem functions like carbon sequestration and water filtration by growing seaweed in degraded coastal areas, while also giving local fishermen a source of income. Direct planting of seagrass shoots, the use of biodegradable seedling substrates, and the transplantation of nursery-grown seaweed fragments onto both natural and artificial substrates are examples of restoration techniques. Sustained local involvement, careful site selection, and water quality management are essential for success.

Southeast Asian case studies show the advantages of connecting restoration to policy support and community well-being. Regenerative seaweed farming has improved household incomes and climate resilience while restoring damaged reef flats in the Philippines. NGO-community collaborations in Johor and Sabah, Malaysia, have combined mangrove and seagrass conservation with ecotourism and education to increase public awareness of the ecological services these ecosystems provide. At the regional level, these initiatives are in line with the Blue Carbon Initiative and the

UN Decade on Ecosystem Restoration, which present seaweed and seagrass restoration as scalable, natural solutions that concurrently address climate change, fisheries decline, and biodiversity loss.

10.5.4 Case Studies and Best Practices

Case Study 1: WildAid Marine Program: Combating Illegal Fishing and Promoting Sustainable Marine Conservation Practices

WildAid's Marine Program was awarded the Earthshot Prize in 2023 for its groundbreaking work in tackling illegal, unreported, and unregulated (IUU) fishing and promoting sustainable marine conservation practices. The organization focuses on strengthening marine enforcement and advocating for the protection of endangered marine species.

Through a combination of community-based solutions, advocacy campaigns, and on-the-ground enforcement, WildAid is fighting illicit fishing operations in vital marine areas. Educating stakeholders and local communities about sustainable fishing methods and making sure conservation initiatives complement the livelihoods of fishing-dependent communities constitute a substantial part of their work.

Additionally, WildAid works with international organizations, regional NGOs, and governments to improve marine governance and guarantee that laws are carried out successfully. In regions like Southeast Asia, where illicit fishing practices have seriously threatened marine biodiversity, their program has already had a positive impact.

Reference: https://marine.wildaid.org/

Case Study 2: Coast 4C: Regenerative Seaweed Farming for Coastal Conservation and Community Resilience

Based in the Philippines, Coast 4C is a ground-breaking project that uses regenerative seaweed farming to combine coastal agriculture and conservation. In addition to helping local communities, seaweed farming, which has long been a sustainable practice in coastal areas, is being used to restore marine ecosystems.

By integrating seaweed farming into conservation strategies, Coast 4C promotes coastal resilience and helps sequester carbon, which is crucial in mitigating climate change. The initiative focuses on using seaweed to restore degraded coastal ecosystems, creating more resilient marine habitats and providing vital ecosystem services, such as coastal protection and water filtration. Furthermore, seaweed farming provides a sustainable source of income for coastal communities, aligning economic growth with environmental conservation.

This strategy could spread throughout the Philippines and other Southeast Asian countries, benefiting society and the environment. Coast 4C serves as an example of how community-driven solutions can be combined with marine conservation by tackling poverty, marine degradation, and climate change.

Reference: https://www.coast4c.com/

Case Study 3: Coral Vita: Restoring Coral Reefs with Resilient, Fast-Growing Coral Species for Ecosystem Revival

The Bahamas-based Coral Vita is well known for its cutting-edge methods of restoring coral. Overfishing, pollution, and climate change are all contributing to the widespread decline of coral reefs, which sustain about 25% of marine species. By creating a scalable model for coral restoration, Coral Vita is tackling this crisis.

Their method entails cultivating robust, quick-growing corals in a controlled setting where growth-promoting conditions can be adjusted. In order to aid in the restoration of damaged reefs, these corals are subsequently moved there. By creating corals that are more resilient to diseases, temperature swings, and other stressors, Coral Vita's methods seek to hasten the natural recovery of coral reefs. They are enhancing coral ecosystems' general health and biodiversity by working with a variety of coral species.

Due to the project's effectiveness, a number of coral reefs have been restored, including previously thought-to-be-lost coral populations. One encouraging example of how science and technology might work together to protect marine biodiversity and restore important ecosystems is Coral Vita's work.

Reference: https://www.coralvita.org/

Case Study 4: Living Seawalls: Enhancing Marine Biodiversity through Innovative Urban Coastal Habitat Restoration

An inventive Australian initiative known for converting urban seawalls into habitats that promote marine biodiversity is called Living Seawalls. The effort, which is headed by the Sydney Institute of Marine Science, tackles the problem of coastal urbanization, which results in "dead zones" for marine life when concrete infrastructure replaces natural ecosystems like mangroves and intertidal zones.

In order to support marine species including barnacles, algae, and molluscs, the Living Seawalls project uses specially made tiles that mimic natural coastal features. These tiles improve biodiversity by providing habitats for tiny fish, crabs, and invertebrates. The system, which is adaptable to different coastal circumstances, offers carbon sequestration, water filtration, biological restoration, and coastal preservation. It also encourages tourism and community involvement in conservation.

Successful implementations of Living Seawalls in Sydney, including at the Sydney Opera House, showcase its potential to restore marine ecosystems in urban areas. The project highlights how environmental conservation can be integrated with urban development, offering a scalable solution to coastal degradation, urbanization, and climate change.

Reference: https://www.livingseawalls.com.au/

10.6 PROPOSED FRAMEWORK FOR MARINE ECOLOGICAL RESTORATION

Marine ecological restoration has long relied upon site-specific, direct interventions—such as coral transplantation, mangrove planting, or artificial reef deployment. Localized strategies will bring early benefits; however, they often are not equipped to address root stressors and lack the integration, scalability, and long-term resilience that are vital for success in a rapidly changing global oceans (Walters, 1986). To progress towards a sustainable, ecosystem-scale recovery, this section will introduce an integrated framework that will incorporate innovative engineering and stable socio-political structures, organized into three important phases.

10.6.1 Foundational Analysis and Integrated Seascape Design

Effective restoration is predicated upon a comprehensive understanding of both the damaged system and its broader context. This requires a multi-disciplinary baseline assessment followed by strategic, seascape-scale planning.

10.6.1.1 The Multi-Disciplinary Baseline

A successful restoration project must move beyond simple ecological surveys to incorporate physical and socio-economic realities.

- **Ecological and physical parameters**: The baseline must quantify not only biodiversity (species richness and functional diversity) but also ecosystem function. This includes measuring primary productivity (oxygen flux measurements) and critical water quality indicators (e.g., nutrient loading), which directly reflect ecosystem health (Basconi et al., 2019). Hydrodynamic models are deployed to map larval dispersal pathways and sediment transport, which are essential for selecting sites where restoration efforts are naturally supported and connected to regional ecological flows (Billionnet, 2021).
- **Socio-economic context**: Recognizing local dependence on the marine environment is essential. By applying tools such as contingent valuation methods, the economic and cultural value of lost ecosystem services can be quantified, ensuring that restoration goals support community well-being. Additionally, integrating TEK helps make interventions both culturally relevant and socially acceptable Mehvar et al., 2018).

10.6.1.2 Strategic Seascape Planning

Effective planning should extend beyond individual projects to encompass the whole seascape. By analysing Cumulative Anthropogenic Pressures, we can identify priority areas and set clear, measurable goals (Madramootoo & Virgil, 2024). For example, instead of only restoring an area, an objective could aim for a 30% rise in targeted fish biomass over 5 years, emphasizing both structural restoration and ecological function.

10.6.2 Implementation with Innovation and Adaptive Management

Nature-Based Solutions (NBS) are becoming increasingly important in ecosystem management and should be a major focus, as they address advanced eco-engineering based on a flexible, evidence-based management plan.

10.6.2.1 Advancing with NBS

The benefits of applying NBS in an ecosystem management plan include ecological recovery through natural processes, especially in coastal vegetated ecosystems such as mangroves and seagrass meadows. Some examples are:

- **Blue carbon and climate mitigation**: They are considered the world's most efficient carbon sinks. Global studies estimate that organic carbon stored in the top metre of soil is around 280 Mg C/ha in mangroves and 140 Mg C/ha in seagrass meadows, which are typically two to four times greater than those of mature tropical forests in the annual sequestration rates. Hence, they are key assets in blue carbon financing schemes (The Blue Carbon Initiative, 2025).
- **Coastal defense**: These habitats demonstrate proven engineering functionality. Mangrove sizes can reach up to 100 m wide and have been recorded to reduce 70% of incoming wave heights (van Wesenbeeck et al., 2025). Meanwhile, seagrass meadows can reduce wave heights by up to 49% compared to adjacent barren areas, helping mitigate erosion and increasing stabilized sedimentation (Reidenbach & Thomas, 2018).

10.6.2.2 Eco-Engineering for Habitat Enhancement

It is critical to apply eco-engineering where the areas required hard infrastructure or accelerated recovery. The process includes deploying materials such as eco-concrete, which are specially designed with optimized surface textures and chemical properties (like pH levels) to promote marine settlement. Monitoring indicates that ecologically-designed structures attract a higher variety and number of organisms compared to conventional concrete. Biofouling, including 57% biogenic carbonate coverage, additionally enhances the structure's long-term durability (Gutiérrez et al., 2022; Yee et al., 2025).

10.6.2.3 Adaptive Management

Adaptive management is crucial because it functions as a vital feedback mechanism in resilient restoration projects, acting as a structured process of "learning by doing" (Walters, 1986). The emphasis in this type of management lies in monitoring, focusing on clear performance indicators (e.g., fisheries Catch Per Unit Effort, and CPUE) and ecological metrics (e.g., resilience to thermal stress). Adaptive management requires the pre-definition of decision triggers (Thom, 2000). For example, if monitoring shows coral survival dropping below a predetermined threshold (e.g., 50%), it activates an immediate, pre-planned intervention such as installing passive shading devices. Long-term monitoring (≥10 years) is crucial for verifying true success and informing future large-scale initiatives, as ecological processes develop gradually (Potharaju & Aruna, 2023).

10.6.3 Sustainable Governance and Financing

10.6.3.1 Inclusive Governance and Stewardship

For a successful restoration, governance should be inclusive because it can bring together many stakeholders from diverse backgrounds, such as local communities, fishers, NGOs, and policymakers, acting as active co-managers (Vasseur, 2021). This type of participatory approach will foster local ownership and strengthen enforcement. One example is the Cabo Pulmo Marine Reserve in Mexico, where a strong community-led enforcement was able to contribute to an increase in total fish biomass by 463% within a decade (Aburto-Oropeza et al., 2011). This action further reinforces many capacity-building initiatives towards local stewardship.

10.6.3.2 Policy and Financial Coherence

Achieving sustainable restoration needs funding sources that extend beyond short-term grants and formal policy support, such as:

- **Diverse financing**: Mechanisms such as blue carbon credits derived from the verifiable sequestration capacity of restored coastal habitats must be included, as well as Public-Private Partnerships (PPPs), which can create market-based incentives for conservation (The Blue Carbon Initiative, 2025).
- **Policy integration**: National governance must adopt restoration as a strategic component in marine ecosystem management. Incorporating project objectives into official documents like Marine Spatial Plans (MSP) and National Adaptation Plans (NAP) helps ensure policy consistency, reduces conflicts with sectors such as coastal development, and establishes essential long-term regulatory frameworks (Stern, 2018).

10.7 FUTURE DIRECTIONS AND RECOMMENDATIONS

Building on the proposed framework, future efforts in marine ecological restoration should focus on scaling up interventions, strengthening governance, and embedding restoration within broader sustainability agendas (Songwe et al., 2022). The following directions are recommended:

10.7.1 Policy Integration and Climate-Resilient Strategy

This direction focuses on ensuring restoration efforts are institutionalized, financially secure, and robust against climate change.

- **Mainstreaming restoration into policy and planning**: National marine spatial plans, biodiversity strategies, and climate adaption policies should all fully incorporate marine restoration (Wright et al., 2021). Long-term dedication and institutional support are ensured by integrating restoration goals into national and regional frameworks.

- **Preparing for climate uncertainty**: Restoration plans need to be climate resilient because maritime stressors are getting worse due to climate change (IPCC, 2022). This entails choosing coral genotypes that can withstand heat, replanting mangroves in regions where sea level rise is anticipated, and designing projects with future climate scenarios in mind. Planning ahead guarantees that interventions continue to be successful in the face of shifting circumstances (Caruso et al., 2021).
- **Innovating financing mechanisms**: Sustainable financing remains a critical bottleneck. New opportunities exist in blue carbon credits, biodiversity offsets, and blended finance models combining public, private, and philanthropic funding (Tirumala & Tiwari, 2022). Establishing long-term trust funds or regional restoration networks could secure continuity beyond project cycles.

10.7.2 Scaling Nature-Based Solutions and Community Stewardship

This focuses on what interventions are implemented and who leads them on the ground.

- **Scaling nature-based solutions**: Because of its numerous co-benefits for biodiversity, fisheries, coastal protection, and carbon sequestration, nature-based approaches—such as mangrove rehabilitation, oyster reef construction, and seagrass meadow restoration—should be prioritized (Reidenbach & Thomas, 2018). Hybrid solutions that combine ecological and engineering components can be expanded in regions where coastal infrastructure already predominates.
- **Strengthening community co-management**: Long-term restoration success relies on the meaningful involvement of local communities, particularly small-scale fishers and coastal residents (Raha et al., 2024). Moving forward, co-management frameworks, benefit-sharing arrangements, and capacity-building programs should be prioritized to empower local stakeholders as enduring stewards.

10.7.3 Data-Driven Adaptive Management

Strong monitoring systems that record ecological, social, and economic results are necessary for future restoration initiatives. Open-access data systems and standardized indicators help speed up adaptive management and enable cross-regional learning. Monitoring capacity can be increased by investing in digital technologies such citizen science applications, AI-driven habitat mapping, and remote sensing (Piazzolla et al., 2024).

10.8 CONCLUSION

Marine biodiversity preservation is a moral, ecological, and socioeconomic necessity rather than just a scientific goal. The importance of marine ecosystems for maintaining human livelihoods, climatic resilience, and the health of the world was highlighted in this chapter. Marine biodiversity provides essential ecosystem services that support billions of people, control global biogeochemical cycles, and spur scientific and cultural growth. These services range from the rich productivity of mangroves and coral reefs to the unrealized promise of deep-sea ecosystems.

However, there is a serious threat to this variety. Marine systems have reached tipping points due to a combination of natural stresses made worse by climate change and anthropogenic pressures such pollution, overfishing, and habitat destruction. The intricacy of marine conservation concerns is highlighted by the synergistic interactions between different stressors, such as how ocean acidification weakens reef regrowth or microplastic contamination exacerbates climate vulnerability. The Malaysian case of tire-derived microplastics highlights the pressing need to address new contaminants and close important research gaps that jeopardize food safety and ecosystems.

But there is also optimism in this chapter. A sustainable future is achievable with strong governance frameworks like UNCLOS and the CBD, inclusive community-based models, scientific advances in eDNA and AI-driven monitoring, and scalable restoration initiatives. Examples like Coral Vita, Coast 4C, and Living Seawalls show how innovation can restore biodiversity in even the most urbanized or degraded regions when combined with local expertize and community involvement.

The future of marine biodiversity depends on three interconnected priorities: developing interdisciplinary cooperation that connects science, policy, and indigenous knowledge; raising public awareness of ocean issues through literacy and education; and utilizing cutting-edge technologies in fair and moral ways. The global momentum behind frameworks like the Kunming-Montreal GBF and the UN Decade on Ecosystem Restoration offers a once-in-a-lifetime chance to match national initiatives with international goals.

In the end, we cannot separate the fate of the ocean from our own. A healthy ocean environment is essential for intergenerational justice, climate adaption, and sustainable development; it is not a luxury. Therefore, protecting marine biodiversity is important for the future of life on Earth as well as for the preservation of individual species.

DECLARATION OF GENERATIVE AI USE

During the preparation of this manuscript, the author(s) used Grammarly and to review the grammar and refine the writing, and Canva to enhance the clarity and appearance of the figures. The author(s) subsequently reviewed and edited the content and take(s) full responsibility for the final version of the published chapter.

REFERENCES

Abd Rahim, N. N., Peng, P. W. Y., Shahrir, N. F., Wan Mahiyuddin, W. R., Sayed Mohamed Zain, S. M., & Ismail, R. (2025). Characteristics, distribution, and sources of atmospheric microplastics in Southeast Asia: A scoping review. *Atmosphere, 16*(5), 515.

Aburto-Oropeza, O., Erisman, B., Galland, G. R., Mascarenas-Osorio, I., Sala, E., & Ezcurra, E. (2011). Large recovery of fish biomass in a no-take marine reserve. *PLoS One, 6*(8), e23601.

Alcala, A. C., & Russ, G. R. (2006). No-take marine reserves and reef fisheries management in the Philippines: A new people power revolution. *AMBIO: A Journal of the Human Environment, 35*(5), 245–254.

Ali, A. M., Khalid, A. A., Abd Razak, N. I., Maulana, N. S. M., Roslan, N. S., Razmi, R. S. B., Ruseli, W. M., Ibrahim, Y. S., Jaafar, M., Shahrudin, R., & Ismail, K. (2024). A review on the presence of microplastics in environmental matrices within Southeast Asia: Elucidating risk information through an analysis of microplastic characteristics such as size, shape, and type. *Water Emerging Contaminants & Nanoplastics, 3*(2), 12.

Allken, V., Handegard, N. O., Rosen, S., Schreyeck, T., Mahiout, T., & Malde, K. (2019). Fish species identification using a convolutional neural network trained on synthetic data. *ICES Journal of Marine Science, 76*(1), 342–349.

Alongi, D. M. (2002). Present state and future of the world's mangrove forests. *Environmental Conservation, 29*(3), 331–349.

Alongi, D. M. (2012). Carbon sequestration in mangrove forests. *Carbon Management, 3*(3), 313–322.

Alongi, D. M. (2018). *Blue Carbon: Coastal Sequestration for Climate Change Mitigation* (pp. 23–36). Cham, Switzerland: Springer.

Alotaibi, E., & Nassif, N. (2024). Artificial intelligence in environmental monitoring: In-depth analysis. *Discover Artificial Intelligence, 4*(1), 84.

Amaral-Zettler, L. A., Zettler, E. R., & Mincer, T. J. (2020). Ecology of the plastisphere. *Nature Reviews Microbiology, 18*(3), 139–151.

Andersen, N. F., Cavan, E. L., Cheung, W. W., Martin, A. H., Saba, G. K., & Sumaila, U. R. (2024). Good fisheries management is good carbon management. *Ocean Sustainability, 3*(1), 17.

Anuar, S. T., Abdullah, N. S., Yahya, N. K. E., Chin, T. T., Yusof, K. M. K. K., Mohamad, Y., Azmi, A. A., Jaafar, M., Mohamad, N., Khalik, W. M. A. W. M., & Ibrahim, Y. S. (2023). A multidimensional approach for microplastics monitoring in two major tropical river basins, Malaysia. *Environmental Research, 227*, 115717. https://doi.org/10.1016/j.envres.2023.115717

Arrieta, J. M., Arnaud-Haond, S., & Duarte, C. M. (2010). What lies underneath: Conserving the oceans' genetic resources. *Proceedings of the National Academy of Sciences, 107*(43), 18318–18324.

Asante, K. A., Pwamang, J. A., Amoyaw-Osei, Y., & Ampofo, J. A. (2016). E-waste interventions in Ghana. *Reviews on Environmental Health, 31*(1), 145–148. https://doi.org/10.1515/reveh-2015-0047

Azman, M. S., Sharma, S., Shaharudin, M. A. M., Hamzah, M. L., Adibah, S. N., Zakaria, R. M., & MacKenzie, R. A. (2021). Stand structure, biomass and dynamics of naturally regenerated and restored mangroves in Malaysia. *Forest Ecology and Management, 482*, 118852.

Ban, N. C., Gurney, G. G., Marshall, N. A., Whitney, C. K., Mills, M., Gelcich, S., Bennett, N. J., Meehan, M. C., Butler, C., Ban, S., & Tran, T. C. (2019). Well-being outcomes of marine protected areas. *Nature Sustainability, 2*(6), 524–532.

Barbier, E. B., Hacker, S. D., Kennedy, C., Koch, E. W., Stier, A. C., & Silliman, B. R. (2011). The value of estuarine and coastal ecosystem services. *Ecological Monographs, 81*(2), 169–193.

Basconi, L., Cadier, C., & Guerrero-Limón, G. (2019). Challenges in marine restoration ecology: How techniques, assessment metrics, and ecosystem valuation can lead to improved restoration success. In *YOUMARES 9—The Oceans: Our Research, Our Future: Proceedings of the 2018 Conference for YOUng MArine RESearchers in Oldenburg, Germany* (pp. 83–99). Springer International Publishing.

Bavington, D. (2011). *Managed Annihilation: An Unnatural History of the Newfoundland Cod Collapse*. Vancouver, British Columbia, Canada: UBC Press.

Bayraktarov, E., Banaszak, A. T., Montoya Maya, P., Kleypas, J., Arias-González, J. E., Blanco, M., Calle-Triviño, J., Charuvi, N., Cortés-Useche, C., Galván, V., & García Salgado, M. A. (2020). Coral reef restoration efforts in Latin American countries and territories. *PLoS One, 15*(8), e0228477.

Bemis, K. E., Girard, M. G., Santos, M. D., Carpenter, K. E., Deeds, J. R., Pitassy, D. E., Flores, N. A. L., Hunter, E. S., Driskell, A. C., Macdonald, K. S., III, & Weigt, L. A. (2023). Biodiversity of Philippine marine fishes: A DNA barcode reference library based on voucher specimens. *Scientific Data, 10*(1), 411.

Bennett, N. J., Di Franco, A., Calò, A., Nethery, E., Niccolini, F., Milazzo, M., & Guidetti, P. (2019). Local support for conservation is associated with perceptions of good governance, social impacts, and ecological effectiveness. *Conservation Letters, 12*(4), e12640.

Berkes, F. (2017). Sacred ecology. Routledge.

Billionnet, A. (2021). *Designing Protected Area Networks*. EDP Sciences.

Blasiak, R., Jouffray, J. B., Wabnitz, C. C. C., Sundström, E., & Österblom, H. (2018). Corporate control and global governance of marine genetic resources. *Science Advances, 4*(6), eaar5237.

Blue Carbon Initiative. (2023). Mitigating Climate Change Through Coastal Conservation. https://www.thebluecarboninitiative.org

Borrelle, S. B., Rochman, C. M., Liboiron, M., Bond, A. L., Lusher, A., Bradshaw, H., & Provencher, J. F. (2017). Opinion: Why we need an international agreement on marine plastic pollution. *Proceedings of the National Academy of Sciences, 114*(38), 9994–9997. https://doi.org/10.1073/pnas.1714450114

Bosire, J. O., Dahdouh-Guebas, F., Walton, M., Crona, B. I., Lewis, R. R. III, Field, C., Kairo, J. G., & Koedam, N. (2008). Functionality of restored mangroves: A review. *Aquatic Botany, 89*(2), 251–259.

Boström-Einarsson, L., Babcock, R. C., Bayraktarov, E., Ceccarelli, D., Cook, N., Ferse, S. C., Hancock, B., Harrison, P., Hein, M., Shaver, E., & Smith, A. (2020). Coral restoration: A systematic review of current methods, successes, failures and future directions. *PLoS One, 15*(1), e0226631.

Bradford, J., Filgueira, R., & Bailey, M. (2020). Exploring community-based marine aquaculture as a coastal resource management opportunity in Nova Scotia, Canada. *FACETS, 5*(1), 26–48.

Brasier, M. J., Wiklund, H., Neal, L., Jeffreys, R., Linse, K., Ruhl, H., & Glover, A. G. (2016). DNA barcoding uncovers cryptic diversity in 50% of deep-sea Antarctic polychaetes. *Royal Society Open Science, 3*(11), 160432.

Buchan, S. J., Duran, M., Rojas, C., Wuth, J., Mahu, R., Stafford, K. M., & Becerra Yoma, N. (2023). An HMM-DNN-based system for the detection and classification of low-frequency acoustic signals from baleen whales, earthquakes, and air guns off Chile. *Remote Sensing, 15*(10), 2554.

Bucklin, A., DiVito, K. R., Smolina, I., Choquet, M., Questel, J. M., Hoarau, G., & O'Neill, R. J. (2020). Population genomics of marine zooplankton. In O. S. Kjesbu, S. Jentoft, & H. S. Tørresen (Eds.), *Population Genomics: Marine Organisms* (pp. 61–102). Springer.

Bucklin, A., Steinke, D., & Blanco-Bercial, L. (2011). DNA barcoding of marine metazoa. *Annual Review of Marine Science, 3*(1), 471–508.

Bulleri, F., Batten, S., Connell, S. D., Benedetti-Cecchi, L., Gibbons, M., Nugues, M. M., & Gribben, P. (2020). Human pressures and the emergence of novel marine ecosystems. In *Oceanography and marine biology* (pp. 456–535). CRC Press.

Bunting, P., Rosenqvist, A., Hilarides, L., Lucas, R. M., Thomas, N., Tadono, T., Worthington, T. A., Spalding, M., Murray, N. J., & Rebelo, L. M. (2022). Global mangrove extent change 1996–2020: Global Mangrove Watch version 3.0. *Remote Sensing, 14*(15), 3657.

Caruso, C., Hughes, K., & Drury, C. (2021). Selecting heat-tolerant corals for proactive reef restoration. *Frontiers in Marine Science, 8,* 632027.

Carvalho, D. V., Pereira, E. M., & Cardoso, J. S. (2019). Machine learning interpretability: A survey on methods and metrics. *Electronics, 8*(8), 832.

Cheung, V., Bell, A., Creencia, L., Fleming, L. E., Goh, H. C., Maharja, C., Morrissey, K., Richter, I., Then, A. H., & Austen, M. C. (2021). Blue Communities in Southeast Asia. *The Environmental Scientist, 30*(1), 96.

Chu, C., Loh, K. H., Ng, C. C., Ooi, A. L., Konishi, Y., Huang, S. P., & Chong, V. C. (2019). Using DNA barcodes to aid the identification of larval fishes in tropical estuarine waters (Malacca Straits, Malaysia). *Zoological Studies, 58,* e30.

Cinner, J. E., Maire, E., Huchery, C., MacNeil, M. A., Graham, N. A., Mora, C., McClanahan, T. R., Barnes, M. L., Kittinger, J. N., Hicks, C. C., & D'Agata, S. (2018). Gravity of human impacts mediates coral reef conservation gains. *Proceedings of the National Academy of Sciences, 115*(27), E6116–E6125.

Collins, R. A., Wangensteen, O. S., O'Gorman, E. J., Mariani, S., Sims, D. W., & Genner, M. J. (2018). Persistence of environmental DNA in marine systems. *Communications Biology, 1*(1), 185.

Convention on Biological Diversity. (2022). COP15: Final Text of the Kunming–Montreal Global Biodiversity Framework (CBD Decision 15/4). Secretariat of the Convention on Biological Diversity. https://www.cbd.int/article/cop15-final-text-kunming-montreal-gbf-221222

Costa, B., Sweeney, E., & Mendez, A. (2022). Leveraging artificial intelligence to annotate marine benthic species and habitats. NOAA Technical Memorandum NOS NCCOS 306. Silver Spring.

Costello, M. J., Coll, M., Danovaro, R., Halpin, P., Ojaveer, H., & Miloslavich, P. (2010). A census of marine biodiversity knowledge, resources, and future challenges. *PLoS One, 5*(8), e12110.

Crain, C. M., Kroeker, K., & Halpern, B. S. (2008). Interactive and cumulative effects of multiple human stressors in marine systems. *Ecology Letters, 11*(12), 1304–1315.

Crosby, A., Orenstein, E. C., Poulton, S. E., Bell, K. L., Woodward, B., Ruhl, H., Katija, K., & Forbes, A. G. (2023, April). Designing Ocean Vision AI: An investigation of community needs for imaging-based ocean conservation. In *Proceedings of the 2023 CHI Conference on Human Factors in Computing Systems* (pp. 1–16). ACM.

Dayton, P. K. (1985). Ecology of kelp communities. *Annual Review of Ecology and Systematics, 16,* 215–245.

Deiner, K., Bik, H. M., Mächler, E., Seymour, M., Lacoursière-Roussel, A., Altermatt, F., Creer, S., Bista, I., Lodge, D. M., De Vere, N., & Pfrender, M. E. (2017). Environmental DNA metabarcoding: Transforming how we survey animal and plant communities. *Molecular Ecology, 26*(21), 5872–5895.

Di Franco, A., Thiriet, P., Di Carlo, G., Dimitriadis, C., Francour, P., Gutiérrez, N. L., Jeudy de Grissac, A., Koutsoubas, D., Milazzo, M., Otero, M. D. M., & Piante, C. (2016). Five key attributes can increase marine protected areas performance for small-scale fisheries management. *Scientific Reports, 6*(1), 38135. https://doi.org/10.1038/srep38135

Dick, G. J. (2019). The microbiomes of deep-sea hydrothermal vents: Distributed globally, shaped locally. *Nature Reviews Microbiology, 17*(5), 271–283.

Donato, D. C., Kauffman, J. B., Murdiyarso, D., Kurnianto, S., Stidham, M., & Kanninen, M. (2011). Mangroves among the most carbon-rich forests in the tropics. *Nature Geoscience, 4*(5), 293–297.

Doney, S. C., Fabry, V. J., Feely, R. A., & Kleypas, J. A. (2009). Ocean acidification: The other CO_2 problem. *Annual Review of Marine Science, 1,* 169–192.

Dong, T., Zeng, Z., Pan, M., Wang, D., Chen, Y., Liang, L., Yang, S., Jin, Y., Luo, S., Liang, S., & Huang, X. (2025). Record-breaking 2023 marine heatwaves. *Science, 389*(6758), 369–374. https://doi.org/10.1126/science.adr0910

Duarte, C. M., Middelburg, J. J., & Caraco, N. (2005). Major role of marine vegetation on the oceanic carbon cycle. *Biogeosciences, 2*(1), 1–8.

Edgar, G. J., Stuart-Smith, R. D., Willis, T. J., Kininmonth, S., Baker, S. C., Banks, S., Barrett, N. S., Becerro, M. A., Bernard, A. T., Berkhout, J., & Buxton, C. D. (2014). Global conservation outcomes depend on marine protected areas with five key features. *Nature, 506*(7487), 216–220. https://doi.org/10.1038/nature13022

Fabry, V. J., Seibel, B. A., Feely, R. A., & Orr, J. C. (2008). Impacts of ocean acidification on marine fauna and ecosystem processes. *ICES Journal of Marine Science, 65*(3), 414–432.

Falkowski, P. G., Fenchel, T., & Delong, E. F. (2008). The microbial engines that drive Earth's biogeochemical cycles. *Science, 320*(5879), 1034–1039.

Ficetola, G. F., Miaud, C., Pompanon, F., & Taberlet, P. (2008). Species detection using environmental DNA from water samples. *Biology Letters, 4*(4), 423–425.

Ficetola, G. F., Taberlet, E., & Coissac, E. (2016). How to limit false positives in environmental DNA and metabarcoding? *Molecular Ecology Resources, 16*(3), 604–607. https://doi.org/10.1111/1755-0998.12508

Figuerola, B., Hancock, A. M., Bax, N., Cummings, V. J., Downey, R., Griffiths, H. J., Smith, J., & Stark, J. S. (2021). A review and meta-analysis of potential impacts of ocean acidification on marine calcifiers from the Southern Ocean. *Frontiers in Marine Science, 8,* 584445.

Fillyaw, R. M., Donnelly, M. J., Litwak, J. W., Rifenberg, J. L., & Walters, L. J. (2021). Strategies for successful mangrove living shoreline stabilizations in shallow water subtropical estuaries. *Sustainability, 13*(21), 11704. https://doi.org/10.3390/su132111704

Filonzi, L., Chiesa, S., Vaghi, M., & Marzano, F. N. (2010). Molecular barcoding reveals mislabelling of commercial fish products in Italy. *Food Research International*, 43, 1383–1388.

Food and Agriculture Organization of the United Nations. (2018). *The State of World Fisheries and Aquaculture 2018: Meeting the Sustainable Development Goals.* FAO. https://www.fao.org/3/i9540en/i9540en.pdf

Friedman, K., Bridgewater, P., Agostini, V., Agardy, T., Arico, S., Biermann, F., Brown, K., Cresswell, I. D., Ellis, E. C., Failler, P., & Kim, R. E. (2022). The CBD post-2020 biodiversity framework: People's place within the rest of nature. *People and Nature, 4*(6), 1475–1484. https://doi.org/10.1002/pan3.10388

Friess, D. A., Thompson, B. S., Brown, B., Amir, A. A., Cameron, C., Koldewey, H. J., Sasmito, S. D., & Sidik, F. (2016). Policy challenges and approaches for the conservation of mangrove forests in Southeast Asia. *Conservation Biology, 30*(5), 933–949. https://doi.org/10.1111/cobi.12784

Gadgil, M., Berkes, F., & Folke, C. (1993). Indigenous knowledge for biodiversity conservation. *AMBIO, 22*(2–3), 151–156.

Galimberti, A., De Mattia, F., Losa, A., Bruni, I., Federici, S., Casiraghi, M., Martellos, S., & Labra, M. (2013). DNA barcoding as a new tool for food traceability. *Food Research International, 50*(1), 55–63. https://doi.org/10.1016/j.foodres.2012.09.036

Gattuso, J. P., Magnan, A., Billé, R., Cheung, W. W. L., Howes, E. L., Joos, F., Allemand, D., Bopp, L., Cooley, S. R., Eakin, C. M., & Hoegh-Guldberg, O. (2015). Contrasting futures for ocean and society from different anthropogenic CO_2 emissions scenarios. *Science, 349*(6243), aac4722. https://doi.org/10.1126/science.aac4722

Gerona-Daga, M. E. B., & Salmo, S. G., III. (2022). A systematic review of mangrove restoration studies in Southeast Asia: Challenges and opportunities for the United Nation's Decade on Ecosystem Restoration. *Frontiers in Marine Science, 9,* 987737. https://doi.org/10.3389/fmars.2022.987737

Geyer, R., Jambeck, J. R., & Law, K. L. (2017). Production, use, and fate of all plastics ever made. *Science Advances, 3*(7), e1700782. https://doi.org/10.1126/sciadv.1700782

Ghayas, S., Siddiquie, J. S., Safdar, S., & Mansoor, A. (2020). Neural network implementations on the coastal water quality of Manora channel for the years 1996 to 2014. *International Journal of Circuits, Systems, and Signal Processing, 14,* 996–1004.

Giakoumi, S., McGowan, J., Mills, M., Beger, M., Bustamante, R. H., Charles, A., Christie, P., Fox, M., Garcia-Borboroglu, P., Gelcich, S., & Guidetti, P. (2018). Revisiting "success" and "failure" of marine protected areas: A conservation scientist perspective. *Frontiers in Marine Science, 5,* 223. https://doi.org/10.3389/fmars.2018.00223

Global Mangrove Watch. (2022). Global Mangrove Extent and Change Dataset. https://www.globalmangrovewatch.org/

Gold, Z., Sprague, J., Kushner, D. J., Zerecero Marin, E., & Barber, P. H. (2021). eDNA metabarcoding as a biomonitoring tool for marine protected areas. *PLoS One, 16*(2), e0238557. https://doi.org/10.1371/journal.pone.0238557

Goldberg, C. S., Strickler, K. M., & Pilliod, D. S. (2015). Moving environmental DNA methods from concept to practice for monitoring aquatic macroorganisms. *Biological Conservation, 183,* 1–3. https://doi.org/10.1016/j.biocon.2014.11.040

Govan, H., Lalavanua, W., & Steenbergen, D. J. (2024). Coastal fisheries governance in the Pacific Islands: The evolution of policy and the progress of management-at-scale. In M. Nakamura, R. Pomeroy, & J. Urquhart (Eds.), *Implementation of the Small-Scale Fisheries Guidelines: A Legal and Policy Scan* (pp. 207–240). Springer Nature Switzerland. https://doi.org/10.1007/978-3-031-56716-2_11

Gutiérrez, J., Tagar-Hadary, T., Rosenberg, Y., Neder, M., & LeBlanc, P. (2022). Eco-engineering solution for nature-based shoreline protection at the Port of San Diego (USA). In *1st Maritime Engineering Meetings.* https://ingenierie-maritime.com/app/uploads/2023/04/393509.pdf

Halpern B. S., McLeod K. L., Rosenberg A. A. & Crowder L. B. (2008). Managing for cumulative impacts in ecosystem-based management through ocean zoning. *Ocean Coast Management*, 51, 203–211.

Han, G., Zhou, Z., Zhang, Y., Martínez-García, M., Peng, Y., & Xie, L. (2021). Sleep-scheduling-based hierarchical data collection algorithm for gliders in underwater acoustic sensor networks. *IEEE Transactions on Vehicular Technology, 70*(9), 9466–9479. https://ieeexplore.ieee.org/document/9501973

Hanner, R., Becker, S., Ivanova, N. V., & Steinke, D. (2011). FISH-BOL and seafood identification: Geographically dispersed case studies reveal systemic market substitution across Canada. *Mitochondrial DNA, 22*(sup1), 106–122. https://doi.org/10.3109/19401736.2011.588217

Hebert, P. D. N., Cywinska, A., Ball, S. L., & deWaard, J. R. (2003). Biological identifications through DNA barcodes. *Proceedings of the Royal Society of London. Series B: Biological Sciences, 270*(1512), 313–321. https://doi.org/10.1098/rspb.2002.2218

Hoegh-Guldberg, O., Mumby, P. J., Hooten, A. J., Steneck, R. S., Greenfield, P., Gomez, E., Harvell, D., Sale, P. F., Edwards, A. J., Caldeira, K., Knowlton, N., Eakin, C. M., Iglesias-Prieto, R., Muthiga, N., Bradbury, R. H., Dubi, A., & Hatziolos, M. E. (2007). Coral reefs under rapid climate change and ocean acidification. *Science, 318*(5857), 1737–1742. https://doi.org/10.1126/science.1152509

Hopkins, F. E., Suntharalingam, P., Gehlen, M., Andrews, O., Archer, S. D., Bopp, L., Buitenhuis, E., Dadou, I., Duce, R., Goris, N., & Jickells, T. (2020). The impacts of ocean acidification on marine trace gases and the implications for atmospheric chemistry and climate. *Proceedings of the Royal Society A, 476*(2237), 20190769. https://doi.org/10.1098/rspa.2019.0769

Huerta Lwanga, E., Mendoza Vega, J., Ku Quej, V., Chi, J. D. L. A., Sanchez del Cid, L., Chi, C., Escalona Segura, G., Gertsen, H., Salánki, T., van der Ploeg, M., & Koelmans, A. A. (2017). Field evidence for transfer of plastic debris along a terrestrial food chain. *Scientific Reports*, 7(1), 14071. https://doi.org/10.1038/s41598-017-14588-2

Hughes, T. P., Kerry, J. T., Álvarez-Noriega, M., Álvarez-Romero, J. G., Anderson, K. D., Baird, A. H., Babcock, R. C., Beger, M., Bellwood, D. R., Berkelmans, R., & Bridge, T. C. (2017). Global warming and recurrent mass bleaching of corals. *Nature*, 543(7645), 373–377. https://doi.org/10.1038/nature21707

Intergovernmental Panel on Climate Change. (2022). *The Ocean and Cryosphere in a Changing Climate: Special Report of the Intergovernmental Panel on Climate Change*. Cambridge University Press.

IPBES. (2019). *Summary for Policy Makers of the Global Assessment Report on Biodiversity and Ecosystem Services of the Intergovernmental Science-Policy Platform on Biodiversity and Ecosystem Services.* IPBES Secretariat.

Issifu, I., Alava, J. J., Lam, V. W., & Sumaila, U. R. (2022). Impact of ocean warming, overfishing and mercury on European fisheries: A risk assessment and policy solution framework. *Frontiers in Marine Science, 8*, 770805.

Jambeck, J. R., Geyer, R., Wilcox, C., Siegler, T. R., Perryman, M., Andrady, A., Narayan, R., & Law, K. L. (2015). Plastic waste inputs from land into the ocean. *Science*, 347(6223), 768–771. https://doi.org/10.1126/science.1260352

Johnson, J. E., Hooper, E., & Welch, D. J. (2020). Community marine monitoring toolkit: A tool developed in the Pacific to inform community-based marine resource management. *Marine Pollution Bulletin, 159*, 111498.

Keskİn, E., & Atar, H. H. (2013). DNA barcoding commercially important fish species of Turkey. *Molecular Ecology Resources, 13*(5), 788–797.

Kikaki, A., Karantzalos, K., Power, C. A., & Raitsos, D. E. (2020). Remotely sensing the source and transport of marine plastic debris in Bay Islands of Honduras (Caribbean Sea). *Remote Sensing, 12*(11), 1727.

Klein, S. G., Roch, C., & Duarte, C. M. (2024). Systematic review of the uncertainty of coral reef futures under climate change. *Nature Communications, 15*(1), 2224.

Knowlton, N., Brainard, R. E., Fisher, R., Moews, M., Plaisance, L., & Caley, M. J. (2010). Coral reef biodiversity. In A. D. McIntyre (Ed.), *Life in the World's Oceans: Diversity, Distribution, and Abundance* (pp. 65–78). Wiley-Blackwell.

Kole, P. J., Löhr, A. J., Van Belleghem, F. G., & Ragas, A. M. (2017). Wear and tear of tyres: A stealthy source of microplastics in the environment. *International Journal of Environmental Research and Public Health, 14*(10), 1265.

Kroeker, K. J., Kordas, R. L., Crim, R. N., & Singh, G. G. (2010). Meta-analysis reveals negative yet variable effects of ocean acidification on marine organisms. *Ecology Letters, 13*(11), 1419–1434.

Lamont, T. A., Razak, T. B., Djohani, R., Janetski, N., Rapi, S., Mars, F., & Smith, D. J. (2022). Multi-dimensional approaches to scaling up coral reef restoration. *Marine Policy, 143*, 105199.

Leal, M. C., Puga, J., Serôdio, J., Gomes, N. C. M., & Calado, R. (2012). Trends in the discovery of new marine natural products from invertebrates over the last two decades: Where and what are we bioprospecting? *PLoS One, 7*(1), e30580.

Lear, W. H. (1998). History of fisheries in the Northwest Atlantic: The 500-year perspective. *Journal of Northwest Atlantic Fishery Science*, *23*, 41–73.

LeBlanc, F., Belliveau, V., Watson, E., Coomber, C., Simard, N., DiBacco, C., Bernier, R., & Gagné, N. (2020). Environmental DNA (eDNA) detection of marine aquatic invasive species (AIS) in Eastern Canada using a targeted species-specific qPCR approach. *Management of Biological Invasions*, *11*(2), 201–218. https://doi.org/10.3391/mbi.2020.11.2.02

Lebreton, L. & Andrady, A. (2019). Future scenarios of global plastic waste generation and disposal. *Palgrave Communications*, *5*(1), pp.1–11.

Leung, J. Y., Zhang, S., & Connell, S. D. (2022). Is ocean acidification really a threat to marine calcifiers? A systematic review and meta-analysis of 980+ studies spanning two decades. *Small*, *18*(35), 2107407.

Li, X., Xu, X., Yan, L., Zhao, H., & Zhang, T. (2020). Energy-efficient data collection using autonomous underwater glider: A reinforcement learning formulation. *Sensors*, *20*(13), 3758.

Liu, B., Li, Y., Li, G., & Liu, A. (2019). A spectral feature based convolutional neural network for classification of sea surface oil spill. *ISPRS International Journal of Geo-Information*, *8*(4), 160.

Lovelock, C. E., Barbier, E., & Duarte, C. M. (2022). Tackling the mangrove restoration challenge. *PLoS Biology*, *20*(10), e3001836.

Lowe, S. C., Misiuk, B., Xu, I., Abdulazizov, S., Baroi, A. R., Bastos, A. C., Best, M., Ferrini, V., Friedman, A., Hart, D., & Hoegh-Guldberg, O. (2025). BenthicNet: A global compilation of seafloor images for deep learning applications. *Scientific Data*, *12*(1), 230.

Lowenstein, J. H., Amato, G., & Kolokotronis, S. O. (2009). The real maccoyii: Identifying tuna sushi with DNA barcodes–contrasting characteristic attributes and genetic distances. *PLoS One*, *4*(11), e7866.

Lubchenco, J., & Grorud-Colvert, K. (2015). Making waves: The science and politics of ocean protection. *Science*, *350*(6259), 382–383.

Luo, H., Wang, X., Xu, Z., Liu, C., & Pan, J. S. (2022). A software-defined multi-modal wireless sensor network for ocean monitoring. *International Journal of Distributed Sensor Networks*, *18*(1), 15501477211068389.

Lusher, A. L., Mchugh, M., & Thompson, R. C. (2013). Occurrence of microplastics in the gastrointestinal tract of pelagic and demersal fish from the English Channel. *Marine Pollution Bulletin*, *67*(1–2), 94–99.

Madramootoo, C. A., & Virgil, C. (2024). Nature-based solutions for coastal protection in the southern Caribbean. *Blue-Green Systems*, *6*(2), 185–197.

Marriott, S. E., Cox, C., Amolo, R. C., Apistar, D., Mancao, R. H., & de Mutsert, K. (2021). Implications of community-based management of marine reserves in the Philippines for reef fish communities and biodiversity. *Frontiers in Marine Science*, *8*, 731675.

Mascia, M. B., Claus, C. A., & Naidoo, R. (2010). Impacts of marine protected areas on fishing communities. *Conservation Biology*, *24*(5), 1424–1429.

Mat Jaafar, T. N. A., Taylor, M. I., Mohd Nor, S. A., de Bruyn, M., & Carvalho, G. R. (2012). DNA barcoding reveals cryptic diversity within commercially exploited Indo-Malay Carangidae (Teleosteii: Perciformes). *PLoS One*, *7*(11), e49623.

Mehvar, S., Filatova, T., Dastgheib, A., De Ruyter van Steveninck, E., & Ranasinghe, R. (2018). Quantifying economic value of coastal ecosystem services: A review. *Journal of Marine Science and Engineering*, *6*(1), 5. https://doi.org/10.3390/jmse6010005

MIT Press Reader. (2024). The Algorithmic Ocean: How AI Is Revolutionizing Marine Conservation. https://thereader.mitpress.mit.edu/the-algorithmic-ocean-how-ai-is-revolutionizing-marine-conservation/

Molari, M., Guilini, K., Lott, C., Weber, M., de Beer, D., Meyer, S., Ramette, A., Wegener, G., Wenzhöfer, F., Martin, D., & Cibic, T. (2018). CO_2 leakage alters biogeochemical and ecological functions of submarine sands. *Science Advances*, *4*(2), eaao2040.

Molina-Molina, J. C., Salhaoui, M., Guerrero-González, A., & Arioua, M. (2021). Autonomous marine robot based on AI recognition for permanent surveillance in marine protected areas. *Sensors*, *21*(8), 2664.

Mora, C., Tittensor, D. P., Adl, S., Simpson, A. G. B., & Worm, B. (2011). How many species are there on Earth and in the ocean? *PLoS Biology*, *9*(8), e1001127.

Morin, P. A., Archer, F. I., Avila, C. D., Balacco, J. R., Bukhman, Y. V., Chow, W., Fedrigo, O., Formenti, G., Fronczek, J. A., Fungtammasan, A., & Gulland, F. M. (2021). Reference genome and demographic history of the most endangered marine mammal, the vaquita. *Molecular Ecology Resources*, *21*(4), 1008–1020.

Moura, C. J., Harris, D. J., Cunha, M. R., & Rogers, A. D. (2008). DNA barcoding reveals cryptic diversity in marine hydroids (Cnidaria, Hydrozoa) from coastal and deep-sea environments. *Zoologica Scripta*, *37*(1), 93–108.

Nieves, V., Ruescas, A., & Sauzède, R. (2023). AI for marine, ocean and climate change monitoring. *Remote Sensing*, *16*(1), 15.

NOAA Office of Response and Restoration. (2025). How Big Is the Great Pacific Garbage Patch: Science vs. Myth. https://response.restoration.noaa.gov/about/media/how-big-great-pacific-garbage-patch-science-vs-myth.html

Noor, N. S., Yahaya, N., Zain, N. N. M., Kamal, N. N. S. N. M., Mansor, M. S., Aziz, M. Y., & Waras, M. N. (2024). Microplastic pollution in Malaysia: Status and challenges: A brief overview. *Malaysian Journal of Analytical Sciences*, *28*(3), 569–585.

Norouzzadeh, M. S., Nguyen, A., Kosmala, M., Swanson, A., Palmer, M. S., Packer, C., & Clune, J. (2018). Automatically identifying, counting, and describing wild animals in camera-trap images with deep learning. *Proceedings of the National Academy of Sciences*, *115*(25), E5716–E5725.

Nur Korkmaz, B., Diamant, R., Danino, G., & Testolin, A. (2023). Automated detection of dolphin whistles with convolutional networks and transfer learning. *Frontiers in Artificial Intelligence*, *6*, 1099022.

O'Garra, T., Mangubhai, S., Jagadish, A., Tabunakawai-Vakalalabure, M., Tawake, A., Govan, H., & Mills, M. (2023). National-level evaluation of a community-based marine management initiative. *Nature Sustainability*, *6*(8), 908–918.

OECD. (2022). Regional Plastics Outlook for Southeast and East Asia. Organisation for Economic Co-operation and Development. https://www.oecd.org/en/publications/regional-plastics-outlook-for-southeast-and-east-asia_5a8ff43c-en/full-report/component-6.html

Olsen, E., Kaplan, I. C., Ainsworth, C., Fay, G., Gaichas, S., Gamble, R., Girardin, R., Eide, C. H., Ihde, T. F., Morzaria-Luna, H. N., & Johnson, K. F. (2018). Ocean futures under ocean acidification, marine protection, and changing fishing pressures explored using a worldwide suite of ecosystem models. *Frontiers in Marine Science*, 5, 64.

Orth, R. J., Carruthers, T. J., Dennison, W. C., Duarte, C. M., Fourqurean, J. W., Heck, K. L., Hughes, A. R., Kendrick, G. A., Kenworthy, W. J., Olyarnik, S., & Short, F. T. (2006). A global crisis for seagrass ecosystems. *BioScience*, *56*(12), 987–996. https://doi.org/10.1641/0006-3568(2006)56[987:AGCFSE]2.0.CO;2

Osathanunkul, M., & Suwannapoom, C. (2023). Sustainable fisheries management through reliable restocking and stock enhancement evaluation with environmental DNA. *Scientific Reports*, *13*(1), 11297.

Pauly, D., Christensen, V., Guénette, S., Pitcher, T. J., Sumaila, U. R., Walters, C. J., Watson, R., & Zeller, D. (2002). Towards sustainability in world fisheries. *Nature*, *418*(6898), 689–695. https://doi.org/10.1038/nature01017

Pauly, D., Watson, R., & Alder, J. (2005). Global trends in world fisheries: Impacts on marine ecosystems and food security. *Philosophical Transactions of the Royal Society B: Biological Sciences*, *360*(1453), 5–12.

Pennacchio, A., Giampaolo, F., Piccialli, F., Cuomo, S., Notomista, E., Spinelli, M., Amoresano, A., Piscitelli, A., & Giardina, P. (2022). A machine learning-enhanced biosensor for mercury detection based on a hydrophobin chimera. *Biosensors and Bioelectronics*, *196*, 113696. https://doi.org/10.1016/j.bios.2021.113696

Pham, C. V., Wang, H. C., Chen, S. H., & Lee, J. M. (2023). The threshold effect of overfishing on global fishery outputs: International evidence from a sustainable fishery perspective. *Fishes*, *8*(2), 71.

Piazzolla, D., Scanu, S., Mancuso, F. P., Bosch-Belmar, M., Bonamano, S., Madonia, A. & Piermattei, V. (2024). An integrated approach for the benthic habitat mapping based on innovative surveying technologies and ecosystem functioning measurements. *Scientific Reports*, *14*(1), 5888.

Pitcher, T. J., & Cheung, W. W. L. (2013). Fisheries: Hope or despair? *Marine Pollution Bulletin*, *74*(2), 506–516.

Pitois, S. G., Blackwell, R. E., Close, H., Eftekhari, N., Giering, S. L., Masoudi, M., Payne, E., Ribeiro, J., & Scott, J. (2025). RAPID: Real-time automated plankton identification dashboard using Edge AI at sea. *Frontiers in Marine Science*, *11*, 1513463. https://doi.org/10.3389/fmars.2024.1513463

Poloczanska, E. S., Brown, C. J., Sydeman, W. J., Kiessling, W., Schoeman, D. S., Moore, P. J., Brander, K., Bruno, J. F., Buckley, L. B., Burrows, M. T., & Duarte, C. M. (2013). Global imprint of climate change on marine life. *Nature Climate Change*, *3*(10), 919–925. https://doi.org/10.1038/nclimate1958

Potharaju, R., & Aruna, M. (2023). Current advancements, future obstacles, recovery and restoration of aquatic ecosystem habitats. *REDVET: Revista Electrónica de Veterinaria*, 24(4), 611–619.

Puig, P., Canals, M., Company, J. B., Martín, J., Amblas, D., Lastras, G., & Calafat, A. M. (2012). Ploughing the deep-sea floor. *Nature*, *489*(7415), 286–289.

Raha, D., Davies-Vollum, K. S., Hemstock, S. L., Boateng, I., Islam, M. T., & Pierce, C. A. E. (2024). We need collaboration and co-creation to address challenges facing coastal communities. *Nature Human Behaviour*, *8*(5), 814–822. https://doi.org/10.1038/s41562-024-01875-y

Ramirez-Llodra, E., Brandt, A., Danovaro, R., De Mol, B., Escobar, E., German, C. R., ... & Vecchione, M. (2010). Deep, diverse and definitely different: unique attributes of the world's largest ecosystem. *Biogeosciences*, *7*(9), 2851–2899.

Raynor, J., Orofino, S., Costello, C., McDonald, G., Mayorga, J., & Sala, E. (2025). Little-to-no industrial fishing occurs in fully and highly protected marine areas. *Science*, 389(6758), 392–395.Razak, T. B., Boström-Einarsson, L., Alisa, C. A. G., Vida, R. T., & Lamont, T. A. (2022). Coral reef restoration in Indonesia: A review of policies and projects. Marine Policy, 137, 104940.

Reidenbach, M. A., & Thomas, E. L. (2018). Influence of the seagrass Zostera marina on wave attenuation and bed shear stress within a shallow coastal bay. *Frontiers in Marine Science*, *5*, 397. https://doi.org/10.3389/fmars.2018.00397

Risch, D., Norris, T., Curnock, M., & Friedlaender, A. (2019). Common and Antarctic minke whales: Conservation status and future research directions. *Frontiers in Marine Science*, *6*, 247.

Roblet, S., Priouzeau, F., Gambini, G., Cottalorda, J. M., Gastaldi, J. M., Pey, A., Raybaud, V., Suarez, G. R., Serre, C., Sabourault, C., & Dérijard, B. (2024). From sight to sequence: Underwater visual census vs environmental DNA metabarcoding for the monitoring of taxonomic and functional fish diversity. *Science of the Total Environment*, *956*, 177250. https://doi.org/10.1016/j.scitotenv.2024.177250

Rochman, C. M., Hoh, E., Kurobe, T., & Teh, S. J. (2013). Ingested plastic transfers hazardous chemicals to fish and induces hepatic stress. *Scientific Reports*, *3*(1), 3263.

Ruppert, K. M., Kline, R. J., & Rahman, M. S. (2019). Past, present, and future perspectives of environmental DNA (eDNA) metabarcoding: A systematic review in methods, monitoring, and applications of global eDNA. *Global Ecology and Conservation*, *17*, e00547.

Sala, E., Mayorga, J., Bradley, D., Cabral, R. B., Atwood, T. B., Auber, A., Cheung, W., Costello, C., Ferretti, F., Friedlander, A. M., & Gaines, S. D. (2021). Protecting the global ocean for biodiversity, food and climate. *Nature*, *592*(7854), 397–402.

Sattar, F. (2023). A new acoustical autonomous method for identifying endangered whale calls: A case study of blue whale and fin whale. *Sensors*, *23*(6), 3048.

Seguin, R., Le Manach, F., Devillers, R., Velez, L., & Mouillot, D. (2025). Global patterns and drivers of untracked industrial fishing in coastal marine protected areas. *Science*, *389*(6758), 396–401.

Shaver, E. C., McLeod, E., Hein, M. Y., Palumbi, S. R., Quigley, K., Vardi, T., Mumby, P. J., Smith, D., Montoya-Maya, P., Muller, E. M., & Banaszak, A. T. (2022). A roadmap to integrating resilience into the practice of coral reef restoration. *Global Change Biology*, *28*(16), 4751–4764.

Solberg, A. H., Brekke, C., & Husoy, P. O. (2007). Oil spill detection in Radarsat and Envisat SAR images. *IEEE Transactions on Geoscience and Remote Sensing*, *45*(3), 746–755.

Songwe, V., Stern, N., & Bhattacharya, A. (2022). Finance for climate action: Scaling up investment for climate and development. London: Grantham Research Institute on Climate Change and the Environment, London School of Economics and Political Science.

Spalding, M. D., Ravilious, C., & Green, E. P. (2001). *World Atlas of Coral Reefs*. University of California Press.

Spear, M. J., Embke, H. S., Krysan, P. J., & Vander Zanden, M. J. (2021). Application of eDNA as a tool for assessing fish population abundance. *Environmental DNA*, *3*(1), 83–91.

Stephens, T. (2023). The kunming–montreal global biodiversity framework. *International Legal Materials*, *62*(5), 868–887.

Stern, M. J. (2018). *Social Science Theory for Environmental Sustainability: A Practical Guide*. Oxford University Press.

Stowell, D. (2022). Computational bioacoustics with deep learning: a review and roadmap. *PeerJ*, *10*, e13152.

Streck, C. (2023). Synergies between the Kunming-Montreal Global Biodiversity Framework and the Paris Agreement: The role of policy milestones, monitoring frameworks and safeguards. *Climate Policy*, *23*(6), 800–811.

Su, J., Friess, D. A., & Gasparatos, A. (2021). A meta-analysis of the ecological and economic outcomes of mangrove restoration. *Nature Communications*, *12*(1), 5050.

Sulaiman, R. N. R., Bakar, A. A., Ngadi, N., Kahar, I. N. S., Nordin, A. H., Ikram, M., & Nabgan, W. (2023). Microplastics in Malaysia's aquatic environment: Current overview and future perspectives. *Global Challenges*, *7*(8), 2300047.

Sun, X. Q., Qu, Y. Q., Yao, H., Zhang, Y. M., Yan, Q. Q., & Hang, Y. Y. (2015). Applying DNA barcodes for identification of economically important species in Brassicaceae. *Genetics and Molecular Research*, *14*(4), 15050–15061.

Suncls, A., & Cai, I. I. (1982). *United Nations Convention on the Law of the Sea*. United Nations.

Tanaka, Y. (2023). *The International Law of the Sea*. Cambridge University Press.

The Blue Carbon Initiative. (2025). What is Blue Carbon? Accessed September 25, 2025. https://www.thebluecarboninitiative.org/about-blue-carbon

Thom, R. M. (2000). Adaptive management of coastal ecosystem restoration projects. *Ecological Engineering*, *15*(3–4), 215–225.

Thurstan, R. H., & Roberts, C. M. (2010). Ecological meltdown in the Firth of Clyde, Scotland: Two centuries of change in a coastal marine ecosystem. *PLoS One*, *5*(7), e11767.

Tirumala, R. D., & Tiwari, P. (2022). Innovative financing mechanism for blue economy projects. *Marine Policy*, *139*, 104194. https://doi.org/10.1016/j.marpol.2020.104194

U.S. Naval Institute. (2025). Sea change: The rise of AI in ocean observing and forecasting. *Proceedings*, *151*(3), 1465. https://www.usni.org/magazines/proceedings/2025/march/sea-change-rise-ai-ocean-observing-and-forecasting

UN. (2023). *Sustainable Development Goals Report 2023: Special Edition*. United Nations.

UNEP. (2020). *Out of the Blue: The Value of Seagrasses to the Environment and to People*. United Nations Environment Programme.

UNEP. (2022). *Ocean Solutions That Benefit People, Nature and the Economy*. United Nations Environment Programme.

UNEP-WCMC & IUCN. (2018). *Protected Planet Report: Marine Protected Areas Covering Almost 7% of the Global Ocean*. Protected Planet. https://pp-digital-report-document.s3.eu-west-2.amazonaws.com/Protected_Planet_Report_2018.pdf

UNEP-WCMC & IUCN. (2021). *Protected Planet Report 2020: State of Protected and Conserved Areas*. UNEP-WCMC and IUCN. https://protectedplanetreport2020.protectedplanet.net/chapter-3

UNEP-WCMC & IUCN. (2024). *Protected Planet Report 2020: Chapter 3: Coverage*. UNEP-WCMC and IUCN. https://digitalreport.protectedplanet.net/?chapter=3#report_chapters-3

UNEP & FAO. (2020). *UN Decade on Ecosystem Restoration Strategy*. United Nations Environment Programme and Food and Agriculture Organization. https://www.decadeonrestoration.org

United Nations. (1982). *United Nations Convention on the Law of the Sea*. United Nations. https://www.un.org/depts/los/convention_agreements/texts/unclos/unclos_e.pdf

Valavi, R., Elith, J., Lahoz-Monfort, J. J., & Guillera-Arroita, G. (2021). Modelling species presence-only data with random forests. *Ecography*, *44*(12), 1731–1742.

Van de Putte, A. P., Griffiths, H. J., Brooks, C., Bricher, P., Sweetlove, M., Halfter, S., & Raymond, B. (2021). From data to marine ecosystem assessments of the Southern Ocean: Achievements, challenges, and lessons for the future. *Frontiers in Marine Science*, *8*, 637063.

Van Horn, G., Mac Aodha, O., Song, Y., Cui, Y., Sun, C., Shepard, A., Adam, H., Perona, P., & Belongie, S. (2018). The iNaturalist species classification and detection dataset. In *Proceedings of the IEEE Conference on Computer Vision and Pattern Recognition* (pp. 8769–8778). https://doi.org/10.1109/CVPR.2018.00914

Van Wesenbeeck, B. K., van Zelst, V. T. M., Antolinez, J. A. A., & de Boer, W. P. (2025). Quantifying uncertainty in wave attenuation by mangroves to inform coastal green belt policies. *Communications Earth & Environment*, *6*(1), 258. https://doi.org/10.1038/s43247-025-02178-4

Vasseur, L. (2021). How ecosystem-based adaptation to climate change can help coastal communities through a participatory approach. *Sustainability*, *13*(4), 2344. https://doi.org/10.3390/su13042344

Wagner, S., Hüffer, T., Klöckner, P., Wehrhahn, M., Hofmann, T., & Reemtsma, T. (2018). Tire wear particles in the aquatic environment: A review on generation, analysis, occurrence, fate and effects. *Water Research*, *139*, 83–100. https://doi.org/10.1016/j.watres.2018.03.051

Wäldchen, J., & Mäder, P. (2018). Machine learning for image-based species identification. *Methods in Ecology and Evolution*, *9*(11), 2216–2225.

Walters, C. J. (1986). *Adaptive Management of Renewable Resources*. New York: MacMillan.

Wang, C., Zhao, J., & Xing, B. (2021). Environmental source, fate, and toxicity of microplastics. *Journal of Hazardous Materials*, *407*, 124357.

Ward, R. D., Zemlak, T. S., Innes, B. H., Last, P. R., & Hebert, P. D. (2005). DNA barcoding Australia's fish species. *Philosophical Transactions of the Royal Society B: Biological Sciences*, *360*(1462), 1847–1857.

Watt-Pringle, R., Razak, T. B., Jompa, J., Ambo-Rappe, R., Kostaman, A. N., & Smith, D. J. (2024). Coral reef restoration in Indonesia: Lessons learnt from the world's largest coral restoration nation. *Biodiversity and Conservation*, *33*(10), 2675–2707.

Weigand, H., Beermann, A. J., Čiampor, F., Costa, F. O., Csabai, Z., Duarte, S., Geiger, M. F., Grabowski, M., Rimet, F., Rulik, B., & Strand, M. (2019). DNA barcode reference libraries for the monitoring of aquatic biota in Europe: Gap-analysis and recommendations for future work. *Science of the Total Environment*, *678*, 499–524. https://doi.org/10.1016/j.scitotenv.2019.04.247

West, C. F., McClenachan, L., Barbeaux, S. J., Spies, I., Addison, J. A., Anderson, B. T., Hofman, C. A., Reedy, K. L., Elliott Smith, E. A., Etnier, M. A., & Helser, T. E. (2025). Integrating marine historical ecology into management of Alaska's Pacific cod fishery for climate readiness. *ICES Journal of Marine Science*, *82*(4), fsaf056. https://doi.org/10.1093/icesjms/fsaf056

Woodall, L. C., Sanchez-Vidal, A., Canals, M., Paterson, G. L., Coppock, R., Sleight, V., Calafat, A., Rogers, A. D., Narayanaswamy, B. E., & Thompson, R. C. (2014). The deep sea is a major sink for microplastic debris. *Royal Society Open Science*, *1*(4), 140317. https://doi.org/10.1098/rsos.140317

World Bank. (2021, August 23). Beyond borders: Collaborative solutions to plastic pollution in Southeast Asia. World Bank Blogs. https://blogs.worldbank.org/en/eastasiapacific/beyond-borders-collaborative-solutions-plastic-pollution-southeast-asia

Worm, B., Barbier, E. B., Beaumont, N., Duffy, J. E., Folke, C., Halpern, B. S., Jackson, J. B., Lotze, H. K., Micheli, F., Palumbi, S. R., & Sala, E. (2006). Impacts of biodiversity loss on ocean ecosystem services. *Science*, *314*(5800), 787–790. https://doi.org/10.1126/science.1132294

Wright, G., Gjerde, K. M., Johnson, D. E., Finkelstein, A., Ferreira, M. A., Dunn, D. C., Rodriguez Chaves, M., & Grehan, A. (2021). Marine spatial planning in areas beyond national jurisdiction. *Marine Policy*, *132*, 103384. https://doi.org/10.1016/j.marpol.2018.12.003

WWF. (2015). *Reviving the Ocean Economy: The Case for Action*. World Wide Fund for Nature.

Yassir, A., Andaloussi, S. J., Ouchetto, O., Mamza, K., & Serghini, M. (2023). Acoustic fish species identification using deep learning and machine learning algorithms: A systematic review. *Fisheries Research*, *266*, 106790.

Yee, J. C., Cheah, C. B., Tan, L. E., Chong, A. K. M., & Chee, S. Y. (2025). Strength and fluid transport properties of GGBS blended cementitious composite with crushed seashells and quarry dust fine aggregate. *Construction and Building Materials*, *465*, 140134. https://doi.org/10.1016/j.conbuildmat.2025.140134

Zaki, M. R. M., Zainuddin, A. H., Saifuddin, M. F. U., Razak, M. R., Isa, N. M., Ash'aari, Z. H., & Aris, A. Z. (2024). Microplastic contamination and ecological risk assessment in tropical riverine surface water of Selangor River Basin, Malaysia. *ACS ES&T Water*, *4*(8), 3225–3234.

Zettler, E. R., Mincer, T. J., & Amaral-Zettler, L. A. (2013). Life in the "plastisphere": Microbial communities on plastic marine debris. *Environmental Science & Technology*, *47*(13), 7137–7146.

Zhang, M., Zou, Y., Xiao, S., & Hou, J. (2023). Environmental DNA metabarcoding serves as a promising method for aquatic species monitoring and management: A review focused on its workflow, applications, challenges and prospects. *Marine Pollution Bulletin*, *194*, 115430.

Zhao, S., Kvale, K. F., & Zhu, L. (2025). The distribution of subsurface microplastics in the ocean. *Nature*, *641*(7945), 51–61. https://doi.org/10.1038/s41586-025-08818-1

Zhou, P., He, X. Q., Tu, Z. Y., Sun, D., Wang, C. S., Shen, H. B., & Pan, X. (2025). Bridging the gap between the public's knowledge and detection of marine non-indigenous species through developing automated image classification applications for marine species. *Frontiers in Marine Science*, *12*, 1508851.

11 Photoreception by Microbial Rhodopsins

An Alternative Solar Energy Pathway Flowing into Marine Ecosystems without Carbon Intermediates

Susumu Yoshizawa

11.1 INTRODUCTION

Microbial rhodopsins have recently emerged as a fundamental component of marine phototrophy, yet their roles remain incompletely understood (Beja et al., 2000; Pinhassi et al., 2016; Olson et al., 2018). While decades of research established oxygenic photosynthesis as the dominant entry point of solar energy into the ocean, it is now clear that rhodopsins constitute a parallel, carbon-independent light-mediated pathway (Gómez-Consarnau et al., 2019; Munson-McGee et al., 2022). Despite major progress in identifying rhodopsin genes across diverse taxa, critical gaps remain: their quantitative contribution to oceanic energy budgets is unresolved, their integration into host physiology is poorly constrained, and their ecological significance in eukaryotic microorganisms is only beginning to be explored.

This review aims to synthesize the current knowledge of microbial rhodopsins and to address these knowledge gaps by examining their discovery, evolutionary spread, physiological diversity, and ecological consequences. The review is structured as follows: first, microbial rhodopsins and their discovery in marine systems are introduced; then, cultivation studies and phylogenetic diversity are examined; next, global distribution patterns are explored; then, functional diversification beyond proton pumps is highlighted; finally, their genomic context, ecological implications, and emerging roles in eukaryotes are discussed. The conclusion outlines the broader significance of this carbon-independent energy flux for marine ecosystems and identifies priorities for future research. Understanding this is essential because accurate knowledge of the light-energy flows that sustain marine ecosystems is fundamental to appreciating and conserving biodiversity.

11.2 MICROBIAL RHODOPSINS ILLUMINATE A NEW LIGHT-ENERGY FLOW IN THE OCEAN

Life on Earth is fundamentally driven by solar energy. For decades, the prevailing paradigm in marine ecology held that the only biological conduit for solar energy into oceanic food webs was oxygenic photosynthesis, carried out by chlorophyll-containing microorganisms such as cyanobacteria and microalgae. In this classical view, light captured by chlorophyll-based photosystems drives water splitting and CO_2 fixation, producing organic matter that fuels heterotrophic life. This organic matter then sinks through the water column and is laterally transported by ocean currents,

DOI: 10.1201/9781003640899-14

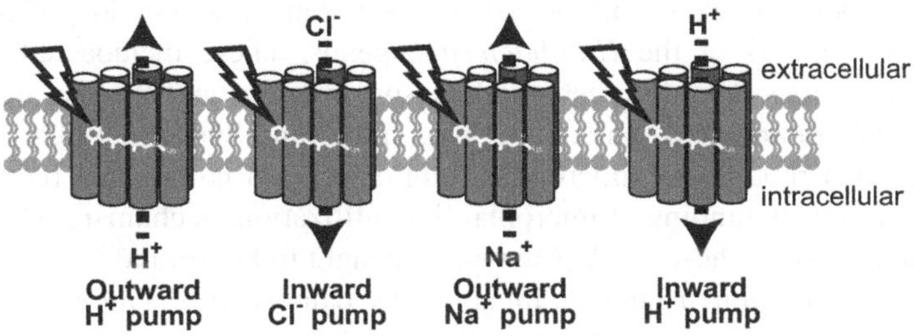

FIGURE 11.1 Functions upon light-induced retinal isomerization of microbial rhodopsins. Schematic representation of representative ion-transport functions of microbial rhodopsins. Upon absorption of a photon, retinal undergoes isomerization, driving conformational changes in the protein that enable ion translocation. Illustrated examples include: outward H^+ pump (e.g., proteorhodopsin and bacteriorhodopsin), inward Cl^- pump (e.g., halorhodopsin and Cl^- pumping rhodopsin), outward Na^+ pump (e.g., Na^+ pumping rhodopsin), and inward H^+ pump (e.g., xenorhodopsin). Arrows indicate the direction of ion movement.

redistributing energy and nutrients throughout the marine ecosystem. Therefore, understanding the dynamics of organic matter flux from the sunlit surface to deeper layers has long been considered central to elucidating the functioning of marine ecosystems.

However, discoveries over the past two decades have revealed a striking omission in this framework. Not all solar energy entering marine ecosystems flows through organic carbon intermediates. Instead, a substantial fraction follows an alternative, carbon-independent light-mediated energy pathway (Beja et al., 2000; DeLong and Beja, 2010). Such ion-motive forces, with the proton-motive force (PMF, the electrochemical gradient of protons across a membrane that drives ATP synthesis, solute transport, and motility) as the most widespread example, function as a universal bioenergetic currency that powers ATP synthesis, solute transport, and cellular motility. This pathway is mediated by microbial rhodopsins—retinal-binding photoproteins that directly convert light into a proton-motive or ion-motive force, bypassing carbon fixation entirely (Figure 11.1). Unlike the complex, multi-gene apparatus of photosynthesis, a single rhodopsin protein with its chromophore can act as a self-contained light-harvesting unit. Through this streamlined mechanism, cells can rapidly exploit light energy for essential processes without generating new organic matter.

First identified in marine bacteria in 2000 (Beja et al., 2000), these type I (microbial) rhodopsins have since been recognized as globally distributed across diverse marine taxa (McCarren and DeLong, 2007; Olson et al., 2018). Their ubiquity suggests that this carbon-independent light-mediated energy pathway is not a biochemical curiosity, but a major, parallel energy channel in the ocean that complements photosynthesis and may account for a substantial portion of the photonic energy budget in surface waters.

11.3 MICROBIAL RHODOPSINS

Rhodopsin is a light-sensitive protein with seven transmembrane domains that functions by forming covalent bonds with retinal pigment. Generally, rhodopsins found in nature are classified into two distinct evolutionary lineages: type I rhodopsins found in microorganisms and type II rhodopsins, which are most well-known in the visual system of vertebrates. These two types are generally considered to have evolved independently and are thought to represent functionally analogous but evolutionarily distinct light-sensing systems (Nagata and Inoue, 2021).

Microbial rhodopsin (hereinafter referred to simply as "rhodopsin") was first discovered in halophilic archaea in 1971. Oesterhelt and Stoeckenius identified a purple membrane protein in *Halobacterium salinarum* that transports protons (H^+) out of the cell in response to light (Oesterhelt and Stoeckenius, 1971). They named this protein bacteriorhodopsin and demonstrated its ability to generate a PMF from light. This function represented the first example of direct light energy utilization by a non-photosynthetic microorganism. Subsequent studies identified additional rhodopsin variants in members of the *Halobacterium* genus. These include halorhodopsins (HR) that transport chloride ions (Cl^-) into cells and sensory rhodopsins (SR) that mediate phototaxis responses (Matsuno-Yagi and Mukohata, 1977; Bogomolni and Spudich, 1982). These discoveries revealed that microbial rhodopsins function as light-driven ion pumps and light sensors, significantly expanding our understanding of microbial light utilization mechanisms (Figure 11.1).

However, for many years, these rhodopsins were thought to be limited to the narrow ecological niche of extreme halophilic environments inhabited by halobacteria. This view was dramatically changed in 2000 with the discovery of a new type of rhodopsin in marine bacteria (Beja et al., 2000). This groundbreaking discovery suggested that rhodopsins are not limited to extreme environment organisms but are widely distributed in sunlit aquatic environments and may constitute an evolutionarily successful light-harvesting strategy. The discovery sparked renewed interest in microbial rhodopsins and laid the foundation for the rapid expansion of this research field over the following decades.

11.4 DISCOVERY OF PROTEORHODOPSIN IN MARINE GAMMAPROTEOBACTERIA (SAR86)

The first compelling evidence for the presence of microbial rhodopsins in the ocean emerged in 2000, identifying a novel rhodopsin gene from the SAR86 lineage—a group of uncultivated Gammaproteobacteria. This gene was named proteorhodopsin (PR) and was shown to function as a light-driven proton pump (Figures 11.1 and 11.2). PR actively exports protons (H^+) across the cell membrane upon illumination, generating a proton motive force. This electrochemical gradient can be harnessed to drive ATP synthesis and solute transport, effectively converting light energy into cellular biochemical energy (Beja et al., 2000).

The initial PR gene was recovered through metagenomic analysis of a bacterial artificial chromosome library, which preserves large fragments of environmental DNA. Heterologous gene expression in *Escherichia coli*, supplemented with the retinal chromophore, demonstrated that the protein could respond to visible light and function as a proton pump. At that point, however, evidence for the *in situ* expression of PR in natural marine microbial communities remained lacking.

This gap was soon filled by spectroscopic studies in 2001, which detected rhodopsin-like absorption signals from microbial biomass collected by filtering large volumes of seawater (Beja et al., 2001). These findings confirmed the widespread presence of PR proteins in the oceanic water column. Notably, spectral variants of PR were also observed: while the first PRs identified absorbed

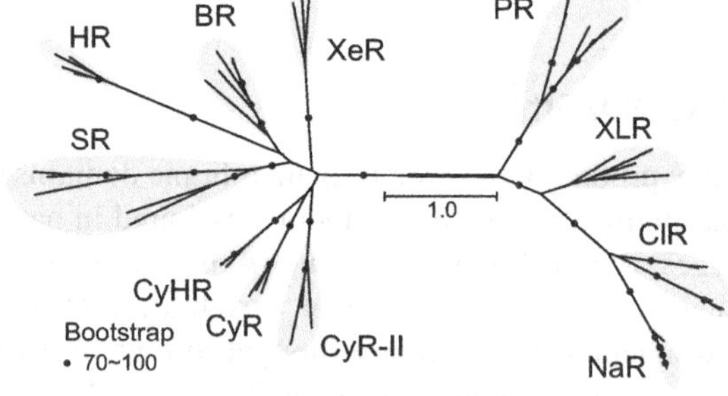

FIGURE 11.2 Phylogenetic relationships of microbial rhodopsins. Major clades are indicated as follows: PR (proteorhodopsin), XLR (xanthorhodopsin-like rhodopsin), ClR (Cl^- pumping rhodopsin), NaR (Na^+ pumping rhodopsin), BR (bacteriorhodopsin), HR (halorhodopsin), SR (sensory rhodopsin I and II), XeR (xenorhodopsin), CyHR (cyanobacterial halorhodopsin), CyR (cyanorhodopsin), and CyR-II (cyanorhodopsin-II). Bootstrap support values (>70%) are shown at the nodes, and the scale bar represents the number of substitutions per site.

green light ($\lambda_{max} \approx 525$ nm), subsequent studies revealed blue-light-absorbing forms ($\lambda_{max} \approx 490$ nm), suggesting that PRs are spectrally tuned to match the prevailing light conditions at different depths in the ocean.

In 2004 and 2006, large-scale environmental metagenomic surveys, including studies of the Sargasso Sea and the Hawaii Ocean Time-series (Venter et al., 2004; DeLong et al., 2006), demonstrated that PR genes are not confined to the SAR86 lineage but are instead widely distributed across a broad range of marine bacterial taxa. These pioneering studies revealed PR as a ubiquitous light-harvesting mechanism in the ocean. They provided a critical foundation for subsequent investigations into its ecological roles, evolutionary diversification, and contributions to marine energy fluxes. To move beyond gene-centric insights and confirm the physiological roles of PR in situ, researchers needed to cultivate PR-containing organisms under controlled conditions—a substantial challenge. It became essential to move from environmental detection to controlled laboratory studies with cultured representatives to elucidate the ecological roles of PR fully.

11.5 CULTIVATION CHALLENGES OF MARINE BACTERIA HARBORING PROTEORHODOPSINS

The initial discovery of PR was made not through classical microbiology, but via environmental genomics. In 2000, metagenomic analysis of DNA fragments retrieved from uncultivated marine bacterioplankton—specifically the SAR86 clade within the Gammaproteobacteria—revealed the presence of a novel gene encoding a rhodopsin-like protein with homology to archaeal proton pumps. This protein was shown to function as a light-driven H^+ pump when heterologously expressed in *Escherichia coli*, marking the beginning of a new paradigm in bacterial phototrophy. However, PR genes remained absent from the genomes of any bacteria cultivated in pure culture for years following this breakthrough. This gap underscored a fundamental limitation in understanding PR function: most PR-bearing marine bacteria are inherently challenging to culture using conventional laboratory techniques.

These "uncultured" bacteria—so designated because they cannot be grown in standard nutrient media—represent a major fraction of marine microbial diversity (Amann et al., 1995). Although metagenomics and structural biology have provided valuable insights into the molecular architecture and photochemistry of PR, they could not answer a fundamental biological question: what is the actual physiological role of PR in its native cellular context? Experimental validation of PR function *in vivo* requires access to pure cultures, where environmental variables such as light, nutrients, and ions can be systematically manipulated. Without this, PR remained largely a genomic and biochemical curiosity, decoupled from ecology and cell physiology.

A turning point arrived with the successful isolation of a member of the SAR11 clade—*Candidatus* Pelagibacter ubique—by Rappe et al. (2002) and subsequent genome sequencing (Giovannoni et al., 2005a). SAR11 is now recognized as the most abundant group of heterotrophic bacteria in surface oceans. Remarkably, the genome of *Ca.* P. ubique revealed a full-length PR gene, marking the first instance of a PR-encoding strain brought into pure culture. This was a landmark finding because it validated the metagenomic predictions and demonstrated that PR is harbored by ecologically dominant bacteria, not just rare or obscure lineages.

Even more compelling was the confirmation of PR protein expression in *Ca.* P. ubique. Using membrane enrichment and mass spectrometry, Giovannoni and colleagues identified PR-derived peptides, directly demonstrating that the gene was transcribed and translated *in vivo*. This firmly established PR as a functional component of the cellular phototrophic machinery in a cultivated marine bacterium. While initial experiments did not reveal a clear light-dependent growth enhancement under nutrient-replete conditions, subsequent studies suggested that PR may improve cellular energy balance during nutrient limitation (Steindler et al., 2011), highlighting the importance of environmental context in assessing PR function.

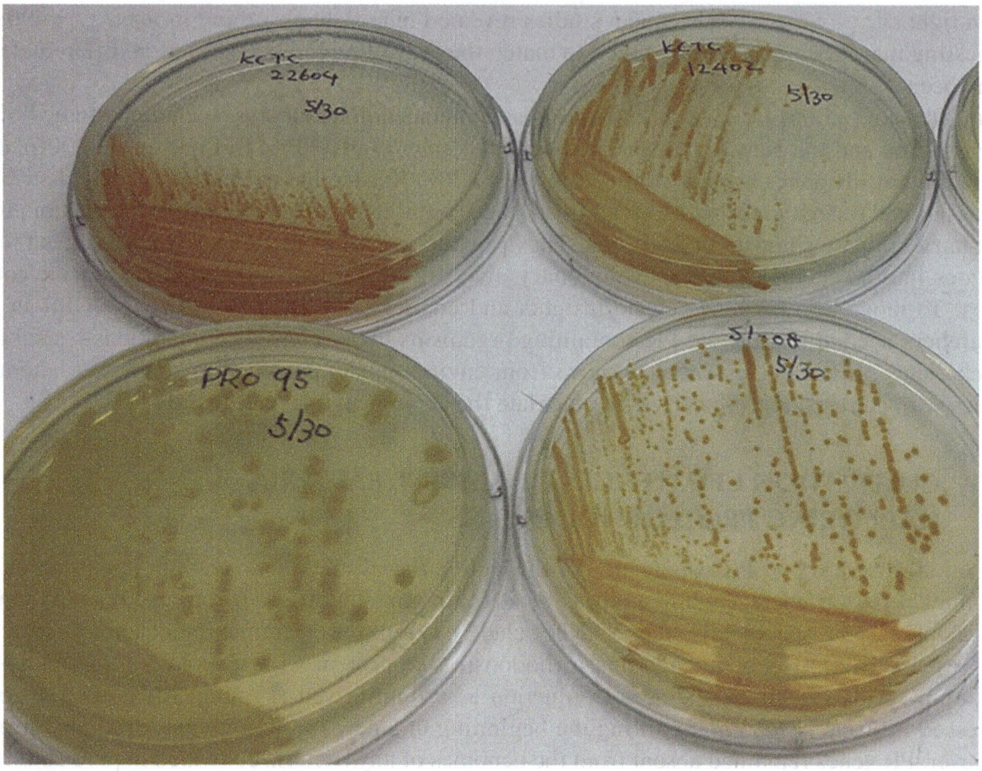

FIGURE 11.3 Pigmented colonies of Flavobacteriaceae strains harboring microbial rhodopsin genes. Photograph of four representative Flavobacteriaceae strains grown on agar plates. All colonies exhibit distinct yellow to orange pigmentation, reflecting carotenoid biosynthesis. These carotenoids serve as precursors for retinal, the chromophore essential for rhodopsin function. (Photo credit: Susumu Yoshizawa.)

Meanwhile, PR genes were also identified in cultured strains of marine Flavobacteria (Figure 11.3), a phylogenetically distinct group within the Bacteroidota phylum. A pivotal study by Gómez-Consarnau et al. (2007) reported that PR-containing Flavobacteria showed significantly higher growth rates under light compared to dark conditions, indicating that PR activity positively influences growth under oligotrophic conditions. This provided the first experimental evidence that PR-based phototrophy could enhance growth in a native host, rather than merely sustaining energy during stress.

Building on these findings, Yoshizawa et al. (2012) successfully cultivated a wide array of marine PR-containing Flavobacteria and performed quantitative assays of light-driven proton transport. Their results showed that illumination induced rapid extracellular H^+ extrusion, consistent with active PR function, across diverse strains. These measurements provided unambiguous functional evidence linking PR expression to photophysiological output in natural isolates.

Together, these studies highlight that pure culture experiments are essential for elucidating the physiological and ecological significance of PR (Figure 11.3). They also demonstrate that PR-bearing bacteria are not limited to Gammaproteobacteria and Alphaproteobacteria but are found across diverse marine taxa, including Bacteroidota, Actinobacteria, and Planctomycetota. The gradual expansion of PR-positive cultured strains has thus been instrumental in moving beyond gene-centric insights toward an integrated understanding of PR's role in microbial survival, growth, and energy balance in the ocean. Building on these cultivation studies, the following section examines how PR genes are distributed across phylogenetically diverse lineages.

11.6 PHYLOGENETIC DIVERSITY AND EVOLUTIONARY SPREAD OF PROTEORHODOPSIN GENES

Following the initial discovery of PR genes in uncultured marine Gammaproteobacteria, extensive metagenomic surveys revealed that these photoproteins are not confined to a single lineage, but are widely distributed across diverse branches of the bacterial domain. As sequencing efforts expanded to new marine environments, PR homologs were successively detected in a broad array of taxonomic groups, including Alphaproteobacteria (notably the SAR11 clade), Bacteroidota (mainly marine Flavobacteria), Actinobacteria, Planctomycetota, and others (Frigaard et al., 2006; Rusch et al., 2007). This unexpectedly broad phylogenetic distribution raises an important question: how did such a relatively compact and functionally specialized gene become so pervasive across microbial taxa?

A key insight arises from the observation that bacterial lineages harboring PR genes are often phylogenetically distant. The same functional gene—encoding a seven-transmembrane, retinal-binding proton pump—has been found in taxonomic groups that otherwise share little genomic similarity. This patchy phylogenetic distribution cannot be explained by vertical inheritance from a common ancestor alone; instead, it strongly suggests that horizontal gene transfer has played a central role in spreading PR genes across marine microbes. PR is often regarded as a "cosmopolitan gene," its broad taxonomic distribution appears to be driven by a combination of strong selective pressures for light utilization and efficient lateral dissemination mechanisms. Genomic signatures such as atypical GC content, codon usage bias, and the frequent presence of transposase genes flanking PR loci provide further evidence for its horizontal mobility. Importantly, PR-based phototrophic systems are genetically compact, typically encoded by only a few genes, and are functionally autonomous, allowing them to be readily integrated into the membranes of diverse microbial hosts. Their simplicity, physiological utility in augmenting cellular energy metabolism, and compatibility across phylogenetic boundaries likely contribute to their propensity for horizontal transfer in the ocean's microbial world (Sabehi et al., 2004; Sabehi et al., 2005; Frigaard et al., 2006).

The possibility of inter-domain gene transfer has also emerged. In a notable case, Frigaard et al. (2006) presented evidence for lateral transfer of PR genes between Bacteria and Archaea, based on phylogenetic incongruence and the presence of PR-like genes in archaeal metagenomic contigs from marine environments. These findings emphasize the extraordinary mobility of PR genes and highlight their potential evolutionary advantage—an advantage compelling enough to overcome major taxonomic barriers. Such mobility implies that PR confers selective benefits across various ecological niches, particularly in the light-penetrated but nutrient-depleted layers of the ocean surface. Once acquired, PR can be rapidly integrated into host energy metabolism because it encodes a self-contained, single-gene phototrophic system that requires only retinal as a chromophore. In contrast to photosynthesis, which entails a substantial genomic investment in complex electron transport machinery, PR-based phototrophy demands minimal cellular and genetic resources. This modular, energetically efficient architecture likely contributes to its evolutionary success and facilitates its horizontal propagation.

The evolutionary plasticity of PR is also reflected in its amino acid sequence diversity, which allows for spectral tuning and adaptation to different light regimes (see Section 11.6). Minor substitutions in the retinal-binding pocket can shift the protein's absorption maximum, enabling microbes to harvest the most abundant wavelengths at their specific depth or geographic location (Man et al., 2003; Mao et al., 2024). This molecular flexibility complements its phylogenetic spread, allowing PR to be co-opted repeatedly by unrelated lineages for similar ecological purposes.

Future research on PR should move beyond merely mapping its taxonomic distribution to investigating its evolutionary trajectories and functional integration within host metabolic networks. An integrative approach combining comparative genomics, synteny analysis, and experimental validation will be essential for elucidating patterns of acquisition, retention, and possible loss. A fundamental challenge is to distinguish ancient vertical inheritance from more recent horizontal acquisitions in reconstructing the evolutionary history of PR.

In summary, the present distribution of PR across both bacterial and archaeal phyla reflects a dynamic interplay between ecological selection pressures and gene mobility (Pereira et al., 2021). The gene's portability, minimal genomic footprint, and immediate energetic benefit position it as an exceptionally successful innovation in marine microbial ecology (Frigaard et al., 2006; Pinhassi et al., 2016). Rather than representing a vertically inherited trait within a single lineage, PR should be understood as a rapidly disseminated and versatile adaptation that has evolved to move, integrate, and function wherever light and oligotrophic conditions create a niche for phototrophic advantage. Phylogenetic analyses reveal how PR genes are shared among diverse lineages; however, understanding their ecological impact requires examining their distribution across oceanic spatial and temporal gradients.

11.7 BIOGEOGRAPHY AND ENVIRONMENTAL DISTRIBUTION OF PROTEORHODOPSINS IN THE OCEAN

Since the initial discovery of PR genes in uncultivated marine bacteria, considerable effort has been devoted to elucidating how these light-driven proton pumps are distributed across the ocean's extensive spatial and temporal gradients (Olson et al., 2018). Early metagenomic surveys swiftly revealed that PRs are not restricted to specific taxonomic groups or localized environments (Venter et al., 2004; DeLong et al., 2006), but instead exhibit a global distribution, characterized by biogeographical and photoadaptive diversity.

A study by Man et al. (2003) demonstrated that PRs can be classified into distinct spectral types based on a single amino acid substitution at position 105 within the retinal-binding pocket. This residue determines the protein's maximum absorption wavelength: green light–absorbing PRs (GPRs; $\lambda_{max} \approx 525\,nm$) and blue light–absorbing PRs (BPRs; $\lambda_{max} \approx 490\,nm$). These spectral types exhibit vertical stratification in the water column: GPRs are predominantly found in the sunlit surface layer, whereas BPRs are more broadly distributed, spanning from the surface to deeper, dimly lit zones. This depth-dependent distribution closely mirrors the spectral attenuation of sunlight in seawater and strongly suggests that PRs have undergone fine-scale adaptive tuning to local light conditions. The fact that such spectral optimization is achieved through a single amino acid change highlights the remarkable plasticity of microbial photoreceptors in responding to subtle ecological gradients.

In addition to vertical stratification, PRs are horizontally widespread across diverse oceanic regions. Metagenomic studies have identified PR genes in microbial communities from various marine habitats—including the Sargasso Sea, Mediterranean Sea, North Pacific Subtropical Gyre, and polar waters—highlighting their ubiquity across ocean basins. These sequences reveal extensive taxonomic and spectral variation, indicating that PRs are encoded by phylogenetically diverse microbes that have independently adapted to distinct environmental conditions (Venter et al., 2004; Rusch et al., 2007; Campbell et al., 2008; Steindler et al., 2011; Boeuf et al., 2016).

Recent advances in high-resolution metagenomic datasets have enabled more quantitative assessments of PR gene abundance and diversity along environmental gradients. For example, Olson et al. (2018) analyzed PR sequence variants in the North Pacific Subtropical Gyre. They found that PR abundance was strongly associated with depth, particularly enriched in the euphotic zone (shallower than 200 m). Interestingly, no significant seasonal correlations were observed between PR abundance and physical-chemical parameters such as temperature, salinity, or oxygen. This is consistent with Station ALOHA's relatively stable, stratified, and oligotrophic nature, where environmental conditions exhibit limited seasonal variability.

Importantly, PRs are recognized as among the most abundant functional genes in surface ocean waters, rivaling those involved in central metabolic processes such as respiration and nutrient uptake. Quantitative PCR, metatranscriptomic analyses, and pigment-based studies have consistently shown that PR-containing cells can comprise approximately half of the bacterial plankton in surface waters, with even greater prevalence reported in oligotrophic gyres (Campbell et al., 2008; DeLong and Beja, 2010; Olson et al., 2018). This widespread occurrence underscores the ecological success of PR-based phototrophy and raises important questions about its role in ocean-scale energy flux and carbon cycling.

These findings indicate that PR functions as a broadly adaptive light-harvesting protein in marine ecosystems. Its environmental distribution reflects evolutionary processes—such as spectral tuning via amino acid substitutions—and ecological selection pressures, including light availability, nutrient limitation, and temperature. Thus, PRs serve as molecular indicators of phototrophic capacity and as markers of microbial adaptation to oceanic light regimes. Continued efforts to integrate PR sequence data with satellite-derived irradiance profiles, ocean circulation models, and biogeochemical measurements hold promise for advancing our understanding of how microbial phototrophy shapes ecosystem function under changing climatic conditions.

11.8 BEYOND PROTON PUMPS: FUNCTIONAL DIVERSIFICATION OF MICROBIAL RHODOPSINS

Marine microbial rhodopsins were initially identified as light-driven proton pumps that generate membrane potential and support ATP synthesis in heterotrophic marine bacteria. This phototrophic mechanism, which requires only a seven-transmembrane protein and the chromophore retinal, offers an efficient strategy for harvesting solar energy in oligotrophic marine environments. However, with the accumulation of genomic data, rhodopsins with functions other than proton transport have also been discovered. This functional diversity is thought to reflect physiological adaptations to various environmental conditions.

A prominent example of this diversification is the discovery of light-driven sodium pumping rhodopsin (NaR). In particular, KR2 (*Krokinobacter eikastus* rhodopsin 2) was first identified by Inoue et al. (2013) from a marine Flavobacterium (Figures 11.1 and 11.2). KR2 shares structural homology with PR, but it selectively transports Na^+ ions in response to light, generating a sodium motive force instead of a proton gradient. Structural and spectroscopic analyses have revealed that key amino acid residues within the ion transport pathway confer Na^+ specificity to KR2. Of particular note, Na^+ ions are abundant in marine environments and, under alkaline conditions, are more readily available than protons due to the low concentration of H^+. These factors suggest that Na^+-translocating rhodopsins may provide a selective ecological advantage in high salinity and alkalinity marine habitats.

Even more unexpected was the discovery of light-driven chloride pumps in marine Flavobacteria. Yoshizawa et al. (2014) characterized a new rhodopsin class in *Nonlabens marinus* that functions as a Cl^- pump, inwardly transporting chloride ions in response to light. While similar in function to archaeal HR, the flavobacterial Cl^- pumps are phylogenetically distinct and belong to a separate lineage within the microbial rhodopsin superfamily (Figures 11.1 and 11.2). These proteins likely play a role in osmoregulation, intracellular pH buffering, or maintaining ionic homeostasis under fluctuating salinity or acid-base stress.

These discoveries point to a broader principle: microbial rhodopsins have evolved as versatile, light-activated ion transporters, with varying ion selectivity depending on ecological context. This diversity is underpinned by subtle yet functionally significant changes in their transmembrane helices, particularly the ion-conducting pathway and retinal-binding pocket (Kato et al., 2015; Hosaka et al., 2016, 2022). Site-directed mutagenesis and structural modeling have revealed that even single amino acid substitutions can switch ion specificity, underscoring the evolutionary plasticity of the rhodopsin scaffold (Sasaki et al., 1995; Inoue et al., 2016). The ability to pump H^+, Na^+, or Cl^-—each with distinct bioenergetic consequences—allows marine bacteria to fine-tune their energy metabolism and ionic balance in response to environmental conditions.

Functionally, these different rhodopsins are thought to support distinct modes of photoheterotrophy. Proton pumps may be optimized for ATP synthesis under standard oceanic conditions, whereas Na^+ pumps could offer a more stable energy source in hypersaline or alkaline niches.

Chloride pumps may play regulatory roles such as cytoplasmic charge compensation or osmotic control while not contributing directly to energy metabolism. These diverse physiological outcomes indicate that rhodopsin-based light utilization is not a monolithic system, but rather a modular and adaptive one—finely tuned to the host organism's metabolic demands and environmental constraints.

Collectively, these findings demonstrate that PR is not a uniform phototrophic mechanism but part of a broader, functionally diverse molecular toolkit that enables microorganisms to harness light under varying environmental conditions. The ongoing expansion of rhodopsin diversity, driven by advances in metagenomics, structural biology, and functional characterization, highlights the likelihood that many light-activated functions remain undiscovered. Elucidating the full spectrum of rhodopsin ion transport mechanisms will be essential for understanding the photophysiology of marine microbes and integrating rhodopsin-based phototrophy into broader models of microbial energy flow, environmental adaptation, and evolutionary dynamics.

11.9 GENOMIC CONTEXT AND ARCHITECTURAL FEATURES OF PROTEORHODOPSIN-CONTAINING MARINE BACTERIA

One of the defining characteristics of PR-based phototrophy is its genetic simplicity. Unlike photosynthetic systems, which require large operons encoding multi-protein complexes, PR phototrophy can be implemented with a minimal gene set—often just a single gene for the rhodopsin protein and a pathway to produce the chromophore retinal. This simplicity facilitates its horizontal transfer (as discussed in Section 11.5) and allows PR phototrophy to be incorporated into compact and streamlined genomes, such as those of oligotrophic marine bacteria.

A striking example is *Ca.* P. ubique (SAR11 clade), whose genome is one of the smallest known among free-living bacteria, at approximately 1.3 Mbp (Giovannoni et al., 2005b). Despite its extremely reduced gene content and metabolic specialization, the genome retains a full-length PR gene, suggesting that PR-mediated phototrophy is maintained under strong selective pressure, even in organisms operating near the minimal threshold for autonomous life (Giovannoni et al., 2005a). This observation supports the view that PR-mediated energy acquisition is an important auxiliary mechanism under oligotrophic conditions, where the ability to harness light can reduce reliance on scarce organic substrates.

The genomic neighborhood of PR often reveals additional clues to its functional integration. In many marine bacteria, PR genes are located adjacent to genes involved in retinal biosynthesis, particularly blh, which encodes β-carotene 15,15′-monooxygenase (Gómez-Consarnau et al., 2007; Martinez et al., 2007). This enzyme catalyzes the cleavage of β-carotene to retinal, the chromophore essential for PR activity. Sabehi et al. (2005) reported that roughly one-third of PR clones obtained from metagenomic libraries contain co-localized blh homologs, suggesting that the PR–retinal module may be functionally and genetically linked, potentially co-transcribed as an operon. This arrangement ensures that PR is coupled with a self-sufficient retinal supply system, particularly in hosts that cannot rely on exogenous retinal from the environment.

Beyond retinal biosynthesis, comparative genomics of PR-containing strains reveals further metabolic and ecological adaptation patterns. For example, Kumagai et al. (2018) described two contrasting photoadaptive strategies in marine Flavobacteriia based on genomic content: a "solar-panel strategy," characterized by the presence of PR and blh genes for light harvesting, and a "parasol strategy," in which bacteria express photoprotective pigments to avoid light stress but lack PR genes (Kumagai et al., 2018). This analysis is primarily based on genomic data from isolated cultures. However, future studies using metagenome-assembled genomes (MAGs) are expected to provide new insights into bacterial adaptation to marine surface environments.

Some PR-containing genomes show signs of genomic modularity, where phototrophic functions are encoded in discrete genomic islands potentially subject to horizontal transfer (Gonzalez et al., 2011). These islands may include PR, blh and genes related to retinal biosynthesis from isoprenoid precursors, carotenoid pathways, or even rhodopsin-associated regulatory elements. Such modular arrangements enable the acquisition of a complete phototrophic system through a single evolutionary event, facilitating the rapid integration and retention of PR phototrophy in newly colonized hosts.

Notably, incorporating PR into bacterial genomes does not necessarily require large-scale rewiring of cellular metabolism. As a membrane protein that directly converts light into an electrochemical gradient, PR can interface with pre-existing ATP synthase complexes, energize nutrient transporters, or indirectly support carbon assimilation without a dedicated photosynthetic apparatus. This high degree of compatibility allows PR phototrophy to be integrated with minimal physiological cost across a wide range of bacterial hosts—from metabolically streamlined oligotrophs like *Pelagibacter* to more versatile heterotrophs like *Dokdonia*.

In conclusion, PR-containing marine bacteria exhibit genomic signatures that reflect both the streamlined efficiency of the PR system and its adaptive flexibility. From compact genomes with minimal auxiliary genes to modular phototrophy islands embedded in larger, more plastic genomes, PR phototrophy exemplifies how light-utilization strategies can evolve and persist under diverse genomic and ecological constraints. As environmental genome datasets and single-cell assemblies continue to grow, comparative genomic studies will further elucidate how PR is integrated into the genomic architecture of marine microbes—and how this light-driven system contributes to the ecological success of bacteria in the sunlit ocean.

11.10 LIGHT ENERGY WITHOUT ORGANIC SUBSTRATES: PHYSIOLOGICAL AND ECOLOGICAL IMPLICATIONS OF PROTEORHODOPSIN PHOTOTROPHY

Unlike chlorophyll-based phototrophy, which relies on carbon fixation or organic substrate oxidation, PR-mediated phototrophy enables marine heterotrophic bacteria to directly convert sunlight into electrochemical energy—bypassing carbon assimilation as an intermediate. This "carbon-independent light-mediated energy pathway" represents a conceptually distinct mode of phototrophy that has profound implications for bacterial physiology, microbial survival strategies, and energy fluxes in the ocean's surface layers.

Experimental studies have shown that PR can contribute to growth or survival depending on the environmental context. For example, Gómez-Consarnau et al. (2007) demonstrated that PR-containing Flavobacteria, such as *Dokdonia* sp., exhibit enhanced growth under light in oligotrophic seawater media. In these conditions, light-driven proton pumping provides an auxiliary energy source that supplements heterotrophic metabolism, reducing the dependency on limited organic carbon and enabling faster biomass accumulation. This represents a rare case in which PR phototrophy contributes to maintenance and light-enhanced growth—a hallmark of functional photometabolism.

However, the role of PR appears more nuanced in other lineages. In *Ca.* P. ubique, a streamlined bacterium from the SAR11 clade, light does not stimulate growth under standard nutrient-replete conditions. Instead, Steindler et al. (2011) showed that PR activity enhances cellular viability by maintaining ATP levels, effectively delaying death under carbon-limited or starvation conditions. In such cases, PR operates not as a growth promoter but as an energy-sparing system—a molecular "battery" that supports survival when organic substrates are depleted. However, these interpretations are based mainly on laboratory studies under simplified conditions and may not fully capture the complexity of natural environments. Contrasting results across lineages highlight unresolved debates regarding how PR contributes to biomass production versus maintenance.

Recent geochemical and metaproteomic analyses suggest that the amount of light energy captured by PRs in oligotrophic ocean regions is ecologically significant and cannot be ignored. Gómez-Consarnau et al. (2019) quantified retinal concentrations in natural bacterioplankton assemblages and estimated PR-driven photon absorption across oceanic gyres. Surprisingly, they found that in some regions of the Mediterranean Sea, PR-mediated light absorption may rival or even exceed chlorophyll-based photosynthesis on a per-cell basis. This challenges the classical paradigm that primary production is the sole entry point of solar energy into marine food webs, revealing PR as a parallel conduit of photonic energy into microbial metabolism.

Moreover, PR activity may influence bacterial growth efficiency (BGE), a key parameter in ocean carbon cycling (Steindler et al., 2011). Heterotrophic bacteria typically oxidize organic carbon to obtain energy, releasing a portion as CO_2 through respiration. If the phototrophic ATP generated by PR activity can offset the cellular energy demand typically met through respiration, a greater proportion of the assimilated carbon may be diverted toward biomass production. This "respiration-load reduction" mechanism may enhance BGE, improving the carbon retention efficiency of microbial communities. PR phototrophy could decouple energy production from carbon oxidation, thereby shifting the balance between respiration and biosynthesis.

Support for this concept comes from diel respiration studies. For instance, SAR86-related bacteria have been observed to exhibit nighttime respiration rates up to six times higher than those during the day (Munson-McGee et al., 2022). This pattern suggests that light availability—and PR activity—suppresses cellular respiration during daylight hours. These observations indicate that PR-based phototrophy actively shapes the temporal dynamics of bacterial metabolism, contributing to diel variation in microbial energy demand and carbon processing. When scaled to the level of entire microbial communities in the euphotic zone, such effects could alter the fate of dissolved organic matter and modulate carbon flux to deeper waters.

PR-mediated phototrophy represents a multifaceted ecological innovation: it enables bacteria to survive starvation, promotes growth under nutrient-poor conditions, reduces dependence on respiratory metabolism, and potentially enhances carbon retention efficiency. Rather than serving as a primary driver of biomass production, PR functions as an auxiliary energy conduit, fine-tuning the metabolic responses of heterotrophic bacteria to fluctuating environmental conditions. Its broad distribution and functional versatility suggest that PR contributes to the resilience of individual bacterial cells and the overall stability and efficiency of microbial communities in the sunlit ocean.

As our understanding of PR phototrophy deepens, it is becoming increasingly evident that this light-dependent mechanism—once underestimated—plays a vital role in the non-photosynthetic energy economy of the ocean. By enabling marine bacteria to utilize solar energy without carbon fixation, PR expands the conceptual framework of phototrophy and reshapes our understanding of energy flow in marine ecosystems. Although research on PR phototrophy has focused mainly on bacteria and archaea, recent discoveries show that similar strategies are also present in eukaryotic microorganisms, expanding this light-harvesting mechanism's evolutionary and ecological scope.

11.11 BEYOND THE PROKARYOTIC REALM: RHODOPSINS IN MARINE EUKARYOTIC MICROORGANISMS

Microbial rhodopsins were long regarded as exclusive to bacteria and archaea, functioning primarily as light-driven ion pumps and sensory proteins. However, recent advances in genomics and transcriptomics have revealed that rhodopsin genes are also present in eukaryotic microorganisms, dramatically broadening this photoreceptive protein family's phylogenetic and functional reach.

The first eukaryotic microbial rhodopsins were discovered in a freshwater alga. One of the earliest functionally characterized examples was the identification of channelrhodopsins (ChR1 and ChR2) in the green alga *Chlamydomonas reinhardtii*. Electrophysiological studies of phototactic behavior in this species revealed submillisecond light-induced membrane currents, suggesting the presence of directly light-gated ion channels. In 2002, Nagel et al. identified ChR1 as a microbial-type rhodopsin that mediates light-dependent proton conductance when expressed in *Xenopus laevis* oocytes (Nagel et al., 2002). This was followed by the characterization of ChR2 as a light-gated cation channel permeable to H^+, Na^+, K^+, and Ca^{2+}, capable of opening within microseconds after photon absorption (Nagel et al., 2003). These findings demonstrated that a single protein could integrate light sensing and ion conductance without relying on second-messenger cascades. Most notably, the ability of ChR2 to induce light-triggered depolarization in animal cells paved the way for optogenetics, marking a paradigm shift in the control of neural activity using light (Boyden et al., 2005).

An important discovery concerning ion-transporting rhodopsins in marine eukaryotes came from the heterotrophic dinoflagellate *Oxyrrhis marina*. Using genomic and expression analyses, Slamovits et al. (2011) identified a rhodopsin gene closely related to bacterial PR, which was highly expressed under light-exposed conditions (Slamovits et al., 2011). Functional assays suggested that this eukaryotic rhodopsin retained proton-pumping activity, indicating that it contributes to the cellular energy budget in a light-dependent manner. Importantly, the gene sequence bears hallmarks of lateral gene transfer, suggesting that eukaryotic rhodopsins may have originated via horizontal acquisition from bacteria, subsequently becoming integrated into eukaryotic bioenergetic systems.

The presence of rhodopsins in photosynthetic eukaryotes, particularly diatoms, adds a new dimension of functional versatility. Multiple rhodopsin homologs have been identified in several diatom genomes, some of which are differentially expressed under environmental stress, notably, under iron-limiting conditions—a common constraint in large regions of the ocean—transcript levels of rhodopsin genes increase, while photosynthetic activity declines. This inverse correlation suggests that rhodopsins may act as auxiliary phototrophic systems, helping to sustain cellular energetics when photosynthesis is impaired (Marchetti et al., 2015).

Further supporting this idea, subcellular localization studies have revealed that certain diatom rhodopsins are targeted to the outermost membrane of the chloroplast complex (Yoshizawa et al., 2023). This strategic positioning implies a functional role in the carbon concentrating mechanism (CCM), where proton pumping across the membrane could acidify the periplastidal space and enhance CO_2 conversion from bicarbonate. If verified, this would represent a remarkable case of microbial-type rhodopsins being repurposed for photoenhanced carbon assimilation in photosynthetic eukaryotes.

The evolutionary and ecological significance of rhodopsins in marine eukaryotes is only beginning to be understood. Some rhodopsins may act primarily as light sensors, modulating cellular responses such as diel vertical migration, phototaxis, or circadian regulation. Others may contribute directly to energy metabolism, particularly in mixotrophic protists that combine photosynthesis, phagotrophy, and photoheterotrophy. Given the metabolic flexibility of many marine protists, rhodopsins may represent an evolutionarily streamlined solution for capturing and utilizing light—compact, efficient, and highly adaptable.

From an ecological perspective, the widespread expression of rhodopsins in both autotrophic and heterotrophic eukaryotic plankton suggests that light-derived energy flow is far more taxonomically and functionally pervasive than previously recognized. Rhodopsin-based phototrophy and photoregulation may buffer microbial eukaryotes against variable light and nutrient conditions, enhancing their resilience and competitiveness in dynamic marine environments.

In summary, the discovery of functional rhodopsins in eukaryotic microorganisms fundamentally reshaped our view of microbial phototrophy. What began as a bacterial adaptation to oligotrophic habitats has evolved into a versatile, lineage-spanning photometabolic strategy, co-opted by eukaryotes for energy harvesting, environmental sensing, and possibly for enhancing carbon fixation. As genomic and experimental tools continue to expand our access to non-model marine protists, eukaryotic rhodopsins are poised to become a central focus in the next generation of microbial photobiology research.

11.12 CONCLUSION

Research on microbial rhodopsins has substantially expanded our understanding of how solar energy enters marine ecosystems. These proteins provide a carbon-independent pathway of light utilization that operates in parallel with photosynthesis. Their broad distribution and functional diversity indicate that rhodopsins play important roles in microbial survival and growth under oligotrophic conditions and can influence carbon cycling by reducing respiratory energy demand and enhancing microbial growth efficiency.

To evaluate their ecological significance more fully, future work needs to integrate molecular mechanisms, cellular physiology, community dynamics, and global biogeochemical modeling. Evidence from both prokaryotes and eukaryotes shows that rhodopsins represent an adaptable strategy for harvesting light energy, but their quantitative contribution to ocean energy budgets remains unresolved. Determining this contribution, clarifying their roles across different taxa, and understanding their interactions with viruses and mobile genetic elements will be important next steps.

Incorporating rhodopsin-mediated processes into ecosystem and climate models will allow a more accurate assessment of how light supports life in the ocean. In this way, the study of microbial rhodopsins can contribute not only to basic knowledge of biodiversity and microbial physiology but also to a broader understanding of the resilience and function of marine ecosystems.

DECLARATION OF GENERATIVE AI USE

During the preparation of this manuscript, the author used ChatGPT to assist with English translation and grammatical refinement. The author subsequently reviewed and edited the content and takes full responsibility for the final version of the published chapter.

REFERENCES

Amann, R.I., Ludwig, W., and Schleifer, K.H. (1995) Phylogenetic identification and in-situ detection of individual microbial-cells without cultivation. *Microbiological Reviews* **59**: 143–169.

Beja, O., Aravind, L., Koonin, E.V., Suzuki, M.T., Hadd, A., Nguyen, L.P. et al. (2000) Bacterial rhodopsin: evidence for a new type of phototrophy in the sea. *Science* **289**: 1902–1906.

Beja, O., Spudich, E.N., Spudich, J.L., Leclerc, M., and DeLong, E.F. (2001) Proteorhodopsin phototrophy in the ocean. *Nature* **411**: 786–789.

Boeuf, D., Lami, R., Cunnington, E., and Jeanthon, C. (2016) Summer abundance and distribution of proteorhodopsin genes in the Western Arctic Ocean. *Frontiers in Microbiology* **7**: 1584.

Bogomolni, R.A., and Spudich, J.L. (1982) Identification of a third rhodopsin-like pigment in phototactic *Halobacterium halobium*. *Proceedings of the National Academy of Sciences of the United States of America* **79**: 6250–6254.

Boyden, E.S., Zhang, F., Bamberg, E., Nagel, G., and Deisseroth, K. (2005) Millisecond-timescale, genetically targeted optical control of neural activity. *Nature Neuroscience* **8**: 1263–1268.

Campbell, B.J., Waidner, L.A., Cottrell, M.T., and Kirchman, D.L. (2008) Abundant proteorhodopsin genes in the North Atlantic Ocean. *Environmental Microbiology* **10**: 99–109.

DeLong, E.F., and Beja, O. (2010) The light-driven proton pump proteorhodopsin enhances bacterial survival during tough times. *Plos Biology* **8**: e1000359.

DeLong, E.F., Preston, C.M., Mincer, T., Rich, V., Hallam, S.J., Frigaard, N.U. et al. (2006) Community genomics among stratified microbial assemblages in the ocean's interior. *Science* **311**: 496–503.

Frigaard, N.U., Martinez, A., Mincer, T.J., and DeLong, E.F. (2006) Proteorhodopsin lateral gene transfer between marine planktonic Bacteria and Archaea. *Nature* **439**: 847–850.

Giovannoni, S.J., Bibbs, L., Cho, J.C., Stapels, M.D., Desiderio, R., Vergin, K.L. et al. (2005a) Proteorhodopsin in the ubiquitous marine bacterium SAR11. *Nature* **438**: 82–85.

Giovannoni, S.J., Tripp, H.J., Givan, S., Podar, M., Vergin, K.L., Baptista, D. et al. (2005b) Genome streamlining in a cosmopolitan oceanic bacterium. *Science* **309**: 1242–1245.

Gómez-Consarnau, L., Gonzalez, J.M., Coll-Llado, M., Gourdon, P., Pascher, T., Neutze, R. et al. (2007) Light stimulates growth of proteorhodopsin-containing marine Flavobacteria. *Nature* **445**: 210–213.

Gómez-Consarnau, L., Raven, J.A., Levine, N.M., Cutter, L.S., Wang, D.L., Seegers, B. et al. (2019) Microbial rhodopsins are major contributors to the solar energy captured in the sea. *Science Advances* **5**: eaaw8855.

Gonzalez, J.M., Pinhassi, J., Fernandez-Gomez, B., Coll-Llado, M., Gonzalez-Velazquez, M., Puigbo, P. et al. (2011) Genomics of the proteorhodopsin-containing marine flavobacterium Dokdonia sp. strain MED134. *Applied and Environmental Microbiology* **77**: 8676–8686.

Hosaka, T., Nomura, T., Kubo, M., Nakane, T., Luo, F.J., Sekine, S.I. et al. (2022) Conformational alterations in unidirectional ion transport of a light-driven chloride pump revealed using X-ray free electron lasers. *Proceedings of the National Academy of Sciences of the United States of America* **119**: e2117433119.

Hosaka, T., Yoshizawa, S., Nakajima, Y., Ohsawa, N., Hato, M., DeLong, E.F. et al. (2016) Structural mechanism for light-driven transport by a new type of chloride ion pump, Nonlabens marinus rhodopsin-3. *Journal of Biological Chemistry* **291**: 17488–17495.

Inoue, K., Nomura, Y., and Kandori, H. (2016) Asymmetric functional conversion of eubacterial light-driven ion pumps. *Journal of Biological Chemistry* **291**: 9883–9893.

Inoue, K., Ono, H., Abe-Yoshizumi, R., Yoshizawa, S., Ito, H., Kogure, K., and Kandori, H. (2013) A light-driven sodium ion pump in marine bacteria. *Nature Communications* **4**: 1678.

Kato, H.E., Inoue, K., Abe-Yoshizumi, R., Kato, Y., Ono, H., Konno, M. et al. (2015) Structural basis for Na^+ transport mechanism by a light-driven Na^+ pump. *Nature* **521**: 48–53.

Kumagai, Y., Yoshizawa, S., Nakajima, Y., Watanabe, M., Fukunaga, T., Ogura, Y. et al. (2018) Solar-panel and parasol strategies shape the proteorhodopsin distribution pattern in marine Flavobacteriia. *ISME Journal* **12**: 1329–1343.

Man, D.L., Wang, W.W., Sabehi, G., Aravind, L., Post, A.F., Massana, R. et al. (2003) Diversification and spectral tuning in marine proteorhodopsins. *EMBO Journal* **22**: 1725–1731.

Mao, J.F., Jin, X.S., Shi, M., Heidenreich, D., Brown, L.J., Brown, R.C.D. et al. (2024) Molecular mechanisms and evolutionary robustness of a color switch in proteorhodopsins. *Science Advances* **10**: eadj0384.

Marchetti, A., Catlett, D., Hopkinson, B.M., Ellis, K., and Cassar, N. (2015) Marine diatom proteorhodopsins and their potential role in coping with low iron availability. *The ISME Journal* **9**: 2745–2748.

Martinez, A., Bradley, A.S., Waldbauer, J.R., Summons, R.E., and DeLong, E.F. (2007) Proteorhodopsin photosystem gene expression enables photophosphorylation in a heterologous host. *Proceedings of the National Academy of Sciences of the United States of America* **104**: 5590–5595.

Matsuno-Yagi, A., and Mukohata, Y. (1977) Two possible roles of bacteriorhodopsin; a comparative study of strains of *Halobacterium halobium* differing in pigmentation. *Biochemical and Biophysical Research Communications* **78**: 237–243.

McCarren, J., and DeLong, E.F. (2007) Proteorhodopsin photosystem gene clusters exhibit co-evolutionary trends and shared ancestry among diverse marine microbial phyla. *Environmental Microbiology* **9**: 846–858.

Munson-McGee, J.H., Lindsay, M.R., Sintes, E., Brown, J.M., D'Angelo, T., Brown, J. et al. (2022) Decoupling of respiration rates and abundance in marine prokaryoplankton. *Nature* **612**: 764–770.

Nagata, T., and Inoue, K. (2021) Rhodopsins at a glance. *Journal of Cell Science* **134**: jcs258989.

Nagel, G., Ollig, D., Fuhrmann, M., Kateriya, S., Mustl, A.M., Bamberg, E., and Hegemann, P. (2002) Channelrhodopsin-1: A light-gated proton channel in green algae. *Science* **296**: 2395–2398.

Nagel, G., Szellas, T., Huhn, W., Kateriya, S., Adeishvili, N., Berthold, P. et al. (2003) Channelrhodopsin-2, a directly light-gated cation-selective membrane channel. *Proceedings of the National Academy of Sciences of the United States of America* **100**: 13940–13945.

Oesterhelt, D., and Stoeckenius, W. (1971) Rhodopsin-like protein from the purple membrane of *Halobacterium halobium*. *Nature New Biology* **233**: 149–152.

Olson, D.K., Yoshizawa, S., Boeuf, D., Iwasaki, W., and DeLong, E.F. (2018) Proteorhodopsin variability and distribution in the North Pacific Subtropical Gyre. *ISME Journal* **12**: 1047–1060.

Pereira, O., Hochart, C., Boeuf, D., Auguet, J.C., Debroas, D., and Galand, P.E. (2021) Seasonality of archaeal proteorhodopsin and associated Marine Group IIb ecotypes (Ca. Poseidoniales) in the North Western Mediterranean Sea. *Isme Journal* **15**: 1302–1316.

Pinhassi, J., DeLong, E.F., Beja, O., Gonzalez, J.M., and Pedros-Alio, C. (2016) Marine bacterial and archaeal ion-pumping rhodopsins: Genetic diversity, physiology, and ecology. *Microbiology and Molecular Biology Reviews* **80**: 929–954.

Rappe, M.S., Connon, S.A., Vergin, K.L., and Giovannoni, S.J. (2002) Cultivation of the ubiquitous SAR11 marine bacterioplankton clade. *Nature* **418**: 630–633.

Rusch, D.B., Halpern, A.L., Sutton, G., Heidelberg, K.B., Williamson, S., Yooseph, S. et al. (2007) The sorcerer II global ocean sampling expedition: Northwest Atlantic through Eastern Tropical Pacific. *PLoS Biology* **5**: 398–431.

Sabehi, G., Beja, O., Suzuki, M.T., Preston, C.M., and DeLong, E.F. (2004) Different SAR86 subgroups harbour divergent proteorhodopsins. *Environmental Microbiology* **6**: 903–910.

Sabehi, G., Loy, A., Jung, K.H., Partha, R., Spudich, J.L., Isaacson, T. et al. (2005) New insights into metabolic properties of marine bacteria encoding proteorhodopsins. *PLoS Biology* **3**: 1409–1417.

Sasaki, J., Brown, L.S., Chon, Y.S., Kandori, H., Maeda, A., Needleman, R., and Lanyi, J.K. (1995) Conversion of bacteriorhodopsin into a chloride-ion pump. *Science* **269**: 73–75.

Slamovits, C.H., Okamoto, N., Burri, L., James, E.R., and Keeling, P.J. (2011) A bacterial proteorhodopsin proton pump in marine eukaryotes. *Nature Communications* **2**: 183.

Steindler, L., Schwalbach, M.S., Smith, D.P., Chan, F., and Giovannoni, S.J. (2011) Energy starved *candidatus* Pelagibacter ubique substitutes light-mediated ATP production for endogenous carbon respiration. *PLoS One* **6**: e19725.

Venter, J.C., Remington, K., Heidelberg, J.F., Halpern, A.L., Rusch, D., Eisen, J.A. et al. (2004) Environmental genome shotgun sequencing of the Sargasso Sea. *Science* **304**: 66–74.

Yoshizawa, S., Azuma, T., Kojima, K., Inomura, K., Hasegawa, M., Nishimura, Y. et al. (2023) Light-driven proton pumps as a potential regulator for carbon fixation in marine diatoms. *Microbes and Environments* **38**: ME23015.

Yoshizawa, S., Kawanabe, A., Ito, H., Kandori, H., and Kogure, K. (2012) Diversity and functional analysis of proteorhodopsin in marine *Flavobacteria*. *Environmental Microbiology* **14**: 1240–1248.

Yoshizawa, S., Kumagai, Y., Kim, H., Ogura, Y., Hayashi, T., Iwasaki, W. et al. (2014) Functional characterization of flavobacteria rhodopsins reveals a unique class of light-driven chloride pump in bacteria. *Proceedings of the National Academy of Sciences of the United States of America* **111**: 6732–6737.

12 Microplastics in Freshwater Environment
Assessing Sources, Pathways, and Effects on Biodiversity

Minhaz Uddin, Md. Rokonuzzaman, Md. Iqram Uddin Al Amran, Md. Kamrul Haque, and Wai Chin Li

12.1 INTRODUCTION

Plastics have become an important part of a modern society because plastic products are being used for packaging, cosmetics production, construction, and in healthcare industries. This reliance, however, has had an immense negative impact on the environment globally. As of 2015, the global plastic waste amounted to more than 6.3 billion metric tons, and only 9% were being recycled, the majority heading to landfills or into the environment (Geyer et al., 2017). These numbers depict how large plastic pollution is in both the land-based and aquatic systems. Plastics are categorized based on size, where macro-plastics are more than 2.5 cm, mesoplastics have 2.5–5 mm, and MPs have less than 5 mm (Jeyasanta et al., 2020). The high prevalence and the high degree of resistance to degradation have drawn some special scientific interest to MPs. They can be split into primary MPs produced at small sizes to be utilized in products such as facial scrubs and cleanings, and secondary MPs produced when large plastic materials were decomposed to the smaller size because of photodegradation, microorganisms, or chemical reactions (Alimi et al., 2018; Hermsen et al., 2018). They resist environmental disintegration, resulting in the fragmentation rather than decomposition, and this makes them all the more persistent in water systems. Because of their resistance, aquatic organisms may consume MPs and transmit them easily across the freshwater (Koelmans et al., 2017; Peiponen et al., 2019). Their small size is beneficial as this enables them to penetrate through the natural filtration barriers, making them reside on soil, water, food webs, and even the atmosphere, where they are able to spend longer periods of time (Aib et al., 2025).

Initial research was devoted to the impact of MPs in the marine environment (Botterell et al., 2019; Everaert et al., 2018; Porter et al., 2018), yet freshwater systems have been identified as major pathways in which MPs move and change. The most important sources of MPs in the loads of these waters are urban runoffs, industrial effluents, agricultural discharge, and treated sewage (Jambeck et al., 2015; Lebreton et al., 2017). This contaminated water finds its way to oceanic ecosystems, especially through the freshwater sources, and this indicates that freshwater sources are channels that connect land-based pollution to the marine accumulation. This increased degree of awareness has resulted in research on the ecological and physiological effects of MPs in freshwater habitats. The articles by Gatidou et al. (2019), Kataoka et al. (2019), Shruti et al. (2019), and Teng et al. (2019) explain the growing evidence of adverse impacts on aquatic life. Two main issues have been raised. First, freshwater environments are susceptible to the biological risks associated with MPs, similar to marine ecosystems, with ingestion reported in invertebrates, crustaceans, and fish alongside oxidative stress, gastrointestinal obstruction, reproductive impairment, and mortality in severe incidences. Second, MPs are carriers of environmental pollutants, which makes it possible to transfer pathogens, heavy metals, and persistent organic

DOI: 10.1201/9781003640899-15

pollutants across the ecological systems (Jambeck et al., 2015). This stance is closely correlated to their chemical composition. As Campanale et al. (2020) note, toxic synthetic additives are frequently employed by MPs, and they might contain dyes, plasticizers, stabilizers, and flame retardants. Some of these compounds interfere with the endocrine processes or are carcinogenic in nature. Once the MPs are deposited into the water bodies, there is a possibility that the chemicals reach the surrounding waters, thus posing long-term health risks. Bisphenol A (BPA) and phthalates are examples of chemicals that have been associated with hormonal issues, reproductive and birth problems, as well as intestinal disturbances (Campanale et al., 2020).

The MPs are also mobile to a significant degree and can be transported over a long distance by wind or surface water flow or even by biological carriers (Arias-Andres et al., 2018). The proximity of freshwater systems to the anthropogenic environment puts them at a disadvantage; their location in direct relation to the anthropogenic environment exposes them to direct anthropogenic pollutant exposure in comparison with other marine systems (Xu et al., 2020). The routes of human exposure are through drinking water, irrigation, and food production. It is also said that the prevalence of MPs has now been documented in a wide range of environments, not only in the most highly urbanized rivers but also in artificial ponds, as well as in isolated areas, such as the Mariana Trench (Ajay et al., 2021; Nayal and Suthar, 2022; Romano et al., 2023; Stang et al., 2022). The global modeling according to Evangeliou et al. (2022) confirms that the quantity of MP emissions may reach up to 9.6 teragrams annually, which implies the level of worldwide spreading. This is confirmed by field research, including the study by Rajan et al. (2023) that has measured the average of 92.85 ± 50.69 particles/m^3 along the Ganga River, one of the most populated and economically significant water basins in the world. Moreover, contrary to popular expectations, groundwater sources were found to possess MPs between 0.1 and more than 6,832 particles/L (Sangkham et al., 2023).

MPs offer appropriate surfaces to enable microbial colonization in water, favoring the growth of biofilms that facilitate the growth of bacteria and viruses. Such biofilms have the ability to modify microbial diversity, disturb ecological balance, and accumulate toxic metals as well as other dangerous chemicals. Upadhyay et al. (2024) noted that these interactions allow MPs to be actively involved in ecological degradation, not acting only as passive pollutants. Its consequences are seen on public health as MPs-contaminated freshwater is popular in irrigation, food preparation, drinking, and domestic purposes like bathing. They also highlighted that the long-term human exposure presents more risks because MPs travel along the food chain-aquatic organisms to agricultural products and eventually to human beings. The detection of MPs in human blood, respiratory tissues, and placental cells highlights their potential long-term public health impacts. This chapter examines the sources, transport, and transformation of MPs in freshwater systems, with a focus on rivers, lakes, and estuaries. It also considers the factors influencing their movement and distribution and concludes by assessing their interactions with freshwater species and the cascading ecological effects that threaten biodiversity.

12.2 METHODOLOGY

This chapter is based on a narrative review of the literature that synthesizes current knowledge on MPs in freshwater ecosystems. Relevant studies were identified primarily through Web of Science Core Collection using keyword combinations such as “microplastic” (MP) OR “microbead*” in combination with “freshwater,” “river,” “lake,” “wetland,” “biodiversity,” “species,” “ecology,” “pathways,” or “sources.” Both laboratory and field-based studies published between 2016 and 2024 were considered. Particular attention was given to research addressing the sources and pathways of MPs, their physical and chemical interactions with freshwater organisms, and their ecological consequences for population dynamics and biodiversity. This review includes studies from rivers, lakes, wetlands, reservoirs, and ponds to show how MPs influence freshwater life, from single species to entire ecosystems.

12.3 OCCURRENCE AND FATE OF MPs IN FRESHWATER ECOSYSTEM

Once regarded primarily as a marine issue, MPs are now recognized as pervasive in rivers, lakes, wetlands, and streams, originating from numerous point and non-point sources. The rapid growth of plastic manufacturing in recent years has substantially increased the presence of MPs in environmental systems (Kurniawan et al., 2023). Plastic waste undergoes degradation through prolonged ultraviolet radiation exposure and sustained mechanical stress, fragmenting into microscopic particles that disperse across marine and freshwater environments, including oceans, rivers, estuaries, and lakes (Mahon et al., 2017a). Rivers are estimated to transport approximately 70%–80% of total plastic waste into aquatic ecosystems, much of it originating from mismanaged plastic during production and consumption stages. Additional sources include agricultural runoff and effluent from wastewater treatment facilities (Kurniawan et al., 2023). MPs are present in all major compartments of freshwater systems, including surface waters, mid-water columns, benthic sediments, and shorelines (Atugoda et al., 2020). Eriksen et al. (2013) have documented downstream MP concentrations reaching more than 466,000 particles/km^2 in the Laurentian Great Lakes of the United States, especially in high population density areas. Similarly, the concentration of MP and mesoplastic in the Ganga River in India was 99–410 particles/km in the sediments (Sarkar et al., 2019). Table 12.1 gives representative concentrations of MPs in some of the world's freshwater systems.

The way in which MPs are accumulated and maintained in aquatic environments is important in the evaluation of their ecological risks, particularly their adsorption and transportation characteristics of metal ions and hydrophobic organic pollutants over extended periods of time. The shape and size of MPs, their density, and external influences connected with water chemistry and the interactions of biological factors influence movement and behavior in freshwater systems (Ding et al., 2021). Another important source of MPs is the wastewater treatment plant (WWTP) where the effluents are disposed in rivers, lakes, and coastal waters (Mahon et al., 2017a; Kurniawan et al., 2023). The disparity in water density is more likely to cause MPs to accelerate the sinkage in freshwater, which is one of the noticeable factors in transport and distribution (Haque and Fan, 2023). Among the processes that are properly documented is the development of biofilms on the MP surfaces, and it involves the growth of different kinds of photosynthetically active microorganisms, such as *Rhodobacteraceae* spp. and cyanobacteria (Oberbeckmann et al., 2015; Haque and Fan, 2023). Examples of environmental factors that may influence biofilm growth are the temperature, light, nutrients, water flow, and hydrodynamics (Chen et al., 2019; Haque and Fan, 2023). The less dense MPs than the freshwater are positively buoyant and can float, travel along water currents, and eventually settle on the lakeshore sediments (Wong et al., 2020). The presence and behavior of MPs in freshwater systems are related to the interaction between MPs and the abiotic and biotic components, and this makes the fate of MPs challenging. Figure 12.1 is the conceptual figure of the sources, transport, and environmental processes of MPs in the interdependent freshwater and marine systems.

There are various ways in which the freshwater organisms will be exposed to MPs, such as direct water column filtration of the water, ingesting the MPs in the form of food particles, or absorbing them through feeding on prey that ingested the particles (Nelms et al., 2018). Due to the discovery of polybrominated diphenyl ether (PBDE) in the tissues of Myctophidae fish, one may predict that MPs can promote the transportation of toxic compounds in the local water to aquatic life (Zalasiewicz et al., 2016; Kurniawan et al., 2023). The distribution and behavior of the MPs in the freshwater environments are required to know in defining the extent of the effects on the entire ecosystem, the source of the pollution, and the development of the most appropriate remedial actions at the ecological levels. There are four main MPs entry routes, including terrestrial runoff, wastewater discharge, agricultural and aquaculture activity, and atmospheric deposition.

TABLE 12.1
Presence of MPs in Global Freshwater Systems

Continent	Location	MP Types	Size Range	Concentration	References
Africa	South African estuaries	PP, PS	0.33–5 mm	745.4±129.7 particles/500 mL	Naidoo et al. (2015)
Africa	South Africa, Bloukrans River	Fragments	0.4–1.00 mm	6.3–160.1/kg sediment	Nel et al. (2018)
Asia	River Saigon, Vietnam	Fragments, fibers	0.05–0.3 mm	10–519,000 particles/m^3	Lahens et al. (2018)
Asia	China, River Wei	Fiber, film, foam, fragment, pellet	0.5–5 mm	3.67–10.7 particles/L	Ding et al. (2019)
Asia	Malaysia, Skudai River	Fragment, fiber, film	0.001–5 mm	200 particles/kg	Sarijan et al. (2018)
Asia	Vembanad Lake, India	PE	<5 mm	252.8 particles/m^2	Sruthy and Ramasamy (2017)
Asia	Wuhan Lakes, China	PET, PP	0.05–5 mm	1,660.0±639.1 n/m^3 to 8,925±1,591 n/m^3	Wang et al. (2017b)
Asia	Three Gorges Reservoir, China	Not specified	<1 mm	1,597–12,611 n/m^3	Di and Wang (2018)
Asia	Taihu Lake, China	Fibres	0.1–1 mm	3.4–25.8 items/L	Su et al. (2016)
North America	Lake Winnipeg, Canada	Fiber, film, fragment, foam	<5 mm	193,420 particles/km^2	Anderson et al. (2017)
South America	Patagonian lakes, Argentina	PET	0.05–5 mm	0.9±0.6 MPs/m^3	Alfonso et al. (2020)
Europe	Elbe River, Germany	PE, PP	0.15–5 mm	5.57±4.33 particles/m^3	Scherer et al. (2020)
Europe	River Kelvin and River Thames, UK	PP, polyarylsulfone, PET	1–5 mm	296.5 particles/L (Kelvin); 33.2±16.1 particles/100 g (Thames sediment)	Blair et al. (2019) and Horton et al. (2017)
Europe	Poland, Carpathian basin	PAC, PES, PP, PS, PE	0.1–0.3 mm	3.52–32.05 particles/m^3	Bordós et al. (2019)
Oceania	Brisbane River	PA, PE, PP	<3 mm	10–520 items/kg	He et al. (2020)
Oceania	New Zealand, Urban stream	Fibres, fragments, beads, PS	0.255–5 mm	44.8 particles/m^3	Mora-Teddy and Matthaei (2020)
North Africa	Bizerte Channel, Tunisia	Fragments, fibers, pellets	0.0003–0.005 mm	3,000–18,000 particles/kg	Abidli et al. (2017)
North America	Atoyac River Basin, Mexico	Films, pellets, fibers, fragments	Not reported	833.33–1,633 particles/kg	Shruti et al. (2019)

12.3.1 Atmospheric Inputs of MPs

The first data of airborne MP pollution reported by Dris et al. (2016) with a rate of 2–355 particles/m^2/day and an average of 110±96 at an urban monitoring location in Paris. Since that period, the concern toward airborne MPs has increased, particularly concerning the tendency to affect respiratory health. Human exposure has now been suggested to be likely by way of inhalation. In addition to health effects, the airborne MPs can even cause further pollution of the environment, which will eventually be transferred to the soils, water, and sediments (Yao et al., 2022; Raju et al., 2023). This makes the atmosphere an effective distributor of MP. Allen et al. (2019) have also found that the MPs do not only travel by air but can even be placed at remote and not industrial locations,

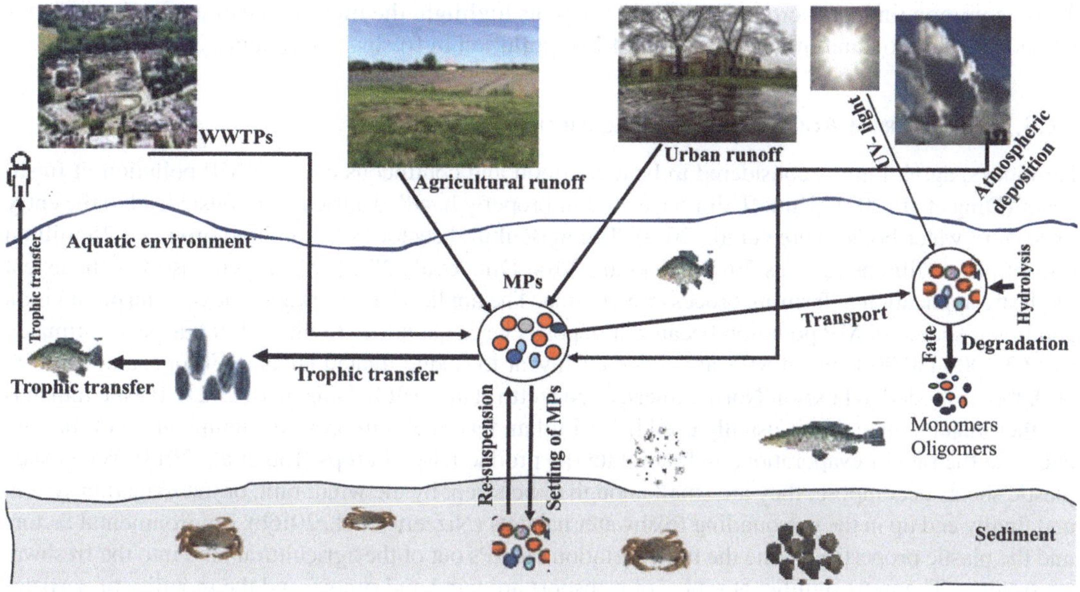

FIGURE 12.1 Occurrence and fate of MPs in Aquatic Ecosystem. (Adapted with permission from Wu et al., 2019, © Elsevier.)

which confirms that the problem of plastic pollution is not limited to the cities. Both the fibers and fragments were observed in a mountain catchment in their study, and thus proved that there was atmospheric transport. Spherical and film-like particles were the most prevalent in street dust, but micro rubbers were fibrous and black fragment particles (Abbasi et al., 2019).

Atmospheric deposition sometimes leads to the entry of MPs into fresh water and industrial discharge, and urban dust is the main source (Abbasi et al., 2019; Ding et al., 2021). In Dongguan, China, it has been reported that researchers have found an average of 36 particles/m²/day and that the most common polymers here are polyethylene terephthalate (PET), polypropylene (PP), and polystyrene (PS) (Cai et al., 2017). The levels were even higher in central London at 575–1,008 particles/m²/day (Wright et al., 2020). Contrarily, previous research discovered lower rates in Paris (110), Dongguan (36), and the Pyrenees (40), which indicated a clear increase in MP pollution in high-density urban locations. Airborne MPs such as artificial fibers, pieces, and particles, have been detected in urban, rural, and even isolated areas, and it is thought that atmospheric distribution is an important vector in the dispersion of MPs. Urban dusts are also the sources and sinks of airborne MPs (Abbasi et al., 2019). Several anthropogenic activities, including industrialization, construction, textile production, tire wear, and the re-suspension of already deposited particles, result in the formation of the atmospheric MPs (Dris et al., 2016; Chen et al., 2020). These particles get into the freshwater systems either by dry deposition of airborne materials settling naturally or by wet deposition like rainfall and snowfall. There is a strong connection between population density and higher atmospheric MPs concentrations.

Rainfall can be a significant mode of transportation, particularly of synthetic microfibers of domestic and industrial sources (Dris et al., 2016; Chen et al., 2021). Synthetic fibers released by household laundry can comprise a large portion of atmospheric fallout among airborne MPs (Cai et al., 2017; Dris et al., 2016). The winds also contribute to the transport of such particles, which ultimately settle on the land or the water in lakes (Chen et al., 2020). According to Dris et al. (2016), the level of precipitation affects the amount of atmospheric deposition, whereby a lower amount of precipitation is associated with lower levels of deposition, and the stream and river loads of MP are lighter. MP in the atmosphere has been measured ranging between 2 and 355 particles/m²/day; similar values have been obtained in China and France (Ding et al., 2021; Dris et al., 2016; Cai et al., 2017).

These common findings across the different regions highlight the importance of atmospheric deposition as a common and influential cause of MP pollution in freshwater systems.

12.3.2 MPs from Agriculture and Aquaculture

Nowadays, agriculture is considered to be a common and continuous cause of MP pollution of freshwater (Ding et al., 2021). Runoff, drainage, and improperly handled agricultural waste lead to the entry of MPs in water bodies (Free et al., 2014). The agricultural sector in China used around 1.25 million tons of plastic film as early as 2011 (Liu et al., 2014; Ding et al., 2021), which demonstrates the extent of plastic application in farming processes in China. The application of sewage sludge to farmland is an additional source of MP pollution because it deposits large quantities to the soil. It has been estimated that 63,000–430,000 tons of MPs are added each year to sludge-treated lands in Europe, and 44,000–300,000 are added to lands in North America (Nizzetto et al., 2016b; Ding et al., 2021). Plastic mulch is another major source that is usually used in peri-urban farming to improve the temperature of the soil, decrease the rate of evaporation, and facilitate the productivity of crops (Liu et al., 2018). When such plastic sheets decompose, they are small enough to be swept by the wind, rain, or the irrigation system and finally end up in the surrounding freshwater habitats (Nizzetto et al., 2016b). Environmental factors and the plastic properties define the transportation of MPs out of the agricultural land into the freshwater systems. The most significant routes of transport are wind and runoff, and the mobility of particles is characterized by features of shape, density, and size (Nizzetto et al., 2016a; Ding et al., 2021).

Biological processes also contribute to MP mobility along with physical vectors (e.g., runoff, wind). Earthworms have been used as an example to enhance the movement of MPs through farm soils and to freshwater sources (Rillig et al., 2017). These findings confirm the idea that cultivated soils can be considered as nonpoint, diffuse sources of MP pollution that can enter rivers and lakes (Ding et al., 2021). This is also augmented by the farming application of sewage sludge. In Europe and North America, about 50% of the generated sludge is applied to the fields, and there is as much as 72% of sludge applied in certain countries (Ireland, Finland) (Mahon et al., 2017b; Raju et al., 2023). Although the processes, such as anaerobic digestion, stabilization, or thermal drying, are typically utilized to treat sludge, they generally do not work in the context of the treatment when it comes to the removal of MPs. Hence, treated sludge is still a significant source of MP in soils (Raju et al., 2023). Fishing is also another serious contributor to the pollution of the freshwater environment by MPs (Ding et al., 2021). Artificial fishing gear such as discarded ropes and nets is harmful over time and is emitting MPs into the water. Nylons have been found in the environment in locations where inland fisheries are very active, which shows that the intensity of fishing is directly correlated with the presence of MPs (Yuan et al., 2019). The use of freshwater sources is also a point of specific interest because the MPs were found in sediment and fish, and posing a threat to human health (Di et al., 2019; Yuan et al., 2019). In China, studies have shown that fish in the Qinghai Lake, Poyang Lake, and Xiangxi River have ingested nylon particles (Xiong et al., 2018; Yuan et al., 2019). This has been documented in fish sampled in Brazilian estuaries (Ferreira et al., 2018) and in Chinese edible bivalves on the seafood markets (Li et al., 2015).

Due to aquaculture activities, there is an increased MPs pollution in freshwater habitats, particularly because plastic-based compounds are extensively used in various production processes (Zhou et al., 2020; Ding et al., 2021). Fish farming infrastructure has breeding enclosures, hatching equipment, storage, synthetic nets, and so on. Wearing and tearing of the plastic material used in the farming equipment's, particularly the nets and gears, through the years are also the key pathways through which the MPs are introduced to the freshwater ecosystems. It has been reported that approximately 4,000 tons of plastic fishing gear is used by aquaculture activities in Norway annually, with a part of it estimated to be 400 tons of equipment disposed of in the surrounding marine environment (Deshpande et al., 2020; Ding et al., 2021). MPs have also been identified in formulated fish feeds, which further raises doubts about pollution in the production cycle. In Malaysia, a study that compared commercial feeds of three brands retrieved 336 MP, which were mostly synthetic polymers (Karbalaei et al., 2020; Ding et al., 2021). In addition, the pharmaceutical residues and veterinary products, including packaging elements, may be used as other sources of MPs, which are produced through aquaculture means. More to the point, the MPs associated with these systems have been found to transmit antibiotic resistance genes, which have greater ecological and human health effects (Dong et al., 2021; Ding et al., 2021). Looking at all these results, it is obvious that MPs pollution due to aquaculture is a compound matter with various risks to freshwater ecosystems.

12.3.3 MPs from WWTPs

The Wastewater systems are a significant source of MPs input into water bodies, which originate from numerous household and industrial sources. These contaminated effluents are sent to WWTPs in urban areas, where they serve a dual purpose as part of the MPs cycle to capture a portion of the particles but release other particles during regular operation (Mahon et al., 2017a; Ding et al., 2021). MPs are able to access freshwater either in a direct or indirect manner. WWTPs emit large volumes of MPs directly into rivers and lakes (Ding et al., 2021). According to the majority of studies, WWTPs have the ability to eliminate over 95% of MPs, although the elimination rate may range between 64.4% and 99.9% (Ding et al., 2021). Cheung and Fok (2017) reported that over 80% of the 209.7 trillion microbeads, amounting to 306.9 metric tons annually, are discharged into aquatic systems by WWTPs in China. Beyond direct effluent discharge, WWTPs also contribute indirectly through the disposal of sludge by-products. When this sludge is repurposed for applications such as green infrastructure or agricultural fertilization, between 69% and 80% of its contained MPs may be introduced into the broader environment (Sun et al., 2019). Rainstorms and/or farming or agricultural practices may allow these transported MPs to enter freshwater habitats. The MPs found in lakes and rivers are probably the same as those found in inland soils in terms of form, shape, size, and color. Sewage sludge has 10–100 times more MPs than freshwater sediment. These findings indicate that MPs may also reach riverine and lacustrine environments via indirect pathways, particularly through effluent discharged from wastewater treatment facilities (Ding et al., 2021). Prata (2018) reported that these plants release about 600 million MP particles into the water every year. In China, over half of the 3,340 treatment plants can clean around 78 million m^3 of sewage each day (Jiang, 2018), and some of that water goes into rivers and lakes. In the United States, WWTPs are estimated to release between 3 and 23 billion MPs into canals on a daily basis, with individual facilities discharging as many as 4 million particles each day (Mason et al., 2016; Raju et al., 2023). Comparable issues have been documented in the Netherlands, where MP concentrations in sewage sludge range from 510 to 760 particles/kg, and treated effluent contains between 51 and 81 particles/L (Leslie et al., 2017; Raju et al., 2023). The reuse of sewage sludge (biosolids) further extends the risk, facilitating MP dispersion into both freshwater bodies and surrounding soils. Even highly efficient plants are not exempt. A Scottish WWTP demonstrated a 98.4% MP removal efficiency, yet measurable levels remained in both influent (15.7±5.23 particles/L) and effluent (0.25±0.04 particles/L). Given the high flow rate, the facility was estimated to release approximately 65 million MPs daily into the environment (Murphy et al., 2016; Raju et al., 2023). In some countries, MPs have been found in waste at concentrations reaching 113 particles/g, as noted by Crespy et al. (2008).

12.3.4 Runoff from Land into Freshwater Environment

Terrestrial runoff constitutes another significant route by which MPs are introduced into freshwater environments. In France, for example, the Seine River is estimated to receive approximately 28,000 kg of floating plastic debris annually (Gasperi et al., 2014). In some regions of Southeast Asia, such as the Saigon River in Vietnam, plastic pollution is severe (receiving ~7,500–13,700 tons of plastic per year) (Van Emmerik et al., 2018). This input is enhanced by rainfall and stormwater occurrences that carry plastics by combined sewer overflows as well as by surface runoff of urbanized areas (Fischer et al., 2016). The effect of flooding could also destabilize sediment and stored MPs to immediate water (Hurley and Nizzetto, 2018). It is important to note that much of these works are specifically about buoyant polymers, so this may not reflect the overall MPs burden (which includes denser types that may settle in transit). Regarding composition, the common polymers such as PP, polyethylene (PE), PS, and PET are mainly driven by the consumer products, such as plastic bags, bottles, and sewage generated materials. These patterns show that human activities have direct effects on the accumulation and persistence of MPs (Haque and Fan, 2023). Primary MPs enter freshwater environments through direct and indirect pathways: industrial emissions, domestic sewage, and surface runoff from land. Sources such as wastewater effluents, personal care products, and packaging materials contribute substantially, specifically through foam-like particles. Moreover, fibers and polymer particles from synthetic textiles are often found in municipal wastewater and urban solid waste (Nayal and Suthar, 2022; Upadhyay et al., 2024). Industrial areas make a significant contribution to MPs load as the management of plastic waste is, in most cases, inefficient and the emissions during their manufacture are not controlled (Horton and Dixon, 2018). In agricultural areas, MPs originating from plastic mulch films and biosolids applied as fertilizers are common in the runoff (Lutz et al., 2021). Among environmental carriers, stormwater is the most significant one. During rainfall events, it mobilizes MPs as well as a wide variety of other pollutants, including sediments, nutrients, and organic matter (Soltaninia et al., 2023). MPs have been reported in stormwater runoff ranging from 0.49 to 22.9 particles/L and are highly dependent on local conditions (Liu et al., 2019a). Urban and highway runoff contribute to MP's release and thus promote their transfer from terrestrial surfaces to the nearby freshwater bodies (Liu et al., 2019a; Raju et al., 2023).

12.4 TRANSPORT AND DISTRIBUTION OF MPs WITHIN FRESHWATER ECOSYSTEMS

12.4.1 Vertical Stratification of MPs

The distribution of MPs vertically in fresh water is determined by processes such as homoaggregation, and heteroaggregation that dictate how MPs move in water. Aggregation typically takes place in the water column, which results in MPs settling in the sediments. Although particle size has a very strong influence on the possibility to homoaggregate, relatively little is known about this in freshwater environments (Besseling et al., 2017). A general overview about MPs distribution from surface to benthic layer is crucial to assess its ecological implications (Chevalier et al., 2023). Despite the intense study of the spatial distribution of MPs in marine ecosystems, paying particular attention to their horizontal accumulation onto the sea surface and in the coastal seabed's (Tsiaras et al., 2022), the information regarding their presence in the intermediate water layers is less developed. This lake of such information is primarily because of the paucity of investigations on vertical trends in the water column (Chevalier et al., 2023). A series of multiple interrelated factors, including the MPs physicochemical properties, hydrodynamic status, bio-interaction, or other environmental drivers, compounding and driving the vertical fate and transport of MPs in the aquatic environment. It is important to know these processes for the evaluation of the environmental risks of MP pollution and the development of possible effective control measures (Kumar et al., 2021).

Despite recent increasing interest in MPs, their vertical distribution in the mid-water column is still poorly understood because there has been little specific study on them. The current studies have concentrated on surface and benthic districts and vertical profile, particularly at the middle depths, has not yet been well analyzed. Therefore, vertical stratification remains a poorly researched issue because MPs within the water column were regarded as temporary and their reservoirs were relatively small as compared to the surface or bottom layers (Kooi et al., 2016; Lenaker et al., 2019). Direct sampling at varying depth intervals is uncommon, and nonloading of other size-fractions can result in the underestimation of total plastic loads in aquatic systems (Lenaker et al., 2019; Van

Sebille et al., 2020). Besides that, vertical distribution is a vital aspect of knowledge in the evaluation of MP interaction with the aquatic biota. An example would be a planktivore preferring to feed on plankton layers of maximum density; thereby, the exposure of animals to the maximum concentration of MPs as well as the risk of ingesting MPs at deeper depths could have ecological implications.

Hydrodynamics or aquatic bioturbation can resuspend MPs deposited in the sediment. Two main processes have been put forward, including hydrological processes that can carry MPs back into the water column. When defecating and excreting, fish can release MPs back to the water column and thus re-expose MPs to an aquatic life. Another more complex route is by ingesting the MPs by soil living organisms that can then be consumed by predator fish. These particles can be excreted into the water column (Ding et al., 2021). Any vertical distribution of MPs would be of interest to determine their effect on aquatic organisms and ultimately the health of the ecosystem, although this has been a challenge experimentally.

The research on the factors that regulate MPs distribution in the water column has indicated that the primary regulating factors are the size of the particles, their shape, and density (Erni-Cassola et al., 2019). Their vertical movement in water is largely dependent on the MP density, and re-suspension of the settled particles can occur with the change of the flow conditions. Hence, under some local conditions, freshwater sediments may act as short-term or long-term storage of MPs. Sediments play a significant role in defining the fate of MPs, and this should be accounted for when modeling the pathways of MPs transport and their environmental persistence (Chen et al., 2024; Elagami et al., 2023). The most crucial point is that the physical properties of MPs do not remain constant but could also be changed during the chemical and biological interaction with the natural environment, causing it to fragment, biofoul, and experience other changes in size, shape, and density over time (Kooi et al., 2016). These changes in MPs can affect their movement in the vertical direction in the water column, as alterations in particle size and density influence their ultimate settling velocity. Also, hydrodynamic forces, especially wind, caused turbulence and velocity gradients to act on them, driving vertical mixing through Langmuir circulation at critical wind speeds of 6–9 ms^{-1} (Wang et al., 2025).

12.4.2 Lateral Movement of MPs across Freshwater Habitats

The main causes of the lateral transportation of MPs between freshwater environments are natural factors such as variation in hydrodynamic characteristics as well as the impact of biological characteristics (Jambeck et al., 2015). Additionally, flow, water depth, and the substrate characteristics are among the main factors that affect the movement of MPs in the water (Kataoka et al., 2019; Lin et al., 2018). Moreover, the distribution of MPs in surface waters is based on the interactions of the physicochemical characteristics of MPs, and a combination of a variety of environmental factors such as flow, rainfall, flood, temperature, wind, terrain, and catchment climates (Xu et al., 2022). These differences in water streams and flows are critical in distribution of MPs in the water environment.

This change will allow migration of MPs out of rivers to larger systems and has the potential of helping MP movement to even further destinations such as marine and polar areas. At a smaller scale, MPs can be transported laterally from tributaries to lakes. On the other hand, relatively stagnant areas with low hydrodynamic activity (such as backwater zones in lakes and rivers) are prone to less vertical mixing, and MPs are prone to accumulate in the sediment in these areas (Ding et al., 2021). The transport and distribution of MPs in freshwater are mainly driven by local environmental conditions. In estuaries, the transition from freshwater to marine water conditions, along with mixing, salinity, and particle characteristics may influence the movement and deposition of MPs (Wong et al., 2020). On the other hand, in open, fluvial systems directly connected to the sea, MPs are expected to be exported to marine water bodies (Leslie et al., 2017).

However, in enclosed or slow-flow hydrological systems, MPs have a higher tendency to be retained and accumulate, causing them to serve as "sinks" for plastics (Eerkes-Medrano et al., 2015). Biological interactions also affect the distribution of MPs. The MPs are also able to be ingested or sorbed to the surfaces of freshwater organisms, and the latter biotas could be releasing the particles via mobilization or direct excretion to re-distribute them within the environment. The distribution of MPs among organisms is dependent on the species and the color, size, and shape of the particles. In particular, although *Daphnia magna* ingests and adsorbs MPs onto their bodies (Zhang et al., 2019), and fish ingest them (Ory et al., 2018), the spread of MPs is different. Also, fish species with varying habitat preferences can also vary in their capability to redistribute MPs in the water column (Ding et al., 2021).

Episodic changes in MPs transport include tides, storm events, floods, and anthropogenic discharges, which affect temporal variations in MPs transport. Particularly turbulent storm events can add sufficient turbulence to re-suspend settled MPs or those living in the biofilm on the sediment. As an example, Fischer et al. (2016) noted that MPs in Lake Bolsena had at least doubled after a wind event because of the lateral transfer of MPs at the lake bottom. Similarly, MPs can also be resuspended and redistributed by external forces acting on stationary standing water bodies (Wong et al., 2020). In general, the MPs present in the water and sediment are of different sizes, shapes, colors, and chemistry that have been used to imply their migration and ecological hazards. Lateral transportation can distribute these particles in the freshwater systems. In this way, the problem of the MP's global distribution in various eco-systems is a matter of concern, and it should be carefully monitored and investigated (Ding et al., 2021).

12.5 ADVERSE IMPACTS OF MPs ON FRESHWATER BIODIVERSITY

Freshwater ecosystems are supportive of biodiversity and ecosystem processes. The concern about the MPs' effects on such systems has been on the rise. MPs' reactions to aquatic life may either be nonexistent or additive or antagonistic, depending upon the interaction of other anthropogenic factors. They could be captured by their physical characteristics, such as size and shape, and biological characteristics, such as feeding habits and life cycle of a species (Castro-Castellon et al., 2022). The MPs could either be directly or indirectly taken by phytoplankton which is the basis of food chain in the aquatic environment. Zooplankton that feed on phytoplankton are also affected as well as they have the potential of accumulating MPs in tissues. As it was suggested by Martínez-Gómez et al. (2017), zooplankton as the main food source of aquatic fauna may have been among the main channels through which plastic pollutants may be deposited in the aquatic food web. The toxicity of MPs may be as a result of ingestion, chemical additives, leaching, and adsorption of external pollutants (Wong et al., 2020). Feeding organs of freshwater species can be physically blocked or functionally impaired if they ingest MPs (Barnes et al., 2009; Wong et al., 2020). Although some species can remove or degrade MPs more quickly than others, some species are not able to remove or degrade MPs leading to the accumulation of MPs within their bodies (Anderson et al., 2016). This chapter considers both direct toxicity from MPs, and their capacity to engage with and amplify the effects of other pollutants. Figure 12.2 depicts the negative effects of MPs on freshwater biodiversity. Although many toxicological studies are confined to individual endpoints, these results have distinct ecological relevance. Lower reproduction, feeding success, and survival reduce population size, differential sensitivity among these taxa changes community structure, and accumulating impacts across trophic levels diminish species richness. These biodiversity-level impacts of MPs are increasingly documented in recent syntheses (Ding et al., 2021; Mithun et al., 2024), highlighting that MP's impacts on organisms must be understood in the context of ecosystem stability. In addition to their organismal toxicity, MPs may have direct effects on biodiversity through changes in species' richness, abundance, and community composition. Population size is reduced as reproduction, feeding, and survival decrease, and species-specific sensitivity can alter community composition through time. These processes eventually decrease species richness and destabilize ecosystem

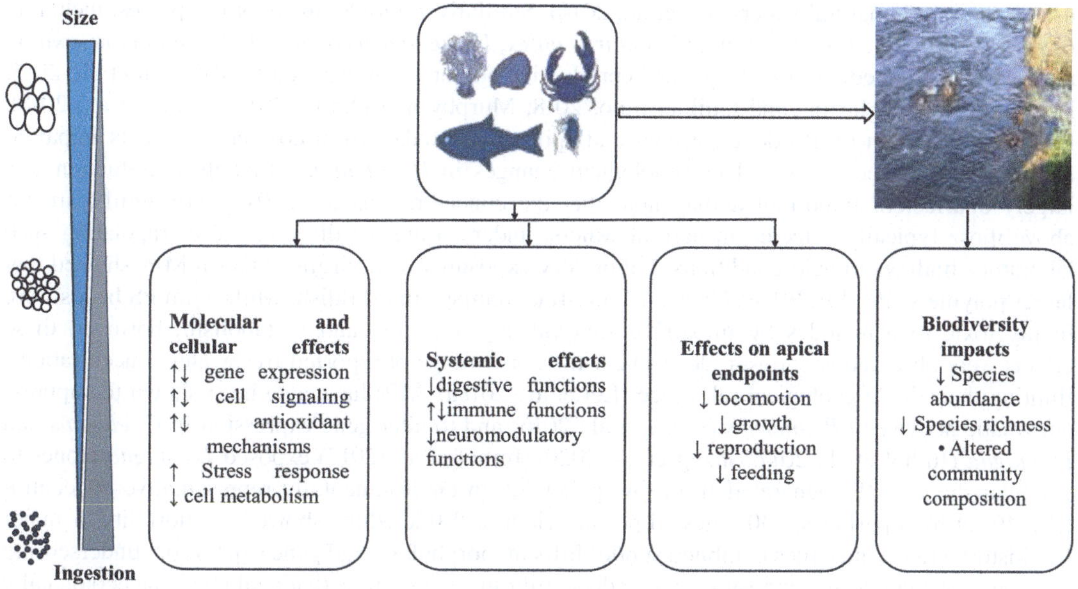

FIGURE 12.2 Pathway showing how MPs exposure disrupts key biological functions in freshwater organisms, contributing to population- and biodiversity-level impacts. (Adapted with permission from Franzellitti et al., 2019, © Elsevier.)

functioning (Ory et al., 2018, Jemec et al., 2016). Evidence from freshwater systems also agrees with this upscaling: MPs can harm both primary producers and consumers, altering eventually the ecosystem size (Ding et al., 2021).

12.5.1 Direct Physical and Biological Harm Caused by MPs

MPs cause a variety of physical and biochemical effects on organisms, such as inflammation, and decrease in energy reserves due to the reduction of nutrient uptake from natural food (Meng et al., 2020). These effects could be due to direct physical injuries, release of toxic chemicals, or transfer of environmental pollutants (Chen et al., 2021). Greven et al. (2016) demonstrated that laboratory-generated PS and polycarbonate nanoplastics (NPs) induced stress responses (i.e., degranulation, oxidative burst, neutrophil extracellular traps release) in fathead minnow neutrophils, suggesting disruption of innate immunity, but these effects were observed in short-term in vitro assays with spike-in particles not previously measured in natural waters. Physiological damages, including laceration and inflammation, have been broadly reported as a result of MP ingestion (Horton et al., 2017; Lei et al., 2018a). According to Sunny et al. (2021) and Mithun et al. (2024), MPs are often found in fish digestive tracts and can be transferred to their tissues, thus causing a decrease in feeding efficiency, impairment of digestion, and potential toxicological impacts due to chemical leachates. Ingestion can also induce false satiation, suppressing feeding behavior and increasing mortality risk (Horton et al., 2017; Chen et al., 2021). Ding et al. (2018) demonstrated that red tilapia exposed to spike-in PS MPs (0.1 µm; 1–100 µg/L for 14 days) accumulated particles across multiple tissues and exhibited biochemical disruptions, including brain acetylcholinesterase inhibition, oxidative stress responses, highlighting potential neurotoxic and metabolic impacts under concentrations closer to, though still exceeding, many field reports. Mollusks, particularly filter-feeding species such as mussels and clams, are highly vulnerable; MPs accumulate in their tissues and disrupt normal biological functions (Mithun et al., 2024).

Experimental exposures with spike-in PS MPs have shown liver inflammation and metabolic disruption in zebrafish, although such effects were detected only under concentrations higher than those usually present in natural waters (Lu et al., 2016). Similarly, multiple invertebrate species, including *Caenorhabditis elegans*, *D. magna*, *Gammarus pulex*, *Hyalella azteca*, and *Hydra attenuata*, exhibit reduced growth, feeding efficiency, and reproductive output when exposed to MPs (Au et al., 2015; Lei et al., 2018a; Martins and Guilhermino, 2018; Murphy and Quinn, 2018; Weber et al., 2018). Murphy and Quinn (2018) demonstrated that spike-in PE flakes from cosmetic products impaired feeding and induced non-lethal morphological changes in *H. attenuata*, though reproduction was largely unaffected. Importantly, they note that the concentrations used (0.01–0.08 g/mL) are far above those typically detected in natural waters, underscoring the difficulty of extrapolating such laboratory findings to field conditions. Laboratory exposures with virgin, spike-in MPs showed that larger polymers (PA, PE, PP, PVC) caused intestinal damage in zebrafish, while 1 µm PS beads were highly toxic to nematodes by impairing survival, reproduction, and metabolism; however, these effects were observed at exposure levels near the upper range of reported freshwater concentrations, limiting their direct ecological relevance (Lei et al., 2018a). MPs have also been shown to suppress enzymatic activity in *P. microps* (Fonte et al., 2016) and to alter gene expression in *C. elegans* and *D. magna* (Imhof et al., 2017; Shang et al., 2020). Imhof et al. (2017) exposed *D. magna* clones to irregular-shaped MPs generated from raw pellets at an environmentally representative concentration (1% of food particles; ~30 ingested per individual). While adults showed no mortality or major life-history effects, juveniles exhibited subtle shifts in morphology and gene expression, underscoring that even low but realistic exposures may induce mild stress responses that could become ecologically relevant under multi-stressor conditions. In *D. rerio*, MPs impair intestinal cell function, disrupt metabolic regulation, and promote hepatic fat accumulation (Lei et al., 2018b; Lu et al., 2016). Lei et al. (2018b) demonstrated that pristine PS beads induced size-dependent neurotoxicity in nematodes, with 1.0 µm particles causing the greatest reductions in survival, growth, and lifespan, as well as damage to cholinergic and GABAergic neurons and strong oxidative stress responses; however, these effects were observed under uniform spike-in exposures of 1 mg/L, a level considerably higher than most field detections. They may also act as vectors for plasticizers and alter gene expression in larval stages (Batel et al., 2018; Le Moine et al., 2018). MPs accumulation has been associated with reduced fecundity and growth rates in *O. latipes* (Chisada et al., 2019). Beyond direct biological impacts, MPs can modify sediment properties and reduce light penetration in water columns, impairing trophic interactions and primary production. These effects are amplified over time by their persistence and capacity for ecosystem-to-ecosystem transfer, raising concerns over long-term biodiversity loss.

12.5.2 Joint Toxic Effects of MPs with Chemical Pollutants

In natural habitats, organisms are rarely exposed to a single contaminant, making it essential to investigate the combined ecotoxicological effects of MPs and other environmental pollutants. Mithun et al. (2024) synthesized recent evidence showing that MPs are pervasive across aquatic taxa and can trigger oxidative stress, reproductive impairment, and neurotoxicity, with toxicity often amplified through interactions with pollutants; however, the review highlights that most available studies rely on laboratory exposures with pristine particles and that impacts on freshwater biodiversity remain comparatively underexplored. The chemical additives that are commonly introduced into plastics to facilitate their performance include metals, phthalates, PBDE colorants, and BPA (Espinosa et al., 2016; Teuten et al., 2009). The release of these substances into the environment not only interferes with the structural and functional integrity of polymers but also has toxicological hazards to the organisms that have been exposed to them. In combination with other pollutants, such as polychlorinated biphenyls (PCBs) and PBDEs, MPs may be even more lethal and may impact organ homeostasis (Chen et al., 2021).

The effectiveness of the MPs in aquatic ecosystems has increased because of the increased use of antibiotics and herbicides (Sazzad et al., 2024). Among antibiotics, triclosan and cephalexin can be adsorbed on MPs and are less persistent in the environment which results in uptake by aquatic organisms. Such contacts are extremely dangerous to the population because of the altered intestinal microbiota and the emergence of antibiotic resistance (Sunny et al., 2021). Herbicides like paraquat are capable of getting into contact with MPs, making them toxic with subsequent effect on the metabolites and oxidative stress. Lu et al. (2016) highlighted that these interactions were an indication that it is important to develop effective measures to address the arising pollutants in the aquatic ecosystems. The endocrine-disrupting chemicals such as BPA and 17α-ethinylestradiol can interfere with hormones in aquatic lives causing alteration in their behavior, reproductive functions, and development. These chemicals are bioavailable to tissues because of MPs which enable the chemicals to penetrate the tissues following ingestion (Lu et al., 2016; Mithun et al., 2024). This can result in endocrine-based diseases in the fish that can involve feminization, reduction in the reproductive capacity, and growth rate of the fish. These are adsorption, chemical bonding and surface-mediated process. MPs are small in size, high in surface area, and selective when it comes to chemical properties, which is why they are efficient vectors of hydrophobic and hydrophilic pollutants (Chen et al., 2017). Once ingested, MPs can release toxins within the body that can cause both long-term exposure and long-term physiological effects (Mattsson et al., 2017). In the case of chemical pollutants, the overall toxicity of the pollutants may be additive, synergistic, or antagonistic, and these effects are usually quite different than those of the individual pollutants.

12.5.2.1 Organic Pollutants Interacting with MPs

Organic pollutants have the ability to move and be concentrated in aquatic environments by MPs. Pollutants that are adsorbed on MPs surfaces can be at very high concentrations than those found in ambient waters. The most prevalent organic pollutants in freshwaters include Polychlorinated biphenyls (PCBs), DDT, PBDEs, and PAHs (Anderson et al., 2016; Wong et al., 2020). The prolonged nature of these legacy pollutants is because they are chemically stable as well as the impact of hydrophobic character and higher surface area of MPs contributing to adsorption. This leads to the occurrence of hazardous materials on the surface of MPs that make them even more harmful to the environment by causing the spread of pollutants in aquatic environments (Kuddus et al., 2020). When absorbed, MPS containing adsorbed pollutants can provide to organism the increased levels of pollutants compared with water only or only sediment. The large surface-area-volume ratio of MPs also ensures that the particles can react with organic pollutants to a large extent, thereby enhancing their adsorption capacity (Wong et al., 2020). After consumption, it has been identified that digestive fluids trigger the desorption of adsorbed chemicals on MPs (Teuten et al., 2009; Wong et al., 2020). Through such a process, contaminated MPs can discharge POPs in aquatic organisms that ultimately result in endocrine disruption, reproductive failure, and eventually, carcinogenesis (Salam et al., 2024; Mithun et al., 2024). In aquatic freshwater ecosystems, the simultaneous determination of MPs and organic pollutants is especially regarded for, because of the proximity to urban and industrial sources. In contrast to the open marine environment, freshwater bodies have limited ability to dilute and disperse. Poor wastewater treatment may cause harmful substances to accumulate in spots, which would pose a long-term risk to aquatic organisms and ecosystems (Wong et al., 2020). While these impacts may not be acutely lethal, chronic exposure may damage animal health, suppress populations and destabilize ecological interactions.

12.5.2.2 Heavy Metal Binding and Toxicity Involving MPs

MPs can be associated with heavy metals due to physical and chemical binding, and form MPs-containing heavy metal complexes that may have higher toxicity. They can bind to clay, silt, and metal ions to assist in the transportation of metals deposited in riverbed sediments (Wang et al., 2017a). Medicated MP surfaces containing charged functional groups allow metal ions to be adsorbed directly via site-specific interactions (Ding et al., 2021). Aged MPs that exhibit more surface oxidation and porosity afford increased abilities to bind metal. The mechanical stress of MPs and the chemical toxicity of the metals could potentially impair growth, reduce reproductive success, and cause long-term physiological harm to aquatic organisms. As an example, the old polyvinyl chloride is more effective in adsorbing copper as opposed to virgin PS. The freshwater environment has the ability to adsorb biofilm and natural organic matter (NOM) onto MPs that increase metal adsorption on the particle surfaces, and this is a concern of local pollution as well as the potential of MP-mediated metal transport (Wang et al., 2017a).

MPs were observed to increase bioavailability and toxicity of chromium (VI) in aquatic organisms, and, therefore, cause oxidative stress, DNA damage, and physiological abnormalities (De Sá et al., 2018; Mithun et al., 2024). In other cases, heavy metals on MPs have not been present in the surrounding environment but in the plastic products, having been introduced during the production (Wang et al., 2017a). The metals attached to MPs may be discharged on ingestion (Khan et al., 2015). The effect of MPs on the toxicity of metals on organisms can be antagonistic or synergistic. In the antagonistic mode, metals adsorb to MPs such as PE beads, reducing their bioavailability to organisms. Conversely, in the "Trojan horse" mechanism, MPs facilitate the transport of metals into organisms, increasing internal exposure. For example, PE MPs have been found to elevate Ag concentrations in the intestines of exposed animals (Khan et al., 2015; Ding et al., 2021). In *Lumbricus terrestris*, MPs increased the availability of Zn (Hodson et al., 2017; Ding et al., 2021), while in seabass, MPs contributed to Hg accumulation, oxidative stress, and behavioral changes (Barboza et al., 2018; Ding et al., 2021). Hodson et al. (2017) demonstrated that fragmented HDPE bag-derived MPs can adsorb Zn at levels comparable to soils, but release far greater amounts in synthetic earthworm gut fluids (40%–60% vs. 2%–15%). In mesocosm experiments, *Lumbricus terrestris* ingested Zn-loaded MPs (236–4,505 mg/kg) without significant bioaccumulation or toxicity, underscoring both the vector potential of MPs and the regulating role of essential metals like Zn.

12.6 ANALYTICAL APPROACHES TO MPs

The detection and characterization of MPs and NPs in environmental and biological samples is very challenging, mainly because MPs and NPs exhibit a diversity of chemical structures, wide size ranges, and environmental transformation processes (e.g., oxidation, degradation) (Fan et al., 2022; Kurniawan et al., 2023). These changes have often resulted in the development of complex particle shapes that make the identification of particle sources difficult. The lack of standardized methods worldwide also hinders direct comparison among studies and the evaluation of ecological and health risks arising from MPs (Zhu et al., 2022; Kurniawan et al., 2023). Even with methodological variations, the majority of studies share a sequence of general steps for MP detection: collection, pre-treatment, filtration, and identification by means of physical or chemical process (Upadhyay et al., 2024). The instruments used in the sample collection from freshwater bodies usually include neuston nets, manta trawls, Ekman grabs, Van Veen grabs, and sludge collectors (Upadhyay et al., 2024). Purification of MPs can be done by the pre-treatment process, where the organic matter is chemically digested, and the density separation is conducted. The degradation of organic substances is usually done in the presence of hydrogen peroxide (H_2O_2), nitric acid (HNO_3), potassium hydroxide (KOH), and Proteinase K (Solasz and Toia, 2020; Kye et al., 2023). The samples are then filtered using various membranes, which include anodisc, cellulose, PTFE, or gold-coated and glass fiber membranes (Zhu et al., 2020; Upadhyay et al., 2024).

One of the most crucial challenges of the accurate determination of MPs and NPs is the limitation of the routine instruments of the particle-characterization linked to the inability to identify smaller particles. One such example is the possibility to observe particles with scanning electron microscopy (SEM) and transmission electron microscopy (TEM) but does not provide any information regarding polymers due to the need to complement it with additional data (chemical) (Faure et al., 2015; Kurniawan et al., 2023). Such techniques fail to isolate polymers with similar functional groups despite using these techniques along with energy-dispersive X-ray spectroscopy (EDX) to identify the elemental composition. Moreover, amorphous NPs are not very visible in TEM images and usually require metal staining to be better seen (Free et al., 2014; Boyle and Örmeci, 2020). The existing analytical techniques can be broadly grouped into two categories, namely, morphological and chemical characterization. The morphological analysis involves visual analysis and microscopic analysis, in which the MPs are described in terms of size, shape, color, and surface texture. Although these methods are typically accessible, they are inaccurate when the primary interest is on particles that are smaller than a size cutoff value (Liu et al., 2019b; Farrell and Nelson, 2013). On the other hand, chemical characterization gives composition information and is generally carried out via more elaborate instrumentation. Main methods used include Fourier-transform infrared (FTIR) spectroscopy, Raman spectroscopy, X-ray diffraction, pyrolysis–gas chromatography–mass spectrometry (Pyr-GC/MS), thermal desorption chromatography and SEM together with EDX analysis (Ha and Yeo, 2018; Da Costa et al., 2018; Kurniawan et al., 2023).

There are benefits and limitations to each of the methods. Economical and easy-to-use methods such as FTIR, chemical staining, and visual sorting are still popular for initial analysis (Onoja et al., 2022; Upadhyay et al., 2024). In order to demonstrate how different analytical tools are used in the real-life environment, Table 12.2 provides a summary of the selected studies reporting on the abundance of MPs, types of polymer types, and identification methods in freshwater ecosystems in different parts of the world (modified from: Mukhopadhyay and Valsalan 2022; Neelavannan and Sen, 2023). These illustrate the variability of the instruments in use, reporting unit, and geographic range, and once again, the importance of standardized protocols in MP research is further emphasized. Due to their time effectiveness, techniques such as FTIR, stereomicroscopy, and atomic force microscopy (AFM) have become a good alternative, while others such as Pyr-GC/MS, inductively coupled plasma (ICP) MS or SEM/EDS offer high analytical accuracy. AFM-based infrared and Raman platforms, along with visual techniques, are valued for ease of handling (Upadhyay et al., 2024). However, trade-offs are inevitable. Advanced methods often demand skilled operators and are labor- and time-intensive. Others, such as visual identification, FTIR, and stereomicroscopy, are limited by particle size resolution, and some – like Raman spectroscopy or AFM – may yield inconsistent accuracy. As a result, researchers often select or combine methods based on the study's scope, available infrastructure, and analytical priorities (Upadhyay et al., 2024; Kurniawan et al., 2023).

12.7 KNOWLEDGE GAPS AND FUTURE RESEARCH DIRECTIONS

The continued accumulation of plastic waste in the environment suggests that MP concentrations in freshwater systems are expected to rise steadily and spread over broader geographic areas in the coming years. While much of the global research to date has focused on marine environments, similar trends are increasingly predicted in freshwater ecosystems such as rivers and lakes. For instance, it has been projected that plastic waste generated by populations living within 50 km of coastlines could increase tenfold between 2010 and 2025 (Jambeck et al., 2015), a trend that may mirror itself in inland waters (Eerkes-Medrano and Thompson, 2018). Freshwater organisms, from microorganisms to fish and amphibians, are becoming more exposed to MPs while the long-term ecological impacts are largely unknown. Laboratory-based research has shown a variety of negative impacts, such as growth inhibition, decreases in feeding habits, oxidative stress, and inflammatory responses. However, many of these tests are performed in MPs concentrations which are well above environmental levels and therefore may not be ecologically relevant (Eerkes-Medrano and Thompson, 2018). To improve real-world applicability, future research should focus on exposure situations that consider environmentally relevant exposures.

Another significant research need is the establishment and integration of standard procedures to be used in the analysis at the international level. The consistency in routine sampling, digestion, filtration, and identification procedures must be such that detection and quantification of MPs in both biological tissues and environmental matrices can be consistent. This harmonization will facilitate strong cross-study comparisons and enhance larger scale ecological risk assessments (Hermsen et al., 2017; Eerkes-Medrano and Thompson, 2018). Although there is an increasing concern, little research on MPs in freshwater environments has been done compared to the marine environment. Many approaches in use are adapted from the marine strategy and may be inappropriate for freshwater conditions with respect to hydrology and geomorphology of the systems. For instance, rivers

TABLE 12.2
Global Studies on MPs Analysis and Identification in Freshwater Bodies

Location & Sample Type	Identification Method	Polymer Type	MPs Abundance	References
Netravathi River (India) – Water & sediment	ATR-FTIR	Polyethylene (PE), PET	288 pieces/m^3 (water); 96 (sediment)	Amrutha and Warrier (2020)
Rhine River (Europe) – Surface water	ATR-FTIR	PE, PP, PS	0.04–9.97 MP/m^3	Mani and Burkhardt-Holm (2020)
Sabarmati River (India) – Sediment	SEM analysis	Plastic debris and fiber	134.53–581.70 mg/kg	Patel et al. (2020)
Donting Lake (China) – Water & sediment	μ-FTIR	Fiber	0.62–4.31 items/m^2 (water); 21–52/100 g (sediment)	Hu et al. (2020)
Adyar River (India) – Water	Microscope and ATR-FTIR	PE, PP, PS	Mean 330 items/m^3	Lechthaler et al. (2021)
Kodaikanal Lake (India) – Water	Microscope, ATR-FTIR, and SEM with EDAX	PE, PP, PS, PET	Mean 24.42 ± 3.22 items/L	Laju et al. (2022)
Zhangjiang River (China) – Catchment	Stereomicroscopy & micro-Raman spectroscopy	PE, PP	50–725 items/m^3 (Mean 246)	Pan et al. (2020)
Qin River (China) – Sediment	μ-FTIR	PE, PP	0–97 items/kg	Zhang et al. (2020)
Maozhou River (China) – Water	ATR-FTIR	PE, PP, PS, PVC	Dry season: 4.0 ± 1.0 to 25.5 ± 3.5 items/L; Wet season: 3.5 ± 1.0 to 10.5 ± 2.5 items/L	Wu et al. (2020)
Manas River Basin (China) – Surface water	EDS & μ-FTIR	PP, PET	21 ± 3 to 49 ± 3 items/L	Wang et al. (2020)
Qiantang River (China) – Surface water	FTIR spectroscopy	Polyamide, polyester, PET	1,183 ± 269 particles/m^3	Zhao et al. (2020)
Victoria River, (Australia) – Surface water	Stereo microscope	Polyester, rayon	0.40 ± 0.27 items/L	Nan et al. (2020)
Renuka Lake (India) – Water	Microscope, ATR-FTIR, and Raman spectroscopy	PP, PS, PE	21 ± 13 particles/L	Ajay et al. (2021)
Veeranam Lake (India) – Water	Microscope and ATR-FTIR	PVC, PE, PP, PS, NY	28 items/km^2	Bharath et al. (2021)
Tuul River (Mongolia) – Sediment	μ-FTIR	Polyester, polyamide	603 ± 251 items/kg	Battulga et al. (2020)
Ganges River (India) – Surface water	FTIR	Rayon, PVC, PS, PET, Nylon	0.038 ± 0.004 MP/L	Napper et al. (2021)
Yongfeng River (China) – Sediment	ATR-FTIR	PE, PP, PET, PS	5–72 items/kg	Rao et al. (2020)
Nakdong River (South Korea) – Water	FTIR	PP, Polyester	293–4760 particles/m^3 (water)	Eo et al. (2019)
Red Hills Lake (India) – Water	Microscope, ATR-FTIR, and SEM with EDS	PE, PP, HDPE, LDPE	Mean 5.9 particles/L	Gopinath et al. (2020)
Poyang Lake (China) – Water	Micro-Raman spectroscopy	PE, PP	5–34 items/L (water)	Yuan et al. (2019)
Three Gorges Reservoir (China) – Water.	Micro-Raman, SEM	PS, PE, PP	1,597–12,611 n/m^3 (water)	Di and Wang (2018)

and lakes have complex flow regime, sediment transport processes, and seasonality affecting the MP load and its fate (Eerkes-Medrano and Thompson, 2018). It is urgently required to develop specific research methodologies that are adapted to freshwater ecosystems.

To overcome these limitations, the following are some recommended directions of future research:

- Exploring the fate of MPs in freshwater sediments and their role as secondary pollution sources in the long term.
- Studying chronic low-dose MP exposure in freshwater species across multiple life stages to assess sub-lethal and cumulative effects.
- Exploring bioaccumulation and biomagnification patterns of MPs within food webs, particularly in key indicator or economically important species.
- Evaluating how MPs interact with other environmental pollutants such as heavy metals or POPs to assess compound toxicity risks.
- Examining trophic transfer and the potential cascading impacts of MPs across different ecological levels in freshwater ecosystems.

Addressing these challenges will require integrated, interdisciplinary efforts involving scientists, policymakers, and industry stakeholders. Enhanced monitoring networks, consistent analytical tools, and freshwater-specific models will be crucial. Ultimately, advancing our understanding of MPs in freshwater systems will depend on cross-sector collaboration and a commitment to filling the critical knowledge gaps that currently limit mitigation and management strategies (Eerkes-Medrano and Thompson, 2018).

12.8 CONCLUSION

MPs have emerged as pervasive and persistent pollutants in freshwater ecosystems, with far-reaching ecological consequences. Unlike earlier studies that predominantly focused on marine environments, this chapter shifts the lens toward freshwater systems, recognizing their critical role not only as recipients of plastic waste but also as transport pathways for MPs into marine ecosystems. These particles are introduced into rivers, lakes, and streams through diverse sources, including domestic and industrial effluents, agricultural runoff, aquaculture operations, atmospheric deposition, and urban stormwater discharge. Once released, MPs interact dynamically with both biotic and abiotic components of the ecosystem. Their small size, stability, and chemical composition make them capable of adsorbing and transporting other pollutants such as heavy metals or POPs. MPs may get entry to aquatic food webs through aquatic organisms after ingestion at various trophic levels. They can cause endocrine disruption, genetic damage, oxidative stress, and physiological disorders that pose dangers to the biodiversity and ecosystem stability. Moreover, MPs with the presence of other pollutant exposes the environment to more risks and make the remediation approach difficult.

The MPs pollution problem must be resolved by international effort. A good action requires good policy measures that are backed by scientific developments. It can be effective to apply strict regulatory policy on the plastic industry and single-use items, better waste disposal and recycling processes, and the development of more eco-friendly alternatives to plastics. Awareness of the population regarding the use of plastics and changing the attitudes on the individual level is necessary. In addition, the intergovernmental, intersectoral, inter-non-governmental, and inter-scientific collaboration will accelerate the development of sustainable materials and scalable solutions. International collaboration is also crucial to coordinate the monitoring guidelines, unify the analysis procedures, and achieve uniform data gathering within the regions. After all, conservation of freshwater ecosystems and the biodiversity they sustain is achieved through a collective action of alleviating the plastic pollution at its source, reinforcing environmental laws, and advancing science and innovation.

With scientific knowledge, governance, and engagement of the people, the future can be resilient and sustainable, where the negative effects of MPs are greatly reduced.

DECLARATION OF GENERATIVE AI USE

AI-assisted tools (ChatGPT and Grammarly) were used solely for grammar corrections, language editing and improve readability. After using these tools, the authors reviewed and edited the content as required.

ACKNOWLEDGMENTS

The authors gratefully acknowledge the financial support received from the Research Grants Council of the Hong Kong Special Administrative Region, China (Project No. EdUHK PDFS2324-8S02), the Research Postgraduate Studentship awarded by The Education University of Hong Kong, and the Belt and Road Scholarship.

REFERENCES

Abbasi, S., Keshavarzi, B., Moore, F., Turner, A., Kelly, F. J., Dominguez, A. O., & Jaafarzadeh, N. (2019). Distribution and potential health impacts of microplastics and microrubbers in air and street dusts from Asaluyeh County, Iran. *Environmental Pollution*, *244*, 153–164.

Abidli, S., Lahbib, Y., & Trigui El Menif, N. (2017). Microplastics in sediments from lagoon of Bizerte (Northern Tunisia): Occurrence and spatial distribution. *Environmental Pollution*, *229*, 117–125. https://doi.org/10.1016/j.envpol.2017.05.050

Aib, H., Parvez, M. S., & Czédli, H. M. (2025). Pharmaceuticals and microplastics in aquatic environments: A comprehensive review of pathways and distribution, toxicological and ecological effects. *International Journal of Environmental Research and Public Health*, *22*(5), Article 5. https://doi.org/10.3390/ijerph22050799

Ajay, K., Behera, D., Bhattacharya, S., Mishra, P. K., Ankit, Y., & Anoop, A. (2021). Distribution and characteristics of microplastics and phthalate esters from a freshwater lake system in Lesser Himalayas. *Chemosphere*, *283*, 131132. https://doi.org/10.1016/j.chemosphere.2021.131132

Alfonso, M. B., Arias, A. H., Ronda, A. C., Piccolo, M. C., & Marcovecchio, J. E. (2020). Evidence of microplastic ingestion by fish from the Patagonian coast (South America). *Marine Pollution Bulletin*, *150*, 110689. https://doi.org/10.1016/j.marpolbul.2019.110689

Alimi, O. S., Farner Budarz, J., Hernandez, L. M., & Tufenkji, N. (2018). Microplastics and nanoplastics in aquatic environments: Aggregation, deposition, and enhanced contaminant transport. *Environmental Science & Technology*, *52*(4), 1704–1724. https://doi.org/10.1021/acs.est.7b05559

Allen, S., Allen, D., Phoenix, V. R., Le Roux, G., Durántez Jiménez, P., Simonneau, A., Binet, S., & Galop, D. (2019). Atmospheric transport and deposition of microplastics in a remote mountain catchment. *Nature Geoscience*, *12*(5), 339–344.

Amrutha, K., & Warrier, A. K. (2020). The first report on the source-to-sink characterization of microplastic pollution from a riverine environment in tropical India. *Science of the Total Environment*, *739*, 140377. https://doi.org/10.1016/j.scitotenv.2020.140377

Anderson, J. C., Park, B. J., & Palace, V. P. (2016). Microplastics in aquatic environments: Implications for Canadian ecosystems. *Environmental Pollution*, *218*, 269–280. https://doi.org/10.1016/j.envpol.2016.06.074

Anderson, P. J., Warrack, S., Langen, V., Challis, J. K., Hanson, M. L., & Rennie, M. D. (2017). Microplastic pollution in Lake Winnipeg, Canada. *Environmental Pollution*, *225*, 223–231. https://doi.org/10.1016/j.envpol.2017.03.054

Arias-Andres, M., Klümper, U., Rojas-Jimenez, K., & Grossart, H.-P. (2018). Microplastic pollution increases gene exchange in aquatic ecosystems. *Environmental Pollution (Barking,* Essex*: 1987)*, *237*, 253–261. https://doi.org/10.1016/j.envpol.2018.02.058

Atugoda, T., Wijesekara, H., Werellagama, D. R. I. B., Jinadasa, K. B. S. N., Bolan, N. S., & Vithanage, M. (2020). Fate and behavior of microplastics in freshwater systems. In T. Rocha-Santos, M. F. Costa, & C. Mouneyrac (Eds.), *Handbook of microplastics in the environment* (pp. 1–31). Springer International Publishing. https://doi.org/10.1007/978-3-030-10618-8_42-1

Au, S. Y., Bruce, T. F., Bridges, W. C., & Klaine, S. J. (2015). Responses of *Hyalella azteca* to acute and chronic microplastic exposures. *Environmental Toxicology and Chemistry, 34*(11), 2564–2572.

Barboza, L. G. A., Vieira, L. R., Branco, V., Figueiredo, N., Carvalho, F., Carvalho, C., & Guilhermino, L. (2018). Microplastics cause neurotoxicity, oxidative damage and energy-related changes and interact with the bioaccumulation of mercury in the European seabass, *Dicentrarchuslabrax* (Linnaeus, 1758). *Aquatic Toxicology, 195*, 49–57.

Barnes, D. K. A., Galgani, F., Thompson, R. C., & Barlaz, M. (2009). Accumulation and fragmentation of plastic debris in global environments. *Philosophical Transactions of the Royal Society B: Biological Sciences, 364*, 1985–1998. https://doi.org/10.1098/rstb.2008.0205

Batel, A., Borchert, F., Reinwald, H., Erdinger, L., & Braunbeck, T. (2018). Microplastic accumulation patterns and transfer of benzo[a]pyrene to adult zebrafish (*Danio rerio*) gills and zebrafish embryos. *Environmental Pollution, 235*, 918–930.

Battulga, B., Kawahigashi, M., & Oyuntsetseg, B. (2020). Behavior and distribution of polystyrene foams on the shore of Tuul River in Mongolia. *Environmental Pollution, 260*, 113979.

Besseling, E., Quik, J. T. K., Sun, M., & Koelmans, A. A. (2017). Fate of nano- and microplastic in freshwater systems: A modeling study. *Environmental Pollution, 220*, 540–548. https://doi.org/10.1016/j.envpol.2016.10.001

Bharath, K. M., Srinivasalu, S., Natesan, U., Ayyamperumal, R., Kalam, S., Anbalagan, S., Sujatha, K., & Alagarasan, C. (2021) Microplastics as an emerging threat to the freshwater ecosystems of Veeranam lake in south India: A multidimensional approach. *Chemosphere, 264*, 128502.

Blair, R. M., Waldron, S., Phoenix, V. R., & Gauchotte-Lindsay, C. (2019). Micro- and nanoplastic pollution in freshwater ecosystems: Linking occurrence, fate, and effects. *Science of the Total Environment, 717*, 135455. https://doi.org/10.1016/j.scitotenv.2019.135455

Bordós, G., Urbányi, B., Micsinai, A., Kriszt, B., Palotai, Z., Szabó, I., Hantosi, Z., & Szoboszlay, S. (2019). Identification of microplastics in fish ponds and natural freshwater environments of the Carpathian basin, Europe. *Chemosphere, 216*, 110–116. https://doi.org/10.1016/j.chemosphere.2018.10.110

Botterell, Z. L. R., Beaumont, N., Dorrington, T., Steinke, M., Thompson, R. C., & Lindeque, P. K. (2019). Bioavailability and effects of microplastics on marine zooplankton: A review. *Environmental Pollution, 245*, 98–110. https://doi.org/10.1016/j.envpol.2018.10.065

Boyle, K., & Örmeci, B., (2020). Microplastics and nanoplastics in freshwater and terrestrial environment: A review. *Water, 12*, 2633. https://doi.org/10.3390/w12092633

Cai, L., Wang, J., Peng, J., Tan, Z., Zhan, Z., Tan, X., & Chen, Q. (2017). Characteristic of microplastics in the atmospheric fallout from Dongguan city, China: Preliminary research and first evidence. *Environmental Science and Pollution Research, 24*(32), 24928–24935.

Campanale, C., Massarelli, C., Savino, I., Locaputo, V., & Uricchio, V. F. (2020). A detailed review study on potential effects of microplastics and additives of concern on human health. *International Journal of Environmental Research and Public Health, 17*(4), Article 4. https://doi.org/10.3390/ijerph17041212

Castro-Castellon, A. T., Horton, A. A., Hughes, J. M. R., Rampley, C., Jeffers, E. S., Bussi, G., & Whitehead, P. (2022). Ecotoxicity of microplastics to freshwater biota: Considering exposure and hazard across trophic levels. *Science of the Total Environment, 816*, 151638. https://doi.org/10.1016/j.scitotenv.2021.151638

Chen, G., Feng, Q., & Wang, J. (2020). Mini review of microplastic in the atmosphere and their risks to humans. *Science of the Total Environment, 703*, 135504. https://doi.org/10.1016/j.scitotenv.2019.135504

Chen, H. L., Selvam, S. B., Ting, K. N., & Gibbins, C. N. (2021). Microplastic pollution in freshwater systems in Southeast Asia: Pollution levels, sources, and ecological impacts. *Environmental Science and Pollution Research, 28*(39), 54222–54237.

Chen, L., Zhou, S., Su, B., Qiu, Y., & Li, Y. (2024). Microplastic pollution in Taihu Lake: Spatial distribution from the lake inlet to the lake centre and vertical stratification in the water column. *Environmental Pollution, 363*, 125102. https://doi.org/10.1016/j.envpol.2024.125102

Chen, Q., Gundlach, M., Yang, S., Jiang, J., Velki, M., Yin, D., & Hollert, H. (2017). Quantitative investigation of the mechanisms of microplastics and nanoplastics toward zebrafish larvae locomotor activity. *Science of the Total Environment, 584*, 1022–1031.

Chen, X., Xiong, X., Jiang, X., Shi, H., & Wu, C. (2019). Sinking of floating plastic debris caused by biofilm development in a freshwater lake. *Chemosphere, 222*, 856–864.

Cheung, P. K., & Fok, L. (2017). Characterisation of plastic microbeads in facial scrubs and their estimated emissions in Mainland China. *Water Research, 122*, 53–61

Chevalier, C., Vandenberghe, M., Pagano, M., Pellet, I., Pinazo, C., TesánOnrubia, J. A., Guilloux, L., & Carlotti, F. (2023). Investigation of dynamic change in microplastics vertical distribution patterns: The seasonal effect on vertical distribution. *Marine Pollution Bulletin, 189*, 114674. https://doi.org/10.1016/j.marpolbul.2023.114674

Chisada, S., Yoshida, M., & Karita, K. (2019). Ingestion of polyethylene microbeads affects the growth and reproduction of medaka, *Oryziaslatipes. Environmental Pollution, 254*, 113094.

Crespy, D., Bozonnet, M., & Meier, M. (2008). 100 years of bakelite, the material of a 1000 uses. *Angewandte Chemie International, 47*, 3322–3328. https://doi.org/10.1002/anie.200704281

Da Costa, J. P., Nunes, A. R., Santos, P. S., Girao, A. V., Duarte, A. C., & Rocha-Santos, T. (2018). Degradation of polyethylene microplastics in seawater: Insights into the environmental degradation of polymers. *Journal of Environmental Science and Health, Part A, Toxic/Hazardous Substances and Environmental Engineering, 53*, 866–875. https://doi.org/10.1080/10934529.2018.1455381

De Sá, L. C., Oliveira, M., Ribeiro, F., Rocha, T. L., & Futter, M. N. (2018). Studies of the effects of microplastics on aquatic organisms: What do we know and where should we focus our efforts in the future? *Science of the Total Environment, 645*, 1029–1039

Deshpande, P. C., Philis, G., Brattebø, H., & Fet, A. M. (2020). Using material flow analysis (MFA) to generate the evidence on plastic waste management from commercial fishing gears in Norway. *Resources, Conservation & Recycling: X, 5*, 100024.

Di, M., Liu, X., Wang, W., & Wang, J. (2019). Pollution in drinking water source areas: Microplastics in the Danjiangkou Reservoir, China. *Environmental Toxicology and Pharmacology, 65*, 82–89.

Di, M., & Wang, J. (2018). Microplastics in surface waters and sediments of the Three Gorges Reservoir, China. *Science of the Total Environment, 616–617*, 1620–1627. https://doi.org/10.1016/j.scitotenv.2017.10.150

Ding, J., Zhang, S., Razanajatovo, R. M., Zou, H., & Zhu, W. (2018). Accumulation, tissue distribution, and biochemical effects of polystyrene microplastics in the freshwater fish red tilapia (*Oreochromis niloticus*). *Environmental Pollution, 238*, 1–9.

Ding, L., Zhang, S., Wang, X., Yang, X., & Wang, J. (2019). The occurrence and distribution characteristics of microplastics in the surface water and sediment of the Wei River, China. *Environmental Pollution, 254*, 113026. https://doi.org/10.1016/j.envpol.2019.113026

Ding, R., Tong, L., & Zhang, W. (2021). Microplastics in freshwater environments: Sources, fates, and toxicity. *Water, Air, & Soil Pollution, 232*, 181. https://doi.org/10.1007/s11270-021-05081-8

Dong, H., Chen, Y., Wang, J., Zhang, Y., Zhang, P., Li, X., Zou, J., & Zhou, A. (2021). Interactions of microplastics and antibiotic resistance genes and their effects on the aquaculture environments. *Journal of Hazardous Materials, 403*, 123961–123961.

Dris, R., Gasperi, J., Saad, M., Mirande, C., & Tassin, B. (2016). Synthetic fibers in atmospheric fallout: A source of microplastic in the environment? *The Marine Pollution Bulletin, 104*(1–2), 290–293. https://doi.org/10.1016/j.marpolbul.2016.01.006

Eerkes-Medrano, D., & Thompson, R. (2018). Occurrence, fate, and effect of microplastics in freshwater systems. In E. M. Foekema & A. A. Koelmans (Eds.), *Microplastic Pollution in Aquatic Environments: An Emerging Matter of Environmental Urgency* (pp. 95–132). Elsevier. https://doi.org/10.1016/B978-0-12-813747-5.00004-7

Eerkes-Medrano, D., Thompson, R. C., & Aldridge, D. C. (2015). Microplastics in freshwater systems: A review of the emerging threats, identification of knowledge gaps and prioritisation of research needs. *Water Research, 75*, 63–82. https://doi.org/10.1016/j.watres.2015.02.012

Elagami, H., Frei, S., Boos, J.-P., Trommer, G., & Gilfedder, B. S. (2023). Quantifying microplastic residence times in lakes using mesocosm experiments and transport modelling. *Water Research, 229*, 119463. https://doi.org/10.1016/j.watres.2022.119463

Eo, S., Hong, S. H., Song, Y. K., Han, G. M., & Shim, W. J. (2019). Spatiotemporal distribution and annual load of microplastics in the Nakdong River, South Korea. *Water Research, 160*, 228–237.

Eriksen, M., Mason, S., Wilson, S., Box, C., Zellers, A., Edwards, W., Farley, H., & Amato, S. (2013). Microplastic pollution in the surface waters of the Laurentian Great Lakes. *Marine Pollution Bulletin, 77*, 177–182. https://doi.org/10.1016/j.marpolbul.2013.10.007

Erni-Cassola, G., Zadjelovic, V., Gibson, M. I., & Christie-Oleza, J. A. (2019). Distribution of plastic polymer types in the marine environment: A meta-analysis. *Journal of Hazardous Materials, 369*, 691–698. https://doi.org/10.1016/j.jhazmat.2019.02.067

Espinosa, C., Esteban, M. Á., & Cuesta, A. (2016). Microplastics in aquatic environments and their toxicological implications for fish. In S. Soloneski & M. L. Larramendy (Eds.), *Toxicology — New aspects to this scientific conundrum* (pp. 113–145). IntechOpen. https://doi.org/10.5772/64815

Evangeliou, N., Tichý, O., Eckhardt, S., Zwaaftink, C. G., & Brahney, J. (2022). Sources and fate of atmospheric microplastics revealed from inverse and dispersion modelling: From global emissions to deposition. *Journal of Hazardous Materials*, *432*, 128585. https://doi.org/10.1016/j.jhazmat.2022.128585

Everaert, G., Van Cauwenberghe, L., De Rijcke, M., Koelmans, A. A., Mees, J., Vandegehuchte, M., & Janssen, C. R. (2018). Risk assessment of microplastics in the ocean: Modelling approach and first conclusions. *Environmental Pollution (Barking,* Essex*: 1987)*, *242*(Pt B), 1930–1938. https://doi.org/10.1016/j.envpol.2018.07.069

Fan, J., Zou, L., Duan, T., Qin, L., Qi, Z., & Sun, J. (2022). Occurrence and distribution of microplastics in surface water and sediments in China's inland water systems. *Journal of Cleaner Production*, *331*, 129968. https://doi.org/10.1016/j.jclepro.2021.129968

Farrell, P., & Nelson, K. (2013). Trophic level transfer of microplastics: Mytilus edulis (L.) to Carcinus maenas (L.). *Environmental Pollution*, *177*, 1–3. https://doi.org/10.1016/j.envpol.2013.01.046

Faure, F., Demars, C., Wieser, O., Kunz, M., & De Alencastro, L. F. (2015). Plastic pollution in Swiss surface waters: Nature and concentrations, interaction with pollutants. *Environmenal Chemistry*, *12*, 582–591. https://doi.org/10.1071/EN14218

Ferreira, G. V. B., Barletta, M., Lima, A. R. A., Morley, S. A., Justino, A. K. S., & Costa, M. F. (2018). High intake rates of microplastics in a Western Atlantic predatory fish, and insights of a direct fishery effect. *Environmental Pollution*, *236*, 706–717.

Fischer, E. K., Paglialonga, L., Czech, E., & Tamminga, M. (2016). Microplastic pollution in lakes and lake shoreline sediments: A case study on Lake Bolsena and Lake Chiusi (central Italy). *Environmental Pollution*, *213*, 648–657. https://doi.org/10.1016/j.envpol.2016.03.012

Fonte, E., Ferreira, P., & Guilhermino, L. (2016). Temperature rise and microplastics interact with the toxicity of the antibiotic cefalexin to juveniles of the common goby (*Pomatoschistus microps*): Post-exposure predatory behaviour, acetylcholinesterase activity and lipid peroxidation. *Aquatic Toxicology*, *180*, 173–185.

Franzellitti, S., Canesi, L., Auguste, M., Wathsala, R. H. G. R., & Fabbri, E. (2019). Microplastic exposure and effects in aquatic organisms: A physiological perspective. *Environmental Toxicology and Pharmacology, 68,* 37–51. https://doi.org/10.1016/j.etap.2019.03.009

Free, C. M., Jensen, O. P., Mason, S. A., Eriksen, M., Williamson, N. J., & Boldgiv, B. (2014). Highlevels of microplastic pollution in a large, remote, mountain lake. *The Marine Pollution Bulletin*, *85*, 156–163. https://doi.org/10.1016/j.marpolbul.2014.06.001

Gasperi, J., Dris, R., Bonin, T., Rocher, V., & Tassin, B. (2014). Assessment of floating plastic debris in surface water along the Seine River. *Environmental Pollution*, *195*(dec.), 163–166.

Gatidou, G., Arvaniti, O. S., & Stasinakis, A. S. (2019). Review on the occurrence and fate of microplastics in Sewage Treatment Plants. *Journal of Hazardous Materials*, *367*, 504–512. https://doi.org/10.1016/j.jhazmat.2018.12.081

Geyer, R., Jambeck, J. R., & Law, K. L. (2017). Production, use, and fate of all plastics ever made. *Science Advances*, *3*(7), e1700782. https://doi.org/10.1126/sciadv.1700782

Gopinath, K., Seshachalam, S., Neelavannan, K., Anburaj, V., Rachel, M., Ravi, S., Bharath, M., & Achyuthan, H. (2020). Quantification of microplastic in Red Hills Lake of Chennai city, Tamil Nadu, India. *Environmental Science and Pollution Research*, *27*, 33297–33306.

Greven, A. C., Merk, T., Karagöz, F., Mohr, K., Klapper, M., Jovanović, B., & Palić, D. (2016). Polycarbonate and polystyrene nanoplastic particles act as stressors to the innate immune system of fathead minnow (Pimephalespromelas). *Environmental Toxicology and Chemistry*, *35*(12), 3093–3100.

Ha, J., & Yeo, M. K. (2018). The environmental effects of microplastics on aquatic ecosystems. *Molecular & Cellular Toxicology*, *14*, 353–359. https://doi.org/10.1007/s13273-018-0039-8

Haque, F., & Fan, C. (2023). Fate of microplastics under the influence of climate change. *iScience, 26*(9), 107649. https://doi.org/10.1016/j.isci.2023.107649

He, D., Luo, Y., Lu, S., Liu, M., Song, Y., & Lei, L. (2020). Microplastics in soils: Analytical methods, pollution characteristics and ecological risks. *TrAC Trends in Analytical Chemistry*, *109*, 163–172. https://doi.org/10.1016/j.trac.2018.10.006

Hermsen, E., Mintenig, S. M., Besseling, E., & Koelmans, A. A. (2018). Quality criteria for the analysis of microplastic in biota samples: A critical review. *Environmental Science & Technology*, *52*(18), 10230–10240. https://doi.org/10.1021/acs.est.8b01611

Hermsen, E., Pompe, R., Besseling, E., & Koelmans, A. A. (2017). Detection of low numbers of microplastics in North Sea fish using strict quality assurance criteria. *The Marine Pollution Bulletin*, *122*(1–2), 253–258.

Hodson, M. E., Duffus-Hodson, C. A., Clark, A., Prendergast Miller, M. T., & Thorpe, K. L. (2017). Plastic bags derived microplastics as a vector for metal exposure in terrestrial invertebrates. *Environmental Science & Technology*, *51*(8), 4714–4721.

Horton, A. A., & Dixon, S. J. (2018). Microplastics: An introduction to environmental transport processes. *Wiley Interdisciplinary Reviews: Water*, *5*(2), e1268.

Horton, A. A., Walton, A., Spurgeon, D. J., Lahive, E., & Svendsen, C. (2017). Microplastics in freshwater and terrestrial environments: Evaluating the current understanding to identify the knowledge gaps and future research priorities. *Science of the Total Environment*, *586*, 127–141. https://doi.org/10.1016/j.scitotenv.2017.01.190

Hu, D., Zhang, Y., & Shen, M. (2020). Investigation on microplastic pollution of Dongting Lake and its affiliated rivers. *Marine Pollution Bulletin*, *160*, 111555. https://doi.org/10.1016/j.marpolbul.2020.111555

Hurley, R. R., & Nizzetto, L. (2018). Fate and occurrence of micro(nano)plastics in soils: Knowledge gaps and possible risks. *Current Opinion in Environmental Science & Health*, *1*, 6–11.

Imhof, H. K., Rusek, J., Thiel, M., Wolinska, J., & Laforsch, C. (2017). Do microplastic particles affect *Daphnia magna* at the morphological, life history and molecular level? *PLoS One*, *12*(11), e0187590.

Jambeck, J. R., Geyer, R., Wilcox, C., Siegler, T. R., Perryman, M., Andrady, A., Narayan, R., & Law, K. L. (2015). Plastic waste inputs from land into the ocean. *Science*, *347*(6223), 768–771. https://doi.org/10.1126/science.1260352

Jemec, A., Horvat, P., Kunej, U., Bele, M., & Krzan, A. (2016). Uptake and effects of microplastic textile fibers on freshwater crustacean Daphnia magna. *Environmental Pollution*, *219*, 201–209.

Jeyasanta, K. I., Sathish, N., Patterson, J., & Edward, J. K. P. (2020). Macro-, meso- and microplastic debris in the beaches of Tuticorin district, Southeast coast of India. *Marine Pollution Bulletin*, *154*, 111055. https://doi.org/10.1016/j.marpolbul.2020.111055

Jiang, J. (2018). Occurrence of microplastics and its pollution in the environment: A review. *Sustainable Production and Consumption*, *13*, 16–23. https://doi.org/10.1016/j.spc.2017.11.003

Karbalaei, S. S., Golieskardi, A., Watt, D. U., Boiret, M., & Karami, A. (2020). Analysis and inorganic composition of microplastics in commercial Malaysian fish meals. *Marine Pollution Bulletin*, *150*, 110687.

Kataoka, T., Nihei, Y., Kudou, K., & Hinata, H. (2019). Assessment of the sources and inflow processes of microplastics in the river environments of Japan. *Environmental Pollution*, *244*, 958–965. https://doi.org/10.1016/j.envpol.2018.10.111

Khan, F. R., Syberg, K., Shashoua, Y., & Bury, N. R. (2015). Influence of polyethylene microplastic beads on the uptake and localization of silver in zebrafish (*Danio rerio*). *Environmental Pollution*, *206*, 73–79.

Koelmans, A. A., Besseling, E., Foekema, E., Kooi, M., Mintenig, S., Ossendorp, B. C., Redondo-Hasselerharm, P. E., Verschoor, A., van Wezel, A. P., & Scheffer, M. (2017). Risks of plastic debris: Unravelling fact, opinion, perception, and belief. *Environmental Science & Technology*, *51*(20), 11513–11519. https://doi.org/10.1021/acs.est.7b02219

Kooi, M., Reisser, J., Slat, B., Ferrari, F. F., Schmid, M. S., Cunsolo, S., Brambini, R., Noble, K., Sirks, L.-A., Linders, T. E. W., Schoeneich-Argent, R. I., & Koelmans, A. A. (2016). The effect of particle properties on the depth profile of buoyant plastics in the ocean. *Scientific Reports*, *6*(1). https://doi.org/10.1038/srep33882

Kuddus, M. A., Datta, G. C., Miah, M. A., Sarker, A. K., Hamid, S. M. A., & Sunny, A. R. (2020). Performance study of selected orange fleshed sweet potato varieties in northeastern Bangladesh. *International Journal of Agriculture Environment and Biotechnology*, *5*, 673–682.

Kumar, R., Sharma, P., Verma, A., Jha, P. K., Singh, P., Gupta, P. K., Chandra, R., & Prasad, P. V. V. (2021). Effect of physical characteristics and hydrodynamic conditions on transport and deposition of microplastics in Riverine ecosystem. *Water*, *13*(19), 2710. https://doi.org/10.3390/w13192710

Kurniawan, T. A., Haider, A., Ahmad, H. M., Mohyuddin, A., Aslam, H. M. U., Nadeem, S., Javed, M., Othman, M. H. D., Goh, H. H., & Chew, K. W. (2023). Source, occurrence, distribution, fate, and implications of microplastic pollutants in freshwater on environment: A critical review and way forward. *Chemosphere*, *325*, 138367.

Kye, H., Kim, J., Ju, S., Lee, J., Lim, C., & Yoon, Y. (2023). Microplastics in water systems: A review of their impacts on the environment and their potential hazards. *Heliyon*, *9*, e14359. https://doi.org/10.1016/J.HELIYON.2023.E14359

Lahens, L., Strady, E., Kieu-Le, T. C., Boukerma, K., Rinnert, E., Gasperi, J., & Tassin, B. (2018). Macroplastic and microplastic pollution assessment of a tropical river (Saigon River, Vietnam) transversed by a developing megacity. *Environmental Pollution*, *236*, 661–671. https://doi.org/10.1016/j.envpol.2018.02.005

Laju, R. L., Jayanthi, M., Jeyasanta, K. I., Patterson, J., Asir, N. G. G., Sathish, M. N., & Edward, J. K. P. (2022). Spatial and vertical distribution of microplastics and their ecological risk in an Indian freshwater lake ecosystem. *Science of the Total Environment*, *820*, 153337. https://doi.org/10.1016/j.scitotenv.2022.153337

Le Moine, C. M., Kelleher, B. M., Lagarde, R., Northam, C., Elebute, O. O., & Cassone, B. J. (2018). Transcriptional effects of polyethylene microplastic ingestion in developing zebrafish (*Danio rerio*). *Environmental Pollution*, *243*, 591–600.

Lebreton, L. C. M., van der Zwet, J., Damsteeg, J.-W., Slat, B., Andrady, A., & Reisser, J. (2017). River plastic emissions to the world's oceans. *Nature Communications*, *8*(1), 15611. https://doi.org/10.1038/ncomms15611

Lechthaler, S., Waldschlager, K., Sandhani, C. G., Sannasiraj, S. A., Sundar, V., Schwarzbauer, J., & Schuttrumpf, H. (2021). Baseline study on microplastics in Indian rivers under different anthropogenic influences. *Water (Switzerland)*, *13*, 1648.

Lei, L., Liu, M., Song, Y., Lu, S., Hu, J., Cao, C., Xie, B., Shi, H., & He, D. (2018a). Polystyrene (nano) microplastics cause size-dependent neurotoxicity, oxidative damage and other adverse effects in Caenorhabditis elegans. *Environmental Science: Nano*, *5*(8), 2009–2020.

Lei, L., Wu, S., Lu, S., Liu, M., Song, Y., Fu, Z., Shi, H., Raley-Susman, K. M., & He, D. (2018b). Microplastic particles cause intestinal damage and other adverse effects in zebrafish Danio rerio and nematode Caenorhabditis elegans. *Science of the Total Environment*, *619*, 1–8.

Lenaker, P. L., Baldwin, A. K., Corsi, S. R., Mason, S. A., Reneau, P. C., & Scott, J. W. (2019). Vertical distribution of microplastics in the water column and surficial sediment from the Milwaukee River Basin to Lake Michigan. *Environmental Science & Technology*, *53*(21), 12227–12237. https://doi.org/10.1021/acs.est.9b03850

Leslie, H. A., Brandsma, S. H., Van Velzen, M. J. M., & Vethaak, A. D. (2017). Microplastics en route: Field measurements in the Dutch river delta and Amsterdam canals, wastewater treatment plants, North Sea sediments and biota. *Environment International*, *101*, 133–142. https://doi.org/10.1016/j.envint.2017.01.018

Li, J., Yang, D., Li, L., Jabeen, K., & Shi, H. (2015). Microplastics in commercial bivalves from China. *Environmental Pollution*, *207*, 190–195.

Lin, L., Zuo, L.-Z., Peng, J.-P., Cai, L.-Q., Fok, L., Yan, Y., Li, H.-X., & Xu, X.-R. (2018). Occurrence and distribution of microplastics in an urban river: A case study in the Pearl River along Guangzhou City, China. *Science of the Total Environment*, *644*, 375–381. https://doi.org/10.1016/j.scitotenv.2018.06.327

Liu, E. K., He, W. Q., & Yan, C. R. (2014). 'White revolution' to 'white pollution': Agricultural plastic film mulch in China. *Environmental Research Letters*, *9*(9), 091001.

Liu, F., Olesen, K. B., Borregaard, A. R., & Vollertsen, J. (2019a). Microplastics in urban and highway stormwater retention ponds. *Science of the Total Environment*, *671*, 992–1000.

Liu, K., Wang, X., Wei, N., Song, Z., & Li, D. (2019b). Accurate quantification and transport estimation of suspended atmospheric microplastics in megacities: Implications for human health. *Environment International*, *132*, 105127. https://doi.org/10.1016/j.envint.2019.105127

Liu, M., Lu, S., Song, Y., Lei, L., Hu, J., Lv, W., Zhou, W., Cao, C., Shi, H., Yang, X., & He, D. (2018). Microplastic and mesoplastic pollution in farmland soils in suburbs of Shanghai, China. *Environmental Pollution, 242,* 855–862. https://doi.org/10.1016/j.envpol.2018.07.051

Lu, Y., Zhang, Y., Deng, Y., Jiang, W., Zhao, Y., Geng, J., Ding, L., & Ren, H. (2016). Uptake and accumulation of polystyrene microplastics in zebrafish (*Danio rerio*) and toxic effects in liver. *Environmental Science & Technology*, *50*(7), 4054–4060.

Lutz, N., Fogarty, J., & Rate, A. (2021). Accumulation and potential for transport of microplastics in stormwater drains into marine environments, Perth region, Western Australia. *Marine Pollution Bulletin*, *168*, 112362.

Mahon, A. M., Nash, R., Andrew, R., & O'Connor, I. (2017a). Scope, fate, risks and impacts of microplastics pollution in Irish freshwater systems. In *Irish Environmental Protection Agency* (pp. 1–57). https://doi.org/10.13140/RG.2.2.13353.60005. ISBN: 978-1-84095-705-1.

Mahon, A. M., O'Connell, B., Healy, M. G., O'Connor, I., Officer, R., Nash, R., & Morrison, L. (2017b). Microplastics in sewage sludge: Effects of treatment. *Environmental Science & Technology*, *51*(2), 810–818.

Mani, T., & Burkhardt-Holm, P. (2020). Seasonal microplastics variation in nival and pluvial stretches of the Rhine River: From the Swiss catchment towards the North Sea. *Science of the Total Environment, 707*, 135579. https://doi.org/10.1016/j.scitotenv.2019.135579

Martínez-Gómez, C., León, V. M., Calles, S., Gomáriz-Olcina, M., & Vethaak, A. D. (2017). The adverse effects of virgin microplastics on the fertilization and larval development of sea urchins. *Marine Environmental Research*, *130*, 69–76.

Martins, A., & Guilhermino, L. (2018). Transgenerational effects and recovery of microplastics exposure in model populations of the freshwater cladoceran *Daphnia magna* Straus. *Science of the Total Environment*, *631*, 421–428.

Mason, S. A., Garneau, D., Sutton, R., Chu, Y., Ehmann, K., Barnes, J., Fink, P., Papazissimos, D., & Rogers, D. L. (2016). Microplastic pollution is widely detected in US municipal wastewater treatment plant effluent. *Environmental Pollution*, *218*,1045–1054.

Mattsson, K., Johnson, E. V., Malmendal, A., Linse, S., Hansson, L. A., &Cedervall, T. (2017). Brain damage and behavioural disorders in fish induced by plastic nanoparticles delivered through the food chain. *Scientific Reports*, *7*(1), 11452.

Meng, Y., Kelly, F. J., & Wright, S. L. (2020). Advances and challenges of microplastic pollution in freshwater ecosystems: A UK perspective. *Environmental Pollution*, *256*, 113445.

Mithun, M. H., Shaikat, M. F. B., Sazzad, S. A., Billah, M., Salehin, S., Foysal, A. M., Jubayer, A., Islam, R., Anzum, A., & Sunny, A. R. (2024). Microplastics in aquatic ecosystems: Sources, impacts, and challenges for biodiversity, food security, and human health: A meta analysis. *Integrative Biomedical Research*, *8*(11), 1–12.

Mora-Teddy, M., & Matthaei, C. D. (2020). Microplastics in urban streams: Abundance, composition, sources, and pathways. *Science of the Total Environment, 720*, 137620. https://doi.org/10.1016/j.scitotenv.2020.137620

Mukhopadhyay, P., & Valsalan, S. A. (2022). Presence of microplastics in freshwater ecosystems: An unheeded emerging concern: A global review. *Pollution*, *8*(1), 69–104.

Murphy, F., Ewins, C., Carbonnier, F., & Quinn, B. (2016). Wastewater treatment works (WwTW) as a source of microplastics in the aquatic environment. *Environmental Science & Technology*, *50*(11), 5800–5808.

Murphy, F., & Quinn, B. (2018). The effects of microplastic on freshwater *Hydra attenuata* feeding, morphology & reproduction. *Environmental Pollution*, *234*, 487–494.

Naidoo, T., Glassom, D., & Smit, A. J. (2015). Plastic pollution in five urban estuaries of KwaZulu-Natal, South Africa. *Marine Pollution Bulletin*, *101*(1), 473–480. https://doi.org/10.1016/j.marpolbul.2015.09.044

Nan, B., Su, L., Kellar, C., Craig, N. J., Keough, M. J., & Pettigrove, V. (2020). Identification of microplastics in surface water and Australian freshwater shrimp *Paratya australiensis* in Victoria, Australia. *Environmental Pollution*, *259*, 113865.

Napper, I. E., Baroth, A., Barrett, A. C., Bhola, S., Chowdhury, G. W., Davies, B. F. R., Duncan, E. M., Kumar, S., Nelms, S. E., Hasan Niloy, M. N., Nishat, B., Maddalene, T., Thompson, R. C., & Coldewey, H. (2021). The abundance and characteristics of microplastics in surface water in the transboundary Ganges River. *Environmental Pollution*, *274*, 116348.

Nayal, R., & Suthar, S. (2022). First report on microplastics in tributaries of the upper Ganga River along Dehradun, India: Quantitative estimation and characterizations. *Journal of Hazardous Materials Advances*, *8*, 100190. https://doi.org/10.1016/j.hazadv.2022.100190

Neelavannan, K., & Sen, I. S. (2023). Microplastics in freshwater ecosystems of India: Current trends and future perspectives. *ACS Omega, 8*(38), 34235–34248. https://doi.org/10.1021/acsomega.3c01214

Nel, H. A., Dalu, T., & Wasserman, R. J. (2018). Sinks and sources: Assessing microplastic abundance in river sediment and deposit feeders in an Austral temperate urban river system. *Science of the Total Environment*, *612*, 950–956. https://doi.org/10.1016/j.scitotenv.2017.08.298

Nelms, S. E., Galloway, T. S., Godley, B. J., Jarvis, D. S., & Lindeque, P. K. (2018). Investigating microplastic trophic transfer in marine top predators. *Environmental Pollution*, *238*, 999–1007. https://doi.org/10.1016/j.envpol.2018.02.016

Nizzetto, L., Bussi, G., Futter, M. N., Butterfield, D., & Whitehead, P. G. (2016a). A theoretical assessment of microplastic transport in river catchments and their retention by soils and river sediments. *Environmental Science: Processes & Impacts*, *18*(8), 1050–1059.

Nizzetto, L., Futter, M., & Langaas, S. (2016b). Are agricultural soils dumps for microplastics of urban origin? *Environmental Science & Technology*, *50*(20), 10777–10779.

Oberbeckmann, S., Löder, M. G. J., & Labrenz, M. (2015). Marine microplasticassociated biofilms: A review. *Environmental Chemistry*, *12*, 551–562.

Onoja, S., Nel, H. A., Abdallah, M. A. E., & Harrad, S. (2022). Microplastics in freshwater sediments: Analytical methods, temporal trends, and risk of associated organophosphate esters as exemplar plastics additives. *Environmental Research*, *203*. https://doi.org/10.1016/j.envres.2021.111830

Ory, N. C., Gallardo, C., Lenz, M., & Thiel, M. (2018). Capture, swallowing, and egestion of microplastics by a planktivorous juvenile fish. *Environmental Pollution*, *240*, 566–573. https://doi.org/10.1016/j.envpol.2018.04.093

Pan, Z., Sun, Y., Liu, Q., Lin, C., Sun, X., He, Q., Zhou, K., & Lin, H. (2020). Riverine microplastic pollution matters: A case study in the Zhangjiang River of Southeastern China. *Marine Pollution Bulletin*, *159*, 111516.

Patel, A. K., Bhagat, C., Taki, K., & Kumar, M. (2020). Microplastic vulnerability in the sediments of the Sabarmati River of India. In M. Kumar, A. H. Ansari, A. Ramanathan, & K. Schneider (Eds.), *Resilience, Response, and Risk in Water Systems* (pp. 127–138). Springer. https://doi.org/10.1007/978-981-15-4668-6_7

Peiponen, K.-E., Räty, J., Ishaq, U., Pélisset, S., & Ali, R. (2019). Outlook on optical identification of micro- and nanoplastics in aquatic environments. *Chemosphere*, *214*, 424–429. https://doi.org/10.1016/j.chemosphere.2018.09.111

Porter, A., Lyons, B. P., Galloway, T. S., & Lewis, C. (2018). Role of Marine Snows in Microplastic Fate and Bioavailability. *Environmental Science & Technology*, *52*(12), 7111–7119. https://doi.org/10.1021/acs.est.8b01000

Prata, J. C. (2018). Airborne microplastics: Consequences to human health? *Environmental Pollution*, *234*, 115–126. https://doi.org/10.1016/j.envpol.2017.11.043

Rajan, K., Khudsar, F. A., & Kumar, R. (2023). Urbanization and population resources affect microplastic concentration in surface water of the River Ganga. *Journal of Hazardous Materials Advances*, *11*, 100342. https://doi.org/10.1016/j.hazadv.2023.100342

Raju, M., Gandhimathi, R., & Nidheesh, P. V. (2023). The cause, fate and effect of microplastics in freshwater ecosystem: Ways to overcome the challenge. *Journal of Water Process Engineering*, *55*, 104199.

Rao, Z., Niu, S., & Zhan, N. (2020). Microplastics in sediments of river Yengdong from Maanshan City, Anhui Province, China. *Bulletin of Environmental Contamination and Toxicology*, *104*, 161–172.

Rillig, M. C., Ziersch, L., & Hempel, S. (2017). Microplastic transport in soil by earthworms. *Scientific Reports*, *7*(1), 1362.

Romano, E., Bergamin, L., Di Bella, L., Baini, M., Berto, D., D'Ambrosi, A., Di Fazio, M., Galli, M., Medeghini, L., Panti, C., Provenzani, C., Rampazzo, F., & Fossi, M. C. (2023). First record of microplastic in the environmental matrices of a Mediterranean marine cave (Bue Marino, Sardinia, Italy). *Marine Pollution Bulletin*, *186*, 114452. https://doi.org/10.1016/j.marpolbul.2022.114452

Salam, M. T., Bari, K. B., Rahman, M. M., Gafur, D. M. M., Faruk, M. O., Akter, K., Irin, F., Ashakin, M. R., Shaikat, T. A., Das, A. C., Tufael, M., Mithun, M. M., & Sunny, A. R. (2024). Emergence of antibiotic-resistant infections in ICU patients. *Journal of Angiotherapy*, *8*(5), 1–9, 9560.

Sangkham, S., Islam, A., Adhikari, S., Kumar, R., Sharma, P., Sakunkoo, P., Bhattacharya, P., & Tiwari, A. (2023). Evidence of microplastics in groundwater: A growing risk for human health. *Groundwater for Sustainable Development*, *23*, 100981. https://doi.org/10.1016/j.gsd.2023.100981

Sarijan, S., Said, M. I. M., Sze, S. H., & Ibrahim, M. H. (2018). Microplastics in Skudai River and the adjacent coastal waters of Johor, Malaysia: Abundance, distribution and potential sources. *Marine Pollution Bulletin*, *136*, 150–162. https://doi.org/10.1016/j.marpolbul.2018.09.022

Sarkar, D. J., Das Sarkar, S., Das, B. K., Manna, R. K., Behera, B. K., & Samanta, S. (2019). Spatial distribution of meso and microplastics in the sediments of river Ganga at eastern India. *Science of the Total Environment*, *694*, 133712. https://doi.org/10.1016/j.scitotenv.2019.133712

Sazzad, S. A., Shawon, R. A. R., Moniruzzaman, Hussain, M., & Zaman, F. (2024). Climate change and socio-economic challenges of fishing communities in the coastal district of Shariatpur in Bangladesh. *Pathfinder of Research*, *2*(1), 74–94.

Scherer, C., Weber, A., Lambert, S., & Wagner, M. (2020). Interactions of microplastics with freshwater biota. In M. Wagner & S. Lambert (Eds.), *Freshwater Microplastics* (pp. 153–180). Springer. https://doi.org/10.1007/978-3-319-61615-5_8

Shang, X., Lu, J., Feng, C., Ying, Y., He, Y., Fang, S., Lin, Y., Dahlgren, R., & Ju, J. (2020). Microplastic (1 and 5 μm) exposure disturbs lifespan and intestine function in the nematode *Caenorhabditis elegans*. *Science of the Total Environment*, *705*, 135837.

Shruti, V. C., Pérez-Guevara, F., Roy, P. D., & Kutralam-Muniasamy, G. (2019). First evidence of microplastics bioaccumulation by estuarine macrofauna in a South American ecosystem. *Science of the Total Environment*, *695*, 133801. https://doi.org/10.1016/j.scitotenv.2019.133801

Solasz, S., & Toia, K. (2020). *Fishing for microplastics: Assessing microplastic ingestion by freshwater fish: A review* [Senior Capstone Paper, Skidmore College]. Skidmore College Digital Collections. https://digitalcoll.skidmore.edu/record/6312/files/fishing_microplastics_2020.pdf

Soltaninia, S., Taghavi, L., Hosseini, S. A., Motamedvaziri, B., & Eslamian, S. (2023). The effects of antecedent dry days and land use types on urban runoff quality in a semi-Arid region. *International Journal of Urban Sciences*, *27*(2), 215–238.

Sruthy, S., & Ramasamy, E. V. (2017). Microplastic pollution in Vembanad Lake, Kerala, India: The first report of a benthic microplastic from an urban estuary. *Environmental Pollution*, *222*, 315–322. https://doi.org/10.1016/j.envpol.2016.12.038

Stang, C., Mohamed, B. A., & Li, L. Y. (2022). Microplastic removal from urban stormwater: Current treatments and research gaps. *Journal of Environmental Management*, *317*, 115510. https://doi.org/10.1016/j.jenvman.2022.115510

Su, L., Xue, Y., Li, L., Yang, D., Kolandhasamy, P., Li, D., & Shi, H. (2016). Microplastics in Taihu Lake, China. *Environmental Pollution*, *216*, 711–719. https://doi.org/10.1016/j.envpol.2016.06.036

Sun, J., Dai, X., Wang, Q., van Loosdrecht, M. C. M., & Ni, B.-J. (2019). Microplastics in wastewater treatment plants: Detection, occurrence and removal. *Water Research*, *152*, 21–37.

Sunny, A. R., Mithun, M. H., Prodhan, S. H., Ashrafuzzaman, M., Rahman, S. M. A., Billah, M. M., Hussain, M., Ahmed, K. J., Sazzad, S. A., Alam, M. T., & Rashid, A. (2021). Fisheries in the context of attaining Sustainable Development Goals (SDGs) in Bangladesh: COVID-19 impacts and future prospects. *Sustainability*, *13*(17), 9912.

Teng, J., Wang, Q., Ran, W., Wu, D., Liu, Y., Sun, S., Liu, H., Cao, R., & Zhao, J. (2019). Microplastic in cultured oysters from different coastal areas of China. *The Science of the Total Environment*, *653*, 1282–1292. https://doi.org/10.1016/j.scitotenv.2018.11.057

Teuten, E. L., Saquing, J. M., Knappe, D. R. U., Rowland, S. J., Barlaz, M. A., Jonsson, S., Bjo, A., Thompson, R. C., Galloway, T. S., Yamashita, R., Ochi, D., Watanuki, Y., Moore, C., Viet, P. H., & Tana, T. S. (2009). Transport and release of chemicals from plastics to the environment and to wildlife. *Philosophical Transactions of the Royal Society B: Biological Sciences*, *364*, 2027–2045. https://doi.org/10.1098/rstb.2008.0284

Tsiaras, K., Costa, E., Morgana, S., Gambardella, C., Piazza, V., Faimali, M., Minetti, R., Zeri, C., Thyssen, M., Ben Ismail, S., Hatzonikolakis, Y., Kalaroni, S., & Garaventa, F. (2022). Microplastics in the Mediterranean: Variability from observations and model analysis. *Frontiers in Marine Science*, *9*. https://doi.org/10.3389/fmars.2022.784937

Upadhyay, S., Sharma, P. K., Dogra, K., Bhattacharya, P., Kumar, M., Tripathi, V., & Karmakar, R. (2024). Microplastics in freshwater: Unveiling sources, fate, and removal strategies. *Groundwater for Sustainable Development*, *26*, 101185. https://doi.org/10.1016/j.gsd.2024.101185

Van Emmerik, T., Kieu-Le, T. C., Loozen, M., Van Oeveren, K., Strady, E., Bui, X. T., Egger, M., Gasperi, J., Lebreton, L., & Nguyen, P. D. (2018). A methodology to characterize riverine macroplastic emission into the ocean. *Frontiers in Marine Science*, *5*, 00372.

van Sebille, E., Aliani, S., Law, K. L., Maximenko, N., Alsina, J. M., Bagaev, A., Bergmann, M., Chapron, B., Chubarenko, I., Cózar, A., Delandmeter, P., Egger, M., Fox-Kemper, B., Garaba, S. P., Goddijn-Murphy, L., Hardesty, B. D., Hoffman, M. J., Isobe, A., Jongedijk, C. E., Kaandorp, M. L. A., Khatmullina, L., Koelmans, A. A., Kukulka, T., Laufkötter, C., Lebreton, L., Lobelle, D., Maes, C., Martinez-Vicente, V., Morales Maqueda, M. A., Poulain-Zarcos, M., Rodríguez, E., Ryan, P. G., Shanks, A. L., Shim, W. J., Suaria, G., Thiel, M., van den Bremer, T. S., & Wichmann, D. (2020). The physical oceanography of the transport of floating marine debris. *Environmental Research Letters*, *15*(2), 023003. https://doi.org/10.1088/1748-9326/ab6d7d

Wang, G., Lu, J., Tong, Y., Liu, Z., Zhou, H., & Xiayihazi, N. (2020). Occurrence and pollution characteristics of microplastics in surface water of the Manas River Basin, China. *Science of the Total Environment*, *710*, 136099.

Wang, J., Peng, J., Tan, Z., Gao, Y., Zhan, Z., Chen, Q., & Cai, L. (2017a). Microplastics in the surface sediments from the Beijiang River littoral zone: Composition, abundance, surface textures and interaction with heavy metals. *Chemosphere*, *171*, 248–258. https://doi.org/10.1016/j.chemosphere.2016.12.074

Wang, W., Ndungu, A. W., Li, Z., & Wang, J. (2017b). Microplastics pollution in inland freshwaters of China: A case study in urban surface waters of Wuhan, China. *Science of the Total Environment*, *575*, 1369–1374. https://doi.org/10.1016/j.scitotenv.2016.09.213

Wang, H., Wu, Y., Deng, Y., Wu, X., Li, X., Xu, H., Zeng, Y., & Yan, Y. (2025). Impacts of wind forcing on microplastics kinematic in a sensitive water area. *Science of the Total Environment*, *958*, 177983. https://doi.org/10.1016/j.scitotenv.2024.177983.

Weber, A., Scherer, C., Brennholt, N., Reifferscheid, G., & Wagner, M. (2018). PET microplastics do not negatively affect the survival, development, metabolism and feeding activity of the freshwater invertebrate *Gammarus pulex*. *Environmental Pollution*, *234*, 181–189.

Wong, J. K. H., Lee, K. K., Tang, K. H. D., & Yap, P.-S. (2020). Microplastics in the freshwater and terrestrial environments: Prevalence, fates, impacts and sustainable solutions. *Science of the Total Environment*, *719*, 137512. https://doi.org/10.1016/j.scitotenv.2020.137512

Wright, S. L., Ulke, J., Font, A., Chan, K. L. A., & Kelly, F. J. (2020). Atmospheric microplastic deposition in an urban environment and an evaluation of transport. *Environment International*, *136*, 105411.

Wu, P., Huang, J., Zheng, Y., Yang, Y., Zhang, Y., He, F., Chen, H., Quan, G., Yan, J., Li, T., & Gao, B. (2019). Environmental occurrences, fate, and impacts of microplastics. *Ecotoxicology and Environmental Safety*, *184*, 109612. https://doi.org/10.1016/j.ecoenv.2019.109612

Wu, P., Tang, Y., Dang, M., Wang, S., Jin, H., Liu, Y., Jing, H., Zheng, C., Yi, S., & Cai, Z. (2020). Spatial-temporal distribution of microplastics in surface water and sediments of Maozhou River within Guangdong-Hong Kong-Macao Greater Bay Area. *Science of the Total Environment*, *717*, 135187.

Xiong, X., Zhang, K., Chen, X., Shi, H., Luo, Z., & Wu, C. (2018). Sources and distribution of microplastics in China's largest inland lake: Qinghai Lake. *Environmental Pollution*, *235*, 899–906.

Xu, D., Gao, B., Wan, X., Peng, W., & Zhang, B. (2022). Influence of catastrophic flood on microplastics organization in surface water of the Three Gorges Reservoir, China. *Water Research*, *211*, Article 118018. https://doi.org/10.1016/j.watres.2021.118018

Xu, Q., Xing, R., Sun, M., Gao, Y., & An, L. (2020). Microplastics in sediment from an interconnected River-Estuary region. *Science of the Total Environment*, *729*, 139025. https://doi.org/10.1016/j.scitotenv.2020.139025

Yao, Y., Glamoclija, M., Murphy, A., & Gao, Y. (2022). Characterization of microplastics in indoor and ambient air in northern New Jersey. *Environmental Research*, *207*, 112142.

Yuan, W., Liu, X., Wang, W., Di, M., & Wang, J. (2019). Microplastic abundance, distribution and composition in water, sediments, and wild fish from Poyang Lake, China. *Ecotoxicology and Environmental Safety*, *170*, 180–187.

Zalasiewicz, J., Waters, C. N., Sul, J. A. I., Corcoran, P. I., Barnosky, A. D., Cearetta, A., Edgeworth, M., Galuszka, A., Jenadel, C., Leinfelder, R., McNeil, J. R., & Yonan, Y. (2016). The geological cycle of plastics and their use as a stratigraphic indicator of the Anthropocene. *Anthropocene*, *13*, 4–17. https://doi.org/10.1016/j.ancene.2016.01.002

Zhang, F., Wang, Z., Wang, S., Fang, H., & Wang, D. (2019). Aquatic behavior and toxicity of polystyrene nanoplastic particles with different functional groups: Complex roles of pH, dissolved organic carbon and divalent cations. *Chemosphere*, *228*, 195–203. https://doi.org/10.1016/j.chemosphere.2019.04.115

Zhang, L., Liu, J., Xie, Y., Zhong, S., Yang, B., Lu, D., & Zhong, Q. (2020). Distribution of microplastics in surface water and sediments of Qin river in Beibu Gulf, China. *Science of the Total Environment*, *708*, 135176.

Zhao, W., Huang, W., Yin, M., Huang, P., Ding, Y., Ni, X., Xia, H., Liu, H., Wang, G., Zheng, H., & Cai, M. (2020). Tributary inflows enhance the microplastic load in the estuary: A case from the Qiantang River. *The Marine Pollution Bulletin*, *156*, 111152.

Zhou, A., Zhang, Y., Xie, S., Chen, Y., Li, X., Wang, J., & Zou, J. (2020). Microplastics and their potential effects on the aquaculture systems: A critical review. *Reviews in Aquaculture*, *13*(1), 719–733.

Zhu, C., Kanaya, Y., Nakajima, R., Tsuchiya, M., Nomaki, H., Kitahashi, T., & Fujikura, K. (2020). Characterization of microplastics on filter substrates based on hyperspectral imaging: Laboratory assessments. *Environmental Pollution*, *263*. https://doi.org/10.1016/j.envpol.2020.114296

Zhu, M., Kurniawan, T. A., Liang, D., Song, Y., Hermanowicz, S. W., & Othman, M. H. D. (2022). Advances in BiO*X*-based ternary photocatalysts for water treatment and energy storage applications: A critical review. *Reviews in Environmental Science and Biotechnology*, *21*, 331–370. https://doi.org/10.1007/s11157-022-09617-0

Part IV

Terrestrial Biodiversity Conservation

13 Community Ecology for Disentangling the Causes and Consequences of Biodiversity

Keita Nishizawa, Xinyu Xu, Dai Saito, and Akira S. Mori

13.1 INTRODUCTION

Biodiversity underpins ecosystem functioning and the ecosystem services that support human well-being. Yet biodiversity is currently declining at an unprecedented rate due to drivers such as deforestation, land-use change, climate change, pollution, and biological invasions (IPBES, 2019). Biodiversity loss is not driven solely by direct destruction or overexploitation. In many regions, rural depopulation and the abandonment of traditional land management are becoming major contributors to biodiversity decline. In East and Southeast Asia, for example, traditionally managed landscapes such as satoyama agricultural–forest mosaics historically supported high levels of biodiversity through practices such as coppicing, controlled burning, grazing, and seasonal harvesting. These practices created heterogeneous habitat structures that allowed diverse species to coexist (Katayama et al., 2014; Takeuchi, 2010). As these practices disappear, many species that depend on such fine-scale habitat diversity are rapidly declining. Importantly, this example shows that human activities can both erode biodiversity and enhance it, depending on how landscapes are managed.

This perspective highlights a critical challenge for conservation and Nature-based Solutions (NbS): It is not biodiversity itself, but how biodiversity is generated, maintained, and structured in real ecosystems that determines ecosystem functioning and service delivery. Research on biodiversity–ecosystem functioning (BEF) has shown that higher biodiversity can enhance ecosystem functioning, stability, and resilience (Cardinale et al., 2012; Loreau et al., 2001; Tilman et al., 1997), and these findings provide a scientific foundation for NbS. Yet BEF research alone is insufficient to guide practical management, because it typically does not address which species can coexist under real environmental conditions, how community composition changes through time and space, or how management actions influence community outcomes. In other words, BEF often tells us that "more diversity is better," but not how to design or sustain such diversity in real landscapes.

To translate BEF insights into effective conservation and management, we must understand the mechanisms that assemble ecological communities (Mori et al., 2018, Figure 13.1). This is the domain of community ecology, which provides a mechanistic framework for linking the generation (causes) and functioning (consequences) of biodiversity.

In this chapter, we first review the ecological processes that generate biodiversity—how communities are assembled—and then examine the consequences of biodiversity for ecosystem functioning and human well-being. Finally, we explore how integrating these perspectives can inform sustainable ecosystem management in rapidly changing socio-ecological landscapes. While the concepts discussed are broadly applicable across ecosystems, this chapter primarily draws on examples from terrestrial systems.

DOI: 10.1201/9781003640899-17

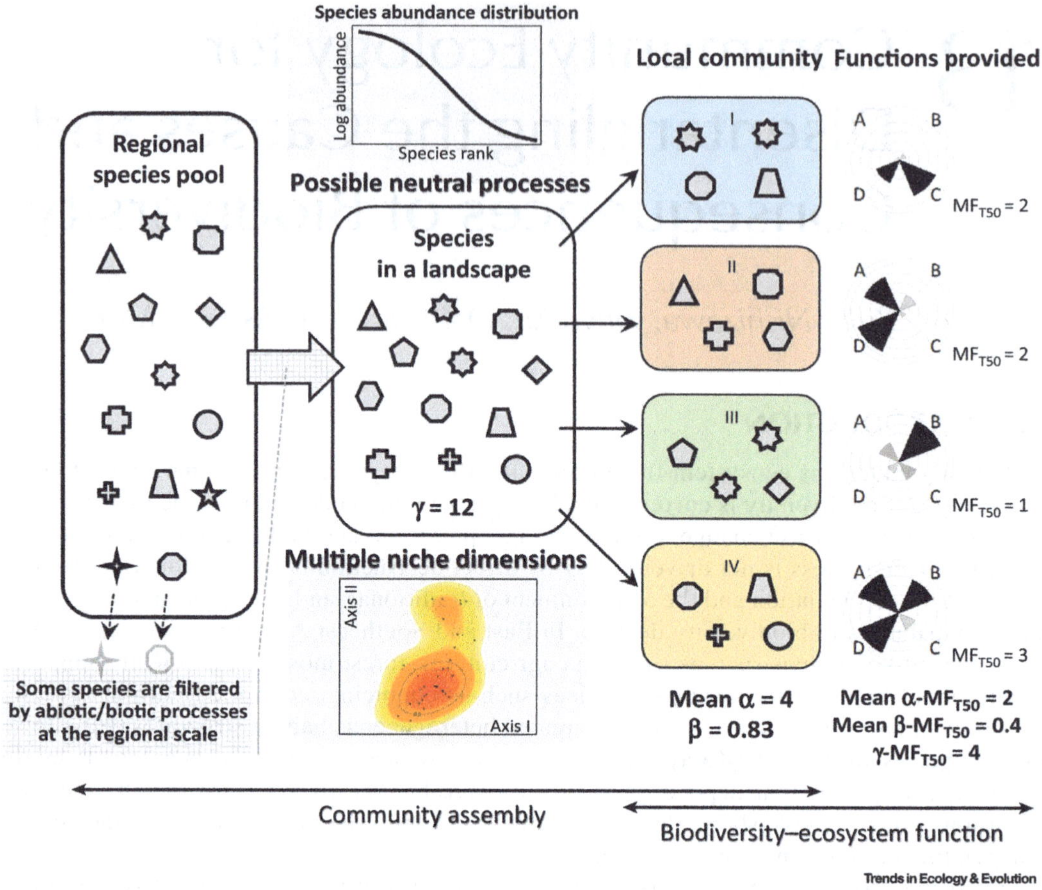

FIGURE 13.1 Schematic illustration of key processes of community assembly and biodiversity–ecosystem functioning operating at different spatial scales. Each icon represents different species, and those shown in gray outline are lost from the meta-community of a landscape. The number of species in each local community, and in the landscape containing these communities, constitutes α- and γ-diversity, respectively. β-Diversity is calculated based on the Sørensen index. To illustrate meta-community structure, the species abundance distribution and kernel density estimation of the trait spectrum are shown. For the latter, contour lines indicate 0.5 and 0.95 quantiles of the occurrence probability of traits. Spatial variation in local environmental conditions is shown by different colors (blue, red, green, and yellow) for the background of local communities [I–IV]. Evaluation of individual functions (A–D) is based on the threshold approach. Levels of individual functions are shown with radial bar diagrams. (Figure from Mori et al. (2018).)

13.2 CAUSE OF BIODIVERSITY

13.2.1 Why Are Living Organisms So Diverse?

The origins of biodiversity lie in a complex interplay of processes that operate across multiple spatial and temporal scales, from global climatic patterns to highly localized interactions, and even to pure chance.

At the broad scale, **climate** is among the primary drivers of biodiversity. Climatic differences across the globe give rise to distinct ecosystems, or biomes, such as tropical rainforests, savannas, deserts, and boreal coniferous forests (Whittaker, 1975). Not only do species compositions differ among regions, but there is also a pronounced gradient in species richness associated with climatic conditions. For instance, the well-known latitudinal diversity gradient shows that species richness generally decreases from the equator toward the poles (Gaston, 2000; Hillebrand, 2004; Willig et al., 2003). In hot and humid tropical rainforests, thousands of tree species coexist within a single landscape, whereas in the cold and dry Arctic, only a few species that are adapted to the environment are able to persist.

Geographic isolation plays a decisive role as a driver of biodiversity. Even under similar climates, separation by oceans, deserts, or mountain ranges can lead to independent evolutionary trajectories, resulting in distinct regional biotas. A classic example is the divergence in tree species composition between the Amazonian and African rainforests, where similar climatic conditions have not prevented extensive taxonomic turnover (Wallace, 1876; Wiens & Donoghue, 2004). Likewise, for many taxa such as oaks, beeches, and birches—though not limited to these—closely related species with similar yet distinct traits are distributed across continents or regions (e.g., Europe, North America, and East Asia). This pattern reflects speciation driven by geographic isolation over evolutionary timescales.

Species differ in their propensity for geographic speciation depending on their biological traits. Taxa with low dispersal ability, such as many terrestrial invertebrates, amphibians, or understory plants, are generally more susceptible to isolation, as physical barriers like mountains or water bodies strongly limit gene flow (Avise, 2000; Bohonak, 1999; Coyne & Orr, 2004). Unlike highly mobile taxa such as birds or wind-dispersed plants, they are more likely to diverge when geographically separated (Claramunt et al., 2012; Jetz et al., 2012).

Biotic interactions also modulate the likelihood of geographic speciation. Taxa that depend strongly on specific mutualistic partners—such as plants with specialized pollinators—often show elevated levels of regional endemism. The coevolution between flowering plants and their pollinators can restrict gene flow when pollinator assemblages differ geographically, ultimately leading to reproductive isolation and speciation (Kay & Sargent, 2009; Schiestl & Johnson, 2013). The presence or absence of particular taxonomic groups can have cascading evolutionary and ecological consequences. For instance, whether certain key taxa—such as large herbivores, pollinators, or seed dispersers—occur in a region can influence the trajectory of community assembly and trait evolution across multiple trophic levels. A particularly interesting example comes from alpine ecosystems. Although the alpine zones of Japan and New Zealand share comparable climatic conditions, their floral color patterns differ sharply. Japanese alpine floras contain many red and purple flowers, whereas New Zealand's alpine flora is dominated by white flowers. This contrast reflects differences in pollinator vision and the geographic isolation of pollinator communities—bumblebees (*Bombus* spp.), which preferentially visit red and purple flowers, are abundant in Japan but absent from New Zealand's native fauna (Ishii et al., 2019). The long-term absence of key pollinators has constrained the evolution of floral traits, illustrating how the regional absence of particular taxa can have cascading effects on community composition and trait evolution.

Yet biodiversity is not determined by macro-scale factors alone. Even within the same climate zone or region, differences abound at fine scales. These local factors include soil heterogeneity, microclimatic variation, biotic interactions such as competition, facilitation, or mutualism, and disturbance, with varying intensities and combinations of these factors generating sharp contrasts in species composition even across short distances. In addition to these deterministic processes linked to species' traits, the accumulation of stochastic events in nature cannot be ignored. Which seeds happen to arrive at a site, which individuals persist or disappear, and the historical sequence of species colonization can all lead to distinct community outcomes even under identical environmental conditions (Hubbell, 2001; Vellend, 2010).

The combination of deterministic and stochastic processes at local scales generates striking variation in community composition even across short distances. This interplay forms one of the most fundamental bases of biodiversity in the world.

Thus, to understand how biodiversity is generated and maintained, it is crucial to consider not only climatic and regional differences but also the processes that create fine-scale local variation.

This represents the central question of community ecology and defines the core focus of the field known as "community assembly".

13.2.2 Community Assembly

Understanding how the composition of biological communities—species diversity, species distributions, and relative abundances—is determined is a central question in community ecology and one of the oldest inquiries in the field. In plant communities, in particular, the diversity of species combinations is remarkable. For instance, if a regional species pool contains 100 species, the number of possible ways to select 20 species for a 1 m^2 plot is roughly on the order of 10^{20}. Of course, community assembly in nature is not completely random, so the actual number of realized combinations is much smaller—but this still illustrates the enormous potential diversity of local communities, and uncovering the mechanisms behind the formation of different local communities is the central goal of community assembly research.

Historically, a classical approach to explaining community formation is the deterministic assembly based on niches. In this view, species are selected according to their ecological traits and interactions with the environment and other species, resulting in a predictable community structure. Deterministic processes suggest that species with traits best suited to local environmental conditions will dominate. For example, plants partition resources such as soil moisture and light, and competitive exclusion can lead to dominance of particular species in local plots (Chesson, 2000; Tilman, 1982). Diamond (1975) famously observed a checkerboard pattern in the distribution of closely related birds across islands, illustrating that interspecific competition can shape community composition. Deterministic, niche-based processes thus highlight the importance of environmental filtering and species interactions in structuring communities.

However, real communities are often too diverse to be fully explained by deterministic mechanisms alone. In many cases, weakly competitive species persist locally, maintaining high levels of diversity. To account for such patterns, stochastic processes have been introduced. Connor and Simberloff (1979) demonstrated that the checkerboard patterns could arise even when species were randomly distributed, indicating that deterministic explanations are insufficient. Furthermore, Hubbell's (2001) unified neutral theory shows that assuming all species are ecologically equivalent can still reproduce observed species abundance distributions and spatial patterns. Neutral theory emphasizes ecological drift, in which stochastic events, such as random deaths and colonization probabilities, shape community dynamics. These processes related to the stochasticity also play a major role in generating and maintaining community structure.

In this regard, priority effects and historical contingencies influence community formation. Early-arriving species may preempt resources, limiting the establishment of later arrivals, so that even under identical environmental conditions, different community compositions can emerge (Fukami, 2015). Such historical and sequence-dependent effects interact with deterministic and stochastic processes to shape local community diversity and structure.

In conclusion, local communities are generated by the overlay of deterministic processes driven by species traits and environmental conditions, stochastic processes governed by chance, and historical contingencies. Understanding the interplay of these factors is essential for explaining why highly diverse local communities arise from the same regional species pool and for elucidating the mechanisms of community assembly in nature.

13.2.3 Methods for Quantifying Community Assembly

Understanding the mechanisms that shape the composition of biological communities, including species diversity, species distributions, and relative abundances, is a central question in community ecology. A central goal of community assembly research is therefore to quantify the relative contributions of deterministic of stochastic processes to observed patterns of biodiversity. Over the past decades, a suite of methodological approaches has been developed to address this challenge, each providing complementary insights into how local communities are structured.

13.2.3.1 Null Model Analysis

Null model analysis is a foundational approach in community ecology for inferring the rules that govern local community assembly. The core idea is to evaluate how much observed community patterns deviate from what would be expected under purely random assembly. This approach allows β-diversity (site-to-site differences in species composition)—which is otherwise strongly influenced by differences in α-diversity (local species richness within sites) and γ-diversity (total species richness across all sites)—to be standardized, making its values more interpretable and comparable across sites or regions (Mori et al., 2014).

A common application of null models in among-community comparisons is the analysis of **β-diversity**. Raw β-diversity is strongly influenced by differences in α- and γ-diversity among sites, which makes direct comparisons across regions difficult. Null models address this by standardizing β-diversity: species occurrences (or abundances) are randomly shuffled across sites according to an algorithm that preserves particular features of the data (for example, the species richness of each site and/or the overall frequency of each species). Repeating this randomization many times generates a null distribution of expected community dissimilarities. From this null distribution, **β-deviation** can be calculated as either the difference between observed and mean expected β (obs − mean(null)) or as a standardized effect size (SES = (obs − mean(null))/sd(null)).

A positive β-deviation indicates that communities are more dissimilar (i.e., show higher turnover) than expected by chance (overdispersion). This pattern suggests that processes accentuating differences among sites such as strong environmental contrasts, dispersal limitation, or priority effects are important. Conversely, a negative β-deviation indicates that communities are more similar (i.e., show lower turnover) than expected by chance (clustering), which can arise from homogenizing forces such as strong regionally consistent environmental filtering, high connectivity and mass effects, or widespread disturbance (Mori et al., 2014). Importantly, β-deviation itself does not uniquely identify a single process; rather, it indicates whether the net effect of assembly processes leads to greater or lesser turnover than a specified random expectation.

Null model approaches are not restricted to taxonomic β-diversity. Functional diversity (FD) and phylogenetic diversity (PD) can be analyzed within similar frameworks. For FD, metrics such as functional richness, functional dispersion, or functional divergence are compared with null expectations obtained by randomizing species identities or trait values, while holding species richness or abundance constant. This allows us to test whether co-occurring species are more functionally similar (functional clustering) or more functionally distinct (functional overdispersion) than expected by chance, thereby providing insight into the roles of environmental filtering and limiting similarity, that is, the constraint on species coexistence imposed by strong niche overlap and competition (MacArthur & Levins, 1967). Likewise, PD metrics such as mean pairwise distance or mean nearest taxon distance can be evaluated against null distributions to detect phylogenetic clustering (co-occurring species more closely related than expected) or phylogenetic overdispersion (co-occurring species more distantly related than expected).

Recent methodological advances have further strengthened null model approaches. SES allow for the comparison of observed versus expected metrics across communities with differing species pools, improving comparability. SES values near zero indicate that observed patterns are consistent with random assembly under the chosen null model, whereas strongly positive or negative SES values indicate overdispersion or clustering, respectively (Spasojevic and Suding, 2012; Webb et al., 2002). The same terminology—clustering vs. overdispersion—is often applied to within-community (α) structure in trait or phylogenetic space, while β-deviation explicitly describes between-community patterns (turnover vs. homogenization).

Despite their power, null models must be used with care. Different randomization algorithms implement different ecological assumptions (e.g., whether species' overall frequencies are fixed, whether site richness is fixed, whether traits or phylogenetic relationships are preserved), and the choice of null model directly influences the interpretation of clustering or overdispersion. Moreover, non-random patterns detected by null models usually reflect the combined effects of multiple processes rather than a single cause.

In summary, null model analysis provides a rigorous statistical framework for detecting non-random patterns in species composition, traits, and phylogenetic relationships. By comparing observed communities with assemblies generated under explicit random expectations, ecologists can infer whether community structure is consistent with environmental filtering, species interactions, dispersal processes, or largely stochastic assembly—and can do so in a way that is comparable across sites, regions, and studies.

13.2.3.2 Variation Partitioning

Variation Partitioning (VP) is a widely used statistical approach in community ecology that aims to quantify the relative contributions of environmental, spatial, and sometimes biotic factors in shaping community composition. The method decomposes the total variation in species composition into distinct fractions attributable to observed environmental variables, spatial structure, and their shared effects (Legendre et al., 2005). By doing so, VP allows us to separate deterministic (niche-based) processes, such as environmental filtering, from stochastic processes related to dispersal limitation or unobserved drivers.

In practice, VP is commonly implemented through multivariate ordination techniques, such as Redundancy Analysis (RDA) or Canonical Correspondence Analysis. Environmental variables (e.g., soil properties, climate, topography) are included as explanatory matrices, while spatial structure is represented using methods such as Principal Coordinates of Neighbor Matrices or Moran's Eigenvector Maps, which capture spatial patterns across scales (Borcard et al., 1992). Variation in community composition is then partitioned into four fractions: the pure environmental fraction, the pure spatial fraction, the shared fraction, and the residual unexplained fraction.

The pure environmental fraction reflects the deterministic influence of measured abiotic variables on community assembly. This component captures niche-based, deterministic processes in which species distributions are explained by suitable environmental conditions, as species maximize their reproductive success and outcompete others only under specific conditions (environmental filtering). In contrast, the pure spatial fraction represents spatial structure in community composition that is independent of measured environmental gradients. Such patterns are consistent with spatially structured stochastic processes, including dispersal limitation, priority effects, and ecological drift. These processes—often emphasized in neutral theory—can generate conspecific aggregation and distance decay in community similarity through chance colonization, stochastic demographic fluctuations, and limited dispersal, even among environmentally similar sites. The shared fraction captures variation that could be attributed to either environmental or spatial effects, highlighting the difficulty of disentangling correlated drivers. Finally, the residual fraction represents unexplained variation, which may arise from unmeasured environmental factors, fine-scale stochasticity, or historical contingencies (Legendre et al., 2005).

Not only for the environment and space, but this method can also be widely applied to other factors that explain the differences in community compositions, such as climate, land use, bed rocks, and host species for the fungi (Ji et al., 2024; Matsuoka et al. 2016)

Recent methodological advances have improved the robustness of VP analyses. Techniques such as partial RDAs with permutation testing allow for formal statistical assessment of each fraction's significance, while approaches integrating phylogenetic or functional traits extend VP beyond species identity to evolutionary and functional dimensions of community assembly (Villéger et al., 2017). Furthermore, multiscale spatial modeling now enables ecologists to partition variation across hierarchical spatial scales, enhancing our understanding of how local and regional processes jointly shape biodiversity patterns.

In summary, Variation Partitioning provides a quantitative framework for disentangling the roles of environmental determinism, spatially structured stochasticity, and unexplained variation in community assembly. By explicitly separating these components, VP helps elucidate the balance between niche-based and dispersal-driven processes, offering mechanistic insight into the formation of local communities across spatial scales.

13.2.4 Known Patterns in Community Assembly

Community assembly is highly context-dependent, and observed patterns can vary even across neighboring sites. For instance, the presence or absence of large herbivores such as deer can dramatically alter local plant composition and the underlying assembly processes—being more deterministic in the absence of deer and more stochastic when deer are present (Nishizawa et al., 2016). Despite this variability, recent research has begun to identify general trends that provide valuable insights into the relative importance of deterministic and stochastic processes across different scales and environments.

13.2.4.1 Scale Dependence

The relative influence of deterministic and stochastic processes on community assembly is strongly scale-dependent and represents one of the few areas in ecology where general patterns have been consistently observed (Chase & Myers, 2011; Vellend, 2010). Understanding these scale-dependent effects is essential for linking local assembly rules to regional biodiversity patterns.

The effects of spatial scale on community assembly processes (i.e., relative influence of deterministic versus stochastic processes) are now widely recognized as a critical factor, and indeed represent one of the few areas in ecology where general patterns have been consistently observed. It is widely recognized that the relative importance of processes structuring ecological communities changes with spatial scale (Chase & Myers, 2011). This scale dependence is a key consideration when asking whether deterministic processes (such as environmental filtering and species interactions) or stochastic processes (such as dispersal limitation and ecological drift) more strongly govern community structure.

A growing body of work shows a general tendency for stochastic processes to dominate at small (local) spatial scales, while deterministic processes become increasingly important as spatial extent increases (Chase, 2014; Garzon-Lopez et al., 2014; Viana & Chase, 2019). At small scales, where environmental conditions are relatively homogeneous, demographic stochasticity and ecological drift can play a dominant role in explaining community compositional differences between sites, leading to the idea that local communities may appear to be structured in a "neutral" way. In contrast, as spatial extent increases, environmental heterogeneity (in climate, topography, soils, vegetation structure, etc.) also increases. This leads to clearer differences in species' distributions and adaptations along environmental gradients, such that community composition changes systematically with the underlying environment and the relative contribution of deterministic processes, particularly environmental filtering, becomes greater (Chase, 2014; Garzon-Lopez et al., 2014).

Analyses of large-scale plant community datasets spanning, for example, tropical to boreal regions have shown that the explanatory power of environmental variables tends to increase with spatial extent. Nishizawa et al. (2022), using a meta-analysis that integrated plant community data from sites worldwide, demonstrated that this scale-dependent pattern has global generality. In their analysis, the proportion of variation in community structure explained by environmental variables increased as the spatial extent of observation widened, whereas the contribution of the spatial component (spatial autocorrelation) tended to decrease (Figure 13.2). These findings suggest that, at broad spatial scales, environmental filtering becomes more influential than dispersal limitation in shaping community composition.

Thus, understanding community assembly requires moving beyond analyses at a single spatial scale and explicitly comparing patterns across multiple scales (local–regional–continental).

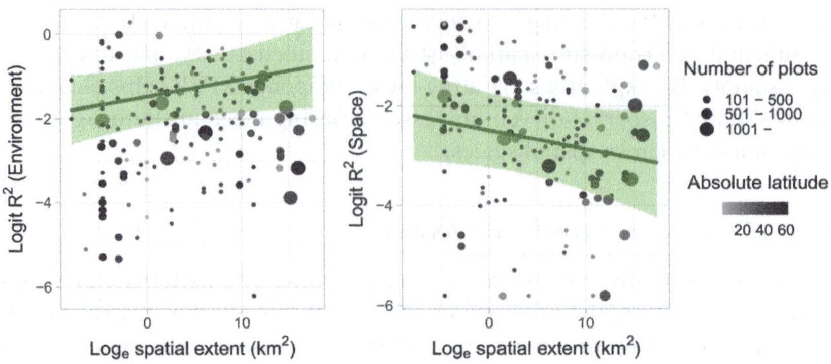

FIGURE 13.2 (a) Relationships between spatial extent (the area [km^2] covered by each study in log scale) and explanatory power for species composition (R^2 value) of environment, space, and spatially structured environment. The regression lines and 95% credible intervals were drawn based on the posterior distributions of the coefficient with the other parameter values fixed as their medians. The size and color of the points represent the number of plots and the absolute latitude of the raw results, respectively. (Figure modified from Nishizawa et al. (2022).)

Even if stochastic factors dominate at local scales, deterministic factors often prevail at broader scales that encompass a greater range of environments, and the superposition of these processes generates the scale-dependent patterns of biodiversity that we observe (Chase, 2014). In this sense, the "scale issue" is now widely recognized as a central topic in community ecology, and spatial scale is one of the relatively few dimensions along which general patterns in community assembly have been consistently identified. Scale dependence is therefore an essential element to consider when evaluating the relative roles of deterministic and stochastic processes in shaping community structure.

13.2.4.2 Environmental Gradients

Environmental harshness is a key factor shaping the balance between deterministic and stochastic assembly processes. In general, the more severe the environment, the more strongly community composition is constrained by abiotic conditions, so that only species with particular traits can persist. Where such constraints are weaker, chance events, dispersal, and species interactions have greater scope to influence which species establish, making assembly more stochastic.

In harsh or extreme environments—such as alpine, arctic, or desert ecosystems—strong abiotic filtering often constrains the functional traits and phylogenetic composition of communities, producing pronounced clustering. Only species that tolerate low temperatures, drought, nutrient limitation, or other stresses can survive, so communities tend to be composed of functionally or phylogenetically similar species. By contrast, in more benign or resource-rich environments, where many species can potentially persist, interspecific interactions and stochastic colonization can lead to greater overdispersion in trait or phylogenetic space and higher variation among communities (Kraft & Ackerly, 2010; Weiher et al., 2011). These patterns illustrate how environmental context modulates the relative importance of niche-based versus stochastic processes.

Latitudinal patterns in community assembly also reflect the interaction between environmental conditions and regional species pools. In tropical regions, high species richness and relatively stable climates favor strong niche differentiation, often producing trait or phylogenetic overdispersion within local communities. In contrast, at higher latitudes, where species pools are smaller and environmental conditions are more variable and seasonal, deterministic processes such as dispersal limitation, ecological drift, and historical contingencies can exert a greater influence on community structure (Jabot et al., 2020; Kraft et al., 2011). These trends highlight how deterministic and stochastic drivers interact along global environmental gradients.

13.2.4.4 Biotic Context and Species Absences

Beyond abiotic factors, the biotic context—the presence, abundance, and interaction strengths of other organisms—can have cascading effects on community assembly. Certain "key taxa," such as large herbivores, apex predators, dominant pollinators, or ecosystem engineers, can strongly modify local environmental conditions and interaction networks. Where these biotic drivers are consistently present and exert strong effects, they impose additional deterministic filters on community composition. Conversely, when such taxa are patchily distributed, declining, or locally absent, assembly outcomes can become more contingent and stochastic, as similar abiotic conditions give rise to different community states depending on which interactions are realized (Paine, 1969; Schmitz et al., 2015).

Herbivores provide a clear example. High densities of large herbivores, such as deer, can selectively remove palatable species, alter light and soil conditions, and suppress regeneration of particular plant functional types (Bugalho et al., 2013; Milne-Rostkowska et al., 2020; Nishizawa et al., 2016). This generates more deterministic assembly patterns, with communities converging toward herbivore-tolerant or chemically defended species (Alberti et al., 2017). Where herbivore presence is spatially heterogeneous—due to hunting, fencing, or landscape configuration—neighboring sites with similar abiotic environments may nonetheless diverge strongly in composition, reflecting differences in browsing pressure rather than underlying environmental gradients (Hodapp et al., 2018). Similarly, dominant pollinators or seed dispersers can structure plant communities by preferentially interacting with certain floral traits or fruit types (Kemp et al., 2019; Koski, 2023; Sargent & Ackerly, 2008). For instance, regional absence of specific pollinators can constrain the evolution of floral traits and limit which plant species can establish and reproduce successfully (Dawson-Glass & Hargreaves, 2022; Kriebel et al., 2020). In such cases, the "missing" interaction partner functions as an invisible filter, shaping community assembly through species absences rather than presences.

Defaunation and functional extinction further illustrate the importance of biotic context (Dirzo et al., 2014; Fricke et al., 2022, 2025; Galetti et al., 2013). Even when a species persists at low density, its ecological role—for example, as a top predator, large frugivore, or habitat-forming engineer—may be effectively lost. The removal or weakening of such roles can relax previously strong deterministic filters, allowing more opportunistic or disturbance-adapted species to proliferate and increasing the apparent importance of stochastic colonization and drift (Albert et al., 2015; Rogers et al., 2021; Vonesh et al., 2009; Wotton & Kelly, 2011). These biotically driven changes often interact with spatial structure and environmental heterogeneity, creating complex, scale-dependent outcomes in community assembly. Thus, understanding the presence, absence, and functional strength of key taxa is essential for interpreting patterns of clustering or overdispersion and for predicting how communities respond to management interventions such as rewilding, herbivore control, or pollinator restoration.

13.3 CONSEQUENCE OF BIODIVERSITY

This section examines the consequences of biodiversity for ecosystem functioning, ecosystem services, and ultimately human well-being. We begin by reviewing key findings from BEF research, focusing on the empirical patterns and the mechanisms. We then highlight the remaining research gap from a management perspective: BEF studies often reveal that "more diversity is better," but they typically do not clarify which species combinations sustain ecosystem functioning, under what environmental or spatial conditions, or how management actions shape community outcomes. Finally, we discuss how insights from community ecology can address this gap by explaining how species assemble and coexist, thereby linking the causes and consequences of biodiversity and informing the design of resilient and functional ecosystems.

13.3.1 Biodiversity–Ecosystem Functioning Relationships

Understanding how biodiversity influences ecosystem functioning has become one of the central themes of modern ecology. Here, "ecosystem functioning" encompasses processes such as primary productivity, decomposition and nutrient cycling, carbon storage, and resilience to disturbances. Since the 1990s, numerous experiments in grasslands, forests, and other ecosystems have demonstrated that plant species richness often has a positive effect on ecosystem functions such as productivity, ecosystem stability, and nutrient cycling (Cardinale et al., 2012; Loreau et al., 2001; Tilman et al., 1997). Communities with higher species richness tend to exhibit greater average biomass or productivity than monocultures, and they often show reduced interannual variability and faster recovery from perturbations such as drought.

Research on BEF can be broadly divided into experimental studies and observational studies under natural conditions. Experimental studies typically manipulate species richness and species composition in small to medium-sized plots and then monitor ecosystem functions through time. This approach has the major advantage of allowing researchers to isolate the effect of biodiversity itself while controlling, as much as possible, for other factors. However, experimental settings are necessarily simplified compared to real landscapes, and experimental durations are often relatively short, so care is needed when generalizing the results.

By contrast, observational studies investigate BEF relationships in naturally assembled communities and landscapes, such as forest stands, agricultural mosaics, or river catchments (van der Plas, 2019). These studies analyze how variation in species richness or functional diversity (FD) correlates with variables such as biomass, carbon stocks, pollination services, or soil fertility under realistic land-use and management conditions (Cao et al., 2025; Fan et al., 2023; Soliveres et al., 2016). The strength of this approach is that it reflects "real-world" ecological and socio-economic contexts, including historical land-use, disturbance regimes, and evolutionary legacies. At the same time, many covarying factors—such as climate, soils, and management practices—can influence both biodiversity and ecosystem functioning, making causal inference more challenging than in controlled experiments.

Syntheses that combine results from both experimental and observational studies generally support a common conclusion: higher biodiversity tends to increase average levels of ecosystem functioning and to stabilize those functions through time (Cardinale et al., 2012; Loreau et al., 2023). This empirical pattern provides the starting point for examining the mechanisms that link biodiversity to ecosystem functioning.

13.3.2 Underlying Mechanisms

Why does biodiversity enhance and stabilize ecosystem functioning? BEF research has identified several key mechanisms, three of which are particularly prominent (Isbell et al., 2011; Loreau et al., 2023; Loreau & Hector, 2001; Naeem et al., 2009). Importantly, Loreau and Hector (2001) developed a statistical framework to partition net biodiversity effects into complementarity and selection components (also see Isbell et al. 2018). Although these components are not fully identical to genuine mechanisms (Loreau et al. 2012), many BEF experiments have applied this approach to disentangle the potential underlying mechanisms of the diversity effects.

- **Complementarity effect**: Complementarity effects arise when species differ in their resource use or functional traits in ways that allow them to collectively exploit resources more completely or efficiently than any single species could alone (Hector, 1998; Loreau, 1998). For example, plant species that differ in rooting depth can access water and nutrients from different soil layers, increasing overall resource uptake when they coexist (Berendse, 1979; de Kroon et al., 2012; Mueller et al., 2013). Similarly, tree species that differ in shade tolerance or canopy architecture can stratify light use vertically, increasing the total photosynthetic capacity of a forest stand (Canham, 1989; H. Ishii & Asano, 2010). Such functional differentiation and niche complementarity can lead to higher productivity, faster nutrient cycling, and more efficient resource use in diverse communities (Hardiman et al., 2011; LaRue et al., 2023).

- **Selection or dominance effect**: Selection (or dominance) effects occur when diverse mixtures are more likely to include one or a few species that have exceptionally high contributions to ecosystem functioning. In such cases, the presence and dominance of these "driver" or "high-performing" species elevate the overall level of ecosystem functioning (Fargione et al., 2007; Loreau & Hector, 2001). For example, in forest ecosystems, the presence of a highly productive or dominant tree species can disproportionately increase stand-level carbon storage or productivity. As species richness increases, the probability of including such species also increases (Huston, 1997; Tilman et al., 1997).
- **Stability mechanisms**: For the temporal stability of ecosystem functioning, asynchronous species responses to environmental variability are especially important (Feng et al. 2025, Gonzalez & Loreau, 2009). Different species often respond differently in timing and magnitude to fluctuations in climate, disturbances, or resource availability. As a result, a year that is unfavorable for some species may be favorable for others, so that declines in one subset of species are compensated for by increases in another. This "portfolio effect" (Tilman et al., 1998) or "insurance effect" (Yachi & Loreau, 1999) leads to more stable aggregate ecosystem properties over time. Higher biodiversity increases the potential for such compensatory dynamics, thereby enhancing the long-term stability of ecosystem functioning.

These mechanisms are not mutually exclusive. In real ecosystems, complementarity, selection, and stability mechanisms typically operate simultaneously, and their relative importance can vary with environmental conditions, temporal scale, and spatial scale. Therefore, a comprehensive understanding of BEF relationships requires integrating evidence from experiments and observations across short and long timescales and from local to regional spatial scales.

13.3.3 Methods for Quantifying the Mechanisms That Link Biodiversity to Ecosystem Functioning

This section reviews methodological approaches for quantifying how biodiversity influences ecosystem functioning. Two complementary families of methods are emphasized: (i) classic additive partitioning, which decomposes net biodiversity effects into complementarity (CE) and selection (SE) components; and (ii) more recent trait-based approaches, which link community assembly and the shape of community trait distributions to functional outcomes to support more mechanistic inference.

13.3.3.1 Additive Partitioning Approach

Additive partitioning (Loreau & Hector, 2001) is a classic framework for separating the net biodiversity effect on plot-level ecosystem functions (e.g., productivity, nutrient retention, decomposition, carbon storage) into two interpretable components using commonly available data from biodiversity experiments—the value of the function in mixtures and in the monoculture of each species. The complementarity effect (CE) is the species-averaged excess of relative performance in mixture over its expectation derived from monocultures, whereas the selection (dominance) effect (SE) is quantified as the covariance between species' monoculture performance and their deviations in mixture—positive when inherently high performers also gain disproportionately in mixtures, negative when they do not. As an illustrative example of this framework's contribution, a recent global meta-analysis of 452 experiments reported an average +15.2% increase in productivity in mixtures (mean richness 2.6), with CE and SE accounting for ~65.6% and ~34.4% of the net effect, respectively; CE was stronger in grassland and forest than in container, cropland, or aquatic systems, increased with PD, with mixtures of N-fixing and non-N-fixing species, and with the FD of leaf N, and tended to increase over time while SE decreased (Chen et al., 2025).

Importantly, CE and SE are statistical summaries of community outcomes, not direct identifications of the underlying biological pathways. For example, what is captured as "complementarity" can arise from resource partitioning, abiotic facilitation, or biotic feedbacks; the additive partitioning approach does not distinguish among these mechanisms (Barry et al., 2019). Thus, while additive partitioning has made a major contribution as a comparative diagnostic across hundreds of experiments, more mechanistic, trait-based approaches (next subsection) are needed to determine which processes generated the observed patterns. In parallel, several studies have extended additive partitioning to resolve mechanisms more finely and to evaluate biodiversity effects across broader spatiotemporal scales. Fox (2005) proposed a further partitioning of the classic selection effect (SE) into a dominance effect (a selection-like component analogous to natural selection, when species with particular traits dominate at the expense of others) and a trait-dependent complementarity effect (when species with particular traits perform better than expected, but not at the expense of others). More recently, Tatsumi & Loreau (2023) decomposed the net, CE, and SE components into additive size vs. density contributions to understand plant diversity-productivity relationships, revealing contrasting demographic pathways: in an annual plant experiment, overyielding was largely explained by selection via increased individual size of highly productive species, whereas in a tree experiment the cause of overyielding shifted over time from enhanced growth (size) to reduced mortality via complementary canopy use (density). At broader spatiotemporal scales, Isbell et al. (2018) generalized the classic additive partitioning approach to quantify insurance effects across time and space due to β-diversity (species turnover), showing in an 18-year grassland experiment that these large-scale insurance effects were positive and substantial (≈ 19% of the net biodiversity effect), driven mainly by temporal asynchrony. Together, these developments illustrate viable paths to relate CE/SE diagnostics more directly to mechanisms and to evaluate biodiversity effects beyond single plots and time steps.

Additive partitioning is also difficult to apply directly to natural communities, where the required monoculture baselines are rarely observable. To bridge this gap, Mori (2018) identified functionally important species in observational data and used their contributions as diagnostics analogous to SE. Using a dataset of natural tree communities spanning four biomes from subtropical to hemiboreal forests in Japan, the study quantified, for each local community, "important-species richness"—the number of species that significantly increased local productivity—and found it to be positively associated with productivity. After statistically accounting for this species-performance effect, the residual diversity signal was weak: species richness no longer explained the remaining variation in productivity, supporting the view that higher tree diversity increased productivity primarily via a selection pathway—greater probability of including high-performance species and their dominance at higher richness (Figure 13.3). In a supplementary analysis using biomass data from the Jena Experiment—a classic biodiversity experiment with monocultures and mixtures—the same metric (important-species richness) correlated with SE estimated from the experimental design. Crucially, the portion of productivity variation not explained by high-performance species was interpreted as arising from other biological interactions—most plausibly niche complementarity and facilitation among coexisting species. More broadly, this kind of flexible reasoning—linking observational diagnostics of functionally-important-species effects to experimental SE estimates—underscores how triangulating across approaches can advance the field: CE/SE should be treated as statistical lenses rather than endpoints, and explicitly relating alternative diagnostics to underlying biological processes is essential for extending BEF insights to natural communities.

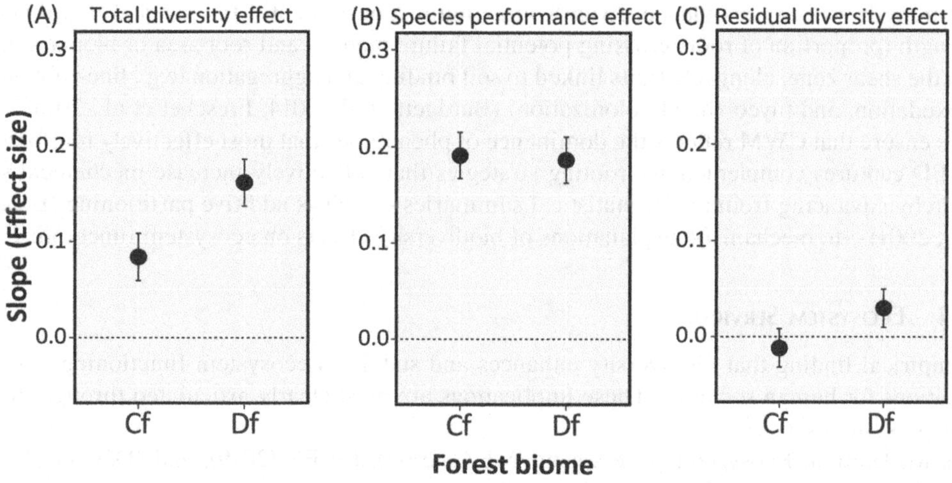

FIGURE 13.3 Standardized effects (*z* scores) of species richness on above-ground productivity (AGP) at the local community scale (0.01 ha). The effect of total basal area within each community was simultaneously considered in the models. (A) Total diversity effect, (B) species performance effect (based on functionally important species richness; Fig. S1), and (C) residual diversity effect (variation in AGP not explained by important-species richness). Points represent slope values (effect sizes) estimated from linear mixed-effects models, and error bars indicate 95% confidence intervals. Results are shown separately for two composite forest biomes: Cf (humid subtropical and maritime temperate climates) and Df (continental and hemiboreal climates). Horizontal dashed lines indicate zero effect. (Figure modified from Mori, 2018.)

13.3.3.2 Trait-based Mechanistic Approaches: Viewing CE/SE through the Lens of Community Trait Distributions

Traits govern both organismal performance and species interactions—and thus the selection-type and complementarity-type processes through which biodiversity influences ecosystem functioning (e.g., Cadotte et al., 2011). Trait-based approaches therefore provide a mechanistic understanding of CE/SE by relating ecosystem functioning to community trait distributions. Within these distributions, the community-weighted mean (CWM) and FD have been most intensively tested under clear hypotheses: the mass-ratio hypothesis (Grime, 1998) predicts that functioning tracks dominant trait values (CWM), conceptually aligning with SE; the niche-complementarity hypothesis (Tilman et al., 1996) predicts that greater trait dissimilarity (FD) enhances joint resource capture or facilitation, aligning with CE. Empirically, both components often matter—for example, a meta-analysis of tree diversity experiments showed that both CWM (trait identity of dominant species) and FD (trait dispersion among species) explain higher mixture performance relative to monocultures, indicating the frequent co-occurrence of selection- and complementarity-type processes (Zheng et al., 2021).

Critically, the interpretation of CWM and FD depends on which traits are chosen. If traits are weakly linked to the targeted process, both CWM–function and FD–function relationships are likely to be diluted; conversely, selecting traits that map directly onto resource acquisition, stress tolerance, or interaction axes sharpens interpretation of selection-versus complementarity-type effects. Because traits covary due to biophysical constraints and ecological–evolutionary trade-offs, single-trait predictors rarely generalize across contexts. A practical strategy is to (1) state the specific mechanism to be tested, (2) select traits with a priori links to that mechanism, and (3) account for trait covariation using multivariate reductions (e.g., PCA), multi-trait indices or distribution moments, and allometric relationships (Chacón-Labella et al., 2023). As a concrete example—matching traits to a belowground function—if the focal ecosystem function is soil reinforcement against shallow landslides, aboveground traits are less informative than root-system traits that directly affect shear strength. Priority should be given to maximum rooting depth, the vertical distribution of root length (proportion of roots crossing potential failure planes), and root area or biomass density within the shear zone, alongside traits linked to soil binding and aggregation (e.g., fine-root production, exudation, and mycorrhizal colonization) (Bardgett et al., 2014; Freschet et al., 2021). These choices ensure that CWM reflects the dominance of phenotypes that most effectively reinforce soil, while FD captures complementary rooting strategies that collectively increase mechanical stability, thereby advancing from purely statistical summaries—such as additive partitioning (Loreau & Hector, 2001)—to mechanistic explanations of biodiversity effects on ecosystem functioning.

13.3.4 Ecosystem Services

The empirical finding that biodiversity enhances and stabilizes ecosystem functioning has direct implications for human societies. These implications are most clearly articulated through the concept of ecosystem services.

The Millennium Ecosystem Assessment (MEA, 2005), IPBES (2019), and (Díaz et al., 2019) classify ecosystem services into several categories:

- **Provisioning services**: such as food, fresh water, timber, and fiber.
- **Regulating services**: such as climate regulation, flood mitigation, disease and pest control, and water purification.
- **Cultural services**: such as recreation, aesthetic and spiritual values, and cultural identity.
- **Supporting or maintenance services**: such as soil formation, nutrient cycling, and primary production.

Findings from BEF research suggest that many of these services are strengthened by biodiversity. For instance, plant diversity in agricultural systems and forests can enhance not only yield and biomass production but also pest suppression, erosion control, and water quality protection (Isbell et al., 2017). Similarly, biodiverse wetlands and coastal ecosystems often provide strong regulating services, including flood attenuation, storm-surge buffering, and carbon sequestration.

The concept of NbS explicitly builds on this understanding by seeking to use biodiversity-supported ecosystem functions to address societal challenges such as climate change mitigation and adaptation, disaster risk reduction, and human health (Cohen-Shacham et al., 2016; Mori, 2020; Mori et al., 2021). Conserving and restoring diverse forests and wetlands, for example, can simultaneously increase carbon sinks, protect biodiversity, and improve local livelihoods and quality of life (Edwards & Cerullo, 2024; Gilroy et al., 2014; Hua et al., 2022). Likewise, emerging policy narratives such as "Nature Positive" call for a transition from merely reducing environmental damage to actively restoring and enhancing natural capital and biodiversity. Scientific insights from BEF research provide a crucial foundation for designing, implementing, and evaluating such strategies (Balvanera et al., 2014; Klaus et al., 2020; Mace, 2014).

In this way, understanding how biodiversity shapes ecosystem functioning, and how those functions translate into ecosystem services, offers a practical answer to the question of why biodiversity matters for conservation and sustainable management.

13.3.5 Human Well-Being

Ultimately, ecosystem services feed into human well-being and health. Clarifying how biodiversity and ecosystem functioning affect human lives is essential for integrating conservation and management into broader societal decision-making.

First, diverse ecosystems support food security and nutrition. A wide range of crops, landraces, and wild food resources (e.g. fish, mushrooms, fruits, and animals) enhance dietary diversity and nutritional balance, while reducing risks associated with dependence on a single crop species, such as susceptibility to pests, diseases, or climatic extremes (Dannenberg et al., 2024). Second, biodiversity plays a key role in disaster risk reduction and climate adaptation. Diverse forests can reduce the risk of landslides and floods, while mangroves and salt marshes act as natural buffers against storm surges and coastal erosion (Akram et al., 2023). In urban areas, diverse street trees, parks, and green spaces can mitigate urban heat island effects, reduce peak runoff during heavy rainfall events, and contribute to climate resilience (Wood & Dupras, 2021). Third, biodiverse natural environments and urban green spaces contribute to mental and physical health. Many studies indicate that visiting species-rich forests, agricultural landscapes, riverscapes, and coastal environments can reduce stress, restore attention, and foster a sense of connection and well-being (Franco et al., 2017; Nisbet et al., 2020). Urban parks and agricultural landscapes that provide diverse floral resources support pollinators while simultaneously offering aesthetic and cultural benefits to people (Ranalli et al., 2025; Ulrich & Sargent, 2025; Westphal et al., 2015).

13.3.6 Why BEF Alone Is Not Enough: Toward Integrating Community Ecology

Despite substantial advances, BEF research alone provides limited guidance for real-world conservation and management. Most BEF experiments are conducted under simplified and short-term conditions where species coexistence is assumed rather than examined, and where environmental heterogeneity, disturbance regimes, and historical contingencies are minimized. As a result, BEF studies often show that "more diversity increases functioning," but they rarely identify which species combinations sustain functions, under what environmental or spatial conditions, or how community composition and functioning change through time and space. For management, however, such a mechanistic understanding is essential.

These limitations become especially apparent when considering the strong scale-dependence of biodiversity–function relationships. For example, Chisholm et al. (2013) showed that the apparent relationship between biodiversity and ecosystem functioning can change markedly depending on spatial scale: positive richness–biomass and richness–productivity relationships are common at small spatial grains (0.04 ha), where neighborhood-scale interactions and complementarity dominate. However, these relationships weakened or even became negative at larger scales (0.25–1 ha). At broader spatial extents, environmental conditions are expected to play a much stronger role in shaping communities. As the authors note, environmental gradients such as soil fertility can cause productivity to increase monotonically, while species richness peaks at intermediate fertility (Grime, 1979). When such unimodal patterns are viewed across broad spatial extents, mixed richness–productivity relationships inevitably emerge.

Importantly, this implies that the same observed level of species richness at broad scales may arise from very different ecological contexts: communities in low-productivity sites may be structured by stress tolerance or facilitation, whereas those in high-productivity sites may be shaped by competitive niche differentiation. As spatial scale increases, environmental heterogeneity expands and these distinct coexistence processes become aggregated—producing the scale-dependent BEF patterns observed empirically. This scale dependence is consistent with well-established findings in community assembly research, which show that stochastic processes often dominate at fine scales, while deterministic processes driven by environmental filtering become increasingly important at larger spatial extents. Consequently, shifts in coexistence mechanisms across scales imply corresponding shifts in the interaction processes (e.g., complementarity, compensatory dynamics, and dominance effects) that generate biodiversity effects on ecosystem functioning—highlighting why BEF experiments alone cannot disentangle these mechanisms.

Recent theoretical work strengthens this perspective. Godoy et al. (2020) showed that complementarity and selection effects—core BEF mechanisms—are tightly linked to niche differences and relative fitness differences, the central axes of coexistence theory. Although these relationships are not one-to-one, they indicate that mechanisms generating coexistence also shape functioning outcomes, pointing toward a unified framework connecting the causes and consequences of biodiversity. Conceptual contributions by Barry et al. (2019) and Wang et al. (2024) emphasize the need to explicitly integrate coexistence processes and ecosystem functioning to build predictive tools relevant to real-world management.

13.3.7 Synthesizing Cause and Consequence through Trait-based Approaches

Building on Section 13.3.5's call to integrate BEF with community ecology, a range of conceptual and theoretical approaches have been proposed to link community assembly with ecosystem functioning. Among these, functional traits provide a particularly practical and widely applicable bridge between assembly processes and functional outcomes. In the response–effect trait perspective (Suding et al., 2008), response traits govern how species' abundances and occurrences shift under environmental or biotic change, whereas effect traits determine how the resulting community state influences ecosystem processes; in some cases, traits act as dual response–effect traits, enabling direct scaling from community change to ecosystem outcomes. When effect traits are concentrated in dominant species, ecosystem functioning tends to track the CWM—linked with selection effects; when effect traits are dispersed among coexisting species, greater FD along those axes expands opportunities for resource partitioning, abiotic facilitation, or biotic feedbacks—linked with complementarity effects.

Within this synthesis, Trait Driver Theory (TDT; Enquist et al., 2015) offers a concise, quantitative view. Building on trait-based theories (including the mass-ratio hypothesis (Grime, 1998)) and the metabolic scaling theory (Enquist et al., 1998, Gillooly et al., 2001, West et al., 1997), TDT posits that each functional trait has an environment-specific performance optimum and that environmental change shifts this optimum. Inference therefore centers on the biomass-weighted trait distribution: (1) the CWM indicates where dominant phenotypes sit relative to the moving optimum; (2) the variance captures spread around that optimum; and (3) the shape of the distribution—its skewness and kurtosis—captures lag and concentration. From these elements, TDT predicts functional responses: states in which trait values are tightly concentrated near the (moving) optimum—CWM close to the optimum, lower variance, higher (positive) kurtosis—should most strongly govern rates such as productivity. This contrasts with the niche-complementarity hypothesis emphasizing benefits of dispersion (high FD) via partitioning or facilitation to predict ecosystem functioning (Tilman et al., 1996). As a concrete field test of TDT, a Peruvian tropical elevation gradient (spanning 3,300 m in elevation with a ~16.2°C difference in mean annual temperature) served as a natural laboratory to assess how climate-related temperature differences shape forest community structure and, in turn, productivity (Enquist et al., 2017). Across 16 1-ha plots, thousands of tree-level trait measurements were paired with GPP and NPP. Consistent with TDT, variation in productivity scaled primarily with stand biomass, and the apparent direct effect of temperature on productivity was much weaker than expected once shifts in community trait distributions were accounted for. Along the gradient, biomass-weighted trait means approached local performance optima and trait distributions were sharply peaked and skewed, indicating strong local filtering and lag. These patterns support the view that the position (CWM) and shape (variance, skewness, kurtosis) of community trait distributions mediate the link from environment-driven community change (cause) to ecosystem functioning (consequence).

Making these trait-based theories testable requires careful design (Dias et al., 2013; Chacón-Labella et al., 2023). Two considerations are paramount. First, because moments of trait distributions (mean, variance, skewness, kurtosis) often covary, studies should, where possible, separate these moments (via targeted assembly, stratified sampling, or factorial manipulations), incorporate intraspecific variation, and select trait axes mechanistically linked to the focal function (e.g., Tobner et al., 2014). Second, because community assembly reflects both deterministic filters and stochastic processes, designs should contrast trait-directed and random scenarios (e.g., trait-driven vs. random species loss, as demonstrated in BEF-China) to distinguish trait-based determinism from chance (Bruelheide et al., 2014; Chen et al., 2020). Indeed, experimental studies demonstrate that stochastic assembly history can strongly shape ecosystem functioning. In wood decomposer communities, variation in the order and timing of species arrival dictated decomposition rates independently of environmental conditions (Fukami et al., 2010), highlighting the functional consequences of priority effects. Framed this way, trait-based approaches—TDT included—provide one concrete route for linking community assembly processes to ecosystem functioning. More broadly, they illustrate how integrating information on species identity, trait distributions, and assembly history can transform BEF relationships from statistical patterns into mechanistic, process-based understanding.

13.3.8 Implications for Ecosystem Management and NbSs

The integration of BEF research with community ecology offers an important shift in perspective for conservation and ecosystem management. Rather than assuming that "more biodiversity automatically enhances functioning," a mechanistic understanding highlights that the identity, functional traits, and coexistence mechanisms of species, as well as spatial heterogeneity and turnover, determine how reliably ecosystems provide functions and services. This perspective emphasizes that biodiversity must be understood as a dynamic process shaped by species interactions, disturbance regimes, land-use practices, and historical contingencies—not just as a static number to be maximized.

From a management perspective, several insights emerge. First, maintaining ecosystem functioning and resilience requires designing communities that sustain desirable interaction mechanisms such as niche complementarity and compensatory dynamics. This suggests strategic use of species mixtures with complementary traits, appropriate spatial configurations, and disturbance regimes that maintain coexistence processes. Second, because ecosystem functioning at larger spatial scales is strongly influenced by species turnover and environmental heterogeneity, effective management must consider landscapes rather than isolated sites. Protecting or restoring mosaics of habitats, including traditional cultural landscapes such as satoyama, can enhance multifunctionality by supporting different species assemblages that perform well under varying local conditions.

Third, management strategies must explicitly address temporal dynamics. As climate change and land-use transformation accelerate, community composition will inevitably shift. Mechanistic understanding of assembly processes can support adaptive management that anticipates future community trajectories rather than reacting after functional losses occur. For example, selecting species combinations that are likely to coexist and perform well under projected climate conditions can enhance long-term ecosystem functioning and resilience.

These insights align with emerging approaches such as NbS and the Nature Positive agenda, which seek to restore and manage ecosystems in ways that support both biodiversity and human well-being (Figure. 13.4). Moving toward predictive, process-based frameworks can improve how NbS projects are designed, evaluated, and monitored—shifting the focus from simple measures of species richness to deeper questions of which species, interacting in which ways, under which environmental and social conditions, sustain critical ecosystem services (Mori et al., 2017). Ultimately, linking the causes and consequences of biodiversity provides a pathway toward evidence-based design of multifunctional, resilient landscapes capable of supporting societies in a rapidly changing world.

13.4 SUMMARY AND FUTURE PERSPECTIVES

Biodiversity enhances ecosystem functioning and human well-being, yet the benefits it provides depend on the ecological processes that generate and maintain diversity. Although BEF research has demonstrated positive effects of biodiversity on ecosystem functioning, stability, and resilience, these insights alone offer limited guidance for practical management because they seldom clarify which species combinations, trait configurations, or spatial arrangements sustain ecosystem functioning under real environmental conditions. Integrating BEF with community

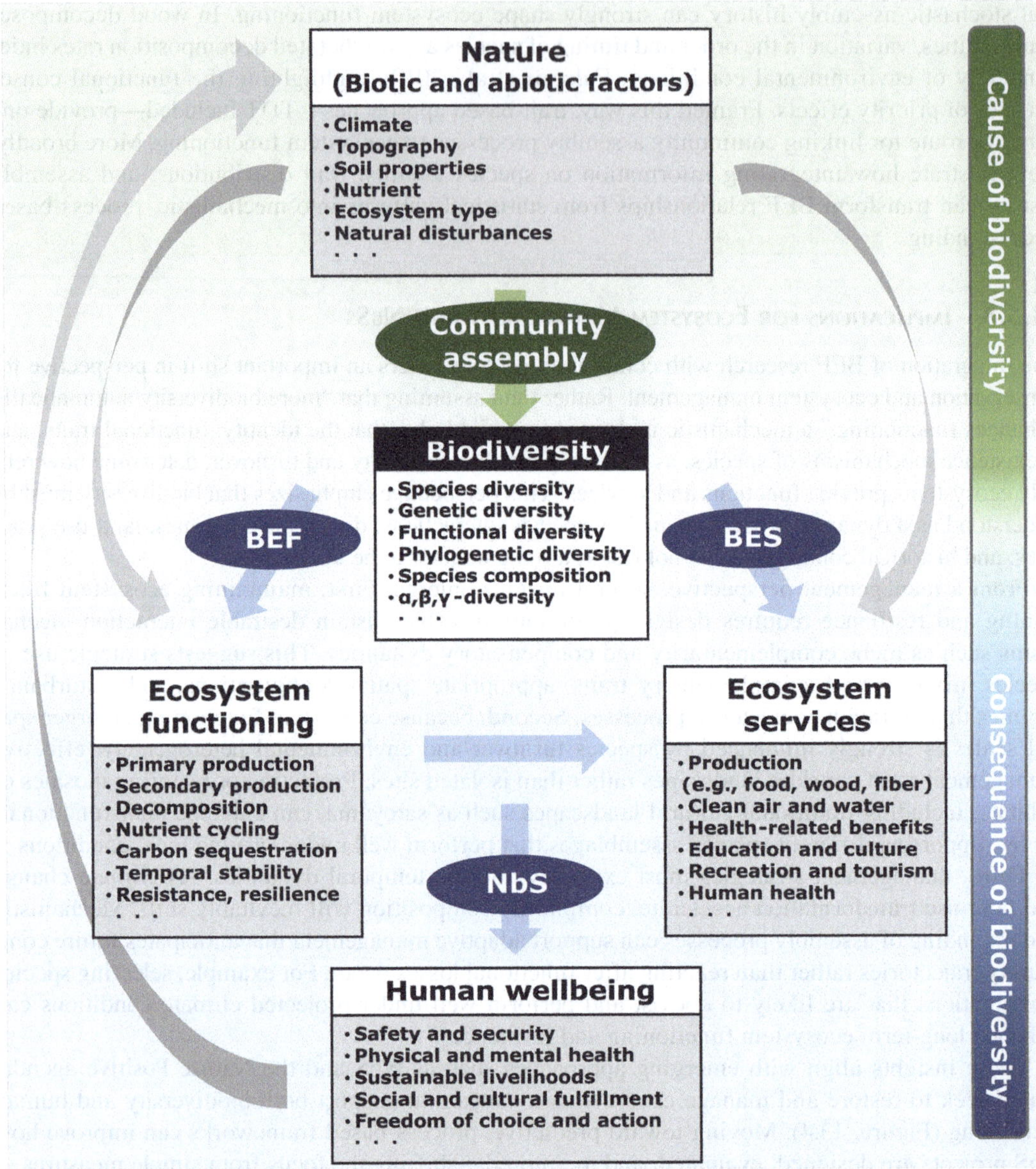

FIGURE 13.4 Conceptual framework of the linkages between nature (Biotic and abiotic factors), biodiversity, ecosystem functions, ecosystem services, and human well-being. The diagram illustrates the sequential flow from nature to societal benefits, mediated by biodiversity. Environmental factors such as climate, topography, and soil properties shape the state of biodiversity—encompassing species, genetic, and FD—through community assembly processes. This biodiversity, in turn, underpins ecosystem functions (e.g., primary production, nutrient cycling) and ecosystem services (e.g., provision of food, clean air, and water) through the pathways of BEF (Biodiversity–Ecosystem Functioning) and BES (Biodiversity–Ecosystem Services), which are collectively framed in the contemporary context as Nature's Contributions to People (NCPs). Finally, these functions and services contribute to human well-being, including safety, health, and sustainable livelihoods. This integrated cascade forms the scientific foundation of NbSs, where protecting and managing biodiversity directly addresses societal challenges by enhancing the delivery of essential ecosystem services.

ecology—linking community assembly processes to functional outcomes—will enable more predictive and process-based ecosystem design.

Importantly, community assembly is shaped not only by deterministic, niche-based processes but also by stochastic forces such as dispersal limitation, ecological drift, and historical contingency. These stochastic processes can generate alternative community states with distinct functional consequences even under similar environmental conditions. In contrast, most BEF mechanisms have been developed primarily under deterministic assumptions, emphasizing complementarity and selection effects, while the functional implications of stochastic assembly processes remain insufficiently explored.

Recognizing the interplay between deterministic and stochastic assembly across spatial and temporal scales is therefore essential for advancing our understanding of BEF relationships and for designing resilient, multifunctional landscapes under accelerating social–ecological change, including the effective implementation of NbSs.

DECLARATION OF GENERATIVE AI USE

During the preparation of this manuscript, the authors used an AI-assisted language editing tool to improve the English expression and readability. The authors subsequently reviewed and edited the content and take full responsibility for the final version of the published chapter.

REFERENCES

Akram, H., Hussain, S., Mazumdar, P., Chua, K. O., Butt, T. E., & Harikrishna, J. A. (2023). Mangrove health: A review of functions, threats, and challenges associated with mangrove management practices. *Forests*, *14*(9), 1698. https://doi.org/10.3390/F14091698

Albert, A., Auffret, A. G., Cosyns, E., Cousins, S. A. O., D'hondt, B., Eichberg, C., … & Baltzinger, C. (2015). Seed dispersal by ungulates as an ecological filter: A trait-based meta-analysis. *Oikos*, 124(9), 1109–1120. https://doi.org/10.1111/OIK.02512

Alberti, J., Bakker, E. S., van Klink, R., Olff, H., & Smit, C. (2017). Herbivore exclusion promotes a more stochastic plant community assembly in a natural grassland. *Ecology*, 98(4), 961–970. https://doi.org/10.1002/ECY.1741

Avise, J. C. (2000). *Phylogeography: The History and Formation of Species*. Harvard University Press.

Balvanera, P., Siddique, I., Dee, L., Paquette, A., Isbell, F., Gonzalez, A., … & Griffin, J. N. (2014). Linking biodiversity and ecosystem services: Current uncertainties and the necessary next steps. *BioScience*, 64(1), 49–57. https://doi.org/10.1093/BIOSCI/BIT003

Bardgett, R. D., Mommer, L., & De Vries, F. T. (2014). Going underground: Root traits as drivers of ecosystem processes. *Trends in Ecology & Evolution*, 29(12), 692–699. https://doi.org/10.1016/j.tree.2014.10.006

Barry, K. E., de Kroon, H., Dietrich, P., Stanley Harpole, W., Roeder, A., Schmid, B., … & Roscher, C. (2019). Linking local species coexistence to ecosystem functioning: A conceptual framework from ecological first principles in grassland ecosystems. In *Advances in Ecological Research* (Vol. 61, pp. 265–296). Academic Press. https://doi.org/10.1016/BS.AECR.2019.06.007

Berendse, F. (1979). Competition between plant populations with different rooting depths - I. Theoretical considerations. *Oecologia*, 43(1), 19–26. https://doi.org/10.1007/BF00346669

Bohonak, A. J. (1999). Dispersal, gene flow, and population structure. *The Quarterly Review of Biology*, 74(1), 21–45. https://doi.org/10.1086/392950

Borcard, D., Legendre, P., & Drapeau, P. (1992). Partialling out the spatial component of ecological variation. *Ecology*, 73(3), 1045–1055. https://doi.org/10.2307/1940179

Bruelheide, H., Nadrowski, K., Assmann, T., Bauhus, J., Both, S., Buscot, F., … & Schmid, B. (2014). Designing forest biodiversity experiments: General considerations illustrated by a new large experiment in subtropical China. *Methods in Ecology and Evolution*, 5, 74–89. https://doi.org/10.1111/2041-210X.12126

Bugalho, M. N., Ibáñez, I., & Clark, J. S. (2013). The effects of deer herbivory and forest type on tree recruitment vary with plant growth stage. *Forest Ecology and Management*, 308, 90–100. https://doi.org/10.1016/J.FORECO.2013.07.036

Cadotte, M. W., Carscadden, K., & Mirotchnick, N. (2011). Beyond species: Functional diversity and the maintenance of ecological processes and services. *Journal of Applied Ecology*, 48(5), 1079–1087. https://doi.org/10.1111/j.1365-2664.2011.02048.x

Canham, C. D. (1989). Different respones to gaps among shade-tollerant tree species. *Ecology*, 70(3), 548–550. https://doi.org/10.2307/1940200

Cao, R., Zhang, Y., Fernández-Martínez, M., Zhang, Z., Lai, G., Ju, W., & Peñuelas, J. (2025). Global evidence for a positive relationship between tree species richness and ecosystem photosynthesis. *Nature Plants*, 11(7), 1429–1440. https://doi.org/10.1038/s41477-025-02046-1

Cardinale, B. J., Duffy, J. E., Gonzalez, A., Hooper, D. U., Perrings, C., Venail, P., ... & Naeem, S. (2012). Biodiversity loss and its impact on humanity. *Nature*, 486(7401), 59–67. https://doi.org/10.1038/nature11148

Chacón-Labella, J., Hinojo-Hinojo, C., Bohner, T., Castorena, M., Violle, C., Vandvik, V., & Enquist, B. J. (2023). How to improve scaling from traits to ecosystem processes. *Trends in Ecology & Evolution*, 38(3), 228–237. https://doi.org/10.1016/j.tree.2022.10.007

Chase, J. M. (2014). Spatial scale resolves the niche versus neutral theory debate. *Journal of Vegetation Science*, 25(2), 319–322. https://doi.org/10.1111/JVS.12159

Chase, J. M., & Myers, J. A. (2011). Disentangling the importance of ecological niches from stochastic processes across scales. *Philosophical Transactions of the Royal Society B: Biological Sciences*, 366(1576), 2351–2363. https://doi.org/10.1098/RSTB.2011.0063

Chen, C., Xiao, W., & Chen, H. Y. (2025). Meta-analysis reveals global variations in plant diversity effects on productivity. *Nature*, 638(8050), 435–440. https://doi.org/10.1038/s41586-024-08407-8

Chen, Y., Huang, Y., Niklaus, P. A., Castro-Izaguirre, N., Clark, A. T., Bruelheide, H., ... & Schmid, B. (2020). Directed species loss reduces community productivity in a subtropical forest biodiversity experiment. *Nature Ecology & Evolution*, 4(4), 550–559. https://doi.org/10.1038/s41559-020-1127-4

Chesson, P. (2000). Mechanisms of maintenance of species diversity. *Annual Review of Ecology and Systematics*, 31, 343–366. https://doi.org/10.1146/annurev.ecolsys.31.1.343

Chisholm, R. A., Muller-Landau, H. C., Abdul Rahman, K., Bebber, D. P., Bin, Y., Bohlman, S. A., ... & Zimmerman, J. K. (2013). Scale-dependent relationships between tree species richness and ecosystem function in forests. *Journal of Ecology*, 101(5), 1214–1224. https://doi.org/10.1111/1365-2745.12132

Claramunt, S., Derryberry, E. P., Remsen, J. V., & Brumfield, R. T. (2012). High dispersal ability inhibits speciation in a continental radiation of passerine birds. *Proceedings of the Royal Society B: Biological Sciences*, *279*(1733), 1567–1574. https://doi.org/10.1098/RSPB.2011.1922

Cohen-Shacham, E., Janzen, C., Maginnis, S., & Walters, G. (2016). Nature-based solutions to address global societal challenges. In *Nature-based Solutions to Address Global Societal Challenges*. IUCN International Union for Conservation of Nature. https://doi.org/10.2305/IUCN.CH.2016.13.EN

Connor, E. F., & Simberloff, D. (1979). The assembly of species communities: Chance or competition? *Ecology*, 60(6), 1132–1140. https://doi.org/10.2307/1936961

Coyne, J. A., & Orr, H. A. (2004). *Speciation/Jerry A. Coyne, H. Allen Orr.* Oxford University Press.

Dannenberg, P., Braun, B., Greiner, C., Follmann, A., Haug, M., Semedi Hargo Yuwono, P., ... & Kopriva, S. (2024). Eight arguments why biodiversity is important to safeguard food security. *Plants People Planet*, 6(3), 604–610. https://doi.org/10.1002/PPP3.10492

Dawson-Glass, E., & Hargreaves, A. L. (2022). Does pollen limitation limit plant ranges? Evidence and implications. *Philosophical Transactions of the Royal Society B: Biological Sciences*, *377*(1846). https://doi.org/10.1098/rstb.2021.0014

de Kroon, H., Hendriks, M., van Ruijven, J., Ravenek, J., Padilla, F. M., Jongejans, E., ... & Mommer, L. (2012). Root responses to nutrients and soil biota: Drivers of species coexistence and ecosystem productivity. *Journal of Ecology*, 100(1), 6–15. https://doi.org/10.1111/J.1365-2745.2011.01906.X

Diamond, J. M. (1975). Assembly of species communities. In J. M. Diamond & M. L. Cody (Eds.), *Ecology and Evolution of Communities*. Harvard University Press. https://www.scirp.org/reference/referencespapers?referenceid=1901245

Dias, A. T., Berg, M. P., de Bello, F., Van Oosten, A. R., Bílá, K., & Moretti, M. (2013). An experimental framework to identify community functional components driving ecosystem processes and services delivery. *Journal of Ecology*, 101(1), 29–37. https://doi.org/10.1111/1365-2745.12024

Díaz, S., Settele, J., Brondízio, E. S., Ngo, H. T., Agard, J., Arneth, A., ... & Zayas, C. N. (2019). Pervasive human-driven decline of life on Earth points to the need for transformative change. *Science*, *366*(6471). https://doi.org/10.1126/SCIENCE.AAX3100

Dirzo, R., Young, H. S., Galetti, M., Ceballos, G., Isaac, N. J. B., & Collen, B. (2014). Defaunation in the anthropocene. *Science*, 345(6195), 401–406. https://doi.org/10.1126/SCIENCE.1251817

Edwards, D. P., & Cerullo, G. R. (2024). Biodiversity is central for restoration. *Current Biology*, 34(9), R371–R379. https://doi.org/10.1016/J.CUB.2024.02.032

Enquist, B. J., Bentley, L. P., Shenkin, A., Maitner, B., Savage, V., Michaletz, S., ... & Malhi, Y. (2017). Assessing trait-based scaling theory in tropical forests spanning a broad temperature gradient. *Global Ecology and Biogeography*, 26(12), 1357–1373. https://doi.org/10.1111/geb.12645

Enquist, B. J., Brown, J. H., & West, G. B. (1998). Allometric scaling of plant energetics and population density. *Nature*, 395(6698), 163–165. https://doi.org/10.1038/25977

Enquist, B. J., Norberg, J., Bonser, S. P., Violle, C., Webb, C. T., Henderson, A., ... & Savage, V. M. (2015). Scaling from traits to ecosystems: Developing a general trait driver theory via integrating trait-based and metabolic scaling theories. In *Advances in Ecological Research* (Vol. 52, pp. 249–318). Academic Press. https://doi.org/10.1016/bs.aecr.2015.02.001

Fan, K., Chu, H., Eldridge, D. J., Gaitan, J. J., Liu, Y. R., Sokoya, B., ... & Delgado-Baquerizo, M. (2023). Soil biodiversity supports the delivery of multiple ecosystem functions in urban greenspaces. *Nature Ecology & Evolution*, 7(1), 113–126. https://doi.org/10.1038/s41559-022-01935-4

Fargione, J., Tilman, D., Dybzinski, R., Lambers, J. H. R., Clark, C., Harpole, W. S., ... & Loreau, M. (2007). From selection to complementarity: Shifts in the causes of biodiversity–productivity relationships in a long-term biodiversity experiment. *Proceedings of the Royal Society B: Biological Sciences*, 274(1611), 871–876. https://doi.org/10.1098/RSPB.2006.0351

Feng, J., Wan, X., Wang, R., Li, S., Wei, M., Dakos, V., ... & Stenseth, N. C. (2025). Mediating role of food web structure in linking diversity to multidimensional stability: Evidence from global marine ecosystems. *Science Advances*, 11(41), eadv3841. https://doi.org/10.1126/sciadv.adv3841

Fox, J. W. (2005). Interpreting the 'selection effect'of biodiversity on ecosystem function. *Ecology Letters*, 8(8), 846–856. https://doi.org/10.1111/j.1461-0248.2005.00795.x

Franco, L. S., Shanahan, D. F., & Fuller, R. A. (2017). A review of the benefits of nature experiences: More than meets the eye. *International Journal of Environmental Research and Public Health*, 14(8), 864. https://doi.org/10.3390/IJERPH14080864

Freschet, G. T., Roumet, C., Comas, L. H., Weemstra, M., Bengough, A. G., Rewald, B., ... & Stokes, A. (2021). Root traits as drivers of plant and ecosystem functioning: Current understanding, pitfalls and future research needs. *New Phytologist*, 232(3), 1123–1158. https://doi.org/10.1111/nph.17072

Fricke, E. C., Bello, C., Chaplin-Kramer, R., Dent, D. H., Feeley, K. J., Galetti, M., ... & Reid, J. L. (2025). Drivers and impacts of global seed disperser decline. *Nature Reviews Biodiversity*, 1(6), 386–400. https://doi.org/10.1038/s44358-025-00053-w

Fricke, E. C., Ordonez, A., Rogers, H. S., & Svenning, J. C. (2022). The effects of defaunation on plants' capacity to track climate change. *Science*, 375(6577), 210–214. https://doi.org/10.1126/SCIENCE.ABK3510

Fukami, T. (2015). Historical contingency in community assembly: Integrating niches, species pools, and priority effects. *Annual Review of Ecology, Evolution, and Systematics*, 46, 1–23. https://doi.org/10.1146/annurev-ecolsys-110411-160340

Fukami, T., Dickie, I. A., Wilkie, J. P., Paulus, B. C., Park, D., Roberts, A., Buchanan, P. K., & Allen, R. B. (2010). Assembly history dictates ecosystem functioning: evidence from wood decomposer communities. *Ecology Letters*, 13(6), 675–684. https://doi.org/10.1111/j.1461-0248.2010.01465.x

Galetti, M., Guevara, R., Côrtes, M. C., Fadini, R., Von Matter, S., Leite, A. B., ... & Jordano, P. (2013). Functional extinction of birds drives rapid evolutionary changes in seed size. *Science*, 340(6136), 1086–1090. https://doi.org/10.1126/SCIENCE.1233774

Garzon-Lopez, C. X., Jansen, P. A., Bohlman, S. A., Ordonez, A., & Olff, H. (2014). Effects of sampling scale on patterns of habitat association in tropical trees. *Journal of Vegetation Science*, 25(2), 349–362. https://doi.org/10.1111/JVS.12090

Gaston, K. J. (2000). Global patterns in biodiversity. *Nature*, 405(6783), 220–227. https://doi.org/10.1038/35012228

Gillooly, J. F., Brown, J. H., West, G. B., Savage, V. M., & Charnov, E. L. (2001). Effects of size and temperature on metabolic rate. *Science*, 293(5538), 2248–2251. https://doi.org/10.1126/science.1061967

Gilroy, J. J., Woodcock, P., Edwards, F. A., Wheeler, C., Baptiste, B. L. G., Medina Uribe, C. A., ... & Edwards, D. P. (2014). Cheap carbon and biodiversity co-benefits from forest regeneration in a hotspot of endemism. *Nature Climate Change*, 4(6), 503–507. https://doi.org/10.1038/nclimate2200

Godoy, O., Gómez-Aparicio, L., Matías, L., Pérez-Ramos, I. M., & Allan, E. (2020). An excess of niche differences maximizes ecosystem functioning. *Nature Communications*, 11(1), 4180. https://doi.org/10.1038/S41467-020-17960-5

Gonzalez, A., & Loreau, M. (2009). The causes and consequences of compensatory dynamics in ecological communities. *Annual Review of Ecology, Evolution, and Systematics*, 40, 393–414. https://doi.org/10.1146/annurev.ecolsys.39.110707.173349

Grime, J. P. (1979). *Plant Strategies and Vegetation Processes*. John Wiley & Sons, Ltd.

Grime, J. P. (1998). Benefits of plant diversity to ecosystems: Immediate, filter and founder effects. *Journal of Ecology*, 86(6), 902–910. https://doi.org/10.1046/j.1365-2745.1998.00306.x

Hardiman, B. S., Bohrer, G., Gough, C. M., Vogel, C. S., & Curtis, P. S. (2011). The role of canopy structural complexity in wood net primary production of a maturing northern deciduous forest. *Ecology*, 92(9), 1818–1827. https://doi.org/10.1890/10-2192.1

Hector, A. (1998). The effect of diversity on productivity: Detecting the role of species complementarity. *Oikos*, 82(3), 597. https://doi.org/10.2307/3546380

Hillebrand, H. (2004). On the generality of the latitudinal diversity gradient. *The American Naturalist*, 163(2), 192–211. https://doi.org/10.1086/381004

Hodapp, D., Borer, E. T., Harpole, W. S., Lind, E. M., Seabloom, E. W., Adler, P. B., … & Hillebrand, H. (2018). Spatial heterogeneity in species composition constrains plant community responses to herbivory and fertilisation. *Ecology Letters*, 21(9), 1364–1371. https://doi.org/10.1111/ELE.13102

Hua, F., Adrian Bruijnzeel, L., Meli, P., Martin, P. A., Zhang, J., Nakagawa, S., … & Balmford, A. (2022). The biodiversity and ecosystem service contributions and trade-offs of forest restoration approaches. *Science*, 376(6595), 839–844. https://doi.org/10.1126/SCIENCE.ABL4649

Hubbell, S. P. (2001). *The Unified Neutral Theory of Biodiversity and Biogeography*. Princeton University Press.

Huston, M. A. (1997). Hidden treatments in ecological experiments: Re-evaluating the ecosystem function of biodiversity. *Oecologia*, 110(4), 449–460. https://doi.org/10.1007/S004420050180

IPBES. (2019). *Global Assessment Report on Biodiversity and Ecosystem Services of the Intergovernmental Science-Policy Platform on Biodiversity and Ecosystem Services*. https://doi.org/10.5281/ZENODO.6417333

Isbell, F., Calcagno, V., Hector, A., Connolly, J., Harpole, W. S., Reich, P. B., … & Loreau, M. (2011). High plant diversity is needed to maintain ecosystem services. *Nature*, 477(7363), 199–202. https://doi.org/10.1038/nature10282

Isbell, F., Cowles, J., Dee, L. E., Loreau, M., Reich, P. B., Gonzalez, A., … & Schmid, B. (2018). Quantifying effects of biodiversity on ecosystem functioning across times and places. *Ecology Letters*, 21(6), 763–778. https://doi.org/10.1111/ele.12928

Isbell, F., Gonzalez, A., Loreau, M., Cowles, J., Díaz, S., Hector, A., … & Larigauderie, A. (2017). Linking the influence and dependence of people on biodiversity across scales. *Nature*, 546(7656), 65–72. https://doi.org/10.1038/nature22899

Ishii, H., & Asano, S. (2010). The role of crown architecture, leaf phenology and photosynthetic activity in promoting complementary use of light among coexisting species in temperate forests. *Ecological Research*, 25(4), 715–722. https://doi.org/10.1007/S11284-009-0668-4

Ishii, H. S., Kubota, M. X., Tsujimoto, S. G., & Kudo, G. (2019). Association between community assemblage of flower colours and pollinator fauna: A comparison between Japanese and New Zealand alpine plant communities. *Annals of Botany*, 123(3), 533–541. https://doi.org/10.1093/AOB/MCY188

Jabot, F., Laroche, F., Massol, F., Arthaud, F., Crabot, J., Dubart, M., … & Datry, T. (2020). Assessing metacommunity processes through signatures in spatiotemporal turnover of community composition. *Ecology Letters*, 23(9), 1330–1339. https://doi.org/10.1111/ELE.13523

Jetz, W., Thomas, G. H., Joy, J. B., Hartmann, K., & Mooers, A. O. (2012). The global diversity of birds in space and time. *Nature*, 491(7424), 444–448. https://doi.org/10.1038/nature11631

Ji, M., Zhou, J., Li, Y., Ma, K., Song, W., Li, Y., … & Tu, Q. (2024). Biodiversity of mudflat intertidal viromes along the Chinese coasts. *Nature Communications*, 15(1), 8611. https://doi.org/10.1038/s41467-024-52996-x

Katayama, N., Amano, T., Naoe, S., Yamakita, T., Komatsu, I., Takagawa, S. I., … & Miyashita, T. (2014). Landscape heterogeneity–biodiversity relationship: Effect of range size. *PLoS One*, *9*(3), e93359. https://doi.org/10.1371/JOURNAL.PONE.0093359

Kay, K. M., & Sargent, R. D. (2009). The role of animal pollination in plant speciation: Integrating ecology, geography, and genetics. *Annual Review of Ecology, Evolution, and Systematics*, 40, 637–656. https://doi.org/10.1146/annurev.ecolsys.110308.120310

Kemp, J. E., Bergh, N. G., Soares, M., & Ellis, A. G. (2019). Dominant pollinators drive non-random community assembly and shared flower colour patterns in daisy communities. *Annals of Botany*, 123(2), 277–288. https://doi.org/10.1093/AOB/MCY126

Klaus, V. H., Whittingham, M. J., Báldi, A., Eggers, S., Francksen, R. M., Hiron, M., … & Buchmann, N. (2020). Do biodiversity-ecosystem functioning experiments inform stakeholders how to simultaneously conserve biodiversity and increase ecosystem service provisioning in grasslands? *Biological Conservation*, 245, 108552. https://doi.org/10.1016/J.BIOCON.2020.108552

Koski, M. H. (2023). Pollinators exert selection on floral traits in a pollen-limited, narrowly endemic spring ephemeral. *American Journal of Botany*, 110(1), e16101. https://doi.org/10.1002/AJB2.16101

Kraft, N. J. B., & Ackerly, D. D. (2010). Functional trait and phylogenetic tests of community assembly across spatial scales in an Amazonian forest. *Ecological Monographs*, 80(3), 401–422. https://doi.org/10.1890/09-1672.1

Kraft, N. J. B., Comita, L. S., Chase, J. M., Sanders, N. J., Swenson, N. G., Crist, T. O., … & Myers, J. A. (2011). Disentangling the drivers of β diversity along latitudinal and elevational gradients. *Science*, 333(6050), 1755–1758. https://doi.org/10.1126/SCIENCE.1208584

Kriebel, R., Drew, B., González-Gallegos, J. G., Celep, F., Heeg, L., Mahdjoub, M. M., & Sytsma, K. J. (2020). Pollinator shifts, contingent evolution, and evolutionary constraint drive floral disparity in Salvia (Lamiaceae): Evidence from morphometrics and phylogenetic comparative methods. *Evolution*, 74(7), 1335–1355. https://doi.org/10.1111/EVO.14030

LaRue, E. A., Knott, J. A., Domke, G. M., Chen, H. Y. H., Guo, Q., Hisano, M., … & Fei, S. (2023). Structural diversity as a reliable and novel predictor for ecosystem productivity. *Frontiers in Ecology and the Environment*, 21(1), 33–39. https://doi.org/10.1002/FEE.2586

Legendre, P., Borcard, D., & Peres-Neto, P. R. (2005). Analyzing beta diversity: Partitioning the spatial variation of community composition data. *Ecological Monographs*, 75(4), 435–450. https://doi.org/10.1890/05-0549

Loreau, M. (1998). Biodiversity and ecosystem functioning: A mechanistic model. *Proceedings of the National Academy of Sciences of the United States of America*, 95(10), 5632–5636. https://doi.org/10.1073/PNAS.95.10.5632

Loreau, M., & Hector, A. (2001). Partitioning selection and complementarity in biodiversity experiments. *Nature*, 412(6842), 72–76. https://doi.org/10.1038/35083573

Loreau, M., Jarne, P., & Martiny, J. B. H. (2023). Opportunities to advance the synthesis of ecology and evolution. *Ecology Letters*, 26(S1), S11–S15. https://doi.org/10.1111/ELE.14175

Loreau, M., Naeem, S., Inchausti, P., Bengtsson, J., Grime, J. P., Hector, A., … & Wardle, D. A. (2001). Ecology: Biodiversity and ecosystem functioning: Current knowledge and future challenges. *Science*, 294(5543), 804–808. https://doi.org/10.1126/SCIENCE.1064088

Loreau, M., Sapijanskas, J., Isbell, F., & Hector, A. (2012). Niche and fitness differences relate the maintenance of diversity to ecosystem function: Comment. *Ecology*, 93(6), 1482–1487. https://doi.org/10.1890/11-0792.1

MacArthur, R., & Levins, R. (1967). The limiting similarity, convergence, and divergence of coexisting species. *The American Naturalist*, 101(921), 377–385.

Mace, G. M. (2014). Whose conservation? *Science*, 345(6204), 1558–1560. https://doi.org/10.1126/SCIENCE.1254704

Matsuoka, S., Mori, A. S., Kawaguchi, E., Hobara, S., & Osono, T. (2016). Disentangling the relative importance of host tree community, abiotic environment and spatial factors on ectomycorrhizal fungal assemblages along an elevation gradient. *FEMS Microbiology Ecology*, 92(5), fiw044.

Millennium Ecosystem Assessment (MEA). (2005). *Ecosystems and Human Well-Being: Biodiversity Synthesis*. https://www.millenniumassessment.org/en/index.html

Milne-Rostkowska, F., Holeksa, J., Bogdziewicz, M., Piechnik, Ł., Seget, B., Kurek, P., … & Żywiec, M. (2020). Where can palatable young trees escape herbivore pressure in a protected forest? *Forest Ecology and Management*, 472, 118221. https://doi.org/10.1016/J.FORECO.2020.118221

Mori, A. S. (2018). Environmental controls on the causes and functional consequences of tree species diversity. *Journal of Ecology*, 106(1), 113–125. https://doi.org/10.1111/1365-2745.12851

Mori, A. S. (2020). Advancing nature-based approaches to address the biodiversity and climate emergency. *Ecology Letters*, 23(12), 1729–1732. https://doi.org/10.1111/ELE.13594

Mori, A. S., Dee, L. E., Gonzalez, A., Ohashi, H., Cowles, J., Wright, A. J., … & Isbell, F. (2021). Biodiversity–productivity relationships are key to nature-based climate solutions. *Nature Climate Change*, 11(6), 543–550. https://doi.org/10.1038/s41558-021-01062-1

Mori, A. S., Fujii, S., Kitagawa, R., & Koide, D. (2014). Null model approaches to evaluating the relative role of different assembly processes in shaping ecological communities. *Oecologia*, 178(1), 261–273. https://doi.org/10.1007/S00442-014-3170-9

Mori, A. S., Isbell, F., & Seidl, R. (2018). β-diversity, community assembly, and ecosystem functioning. *Trends in Ecology & Evolution*, 33(7), 549–564. https://doi.org/10.1016/j.tree.2018.04.012

Mori, A. S., Lertzman, K. P., & Gustafsson, L. (2017). Biodiversity and ecosystem services in forest ecosystems: A research agenda for applied forest ecology. *Journal of Applied Ecology*, 54(1), 12–27. https://doi.org/10.1111/1365-2664.12669

Mueller, K. E., Tilman, D., Fornara, D. A., & Hobbie, S. E. (2013). Root depth distribution and the diversity–productivity relationship in a long-term grassland experiment. *Ecology*, 94(4), 787–793. https://doi.org/10.1890/12-1399.1

Naeem, S., Bunker, D. E., Hector, A., Loreau, M., & Perrings, C. (2009). Biodiversity, ecosystem functioning, and human wellbeing: An ecological and economic perspective. In *Biodiversity, Ecosystem Functioning, and Human Wellbeing: An Ecological and Economic Perspective* (Vol. 9780199547951). Oxford University Press. https://doi.org/10.1093/ACPROF:OSO/9780199547951.001.0001

Nisbet, E. K., Shaw, D. W., & Lachance, D. G. (2020). Connectedness with nearby nature and well-being. *Frontiers in Sustainable Cities*, 2, 523967. https://doi.org/10.3389/frsc.2020.00018

Nishizawa, K., Shinohara, N., Cadotte, M. W., & Mori, A. S. (2022). The latitudinal gradient in plant community assembly processes: A meta-analysis. *Ecology Letters*, 25(7), 1711–1724. https://doi.org/10.1111/ELE.14019

Nishizawa, K., Tatsumi, S., Kitagawa, R., & Mori, A. S. (2016). Deer herbivory affects the functional diversity of forest floor plants via changes in competition-mediated assembly rules. *Ecological Research*, 31(4), 569–578. https://doi.org/10.1007/S11284-016-1367-6

Paine, R. T. (1969). A note on trophic complexity and community stability. *The American Naturalist*, 103(929), 91–93. https://doi.org/10.1086/282586

Ranalli, R., Borghesan, S., Labra, M., & Biella, P. (2025). From cities to refuges for pollinators: Urban practices for enhancing pollinator habitats in changing landscapes. *Journal of Applied Ecology*, 62(11), 2874–2881. https://doi.org/10.1111/1365-2664.70005

Rogers, H. S., Donoso, I., Traveset, A., & Fricke, E. C. (2021). Cascading impacts of seed disperser loss on plant communities and ecosystems. *Annual Review of Ecology, Evolution, and Systematics*, 52, 641–666. https://doi.org/10.1146/annurev-ecolsys-012221-111742

Sargent, R. D., & Ackerly, D. D. (2008). Plant–pollinator interactions and the assembly of plant communities. *Trends in Ecology & Evolution*, 23(3), 123–130. https://doi.org/10.1016/J.TREE.2007.11.003

Schiestl, F. P., & Johnson, S. D. (2013). Pollinator-mediated evolution of floral signals. *Trends in Ecology & Evolution*, 28(5), 307–315. https://doi.org/10.1016/J.TREE.2013.01.019

Schmitz, O. J., Buchkowski, R. W., Burghardt, K. T., & Donihue, C. M. (2015). Functional traits and trait-mediated interactions: Connecting community-level interactions with ecosystem functioning. In *Advances in Ecological Research* (Vol. 52, pp. 319–343). Academic Press. https://doi.org/10.1016/BS.AECR.2015.01.003

Soliveres, S., Van Der Plas, F., Manning, P., Prati, D., Gossner, M. M., Renner, S. C., … & Allan, E. (2016). Biodiversity at multiple trophic levels is needed for ecosystem multifunctionality. *Nature*, 536(7617), 456–459. https://doi.org/10.1038/nature19092

Spasojevic, M. J., & Suding, K. N. (2012). Inferring community assembly mechanisms from functional diversity patterns: the importance of multiple assembly processes. *Journal of Ecology*, 100(3), 652–661. https://doi.org/10.1111/j.1365-2745.2011.01945.x

Suding, K. N., Lavorel, S., Chapin Iii, F. S., Cornelissen, J. H., DIAz, S., Garnier, E., ... & Navas, M. L. (2008). Scaling environmental change through the community-level: A trait-based response-and-effect framework for plants. *Global Change Biology*, 14(5), 1125–1140. https://doi.org/10.1111/j.1365-2486.2008.01557.x

Takeuchi, K. (2010). Rebuilding the relationship between people and nature: The Satoyama initiative. *Ecological Research*, 25(5), 891–897. https://doi.org/10.1007/S11284-010-0745-8

Tatsumi, S., & Loreau, M. (2023). Partitioning the biodiversity effects on productivity into density and size components. *Ecology Letters*, 26(11), 1963–1973. https://doi.org/10.1111/ele.14300

Tilman, D. (1982). Resource competition and community structure. In *Resource Competition and Community Structure. (M*PB-17*), V*olume 17. Princeton University Press. https://doi.org/10.2307/J.CTVX5WB72

Tilman, D., Knops, J., Wedin, D., Reich, P., Ritchie, M., & Siemann, E. (1997). The influence of functional diversity and composition on ecosystem processes. *Science*, 277(5330), 1300–1302. https://doi.org/10.1126/SCIENCE.277.5330.1300

Tilman, D., Lehman, C. L., & Bristow, C. E. (1998). Diversity-stability relationships: Statistical inevitability or ecological consequence? *The American Naturalist*, 151(3), 277–282. https://doi.org/10.1086/286118

Tilman, D., Wedin, D., & Knops, J. (1996). Productivity and sustainability influenced by biodiversity in grassland ecosystems. *Nature*, 379(6567), 718–720. https://doi.org/10.1038/379718a0

Tobner, C. M., Paquette, A., Reich, P. B., Gravel, D., & Messier, C. (2014). Advancing biodiversity–ecosystem functioning science using high-density tree-based experiments over functional diversity gradients. *Oecologia*, 174(3), 609–621. https://doi.org/10.1007/s00442-013-2815-4

Ulrich, J., & Sargent, R. D. (2025). Urban landscapes with more natural greenspace support higher pollinator diversity. *Ecological Applications*, 35(1), e70005. https://doi.org/10.1002/EAP.70005

van der Plas, F. (2019). Biodiversity and ecosystem functioning in naturally assembled communities. *Biological Reviews*, 94(4), 1220–1245. https://doi.org/10.1111/brv.12499

Vellend, M. (2010). Conceptual synthesis in community ecology. *The Quarterly Review of Biology*, 85(2), 183–206. https://doi.org/10.1086/652373

Viana, D. S., & Chase, J. M. (2019). Spatial scale modulates the inference of metacommunity assembly processes. *Ecology*, 100(2), e02576. https://doi.org/10.1002/ECY.2576

Villéger, S., Maire, E., & Leprieur, F. (2017). On the risks of using dendrograms to measure functional diversity and multidimensional spaces to measure phylogenetic diversity: A comment on Sobral et al. (2016). *Ecology Letters*, 20(4), 554–557. https://doi.org/10.1111/ELE.12750

Vonesh, J. R., Kraus, J. M., Rosenberg, J. S., & Chase, J. M. (2009). Predator effects on aquatic community assembly: Disentangling the roles of habitat selection and post-colonization processes. *Oikos*, 118(8), 1219–1229. https://doi.org/10.1111/j.1600-0706.2009.17369.x

Wallace, A. R. (1876). *The Geographical Distribution of Animals; with a Study of the Relations of Living and Extinct Faunas as Elucidating the Past Changes of the Earth's Surface* (Vol. 1). Harper & Brothers.

Wang, S., Hong, P., Adler, P. B., Allan, E., Hautier, Y., Schmid, B., Spaak, J. W., & Feng, Y. (2024). Towards mechanistic integration of the causes and consequences of biodiversity. *Trends in Ecology & Evolution*, 39(7), 689–700. https://doi.org/10.1016/J.TREE.2024.02.008

Webb, C. O., Ackerly, D. D., McPeek, M. A., & Donoghue, M. J. (2002). Phylogenies and community ecology. *Annual Review of Ecology and Systematics*, 33, 475–505. https://doi.org/10.1146/annurev.ecolsys.33.010802.150448

Weiher, E., Freund, D., Bunton, T., Stefanski, A., Lee, T., & Bentivenga, S. (2011). Advances, challenges and a developing synthesis of ecological community assembly theory. *Philosophical Transactions of the Royal Society B: Biological Sciences*, 366(1576), 2403–2413. https://doi.org/10.1098/RSTB.2011.0056

West, G. B., Brown, J. H., & Enquist, B. J. (1997). A general model for the origin of allometric scaling laws in biology. *Science*, 276(5309), 122–126. https://doi.org/10.1126/science.276.5309.122

Westphal, C., Vidal, S., Horgan, F. G., Gurr, G. M., Escalada, M., Van Chien, H., … & Settele, J. (2015). Promoting multiple ecosystem services with flower strips and participatory approaches in rice production landscapes. *Basic and Applied Ecology*, 16(8), 681–689. https://doi.org/10.1016/J.BAAE.2015.10.004

Whittaker, R. H. (1975). *Communities and Ecosystems* (2nd Revise Ed.). MacMillan Publishing Co.

Wiens, J. J., & Donoghue, M. J. (2004). Historical biogeography, ecology and species richness. *Trends in Ecology & Evolution*, 19(12), 639–644. https://doi.org/10.1016/J.TREE.2004.09.011

Willig, M. R., Kaufman, D. M., & Stevens, R. D. (2003). Latitudinal gradients of biodiversity: Pattern, process, scale, and synthesis. *Annual Review of Ecology, Evolution, and Systematics*, 34, 273–309. https://doi.org/10.1146/annurev.ecolsys.34.012103.144032

Wood, S. L. R., & Dupras, J. (2021). Increasing functional diversity of the urban canopy for climate resilience: Potential tradeoffs with ecosystem services? *Urban Forestry & Urban Greening*, 58, 126972. https://doi.org/10.1016/J.UFUG.2020.126972

Wotton, D. M., & Kelly, D. (2011). Frugivore loss limits recruitment of large-seeded trees. *Proceedings of the Royal Society B: Biological Sciences*, 278(1723), 3345–3354. https://doi.org/10.1098/RSPB.2011.0185

Yachi, S., & Loreau, M. (1999). Biodiversity and ecosystem productivity in a fluctuating environment: The insurance hypothesis. *Proceedings of the National Academy of Sciences of the United States of America*, 96(4), 1463–1468. https://doi.org/10.1073/PNAS.96.4.1463

Zheng, L. T., Chen, H. Y., Biswas, S. R., Bao, D. F., Fang, X. C., Abdullah, M., & Yan, E. R. (2021). Diversity and identity of economics traits determine the extent of tree mixture effects on ecosystem productivity. *Journal of Ecology*, 109(4), 1898–1908. https://doi.org/10.1111/1365-2745.13614

14 Incorporating Biodiversity Conservation in the Restoration of Fully Functioning Subtropical Evergreen Forest at Kadoorie Farm and Botanic Garden in Hong Kong

An Overview of Principles and Practices

Jinlong Zhang, Fernanda C. G. Cardoso, Huiling Zhu, Mang Lung Cheuk, and Stephan W. Gale

14.1 INTRODUCTION

14.1.1 Background and Aims: Deforestation in South China and Hong Kong

The climate of South China is warm and humid, especially in the Pearl River Delta region, which receives abundant rainfall, with a mean annual rainfall of over 1,600 mm (Zhang et al., 2019), rising to 2,600–3,300 mm in mountainous areas (Hong Kong Observatory, 2023). Accordingly, tall, subtropical evergreen forest is thought to have covered the entire region before human impact (Ashton, 2014; Zheng et al., 2003). However, due to centuries of slash-and-burn agriculture and logging for timber and firewood, the region now has only comparatively small and scattered remaining patches of high diversity natural forest (Dudgeon & Corlett, 2011). It is believed that by at least the Ming (1368–1644) and Qing (1644–1912) Dynasties, the forests of the Pearl River Delta had already been largely depleted (Dudgeon & Corlett, 2011).

During the Qing dynasty, Hong Kong (Figure 14.1) was part of Xin'an County, which was under the jurisdiction of Guangzhou Prefecture (Ng, 1983). In 1841, after Hong Kong was ceded to Britain, the British Foreign Secretary Lord Palmerston (1784–1865) described it as "A barren island with hardly a house upon it" (Owen & Shaw, 2007). Of course, this was inaccurate—at that time, about 7,000 people lived on Hong Kong Island, in 16 villages (Hong Kong Chronicles Institute, 2022). Agarwood (*Aquilaria sinensis*) was widely planted, and Hong Kong itself had been an important transit port for the agarwood trade for several hundred years previously, which is assumed to be how the island got its name (the vernacular for agarwood being *heung* and *gong* meaning harbor in

DOI: 10.1201/9781003640899-18

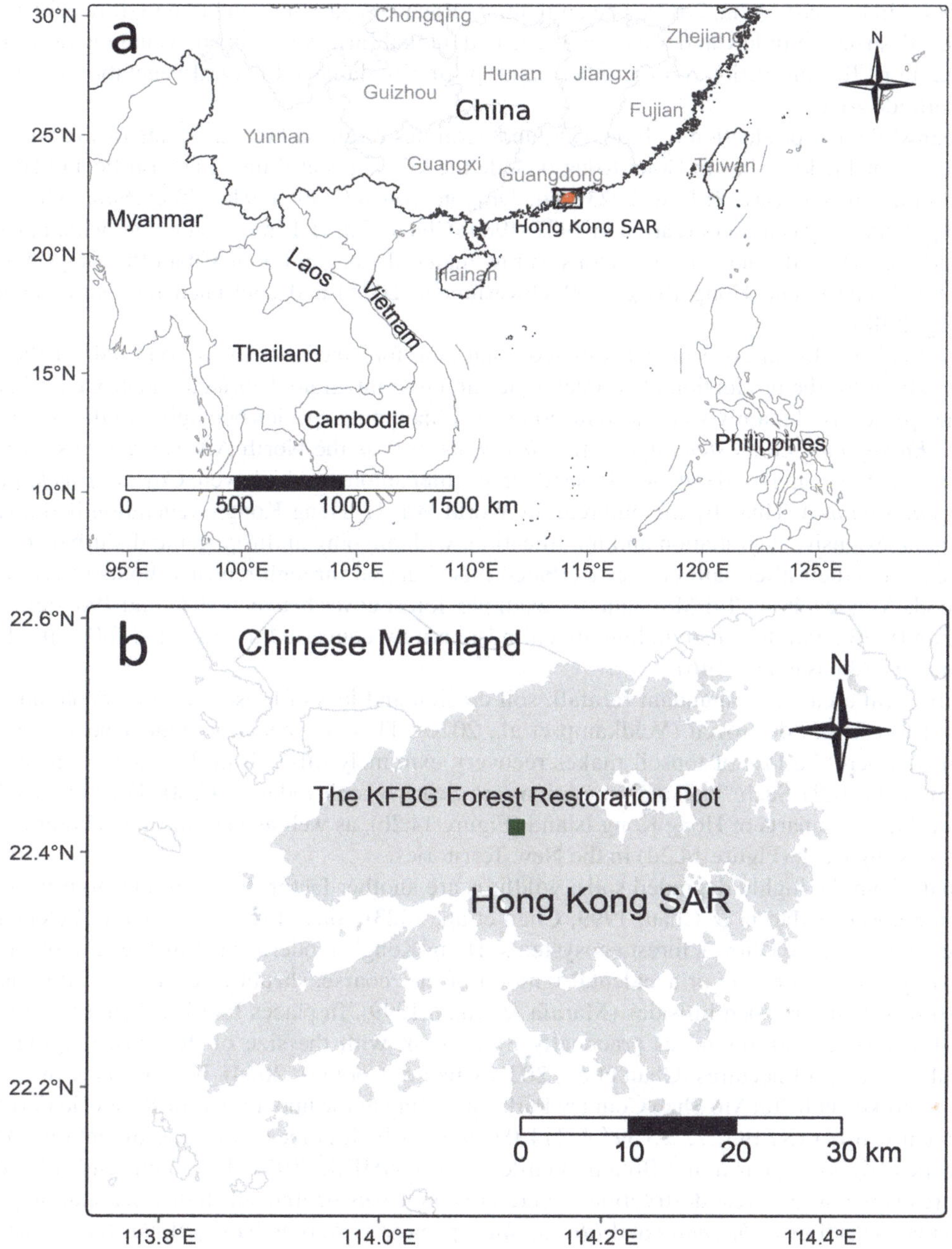

FIGURE 14.1 Location of Hong Kong Special Administrative Region (a) and Kadoorie Farm and Botanic Garden (KFBG), where its forest restoration plot is situated (b).

Cantonese; Jim, 2015). In addition to plantations of agarwood, Hong Kong also produced tea, salt, porcelain, and lime during the Ming and Qing Dynasties (Hong Kong Chronicles Institute, 2022).

Tea was typically grown at elevations above 500 m (Planning Department, Hong Kong SAR Government, 2025). Even today, at mid- to high-altitude areas of Tai Mo Shan (957 m a.s.l., Hong Kong's highest peak; 22°24′42″N, 114°7′23″E) and Tai To Yan (566 m a.s.l., a ridge north of Tai Mo Shan; 22°27′1″N, 114°7′5″E) in the New Territories, one can still clearly see ancient terraced fields built from piled stones, believed to have been used for tea cultivation. Rice farming was also a vital livelihood in the past (Atha, 2012). The cultivation of native agarwood and rice mainly contributed to forest destruction in lowland areas, and repeated annual farming cycles prevented natural forest regeneration. Tea cultivation, on the other hand, primarily damaged natural vegetation in mid- to high-altitude zones.

Meanwhile, the production of lime, salt, and ceramics required large amounts of firewood and charcoal. Notable kiln sites in Hong Kong include the kiln ruins at Wun Yiu in Tai Po and Ha Law Wan on Lantau Island (Ho & Law, 2023). The Dragon Kiln at Castle Peak in Tuen Mun, which produced pottery, only ceased operation in 1982–1984 (Chuen, 2019). Lime production, which primarily used marine shells and coral skeletons as raw materials, began no later than the Tang Dynasty (Planning Department, Hong Kong SAR Government, 2025) and continued for many centuries (Delang, 2018).

The British colonial government initiated plantation forestry from the second half of the 19th century, both for the production of wooden poles and to reintegrate degraded and eroded hillsides, but the species used mainly were introduced exotics (including species belonging to the Australian genera *Eucalyptus*, *Lophostemon* and *Melaleuca*, as well as the North American *Pinus elliottii*), with native trees being used only to a lesser degree (chief amongst which were *Castanopsis fissa* and *Pinus massoniana*). Thus, by the outbreak of World War II, Hong Kong's vegetation had already undergone extensive degradation and modification, with ongoing anthropogenic disturbances. The war years and their aftermath only exacerbated this situation through repeated bouts of felling for firewood. As a result, on Tai Mo Shan, for example, forest cover between 1945 and 1963 remained less than 0.17%, with the surrounding area almost entirely converted to grassland following forest destruction (Abbas et al., 2016).

In tropical areas with abundant rainfall, soil erosion and loss of topsoil are inevitable once the original forests are destroyed (Veldkamp et al., 2020). This is especially evident near mountain summits, where the loss of topsoil makes recovery extremely difficult in the short term (but see Poorter et al., 2021). Such "badland" conditions are clearly visible today on Castle Peak in Tuen Mun (Figure 14.2a), on parts of Hong Kong Island (Figure 14.2b), as well as Pat Sin Leng (Figure 14.2c) and Kai Kung Leng (Figure 14.2d) in the New Territories.

Apart from the highly degraded soils, wildfires are another factor that continues to hinder vegetation recovery (Marafa & Chau, 1999; Chan et al., 2023). Since fire is not a natural element in the ecology of South China's forest ecosystems, Hong Kong's modern-day hill fires are invariably caused by either deliberate or accidental ignition of dry, coarse, shrubby vegetation, often during ancestral worship on open hillsides (Marafa & Chau, 1999). In places like Kai Kung Leng in the New Territories, wildfires occur practically every year, with the size of the burn ranging from several to dozens of hectares (Chan et al., 2023; Chan & Coomes, 2024). For instance, in 2004, a wildfire broke out in Tai Mo Shan Country Park, spreading to the upper areas of Kadoorie Farm and Botanical Garden (KFBG; 22°25′39″N, 114°07′19″E), which is located on the mountain's northern slopes (Kadoorie Farm and Botanic Garden, 2006; NHER, 2017). The main reason for these wildfires is that after forest destruction, severe erosion leaves behind substrates that can only support grasses such as *Ischaemum* and *Miscanthus* species, as well as hardy, clonal ferns including *Dicranopteris pedata* and *Blechnum orientale*, with only scattered individuals of common woody pioneers such as *Baeckia frutescens*, *Polyspora axillaris*, and *Rhodomyrtus tomentosa* (Zhang et al., 2025). Due to the effective absence of diverse, active species pools at the landscape scale, as well as the persistent extreme soil degradation compounded by repeated fire, these areas often cannot independently recover into forest, entering what has been referred to as an "alternative stable state" (Beisner et al., 2003) typified by "arrested succession" (Mata et al., 2022).

This chapter aims to highlight the challenges associated with the recovery of natural forests. It provides a concise overview of the restoration practices currently adopted by Kadoorie Farm and Botanic Garden (KFBG), a non-governmental organization that has been engaged in native species-based forest restoration since 1997. Science-based forest restoration experiments aiming at restoring the whole forest ecosystem and its functions using a diverse set of native plant species

FIGURE 14.2 Various degrees of forest degradation in Hong Kong. (a) Castle Peak, Tuen Mun. The vegetation is severely degraded, with topsoil erosion exposing the sandy substrate. Only a limited number of species can survive in such conditions. (b) Shrubland dominated by *Lepidosperma chinense* (Cyperaceae) at Mount Parker, Hong Kong Island. (c) Woodland showing partial recovery at Pat Sin Leng, New Territories. (d) Kai Kung Leng, located in the north-west New Territories. Although vegetation has recovered over the years through active restoration, forest regeneration on the upper hillside has been hindered by frequent hill fire. ((a–c) Photo by Zhang, 2014; (d) photo by Zhang, 2023.)

started in 2013. We here provide information on the principles considered and practices adopted. In particular, we emphasize the importance of incorporating biodiversity conservation into the restoration of forest ecosystems and demonstrate how key facilities—such as herbaria and nurseries—can support the accumulation of botanical and ecological knowledge and skillsets that are key to forest restoration. Additionally, we introduce the Global Biodiversity Standard (TGBS), launched in 2024 by Botanic Gardens Conservation International (BGCI) and the Society for Ecological Restoration (SER), and light the significant role botanical gardens play in meeting the criteria outlined by the TGBS framework.

14.2 CHALLENGES AND OPPORTUNITIES IN FOREST RESTORATION IN HONG KONG

14.2.1 Hong Kong's Remaining Biodiversity Hotspots and the Initiation of Active Forest Restoration

Currently, approximately 40% (44,842 ha) of Hong Kong's land area is designated as protected areas, referred to as Country Parks (Agriculture, Fisheries and Conservation Department, Hong Kong SAR Government, 2025; Lands Department of Hong Kong SAR Government, 2025). Since the mid-1970s, plant harvesting and logging have been strictly prohibited and regulated within these parks. Through a gradual process of natural succession and in some cases plantation forestry, formerly grassland-dominated slopes have, where hill fires have been effectively reduced or eliminated, regained some canopy cover, generally of limited and skewed taxonomic diversity (Abbas et al., 2016; Corlett, 1999). Before the 1970s, Hong Kong saw extensive planting of the native pine species, *Pinus massoniana* (Masson pine), as well as exotic species from Australia and North America, such as *Eucalyptus camaldulensis, Lophostemon confertus, Melaleuca quinquenervia, Casuarina equisetifolia* and *Pinus elliottii* (slash pine) (Corlett, 1999; Nicholson, 1996). Other commonly planted exotic species included *Acacia confusa, A. auriculiformis* and *A. mangium* (Hu & Wu, 2008; Lee, Williams, et al., 2023). In recent years, many of these species have declined and spontaneously regenerating forests are increasingly composed of common native species, such as *Machilus chekiangensis* and *Psychotria asiatica* (Abbas et al., 2016). However, the community structure of these associations remains simple and homogenic, with few discernible strata and a striking absence of understorey herbs, lianas and epiphytes, in marked contrast to the multi-layered and far richer *feng shui* woodlands that are found scattered across Hong Kong, though predominantly at lower elevation and generally in close proximity to New Territories villages. These woodlands are believed to constitute semi-natural pockets of original forest cover that were protected for folkloric or geomantic reasons (Lee et al., 2023). Indeed, analysis of species composition of naturally regenerating woody communities on Tai Mo Shan has revealed that they do not recover the taxonomic or functional diversity, nor the structural complexity, that might be expected of the mature natural forests of the region, even after decades of regrowth, implicating "arrested succession" and underscoring the inference that canopy cover alone may be a poor indicator of ecological functioning (Abbas et al., 2019; Zhu et al., 2023b).

The isolated *feng shui* woodlands encountered near villages at lower elevations are typically very small—usually less than 1 ha in land area—and they continue to suffer varying degrees of disturbance and degradation (Chu, 1998; Lee et al., 2023; Zhuang & Corlett, 1997). In recent years, the transmission of traditional culture and beliefs has weakened, and villagers' awareness of the value of *feng shui* woodland conservation has declined (Lee et al., 2023). The understorey of some *feng shui* woodlands has even been cleared, and the remaining patches are increasingly encroached upon and converted to other land uses, leading to a decline in both area and biodiversity.

Across Hong Kong's modern, highly modified landscape matrix, the areas with the highest remaining plant species diversity include its *feng shui* woodlands (as mainly reported in Chu (1998)) as well as many of the more remote streams and ravines of the numerous rugged mountain blocks, with notable examples including Wong Lung Hang and the vicinity of Lantau Peak on Lantau Island, and Ma On Shan, Ng Tung Chai in Lam Tsuen, Tai Shing Rocky Stream and Tai Lui Rocky Stream on Tai Mo Shan, New Territories. In many of these locations, in excess of 100–200 tree species can often be found within just a small patch (1–2 ha). Common plant families include Moraceae, Lauraceae, Magnoliaceae, Fagaceae, Rubiaceae, Hamamelidaceae, Sapindaceae, Phyllanthaceae, Meliaceae, Aquifoliaceae, Symplocaceae, and Euphorbiaceae (Hu & Wu, 2008). Frequently encountered genera include *Ficus, Beilschmiedia, Litsea, Cinnamomum, Machilus, Litsea, Psychotria, Syzygium, Castanopsis, Lithocarpus, Quercus, Psychotria, Camellia, Ilex, Symplocos, Croton*, and *Sapindus*. Additionally, lianas belonging to Fabaceae (e.g., *Millettia* and *Bauhinia*) and Annonaceae (e.g., *Desmos, Fissistigma*, and *Uvaria*) may become locally abundant, as can epiphytic or lithophytic orchids, reflecting the tropical nature of the flora (Hu & Wu, 2008; Gale et al., 2014; Zhu et al., 2023b).

14.2.2 Plant Diversity and Recent Botanical Discoveries in Hong Kong

Hong Kong was one of the earliest places in East Asia to be explored and studied by Western plant collectors and botanists. George Bentham's *Flora Hongkongensis* published in 1861, is also the earliest flora published for the British colonies (Bentham, 1861). Over the past century, numerous botanists, ecologists, and plant enthusiasts have collected and studied plants in Hong Kong. Despite this long history, new species and new distribution records continue to be reported (Figure 14.3). For example, the new species *Syzygium impressum* N.H.Xia, Y.F.Deng & K.L.Yip (2008), *Carpinus*

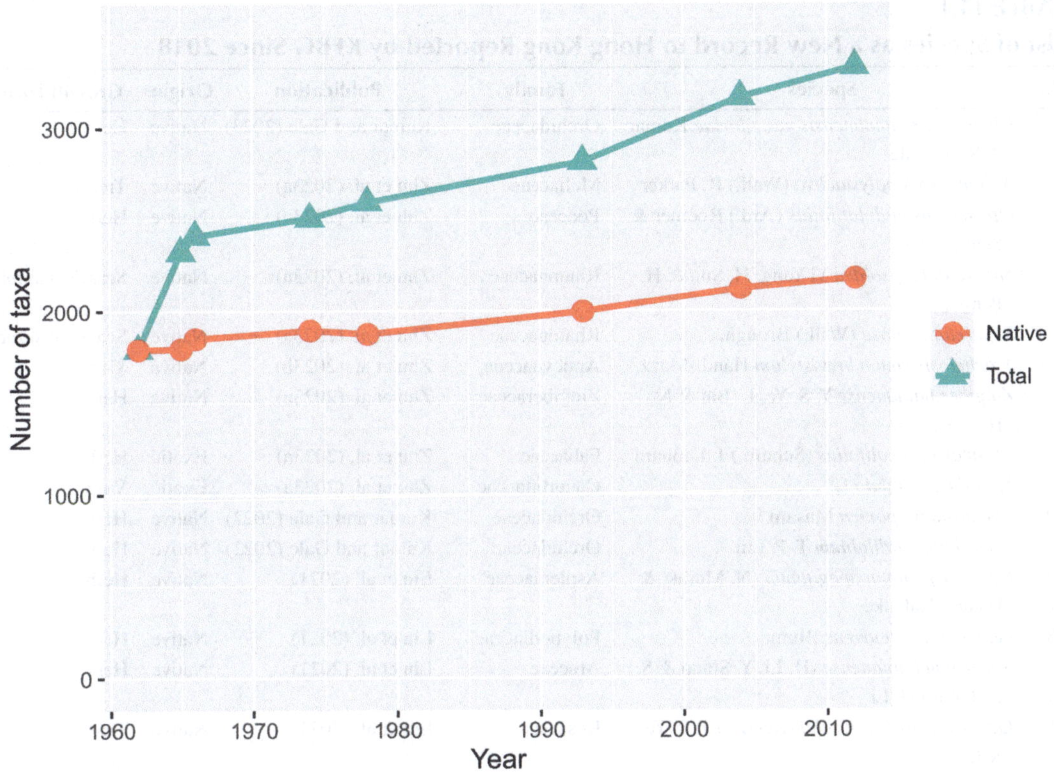

FIGURE 14.3 Number of plant taxa recorded in Hong Kong. (Source: *Check List of Hong Kong Plants 2012* Hong Kong Herbarium, 2012.)

insularis N.H.Xia, K.S.Pang & Y.H.Tong (2014), *Gastrochilus kadooriei* Kumar, S.W.Gale, Kocyan, G.A.Fisch. & Aver. (2014), and *Thismia hongkongensis* Mar & R.M.K.Saunders (2015) were all described within the past two decades. Meanwhile, new distribution records have accumulated at a rate of several species per year, with more than 40 plant species being reported by the Flora Conservation Department of KFBG alone since 2018 (Table 14.1). While some of these represent newly colonized exotic species, the vast majority are native to Hong Kong, reflecting the region's rich biodiversity and the ongoing need for exploration. These findings have contributed to the total number of native higher plant species in Hong Kong, which now exceeds 2,200, including more than 600 woody taxa.

However, many species have minimal populations. For instance, the strict endemics *Sinopora hongkongensis* (N.H.Xia, Y.F.Deng & K.L.Yip) J.Li, N.H.Xia & H.W.Li (2008) and *Carpinus insularis* are, respectively, only known from a single mature individual on Tai Mo Shan and from two small populations on Violet Hill on Hong Kong Island and Ma On Shan, amounting to an estimated population size of <50 individuals (Qin et al., 2018). Some species that are common outside Hong Kong, such as *Xanthophyllum hainanense*, can only be found in a few *feng shui* woodlands in the New Territories (Chu, 1998).

Understanding species occurrences, especially for rare and threatened taxa, is central to compiling local checklists and floras, which are in turn fundamental to ecological research and the development of effective biodiversity conservation strategies—particularly in the context of restoration ecology. For all of these purposes, herbaria play a key role in the discovery, documentation, and analysis of species and their biotic and abiotic interactions (see Section 14.3.1).

TABLE 14.1
List of Species as a New Record to Hong Kong Reported by KFBG Since 2018

No.	Species	Family	Publication	Origin[a]	Growth Form
1	*Cheirostylis yunnanensis* var. *gloriae* Kumar & S. W. Gale	Orchidaceae	Kumar and Gale (2023)	Native	Herb
2	*Aphanamixis polystachya* (Wall.) R. Parker	Meliaceae	Zhu et al. (2023a)	Native	Tree
3	*Oplismenus undulatifolius* (Ard.) Roemer & Schuit.	Poaceae	Zhu et al. (2023a)	Native	Herb
4	*Sageretia ellipsoidea* Yi Yang, H. Sun & H. Peng	Rhamnaceae	Zhu et al. (2023a)	Native	Scandent shrub
5	*Sageretia hamosa* (Wall.) Brongn.	Rhamnaceae	Zhu et al. (2023a)	Native	Scandent shrub
6	*Trachelospermum brevistylum* Hand.-Mazz.	Apocynaceae	Zhu et al. (2023b)	Native	Vine
7	*Zingiber hainanense* Y. S. Ye, L. Bai & N. H. Xia	Zingiberaceae	Zhu et al. (2023a)	Native	Herb
8	*Alysicarpus ovalifolius* (Schum.) J. Léonard	Fabaceae	Zhu et al. (2023a)	Exotic	Herb
9	*Melothria pendula* L.	Cucurbitaceae	Zhu et al. (2023a)	Exotic	Vine
10	*Apostasia nipponica* Masam.	Orchidaceae	Kumar and Gale (2022)	Native	Herb
11	*Crepidium cordilabium* T. P. Lin	Orchidaceae	Kumar and Gale (2022)	Native	Herb
12	*Hymenasplenium apogamum* (N. Murak. & Hatan.) Nakaike	Aspleniaceae	Liu et al. (2021)	Native	Herb
13	*Leptochilus decurrens* Blume	Polypodiaceae	Liu et al. (2021)	Native	Herb
14	*Amydrium hainanense* (H. Li, Y. Shiao & S. L. Tseng) H. Li	Araceae	Liu et al. (2021)	Native	Herb
15	*Laurocerasus hypotricha* (Rehder) T. T. Yu & L. T. Lu	Rosaceae	Liu et al. (2021)	Native	Tree
16	*Rubus buergeri* Miq.	Rosaceae	Liu et al. (2021)	Native	Shrub
17	*Quercus pachyloma* Seemen	Fagaceae	Liu et al. (2021)	Native	Tree
18	*Lithocarpus glaucus* Chun & C. C. Huang ex H. G. Ye	Fagaceae	Liu et al. (2021)	Native	Tree
19	*Lithocarpus longanoides* C. C. Huang & Y. T. Chang	Fagaceae	Liu et al. (2021)	Native	Tree
20	*Murraya euchrestifolia* Hayata	Rutaceae	Liu et al. (2021)	Native	Shrub
21	*Diospyros chunii* F. P. Metcalf & L. Chen	Ebenaceae	Liu et al. (2021)	Native	Tree
22	*Myrsine linearis* (Lour.) Poir.	Primulaceae	Liu et al. (2021)	Native	Tree
23	*Myrsine semiserrata* Wall.	Primulaceae	Liu et al. (2021)	Native	Shrub
24	*Schima remotiserrata* Hung T. Chang	Theaceae	Liu et al. (2021)	Native	Tree
25	*Heterostemma brownii* Hayata	Apocynaceae	Liu et al. (2021)	Native	Vine
26	*Urceola napeensis* (Quint.) D. J. Middleton	Apocynaceae	Liu et al. (2021)	Native	Herb
27	*Utricularia minutissima* Vahl	Lentibulariaceae	Liu et al. (2021)	Native	Herb
28	*Blumea aromatica* DC.	Asteraceae	Liu et al. (2021)	Native	Herb
29	*Calanthe × dominyi* Lindl.	Orchidaceae	Gale et al. (2020)	Native	Herb
30	*Chrysoglossum assamicum* Hook. f.	Orchidaceae	Gale et al. (2020)	Native	Herb
31	*Gastrodia clausa* T. C. Hsu, S. W. Chung & C. M. Kuo	Orchidaceae	Gale et al. (2020)	Native	Herb
32	*Liparis sootenzanensis* Fukuy.	Orchidaceae	Gale et al. (2020)	Native	Herb
33	*Anoectochilus formosanus* Hayata	Orchidaceae	Kumar and Gale (2020)	Native	Herb
34	*Haplopteris elongata* (Swartz) E. H. Crane	Pteridaceae	Zhu et al. (2018)	Native	Herb
35	*Amydrium sinense* (Engl.) H. Li	Araceae	Zhu et al. (2018)	Native	Vine
36	*Pothos repens* (Lour.) Druce	Araceae	Zhu et al. (2018)	Native	Vine
37	*Cyclea sutchuenensis* Gagnep.	Menispermaceae	Zhu et al. (2018)	Native	Vine
38	*Dumasia truncata* Siebold & Zucc.	Fabaceae	Zhu et al. (2018)	Native	Vine

(*Continued*)

TABLE 14.1 (*Continued*)
List of Species as a New Record to Hong Kong Reported by KFBG Since 2018

No.	Species	Family	Publication	Origin[a]	Growth Form
39	*Aucuba obcordata* (Rehder) Fu ex W. K. Hu et Soong	Garryaceae	Zhu et al. (2018)	Native	Tree
40	*Ficus stenophylla* Hemsl.	Moraceae	Zhu et al. (2018)	Native	Shrub
41	*Ficus gasparriniana* var. *esquirolii* (H. Lév. & Vaniot) Corner	Moraceae	Zhu et al. (2018)	Native	Shrub
42	*Ilex ficifolia* C. J. Tseng ex S. K. Chen et Y. X. Feng	Aquifoliaceae	Zhu et al. (2018)	Native	Tree
43	*Paraprenanthes sororia* (Miq.) C. Shih.	Asteraceae	Zhu et al. (2018)	Native	Herb

[a] Whether the species is native to Hong Kong SAR, China.

14.2.3 Overcoming Extinction Debt

For most rare species, small population sizes and biological or geographic isolation pose significant risks of extinction. Small populations often exhibit low genetic diversity (Vrijenhoek, 1994), and if there is limited effective gene flow with external populations, population decline may be inevitable (Bohonak, 1999). Due to the long life cycles of certain trees and other long-lived plants, such declines may not be immediately noticeable. Without appropriate intervention to address the depauperate genetic diversity of these populations, reproductive failure, inbreeding, and recruitment failure may persist (Zeng & Fischer, 2021). In other words, even if extinction does not occur immediately, it may be a danger in the future—which is known as ***extinction debt*** (Hylander & Ehrlén, 2013; Kuussaari et al., 2009). To reduce the risk of extinction, it is important to both study the causes of potential extinction and start conservation actions, which may include collecting seeds or pollen from different populations to facilitate cross-pollination or introducing new individuals into existing populations (Bohonak, 1999).

For tree species, such interventions are rarely reported. However, an applied study involving genetic profiling, hand pollination, and propagation of the extremely rare and self-incompatible orchid *Bulbophyllum bicolor* (Lindl.) Hook.f. in Hong Kong and surrounding regions, led by KFBG's botanists, has enabled the successful reintroduction of micropropagated progeny with high genetic diversity (Hu et al., 2017; Gale et al., 2019). This orchid is critically endangered and grows epiphytically on tree trunks and cliffs along streams. Due to vegetation destruction and illegal harvesting, it is now sparsely distributed across Hong Kong, Shenzhen (in Guangdong), and Guangxi, with distances between remnant monoclonal populations often exceeding several dozen kilometers, precluding the possibility of natural gene flow by pollen for successful fruit set. The species relies on fly pollination and is self-sterile, meaning that although individual genets may flower normally, they will be unable to produce viable seeds without cross-pollination between genetically distinct clones. By collecting pollen from different populations and conducting hand pollinations, KFBG researchers successfully obtained fertile seeds (Flora Conservation Department, Kadoorie Farm and Botanic Garden, 2025). These were propagated into seedlings and reintroduced into suitable habitats in the wild.

14.2.4 Challenging Conventional Approaches to Reforestation: Alien Invasive Species, Winner Species, and Arrested Succession

Traditionally, reforestation has often placed emphasis on practical outcomes, such as seedling survival rates and canopy coverage, whilst overlooking impacts in terms of biodiversity and ecosystem function (Wortley et al., 2013). This has led to the selection of species deemed most suitable for rapid regeneration, including many non-native species such as (in a Hong Kong context) *Acacia confusa*, *Eucalyptus* spp. and *Pinus elliottii*—species known for their resilience and hardiness (Reid et al., 2018). In the early days of post-disturbance land management, when survival and coverage were paramount, planting these species was understandable. However, recent studies and observations have revealed that many of the seedlings planted decades ago have now matured into large trees, with some proving to be strong colonizers of an invasive character, whilst others have aged and show a high risk of collapsing (Abbas et al., 2016). Not surprisingly, therefore, these planted forests are increasingly leading to the proliferation of non-native species in natural areas, whilst in other cases they are progressively being replaced by robust, native "winner" species of high reproductive potential, such as *Machilus chekiangensis* and *Psychotria asiatica*, that are able to recruit into the species-poor, drought-prone and low light environments that these aging stands present (Albrecht et al., 2022; Zhu et al., 2023b). Once the "winner" species establishes itself and forms skewed communities, the resulting vegetation is typically very simple in structure and species composition, but also more or less stable. On the upper slopes of many New Territories mountains, including at KFBG, sites formerly characterized by fire-maintained grassland or shrubland have become dominated by *M. chekiangensis*, which, over time, forms more or less monospecific stands with a very shaded understorey, making it difficult for other, rarer, native tree seedlings to establish and grow. Because these winning species dominate early on, the "priority effect" takes hold (Fukami, 2015; Fukami et al., 2016), whereby early colonizers influence the trajectory of community development, resulting in a relatively locked-in structure and composition typical of long-term, alternative stable ecological states (Mata et al., 2022).

14.2.5 Forest Dynamics Plots as Windows onto Forest Diversity and Ecological Functioning

The restoration of high-quality forest is, in fact, an exercise in restoring species composition and ecosystem functions (Bartholomew et al., 2024). A healthy forest provides a wide range of ecological services, including providing clean water, reduction of soil erosion, resistance to droughts and floods, providing suitable habitat for flora and fauna, mitigation of wind speed and sand damage, climate regulation, and carbon sequestration (Aerts & Honnay, 2011; Mori et al., 2017; Zhu et al., 2024b). Species diversity, structure and associated distribution patterns—including abundance, spatial arrangement, and population age structure—are crucial for securing these ecosystem functions and ensuring a forest's stability and resilience (Chen et al., 2025).

Understanding the process of forest recovery and the mechanisms that maintain biodiversity is essential to ecology. Traditional phytosociological methods typically involve establishing smaller plots (e.g., 20 m×20 m) and surveying them by layers—canopy, shrub and herbaceous—to understand species composition and structure (for example, Fang et al., 2004b; Xiao et al., 2022). However, such plots usually only reflect the status at the time of survey, often fail to capture seasonal and long-term changes.

In the 1980s, American scientists Stephen Hubbell, Robin Foster and colleagues established the world's first forest monitoring permanent plot—the BCI (Barro Colorado Island) plot in Panama (Condit, 1998; Hubbell et al., 2024). Based on their survey protocol, over 80 large forest monitoring plots have been established across continents, resulting in the documentation of more than 12,000 species and the formation of the ForestGEO research network in 2025 (Anderson-Teixeira et al., 2015; Davies et al., 2021; *ForestGEO*, 2025). Between 2012 and 2015, Billy Hau, an ecologist of the University of Hong Kong, in collaboration with the Flora Conservation Department of KFBG, established a 20-ha forest monitoring plot in Hong Kong's Tai Po Kau Nature Reserve (Zhu et al., 2024a). The plot measures 500 m north-south and 400 m east-west, within which over 81,100 woody plants belonging to 172 species, each with a diameter at breast height (DBH) ≥1 cm, were mapped, identified and measured. This is the first forest dynamics plot in Hong Kong, and is the only one explicitly established in secondary forest globally (Zhu et al., 2023b).

Historically, the Tai Po Kau site was heavily planted with the native *Pinus massoniana*, as well as other exotic species, in the 1950s. At that time, native woody plants mainly were confined to streamside areas. By comparing the species distribution identified in aerial photographs taken in the 1950s and 1960s with recent survey data, researchers found that areas previously planted with *P. massoniana* still exhibit significantly lower species diversity than valleys with a more natural signature of passive recovery (Zhu et al., 2023b). This suggests that many native species have been unable to spread into low-diversity areas dominated by historical plantations, even after canopy formation during the 1980s. This finding highlights the importance of identifying and protecting remnant native vegetation in stream valleys and illustrates the priority effect where "winner species" (Albrecht et al., 2022) rapidly establish communities in suitable habitats, precluding the natural recovery of a wider cross-section of the original forest assemblage. Once such communities are formed, they may enter a long-term "alternative stable state" (Beisner et al., 2003). Observations show that in species-poor surrounding areas, this stable state can persist for decades with little change in species diversity or composition (Zhu et al., 2023b). Even when adjacent to species-rich forest patches, indicator species of mature forests rarely succeed in colonizing these locked-in communities. This phenomenon challenges the traditional textbook model of forest succession, offering valuable insights into the restoration of forest biodiversity, structure, and function. The results also indicate that assisted restoration may be needed in areas dominated by common "winner" species through enhancement or enrichment planting (Mangueira et al., 2019).

14.3 PRINCIPLES AND PRACTICES OF FOREST RESTORATION AT KFBG

14.3.1 Facilities Facilitating Forest Restoration at KFBG

14.3.1.1 Herbarium

Herbarium specimens usually considered the foundation of plant taxonomy and biodiversity (Davies et al., 2023). Collectors gather specimens from specific regions or representing specific taxonomic groups, and taxonomists study these specimens to describe new species, compile regional floras. Recently, herbarium specimens have been shown to be useful in restoration (Bowles et al., 1993; Davis, 2023). For example, the distribution and phenology of a species, which are critical for forest restoration, can be gleaned from well-curated herbarium specimens. Restoration practitioners can also use herbarium records to identify historical sites and list the suitable species for restoration (Hoffmann et al., 2015). Additionally, herbarium records and phenology data can guide seed collection (Vander Mijnsbrugge et al., 2010). Herbaria are also an ideal place to deposit voucher specimens arising from surveys of natural or planted areas, since it will be essential to collect herbarium specimens throughout the restoration cycle (Figure 14.4).

14.3.1.2 Native Tree Nursery

KFBG's Native Tree Nursery (Figure 14.5) was established in 1997 (Hau, 1999) and produces more than 30,000 native tree seedlings each year. KFBG staff collect seeds of native tree species from Hong Kong's Country Parks with the permission of the Agriculture, Fisheries and Conservation Department of the Hong Kong SAR Government. These seeds are cleaned, sew, germinating, and managed following tailored protocols (Wilkinson et al., 2014). Similar to herbarium specimen collection, seeds gathered from the wild must be accompanied by detailed records, including species name, collector, collection number, location, GPS coordinates, elevation, and a brief description. All this information is recorded in a field collection book (Figure 14.6), which is automatically generated by a database.

After reaching a specific size, seedlings are transplanted into individual seedling bags. Air-pruning pots (Figure 14.5d) are used for some species that are grown to larger sizes or tend to develop taproots at an early stage, such as oak species. Seedlings typically remain in the nursery for 12–18 months after germination and are usually planted when they reach 25–30 cm in height, to

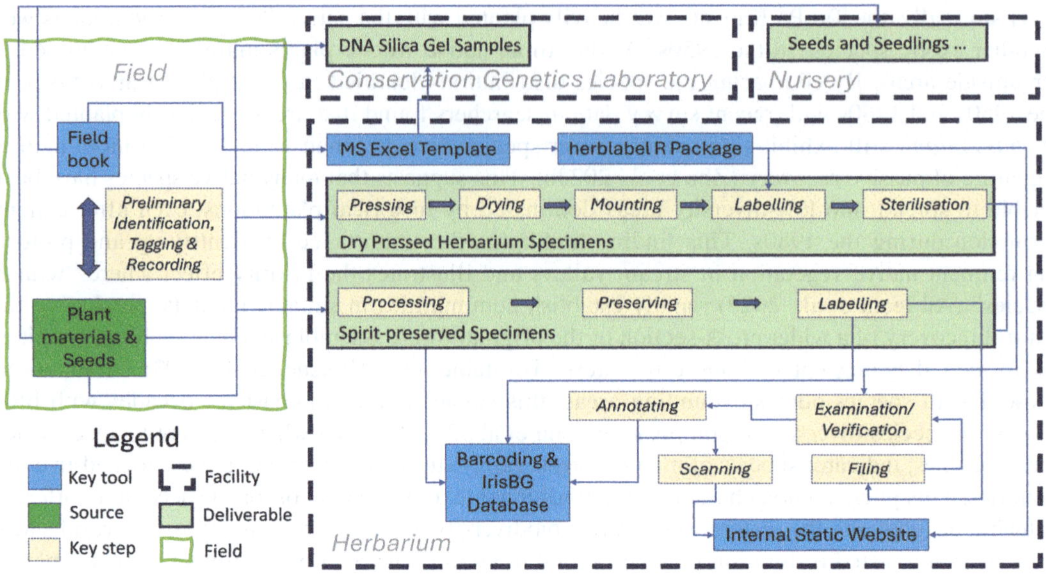

FIGURE 14.4 Role of the herbarium at KFBG. In the field, collectors gather plant specimens and record relevant information, including collector's name, date of collection, location, species identity and ecological notes on both the species and its environment. This information is entered into a standardized field book, with pages generated using Microsoft Access. Collected plant material is sent to the herbarium for processing—including drying, mounting and preparation of herbarium sheets or spirit specimens. Seeds or fruits are passed to the nursery for propagation, while DNA samples preserved in silica gel may be processed and stored in the Conservation Genetics Laboratory. Data recorded in the field book are entered into a standardized template based on the Darwin Core data exchange standard. These records can be used to generate specimen labels using the R package herblabel (Zhang et al., 2016) and to produce static websites. This diagram illustrates the deliverables and workflow (indicated by blue arrows).

prevent them from developing girdled roots due to prolonged pot-bound conditions (Zhang et al., 2025). Every seedling is tagged with a unique ID, and the species name, accession number, collector, collection number, collection site, and collection date are printed on the tag (Figure 14.5c). The number is typically prefixed with initials indicating a particular planting project. For example, starting in 2022, KFBG initiated a restoration project with funding from CLP Holdings Limited (CLP), a listed energy company. All seedlings planted under this project were labeled with a number denoted by the prefix "CLP" (see Section 14.3.3).

Propagation from seeds is effective for most native tree species. However, for a few rare species that do not flower or fruit and which may therefore be suffering from extinction debt (Hylander & Ehrlén, 2013; Kuussaari et al., 2009), vegetative propagation via cuttings becomes a viable alternative. To encourage rooting, these cuttings often require rooting hormones and are placed in a mist house (Figure 14.7) until new roots develop. Rooting cuttings is a last-resort method for propagating endangered species, because cuttings are genetically identical to the parent plant, and even if large numbers of seedlings are propagated, they do not increase the species' genetic diversity (Bonga, 1987).

14.3.2 Standard-based Restoration

In 2024, BGCI and the SER launched TGBS, a site-based land-management certification scheme driven by explicit targets for the uplifting of biodiversity, ecological function, habitat protection, long-term monitoring, and social benefits (Bartholomew et al., 2024). Within the TGBS framework, the success of forest restoration projects is evaluated using a comprehensive set of eight site-level criteria, each with its own scoring system:

FIGURE 14.5 The Native Tree Nursery (NTN) (a) and Native Herbs, Shrubs and Climbers Nursery (b) at KFBG. Established in 1997, the NTN produces over 30,000 seedlings of native trees annually. Its operations follow standard protocols for tropical nurseries, including the use of shade nets and regular watering. Unlike commercial nurseries, nearly all seedlings are raised from seeds collected from natural populations in Hong Kong. Collection details are recorded in field books, and voucher specimens are deposited in the KFBG herbarium. Once seedlings reach a suitable size, each is tagged (c) with an accession number, collector name, collection location and date, sowing date, and other relevant information. This approach maximizes the genetic diversity of planted species and contributes to mitigating extinction debt. For species that develop taproots shortly after germination—particularly oaks such as *Quercus*, *Castanopsis*, and *Lithocarpus*—air-pruning pots may be used (d). ((a) Photo by Gale, 2023; (b–d), photo by Zhang, 2025.)

1. *Select appropriate sites to enhance native biodiversity*
2. *Enhance protection of existing habitats and biodiversity*
3. *Protect, restore, and manage biodiversity in consultation and partnership with local communities and other stakeholders*
4. *Aim to maximize biodiversity recovery through ecosystem restoration*
5. *Avoid and reduce invasive or potentially invasive species*
6. *Prioritize the use of native, threatened, and rare species*
7. *Promote biodiversity and adaptive capacity*
8. *Implement robust monitoring, evaluation, and adaptive management of biodiversity*

For ecosystem integrity, criteria 1, 4, 5, 6, and 7 are assessed using the Ecological Recovery Wheel and the Ecosystem Integrity 5-star system, adapted from the SER (Gann et al., 2019). Six attributes and 21 sub-attributes are considered, as follows:

JL 2401 Date:

Jinlong Zhang

Field Identification

Plant Description

Habitat Description (including abundance and associated species)

Locality

Alt. (m)

Coordinates

N E

GPS Accuracy

Material ☐ Herbarium ☐ Seeds ☐ Whole Plant ☐ Cutting ☐ Silica

Remarks

☐ Population Sampling

☐ Photo

IrisBG No.:

JL 2401	JL 2401	JL 2401	JL 2401	JL 2401
JL 2401	JL 2401	JL 2401	JL 2401	JL 2401

FIGURE 14.6 A blank page from the computer-generated field book record sheet used at Kadoorie Farm and Botanic Garden. (Image by Zhang, 2025.)

1. **Absence of threats**: Over-utilization; Invasive species; Contamination
2. **Physical conditions**: Substrate physical; Substrate chemical; Water chemo-physical
3. **Species composition**: Desirable plants; Desirable animals; No undesirable species
4. **Structural diversity**: All strata present; All trophic levels; Spatial mosaic
5. **Ecosystem function**: Productivity, cycling; Habitat interactions; Resilience, recruitment
6. **External exchanges**: Landscape flows; Gene flows; Habitat links

The Global Biodiversity Standard, which evaluates ecosystem integrity, involves examining the present condition of an ecosystem in relation to two benchmarks: a baseline and a reference model (Comer et al., 2025). The baseline reflects the initial or pre-intervention state of the ecosystem. At the same time, the reference model illustrates the ideal state the ecosystem would have maintained if it had not experienced degradation or transformation. This reference serves as the target against which restoration efforts are assessed.

FIGURE 14.7 KFBG's mist house. For species with extremely small populations and limited seed production, a small number of cuttings may be collected with permission from the Agriculture, Fisheries and Conservation Department (AFCD) of the Hong Kong Government. These cuttings are planted in pots and kept in the mist house for a period determined by the development of their root systems. The mist house is a key component of the nursery, primarily used for producing clonal seedlings. (Photo by Zhang, 2025.)

In criterion 3, TGBS emphasizes the social responsibility and equitable distribution of benefits during project implementation, particularly to local communities and vulnerable groups. In criterion 2, the site's level of protection is evaluated through a different set of tools, including the 5-star level of protection ratings, adapted from the IUCN Protected Areas Categories and the IUCN Red List of Ecosystems. Finally, criterion 8 is assessed through the Robust Monitoring, Evaluation, and Adaptive Management List, adapted from the Restoration Project Information Sharing Framework (Gann et al., 2022) and the Standards of Practice for guiding ecosystem restoration (Nelson et al., 2024).

As of 2026, TGBS has established 16 regional hubs worldwide, and the methodology has been rigorously tested in over 100 projects in various land uses. KFBG is honored to be the first and, as of the time of writing, the only TGBS hub in China. Moreover, its forest restoration project was evaluated at the end of 2024 by two third-party assessors from the Auroville Botanical Gardens in India. The outcome of this assessment was the recognition of KFBG's site management with a Premium-tier certification, the first of its kind globally. Meanwhile, over a dozen ecologists in Hong Kong have received training in TBGS methodology, all of whom successfully qualified as TGBS assessors (Figure 14.8).

The first step in the TGBS process requires project proponents to establish a reference model to determine the intended long-term outcome for the restoration process. In the case of Hong Kong, this involved comprehensive data collection on forest species composition and structure in intact primary and mature secondary forest sites across the Pearl River Delta and even the broader Lingnan region (i.e., Guangdong, Guangxi, Hong Kong, Macau, and Hainan). Large forest dynamics plots—such as the Tai Po Kau plot in Hong Kong (Zhu et al., 2024a), the Dinghushan Plot in

FIGURE 14.8 TGBS training workshop held from 9 to 13 December 2024 at Kadoorie Farm and Botanic Garden. Kiran Baldwin (speaker) from Auroville Botanical Gardens, India, introduced the participants to the assessment criteria and scoring system. Attendees included representatives from KFBG, The University of Hong Kong, and The Conservancy Association, Hong Kong. All participants successfully passed the examination and were certified as TGBS assessors. (Photo by Zhang, 2024.)

Zhaoqing (Cao et al., 2013), the Heishiding Plot in Fengkai County (Bin et al., 2024), Guangdong, and the Jianfengling Plot in Hainan (Fang et al., 2004a; Xu et al., 2015)—served as important references in this regard.

14.3.3 Tree Planting Activities in KFBG

KFBG began its native tree planting activities shortly after establishing its Native Tree Nursery in 1997, with participation from local schools being a unique feature at the time (Kadoorie Farm and Botanic Garden, 2004). In 2004, the wildfire affecting Tai Mo Shan Country Park and the upper areas of KFBG above 550 m, reverted the woodland back to grassland (Kadoorie Farm and Botanic Garden, 2006). Since then, KFBG has organized native tree planting activities annually on Earth Day (22nd April). Unfortunately, these seedlings were not systematically monitored, nor were any management measures taken. As a result, many were outcompeted and shaded out by grasses, and due to a lack of data, the outcomes were never formally reported.

Starting in 2013, KFBG's tree planting programmer transitioned to an experimental approach, featuring long-term monitoring of seedling growth (Fischer & Zhang, 2016) (Figures 14.9 and 14.10a). Each year, more than 3,000 seedlings are planted in the forest restoration area, using subplots measuring 5 m × 5 m and different species combinations (Figure 14.9). Each seedling is tagged with a unique ID (Figure 14.10d), labeled with species name, batch number, collector, collection number, location where the seed was collected and date of collection. Restoration interventions we tested include tree guards, weeding mats, manual and mechanized weeding, structural pruning, application of mulch, and the application of organic fertilizer (Figure 14.10b and c). The current

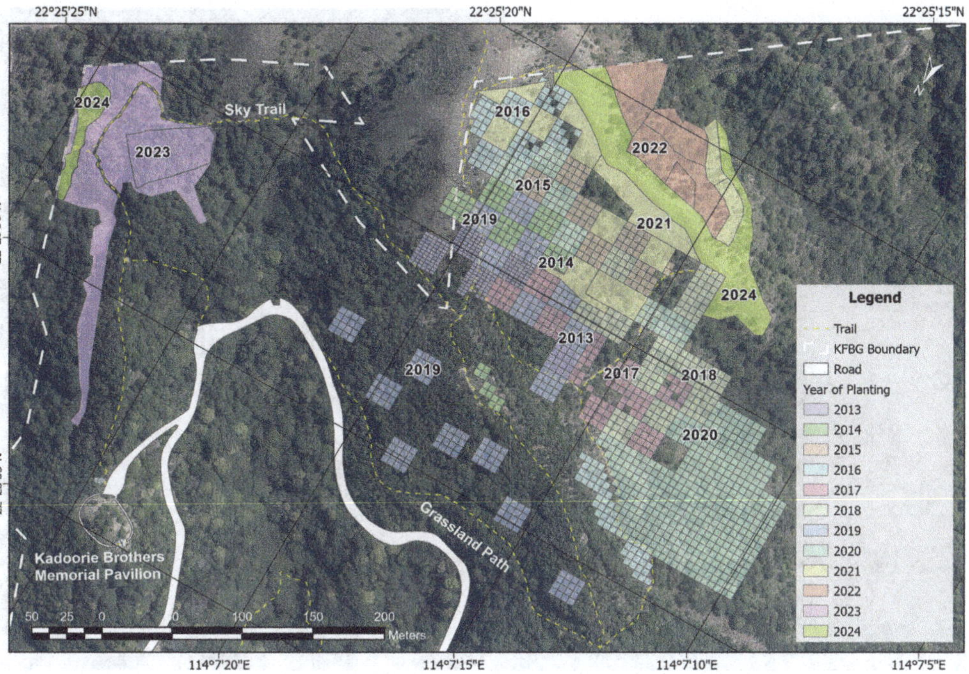

FIGURE 14.9 Forest restoration site at Kadoorie Farm and Botanic Garden. The site is situated on the northern slope of Tai Mo Shan, Hong Kong's highest peak (957 m a.s.l.). The restoration area spans elevations from approximately 550 to 650 m. From 2013 to 2020, plantings were conducted in a plot-based manner, whereby each year's planting mixture—comprising an average of 20–40 species—was replicated multiple times in 5 m×5 m quadrats within 20 m×20 m plots. Subsequently, a more diverse mixture of species was introduced to test the feasibility of enhancement planting and to examine broader ecosystem processes. From 2021 to 2024, planting areas were selected based on site availability, ecological suitability and to test particular methodological approaches. Species selection has been guided by the expertise of KFBG's botanists and ecologists. All planted individuals are protected with tree guards; some were also fitted with weed mats, or fertilizer in the early years (2013). Height and basal diameter were recorded for all seedlings at planting, with average heights ranging from 20 to 40 cm depending on species. Seedlings planted between 2013 and 2019 were censused 2 years after planting, following a protocol developed by the ForestGEO Network. Long-term monitoring was conducted for selected plots at 5-year intervals. Areas planted from 2021 to 2024 were also monitored using 20 m×20 m plots, following the same census protocol. In total, over 100,000 individuals have been planted since 2013, representing in excess of 300 tree and shrub species.

practices include installation of tree guards, application of mulch, manual weeding and structural pruning. Since 2013, seedling height, basal diameter, DBH (if applicable), and x–y coordinates have been measured for selected plots to enable long-term monitoring.

This approach is suitable for testing how the effects of environmental factors and restoration interventions affect seedling survival and growth. For example, in 2013, 12 native tree species were planted, and tree height and basal diameter were measured in 2014, 2015, and 2017 (Zhang et al., 2025). Growth rates and mortality were calculated based on height and basal diameter. The results showed that different species respond differently to different environmental factors (e.g., elevation, slope, and aspect) and restoration treatments (e.g., tree guards, weeding mats, and fertilizers) and varied in different growth periods. The results also highlight the importance of protective measures, especially tree guards, which mitigate damage by large mammalian herbivores (including wild boar, barking deer, and porcupine) (Zhang et al., 2025).

In addition to planting on grasslands, KFBG has also initiated enhancement planting in species-poor secondary forests dominated by a few winner species (Albrecht et al., 2022). Specifically, in forests dominated by *Machilus chekiangensis*, selective thinning of canopy trees

FIGURE 14.10 KFBG's Forest Restoration Plot. (a) Overview of plots planted in 2013; (b) section of the area planted in 2022. From 2022 onward, forest restoration activities were supported by CLP.; (c) A *Reevesia thyrsoidea* seedling protected by a tree guard.; (d) a permanent tree tag with a unique ID, made of stainless steel used in the CLP forest restoration project. ((a–d) Photo by Zhang, 2023.)

and crown reductions were implemented to create canopy gaps ranging from a few to over 100 square meters, thereby increasing light availability on the forest floor. Native seedlings were then planted in these gaps. This project commenced in 2019 (Figure 14.9), and surveys of the planted seedlings were conducted through 2024. Preliminary results suggest that it is indeed possible to overcome arrested succession. Still, tree performance varies from species to species, highlighting the need for a better understanding of species' requirements and suitability. The performance of the seedlings is influenced not only by restoration treatments but also by the existing trees in an enrichment planting (Zhang et al., 2025). To understand the barriers to enriching Hong Kong's secondary forests and to evaluate the potential effect of the dominant tree species, *Machilus chekiangensis,* on forest regeneration, we have also set up an experiment to explore the impact of neighborhood trees in the thinned and pruned secondary forest dominated by this species (Figure 14.11a). The experimental area, part of the plantation in 2023, was fenced off (Figure 14.11b). Water tanks for collecting rainwater were installed to help planted seedlings overcome drought in the spring (Figure 14.11c). The experiment area is manually weeded two to three times a year, depending on the status (Figure 14.11d).

14.3.4 Pollination Studies

Degraded natural forests face multiple barriers to regeneration and recovery. In contrast to the widely occurring "winner species", the majority of native species exist in small, isolated populations, which may affect their ability to produce viable seeds due to inbreeding and other legacy effects (Severns, 2003). Meanwhile, pollinators (Borges et al., 2025; Ulyshen et al., 2023) and seed dispersal agents (Corlett, 2009, 2011) are also limiting factors hindering forest recovery. About 90% of flowering plants rely on faunal pollination (Tong et al., 2023), and many rely on specific pollinators (Borges et al., 2025). Bees, butterflies, and moths are widely known pollinators (Stephens

FIGURE 14.11 Enhancement planting plots. (a) Crowns of the dominant species *Machilus chekiangensis* were heavily pruned to allow increased light penetration to the understorey, thereby creating suitable conditions for enhancement planting. (b) Selected plots were fenced to exclude animals, particularly feral cattle, wild boar, porcupines, and barking deer. (c) Rainwater catchment tanks were installed to support plant survival during the dry season. (d) The plot is manually weeded two to three times a year, depending on the condition ((a–c) Photo by Zhang (2023); (d) photo by Cardoso, 2024.)

et al., 2023), but flies, birds, lizards, and even small mammals can also be active pollinators. Many tropical plants have specialized pollinators. For example, the majority of Annonaceae species are pollinated by small beetles (Gottsberger, 1999; Liu et al., 2025; Saunders, 2012), but some can be pollinated by large beetles, thrips, flies, and cockroaches. A study conducted by Japanese ecologists in collaboration with KFBG reported that *Mucuna birdwoodiana* (Fabaceae) is pollinated by masked palm civets (Kobayashi et al., 2019). What chemical compounds are used to attract pollinators? What rewards do these compounds offer in return, and how critical are these interactions for the survival of the pollinators? Although these questions are critical to understand ecosystem stability, the maintenance, and the recovery of biodiversity, they remain poorly examined, especially in the Asian tropics. Understanding how the pollination of the rare species, no matter whether an orchid or tree species, provides better opportunities for conservation and restoration. Nevertheless, as faunal species serve key ecological functions in the forest ecosystem, in particular, in pollination (Kobayashi et al., 2019; Ulyshen et al., 2023) and seed dispersal (Corlett, 2011), monitoring of the dynamics of birds, mammals, and insects has been considered in the long-term restoration monitoring plan in KFBG.

14.4 FUTURE DIRECTIONS

Forestation restoration has focused on canopy coverage and seedling survival for decades. In recent years, more attention has been given to species composition, carbon sequestration, and other ecosystem processes (Chazdon et al., 2016; Zhu et al., 2024a). However, attempts to restore a highly diverse, fully ecological functioning forest remain scarce (Chen et al., 2025). The release of the Global Biodiversity Standard (TGBS) may shift the focus toward these goals (Bartholomew et al., 2024).

To comprehensively restore the forest's species composition, structure, and function, there is a need for a deeper understanding of biodiversity. For instance, expertise in botany, species identification, and ecological knowledge, as well as generating, managing, and utilizing species-occurrence records is essential. In this regard, botanical gardens are ideally positioned to facilitate the accumulation of relevant knowledge and skills among staff and stakeholders, through facilities such as herbaria, nurseries, and specialized laboratories (Westwood et al., 2021).

We also argue that it is time to embrace new technologies: with the advent of artificial intelligence, species monitoring, and management are undergoing significant changes. For example, trees can now be scanned and mapped using terrestrial LiDAR and drones equipped with LiDAR (Almeida et al., 2019; Francis et al., 2023), as well as hyperspectral sensors (Ceriani et al., 2025; Kayet, 2021; Schratz et al., 2021). Species identification and forest carbon stock assessments are increasingly being conducted using machine learning and AI-based technologies (Liang & Gadow, 2023; Yang et al., 2025). Ultimately, we argue that restoration projects must not only focus on identifying and making use of suitable species, but also on long-term stewardship towards the recreation of diverse, functioning ecosystems (Chen et al., 2025), analogous to those created by nature (Rozendaal et al., 2019; Leblanc et al., 2025). Within this framework, rare species can be protected and conserved by ensuring their inclusion in plantings and restored ecosystems.

14.5 CONCLUSIONS

Though rich in biodiversity, Hong Kong's natural vegetation has been subject to multiple degradation factors for millennia, some of which have compounded and intensified in recent times, leading to intractable challenges to natural recovery. This included, but is not limited to, arrested succession, extinction debt and alternative stable states. Forest restoration has reached a crossroads where biodiversity must become a guiding principle. Forest restoration under the Global Biodiversity Standard (TGBS) framework will be extremely useful in resolving these issues.

Botanical gardens are ideal places to fulfill this role through their facilities and skill sets. Ecological restoration that targets biodiversity and ecosystem functions ultimately integrates forest restoration, biodiversity conservation, and ecosystem processes. This requires multifaceted collaboration and long-term support, which can be achieved through botanical gardens in partnership with broader societal stakeholders, including government, the corporate sector, and the general public. We argue that it is time to incorporate biodiversity and conservation into restoration, and this seems achievable under the framework of the Global Biodiversity Standard.

ACKNOWLEDGMENTS

We gratefully acknowledge present and former colleagues of KFBG who have contributed to the forest restoration project since 2013. We would like to thank Prof. Wong Ming Hung, Dr. Chen Xunwen and Dr. Masayuki Ushio for their comments and edits. The financial support of CLP Holdings Limited is acknowledged for supporting KFBG's ongoing applied forest restoration research.

DECLARATION OF GENERATIVE AI USE

During the preparation of this manuscript, the authors used Copilot (https://copilot.microsoft.com/) to refine the English language. All authors subsequently reviewed and edited the content and take full responsibility for the final version of the published chapter.

REFERENCES

Abbas, S., Nichol, J. E., & Fischer, G. A. (2016). A 70-year perspective on tropical forest regeneration. *Science of the Total Environment*, *544*, 544–552. https://doi.org/10.1016/j.scitotenv.2015.11.171

Abbas, S., Nichol, J. E., Zhang, J., & Fischer, G. A. (2019). The accumulation of species and recovery of species composition along a 70 year succession in a tropical secondary forest. *Ecological Indicators, 106*, 105524. https://doi.org/10.1016/j.ecolind.2019.105524

Aerts, R., & Honnay, O. (2011). Forest restoration, biodiversity and ecosystem functioning. *BMC Ecology, 11*(1), 29. https://doi.org/10.1186/1472-6785-11-29

Agriculture, Fisheries and Conservation Department, Hong Kong SAR Government. (2025). *Hong Kong: The Facts - Country Parks and Special Areas*. https://www.afcd.gov.hk/english/country/cou_lea/the_facts.html

Albrecht, C., Contreras, Z., Wahl, K., Sternberg, M., & Christoffersen, B. (2022). Winners and losers in dryland reforestation: Species survival, growth, and recruitment along a 33-year planting chronosequence. *Restoration Ecology, 30*(4), e13559. https://doi.org/10.1111/rec.13559

Almeida, D. R. A., Broadbent, E. N., Zambrano, A. M. A., Wilkinson, B. E., Ferreira, M. E., Chazdon, R., Meli, P., Gorgens, E. B., Silva, C. A., Stark, S. C., Valbuena, R., Papa, D. A., & Brancalion, P. H. S. (2019). Monitoring the structure of forest restoration plantations with a drone-lidar system. *International Journal of Applied Earth Observation and Geoinformation, 79*, 192–198. https://doi.org/10.1016/j.jag.2019.03.014

Anderson-Teixeira, K. J., Davies, S. J., Bennett, A. C., Gonzalez-Akre, E. B., Muller-Landau, H. C., Joseph Wright, S., Abu Salim, K., Almeyda Zambrano, A. M., Alonso, A., Baltzer, J. L., Basset, Y., Bourg, N. A., Broadbent, E. N., Brockelman, W. Y., Bunyavejchewin, S., Burslem, D. F. R. P., Butt, N., Cao, M., Cardenas, D., … & Zimmerman, J. (2015). CTFS-ForestGEO: A worldwide network monitoring forests in an era of global change. *Global Change Biology, 21*(2), 528–549. https://doi.org/10.1111/gcb.12712

Ashton, P. (2014). *On the Forests of Tropical Asia Lest the Memory Fade*. The University of Chicaco Press.

Atha, M. (2012). A neglected heritage: Towards a fuller appreciation of the landscapes and lifeways of Hong Kong's rice farming past. *Asian Anthropology, 11*(1), 129–156. https://doi.org/10.1080/1683478X.2012.10600860

Bartholomew, D., Mosyaftiani, A., Morgan, B., Shah, T., Shaw, K., Stillman, C., Baldwin, K., Baqueiro, L., Birkinshaw, C., Breman, E., Diniz, G., Fereira, M., Flowers, C., Gichira, A., Jenkins, J., Kindt, R., Molin, P., Mungai, T., Said Musa, M., & Gann, G. (2024). *The Global Biodiversity Standard: Manual for Assessment and Best Practices*. Botanic Gardens Conservation International & Society for Ecological Restoration. https://www.biodiversitystandard.org/our-method/

Beisner, B. E., Haydon, D. T., & Cuddington, K. (2003). Alternative stable states in ecology. *Frontiers in Ecology and the Environment, 1*(7), 376–382. https://doi.org/10.1890/1540-9295(2003)001[0376:ASSIE]2.0.CO;2

Bentham, G. (1861). *Flora Hongkongensis: A Description of the Flowering Plants and Ferns of the Island of Hongkong*. L. Reeve.

Bin, Y., Cao, H., Rodríguez-Hernández, D. I., Wang, Z., Chu, C., & Li, B. (2024). Habitat association and demographic rates for large tree species: Implications for native tree species used in forestry. *Journal of Forestry Research, 36*(1), 8. https://doi.org/10.1007/s11676-024-01801-y

Bohonak, A. J. (1999). Dispersal, gene flow, and population structure. *The Quarterly Review of Biology, 74*(1), 21–45. https://doi.org/10.1086/392950

Bonga, J. (1987). Clonal propagation of mature trees: Problems and possible solutions. In J. M. Bonga & D. J. Durzan (Eds.), *Cell and Tissue Culture in Forestry: General Principles and Biotechnology* (pp. 249–271). Springer Netherlands. https://doi.org/10.1007/978-94-017-0994-1_15

Borges, R. C., Cunha, E. J., dos Santos Júnior, J. E., Silveira, O. T., de Oliveira Andrino, C., Caldeira Jr, C. F., Silva, D. F., Tyski, L., & Giannini, T. C. (2025). Using plant–pollinator interactions to identify key species for planning tropical forest restoration. *Restoration Ecology, 33*(3), e14334. https://doi.org/10.1111/rec.14334

Bowles, M. L., Betz, R. F., & DeMauro, M. M. (1993). Propagation of rare plants from historic seed collections: Implications for species restoration and herbarium management. *Restoration Ecology, 1*(2), 101–106. https://doi.org/10.1111/j.1526-100X.1993.tb00015.x

Cao, H., Wu, L., Wang, Z., Huang, Z., Li, L., Wei, S., Lian, J., & Ye, W. (2013). *Dinghushan Lower Subtropical Forest Dynamics Plot: Tree Species and Their Distribution Patterns*. China Forestry Publishing House.

Ceriani, R., Brocco, S., Pepe, M., Oggioni, S., Vacchiano, G., Motta, R., Berretti, R., Ascoli, D., Garbarino, M., Morresi, D., Bassi, F., & Fava, F. (2025). Hyperspectral and LiDAR space-borne data for assessing mountain forest volume and biomass. *International Journal of Applied Earth Observation and Geoinformation, 141*, 104614. https://doi.org/10.1016/j.jag.2025.104614

Chan, A. H. Y., & Coomes, D. A. (2024). Fire traps in the wet subtropics: New perspectives from Hong Kong. *Journal of Applied Ecology, 61*(4), 884–898. https://doi.org/10.1111/1365-2664.14575

Chan, A. H. Y., Guizar-Coutiño, A., Kalamandeen, M., & Coomes, D. A. (2023). Reconstructing 34 Years of Fire History in the Wet, Subtropical Vegetation of Hong Kong Using Landsat. *Remote Sensing*, *15*(6), 1489. https://doi.org/10.3390/rs15061489

Chazdon, R. L., Broadbent, E. N., Rozendaal, D. M. A., Bongers, F., Zambrano, A. M. A., Aide, T. M., Balvanera, P., Becknell, J. M., Boukili, V., Brancalion, P. H. S., Craven, D., Almeida-Cortez, J. S., Cabral, G. A. L., de Jong, B., Denslow, J. S., Dent, D. H., DeWalt, S. J., Dupuy, J. M., Durán, S. M., … & Poorter, L. (2016). Carbon sequestration potential of second-growth forest regeneration in the Latin American tropics. *Science Advances*, *2*(5), e1501639. https://doi.org/10.1126/sciadv.1501639

Chen, C., Bongers, F. J., Schmid, B., Ma, K., & Liu, X. (2025). Ecosystem consequences of functional diversity in forests and implications for restoration. *New Phytologist*, *247*(3). https://doi.org/10.1111/nph.70247

Chu, W. H. (1998). *Conservation of Terrestrial Biodiversity in Hong Kong* [The University of Hong Kong]. https://hub.hku.hk/handle/10722/33375

Chuen, M. L. P. (2019). *Proposed Public Housing Development at Area 39*, Hin Fat Lane, Tuen Mun. https://www.aab.gov.hk/filemanager/aab/common/hia-report/HIA_Report_HFLPHD.pdf

Comer, P. J., Eckert, G. E., & Gann, G. D. (2025). Establishing Reference Models for Ecological Restoration—Case Study from Colorado National Monument, USA. *Land*, 14(9), 1871. https://doi.org/10.3390/land14091871

Condit, R. (1998). *Tropical Forest Census Plots: Methods and Results from Barro Colorado Island, Panama and a Comparison with Other Plots*. Springer. https://doi.org/10.1007/978-3-662-03664-8

Corlett, R. T. (1999). Environmental Forestry in Hong Kong: 1871–1997. *Forest Ecology and Management*, *116*(1–3), 93–105. https://doi.org/10.1016/S0378-1127(98)00443-5

Corlett, R. T. (2009). Seed dispersal distances and plant migration potential in Tropical East Asia. *Biotropica*, *41*(5), 592–598. https://doi.org/10.1111/j.1744-7429.2009.00503.x

Corlett, R. T. (2011). Seed dispersal in Hong Kong, China: Past, present and possible futures. Integrative Zoology, *6*(2), 97–109. https://doi.org/10.1111/j.1749-4877.2011.00235.x

Davis, C. C. (2023). The herbarium of the future. *Trends in Ecology & Evolution*, *38*(5), 412–423. https://doi.org/10.1016/j.tree.2022.11.015

Davies, N. M. J., Drinkell, C., & Utteridge, T. M. A. (2023). *The Herbarium Handbook: Sharing Best Practice from across the Globe*. Kew Publishing.

Davies, S. J., Abiem, I., Abu Salim, K., Aguilar, S., Allen, D., Alonso, A., Anderson-Teixeira, K., Andrade, A., Arellano, G., Ashton, P. S., Baker, P. J., Baker, M. E., Baltzer, J. L., Basset, Y., Bissiengou, P., Bohlman, S., Bourg, N. A., Brockelman, W. Y., Bunyavejchewin, S., … & Zuleta, D. (2021). ForestGEO: Understanding forest diversity and dynamics through a global observatory network. *Biological Conservation*, *253*, 108907. https://doi.org/10.1016/j.biocon.2020.108907

Delang, C. O. (2018). Local livelihoods and global process: Complex causalities in Hong Kong's Sai Kung Peninsula. *Miscellanea Geographica*, *22*(1), 31–39. https://doi.org/10.2478/mgrsd-2018-0003

Dudgeon, D., & Corlett, R. T. (2011). *The Ecology and Biodiversity of Hong Kong* (Revised Edition, July, 2011). Lions Nature Education Foundation; Cosmos Books Ltd.

Fang, J.-Y., Li, Y.-D., Zhu, B., Liu, G.-H., & Zhou, G.-Y. (2004a). Community structures and species richness in the montane rain forest of Jianfengling, Hainan Island, China. *Biodiversity Science*, *12*(1), 29. https://doi.org/10.17520/biods.2004005

Fang, J.-Y., Shen, Z.-H., Tang, Z.-Y., & Wang, Z.-H. (2004b). The protocol for the survey plan for plant species diversity of China's mountains. *Biodiversity Science*, *12*(1), 5.

Fischer, G. A., & Zhang, J. (2016). Restoring Hong Kong's montane forest: Challenges and perspectives from Kadoorie Farm and Botanic Garden. *BGjournal*, *13*(2), 32–34.

Flora Conservation Department, Kadoorie Farm and Botanic Garden. (2025). *Ecology and Conservation of Bulbophyllum bicolor in Hong Kong*. Kadoorie Farm and Botanic Garden. https://www.kfbg.org/en/flora-conservation/ecology-and-conservation-of-bulbophyllum-bicolor-in-hong-kong

ForestGEO. (2025). ForestGEO. https://forestgeo.si.edu/

Francis, J., Disney, M., & Law, S. (2023). Monitoring canopy quality and improving equitable outcomes of urban tree planting using LiDAR and machine learning. *Urban Forestry & Urban Greening*, *89*, 128115. https://doi.org/10.1016/j.ufug.2023.128115

Fukami, T. (2015). Historical contingency in community assembly: Integrating niches, species pools, and priority effects. *Annual Review of Ecology, Evolution, and Systematics*, *46*(1), 1–23. https://doi.org/10.1146/annurev-ecolsys-110411-160340

Fukami, T., Mordecai, E. A., & Ostling, A. (2016). A framework for priority effects. *Journal of Vegetation Science*, *27*(4), Article 4. https://doi.org/10.1111/jvs.12434

Gale, S. W., Bizid, A., Wai, L.K., Chan, K. (2014). A Guide to Orchids of Hong Kong. *Natural History Publications* (Borneo), Kota Kinabalu.

Gale, S. W., Kumar, P., Hu, A.-Q., Tai, C. P., Fischer, G. A. (2020). *Species Action Plan: Bulbophyllum bicolor (Orchidaceae), Towards restoration of viable wild populations in Hong Kong*. Kadoorie Farm and Botanic Garden, Hong Kong.

Gale, S. W., Yeung, W. K., Ho, N., Chan, C. S., & Kumar, P. (2020). Four new additions and a taxonomic amendment to the orchid flora of Hong Kong. *Taiwania*, *65*(4). https://doi.org/10.6165/tai.2020.65.575

Gann, G. D., McDonald, T., Walder, B., Aronson, J., Nelson, C. R., Jonson, J., Hallett, J. G., Eisenberg, C., Guariguata, M. R., Liu, J., Hua, F., Echeverría, C., Gonzales, E., Shaw, N., Decleer, K., & Dixon, K. W. (2019). International principles and standards for the practice of ecological restoration. Second edition. *Restoration Ecology*, *27*(S1), S1–S46. https://doi.org/10.1111/rec.13035

Gann, G. D., Walder, B., Gladstone, J., Manirajah, S. M., Roe, S. (2022). Restoration Project Information Sharing Framework. Society for Ecological Restoration and Climate Focus. Washington, D.C. https://www.ser.org/news/602976/New-Release-Restoration-Project-Information-Sharing-Framework.htm

Gottsberger, G. (1999). Pollination and evolution in neotropical Annonaceae. *Plant Species Biology*, *14*(2), 143–152. https://doi.org/10.1046/j.1442-1984.1999.00018.x

Hau, C. (1999). The establishment and survival of native trees on degraded hillsides in Hong Kong. (Thesis). University of Hong Kong, Pokfulam, Hong Kong SAR. http://dx.doi.org/10.5353/th_b2979961.

Ho, J. W. I., & Law, L.-Y. (2023). Customary and customised: Nation-building at the Hong Kong Palace Museum. *International Journal of Heritage Studies*, *29*(3), 122–144. https://doi.org/10.1080/13527258.2023.2176349

Hoffmann, P. M., Blum, C. T., Velazco, S. J. E., Gill, D. J. C., & Borgo, M. (2015). Identifying target species and seed sources for the restoration of threatened trees in southern Brazil. *Oryx*, *49*(3), 425–430. https://doi.org/10.1017/S0030605314001069

Hong Kong Chronicles Institute. (2022). *Hong Kong Chronicles: Nature* (First edition). Chung Hwa Book Co., (H.K.) Ltd.

Hong Kong Herbarium. (2012). *Check list of Hong Kong plants 2012*. Agriculture, Fisheries and Conservation Department, Hong Kong SAR. Government.

Hong Kong Observatory. (2023). *Climate of Hong Kong*. https://www.hko.gov.hk/en/cis/climahk.htm

Hu, A., Gale, S. W., Kumar, P., Saunders, R. M., Sun, M., & Fischer, G. A. (2017). Preponderance of clonality triggers loss of sex in *Bulbophyllum bicolor*, an obligately outcrossing epiphytic orchid. *Molecular Ecology*, *26*(13), 3358–3372. https://doi.org/10.1111/mec.14139

Hu, Q., & Wu, D. (Eds.). (2008). *Flora of Hong Kong*. Agriculture, Fisheries and Conservation Department, Government of the Hong Kong Special Administrative Region.

Hubbell, S. P., Foster, R. B., & Rubinoff, I. (2024). Early history of the Barro Colorado Island Forest Dynamics Project. In *The First 100 Years of Research on* Barro Colorado*: Plant and Ecosystem Science*. Smithsonian Institution Scholarly Press. https://doi.org/10.5479/si.26809612

Hylander, K., & Ehrlén, J. (2013). The mechanisms causing extinction debts. *Trends in Ecology & Evolution*, *28*(6), 341–346. https://doi.org/10.1016/j.tree.2013.01.010

Jim, C. Y. (2015). Cross-border itinerant poaching of agarwood in Hong Kong's peri-urban forests. *Urban Forestry & Urban Greening*, *14*(2), 420–431. https://doi.org/10.1016/j.ufug.2015.04.007

Kadoorie Farm and Botanic Garden. (2004). *Kadoorie Farm and Botanic Garden Annual Report 2003/2004* (p. 22). Kadoorie Farm and Botanic Garden Corporation. https://www.kfbg.org/images/ar/report/2003-2004-Eng.pdf

Kadoorie Farm and Botanic Garden. (2006). *Kadoorie Farm and Botanic Garden Annual Report 2004/2005* (p. 29). Kadoorie Farm and Botanic Garden Corporation. https://www.kfbg.org/images/ar/report/2004-2005-Eng.pdf

Kayet, N. (2021). Forest health monitoring using hyperspectral remote sensing techniques. In P. K. Shit, H. R. Pourghasemi, P. Das, & G. S. Bhunia (Eds.), *Spatial Modeling in Forest Resources Management: Rural Livelihood and Sustainable Development* (pp. 239–257). Springer International Publishing. https://doi.org/10.1007/978-3-030-56542-8_10

Kobayashi, S., Gale, S. W., Denda, T., & Izawa, M. (2019). Civet pollination in *Mucuna birdwoodiana* (Fabaceae: Papilionoideae). *Plant Ecology*, *220*(4), 457–466. https://doi.org/10.1007/s11258-019-00927-y

Kumar, P., & Gale, S. (2022). Taxonomic notes on *Apostasia nipponica* and *Crepidium cordilabium* (Orchidaceae), two species newly recorded from Hong Kong. *Feddes Repert*, fedr.202200002. https://doi.org/10.1002/fedr.202200002

Kumar, P., & Gale, S. W. (2020). *Anoectochilus formosanus* (Orchidaceae), a new record for Hong Kong. *Rheedea*, *30*(2), 293–298. https://doi.org/10.22244/rheedea.2020.30.02.06

Kumar, P., & Gale, S. W. (2023). *Cheirostylis yunnanensis* var. *gloriae*, an interesting new caterpillar orchid from Hong Kong, China. *Feddes Repertorium*, *134*(3), 181–190. https://doi.org/10.1002/fedr.202200052

Kumar, P., Gale, S. W., Kocyan, A., Fischer, G. A., Averyanov, L., Borosova, R., Bhattacharjee, A., Li, J.-H., & Pang, K. S. (2014). *Gastrochilus kadooriei* (Orchidaceae), a new species from Hong Kong, with notes on allied taxa in section Microphyllae found in the region. *Phytotaxa*, *164*(2), 091–103. https://doi.org/10.11646/phytotaxa.164.2.3

Kuussaari, M., Bommarco, R., Heikkinen, R. K., Helm, A., Krauss, J., Lindborg, R., Öckinger, E., Pärtel, M., Pino, J., Rodà, F., Stefanescu, C., Teder, T., Zobel, M., & Steffan-Dewenter, I. (2009). Extinction debt: A challenge for biodiversity conservation. *Trends in Ecology & Evolution*, *24*(10), 564–571. https://doi.org/10.1016/j.tree.2009.04.011

Lands Department of Hong Kong SAR Government. (2025). *Hong Kong Geographic Data 2025*. https://www.landsd.gov.hk/en/resources/mapping-information/hk-geographic-data.html

Leblanc, C., Bonnet, P., Servajean, M., Thuiller, W., Chytrý, M., Aćić, S., Argagnon, O., Biurrun, I., Bonari, G., Bruelheide, H., Campos, J. A., Čarni, A., Ćušterevska, R., De Sanctis, M., Dengler, J., Dziuba, T., Garbolino, E., Jandt, U., Jansen, F., … & Joly, A. (2025). Learning the syntax of plant assemblages. *Nature Plants*, 11(10), 2026–2040. https://doi.org/10.1038/s41477-025-02105-7

Lee, K. W. K., Cheuk, M. L., Fischer, G. A., & Gale, S. W. (2023). Can disparate shared social values benefit the conservation of biodiversity in Hong Kong's sacred groves? *Human Ecology*, 1–12. https://doi.org/10.1007/s10745-023-00443-8

Lee, K. W. K., Williams, C., & Gale, S. (2023). *30 of Hong Kong's Worst Weeds* (p. 76). Kadoorie Farm and Botanic Garden. https://www.kfbg.org/images/download/30_of_hk_worst_weeds_eversion_higha-compressed_compressed.pdf

Li, J., Xia, N., & Li, X. (2008). *Sinopora*, a new genus of Lauraceae from South China. *Novon: A Journal for Botanical Nomenclature*, *18*(2), 199–201. https://doi.org/10.3417/2006126

Liang, J., & Gadow, K. (2023). *Applications of Artificial Intelligence in Forest Research and Management*. Association of Consulting Foresters.

Liu, J., Hang, K. Y., Wong, C. K., Hon, C. H., Williams, C., Mar, S. S., Lee, Y. L., Zhang, J., & Fischer, G. A. (2021). Seventeen newly recorded species to the flora of Hong Kong, China. *Journal of Tropical and Subtropical Botany*, *29*(2), 123–131. https://doi.org/10.11926/jtsb.4272

Liu, M., Chen, J., Pang, C., Scharaschkin, T., & Saunders, R. M. K. (2025). More than fruity scents: Pollination biology, scent, and spectral reflectance of Annonaceae species. *Plant Species Biology*, *40*(5), 403–425. https://doi.org/10.1111/1442-1984.70014

Mangueira, J. R. S. A., Holl, K. D., & Rodrigues, R. R. (2019). Enrichment planting to restore degraded tropical forest fragments in Brazil. *Ecosystems and People*, *15*(1), 3–10. https://doi.org/10.1080/21513732.2018.1529707

Mar, S. S., & Saunders, R. (2015). *Thismia hongkongensis* (Thismiaceae): A new mycoheterotrophic species from Hong Kong, China, with observations on floral visitors and seed dispersal. *PhytoKeys*, *46*, 21–33. https://doi.org/10.3897/phytokeys.46.8963

Marafa, L. M., & Chau, K. (1999). Effect of hill fire on upland soil in Hong Kong. *Forest Ecology and Management*, *120*(1–3), 97–104. https://doi.org/10.1016/S0378-1127(98)00528-3

Mata, S., Braga, J. M. A., Moser, P., Sartori, R. A., Sánchez-Tapia, A., & Sansevero, J. B. B. (2022). Forever young: Arrested succession in communities subjected to recurrent fires in a lowland tropical forest. *Plant Ecology*, *223*(6), 659–670. https://doi.org/10.1007/s11258-022-01239-4

Mori, A. S., Lertzman, K. P., & Gustafsson, L. (2017). Biodiversity and ecosystem services in forest ecosystems: A research agenda for applied forest ecology. *Journal of Applied Ecology*, *54*(1), 12–27

Nelson, C.R., Hallett, J.G., Romero Montoya, A.E., Andrade, A., Besacier, C., Boerger, V., Bouazza, K., Chazdon, R.,... & Weidlich, E.W.A. (2024) *Standards of practice to guide ecosystem restoration – A contribution to the United Nations Decade on Ecosystem Restoration 2021-2030*. Rome, FAO, Washington, DC, SER & Gland, Switzerland, IUCN CEM.

Ng, P. Y. (1983). *New Peace County: A Chinese gazetteer of the Hong Kong Region* (Vol. 1). Hong Kong University Press.

NHER. (2017, February 10). Little known side of Hong Kong: Conservation and Restoration work at Kadoorie Farm and Botanic Garden (KFBG). *Natural History of Ecological Restoration*. https://mbgecological-restoration.wordpress.com/2017/02/10/little-known-side-of-hong-kong-conservation-and-restoration-work-at-kadoorie-farm-and-botanic-garden-kfbg/

Nicholson, B. (1996). Tai Po Kau nature reserve, new territories, Hong Kong: A reafforestation history. *Asian Journal of Environmental Management*, *4*(2), 103–119.

Owen, B., & Shaw, R. (2007). *Hong Kong Landscapes: Shaping the Barren Rock* (Vol. 1). Hong Kong University Press.

Planing Department, Hong Kong SAR Government. (2025). *Landscape Value Mapping of Hong Kong Final Report*. https://www.pland.gov.hk/studies/landscape/landscape_final/

Poorter, L., Craven, D., Jakovac, C. C., van der Sande, M. T., Amissah, L., Bongers, F., Chazdon, R. L., Farrior, C. E., Kambach, S., Meave, J. A., Muñoz, R., Norden, N., Rüger, N., van Breugel, M., Almeyda Zambrano, A. M., Amani, B., Andrade, J. L., Brancalion, P. H. S., Broadbent, E. N., ... & Hérault, B. (2021). Multidimensional tropical forest recovery. *Science*, *374*(6573), 1370–1376. https://doi.org/10.1126/science.abh3629

Qin, Q., Tong, Y., Pang, K., Au, W., Tong, H., Cheung, W., Wu, L., & Ye, X. (2018). Community characteristics of *Carpinus insularis* (Betulaceae). *Journal of Tropical and Subtropical Botany*, *26*, 207–211.

Reid, J. L., Fagan, M. E., & Zahawi, R. A. (2018). Positive site selection bias in meta-analyses comparing natural regeneration to active forest restoration. *Science Advances*, *4*(5), eaas9143. https://doi.org/10.1126/sciadv.aas9143

Rozendaal, D. M. A., Bongers, F., Aide, T. M., Alvarez-Dávila, E., Ascarrunz, N., Balvanera, P., Becknell, J. M., Bentos, T. V., Brancalion, P. H. S., Cabral, G. A. L., Calvo-Rodriguez, S., Chave, J., César, R. G., Chazdon, R. L., Condit, R., Dallinga, J. S., De Almeida-Cortez, J. S., De Jong, B., De Oliveira, A., ... & Poorter, L. (2019). Biodiversity recovery of Neotropical secondary forests. *Science Advances*, *5*(3), eaau3114. https://doi.org/10.1126/sciadv.aau3114

Saunders, R. M. K. (2012). The diversity and evolution of pollination systems in Annonaceae: Pollination systems in Annonaceae. *Botanical Journal of the Linnean Society*, *169*(1), 222–244. https://doi.org/10.1111/j.1095-8339.2011.01208.x

Schratz, P., Muenchow, J., Iturritxa, E., Cortés, J., Bischl, B., & Brenning, A. (2021). Monitoring forest health using hyperspectral imagery: Does feature selection improve the performance of machine-learning techniques? *Remote Sensing*, *13*(23), 4832. https://doi.org/10.3390/rs13234832

Severns, P. (2003). Inbreeding and small population size reduce seed set in a threatened and fragmented plant species, *Lupinus sulphureus* ssp. *kincaidii* (Fabaceae). *Biological Conservation*, *110*(2), 221–229. https://doi.org/10.1016/S0006-3207(02)00191-X

Stephens, R. E., Gallagher, R. V., Dun, L., Cornwell, W., & Sauquet, H. (2023). Insect pollination for most of angiosperm evolutionary history. *New Phytologist*, *240*(2), 880–891. https://doi.org/10.1111/nph.18993

Tong, Z.-Y., Wu, L.-Y., Feng, H.-H., Zhang, M., Armbruster, W. S., Renner, S. S., & Huang, S.-Q. (2023). New calculations indicate that 90% of flowering plant species are animal-pollinated. *National Science Review*, *10*(10), nwad219. https://doi.org/10.1093/nsr/nwad219

Ulyshen, M., Urban-Mead, K. R., Dorey, J. B., & Rivers, J. W. (2023). Forests are critically important to global pollinator diversity and enhance pollination in adjacent crops. *Biological Reviews*, *98*(4), 1118–1141. https://doi.org/10.1111/brv.12947

Vander Mijnsbrugge, K., Bischoff, A., & Smith, B. (2010). A question of origin: Where and how to collect seed for ecological restoration. *Basic and Applied Ecology*, *11*(4), 300–311. https://doi.org/10.1016/j.baae.2009.09.002

Veldkamp, E., Schmidt, M., Powers, J. S., & Corre, M. D. (2020). Deforestation and reforestation impacts on soils in the tropics. *Nature Reviews Earth & Environment*, *1*(11), 590–605. https://doi.org/10.1038/s43017-020-0091-5

Vrijenhoek, R. C. (1994). Genetic diversity and fitness in small populations. In V. Loeschcke, S. K. Jain, & J. Tomiuk (Eds.), *Conservation Genetics* (pp. 37–53). Birkhäuser Basel. https://doi.org/10.1007/978-3-0348-8510-2_5

Westwood, M., Cavender, N., Meyer, A., & Smith, P. (2021). Botanic garden solutions to the plant extinction crisis. Plants, People, Planet, *3*(1), 22–32. https://doi.org/10.1002/ppp3.10134

Wilkinson, K., Landis, T., Haase, D., Daley, B., & Dumroese, R. K. (2014). *Tropical Nursery Manual: A Guide to Starting and Operating a Nursery for Native & Traditional Plants*. U.S. Department of Agriculture, Forest Service.

Wortley, L., Hero, J.-M., & Howes, M. (2013). Evaluating ecological restoration success: A review of the literature. *Restoration Ecology*, *21*(5), 537–543. https://doi.org/10.1111/rec.12028

Xia, N., Deng, Y., & Yip, K. (2008). *Syzygium impressum* (Myrtaceae), a new species from Hong Kong. *Journal of Tropical and Subtropical Botany*, *16*(1), 19–22.

Xiao, N., Zhao, Z., Li, G., Gao, X., Ji, S., Xu, J., & Li, J. (2022). Biodiversity survey and assessment methods in biodiversity conservation priority areas in China. *Acta Ecologica Sinica*, *42*(7), 2523–2531.

Xu, H., Li, Y., Lin, M., Wu, J., Luo, T., Zhou, Z., Chen, D., Yang, H., Li, G., & Liu, S. (2015). Community characteristics of a 60 ha dynamics plot in the tropical montane rain forest in Jianfengling, Hainan Island. *Biodiversity Science*, *23*(2), 192–201. https://doi.org/10.17520/biods.2014157

Yang, Z., Ren, W., Liu, S., & Wei, H. (2025). Chapter 25—Artificial intelligence in forest management and tree health protection. In F. O. Asiegbu & A. Kovalchuk (Eds.), *Forest Microbiology* (Vol. 4, pp. 433–446). Academic Press. https://doi.org/10.1016/B978-0-443-21903-0.00026-6

Zeng, X., & Fischer, G. A. (2021). Using multiple seedlots in restoration planting enhances genetic diversity compared to natural regeneration in fragmented tropical forests. *Forest Ecology and Management*, *482*, 118819. https://doi.org/10.1016/j.foreco.2020.118819

Zhang, H., Wu, C., Chen, W., & Huang, G. (2019). Effect of urban expansion on summer rainfall in the Pearl River Delta, South China. *Journal of Hydrology*, *568*, 747–757. https://doi.org/10.1016/j.jhydrol.2018.11.036

Zhang, J., Cardoso, F. C. G., Zhu, H., Cheuk, M. L., Fischer, G. A., & Gale, S. W. (2025). Temporal shifts in the importance of environmental factors and management interventions among species in the early stages of forest restoration. *Journal of Forestry Research*. https://doi.org/10.1007/s11676-025-01857-4

Zhang, J., Zhu, H., Liu, J., & Fischer, G. A. (2016). Principles behind designing herbarium specimen labels and the R package 'herblabel'. *Biodiversity Science*, *24*(12), 1345–1352. https://doi.org/10.17520/biods.2016230

Zheng, Z., Deng, Y., Zhang, H., Yu, R., & Chen, Z. (2003). Holocene environmental changes in the tropical and subtropical areas of the south China and the relation to human activities. *Quaternary Sciences*, *24*, 387–393.

Zhu, H., Liu, J., Wong, C., Williams, C., Hang, K., Zhang, J., & Gale, S. W. (2023a). Additions to the flora of Hong Kong, China. *Subtropical Plant Science*, *52*(1), 70–77. https://doi.org/10.3969/j.issn.1009-7791.2023.01.011

Zhu, H., Liu, J., Zhang, J., Hang, K. Y., Yeung, W. K., & Fischer, G. A. (2018). Ten newly recorded taxa to the flora of Hong Kong, China. *Guihaia*, *22*, 21. https://doi.org/10.11931/guihaia.gxzw201710024

Zhu, H., Zhang, J., Cheuk, M. L., Hau, B. C. H., Fischer, G. A., & Gale, S. W. (2023b). Monoculture plantations impede forest recovery: Evidence from the regeneration of lowland subtropical forest in Hong Kong. *Frontiers in Forests and Global Change*, *6*, 1098666. https://doi.org/10.3389/ffgc.2023.1098666

Zhu, H., Zhang, J., Hau, B. C. H., Shum, B. T. W., Ma, X. K. K., Lo, J. P. L., Fischer, G. A., & Gale, S. W. (2024a). *Tai Po Kau ForestGEO Forest Dynamics Plot: Species Composition and Community Structure*. Kadoorie Farm and Botanic Garden.

Zhu, Y., Wang, L., ma, J., Hua, Z., Yang, Y., & Chen, F. (2024b). Assessment of carbon sequestration potential of mining areas under ecological restoration in China. *Science of the Total Environment*, *921*, 171179. https://doi.org/10.1016/j.scitotenv.2024.171179

Zhuang, X. Y., & Corlett, R. T. (1997). Forest and forest succession in Hong Kong, China. *Journal of Tropical Ecology*, *13*(6), 857–866. https://doi.org/10.1017/S0266467400011032

15 Invasive Alien Species Management in Vulnerable Forest Ecosystems

Katharina Lapin, Marcin K. Dyderski, and Cornelia Amon

15.1 INTRODUCTION

Forest ecosystems represent one of the most biologically rich and ecologically significant habitats on Earth (FAO, 2020; Oslo, 2011; Thuiller et al., 2019). They play fundamental roles in regulating climate, supporting biodiversity, cycling nutrients, and providing a wide range of ecosystem services that sustain human well-being (Mauri et al., 2023). Forest ecosystems, however, are increasingly threatened by a range of anthropogenic pressures. Invasive alien species (IAS) have emerged as one of the most critical challenges (Essl et al., 2020; Brondizio, 2019).

IAS are plants, animals, pathogens, or other organisms that are introduced—intentionally or unintentionally—outside their natural range, and whose establishment and spread cause significant ecological, economic, or social harm (Essl et al., 2020). Their impacts on biodiversity are leading to the decline of native species populations through mechanisms such as competition, predation, and pathogen transmission, as well as the disruption of ecosystem structure and function (Rai & Singh, 2020; Vilà et al., 2011).

The severity of this threat has been well-documented (Roy et al., 2023a). According to the IPBES Global Assessment Report on Biodiversity and Ecosystem Services, IAS is one of the top direct drivers of biodiversity loss globally. Since the 17th century, invasive alien species have been associated with approximately 40% of all known animal extinctions, underscoring their transformative and often irreversible impacts on ecosystems (Essl et al., 2020; Roy 2023). In Europe, IAS are considered among the most urgent environmental challenges, threatening biodiversity, ecosystem functioning, and socioeconomic stability (Pérez et al., 2022; Roy et al., 2023a, b; Vilà et al., 2011).

IAS affects nearly every type of ecosystem worldwide. Their ability to outcompete native vegetation, disrupt ecological networks, and alter key ecosystem processes makes them powerful agents of landscape-scale change (Hejda et al., 2009; Pyšek & Richardson, 2010; Van Kleunen et al., 2010). Forests—particularly those that are already vulnerable due to fragmentation, climate change, or other disturbances—are especially susceptible to invasion and its cascading effects (De Groot et al., 2022; Havrdová et al., 2023).

This chapter aims to discuss the impacts of invasive alien species in vulnerable forest ecosystems, focusing on understanding their mechanisms, monitoring and management strategies, governance frameworks, and emerging challenges. By integrating ecological theory with practical case studies, this chapter aims to provide a comprehensive overview to support effective management and conservation planning.

Subsequent sections will detail the main mechanisms and ecological consequences of IAS in forest ecosystems. We will then discuss how these species are introduced, and the vulnerabilities involved and outline surveillance and early detection methods. Current management strategies, such as prevention, control, and restoration, will be presented. Finally, we will discuss emerging challenges and future considerations. Collectively, these sections aim to provide a comprehensive viewpoint on the management of IAS in susceptible forest ecosystems in the context of a changing climate. The focus will mainly be on Europe, where IAS present significant challenges to ecological integrity.

DOI: 10.1201/9781003640899-19

15.2 UNDERSTANDING IAS IN FOREST ECOSYSTEMS

Forest ecosystems are exposed to a wide range of risk factors that can influence their health, composition, and long-term stability. These risks are often classified as abiotic and biotic. Abiotic risks include forest fires, storm damage, drought events, declining precipitation and changes in its distribution, as well as extreme temperature events (Seidl et al., 2014). Biotic risks include pests, diseases, and invasive alien species (Lindenmayer et al., 2000; Oslo, 2011; Rustad et al., 2012; Sierota et al., 2019). These factors can lead to changes in growth rates, phenology, species composition, and shifts in species distribution. Biotic impacts are often closely linked to abiotic influences, and in Europe, forest damage is most frequently caused by biotic factors. Forest ecosystem dynamics are shaped by a complex interplay of multiple factors (Bartz & Kowarik, 2019).

IAS in forest ecosystems presents particular challenges, mainly as they often occur in already disturbed or fragmented environments (Lapin, Oettel, et al., 2025). Their impacts extend across structural, compositional, and functional aspects of forest biodiversity. In certain conservation contexts, species-poor habitats are often undervalued, despite their significant environmental importance. Biological invasions can alter ecosystem functions and disrupt the provision of ecosystem services (Castro-Díez et al., 2019).

Vulnerability in forest ecosystems arises when ecological stability, structural diversity, or regenerative capacity is reduced by environmental or anthropogenic stressors. Forests that are fragmented, located along potential pathways, near urban infrastructure, or already weakened by drought, pests, diseases, or soil degradation tend to be more likely to be exposed to biological invasions (Figure 15.1). Edge effects—such as increased light availability and soil disturbance—further facilitate the establishment of light-demanding or disturbance-adapted IAS, such as the Tree of

FIGURE 15.1 Massive invasion of the invasive plant Reynoutria japonica in a temperate riparian forest ecosystem in southern Austria. Forest edges and riparian zones often serve are particularly vulnerable to biological invasions, particularly where habitat degradation and warmer microclimatic conditions facilitate the establishment and spread of IAS. (Photo: Lapin (2025).)

Heaven (*Ailanthus altissima*). A better understanding of these vulnerability drivers is crucial for prioritizing management efforts and designing effective early warning measures. Furthermore, IAS can undermine forest resilience, which is the capacity of forests to absorb disturbances such as climate change, wildfires, pest outbreaks, or human activities while maintaining their core functions and biodiversity (Brundu et al., 2025; De La Fuente & Beck, 2018; Minor et al., 2009).

Certain traits make IAS particularly successful and ecologically disruptive (Pyšek et al., 2014). These include rapid growth, adaptability to a wide range of conditions, lack of natural predators, clonal reproduction, high seed output, and tolerance of disturbance. Species, such as *Acacia* spp., not only degrade habitat quality and reduce native biodiversity, but they can also significantly alter fire regimes and the success of nutrient cycling (Richardson & Rejmánek, 2011). The United Kingdom spends over €200 million annually managing (Jovanović et al., 2018; Mandák et al., 2004; Williams et al., 2010) spp., and the costs across the European Union (EU) for managing invasive plants and restoring ecosystems continue to rise (Caffrey et al., 2014; Scalera, 2010). Beyond direct management expenses, IAS impair critical ecosystem services, including pollination, soil stability, water regulation, and carbon storage (Castro-Díez et al., 2019).

15.3 PATHWAYS OF INTRODUCTION

IAS spreads through both natural and human-mediated pathways. Natural dispersal can occur through wind, water, or animal movement. Human-mediated pathways, however, are often more significant, involving trade, tourism, transport infrastructure, and land-use changes, e.g., for Austria, located in Central Europe, stowaway and escape have been identified as the primary mechanisms for the introduction of neophytic vascular plants, with *Asteraceae* emerging as the most successful family (Glaser et al., 2025). In cultural landscapes, IAS can spread rapidly through machinery, along roads, railways, and field edges, as well as through intensive land use. Landscape elements are particularly exposed to edge effects, making them more vulnerable to invasion and leading to habitat degradation (Faulkner et al., 2020). Disturbance events, including those driven by intensive land use and changing climate conditions, often give IAS a competitive advantage over specialized, site-adapted native species (Dudney et al., 2025; Sexton & Lawhorn, 2025). In agriculturally dominated cultural landscapes, these pathways frequently facilitate the movement of IAS into adjacent forest areas.

15.4 ECOLOGICAL IMPACTS

IAS has wide-ranging effects on native biodiversity, ecosystem services, forest structure, and local livelihoods. Forest ecosystems are particularly vulnerable to threats from invasive alien pests and pathogens. An example is the American oak lace bug (*Corythucha arcuata*) (Figure 15.2), which causes defoliation and reduced tree vitality; *Dothistroma* needle blight (*Dothistroma septosporum*), a fungal disease affecting pine forests; and chestnut blight (*Cryphonectria parasitica*), which has had devastating effects on chestnut populations. Such IAS expansions can have severe economic impacts on the forest-related industries (Seidl et al., 2018).

One of the most common mechanisms is competition (Lapin et al., 2021), where invasive plants compete with native vegetation for essential resources such as light, water, nutrients, and physical space. A clear example is *Humulus scandens* (Japanese hop), a rapidly growing vine that forms dense mats over other plants. Shading the understory and suppressing seedlings, this species alters forest structure and reduces native biodiversity. Another mechanism is parasitism, in which invasive organisms act as parasites or pathogens, causing direct harm to native species. *Ophiostoma novo-ulmi*, the fungal pathogen behind Dutch elm disease, infects elm trees by blocking their water transport system. This results in extensive tree mortality and significant changes in forest composition and landscape structure. Structural impacts occur when IAS modify the physical characteristics of ecosystems. For instance, *Asclepias syriaca* (common milkweed) can form dense monocultures

FIGURE 15.2 An illustrative example of the consequences of such introductions is the case of the American oak lace bug (Corythucha arcuata). This leaf pest has been recorded in parts of South-Eastern Europe along highways. The insect damages trees by sucking sap from the underside of leaves, leading to leaf discoloration, tissue damage, and reduced photosynthetic capacity. Over time, repeated infestations weaken trees, making them more susceptible to other stressors. In addition to the direct physiological damage, the presence of Corythucha arcuata creates indirect competition with native, specialized herbivores, potentially altering local trophic interactions and forest ecosystem dynamics. (Photo: Lapin & Dyderski, 2024; Vienna, Austria.)

in disturbed areas, changing the vegetation structure and reducing habitat diversity. Such alterations disrupt pollination networks, soil processes, and the regeneration capacity of native plants. Some IAS act as vectors of disease, transmitting pathogens to which native species have no natural resistance. The leafhopper *Orientus ishidae* is a notable example, as it carries phytoplasmas—plant pathogenic bacteria—that affect forest and fruit trees. Through disease transmission, such insects accelerate the decline of forest health and productivity. Finally, alien species can cause indirect effects, which often manifest through ecological cascades. The chestnut gall wasp (*Dryocosmus kuriphilus*) induces gall formation on chestnut trees, weakening them and increasing their vulnerability to secondary pests and diseases. These indirect stresses reduce tree vitality, affect growth and reproduction, and ultimately alter forest composition and ecosystem functions.

15.5 EARLY DETECTION AND MONITORING

Effective management of IAS in forest ecosystems depends on early detection, accurate monitoring, and rapid response. Monitoring methods typically include field surveys, remote sensing, citizen science initiatives, and biosecurity checkpoints. Traditional field surveys remain essential for identifying new invasions, verifying species presence, and collecting ecological data. However, they require trained specialists and a considerable amount of time. Remote sensing technologies, including satellite imagery and aerial surveys, enable large-scale monitoring of forest structure and vegetation changes that may indicate invasion. Biosecurity checkpoints at borders, transport hubs, and forest entry points are also crucial for preventing the introduction and spread of invasive organisms. The integration of advanced technologies has significantly improved monitoring efficiency and predictive capacity. Machine learning algorithms can be applied to detect unusual vegetation signatures or species occurrences, supporting rapid and automated recognition of IAS in complex landscapes (Große-Stoltenberg et al., 2018; Kopeć et al., 2020).

Species distribution models (SDMs) have become indispensable tools for understanding and anticipating invasion dynamics under changing environmental conditions. Climate change is intensifying invasion risks by facilitating the northward and altitudinal spread of thermophilic species, thereby expanding invasion fronts into previously unsuitable regions (Bellard et al., 2013; Puchałka et al., 2023). SDMs correlate species occurrence data with climatic and environmental predictors to project potential range expansions and identify areas that are vulnerable. When applied to invasive alien species, SDMs help translate ecological risks into regulatory and management priorities (Gallien et al., 2012; Lapin & Dyderski, 2024). Tanner et al. (2017) emphasized the importance of horizon scanning and climate-aware risk assessment in updating SDMs to assess current and future climatic suitability under multiple climate change scenarios across Europe.

The SSPs (Shared Socioeconomic Pathways) represent future climate change scenarios developed by the Intergovernmental Panel on Climate Change (IPCC). They combine assumptions about greenhouse gas emissions, socioeconomic development, and land-use change. For example, projections have shown that range expansion is most pronounced for *Reynoutria*×*bohemica* and *R. japonica*, with increases in suitable habitat of up to 40% relative to current distributions, the SSP370 (moderately pessimistic) and SSP585 (worst-case) scenarios. These species are expected to shift their ranges northward and eastward, particularly into Scandinavia, the Baltic region, and higher altitudes in the Carpathians and Alps. Range expansion was also observed for *R. sachalinensis* and the general *Reynoutria* spp. ensemble model, although to a slightly lesser degree. These findings align with previous risk assessments (Jovanović et al., 2018), demonstrating how SDMs provide valuable foresight for targeted surveillance and management planning (Figure 15.3).

FIGURE 15.3 Current and projected climatic suitability maps for Reynoutria spp., illustrate substantial variation in future distribution patterns under the SSP245 (moderate) and SSP585 (worst-case) scenarios for the periods 2041–2060 and 2061–2080, as per the 6th IPCC report. In the maps, the combination of green and orange colours, which indicates the current potential distribution. Green areas represent climatic persistence, orange areas show projected range contraction, and blue areas highlight zones of future expansion. This shift has direct management implications: preventive action becomes more valuable than reactive control. In practice, this means allocating resources to horizon scanning, updating regional watch lists, increasing monitoring along disturbance-prone habitats (such as ski tourism infrastructure, avalanche corridors, river valleys, and forest road networks), and preparing rapid-response protocols before Reynoutria spp. become established. (Photo: Lapin, 2025; Lower Austria.)

Citizen science plays an increasingly important role in the detection and monitoring of invasive species (De Groot et al., 2023). One of the central challenges in managing invasions is their cross-sectoral nature, as IAS often appear and spread through multiple pathways, including forestry, transport, trade, and recreation. Engaging the public is essential, as people may unintentionally contribute to species introductions or, conversely, act as a valuable early warning network. Citizen science initiatives expand the spatial and temporal coverage of monitoring efforts, increase public awareness, and often lead to earlier detections of invasive species. Several case studies have demonstrated the effectiveness of citizen observations in identifying both insect pests and invasive plant species at early stages of invasion. However, successful citizen science programs require clear communication, standardized protocols, and strong scientific coordination to ensure data quality and meaningful integration into management frameworks (De Groot et al., 2023; H. Roy et al., 2018).

15.6 MANAGEMENT STRATEGIES

Over the past two decades, research on biological invasions in forests has expanded substantially, providing a growing foundation for effective management. However, successful control depends on selecting strategies that are appropriate to the specific ecological context, invasion stage, and management goals. In forest ecosystems, this typically involves combining prevention and regulation with mechanical, chemical, biological, and restoration-based measures. An integrated approach enables managers to adapt to changing environmental conditions and evolving invasion fronts, particularly in areas where climate change increases the likelihood of new establishment events.

15.6.1 Prevention

Prevention remains the most effective and cost-efficient strategy for managing invasive alien species. Policy instruments, quarantine measures, and trade regulations play a crucial role in minimizing introductions. One prominent example is the EU's Regulation No 1143/2014, which aims to prevent and reduce the adverse impacts of invasive species. A key element of this regulation is the Union List, a dynamic inventory of species considered to be of concern to the EU. Inclusion on this list triggers strict legal restrictions on the introduction, trade, breeding, and release of these species. In addition to IAS regulations, plant health legislation requires Member States to take immediate eradication measures when listed organisms are detected, ensuring a coordinated legal response across the EU.

15.6.2 Control Measures

Mechanical and physical methods include manual removal, controlled burning, and habitat restoration. These approaches are often used for invasive plant species that dominate forest edges, riparian zones, or disturbed sites (Weidlich et al., 2020; Wilson et al., 2011). Chemical control involves the use of herbicides or pesticides to suppress or eradicate invasive species. While these methods can be effective, they carry ecological risks, including non-target effects, soil and water contamination, and long-term impacts on native flora and fauna. Chemical treatments are therefore typically applied in combination with other methods and under strict regulatory conditions to minimize unintended consequences. Biological control uses natural enemies, such as herbivores, pathogens, or predators, to manage IAS. This approach requires careful ecological risk assessment to avoid introducing new problems. When implemented responsibly, biological control can provide sustainable long-term suppression of invasive populations, particularly in cases where mechanical or chemical methods are insufficient.

15.6.3 Integrated Management Approaches

Integrated management combines multiple methods—prevention, mechanical removal, chemical control, biological control, and restoration—within an adaptive management framework. Stakeholder participation is essential, involving forestry authorities, landowners, scientists, policymakers, and the public (Hazarika et al., 2024; Novoa et al., 2024). Adaptive management is a flexible strategy that can help to adjust as ecological conditions and invasion dynamics change over time. Furthermore, management strategies that promote species diversity can reduce the risks of pests and IAS. Mixed-species stands tend to support more natural enemies and exhibit lower pest damage than monocultures (Lapin, Hoffmann, et al., 2025; Menéndez-Miguélez et al., 2024). Greater host diversity makes it more difficult for specialized pests and pathogens to spread, providing a natural buffer against large-scale damage.

The integrated management approach also requires active monitoring in vulnerable forest sites, such as nature reserves or unmanaged forest patches (Lapin, Oettel, et al., 2025). Monitoring IAS in unmanaged forests presents challenges due to inaccessibility of these areas. Regular surveys in these vulnerable forest ecosystem sites support the detection of new inaccessible areas (Lapin et al., 2019). For example, a recent survey revealed the spread of non-native Prunus cerasifera in the Strict Reserve of Białowieża Primaeval Forest (Poland), approximately 50 years after its introduction in the surroundings. Regular surveys support the detection of new invasive alien plant species, track changes in abundance of existing ones, and offer opportunities to evaluate the effectiveness of management measures implemented in surrounding areas.

15.6.4 Ecological Restoration Measures

Successful eradication of IAS in forests requires more than simply removing the target organism (Lapin, Hoffmann, et al., 2025; Menéndez-Miguélez et al., 2024). Without ecological restoration, cleared areas remain vulnerable to reinvasion. Restoration involves reinforcing native species, re-establishing natural ecosystem dynamics, and addressing the underlying drivers of invasion. Without these measures, eradication efforts are often temporary and ineffective. Restoration, therefore, forms the ecological backbone of long-term IAS management.

A relevant example is the case of ash dieback caused by *Hymenoscyphus fraxineus* (Figure 15.3), a pathogen introduced from East Asia. The disease affects twigs, crowns, and root collars, often leading to tree mortality through secondary wood-decaying fungi. In Europe, the disease remains active across the country. Breeding programs have identified a small proportion (2%–5%) of ash trees with natural disease tolerance, indicating a genetic basis for resistance. Since 2015, the initiative (George et al., 2022, 2024) has focused on selecting resistant genotypes, collecting seeds, raising plants, and evaluating traits. By 2024, clone selection for orchard establishment was completed. Results of redundancy analyses revealed significant genotype–environment associations related to temperature and elevation, supporting breeding efforts for resistant ash populations to maintain the ecological and economic roles of ash in European forests (Runge et al., 2019) (Figure 15.4).

15.6.5 Non-native Trees and Risk Assessment

Finally, the usage of non-native trees needs to be addressed when discussing the risks of biological invasion. Non-native tree species are widely used in European forestry for climate change adaptation and economic purposes. Some, such as *Robinia pseudoacacia*, have been cultivated outside their native range for centuries (Figure 15.5). In contrast, others, like *Eucalyptus* spp., have been introduced more (Brus et al., 2019) recently (Marinšek et al., 2022; Brundu et al., 2020). Global guidelines and regional strategies recommend prioritizing native tree species for afforestation due to the invasive potential of particular non-native species under changing climatic conditions (Lapinet al., 2023). When non-native trees are considered, site-specific risk assessments are essential (Bindewald et al., 2021). These assessments should account for the proximity of conservation areas and evaluate the invasive potential of the species in question to ensure that adaptation measures do not inadvertently introduce new ecological problems (Bindewald et al., 2021; Hazarika et al., 2024; Konic et al., 2024).

FIGURE 15.4 (a) Healthy, tolerant individual in comparison to (b) symptomatic individual displaying characteristic leaf wilting, necrosis, and shoot dieback indicative of progressive infection. Symptoms of ash dieback (Hymenoscyphus fraxineus) on young Fraxinus excelsior trees in the experimental nursery of the Austrian Forest Research Centre, Tulln, in Lower Austria. Visible leaf wilting, necrosis, and shoot dieback indicate progressive infection in nursery-grown germinants. Such symptomatic individuals contrast with naturally tolerant genotypes (2%–5% of the population), which maintain crown vitality and serve as the genetic basis for breeding-for-resilience and biotechnological restoration programs aimed at sustaining ash populations under ongoing disease pressure. (Photo: (a) Konrad, (b) Unger/BFW.)

FIGURE 15.5 A forest road in south-eastern Austria with multiple invasive alien plant species, such as Robinia pseudoacacia and Solidago ssp. Forest roads serve as crucial pathways for biological invasions, particularly in degraded forest ecosystems where disturbed habitats and warmer microclimatic conditions favour the establishment and spread of alien species. (Photo: Amon, 2025; Burgenland, Lower Austria.)

15.7 CHALLENGES AND FUTURE DIRECTIONS

Despite growing recognition of the ecological and socio-economic impacts of invasive alien species in forest ecosystems, significant challenges remain in translating knowledge into effective and sustainable management practices. One of the most pressing challenges is the accelerating influence of climate change on invasion dynamics. Across all climate scenarios, most IAS exhibit clear patterns of climate-sensitive range shifts, with trends toward geographic expansion becoming especially pronounced under high-emission trajectories. This includes northward and altitudinal range expansions of thermophilic species into areas that were previously unsuitable (Puchałka, Paź-Dyderska, Woziwoda, et al., 2023). These patterns highlight the importance of integrating projections of future climatic suitability into risk assessments, thereby strengthening the scientific basis for adaptive and forward-looking management strategies (Essl et al., 2020; Essl & Rabitsch, 2013; Seidl et al., 2018).

A critical limitation remains insufficient funding and resources, which constrain the capacity to implement prevention, monitoring, and management measures effectively. This is particularly problematic for vulnerable forest ecosystems, which require long-term investments in surveillance, restoration, and adaptive management to mitigate the impacts of invasions. Limited financial support often results in fragmented actions, with priority given to short-term control measures rather than comprehensive, ecosystem-based strategies. Although resource limitations and fragmented coordination can hinder management actions, these same challenges reinforce the importance of prevention, early detection, and prioritization. Cost-effective tools such as SDMs, citizen science monitoring, and standardized communication frameworks can help expand surveillance capacity and support proactive management, even under constrained budgets.

Another persistent challenge is the lack of awareness, capacity, and coordination across sectors and governance levels. IAS management often involves multiple stakeholders, including forestry agencies, conservation organizations, policymakers, landowners, and the general public. Inconsistent communication and weak institutional integration frequently hinder early detection and rapid response. Building institutional capacity and fostering coordinated cross-sectoral networks are essential to overcome these barriers.

There are also substantial research gaps that need to be addressed. These include improving predictive modelling to better anticipate invasion dynamics under changing environmental conditions, advancing ecological restoration approaches following species removal, and enhancing understanding of how multiple stressors interact to shape invasion outcomes. Research into the sustainability and effectiveness of control measures is essential, and innovative new methods—such as electric treatment (Lehnhoff et al., 2022)—need to be tested for their efficiency on different species.

DECLARATION OF GENERATIVE AI USE

During the preparation of this manuscript, the author used ChatGPT to edit language. The author subsequently reviewed and edited the content and takes full responsibility for the final version of the published chapter.

REFERENCES

Bartz, R., & Kowarik, I. (2019). Assessing the environmental impacts of invasive alien plants: A review of assessment approaches. In *NeoBiota* (Vol. 43, pp. 69–99). Pensoft Publishers. https://doi.org/10.3897/NEOBIOTA.43.30122

Bellard, C., Thuiller, W., Leroy, B., Genovesi, P., Bakkenes, M., & Courchamp, F. (2013). Will climate change promote future invasions? *Global Change Biology*, *19*(12), 3740–3748. https://doi.org/10.1111/gcb.12344

Bindewald, A., Brundu, G., Schueler, S., Starfinger, U., Bauhus, J., & Lapin, K. (2021). Site-specific risk assessment enables trade-off analysis of non-native tree species in European forests. *Ecology and Evolution*, *11*(24), 18089–18110.

Brondizio ES. (2019). Global assessment report on biodiversity and ecosystem services of the Intergovernmental Science-Policy Platform on Biodiversity and Ecosystem Services. *IPBES secretariat*, 1148.

Brundu, G., De Groot, M., Kumschick, S., Pergl, J., & Lapin, K. (2025). Managing invasive alien species in forest corridors and stepping stones. In *Ecological Connectivity of Forest Ecosystems* (pp. 347–364). Cham: Springer Nature Switzerland.

Brundu, G., Pauchard, A., Pyšek, P., Pergl, J., Bindewald, A. M., Brunori, A., Canavan, S., Campagnaro, T., Celesti-Grapow, L., & Dechoum, M. de S. (2020). Global guidelines for the sustainable use of non-native trees to prevent tree invasions and mitigate their negative impacts. *NeoBiota*, *61*, 65–116.

Brus, R., Pötzelsberger, E., Lapin, K., Brundu, G., Orazio, C., Straigyte, L., & Hasenauer, H. (2019). Extent, distribution and origin of non-native forest tree species in Europe. *Scandinavian Journal of Forest Research*, *34*(7), 533–544.

Caffrey, J. M., Baars, J.-R., Barbour, J. H., Boets, P., Boon, P., Davenport, K., Dick, J. T. A., Early, J., Edsman, L., & Gallagher, C. (2014). Tackling invasive alien species in Europe: The top 20 issues. *Management of Biological Invasions*, *5*(1), 1.

Castro-Díez, P., Vaz, A. S., Silva, J. S., Van Loo, M., Alonso, Á., Aponte, C., Bayón, Á., Bellingham, P. J., Chiuffo, M. C., & DiManno, N. (2019). Global effects of non-native tree species on multiple ecosystem services. *Biological Reviews*, *94*(4), 1477–1501.

De Groot, M., Pocock, M. J. O., Bonte, J., Fernandez-Conradi, P., & Valdés-Correcher, E. (2023). Citizen science and monitoring forest pests: A beneficial alliance? *Current Forestry Reports*, *9*(1), 15–32.

De Groot, M., Schueler, S., Sallmannshofer, M., Virgillito, C., Kovacs, G., Cech, T., Božič, G., Ogris, N., Hoch, G., & Kavčič, A. (2022). Forest management, site characteristics and climate change affect multiple biotic threats in riparian forests. *Forest Ecology and Management*, *508*, 120041.

De La Fuente, B., & Beck, P. S. A. (2018). IAS may disrupt protected area networks: Insights from the pine wood nematode spread in Portugal. *Forests*, *9*(5), 282.

Dudney, J., Edwards, J., Harvey, B. J., & Seidl, R. (2025). Climate change effects on interacting disturbances in forest ecosystems. *Annual Review of Ecology, Evolution, and Systematics*, *56*, 393–420.

Essl, F., Lenzner, B., Bacher, S., Bailey, S., Capinha, C., Daehler, C., Dullinger, S., Genovesi, P., Hui, C., & Hulme, P. E. (2020). Drivers of future alien species impacts: An expert-based assessment. *Global Change Biology*, *26*(9), 4880–4893.

Essl, F., & Rabitsch, W. (Eds.). (2013). *Biodiversität und Klimawandel: Auswirkungen und Handlungsoptionen für den Naturschutz in Mitteleuropa*. Springer Berlin Heidelberg. https://doi.org/10.1017/CBO9781107415324.004

Essl, F., et al. (2020). The Convention on Biological Diversity (CBD)'s Post-2020 target on invasive alien species–what should it include and how should it be monitored?" *NeoBiota*, 62, 99–121.

FAO. (2020). *Sustainable forest management: Lessons of history and recent developments*. https://www.fao.org/3/v6585e/V6585e07.htm

Faulkner, K. T., Hulme, P. E., Pagad, S., Wilson, J. R. U., & Robertson, M. P. (2020). Classifying the introduction pathways of alien species: Are we moving in the right direction? *NeoBiota*, *62*, 143–159.

Gallien, L., Douzet, R., Pratte, S., Zimmermann, N. E., & Thuiller, W. (2012). IAS distribution models–how violating the equilibrium assumption can create new insights. *Global Ecology and Biogeography*, *21*(11), 1126–1136.

George, J. P., Rusanen, M., Beuker, E., Yrjänä, L., Timmermann, V., Potočić, N., Välimäki, S., & Konrad, H. (2024). Lessons to learn for better safeguarding of genetic resources during tree pandemics: The case of ash dieback in Europe. In *Biological Conservation* (Vol. 299). Elsevier Ltd. https://doi.org/10.1016/j.biocon.2024.110802

George, J. P., Sanders, T. G. M., Timmermann, V., Potočić, N., & Lang, M. (2022). European-wide forest monitoring substantiate the neccessity for a joint conservation strategy to rescue European ash species (Fraxinus spp.). *Scientific Reports*, *12*(1). https://doi.org/10.1038/s41598-022-08825-6

Glaser, M., Gilli, C., Griebl, N., Hohla, M., Pflugbeil, G., Stöhr, O., Pilsl, P., Ehrendorfer-Schratt, L., Niklfeld, H., & Walter, J. (2025). Checklist of Austrian neophytes. *Preslia*, *97*(3), 413–539.

Große-Stoltenberg, A., Hellmann, C., Thiele, J., Werner, C., & Oldeland, J. (2018). Early detection of GPP-related regime shifts after plant invasion by integrating imaging spectroscopy with airborne LiDAR. *Remote Sensing of Environment*, *209*, 780–792.

Havrdová, A., Douda, J., & Doudová, J. (2023). Threats, biodiversity drivers and restoration in temperate floodplain forests related to spatial scales. *Science of the Total Environment*, *854*, 158743.

Hazarika, R., Lapin, K., Bindewald, A., Vaz, A. S., Marinšek, A., La Porta, N., Detry, P., Berger, F., Barič, D., & Simčič, A. (2024). Balancing risks and benefits: Stakeholder perspective on managing non-native tree species in the European Alpine Space. *Mitigation and Adaptation Strategies for Global Change*, *29*(6), 55.

Hejda, M., Pyšek, P., & Jarošík, V. (2009). Impact of invasive plants on the species richness, diversity and composition of invaded communities. *Journal of Ecology*, *97*(3), 393–403.

Jovanović, S., Hlavati-Širka, V., Lakušić, D., Jogan, N., Nikolić, T., Anastasiu, P., Vladimirov, V., & Šinžar-Sekulić, J. (2018). Reynoutria niche modelling and protected area prioritization for restoration and protection from invasion: A Southeastern Europe case study. *Journal for Nature Conservation*, *41*, 1–15.

Konic, J., Heiling, C., Haeler, E., Chakraborty, D., Lapin, K., & Schueler, S. (2024). The potential of non-native tree species to provide major ecosystem services in Austrian forests. *Frontiers in Plant Science*, *15*, 1402601.

Kopeć, D., Sabat-Tomala, A., Michalska-Hejduk, D., Jarocińska, A., & Niedzielko, J. (2020). Application of airborne hyperspectral data for mapping of invasive alien Spiraea tomentosa L.: A serious threat to peat bog plant communities. *Wetlands Ecology and Management*, *28*(2), 357–373.

Lapin, K., Bacher, S., Cech, T., Essl, F., Georges, F.-I., Hoch, G., Kavčič, A., Koltay, A., Kostić, S., & Lukić, I. (2021). Comparing environmental impacts of alien plants, insects and pathogens in protected riparian forests. *NeoBiota*, *69*, 1–28.

Lapin, K., Bindewald, A., Brundu, G., Marinšek, A., Chakraborty, D., Oettel, J., Konrad, H., Porta, N., Alagić, A., & Berger, F. (2023). Transnational strategy on the sustainable management and responsible use of non-native trees in the Alpine Space. *Central European Forestry Journal*, *69*(3), 142–151.

Lapin, K., & Dyderski, M. K. (2024). Expanding range of the invasive shrub Amorpha fruticosa under changing climate. *Regional Environmental Change*, *24*(4), 152.

Lapin, K., Hoffmann, J., Dyderski, M. K., Roche, J., Lakatos, F., Menéndez-Miguélez, M., de -Miguel, S., Erdozain, M., Svensson, J., & Werden, L. K. (2025). Challenges, motivations, and perspectives of practitioners on forest restoration across seven European countries. *Restoration Ecology*, *33*, e70160.

Lapin, K., Oettel, J., Braun, M., & Konrad, H. (2025). *Ecological Connectivity of Forest Ecosystems*. Springer.

Lapin, K., Oettel, J., Steiner, H., Langmaier, M., Sustic, D., Starlinger, F., Kindermann, G., & Frank, G. (2019). Invasive alien plant species in unmanaged forest reserves, Austria. *NeoBiota*, *48*, 71–96.

Lehnhoff, E. A., Neher, P., Indacochea, A., & Beck, L. (2022). Electricity as an effective weed control tool in non-crop areas. *Weed Research*, *62*(2), 149–159.

Lindenmayer, D. B., Margules, C. R., & Botkin, D. B. (2000). Indicators of biodiversity for ecologically sustainable forest management. *Conservation Biology*, *14*(4), 941–950.

Mandák, B., Pyšek, P., & Bímová, K. (2004). History of the invasion and distribution of Reynoutria taxa in the Czech Republic: A hybrid spreading faster than its parents. *Presila*, *76*, 15–64.

Marinšek, A., Bindewald, A., Kraxner, F., La Porta, N., Meisel, P., Stojnić, S., Cocozza, C., & Lapin, K. (2022). *Management of Non-native Tree Species in Urban Areas of the Alpine Space*. BFW (Bundesforschungszentrum für Wald).

Mauri, A., Girardello, M., Forzieri, G., Manca, F., Beck, P. S. A., Cescatti, A., & Strona, G. (2023). Assisted tree migration can reduce but not avert the decline of forest ecosystem services in Europe. *Global Environmental Change*, *80*, 102676.

Menéndez-Miguélez, M., Rubio-Cuadrado, Á., Cañellas, I., Erdozain, M., de Miguel, S., Lapin, K., Hoffmann, J., Werden, L., & Alberdi, I. (2024). How to measure outcomes in forest restoration? A European review of success and failure indicators. *Frontiers in Forests and Global Change*, *7*, 1420127.

Minor, E. S., Tessel, S. M., Engelhardt, K. A. M., & Lookingbill, T. R. (2009). The role of landscape connectivity in assembling exotic plant communities: A network analysis. *Ecology*, *90*(7), 1802–1809.

Novoa, A., Vimercati, G., Brundu, G., Richardson, D. M., Schaffner, U., Brunori, A., Campagnaro, T., Canavan, S., Celesti-Grapow, L., & Dechoum, M. (2024). Stakeholders' views on the global guidelines for the sustainable use of non-native trees. *People and Nature*, *6*(4), 1640–1654.

Oslo, F. E. (Organization). L. U. (2011). *State of Europe's Forests, 2011: Status & Trends in Sustainable Forest Management in Europe*. https://foresteurope.org/publications_type/state-of-europes-forests-2011/

Pérez, G., Vila, M., & Gallardo, B. (2022). Potential impact of four invasive alien plants on the provision of ecosystem services in Europe under present and future climatic scenarios. *Ecosystem Services*, *56*, 101459.

Puchałka, R., Paź-Dyderska, S., Jagodziński, A. M., Sádlo, J., Vítková, M., Klisz, M., Koniakin, S., Prokopuk, Y., Netsvetov, M., & Nicolescu, V.-N. (2023). Predicted range shifts of alien tree species in Europe. *Agricultural and Forest Meteorology*, *341*, 109650.

Puchałka, R., Paź-Dyderska, S., Woziwoda, B., & Dyderski, M. K. (2023). Climate change will cause climatic niche contraction of Vaccinium myrtillus L. and V. vitis-idaea L. in Europe. *Science of the Total Environment, 892*, 164483.

Pyšek, P., Jarošík, V., Pergl, J., Moravcová, L., Chytrý, M., & Kühn, I. (2014). Temperate trees and shrubs as global invaders: the relationship between invasiveness and native distribution depends on biological traits. *Biological Invasions, 16*(3), 577–589.

Pyšek, P., & Richardson, D. M. (2010). Invasive species, environmental change and management, and health. *Annual Review of Environment and Resources, 35*(1), 25–55.

Rai, P. K., & Singh, J. S. (2020). Invasive alien plant species: Their impact on environment, ecosystem services and human health. *Ecological Indicators, 111*, 106020.

Richardson, D. M., & Rejmánek, M. (2011). Trees and shrubs as invasive alien species: A global review. *Diversity and Distributions, 17*(5), 788–809.

Roy, H., Groom, Q., Adriaens, T., Agnello, G., Antic, M., Archambeau, A.-S., Bacher, S., Bonn, A., Brown, P., & Brundu, G. (2018). Increasing understanding of alien species through citizen science (Alien-CSI). *Research Ideas and Outcomes, 4*, e31412.

Runge, C. A., Withey, J. C., Naugle, D. E., Fargione, J. E., Helmstedt, K. J., Larsen, A. E., Martinuzzi, S., & Tack, J. D. (2019). Single species conservation as an umbrella for management of landscape threats. *PloS One, 14*(1), e0209619.

Rustad, L., Campbell, J., Dukes, J. S., Huntington, T., Lambert, K. F., Mohan, J., & Rodenhouse, N. (2012). Changing climate, changing forests: The impacts of climate change on forests of the northeastern United States and eastern Canada. *General Technical Report NRS-99*. Newtown Square, PA: US Department of Agriculture, Forest Service, Northern Research Station. 48 p., 99, 1–48.

Scalera, R. (2010). How much is Europe spending on invasive alien species? *Biological Invasions, 12*(1), 173–177.

Seidl, R., Klonner, G., Rammer, W., Essl, F., Moreno, A., Neumann, M., & Dullinger, S. (2018). Invasive alien pests threaten the carbon stored in Europe's forests. *Nature Communications, 9*(1). https://doi.org/10.1038/s41467-018-04096-w

Seidl, R., Schelhaas, M.-J., Rammer, W., & Verkerk, P. J. (2014). Increasing forest disturbances in Europe and their impact on carbon storage. *Nature Climate Change, 4*(9), 806–810.

Sexton, A. N., & Lawhorn, K. A. (2025). Best practices for designing resilient urban ecosystems through native species restoration. *Urban Forestry & Urban Greening, 104*, 128657.

Sierota, Z., Grodzki, W., & Szczepkowski, A. (2019). Abiotic and biotic disturbances affecting forest health in Poland over the past 30 years: Impacts of climate and forest management. In *Forests* (Vol. 10, Issue 1). MDPI AG. https://doi.org/10.3390/f10010075

Tanner, R., Branquart, E., Brundu, G., Buholzer, S., Chapman, D., Ehret, P., Fried, G., Starfinger, U., & van Valkenburg, J. (2017). The prioritisation of a short list of alien plants for risk analysis within the framework of the Regulation (EU) No. 1143/2014. *NeoBiota, 35*, 87–118.

Thuiller, W., Guéguen, M., Renaud, J., Karger, D. N., & Zimmermann, N. E. (2019). Uncertainty in ensembles of global biodiversity scenarios. *Nature Communications, 10*(1), 1446.

Van Kleunen, M., Weber, E., & Fischer, M. (2010). A meta-analysis of trait differences between invasive and non-invasive plant species. *Ecology Letters, 13*(2), 235–245.

Vilà, M., Espinar, J. L., Hejda, M., Hulme, P. E., Jarošík, V., Maron, J. L., Pergl, J., Schaffner, U., Sun, Y., & Pyšek, P. (2011). Ecological impacts of invasive alien plants: A meta-analysis of their effects on species, communities and ecosystems. *Ecology Letters, 14*(7), 702–708.

Weidlich, E. W. A., Flórido, F. G., Sorrini, T. B., & Brancalion, P. H. S. (2020). Controlling invasive plant species in ecological restoration: A global review. *Journal of Applied Ecology, 57*(9), 1806–1817.

Williams, F., Eschen, R., Harris, A., Djeddour, D., Pratt, C., Shaw, R. S., Varia, S., Lamontagne-Godwin, J., Thomas, S. E., & Murphy, S. T. (2010). The economic cost of invasive non-native species on Great Britain. *CABI Proj No VM10066, 199.*

Wilson, J. R. U., Gairifo, C., Gibson, M. R., Arianoutsou, M., Bakar, B. B., Baret, S., Celesti-Grapow, L., DiTomaso, J. M., Dufour-Dror, J., Kueffer, C., Kull, C. A., Hoffmann, J. H., Impson, F. A. C., Loope, L. L., Marchante, E., Marchante, H., Moore, J. L., Murphy, D. J., Tassin, J., … & Richardson, D. M. (2011). Risk assessment, eradication, and biological control: Global efforts to limit Australian acacia invasions. *Diversity and Distributions, 17*(5), 1030–1046. https://doi.org/10.1111/j.1472-4642.2011.00815.x

16 Changes in the Structure and Functions of Soil Bacterial Communities and Driving Factors in a Restored Forest Soil Chronosequence

Feifei Wang and Wenzhi Cao

16.1 INTRODUCTION

Soil erosion, one of the severe ecological degradations in China, has caused a large reduction in crop yields by soil nutrient loss, infertility, and adverse impacts on soil microbial diversity (Higgitt, 1992; Huang et al., 2013; Shi et al., 2005), frequent flooding by river channel siltation, and living environmental deterioration 20 years ago. The afforestation, as an effective way for vegetation restoration, enhances water conservation, improves soil physicochemical environment, and increases soil biodiversity (Boulmane et al., 2017; Ding et al., 2023; Gellie et al., 2017). Soil microbial communities regulate the responses of plant-soil ecosystems to anthropogenic disturbances and environmental changes by modulating the soil environment and nutrient cycling (Banning et al., 2011; Lauber et al., 2008; Roesch et al., 2007). In eroded areas, owing to the imbalanced nutrient (C, N, and P) supply, the plant growth and local soil microbial functions are often negatively influenced (Liu et al., 2023).

Soil nutrient cycling is dominantly regulated by the bacterial community (Wang et al., 2023). Functional Annotation of Prokaryotic Taxa (FAPROTAX) is a database and annotation tool for predicting the metabolic functions encoded by bacterial taxa in the soil microbiome (Louca et al., 2016). It has been used in a broad spectrum of environments ranging from terrestrial to coastal systems. The structure and function of soil microbial communities are regulated by abiotic factors (pH, nutrients, and moisture) and soil properties that are introduced by plant alterations (Cui et al., 2020). Owing to the vulnerability of abiotic factors to environmental changes, the plant changes can have subsequent effects on nutrient availability, turnover, and other soil functions. This finally can result in changes in the structure and functions of microbial communities in eroded areas.

Changting County is one of the severely eroded areas in Southeast China (Ding et al., 2023). The vegetation restoration of local soil erosion began with establishing a Masson pine monoculture (*Pinus massoniana*) as a pioneer stand, which was later brought through the sequential introduction of *Schima superba*, *Fagus sylvatica*, and other diverse broadleaved species. Research in Changting County indicated that such mixed forests (MF) positively impacted soil physicochemical traits and increased species diversity, thereby generating a range of plant communities distinct in their composition and restoration time (Wang, 2020).

Therefore, this study selected different restored forest chronosequences to study the characteristics of changes in plant nutrients, soil physicochemical properties, and community composition and metabolic function of bacteria (Chen et al., 2024; Guo et al., 2024). This study was designed to: (1) characterize the nutrient levels and stoichiometry of inputs derived from plants and litter to

DOI: 10.1201/9781003640899-20

the eroded soil; (2) assess variations in soil C, N, and P contents and bacterial communities across vegetation restoration types and restoration periods; (3) examine the effect of vegetation types and restoration duration on nutrients inputs from plant-litter-soil system and the turnover rates of soil organic C (SOC); and (4) identify key factors influencing the shifts of bacterial structure and function during vegetation restoration. The research will facilitate ecological restoration policies in subtropical red soil eroded areas.

16.2 MATERIALS AND METHODS

16.2.1 Study Areas

Changting County has a typical subtropical marine monsoon climate, with the mean annual temperature of 18°C and the mean annual precipitation of 1,650 mm. The red soil developed from local coarse crystalline granites, and the surface soil layer has high sand content and is easily eroded by water. More than 30 years ago, the county had experienced severe erosion and land degradation. After more than 30 years of afforestation, the rate of soil and water loss has decreased from 31.5% (1985) to 6.78% (2020), with a reduction in area for soil erosion control by 745.33 km^2. Most of the current forest includes planted *P. massoniniana*, and the understory shrub layer is dominated by shrubby bush clover, *Adinandra millettii*, and cape jasmine (Table 16.1). The herbaceous layer is mainly occupied by *Dicanopteris dichotoma*. The primary evergreen broadleaved forest is the regional climax community in Changting County and is exclusively distributed in the Tingjiang River Source Protection Area.

16.2.2 Site Description and Sampling

Five coniferous forests (CF) and two MF in vegetation restoration areas, and two broadleaf forests (BF) in a nature reserve were selected in Changting County. CF, MF, and BF were dominated by the sole species *P. massoniana*, *P. massoniana* with some broadleaf trees, and local climax communities (age > 50 years), respectively (Table 16.1). Among them, the restoration years of CF and MF included 3, 10, 20, and 30 years, and 10 and 30 years, respectively. Bare land (BL) without any plants was selected as the control.

Three replicate 10 × 10 m quadrats in each vegetation area were selected to investigate plant communities and collect leaf and litter samples. In different vegetation types, we collected and mixed five surface soil samples randomly. One portion of the mixed soils was transported to the laboratory for analyzing the soil physicochemical parameters and nutrient contents. The other portion was stored at −80°C for 16S rRNA analysis.

16.2.3 Soil Physicochemical Analysis

Soil moisture and bulk density (BD) were determined by the oven-drying method and the cutting ring method, respectively. Soil pH and texture were measured using a water quality detector (WTW, Germany) and a laser particle size meter (Malvern Panalytical, USA), respectively. Total N (TN) and total C (TC) were determined using a TOC elemental analyzer (Elementar). The concentrations of SOC, total P (TP), microbial biomass P (MBP), and available P (AP) were measured according to standardized procedures (Guo et al., 2024). SOC fractions, including dissolved organic C (DOC), microbial biomass C (MBC), easily oxidized organic C (EOC), labile organic C (LOC), and recalcitrant organic C (ROC), were determined and calculated according to established methods (Wang et al., 2024b). SOC storage capacity (SOCs) was calculated based on the SOC content, BD, and sampling depth. BD, moisture, DOC, MBC, SOC, TC, TN, AP, and TP were selected and standardized for the calculation of soil multifunctionality using mean value method (Maestre et al., 2012).

TABLE 16.1
Plant Community Compositions and Restoration Years in Different Vegetation Types

Sites	Tree Layers			Herb Layers			Years	Group
	Dominant Species	Minor Species	Shannon-Wiener Index	Dominant Species	Minor Species	Shannon-Wiener Index		
1	—	—	0.00	—	—	0.00	—	BL
2	*Pinus massoniana*	—	0.68	*Dicranopteris pedata*	—	0.17	10	CF
3	*Pinus massoniana*	—	0.36	*Dicranopteris pedata*	—	0.07	20	CF
4	*Pinus massoniana* *Castanopsis fissa*	—	0.83	*Dicranopteris pedata*	—	0.34	10	MF
5	*Pinus massoniana*	*Ilex pubescens*	0.93	*Dicranopteris pedata*	—	0.00	30	CF
6	*Pinus massoniana*	—	0.50	*Dicranopteris pedata*	—	0.16	20	CF
7	*Pinus massoniana* *Schima superba*	*Adinandra millettii*	1.00	*Dicranopteris pedata*	—	0.71	30	MF
8	*Pinus massoniana*	*Schima superba*	0.45	*Dicranopteris pedata*	—	0.35	3	CF
9	*Fagus sylvatica* *Castanopsis fissa* *Fokienia hodginsii*	*Schima superba* *Liquidambar formosana*	2.88	*Gahnia tristis* *Woodwardia japonica* *Selaginella trichoclada*	*Lophatherumgracile* Brongn. —	2.72	> 50	BF

The activities of sucrase (SC), urease (UE), and acid phosphatase (ACP) were determined by an EnSpire Multifunctional Plate Reader (PerkinElmer, USA). Microbial C and N/P limitation were quantified by vector lengths and angles, respectively (Moorhead et al., 2013).

The $\delta^{13}C$ values were detected using stable isotope ratio mass spectrometry. The linear regression coefficient of $\delta^{13}C$ values and SOC contents (log-transformed) represents the turnover rate of SOC (Acton et al., 2013). The functional groups of SOC (alkyl C, O-alkyl C, aromatic C, and carbonyl C) were followed by established protocols (Wang et al., 2024a).

16.2.4 Amplicon Sequencing and Functional Annotation of Soil Bacteria

Amplicon sequencing analyses followed established protocols, including DNA extraction and subsequent sequencing (Dai et al., 2025b; Wang et al., 2023). Bacterial communities were characterized through amplification of the 16S rRNA gene V3–V4 regions using the primer pairs 341F (5′-CCTAYGGGRBGCASCAG-3′) and 806R (5′-GGACTACNNGGGTATCTAAT-3′). The GreenGene Database (v 13.8) was used to annotate the taxonomic information for each representative sequence. Raw sequence data were uploaded to the National Microbiology Data Center (NMDC) under accession number NMDC10018309.

The FAPROTAX database, supplemented with the bacterial taxa of phosphate solubilization, was used to annotate the metabolic functions of C, N, and P (Keshavarz Zarjani et al., 2013; Louca et al., 2016).

16.2.5 Statistical Analysis

A one-way analysis of variance and Tukey's Honestly Significant Difference was used to calculate differences among different vegetation types and restoration years ($p<0.05$). The "vegan" package in R was used to calculate the bacterial Shannon index. Principal coordinate analysis based on the Bray-Curtis dissimilarity and the similarities method was used to examine the differences in β-diversity among different communities. The "Carrplot" and "linkET" packages in R were used to correlate environmental variables and bacterial taxa ($p<0.001$) and perform the Mantel test for physicochemical factors and the communities and functions of bacteria, respectively. Redundancy analysis and random forest models were used to identify the importance of physicochemical factors on the bacteria. The visualization was performed using "ggplot2" package in R.

16.3 RESULTS

16.3.1 C, N, and P Contents of Plant and Litter

For BF, TC content in plant was significantly lower than that in both CF and MF ($p<0.05$), while TC content in litter showed no significant variation in different vegetation types (Figure 16.1). In contrast, TN and TP contents for both plant and litter were significantly higher in BF compared to CF and MF. Additionally, TP content in plants exceeded that across all plant types ($p<0.05$). The C/N and C/P ratios were highest in CF for both plants and litter, whereas N/P ratios did not differ significantly among vegetation types in either plants or litter (Figure 16.1).

16.3.2 Soil Physicochemical Properties and Multifunctionality

Soil texture gradually changed from sandy (BL) to silt (BF) with vegetation restoration (Table 16.2). Most soil nutrient contents in CF and MF did not show a significant difference within the same restoration year, and were higher than those in BL but lower than those in BF. Soil pH of CF and MF

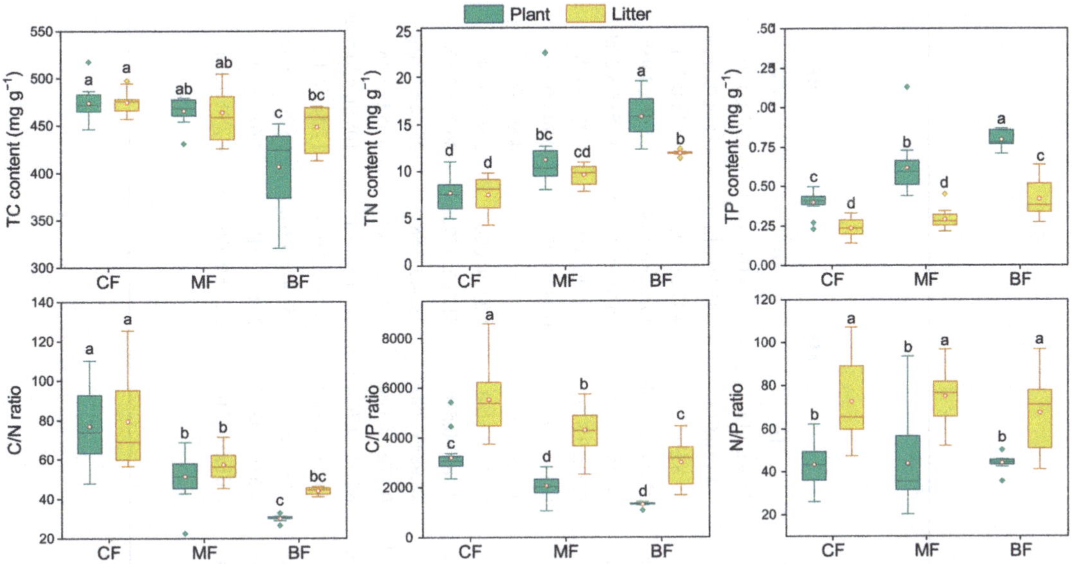

FIGURE 16.1 The contents of TC, TN, and TP and the ratios of C/N, C/P, and N/P of the plant and litter. Different letters indicate significant differences in different vegetation types ($p<0.05$).

was significantly lower compared to the other vegetation types ($p<0.05$). BD, TP content, and sand proportion remained stable across all vegetation types and restoration years, whereas C/N ratios, AP, and TC varied significantly ($p<0.05$). The soil moisture only differed across vegetation types, and the contents of soil TN, DOC, and silt were affected by the restoration years. Both vegetation types and restoration years had significant impacts on soil pH, SOC, and the C/P ratios (Table 16.2).

The soil multifunctionality varied from -1.06 ± 0.10 to 0.88 ± 0.25, and increased along the vegetation restoration stages (Table 16.2). However, the soil multifunctionality showed no obvious difference among restoration years in the same vegetation types.

16.3.3 Soil Nutrient Limitations and SOC Turnover Rates

Enzyme activities showed an increasing trend across the vegetation restoration stages (Figure 16.2a). Among them, UE and SC activities varied significantly among vegetation types. C limitation was most pronounced in BL soils and diminished as restoration progressed (Figure 16.2b). Based on the UE:ACP ratio, the dominant nutrient limitation shifted from P to N, and the extent of N limitation was gradually intensifying as the restoration advanced.

SOC contents exhibited positive correlations with their fractions, with the strongest association observed between SOC and ROC (Figure 16.3a). Among these fractions, ROC constituted the largest portion of SOC, followed by EOC. The proportion of SOC fractions reached a maximum in CF (83.07%) and was lowest in BL (75.10%). Additionally, BF soils had the highest EOC proportion ($p<0.01$). According to ^{13}C-NMR analysis, O-alkyl C was the predominant chemical composition of SOC (45.55%±10.88%), succeeded by alkyl C (33.48%±4.91%), aromatic C (14.45%±3.35%), and carboxyl C (Figure 16.3b). O-alkyl C content rose steadily following the vegetation restoration, whereas aromatic C declined progressively, and alkyl C displayed an initial increase followed by a decrease.

TABLE 16.2
Physicochemical Properties and Functional Diversity in Different Vegetation Types

Type	BL	CF				MF		BF
Years	**0**	**3**	**10**	**20**	**30**	**10**	**30**	**>50**
BD (g/cm^3)	1.35±0.08ab	1.33±0.15ab	1.39±0.17a	1.38±0.18a	1.49±0.04a	1.39±0.1a	1.52±0.06a	1.02±0.29b
Moisture (%)	0.13±0.002bc	0.14±0.03abc	0.13±0.01abc	0.14±0.03abc	0.11±0.01c	0.16±0.01a	0.15±0.01a	0.14±0.01abc
pH	4.93±0.04a	4.15±0.05c	4.61±0.01ab	4.36±0.05bc	4.26±0.01c	4.18±0.02c	4.12±0.29c	4.83±0.26a
Clay (%)	39.95±6.03a	15.62±7.71b	18.02±8.30ab	19.83±7.71ab	28.55±8.3ab	32.04±2.93ab	19.98±6.15ab	15.68±4.98b
Silt (%)	28.28±4.37c	23.83±9.90c	29.35±9.86ab	26.79±9.9bc	46.38±9.86a	26.35±2.86bc	36.81±9.14abc	43.53±3.29ab
Sand (%)	4.08±1.98b	1.32±8.19b	7.80±2.64b	4.59±2.92b	2.60±2.64b	4.93±0.81b	5.43±2.45b	1.75±1.08b
TP (g/kg)	0.43±0.01b	0.45±0.01b	0.48±0.02b	0.43±0.01b	0.48±0.02b	0.43±0.01b	0.48±0.03b	0.74±0.16a
AP (mg/kg)	0.70±0.19e	1.44±0.34e	3.26±0.03c	1.32±0.34e	1.20±0.03e	1.0±0.15e	2.31±0.14d	4.86±0.33b
TN (g/kg)	0.03±0.01d	0.27±0.16d	0.10±0.04d	0.60±0.16cd	1.47±0.04b	0.62±0.01cd	1.23±0.02b	2.89±0.79a
TC (g/kg)	2.67±0.02c	7.71±7.53c	2.80±0.46c	21.99±7.53b	30.05±0.46ab	20.83±0.55b	31.79±0.57ab	36.74±9.52a
SOC (g/kg)	2.49±0.04d	6.04±7.78d	6.66±1.00d	20.85±7.78abc	14.79±1.0c	18.98±0.38bc	28.34±0.68a	25.32±3.30ab
MBC (g/kg)	0.14±0.01f	0.49±0.11ab	0.36±0.05cde	0.34±0.11de	0.40±0.05bcd	0.47±0.06abc	0.25±0.03ef	0.55±0.04a
DOC (mg/kg)	62.56±17.79d	114.75±22.03a	107.91±7.14ab	93.36±22.03abc	80.98±7.14bc	87.77±9.76abc	61.41±4.3cd	108.96±20.58ab
C/N ratio	107.46±30.53a	26.06±10.45ab	140.61±6.48a	43.51±10.45b	45.41±6.48b	35.78±0.88ab	26.99±0.92ab	11.14±4.10c
C/P ratio	15.08±0.01d	34.70±44.17cd	36.29±5.00cd	123.13±44.17ab	80.12±5.0bc	113.12±2.17ab	151.42±7.69a	96.29±45.87ab
N/P ratio	0.05±0.003d	0.70±0.39cd	0.24±0.28d	1.60±0.39c	3.60±0.28ab	1.67±0.03c	2.96±0.21b	4.67±1.24a
Soil multifunctionality	−1.06±0.10c	−0.26±0.20b	−0.30±0.15b	−0.11±0.46b	−0.11±0.14b	0.01±0.07 b	0.17±0.06b	0.88±0.25a

Note: Different letters indicate significant differences in different vegetation types and restoration years ($p<0.05$).

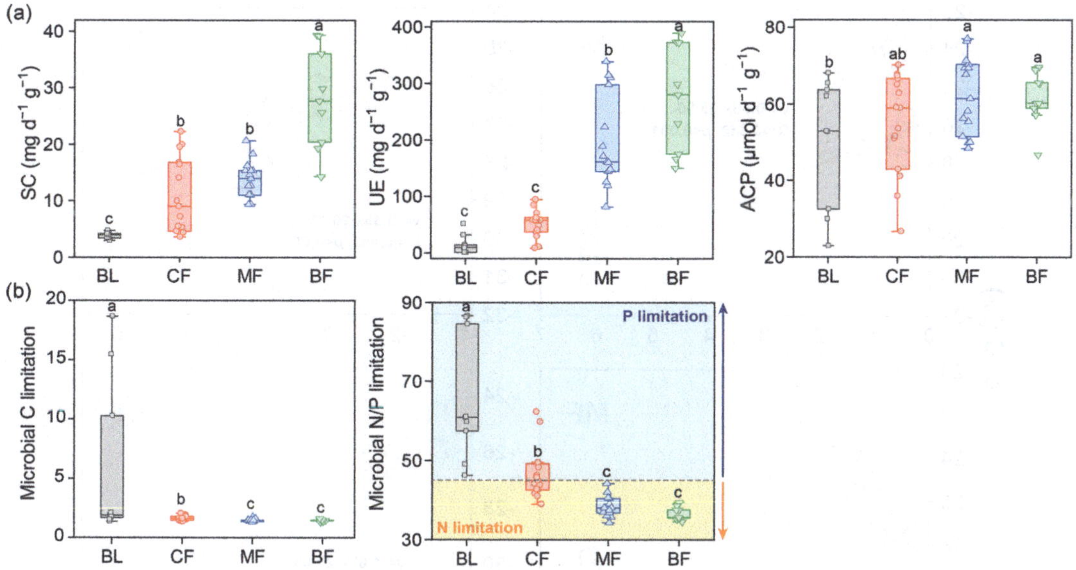

FIGURE 16.2 Enzyme activities (a) and microbial nutrient limitations (b) in different soils of different vegetation types. SC, sucrase; UE, urease; ACP, acid phosphatase. Different letters indicate significant differences in different vegetation types ($p<0.05$).

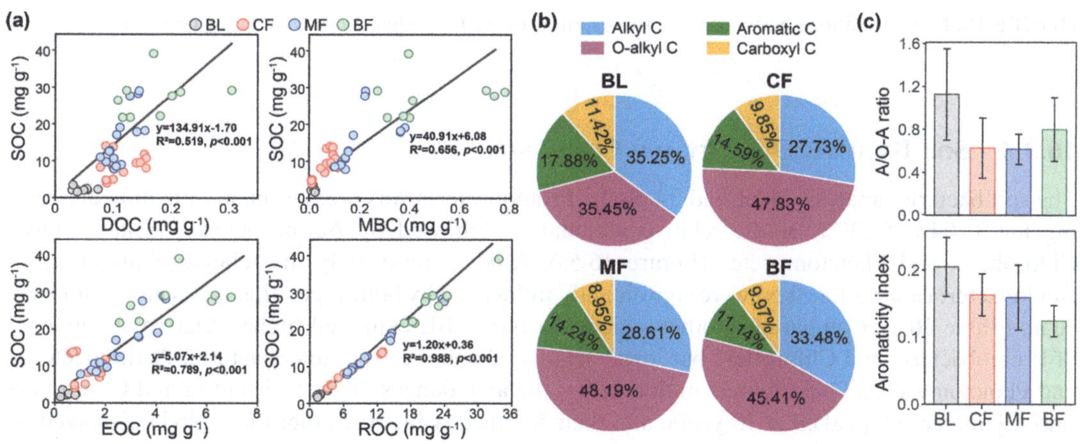

FIGURE 16.3 Correlations between fractions and concentrations of SOC (a) and the contents of chemical composition of SOC in different vegetation types (b). DOC, dissolved organic C; MBC, microbial biomass C; EOC, easily oxidized organic C; ROC, recalcitrant organic C.

Among all the soils, the $\delta^{13}C$ values (–28.55‰ to –22.56‰) were significantly and negatively correlated with SOC contents (Figure 16.4). The slopes of the regression lines, an indicator of C turnover rate, were lowest in BF and peaked in CF. Consequently, the overall turnover rates ranked in the following order BF>MF>BL>CF.

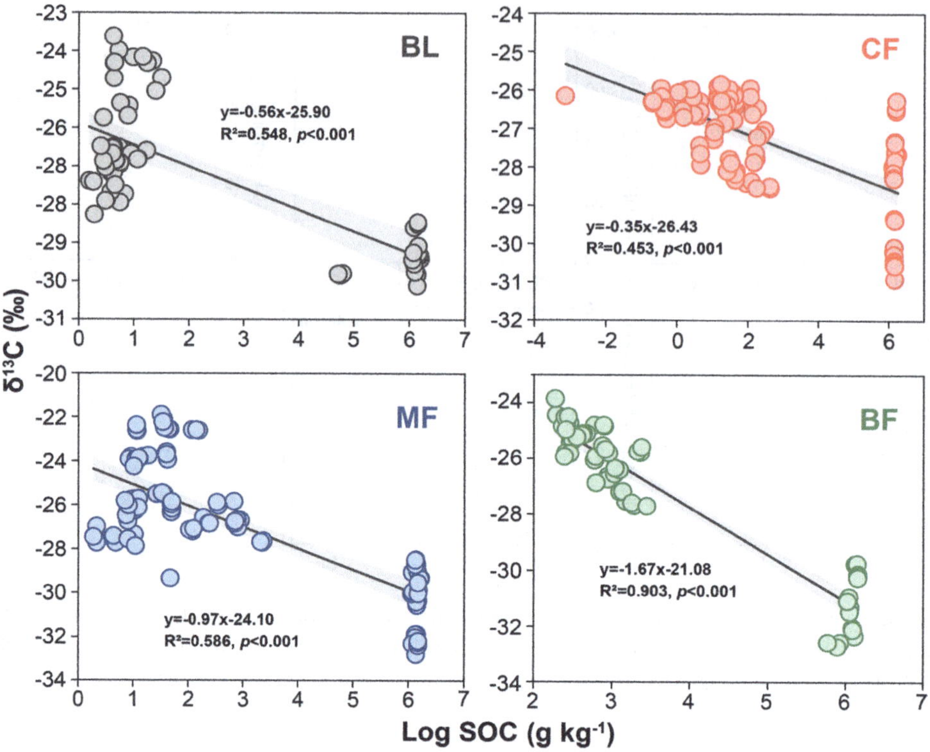

FIGURE 16.4 Correlations between $\delta^{13}C$ values and Log SOC to identify the soil C turnover rates.

16.3.4 Soil Bacterial Diversity and Community

The soil bacterial analysis identified 34 phyla. Proteobacteria was predominantly phylum, accounting for 33.04%–55.30% of the relative abundance, followed by Actinobacteria, Acidobacteria, Chloroflexi, and Planctomycetes (Figure 16.5a). Among these, only the relative abundances of Acidobacteria and Chloroflexi were significantly influenced by both vegetation types and restoration years (Table 16.3). Compared to other vegetation types, BL exhibited higher relative abundances of Proteobacteria and Chloroflexi but lower relative abundances of Acidobacteria, Actinobacteria, and Planctomycetes. Over the restoration years, the abundances of Proteobacteria and Chloroflexi initially increased, peaked at 10 years, and then declined, while the other three phyla displayed an inverse trend.

Koribacteraceae, *Rhodospirillaceae*, and *Sinobacteraceae* were the dominant families in all vegetation types. The relative abundance of *Koribacteraceae* varied significantly across all the restoration years, and the remaining seven families varied significantly in different vegetation types and restoration years (Table 16.3). Across the vegetation types, the abundances of *Koribacteraceae* and *Hyphomicrobiaceae* correlated positively with the increasing of broadleaf species, whereas *Bradyrhizobiaceae* exhibited an inverse relationship. The abundances of *Sinobacteraceae*, *Burkholderiaceae*, *Acidobacteriaceae*, *Thermogemmatisporaceae*, *Bradyrhizobiaceae*, and *Rhodospirillaceae* followed a unimodal pattern throughout the vegetation restoration chronosequence, peaking at intermediate stages (10 or 20 years), while *Koribacteraceae* was highest in the latest stage (>30 years).

The soils with plant cover generally showed higher bacterial diversities than those in unvegetated soils (BL), but there were no significant differences among different restoration years in same

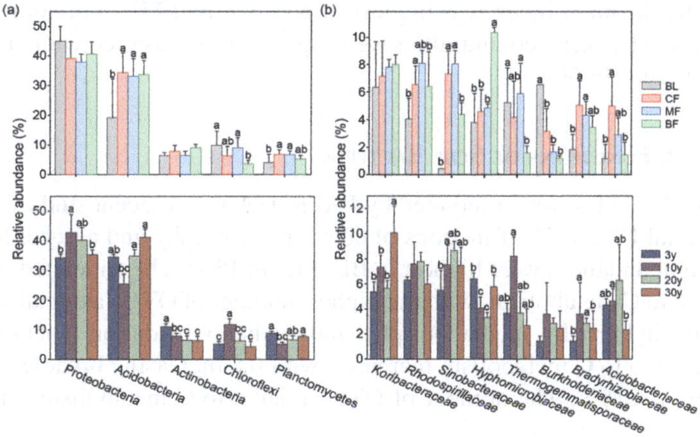

FIGURE 16.5 The relative abundance of dominant phylum (a) and family (b) in different vegetation types and restoration years. Different letters indicate significant differences among the vegetation types and restoration years ($p<0.05$).

TABLE 16.3
Kruskal-Wallis Tests of the Relative Abundance of Taxa at Phylum and Family Levels in Different Vegetation Types and Restoration Years

Taxonomy	Taxa	Type		Year	
		Chi-sq	*p*	Chi-sq	*p*
Phylum	AD3	4.545	0.208	8.359	0.138
	Acidobacteria	2.259	0.52	19.677	**0.001**
	Actinobacteria	7.859	**0.049**	17.381	**0.004**
	Bacteroidetes	2.966	0.397	12.267	**0.031**
	Chloroflexi	8.814	**0.042**	16.123	**0.007**
	Planctomycetes	4.889	0.18	16.31	**0.006**
	Proteobacteria	4.098	0.251	15.929	**0.007**
	WPS_2	15.262	**0.002**	24.755	**<0.001**
Family	*Acidobacteriaceae*	17.359	**0.001**	22.735	**<0.001**
	Koribacteraceae	4.241	0.237	17.785	**0.003**
	Solibacteraceae	4.091	0.252	9.151	0.103
	Ktedonobacteraceae	12.275	**0.006**	21.662	**0.001**
	Thermogemmatisporaceae	10.343	**0.016**	19.869	**0.001**
	Gemmataceae	6.573	0.087	19.292	**0.002**
	Isosphaeraceae	7.025	0.071	7.834	0.166
	Caulobacteraceae	15.549	**0.001**	19.219	**0.002**
	Bradyrhizobiaceae	9.272	**0.026**	22.57	**<0.001**
	Hyphomicrobiaceae	14.917	**0.002**	22.316	**<0.001**
	Methylocystaceae	14.388	**0.002**	15.804	**0.007**
	Acetobacteraceae	16.869	**0.001**	23.348	**<0.001**
	Rhodospirillaceae	7.876	**0.049**	8.955	**0.111**
	Burkholderiaceae	18.029	**0**	18.234	**0.003**
	Syntrophobacteraceae	9.218	**0.027**	19.712	**0.001**
	Sinobacteraceae	15.323	**0.002**	22.955	**0.016**

vegetable types. The Shannon index was highest in both CF and MF communities after 10 years (Figure 16.6a). The soil bacterial communities were significantly clustered with different restoration years (<50 years) (Figure 16.6b).

16.3.5 Metabolic Functions of Soil Bacteria

51 functions and 1,019 OTUs were annotated by FAPROTAX and recent studies, which accounted for 12.57% of the total OTUs. The functions of chemoheterotrophy and aerobic heterotrophy were dominant, and their abundances were highest in BL (Figure 16.7). The annotated functions involved ten functions related to C metabolism, with the highest number of OTUs associating with the phototroph. Except for methylotrophy, methanotrophy, and methanol oxidation functions, the number of functional OTUs related to C metabolism increased with the increasing broadleaf species, and was highest in BF soils ($p<0.05$). The number of OTUs related to C metabolism functions decreased

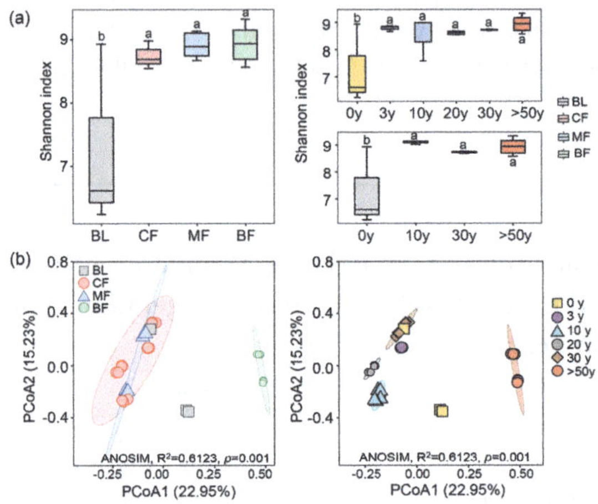

FIGURE 16.6 Shannon index of the soil bacterial communities (a) and PCoA analysis in different vegetation types and restoration years (b).

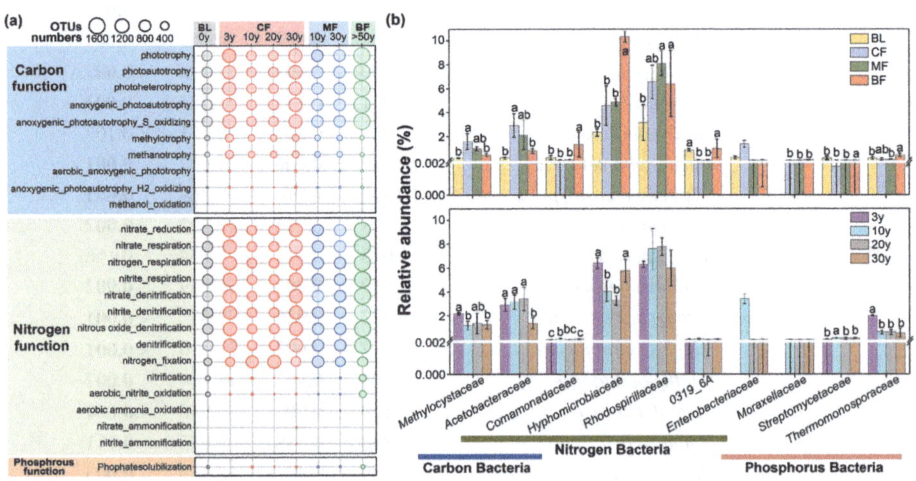

FIGURE 16.7 Metabolic functions of C, N, and P (a) and relative abundance (> 0.015%) of dominant functional bacteria (b) in different vegetation types and restoration years. OTUs, operational taxonomic units.

first and then increased along with restoration years and was highest in early stages (3 years) of different vegetation types.

For N metabolism, 14 functions were identified, with nitrate reduction being the dominant (Figure 16.7a). Denitrification-associated OTUs increased with broadleaf species coverage. Across restoration years, the number of N-related OTUs generally declined initially before rising again. The number of OTUs related to N fixation and nitrification was highest in BF and intermediate stages (20 years in CF), respectively. The OTU numbers for P-solubilization were significantly lower than those for C-cycling and N-cycling functions but markedly higher in BF than other soils ($p<0.05$). In total, the annotated functional taxa comprised 10, 13, and 9 families involved in the C-, N-, and P-cycling, respectively (Figure 16.7b).

16.3.6 Soil Environmental Factors Influencing Bacterial Communities and Functions

Soil BD, moisture, texture, and C/P ratios were not correlated with bacterial metabolic functions, while BD and C/P ratios significantly correlated with bacterial communities (Figure 16.8a).

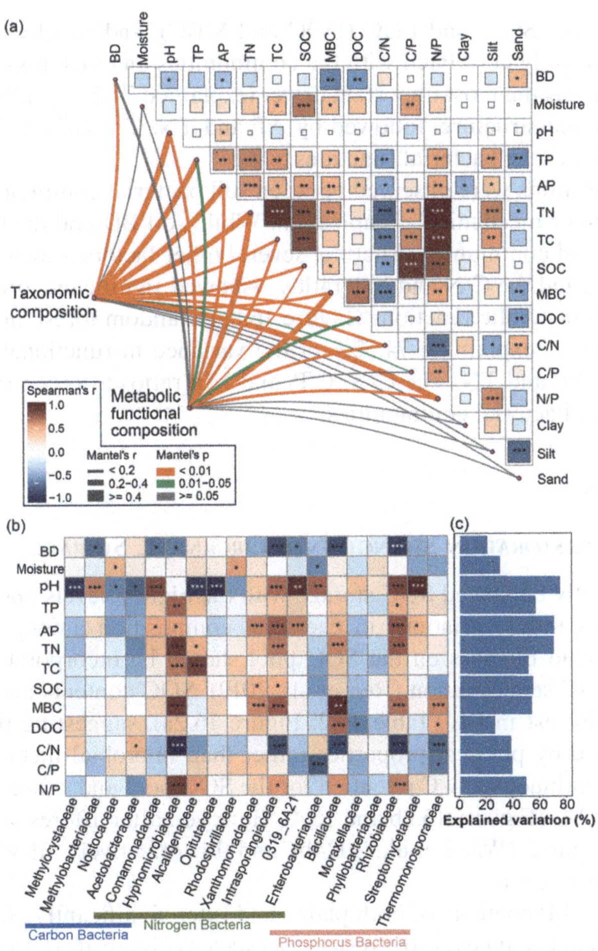

FIGURE 16.8 Correlations of the physicochemical factors with the communities and metabolic functions of bacteria (a) and bacteria (family level) related to metabolic functions of C, N, and P (b), and random forest analysis showing key drivers of the bacterial communities (c). *, **, and *** represent $p<0.05$, 0.01, and 0.001, respectively.

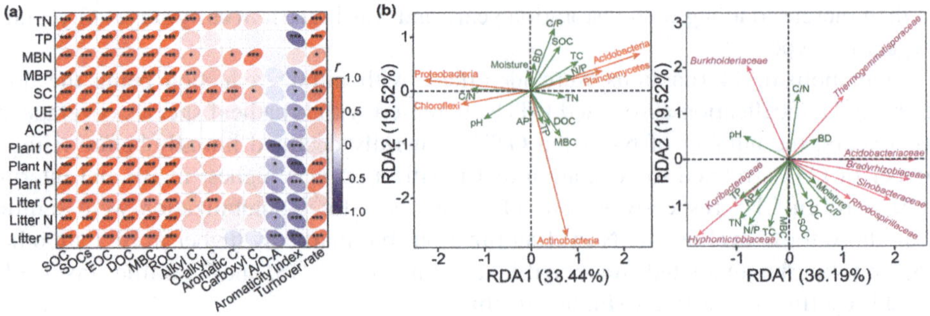

FIGURE 16.9 Heatmap of community C, N, and P stoichiometry correlations with SOC and its fractions, functional groups, and chemical properties (a), and RDA analysis of the relationships between the relative abundance of dominant bacterial phyla and families and soil properties (b). A/O-A, ratio of alkyl C to O-alkyl C; Aromaticity index, the ratio of aromatic C to the sum of alkyl C, O-alkyl C, and aromatic C. *, $p<0.05$; ***, $p<0.001$.

Soil pH, AP, TP, TN, TC, SOC, and LOC (DOC and MBC), and stoichiometry (C/N ratios and N/P ratios) were significantly correlated with the communities and metabolic functions of bacteria (Figure 16.8a). The pH was the most important driver (explained 74.52%) influencing the functional bacterial communities and structure, followed by AP and TN. The C/N, C/P, and N/P ratios also contributed significantly to the bacterial functions.

No significant correlations were observed between soil bacterial communities or their functions and either soil moisture or mechanical composition. While soil BD and the C/P ratios were correlated solely with bacterial community structure, several other factors—including pH, TP, AP, TN, MBC, DOC, SOC, TC, and the C/N and N/P ratios—showed significant correlations with both the communities and their metabolic functions (Figure 16.8a). Random forest analysis identified pH as the most influential driver, explaining 74.52% of the variance in functional bacterial community structure, followed by AP and TN. The C/N, C/P, and N/P ratios also contributed significantly to shape these functions of bacterial communities.

16.4 DISCUSSION

16.4.1 Vegetation Restoration Strengthened Organic C Storage

Afforestation significantly enhanced SOC storage with the highest levels observed in broad-leaved forests. Given the substantial variations in chemical composition among vegetation types, the influence of above-ground biomass on the SOC pool should be incorporated into evaluations of afforestation-driven SOC sequestration (Ren et al., 2019). SOC contents and soil microbial activity were significantly lowest in BL (Table 16.2, Figure 16.2a), suggesting that SOC contents are predominantly governed by physical properties rather than microbial metabolism in unvegetated soils. This leads to a continuous low C retention for the SOC pool and consequent C limitation and low C turnover rates in BL (Figures 16.2b and 16.4). However, after afforestation, the plant litterfall became a major SOC source (Wang et al., 2024b), and this effectively alleviated soil C limitation and accelerated soil C turnover.

Meanwhile, the N and P contents of both plants and litter significantly changed, but the C contents showed no significant changes across restoration stages (Figure 16.1). Both N and P contents of plant and litter showed positive correlation with SOC contents and C turnover rates (Figure 16.9a), suggesting that N and P of plant and litter contributed to SOC accumulation. During litter decomposition, the nutrient stoichiometric ratio is a key driver of decomposition and ultimately influences SOC accumulation (Yeasmin et al., 2020). The litter with lower C/N and C/P ratios is more susceptible to being involved into soil biogeochemical processes and decomposition (Lu et al., 2021). Therefore, after microbial utilization of soil N and P, plant and litter C were preserved in the soil due to their high C but low N and P contents. This residual C of plant and litter with less microbial bioavailability makes large contribution to the ROC pool (Sun et al., 2022). Moreover, soil MBC was significantly positively correlated with plant nutrient inputs, indicating that microbial growth was supported by the plant litter nutrient (Li et al., 2020). The strong microbial activity also accelerates SOC content through more microbial necromass C, which accounts for over 35% of SOC (Gunina and Kuzyakov, 2022; Wang et al., 2021). Therefore, vegetation restoration promotes the SOC storage through contributions from the plant and litter inputs and microbial biomass accumulation.

16.4.2 Vegetation Restoration Enhanced SOC Storage and Stabilizes C Pool

The contents of DOC, MBC, and EOC increased with the SOC accumulation during restoration (Figure 16.3). These LOC fractions represented only minor proportion of the total SOC, but dominantly derived soil microbial activity (Hu et al., 2022). The significant positive correlations between soil N, P, and three LOC fractions revealed that soil nutrient availability regulated the above labile C pool (Chen et al., 2019). ROC is difficult for microbial decomposition and represents a stable C pool in soils owing to its complex structure (Kögel-Knabner, 2002). Despite an increase in ROC contents after vegetation restoration, the percentages of ROC in the whole SOC pool were lower in afforested areas than that in BL. This indicated that vegetation restoration enhanced the activity of the SOC pool and accelerated SOC turnover.

The changes in SOC composition after vegetation restoration were significant based on ^{13}C-NMR analysis. Alkyl C and O-alkyl C were the dominant SOC functional groups, with a significant increase in O-alkyl C content (Figure 16.3b). Previous studies showed that the plant-derived materials are closely related to O-alkyl C (Wang et al., 2024b). O-alkyl C originates primarily from carbohydrates such as cellulose and hemicellulose, it is more easily consumed by the soil microbials than alkyl C (Kögel-Knabner, 2002). Therefore, the alkyl-C to O-alkyl-C (A/O-A) ratio can indicate the degree of SOC decomposition (Wang et al., 2024b).

For BL, no plant litter input and poor microbial activity directly led to a higher A/O-A ratio (Figure 16.3c). Vegetation restoration increased plant-derived litter inputs and microbial activity, accelerated soil C turnover, and enhanced the labile C pool (Figure 16.3b and c). The A/O-A ratio increased first in BF, suggesting an increase in labile C pool during the early restoration stage due to high nutrient availability. Aromatic C and alkyl C were the main function groups in ROC, which is a stable product of microbial metabolism in soils (Knorr et al., 2005). The growing proportions of these function components indicated that long-term afforestation increased SOC storage, and simultaneously enlarged the stable C pool.

16.4.3 Vegetation Restoration Improved Soil Multifunctionality

The gradually introducing indigenous broadleaf species transitioned from CF to MF and finally to evergreen broadleaved forests. This increase in diversity of broadleaf species enhanced more litter and root biomass inputs to the soils, thereby elevating soil organic matter, TC, and silt contents. Silt content significantly and positively correlated with TP, TN, and TC (Figure 16.8a), which demonstrated that the soil texture was significantly improved and the soil nutrient level was increased. Therefore, the restoration strategy, with the pioneer species *P. massoniana* planted first was effectively in severely eroded areas and gradually introducing more indigenous broadleaf species, effectively and efficiently increased ecosystem multifunctionality (Table 16.2, Figure 16.10).

The composition of soil bacterial communities in CF and MF differed significantly with BL and BF (Figure 16.6b). Bacterial diversity of different vegetation types increased significantly during the beginning restoration stage and stabilized after 3 years. Generally, soil multifunctionality positively correlated with bacterial diversity (Figure 16.10), which indicated the positive influence of vegetation restoration on the diversity of soil bacterial communities. It is consistent with a similar increase in bacterial abundance and diversity along a vegetation restoration chronosequence in an oasis-desert ecotone (An et al., 2022). Proteobacteria and Acidobacteria were important bacterial phyla to soil nutrient cycling (Lee et al., 2008; Wang et al., 2023; Zeng et al., 2016). The significant negative correlation between the relative abundance of Acidobacteria and pH also suggest the taxa's preference to acid soil condition, and this was intensified with restoration years by single planting *P. massoniana*, root exudation of organic acids, uptake of alkali cations (Table 16.2, Figure 16.9b), as well as the more resins and tannins in litter composition and poor in base ions (Hou et al., 2019; Jiang et al., 2018). Thus, the relative abundance of Acidobacteria increased progressively with restoration ages (Figure 16.5). In contrast, Actinobacteria and Chloroflexi remained relatively abundant in early stages, owing to their resilience in harsh, nutrient-limited environments (Delgado-Baquerizo et al., 2016). In summary, introducing plant communities with evergreen broadleaved species in the original *P. massoniana* forest effectively increased community diversity, improved soil texture, nutrient availability, and ultimately increased soil bacterial diversity and soil multifunctionality across restoration stages (Maestre et al., 2012).

16.4.4 Vegetation Restoration Regulated Nutrient Metabolic Functions

Significant shifts in C, N, and P metabolic functions were observed based on FAPROTAX database across the restoration stages (Figure 16.7) (Fan et al., 2024; Sansupa et al., 2021). The relative abundance of OTUs associated with phototrophic C metabolism increased to peak in climax broadleaf forest. However, OTUs associated with methylotrophic functions declined with restoration stages (Figure 16.7a), and the relative abundance of *Methylocystaceae*, the taxa of methylotrophic family,

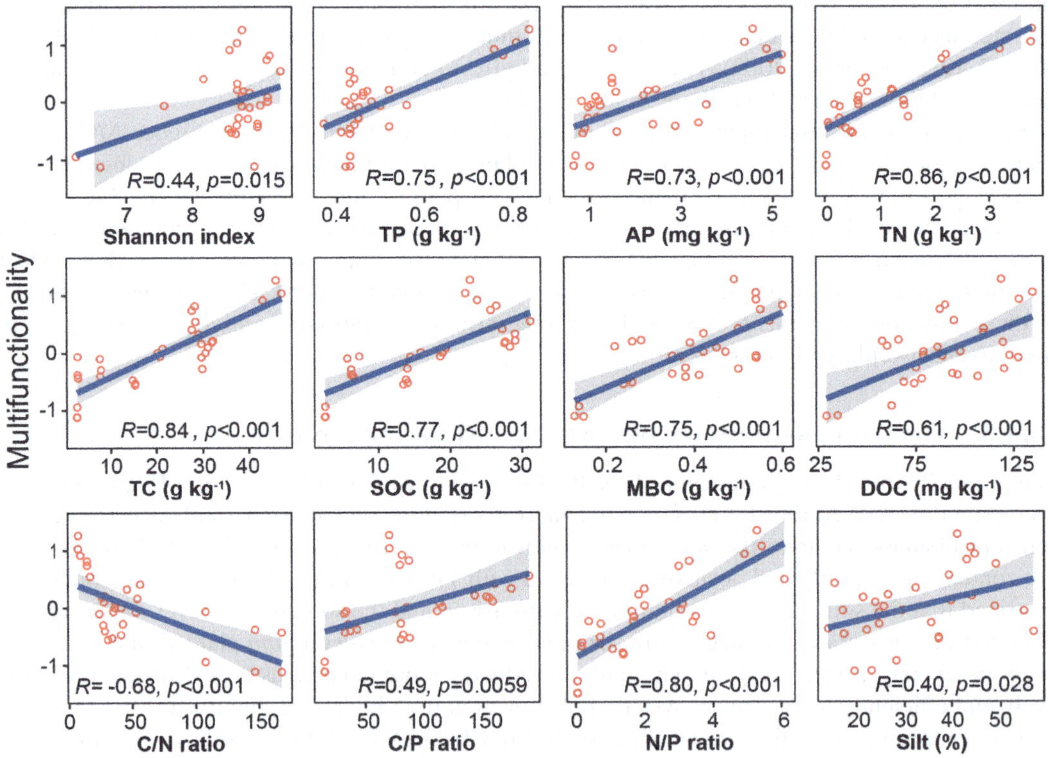

FIGURE 16.10 Relationships among the soil physicochemical properties and bacterial Shannon index and soil multifunctionality.

followed a similar trend (Figure 16.7b). These patterns indicated that phototrophy is the main C sequestration pathway during vegetation restoration, while methylotrophic activity was also influenced by restoration age. The methylotrophic functional intensity was higher in early and lower in later stages, suggesting the nutrient demand from organic matter degradation reduced with restoration advancing (Dai et al., 2025a; Zhong et al., 2018). Together with the increasing soil TC contents in restored vegetation communities (Table 16.2), these functional shifts showed increasing capacity in bacterial C sequestration with the increasing diversity of broadleaf species in the original *P. massoniana* forest, with declined decomposition of SOC. This vegetation restoration improved soil C sequestration (Xue et al., 2022).

Both the OTUs number and the abundance of N fixation taxa (*Acetobacteraceae* and *Rhodospirillaceae*) peaked in the initial restoration stage but declined in later stages (Figure 16.7b). This pattern suggested that biological N fixation is essential during early to mid-restoration stages under conditions of soil nutrient limitation (Li et al., 2026). The nutrient limitation was finally relieved with the improvement of soil nutrient and soil properties after more litter inputs in the later restoration stages (Batterman et al., 2013).

P-solubilizing bacteria, including *Bacillaceae*, *Moraxellaceae*, and *Rhizobiaceae*, are associated with converting organic or particulate P into bioavailable forms (Liu et al., 2023). Their abundance was significantly lower than those of C and N cycling functional groups. (Figure 16.7b), The AP contents decreased with the restoration years and were significantly lower in CF compared to BF due mainly to the transformation of AP into stable complexes with iron and aluminum oxides in acidifying soils (Table 16.2), increased plant P uptake with greater plant diversity in later stages, and the absence of artificial P fertilization. However, litterfall was significantly lower in CF than that in BF (Guo et al., 2016), and this yielded lower soil AP in CF. The positive correlation between soil P content and the abundance of P-solubilizing bacteria suggested an important role of litter inputs (Figure 16.8).

In summary, key microbial functional groups governing C (e.g., methylotrophy), N (e.g., fixation), and P (e.g., solubilization) cycling shifted significantly with restoration age. These microbial changes were closely correlated with soil physicochemical properties (Figure 16.8b), indicating that the microbial community structure is a strong determinant of nutrient cycling dynamics across the restoration areas in Changting County (Zechmeister-Boltenstern et al., 2015).

16.4.5 Vegetation Restoration Governed Bacterial Community Structure and Function

The soil bacterial community structure and function during vegetation restoration are primarily regulated by soil pH and nutrient stoichiometry and other abiotic factors, including soil moisture, pH, temperature, and the C/N ratio (Figure 16.8a). Soil pH variation was significant across restoration years and had a strong influence on the soil bacterial community structure as indicated by a significant negative correlation between pH and Acidobacteria abundance (Figure 16.9b). Meanwhile, soil C:N:P stoichiometry is a critical indicator of soil quality and nutrient supply capacity. Once the balance of soil C:N:P stoichiometry is lost, soil nutrient limitations occur and soil microbial composition is altered.

The nutrient limitations shifted across the restored forest soil chronosequence. Due to the negative relationship between the high C/N ratios and soil mineralization rates, as well as resource availability, increased soil C/N ratios in initial stages indicated that N was a limiting factor for plant productivity and the bacterial community structure (Table 16.2, Figure 16.8b). For the later stages, the subsequent more litter input increased soil N content, gradually relieving the N limitation (Table 16.1). Consequently, this shift from the N limitation to a complex interaction about TP, AP, TN, and the N/P and C/P ratios (Figure 16.8b). Although the soil N/P ratios increased with the succession stages, with a peak in BF, the ratios remained below the national average of 5.2 for all the stages (Table 16.1). This stoichiometric transition significantly shifted from initial N limitation to increasing P limitation and finally the N and P co-limitation. These shifts ultimately shaped microbial communities during vegetation restoration.

16.5 CONCLUSIONS

P. massoniana, with its high tolerance to heat and drought weather, as well as its high adaptation to poor and acidic soil, is commonly used to be typical plant species in severely eroded areas in South China. In Changting county, as a successively vegetation restoration, the plant has successfully established vegetation coverage in most areas with soil erosion, and effectively reduced soil erosion since the 1980s.

However, due to the obvious drawbacks of pure stands of *P. massoniana* in the perspective of soil improvement and the formation of effective SOC and other soil nutrients, people have to continuously enhance these artificial pure forests through mixed planting and increasing biodiversity following the ecological principles of local succession. It is proven that the soil physicochemical properties and soil C, N, and P pools were substantially improved with increasing plant diversity and restoration years. These findings provide valuable insights into ecological restoration in subtropical China with soil erosion.

ACKNOWLEDGMENTS

This study was supported by the National Natural Science Foundation of China (42230407, 42506235), the China Postdoctoral Science Foundation (2025M771276), and projects from Changting County and the Fujian Water Resources Department.

REFERENCES

Acton, P., Fox, J., Campbell, E., Rowe, H., Wilkinson, M., 2013. Carbon isotopes for estimating soil decomposition and physical mixing in well-drained forest soils. *Journal of Geophysical Research: Biogeosciences* 118, 1532–1545.

An, F., Niu, Z., Liu, T., Su, Y., 2022. Succession of soil bacterial community along a 46-year choronsequence artificial revegetation in an arid oasis-desert ecotone. *Science of the Total Environment* 814, 152496.

Banning, N. C., Gleeson, D. B., Grigg, A. H., Grant, C. D., Andersen, G. L., Brodie, E. L., Murphy, D. V., 2011. Soil microbial community successional patterns during forest ecosystem restoration. *Applied and Environmental Microbiology* 77, 6158–6164.

Batterman, S. A., Hedin, L. O., van Breugel, M., Ransijn, J., Craven, D. J., Hall, J. S., 2013. Key role of symbiotic dinitrogen fixation in tropical forest secondary succession. *Nature* 502(7470), 224–227.

Boulmane, M., Oubrahim, H., Halim, M., Bakker, M. R., Augusto, L., 2017. The potential of Eucalyptus plantations to restore degraded soils in semi-arid Morocco (NW Africa). *Annals of Forest Science* 74, 57.

Chen, J., Seven, J., Zilla, T., Dippold, M. A., Blagodatskaya, E., Kuzyakov, Y., 2019. Microbial C:N:P stoichiometry and turnover depend on nutrients availability in soil: A ^{14}C, ^{15}N and ^{33}P triple labelling study. *Soil Biology and Biochemistry* 131, 206–216.

Chen, Y., Dai, Z., Yang, S., Wang, F., Yue, H., Peng, S., Cao, W., 2024. Effects of different restored vegetation on soil organic carbon pools in subtropic erosive lands: Insights from stable carbon isotopes. *Forest Ecology and Management* 564, 122040.

Cui, J., Zhu, Z., Xu, X., Liu, S., Jones, D. L., Kuzyakov, Y., Shibistova, O., Wu, J., Ge, T., 2020. Carbon and nitrogen recycling from microbial necromass to cope with C:N stoichiometric imbalance by priming. *Soil Biology and Biochemistry* 142, 107720.

Dai, Z., Fan, Y., Li, Y., Huang, Y., Mao, X., Xiang, T., Wang, F., Yang, S., Cao, W., 2025a. Global patterns and environmental drives of nitrite/nitrate-dependent anaerobic methane oxidation across natural ecosystems. *Global Change Biology* 31, e70350.

Dai, Z., Zhang, Y., Li, Y., Xiang, T., Wang, F., Peng, J., Yang, S., Cao, W., 2025b. Nitrification regulates the responses of soil nitrous oxide emissions to nitrogen addition in China: A meta-analysis from a gene perspective. *Environmental Science: Processes & Impacts* 27, 2367–2378.

Delgado-Baquerizo, M., Maestre, F. T., Reich, P. B., Trivedi, P., Osanai, Y., Liu, Y. R., Hamonts, K., Jeffries, T. C., Singh, B. K., 2016. Carbon content and climate variability drive global soil bacterial diversity patterns. *Ecological Monographs* 86, 373–390.

Ding, S., Wang, F., Yue, H., Peng, S., Ruan, Q., Lin, J., Cao, W., 2023. Downward trends in streamflow and sediment yield associated with soil and water conservation in the Tingjiang River Watershed, Southeast China. *Water* 15, 212.

Fan, Y., Chen, K., Dai, Z., Peng, J., Wang, F., Liu, H., Xu, W., Huang, Q., Yang, S., Cao, W., 2024. Land use/cover drive functional patterns of bacterial communities in sediments of a subtropical river, China. *Science of the Total Environment* 947, 174564.

Gellie, N. J. C., Mills, J. G., Breed, M. F., Lowe, A. J., 2017. Revegetation rewilds the soil bacterial microbiome of an old field. *Molecular Ecology* 26, 2895–2904.

Gunina, A., Kuzyakov, Y., 2022. From energy to (soil organic) matter. *Global Change Biology* 28, 2169–2182.

Guo, J., Yang, Z., Lin, C., Liu, X., Chen, G., Yang, Y., 2016. Conversion of a natural evergreen broadleaved forest into coniferous plantations in a subtropical area: Effects on composition of soil microbial communities and soil respiration. *Biology and Fertility of Soils* 52, 799–809.

Guo, R., Chen, Y., Xiang, M., Yang, S., Wang, F., Cao, W., Yue, H., Peng, S., 2024. Soil nutrients drive changes in the structure and functions of soil bacterial communities in a restored forest soil chronosequence. *Applied Soil Ecology* 195, 105247.

Higgitt, D., 1992. Soil-erosion and soil problems. *Progress in Physical Geography-Earth and Environment* 16, 230–238.

Hou, X. L., Han, H., Tigabu, M., Cai, L. P., Meng, F. R., Liu, A. Q., Ma, X. Q., 2019. Changes in soil physico-chemical properties following vegetation restoration mediate bacterial community composition and diversity in Changting, China. *Ecological Engineering* 138, 171–179.

Hu, L., Li, Q., Yan, J., Liu, C., Zhong, J., 2022. Vegetation restoration facilitates belowground microbial network complexity and recalcitrant soil organic carbon storage in southwest China karst region. *Science of the Total Environment* 820, 153137.

Huang, J., Li, Z., Zeng, G., Zhang, J., Li, J., Nie, X., Ma, W., Zhang, X., 2013. Microbial responses to simulated water erosion in relation to organic carbon dynamics on a hilly cropland in subtropical China. *Ecological Engineering* 60, 67–75.

Jiang, J., Wang, Y.-P., Yu, M., Cao, N., Yan, J., 2018. Soil organic matter is important for acid buffering and reducing aluminum leaching from acidic forest soils. *Chemical Geology* 501, 86–94.

Keshavarz Zarjani, J., Aliasgharzad, N., Oustan, S., Emadi, M., Ahmadi, A., 2013. Isolation and characterization of potassium solubilizing bacteria in some Iranian soils. *Archives of Agronomy and Soil Science* 59, 1713–1723.

Knorr, W., Prentice, I. C., House, J. I., Holland, E. A., 2005. Long-term sensitivity of soil carbon turnover to warming. *Nature* 433, 298–301.

Kögel-Knabner, I., 2002. The macromolecular organic composition of plant and microbial residues as inputs to soil organic matter. *Soil Biology and Biochemistry* 34, 139–162.

Lauber, C. L., Strickland, M. S., Bradford, M. A., Fierer, N., 2008. The influence of soil properties on the structure of bacterial and fungal communities across land-use types. *Soil Biology & Biochemistry* 40, 2407–2415.

Lee, S., Ka, J., Cho, J., 2008. Members of the phylum Acidobacteria are dominant and metabolically active in rhizosphere soil. *Fems Microbiology Letters* 285, 263–269.

Li, J., Zhang, X., Luo, J., Lindsey, S., Zhou, F., Xie, H., Li, Y., Zhu, P., Wang, L., Shi, Y., He, H., Zhang, X., 2020. Differential accumulation of microbial necromass and plant lignin in synthetic versus organic fertilizer-amended soil. *Soil Biology and Biochemistry* 149, 107967.

Li, Y., Dai, Z., Fan, Y., Wang, F., Xiang, T., Huang, Y., Mao, X., Yang, S., Cao, W., 2026. Aquaculture wastewater input enhances biological nitrogen fixation by altering the physicochemical properties and reshaping the diazotrophic community in mangrove sediments. *Applied Soil Ecology* 217, 106664.

Liu, J., Fang, L. C., Qiu, T. Y., Bing, H. J., Cui, Y. X., Sardans, J., Du, E. Z., Chen, J., Tan, W. F., Delgado-Baquerizo, M., Zhou, G. Y., Cui, Q. L., Penuelas, J., 2023. Disconnection between plant-microbial nutrient limitation across forest biomes. *Functional Ecology* 37, 2271–2281.

Louca, S., Parfrey, L. W., Doebeli, M., 2016. Decoupling function and taxonomy in the global ocean microbiome. *Science* 353, 1272–1277.

Lu, X., Hou, E., Guo, J., Gilliam, F. S., Li, J., Tang, S., Kuang, Y., 2021. Nitrogen addition stimulates soil aggregation and enhances carbon storage in terrestrial ecosystems of China: A meta-analysis. *Global Change Biology* 27, 2780–2792.

Maestre, F. T., Quero, J. L., Gotelli, N. J., Escudero, A., Ochoa, V., Delgado-Baquerizo, M., Garcia-Gomez, M., Bowker, M. A., Soliveres, S., Escolar, C., Garcia-Palacios, P., Berdugo, M., Valencia, E., Gozalo, B., Gallardo, A., Aguilera, L., Arredondo, T., Blones, J., Boeken, B., Bran, D., Conceicao, A. A., Cabrera, O., Chaieb, M., Derak, M., Eldridge, D. J., Espinosa, C. I., Florentino, A., Gaitan, J., Gatica, M. G., Ghiloufi, W., Gomez-Gonzalez, S., Gutierrez, J. R., Hernandez, R. M., Huang, X., Huber-Sannwald, E., Jankju, M., Miriti, M., Monerris, J., Mau, R. L., Morici, E., Naseri, K., Ospina, A., Polo, V., Prina, A., Pucheta, E., Ramirez-Collantes, D. A., Romao, R., Tighe, M., Torres-Diaz, C., Val, J., Veiga, J. P., Wang, D., Zaady, E., 2012. Plant species richness and ecosystem multifunctionality in global drylands. *Science* 335, 214–218.

Moorhead, D. L., Rinkes, Z. L., Sinsabaugh, R. L., Weintraub, M. N., 2013. Dynamic relationships between microbial biomass, respiration, inorganic nutrients and enzyme activities: Informing enzyme-based decomposition models. *Frontiers in Microbiology*, 4, 223.

Ren, C. J., Liu, W. C., Zhao, F. Z., Zhong, Z. K., Deng, J., Han, X. H., Yang, G. H., Feng, Y. Z., Ren, G. X., 2019. Soil bacterial and fungal diversity and compositions respond differently to forest development. *Catena* 181, 104071.

Roesch, L. F., Fulthorpe, R. R., Riva, A., Casella, G., Hadwin, A. K. M., Kent, A. D., Daroub, S. H., Camargo, F. A. O., Farmerie, W. G., Triplett, E. W., 2007. Pyrosequencing enumerates and contrasts soil microbial diversity. *ISME Journal* 1, 283–290.

Sansupa, C., Wahdan, S. F. M., Hossen, S., Disayathanoowat, T., Wubet, T., Purahong, W., 2021. Can we use Functional Annotation of Prokaryotic Taxa (FAPROTAX) to assign the ecological functions of soil bacteria? *Applied Sciences-Basel* 11, 688.

Shi, D., Shi, X., Liang, Y., Yu, D., 2005. Dynamic change of soil erosion in different spatial scale of China. *Bulletin of Soil and Water Conservation* 5, 89–93. 10.13961/j.cnki.stbctb.2005.05.021

Sun, G., Cao, C., Wang, M., Li, H., Wang, Y., Deng, G., Wang, J., 2022. Organic matter biomarker and ^{13}C NMR characteristics of soil and sediment standard reference materials from China. *Science of the Total Environment* 836, 155661.

Wang, F., Guo, R., Zhang, N., Yang, S., Cao, W., 2023. Soil organic carbon storages and bacterial communities along a restored mangrove soil chronosequence in the Jiulong River Estuary: From tidal flats to mangrove afforestation. *Fundamental Research* 3, 880–889.

Wang, F., Tao, Y., Yang, S., Cao, W., 2024a. Warming and flooding have different effects on organic carbon stability in mangrove soils. *Journal of Soils and Sediments* 24, 60–69.

Wang, F., Zhang, N., Yang, S., Li, Y., Yang, L., Cao, W., 2024b. Source and stability of soil organic carbon jointly regulate soil carbon pool, but source alteration is more effective in mangrove ecosystem following Spartina alterniflora invasion. *Catena* 235, 107681.

Wang, X., Teng, Y., Ren, W., Han, Y., Wang, X., Li, X., 2021. Soil bacterial diversity and functionality are driven by plant species for enhancing polycyclic aromatic hydrocarbons dissipation in soils. *Science of the Total Environment* 797, 149204.

Wang, Y., 2020. Analysis on effect of inter-planting broad-leaved trees on quality improvement of Pinus massoniana Stand in soil erosion area: A case study in Changting, Fujian Province. *Forest Resources Management* 6, 116–122.

Xue, Y., Kang, H., Cui, Y., Lu, S., Yang, H., Zhu, J., Fu, Z., Yan, C., Wang, D., 2022. Consistent plant and microbe nutrient limitation patterns during natural vegetation restoration. *Frontiers in Plant Science* 13, 885984.

Yeasmin, S., Singh, B., Smernik, R. J., Johnston, C. T., 2020. Effect of land use on organic matter composition in density fractions of contrasting soils: A comparative study using ^{13}C NMR and DRIFT spectroscopy. *Science of the Total Environment* 726, 138395.

Zechmeister-Boltenstern, S., Keiblinger, K. M., Mooshammer, M., Penuelas, J., Richter, A., Sardans, J., Wanek, W., 2015. The application of ecological stoichiometry to plant-microbial-soil organic matter transformations. *Ecological Monographs* 85, 133–155.

Zeng, Q., Dong, Y., An, S., 2016. Bacterial community responses to soils along a latitudinal and vegetation gradient on the Loess Plateau, China. *PLoS One* 11, e0152894.

Zhong, Y., Yan, W., Wang, R., Wang, W., Shangguan, Z., 2018. Decreased occurrence of carbon cycle functions in microbial communities along with long-term secondary succession. *Soil Biology & Biochemistry* 123, 207–217.

17 Biodiversity Conservation of Arbuscular Mycorrhizal Fungi under Global Change

Tang Rong Zhou, Xun Wen Chen, Miao Miao Zhang, and Hui Li

17.1 INTRODUCTION

More than 70% of the Earth's land area is covered by vegetation, including forests, grasslands, savannas, shrublands, and croplands (Latham et al. 2014). These terrestrial ecosystems provide us with living space and resources. Plants are not alone; they are closely teaming up with microorganisms (also called microbes) to enhance their fitness (Bonfante and Anca 2009; Vorholt 2012; Sikes et al. 2016; Hassani et al. 2018; Toju et al. 2018). Microbes and plants have a long co-evolutionary history. Fossilized plant samples from the Rhynie Chert (~407 million years ago) were discovered with microbial structures preserved within plant cells, suggesting the dawn of plant-fungal symbioses (Remy et al. 1994; Bidartondo et al. 2011). Such successful symbioses helped plants resist stress and facilitated the transition of early plants from aquatic environments to challenging upland habitats (Martin et al. 2017).

In terrestrial ecosystems, there are four types of fungi forming symbiotic relationships with plant roots, including ectomycorrhizal (ECM) fungi, ericoid mycorrhizal (ERM) fungi, orchid mycorrhizal (ORM) fungi, and arbuscular mycorrhizal (AM) fungi (Smith and Read 2008). The first three (ECM, ERM, and ORM fungi) are associated with woody plants (e.g., pines), Ericaceae plants (e.g., blueberries), and green orchids, respectively. In comparison, AM fungi can colonize most terrestrial plants (i.e., trees, herbs, grasses, and many crops), making them distributed on six continents, much wider than the other three types (Davison et al. 2015; Brundrett and Tedersoo 2018; Martin and van der Heijden 2024). AM fungi have also been actively used in agriculture to promote crop yield (Yang et al. 2014; Ortaş et al. 2017; Lutz et al. 2023).

As the unseen majority (van der Heijden et al. 2008), AM fungi explore soils with their extensive hyphal networks (mycelia) and absorb nutrients for host plants; in return, host plants provide AM fungi with carbohydrates (Wipf et al. 2019). AM fungi help plants acquire more nitrogen (+50%) and phosphorus (+70%), and accumulate more biomass (+105%) than the case without AM fungi (synthesizing 187 studies) (Wu et al. 2024). In addition to these nutritional benefits, AM fungi directly and/or indirectly provide other important functions, such as soil aggregate formation (Rillig and Mummey 2006; Wilson et al. 2009), soil nutrient retention (Morrison et al. 2017; Xiao and Chen 2022), abiotic and biotic stress resistance (Querejeta et al. 2006; Hoang et al. 2022; Lutz et al. 2023), soil carbon sequestration (Treseder and Allen 2000; Clemmensen et al. 2013; Tao and Liu 2024), organic matter decomposition (Choreño-Parra and Treseder 2024), heavy metal fixation/transformation (Wu et al. 2014, 2015; Chen et al. 2018), and plant biodiversity regulation (van der Heijden, Klironomos, et al. 1998; Klironomos et al. 2000; Hart et al. 2003).

Notably, because AM fungi typically lack lytic enzymes (genetically) (Tisserant et al. 2013), they rely heavily on soil bacteria for organic matter decomposition (Wang et al. 2022). Then, AM fungi can take up the mobilized nutrients for plants and obtain carbon from plants in return (van der Heijden et al. 2015; Wang et al. 2022; Duan et al. 2024; Jin et al. 2024). Such a genetic "constraint"

DOI: 10.1201/9781003640899-21

(lacking genes encoding lytic enzymes) shapes soil bacterial communities differently, compared with those associated with ECM fungi, which can release lytic enzymes that decompose organic matter (Kohler et al. 2015; Bunn et al. 2019). In this way, AM fungi reinforce the *tripartite interactions* of plants, AM fungi, and soil bacteria, forming the plants-fungi-bacteria continuum.

Besides, we note that without bacteria, AM fungi also interact with saprotrophic fungi (Fang et al. 2020) in that the co-existence of AM fungi and saprotrophic fungi decomposes more organic matter than either alone (Cao et al. 2022). Adding saprotrophic fungus (*Fusarium oxysporum*) has been shown to increase AM fungal colonization rates in plants and the number of propagules (units for dispersal and reproduction) in the field (Fracchia et al. 2000). Competitive interaction between saprotrophic fungi and ECM fungi (as they can decompose organic matter as well) has also been noted (Leake et al. 2003; Bödeker et al. 2016), but it is beyond the scope of this chapter.

Given their essential roles and *multifunctionality*, AM fungi are critical to multiple targets in the United Nations' Sustainable Development Goals (SDGs), particularly SDG 15 "Life on Land" (Stewart et al. 2025). Studies generally suggest that increasing AM fungal diversity promotes ecological functions, such as increases in plant diversity or productivity, supporting the *positive biodiversity-ecosystem function* (BEF) relationship, albeit not universal (van der Heijden, Boller, et al. 1998, Vogelsang et al. 2006; Maherali and Klironomos 2007; Wagg et al. 2011; Powell and Rillig 2018; Janoušková and Jansa 2025; Runnel et al. 2025; Weber et al. 2025). Conserving AM fungal diversity is crucial for sustaining multiple ecosystem functions (Tilman et al. 2014; Soliveres et al. 2016; Hong et al. 2022), particularly under global change (Hong et al. 2022). However, the emerging global change factors, such as rising temperatures (i.e., heat), altered precipitation patterns (e.g., drought), elevated atmospheric carbon dioxide (eCO_2), soil pollution, intended/unintended fertilization, land use change, and commercialization, are bringing complex influences on AM fungal diversity, composition, and distribution (Staddon et al. 2003; Staddon 2005; Bennett and Classen 2020). But note that for soil nutrients/fertilization, Tedersoo et al. (2014, 2022) discussed that soil nutrients do not affect fungal diversity at the global scale, suggesting nutrient availability may not limit fungal growth (Tedersoo et al. 2014, 2022). The discrepancy in nutrient effects on AM fungal diversity could be due to the scale-dependent manner and research methodology. That is, at the global scale, other dominant factors such as precipitation and latitude statistically outweigh the effects of nutrients. Collectively, AM fungi are considered *the most vulnerable fungi* under drought, heat, and land use change (Tedersoo et al. 2022). To conserve AM fungal diversity, we must understand how these factors affect AM fungal diversity and distribution.

In this chapter, we begin with a brief view of the current global status of AM fungal diversity and distribution, followed by our current understanding of the assembly and coexistence of different AM fungal species and the primary governing factors. Then, we summarize and evaluate primary emerging factors that compromise AM fungal biodiversity and their ecosystem functions. Next, we provide possible conservation strategies informed by current knowledge. Finally, by acknowledging the complexity and limitations, we attempt to propose future research and practical needs for conserving AM fungal diversity against the backdrop of global change.

17.2 GLOBAL STATUS OF AM FUNGAL DIVERSITY

As of September 2024, approximately 370 AM fungal species (phylum Glomeromycota) have been described, and they are distributed in three classes, six orders, 17 families, and 49 genera (Větrovský et al. 2023; Goto et al. 2024; Hyde et al. 2024). Historically, identifying AM fungal taxa was based on the morphology of spores; however, this approach has apparent limitations, such as the similarity of spores among different species and difficulties in obtaining spores. Modern AM fungal taxonomy integrates DNA-based methods. Data analysis on environmental DNA (eDNA) sequences included both Glomeromycota and Mucoromycota (putative AM fungi) (Chang et al. 2019; Bonfante and Venice 2020), and it has been extended to 9,117 AM fungal "species" (Tedersoo et al. 2022, 2024; Lutz et al. 2025). The "species" are approximated using operational taxonomic units (OTUs) or virtual taxa (VT) in that similar DNA sequences are clustered to the same OTU/VT (a "species"), albeit still with no assigned taxonomic name (Öpik et al. 2009; Lutz et al. 2025).

Geographically, unlike the species (e.g., giant panda and kangaroo) that are limited to certain locations (known as high *endemism*) (Kier et al. 2009; Maciá-Vicente and Popa 2021), AM fungi are globally distributed on six continents (Antarctica has not been surveyed), and with low endemism (Davison et al. 2015; Tedersoo et al. 2022; Lutz et al. 2025). The diversity is higher in *hotter* areas, such as tropical regions of Africa and Southeast Asia, than in *colder* areas (Davison et al. 2015; Toussaint et al. 2020; Ma et al. 2023; van Galen et al. 2025; Van Nuland et al. 2025). Additionally, AM fungal diversity highly depends on the type of biome, in that the trend of OTU/VT richness is grassland > cropland = desert > shrubland > urban > woodland > bareland > tundra (Davison et al. 2015; Lutz et al. 2025). Because eDNA-based studies may not capture the AM fungal biomass, an early study also shows that the grasslands in savannas and temperate regions tend to have more AM fungi, in terms of abundances and root length colonized, than biomes like boreal forests (based on the data covering nine biomes in 151 geographical locations) (Treseder and Cross 2006).

To provide a high-resolution map of AM fungal diversity hotspots that facilitates conservation, extensive environmental sampling is needed. Thanks to the recent development of machine learning approaches, researchers have leveraged existing eDNA data to enhance resolution (van Galen et al. 2025, Van Nuland et al. 2025). It was predicted that the major hotspots of AM fungal richness and endemicity are *sub/tropical regions* of *Africa* (the Guinean forests of West Africa), *South America* (savannas of the Brazilian Cerrado), and *Southeast Asia* (tropical forests) (van Galen et al. 2025, Van Nuland et al. 2025).

17.3 COMMUNITY ASSEMBLY OF AM FUNGI

Various AM fungi assemble, coexist, and form a community. Assembly can occur at *multiple scales*, including single root, root system, monospecific/mixed-species plant community, landscape, region, continent, and the globe. Although multiple factors (e.g., temperature and soil nutrients) can simultaneously drive community assembly at all scales, the relative importance of assembly factors has been considered *scale-dependent* (Vályi et al. 2016). Contrasting phenomena could be observed at different scales. In this chapter, we adopt an overarching framework that emphasizes spatial and temporal dependency (Vellend 2016; Hewitt et al. 2017; Chase et al. 2018) to illustrate how AM fungi assemble and coexist. Such an approach also logically guides us in formulating conservation strategies, which are often at multiple spatial and temporal scales (Poiani et al. 2000; Monroe et al. 2022; Zurell et al. 2025).

17.3.1 Single Host

Different AM fungal species can coexist within the root system of an individual plant (Maherali and Klironomos 2007; Wolfe et al. 2009) (Figure 17.1). For example, when three AM fungal species (*Glomus mosseae*, *G. claroideum,* and *G. intraradices*; note these names have been updated) or their pair-wise combinations were inoculated to two plant species, two or three AM fungal species could colonize both plant species (Jansa et al. 2008). It clearly shows that various AM fungi can coexist in a single host, regardless of the host's identity (Figure 17.1). Additionally, in the case of these three *Glomus* species, *G. mosseae* was always the dominant one, suggesting it is the most competitive one (Figure 17.1), although its relative dominance decreased over time (Jansa et al. 2008). A competitive AM fungus obtains growth-limiting resources (i.e., carbon from the host) more efficiently than others (Olsson et al. 2003). Also, the amount of carbon obtained by an AM fungus can be proportional to the nutrient (phosphorus) that the fungus returns to its host plant (Kiers et al. 2011).

These studies collectively highlight that the trade-off traits of AM fungi determine their growth and competitiveness (Chagnon et al. 2013), which has been explained using the Biological Market Models (BMMs) (Schwartz and Hoeksema 1998; Wyatt et al. 2014). However, a later synthesizing

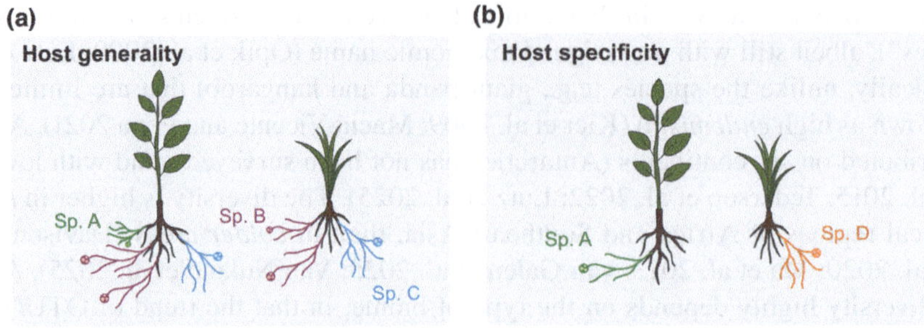

FIGURE 17.1 Host generality and specificity of AM fungi in colonizing plant hosts. (a) Different AM fungal species can colonize a single plant, and a single AM fungal species can colonize different plants. There could be dominant AM fungal species (Sp. B) (Jansa et al. 2008). (b) Different AM fungal species prefer to colonize a specific host (Gosling et al. 2013; Ramana et al. 2023) (see Section 17.3.2). "Sp." denotes species.

work shows that AM fungal growth is not related to the ability of AM fungi in delivering nutrients, but is more dependent on the *surplus carbon* in host plants (Bunn et al. 2024). Surplus carbon is photosynthates that are produced by plants under stress, and it is beyond what the plants can use for any function (Prescott et al. 2020); transferring such excessive carbon to AM fungi is never costly (Bunn et al. 2024). Therefore, temporal fluctuations in the surplus carbon stock and its allocation to fungi probably govern the coexistence of different AM fungi over time. Surplus carbon (or non-structural carbohydrates, NSCs) in plants has been shown to fluctuate over time and under stress (Oberhuber et al. 2011; Blumstein et al. 2024).

17.3.2 Host Population and Community

17.3.2.1 Root Diameter Distribution

In a mixed-species plant community, AM fungi tend to colonize fine roots rather than coarse roots (Ramana et al. 2023) (Figure 17.2a). This may partially explain why grasslands, dominated by plants with fibrous root systems, harbor more AM fungal species and biomass (Treseder and Cross 2006; van Galen et al. 2025; Van Nuland et al. 2025). A study provides a molecular explanation that increased physical engagement between fine roots and AM fungi enhances gene expression related to mineral transport (Gutjahr et al. 2015). It may suggest that absorptive roots (or fine roots < 0.2 mm), which seek minerals, possibly release more strigolactones (SLs) than transportive roots (or coarse roots > 0.2 mm), thereby triggering AM fungal colonization (Xie et al. 2010; Kapulnik and Koltai 2014; Lanfranco et al. 2018). SLs are a group of sesquiterpene lactones found in plants, and have been shown to induce hyphal branching of germinating spores (Akiyama et al. 2005).

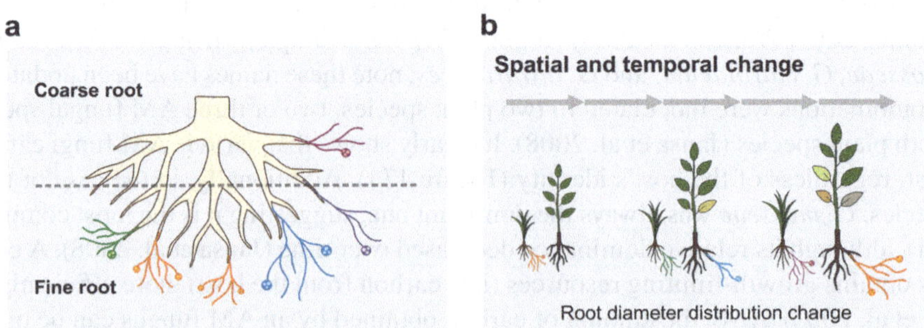

FIGURE 17.2 Effect of root diameter on harboring various AM fungal species. (a) More AM fungal species are observed in the fine root area (Ramana et al. 2023). (b) This phenomenon may also apply to mixed plant communities with varied root-diameter distributions across space and time (Erktan et al. 2023).

Specifically, the plants with fine roots (mean root diameter = 0.16–0.18 mm), such as *Leptospermum scoparium* and *Raoulia subsericea,* had higher AM fungal diversity than those with coarse roots (mean root diameter = 0.20–0.23 mm), such as *Hypericum perforatum* and *Anthoxanthum odoratum* (Ramana et al. 2023). Because the root diameter distribution of a monospecific or mixed-species plant community is different from one another and could vary temporally (i.e., succession) (Erktan et al. 2023), it is plausible that the change of AM fungal diversity and community structure would correlate with the dynamics of root distribution along the spatial (location/community change) and temporal (succession) axes (Figure 17.2b). Note that such an assembly process can also be influenced by the growth stage (Gao et al. 2019) and environmental conditions, such as nutrient- or stress-induced surplus carbon (Prescott et al. 2020). Also, host specificity has been reported, with different AM fungal species preferring to colonize a specific host (Gosling et al. 2013; Ramana et al. 2023) (Figure 17.1b).

17.3.2.2 Plant Diversity

In highly disturbed lands such as agricultural fields, increasing plant diversity, such as polyculture of different crops, can promote AM fungal diversity and compensate for the negative impacts of agricultural intensification (Guzman et al. 2021). In other studies, positive correlations between plant diversity and AM fungal diversity have been observed (van der Heijden, Boller, et al. 1998; Wagg et al. 2011), collectively suggesting that promoting plant diversity is a considerable approach to increase AM fungal diversity.

17.3.2.3 Native and Exotic Plants

A common impression is that AM fungi can colonize a wide range of plant species. However, AM fungi should be divided into generalists and specialists (van der Heijden et al. 2015; Bebber and Chaloner 2022; Ramana et al. 2023, 2025; Jin et al. 2025). Generalists are dominant and widely distributed species, while specialists are minor and relatively restricted to certain places. For a plant community that consists of native and exotic plants, native plants harbor more AM fungal species (both generalists and specialists), while exotic plants harbor fewer (mostly generalists) (Hawkes et al. 2006; Bunn et al. 2015; Ramana et al. 2025). Exotic plants form symbiotic relationships with a subset of AM fungal species, such as the generalists *Glomeraceae* (Waller et al. 2020). In addition, exotic plants can enrich gram-negative bacteria, known as antagonists, suppressing specialist AM fungi but not the generalist ones (Waller et al. 2020). These lead to a substantial reduction of AM fungal diversity in biomes occupied by exotic plants. The invasion of exotic plants has been considered facilitated by AM fungi (Chen et al. 2024). The above results imply that, in degraded lands, growing diverse native plants with more fine roots can promote conservation of AM fungal diversity even in the agricultural context.

17.3.3 Landscape and Beyond

17.3.3.1 Biogeographic Drivers

At broad spatial scales such as landscapes, ecosystems, and biomes, AM fungal community structure is primarily shaped by geographic distance, soil temperature, soil moisture, soil pH, and the composition of the plant community (Kivlin et al. 2011; Zhou et al. 2020; Davison et al. 2021). Such location-dependent biogeographic drivers (Figure 17.3) create habitat filtering and dispersal limitation, which restrict the transport of AM fungi, resulting in relatively unique communities at different sites (Kivlin et al. 2011). In warm and humid regions, rapid decomposition of organic matter increases the supply of inorganic nutrients, which enhances AM fungal diversity (Ma et al. 2023). By contrast, in colder regions, the abundance of AM host plants is lower, carbon allocation to fungi is reduced, and microbial activity is suppressed, all of which limit AM fungal growth and diversity (Brown et al. 2004; Tylianakis et al. 2008; Bueno et al. 2017; Barceló et al. 2019).

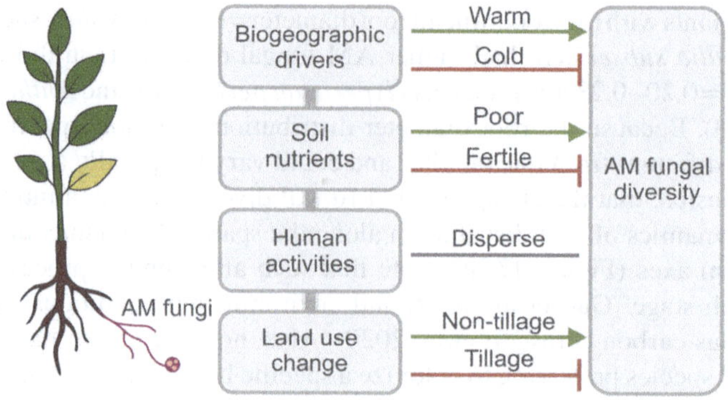

FIGURE 17.3 Key factors influencing AM fungal diversity at large scales. AM fungal diversity is affected by multiple intertwined factors. Biogeographic drivers encompass the temperature and precipitation patterns in various regions. Soil properties are also included. Soil nutrients, particularly nitrogen and phosphorus, are among the best predictors of AM fungal diversity. Human activities, such as the application of commercialized AM fungal products, pose uncertain effects on AM fungal diversity. Land use change poses strong soil disturbance, including tillage in agricultural fields, which negatively impacts AM fungal diversity.

Plant communities are consistently among the best predictors of AM fungal composition (Kivlin et al. 2011; Davison et al. 2015). The C-S-R framework (competitor-stress tolerator-ruderal) provides a useful lens for understanding these relationships, as plant life-history strategies often align with those of their fungal partners: competitor plants tend to associate with competitor fungi, while ruderal plants are more often linked with ruderal fungi (Grime 2006; Chagnon et al. 2013; Koyama et al. 2019). Past plant communities can also leave legacy effects on AM fungal composition (Bittebiere et al. 2020), and mycorrhizal types influence broader patterns such as the island biogeography of plants (Delavaux et al. 2017, 2019, 2021).

17.3.3.2 Soil Nutrients

Although soil nutrients are factors included in biogeographic drivers, i.e., soil properties change across locations, soil nutrients (particularly nitrogen and phosphorus) are very important factors affecting AM fungal diversity. This is because plant-AM fungal symbiosis typically aims at nutritional benefits (Wipf et al. 2019). Nutrient enrichment often reduces both the abundance and diversity of AM fungi (Han et al. 2020; Ma, Geng, et al. 2021). High levels of nitrogen addition, above 50 kg/ha/year, significantly reduce AM fungal diversity in tropical forests, whereas lower levels of nitrogen input cause little change (Li et al. 2025). These effects are linked to plant nutrient allocation strategies: when nitrogen is abundant, plants invest less carbon in mycorrhizal symbionts and rely more heavily on fine roots for nutrient uptake, leading to a decline in fungal diversity (Ma, Zhu, et al. 2021). However, the global trend of nitrogen deposition (Chen et al. 2017; Li et al. 2025) (but see a study on nitrogen decline (Mason et al. 2022) and phosphorus limitation (Zhu et al. 2024) can have uncertain effects on AM fungal diversity.

17.3.3.3 Human Activities

Human activities, such as dispersal and agricultural practices, add another layer of complexity. Global dispersal of AM fungi is increasingly facilitated by human-mediated movement of soils and plants (Davison et al. 2015). In agriculture, commercial AM fungal products are widely applied, often containing species such as *Rhizophagus irregularis* and *Funneliformis mosseae* due to their effectiveness in promoting plant growth (Biel et al. 2021). These inoculants can enhance crop yields and reduce the need for irrigation and fertilizer, as demonstrated by Ziane et al. (2017), who showed that inoculating tomato fields with a mixed inoculum reduced fertilizer use by 50% without lowering yields. However, the long-term ecological consequences of such applications remain uncertain.

Introduced fungi may inhibit the growth of some native plants (Klironomos 2003) or interact unpredictably with local microbial communities, and their spread beyond application sites could have unrecognized impacts (Nayak et al. 2022; Salomon, Demarmels, et al. 2022; Martin and van der Heijden 2024). Monitoring these effects is therefore essential.

Land use and management practices are critical determinants of AM fungal diversity (Figure 17.3). Oehl et al. (2010) found that more than half of the AM fungal species identified in Central Europe were restricted to specific land use types or intensities (Oehl et al. 2010). Agricultural practices strongly influence fungal diversity: permanent pastures tend to support the highest diversity, while intensive farming systems reduce it (Manoharan et al. 2017). Conversely, less intensive practices such as reduced tillage and winter cover cropping can increase AM fungal colonization and improve crop yields (Bowles et al. 2017). Land use change and habitat fragmentation further alter AM fungal communities by reshaping soil and plant environments (Balami et al. 2020).

17.4 EMERGING FACTORS THREATENING AM FUNGAL DIVERSITY

Habitat destruction, environmental pollution, climate change, and the introduction of invasive species are major emerging factors threatening AM fungal biodiversity (García de León et al. 2018; Bennett and Classen 2020). These threats can reduce AM fungal diversity and disrupt the symbiotic relationships between fungi and plants, ultimately affecting ecosystem stability and productivity.

17.4.1 COMMERCIALIZATION OF AM FUNGI

Habitat filtering and dispersal limitation prevent homogeneity of AM fungal communities and diversify community structures (Kivlin et al. 2011). However, commercialized AM fungal products disperse 1–2 AM fungal species over long distances (sometimes continentally) and pose uncertain ecological consequences. Globally, a very low endemism of AM fungi (i.e., not restricted to geographic locations) has been observed (Davison et al. 2015), suggesting efficient dispersal and establishment of certain AM fungi in their non-native range (but see Bruns and Taylor (2016), Öpik et al. (2016), and Bruns et al. (2018)).

In the past few decades, the commercialization of fungi has played a critical role in modern agriculture, particularly in enhancing plant growth, nutrient uptake, and soil health (Rosendahl et al. 2009). The AM fungal inoculants are now widely used in agricultural production to improve crop productivity, save fertilizer, and promote sustainable agriculture practices. However, this growing commercialization of AM fungi has introduced several ecological risks, possibly affecting the native AM fungal diversity (Hart et al. 2018). Also, it has long been documented that AM fungi can be *parasitic*, rather than always mutualistic, to many plants, depending on the plant-AM fungus-soil combination (Johnson et al. 1997; Klironomos 2003; Jones and Smith 2004). It raises significant concerns about whether an AM fungal species beneficial to crops could escape into adjacent ecosystems and be detrimental to non-crop species.

The commercialization of AM fungi is focused on a limited number of generalist species that are easy to cultivate, commercially viable, and known for their effective symbiosis with a broad range of crops. For instance, species like *Rhizophagus intraradices* and *Glomus aggregatum* are formulated as inoculants and are commonly used in agricultural practices (Colombo et al. 2013; Faye et al. 2013). While these species may promote short-term crop growth, their wide use poses a threat to the native AM fungal community. For example, commercial AM fungal inoculants colonize more rapidly than the native strains, resulting in the dominance of a few exotic species (Schwartz et al. 2006). Because the belowground AM fungal community is correlated with the aboveground plant diversity, altering the native AM fungal community can pose uncertain risks, such as facilitating the spread of invasive plants (Sielaff et al. 2019; Chen et al. 2024).

Another major threat arises from the genetic homogenization of AM fungi applications. Commercial AM fungi inoculants are typically mass-produced in controlled environments, which may limit their genetic diversity. The commercial strains often represent only a small subset of actual fungal diversity found in natural ecosystems. Inoculating agricultural soils with such genetically uniform strains may decrease the resilience of the soil microbiome, making it less adaptable to environmental changes (Ambrosini et al. 2016). This lake of genetic diversity can reduce the functional redundancy of the mycorrhizal community, making ecosystems more vulnerable to disturbances (Gianinazzi et al. 2010). For example, the applications of commercial inoculants may damage the native fungi with specialized traits, diminishing the ecosystem services they provide (Basiru and Hijri 2022).

Additionally, commercial AM fungal inoculants are often marketed without sufficient consideration of ecological compatibility with local soil and climate conditions. For example, the commercial AM fungal strains may not perform well in agricultural or natural ecosystems, potentially leading to ecological mismatches that harm native plants and animals (Schwartz et al. 2006). Invasive commercial AM fungal strains may disrupt symbiotic relationships between native fungi and plants, which may alter plant community composition and reduce ecosystem stability (Schwartz et al. 2006). For instance, the AM fungi can enhance the competitive ability of Asteraceae species and reduce the symbiosis between local plants and AM fungi (Sokornova et al. 2022). This may decrease the diversity of the local ecosystem.

Therefore, while the commercialization of AM fungi provides various benefits for agriculture, it also comes with significant risks to fungal biodiversity. The main reasons for this damage are as follows: genetic homogenization, ecological mismatches, lack of regulatory oversight, and potential biopiracy all contribute to the erosion of the native AM fungal community. It is essential to adopt a more cautious and ecologically informed approach to AM fungal commercialization. This includes promoting the use of locally adapted strains, implementing stricter regulatory frameworks, and understanding the potential impacts on soil biodiversity and ecosystem functionality.

17.4.2 Soil Pollution

The soil pollution caused by the accumulation of heavy metals, pesticides, industrial chemicals, and other contaminants significantly threatens the diversity and functioning of soil microbial communities, including the AM fungal community. The AM fungi play a fundamental role in the diversity and functioning of ecosystems, plant growth, and crop productivity. However, soil pollution can severely disrupt these symbiotic relationships and negatively impact AM fungal diversity, community structure, and ecological function.

Metals such as lead (Pb), cadmium (Cd), Zinc (Zn), copper (Cu), and arsenic (As) are common pollutants found in industrial and agricultural soils. These metal pollutants can directly affect AM fungal growth, spore production, and overall fungal biodiversity (Entry et al. 2002). For example, a previous study suggested that the AM fungal species richness of noncontaminated soils was significantly higher than that of metal-contaminated soils adjoining Kanpur Tanneries, Uttar Pradesh, India (Khade and Adholeya 2009). Ruotsalainen et al. (2007) also demonstrated that the heavy metals nickel and copper decreased the colonization of AM fungi (Ruotsalainen et al. 2007).

The widespread use of chemical pesticides and herbicides in agriculture has raised concerns about their effects on soil ecosystems, including AM fungi. Pesticides, particularly fungicides and herbicides, can be toxic to AM fungi by disrupting their metabolic processes or altering their microbial environment. Fungicides are often designed to target fungi, but they may also harm beneficial AM fungal species. For instance, the fungicide carbendazim has been shown to reduce AMF spore density and mycorrhizal colonization in crops, leading to decreased plant growth and nutrient uptake (Li et al. 2015). Benomyl, a broad-spectrum fungicide, is known for its deleterious effect on AM fungi (Kjøller and Rosendahl 2000). Herbicides such as glyphosate, which is widely used to control weeds, have also been implicated in harming the AM fungal population (Savin et al. 2009). Studies have found that glyphosate exposure can reduce the effectiveness of AM fungal symbiosis and decrease plant growth (Helander et al. 2018). This disruption of AM fungal function can have cascading effects on plant productivity and soil health.

Despite the detrimental effects of pollution on AM fungi, recent research has also explored the potential of AM fungi to mitigate the impacts of soil pollution. One of the strategies that AM fungi employ to help plants remediate contaminated soil is plant extraction (which increases the plant absorption of heavy metals and reduces soil pollution) and plant stabilization processes (which reduce the stability of heavy metals). The mechanisms of AM fungal symbionts restore contaminated soil include enhancing plant nutrient absorption, alleviating oxidative stress in plants, improving drought resistance and metal tolerance in plants, and altering the soil microbial composition (Ai et al. 2022). The AM fungi can also release glomalin-related soil protein to adsorb or chelate the heavy metals before the metals enter the fungal-plant symbiosis (Khan 2006).

Soil pollution (heavy metals, pesticides, and herbicides) presents multifaceted threats to AM fungi by reducing their numbers, impairing their symbiotic functions, and altering the soil microbial community. The AM fungi also hold promise for bioremediation strategies, where they may help mitigate the effects of soil pollution.

17.4.3 Climate Change

Climate change is one of the most pressing global challenges of the 21st century. Climate changes, such as rising temperatures, altered precipitation patterns, and eCO_2 levels, have been proven to affect ecosystems, biodiversity, and soil microbial communities.

Increased temperatures can influence the AM fungi in both direct and indirect ways. In general, the increase can enhance the richness of AM fungi, but some species may exhibit different results because they are more sensitive than other species. For example, a global meta-analysis suggested that entire-year and growing-season warming experiments increased the richness of AM fungi by 22.7% and 16.6%, respectively (Hu et al. 2022). However, Mao et al. (2024) demonstrated that warming decreased the diversity and evenness of AM fungi, which may induce changes in ecosystem multifunctionality (Mao et al. 2024).

Precipitation is also a fundamental factor that affects the composition and diversity of the fungal community, including the AM fungi. The precipitation pattern changes are predicted to become more frequent under climate change scenarios, which stress AM fungi by creating unpredictable soil moisture conditions. Water availability has already been a limiting factor for plant growth. Drought conditions can severely affect the symbiotic relationship between plants and AM fungi. For example, Quiroga et al. (2024) demonstrated that drought can negatively affect the rhizosphere microorganisms, including AM fungi (Quiroga et al. 2024). Bittlingmaier et al. (2025) further pointed out that the drought significantly decreases the AM fungal diversity, especially in low soil and plant biodiversity ecosystems (Bittlingmaier et al. 2025). On the other hand, excessive rainfall or flooding is also detrimental to AM fungi, due to anaerobic conditions in the soil. The flooding can decrease the AM fungal colonization and extraradical hyphal density, and induce negative effects on rice production (Wang et al. 2021).

Under eCO_2 conditions, plants should allocate more photosynthetic products to absorb underground resources. Short-term experiments proved that root colonization, hyphal length, and protein glomalin increased linearly along a CO_2 gradient (Rillig et al. 2001). A study also suggests that the eCO_2 promoted the allocation of AM fungal biomass to the external hyphae. In this study, compared to the natural conditions, the abundance and length of external hyphae of AM fungi increased 2 and 5 times, respectively, under eCO_2 (Sanders et al. 1998). The relationship between eCO_2 and AM fungi is multifaceted. The effects of eCO_2 on soil microbial biomass, including AM fungi, are regulated by maintaining time. For example, the eCO_2 has a significant positive effect on soil microbial biomass in a short time (<3 years), but a negative effect in a long time (>3 years). The long-term ecological implications of eCO_2 remain ongoing research.

Climate change is likely to have profound effects on AM fungi, altering their biomass, colonization, diversity, distribution, and functionality. Rising temperatures, altered precipitation patterns, and eCO_2 all interact to influence AM fungal communities. Understanding the response mechanism of AM fungi to climate change, especially the response of AM fungi to various climate conditions, is of vital importance for maintaining healthy soil ecosystems and plant productivity.

17.5 CONSERVATION STRATEGIES

The conservation of AM fungi is critical for the preservation of ecosystem services such as nutrient cycling, soil health, and plant growth. The conservation strategies should focus on both *in-situ* and *ex-situ* conservation efforts, with an emphasis on protecting their natural habitats, promoting sustainable land use practices, and mitigating the impacts of climate change.

17.5.1 *In-Situ* Conservation

In-situ conservation refers to the preservation of organisms in their natural habitats, ensuring the protection of the ecosystems in which they thrive. Protecting natural habitats, such as forests, grasslands, and wetlands, is fundamental for the conservation of AM fungi (Figure 17.4). These ecosystems are not only home to diverse AM fungal species but also provide the necessary environmental conditions, such as soil structure, moisture, and nutrient availability (Zhou et al. 2022). A high level

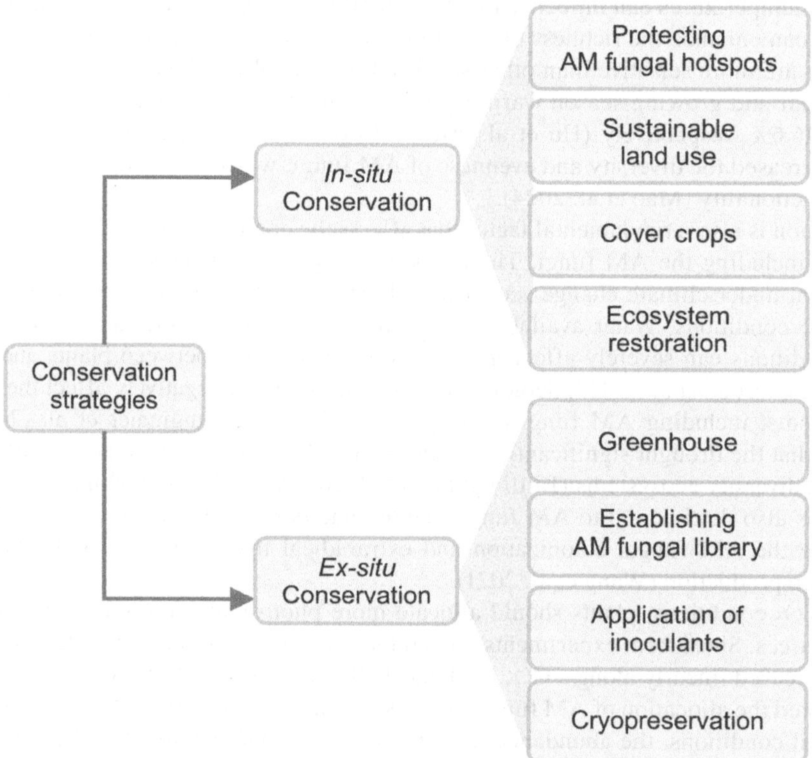

FIGURE 17.4 Scale-dependent conservation strategies of AM fungal diversity. *In-situ* approaches include protecting AM fungal hotspots, promoting fine-root and native plant establishment for sustainable land use, such as using diverse cover crops/native plants/polyculture for conservation, and mitigating negative factors, especially pollution. *Ex-situ* approaches include propagating AM fungal species in greenhouses, isolating more native AM fungal species, and constructing local species libraries, such as using cryopreservation.

of functional diversity (FD) can be observed in an AM fungal community with a lower level of phylogenetic diversity (PD) (Munkvold et al. 2004). In other words, PD is not a suitable predictor of FD. In this case, to play it safe, protecting PD could be a priority rather than FD, as a high PD probably provides the basis for a high PD. Whether there would be a high PD depends on other factors, such as species combinations and environmental conditions, that influence the trade-off traits between plants and AM fungi (Chagnon et al. 2013) or the surplus carbon from plants to be supplied to AM fungi (Bunn et al. 2024).

Protected areas should be expanded and effectively managed to preserve the integrity of these habitats. Meanwhile, the establishment of dedicated conservation areas specifically designed to protect AM fungi can help maintain their diversity and ecological function (Turrini and Giovannetti 2012). Mycorrhizal sanctuaries could be set up in regions that are rich in mycorrhizal diversity and vulnerable to degradation. These regions should be protected from direct threats, such as logging, mining, or agricultural expansion, and indirect threats like invasive species and soil pollution (Turrini and Giovannetti 2012). For example, the first AM fungal *in-situ* conservation area in the world is located in Italy, which is important for the conservation of rare and endangered AM fungi (Turrini et al. 2008). Additionally, encouraging sustainable land-use practices in agricultural and forested areas is also essential for mitigating the impacts of land-use change on AM fungi. It was suggested that agroforestry, no-till agriculture, and organic farming can promote healthy mycorrhizal populations by maintaining soil health and reducing the harmful chemical use (Gosling et al. 2006; Yang et al. 2012; da Silva Maia et al. 2021). Cover crops can also promote soil health by increasing the AM fungal colonization and altering soil microbial community structure (Bowles et al. 2017).

Because the commercialization of AM fungi can pose uncertain ecological risks (Section 17.3.3.3), regulatory measures are necessary to ensure a safe use of AM fungal products. A majority of work highlights the quality assurance and standardization of AM fungal products (Salomon, Watts-Williams, et al. 2022; Ghorui et al. 2024; Koziol et al. 2025). However, the ecological risks, particularly AM fungal invasion, have not been effectively managed. So, a safer approach could be using native AM fungi strains, as they are better adapted to the local ecosystem and may reduce damage to the local microbial communities (Ghorui et al. 2024). At the same time, diversified soil management practices, such as crop rotation and organic farming, can also help support a more balanced and resilient microbial community.

In addition, restoration of degraded ecosystems is also a critical method to protect the AM fungi. Restoring the degraded ecosystems can provide new opportunities to re-establish the AM fungal community. Reforestation and afforestation projects, especially those that use native plant species, can help restore the AM fungal community by creating environments for AM fungi to colonize.

17.5.2 *Ex-Situ* Conservation

Ex-situ conservation involves the conservation of species outside their natural habitats, typically in controlled environments (Figure 17.4). *Ex-situ* conservation plays a vital role in preserving AM fungal genetic diversity. The cultivation of AM fungi in laboratory or greenhouse conditions is one of the most used *ex-situ* conservation strategies (IJdo et al. 2011; Pandit et al. 2025).

The produced AM fungal strains can be used for preservation, restoration projects, and scientific research. The application of AM fungal inoculants in ecosystem restoration and agriculture is also a considerable *ex-situ* conservation strategy (Hart and Trevors 2005; Fester and Sawers 2011). For instance, the AM fungal inoculants have been used in heavy metal-contaminated soils, and improved plant growth and soil nutrient cycling (Riaz et al. 2021). For rare or endangered fungal species, cryopreservation is an emerging *ex-situ* technique used to store AM fungal species in a dormant state at low temperatures, preserving their genetic material for future research or restoration efforts (Lalaymia et al. 2014). Freezing AM fungal spores allows researchers to preserve fungal diversity over time, preventing loss of valuable genetic material due to environmental changes or human impact.

The conservation of AM fungi is crucial for maintaining soil health, plant productivity, and overall ecosystem stability. Effective conservation strategies should include both *in-situ* and *ex-situ* methods. It is vital to combine the two conservation techniques to ensure the long-term survival and functionality of these indispensable soil organisms.

17.6 PROSPECTS

The future research must explore several key areas, such as AM fungal functional diversity, effects of climate change on AM fungi, and application of advanced technologies, to ensure the effective conservation and sustainability of AM fungi. The practice needs to adapt to emerging environmental and societal changes.

Understanding the functional diversity of AM fungi is conducive to protecting ecological functions and genetic diversity of AM fungi, thereby promoting ecosystem stability and agricultural sustainability. However, significant knowledge gaps remain in how different species contribute to ecosystem functions under various environmental conditions. The future research should focus on functional trait diversity, fungal functional groups, and interdependence with other soil organisms, which can help us understand the roles of AM fungi in soil fertility, plant resistance, and drought tolerance (Van Der Heijden and Scheublin 2007; Davison et al. 2020; Bagyaraj et al. 2022). Climate change induces complex positive or negative effects on the AM fungi, so predictive models are needed to investigate the effects and conservation strategies (García and Mendoza 2009; Heuck et al. 2024). Climate change may also induce unknown effects on the conservation of AM fungal diversity. Therefore, monitoring and predicting the impact of climate change on the diversity of AM fungi is an important guarantee for formulating dynamic conservation strategies. Advanced technologies, such as metagenomics, stable isotope labeling, and remote sensing, can help us to monitor AM fungal communities in situ, track nutrient cycling, and assess the impact of climate change (Rowe and Pringle 2005; Peršoh 2015; Lang et al. 2023). These can support the conservation and sustainable management of AM fungi diversity.

17.7 CONCLUSIONS

The biodiversity of AM fungi is crucial for maintaining soil health, plant productivity, and overall ecosystem functioning. However, the AM fungal communities face significant threats from environmental degradation, soil pollution, and climate change. Additionally, the commercialization of AM fungi poses risks to native fungal populations and their ecological functions. Addressing these challenges requires an integrated approach that combines scientific research, sustainable land-use practices, and effective conservation strategies. Both *in-situ* and *ex-situ* conservation methods should be implemented to protect AM fungal diversity. Crucial conservation strategies include promotion of native and fine-root plant species, construction of native AM fungal species libraries, identification of AM fungal diversity hotspots for protection, and projection of overlapping areas of diversity hotspots and intensive global change factors. Future research should focus on understanding the functional diversity of AM fungi and its responses to multiple global change factors to inform better management and conservation practices.

DECLARATION OF GENERATIVE AI USE

During the preparation of this manuscript, the authors used Copilot to search relevant literature. The authors further manually reviewed, screened, and edited the content. The manuscript was then manually drafted, and Grammarly was used to check typos. Some suggested rephrasing was used. The authors take full responsibility for the final version of the published chapter.

ACKNOWLEDGMENTS

We gratefully acknowledge the support from the National Natural Science Foundation of China (42322711, 42277211, and 42477118), the Scientific Research Innovation Capability Support Project for Young Faculty (SRICSPYF-ZY2025136), the Guangdong Basic and Applied Basic Research Foundation (2021B1515020014), and the Guangzhou Science and Technology Program (2024A04J6285).

REFERENCES

Ai, Y.-J., Li, F.-P., Yang, J.-Q., Lu, S., and Gu, H.-H., 2022. Research progress and potential functions of AMF and GRSP in the ecological remediation of metal tailings. *Sustainability*, 14(15), 9611.

Akiyama, K., Matsuzaki, K., and Hayashi, H., 2005. Plant sesquiterpenes induce hyphal branching in arbuscular mycorrhizal fungi. *Nature*, 435(7043), 824–827.

Ambrosini, A., de Souza, R., and Passaglia, L.M.P., 2016. Ecological role of bacterial inoculants and their potential impact on soil microbial diversity. *Plant and Soil*, 400(1), 193–207.

Bagyaraj, D.J., Sridhar, K.R., and Revanna, A., 2022. Arbuscular mycorrhizal fungi influence crop productivity, plant diversity, and ecosystem services. *In*: V.R. Rajpal, I. Singh, and S.S. Navi, eds. *Fungal Diversity, Ecology and Control Management*. Singapore: Springer Nature, 345–362.

Balami, S., Vašutová, M., Godbold, D., Kotas, P., and Cudlín, P., 2020. Soil fungal communities across land use types. *iForest - Biogeosciences and Forestry*, 13(6), 548.

Barceló, M., van Bodegom, P.M., and Soudzilovskaia, N.A., 2019. Climate drives the spatial distribution of mycorrhizal host plants in terrestrial ecosystems. *Journal of Ecology*, 107(6), 2564–2573.

Basiru, S. and Hijri, M., 2022. The potential applications of commercial arbuscular mycorrhizal fungal inoculants and their ecological consequences. *Microorganisms*, 10(10), 1897.

Bebber, D.P. and Chaloner, T.M., 2022. Specialists, generalists and the shape of the ecological niche in fungi. *New Phytologist*, 234(2), 345–349.

Bennett, A.E. and Classen, A.T., 2020. Climate change influences mycorrhizal fungal–plant interactions, but conclusions are limited by geographical study bias. *Ecology*, 101(4), e02978.

Bidartondo, M.I., Read, D.J., Trappe, J.M., Merckx, V., Ligrone, R., and Duckett, J.G., 2011. The dawn of symbiosis between plants and fungi. *Biology Letters*, 7(4), 574–577.

Biel, C., Camprubí, A., Lovato, P.E., and Calvet, C., 2021. On-farm reduced irrigation and fertilizer doses, and arbuscular mycorrhizal fungal inoculation improve water productivity in tomato production. *Scientia Horticulturae*, 288, 110337.

Bittebiere, A.-K., Vandenkoornhuyse, P., Maluenda, E., Gareil, A., Dheilly, A., Coudouel, S., Bahin, M., and Mony, C., 2020. Past spatial structure of plant communities determines arbuscular mycorrhizal fungal community assembly. *Journal of Ecology*, 108(2), 546–560.

Bittlingmaier, M., Séjalon-Delmas, N., Goldmann, K., Johnson, D., Huys, R., and Freschet, G.T., 2025. Plant and soil biodiversity sustain root mycorrhizal fungal richness under drought stress. *The ISME Journal*, 20, wraf102.

Blumstein, M., Oseguera, M., Caso-McHugh, T., and Des Marais, D.L., 2024. Nonstructural carbohydrate dynamics' relationship to leaf development under varying environments. *New Phytologist*, 241(1), 102–113.

Bödeker, I.T.M., Lindahl, B.D., Olson, Å., and Clemmensen, K.E., 2016. Mycorrhizal and saprotrophic fungal guilds compete for the same organic substrates but affect decomposition differently. *Functional Ecology*, 30(12), 1967–1978.

Bonfante, P. and Anca, I.-A., 2009. Plants, mycorrhizal fungi, and bacteria: A network of interactions. *Annual Review of Microbiology*, 63(1), 363–383.

Bonfante, P. and Venice, F., 2020. Mucoromycota: going to the roots of plant-interacting fungi. *Fungal Biology Reviews*, 34(2), 100–113.

Bowles, T.M., Jackson, L.E., Loeher, M., and Cavagnaro, T.R., 2017. Ecological intensification and arbuscular mycorrhizas: A meta-analysis of tillage and cover crop effects. *Journal of Applied Ecology*, 54(6), 1785–1793.

Brown, J.H., Gillooly, J.F., Allen, A.P., Savage, V.M., and West, G.B., 2004. Toward a metabolic theory of ecology. *Ecology*, 85(7), 1771–1789.

Brundrett, M.C. and Tedersoo, L., 2018. Evolutionary history of mycorrhizal symbioses and global host plant diversity. *New Phytologist*, 220(4), 1108–1115.

Bruns, T.D., Corradi, N., Redecker, D., Taylor, J.W., and Öpik, M., 2018. Glomeromycotina: What is a species and why should we care? *New Phytologist*, 220(4), 963–967.

Bruns, T.D. and Taylor, J.W., 2016. Comment on "Global assessment of arbuscular mycorrhizal fungus diversity reveals very low endemism". *Science*, 351(6275), 826–826.

Bueno, C.G., Moora, M., Gerz, M., Davison, J., Öpik, M., Pärtel, M., Helm, A., Ronk, A., Kühn, I., and Zobel, M., 2017. Plant mycorrhizal status, but not type, shifts with latitude and elevation in Europe. *Global Ecology and Biogeography*, 26(6), 690–699.

Bunn, R.A., Corrêa, A., Joshi, J., Kaiser, C., Lekberg, Y., Prescott, C.E., Sala, A., and Karst, J., 2024. What determines transfer of carbon from plants to mycorrhizal fungi? *New Phytologist*, 244(4), 1199–1215.

Bunn, R.A., Ramsey, P.W., and Lekberg, Y., 2015. Do native and invasive plants differ in their interactions with arbuscular mycorrhizal fungi? A meta-analysis. *Journal of Ecology*, 103(6), 1547–1556.

Bunn, R.A., Simpson, D.T., Bullington, L.S., Lekberg, Y., and Janos, D.P., 2019. Revisiting the 'direct mineral cycling' hypothesis: Arbuscular mycorrhizal fungi colonize leaf litter, but why? *The ISME Journal*, 13(8), 1891–1898.

Cao, T., Fang, Y., Chen, Y., Kong, X., Yang, J., Alharbi, H., Kuzyakov, Y., and Tian, X., 2022. Synergy of saprotrophs with mycorrhiza for litter decomposition and hotspot formation depends on nutrient availability in the rhizosphere. *Geoderma*, 410, 115662.

Chagnon, P.-L., Bradley, R.L., Maherali, H., and Klironomos, J.N., 2013. A trait-based framework to understand life history of mycorrhizal fungi. *Trends in Plant Science*, 18(9), 484–491.

Chang, Y., Desirò, A., Na, H., Sandor, L., Lipzen, A., Clum, A., Barry, K., Grigoriev, I.V., Martin, F.M., Stajich, J.E., Smith, M.E., Bonito, G., and Spatafora, J.W., 2019. Phylogenomics of Endogonaceae and evolution of mycorrhizas within Mucoromycota. *New Phytologist*, 222(1), 511–525.

Chase, J.M., McGill, B.J., McGlinn, D.J., May, F., Blowes, S.A., Xiao, X., Knight, T.M., Purschke, O., and Gotelli, N.J., 2018. Embracing scale-dependence to achieve a deeper understanding of biodiversity and its change across communities. *Ecology Letters*, 21(11), 1737–1751.

Chen, B., Nayuki, K., Kuga, Y., Zhang, X., Wu, S., and Ohtomo, R., 2018. Uptake and intraradical immobilization of cadmium by arbuscular mycorrhizal fungi as revealed by a stable isotope tracer and synchrotron radiation μX-ray fluorescence analysis. *Microbes and Environments*, 33(3), 257–263.

Chen, X.W., Pan, M., and Louis, S.H.L., 2024. Microbe-assisted plant invasion. In: Wong, M.H. and Seastedt, T.R. (Eds.) *Sustainable Management of Invasive Species*. World Scientific (Europe), 159–185.

Chen, Y.-L., Xu, Z.-W., Xu, T.-L., Veresoglou, S.D., Yang, G.-W., and Chen, B.-D., 2017. Nitrogen deposition and precipitation induced phylogenetic clustering of arbuscular mycorrhizal fungal communities. *Soil Biology and Biochemistry*, 115, 233–242.

Choreño-Parra, E.M. and Treseder, K.K., 2024. Mycorrhizal fungi modify decomposition: A meta-analysis. *New Phytologist*, 242(6), 2763–2774.

Clemmensen, K.E., Bahr, A., Ovaskainen, O., Dahlberg, A., Ekblad, A., Wallander, H., Stenlid, J., Finlay, R.D., Wardle, D.A., and Lindahl, B.D., 2013. Roots and associated fungi drive long-term carbon sequestration in boreal forest. *Science*, 339(6127), 1615–1618.

Colombo, R.P., Martínez, A.E., di Pardo, A.F., Bidondo, L.F., van Baren, C., Lira, P. di L., and Godeas, A.M., 2013. Differential effects of two strains of *Rhizophagus intraradices* on dry biomass and essential oil yield and composition in *Calamintha nepeta*. *Revista Argentina de Microbiología*, 45(2), 114–118.

Davison, J., García de León, D., Zobel, M., Moora, M., Bueno, C.G., Barceló, M., Gerz, M., León, D., Meng, Y., Pillar, V.D., Sepp, S.-K., Soudzilovaskaia, N.A., Tedersoo, L., Vaessen, S., Vahter, T., Winck, B., and Öpik, M., 2020. Plant functional groups associate with distinct arbuscular mycorrhizal fungal communities. *New Phytologist*, 226(4), 1117–1128.

Davison, J., Moora, M., Öpik, M., Adholeya, A., Ainsaar, L., Bâ, A., Burla, S., Diedhiou, A.G., Hiiesalu, I., Jairus, T., Johnson, N.C., Kane, A., Koorem, K., Kochar, M., Ndiaye, C., Pärtel, M., Reier, Ü., Saks, Ü., Singh, R., Vasar, M., and Zobel, M., 2015. Global assessment of arbuscular mycorrhizal fungus diversity reveals very low endemism. *Science*, 349(6251), 970–973.

Davison, J., Moora, M., Semchenko, M., Adenan, S.B., Ahmed, T., Akhmetzhanova, A.A., Alatalo, J.M., Al-Quraishy, S., Andriyanova, E., Anslan, S., Bahram, M., Batbaatar, A., Brown, C., Bueno, C.G., Cahill, J., Cantero, J.J., Casper, B.B., Cherosov, M., Chideh, S., Coelho, A.P., Coghill, M., Decocq, G., Dudov, S., Fabiano, E.C., Fedosov, V.E., Fraser, L., Glassman, S.I., Helm, A., Henry, H.A.L., Hérault, B., Hiiesalu, I., Hiiesalu, I., Hozzein, W.N., Kohout, P., Kõljalg, U., Koorem, K., Laanisto, L., Mander, Ü., Mucina, L., Munyampundu, J.-P., Neuenkamp, L., Niinemets, Ü., Nyamukondiwa, C., Oja, J., Onipchenko, V., Pärtel, M., Phosri, C., Põlme, S., Püssa, K., Ronk, A., Saitta, A., Semboli, O., Sepp, S.-K., Seregin, A.,

Sudheer, S., Peña-Venegas, C.P., Paz, C., Vahter, T., Vasar, M., Veraart, A.J., Tedersoo, L., Zobel, M., and Öpik, M., 2021. Temperature and pH define the realised niche space of arbuscular mycorrhizal fungi. *New Phytologist*, 231(2), 763–776.

Delavaux, C.S., Smith-Ramesh, L.M., and Kuebbing, S.E., 2017. Beyond nutrients: A meta-analysis of the diverse effects of arbuscular mycorrhizal fungi on plants and soils. *Ecology*, 98(8), 2111–2119.

Delavaux, C.S., Weigelt, P., Dawson, W., Duchicela, J., Essl, F., van Kleunen, M., König, C., Pergl, J., Pyšek, P., Stein, A., Winter, M., Schultz, P., Kreft, H., and Bever, J.D., 2019. Mycorrhizal fungi influence global plant biogeography. *Nature Ecology & Evolution*, 3(3), 424.

Delavaux, C.S., Weigelt, P., Dawson, W., Essl, F., van Kleunen, M., König, C., Pergl, J., Pyšek, P., Stein, A., Winter, M., Taylor, A., Schultz, P.A., Whittaker, R.J., Kreft, H., and Bever, J.D., 2021. Mycorrhizal types influence island biogeography of plants. *Communications Biology*, 4(1), 1–8.

Duan, S., Feng, G., Limpens, E., Bonfante, P., Xie, X., and Zhang, L., 2024. Cross-kingdom nutrient exchange in the plant–arbuscular mycorrhizal fungus–bacterium continuum. *Nature Reviews Microbiology*, 22, 773–790.

Entry, J.A., Rygiewicz, P.T., Watrud, L.S., and Donnelly, P.K., 2002. Influence of adverse soil conditions on the formation and function of *Arbuscular mycorrhizas*. *Advances in Environmental Research*, 7(1), 123–138.

Erktan, A., Roumet, C., and Munoz, F., 2023. Dissecting fine root diameter distribution at the community level captures root morphological diversity. *Oikos*, 2023(1), e08907.

Fang, M., Liang, M., Liu, X., Li, W., Huang, E., and Yu, S., 2020. Abundance of saprotrophic fungi determines decomposition rates of leaf litter from arbuscular mycorrhizal and ectomycorrhizal trees in a subtropical forest. *Soil Biology and Biochemistry*, 149, 107966.

Faye, A., Dalpé, Y., Ndung'u-Magiroi, K., Jefwa, J., Ndoye, I., Diouf, M., and Lesueur, D., 2013. Evaluation of commercial arbuscular mycorrhizal inoculants. *Canadian Journal of Plant Science*, 93(6), 1201–1208.

Fester, T. and Sawers, R., 2011. Progress and challenges in agricultural applications of arbuscular mycorrhizal fungi. *Critical Reviews in Plant Sciences*, 30(5), 459–470.

Fracchia, S., Garcia-Romera, I., Godeas, A., and Ocampo, J.A., 2000. Effect of the saprophytic fungus *Fusarium oxysporum* on arbuscular mycorrhizal colonization and growth of plants in greenhouse and field trials. *Plant and Soil*, 223(1), 177–186.

van Galen, L.G., Stewart, J.D., Qin, C., Corrales, A., Manley, B.F., Kiers, E.T., Crowther, T.W., and Van Nuland, M.E., 2025. Global divergence in plant and mycorrhizal fungal diversity hotspots. *Nature Communications*, 16(1), 6702.

Gao, C., Montoya, L., Xu, L., Madera, M., Hollingsworth, J., Purdom, E., Hutmacher, R.B., Dahlberg, J.A., Coleman-Derr, D., Lemaux, P.G., and Taylor, J.W., 2019. Strong succession in arbuscular mycorrhizal fungal communities. *The ISME Journal*, 13(1), 214–226.

García, I.V. and Mendoza, R.E., 2009. Arbuscular mycorrhizal fungi and plant symbiosis under stress conditions: Ecological implications of drought, flooding and salinity. In: Rai, M. and Bridge, P.D. (Eds.) *Applied Mycology.* CAB International (Wallingford, UK). 17–37.

García de León, D., Davison, J., Moora, M., Öpik, M., Feng, H., Hiiesalu, I., Jairus, T., Koorem, K., Liu, Y., Phosri, C., Sepp, S.-K., Vasar, M., and Zobel, M., 2018. Anthropogenic disturbance equalizes diversity levels in arbuscular mycorrhizal fungal communities. *Global Change Biology*, 24(6), 2649–2659.

Ghorui, M., Chowdhury, S., Balu, P., and Burla, S., 2024. Arbuscular Mycorrhizal inoculants and its regulatory landscape. *Heliyon*, 10(9), e30359.

Gianinazzi, S., Gollotte, A., Binet, M.-N., van Tuinen, D., Redecker, D., and Wipf, D., 2010. Agroecology: The key role of arbuscular mycorrhizas in ecosystem services. *Mycorrhiza*, 20(8), 519–530.

Gosling, P., Hodge, A., Goodlass, G., and Bending, G.D., 2006. Arbuscular mycorrhizal fungi and organic farming. *Agriculture, Ecosystems & Environment*, 113(1), 17–35.

Gosling, P., Mead, A., Proctor, M., Hammond, J.P., and Bending, G.D., 2013. Contrasting arbuscular mycorrhizal communities colonizing different host plants show a similar response to a soil phosphorus concentration gradient. *New Phytologist*, 198(2), 546–556.

Goto, B.T., Queiroz, M., de Souza, F., Magurno, F., and Błaszkowski, J., 2024. How far have we progressed in Glomeromycota taxonomy and systematic? *IMS Newsletter*, 5(2), 18–24.

Grime, J.P., 2006. *Plant Strategies, Vegetation Processes, and Ecosystem Properties*. Chichester: Wiley.

Gutjahr, C., Sawers, R.J.H., Marti, G., Andrés-Hernández, L., Yang, S.-Y., Casieri, L., Angliker, H., Oakeley, E.J., Wolfender, J.-L., Abreu-Goodger, C., and Paszkowski, U., 2015. Transcriptome diversity among rice root types during asymbiosis and interaction with arbuscular mycorrhizal fungi. *Proceedings of the National Academy of Sciences of the United States of America*, 112(21), 6754–6759.

Guzman, A., Montes, M., Hutchins, L., DeLaCerda, G., Yang, P., Kakouridis, A., Dahlquist-Willard, R.M., Firestone, M.K., Bowles, T., and Kremen, C., 2021. Crop diversity enriches arbuscular mycorrhizal fungal communities in an intensive agricultural landscape. *New Phytologist*, 231(1), 447–459.

Han, Y., Feng, J., Han, M., and Zhu, B., 2020. Responses of arbuscular mycorrhizal fungi to nitrogen addition: A meta-analysis. *Global Change Biology*, 26(12), 7229–7241.

Hart, M.M., Antunes, P.M., Chaudhary, V.B., and Abbott, L.K., 2018. Fungal inoculants in the field: Is the reward greater than the risk? *Functional Ecology*, 32(1), 126–135.

Hart, M.M., Reader, R.J., and Klironomos, J.N., 2003. Plant coexistence mediated by arbuscular mycorrhizal fungi. *Trends in Ecology & Evolution*, 18(8), 418–423.

Hart, M.M. and Trevors, J.T., 2005. Microbe management: Application of mycorrhyzal fungi in sustainable agriculture. *Frontiers in Ecology and the Environment*, 3(10), 533–539.

Hassani, M.A., Durán, P., and Hacquard, S., 2018. Microbial interactions within the plant holobiont. *Microbiome*, 6(1), 58.

Hawkes, C.V., Belnap, J., D'Antonio, C., and Firestone, M.K., 2006. Arbuscular mycorrhizal assemblages in native plant roots change in the presence of invasive exotic grasses. *Plant and Soil*, 281(1), 369–380.

van der Heijden, M.G.A., Bardgett, R.D., and van Straalen, N.M., 2008. The unseen majority: Soil microbes as drivers of plant diversity and productivity in terrestrial ecosystems. *Ecology Letters*, 11(3), 296–310.

van der Heijden, M.G.A., Boller, T., Wiemken, A., and Sanders, I.R., 1998. Different arbuscular mycorrhizal fungal species are potential determinants of plant community structure. *Ecology*, 79(6), 2082–2091.

van der Heijden, M.G.A., Klironomos, J.N., Ursic, M., Moutoglis, P., Streitwolf-Engel, R., Boller, T., Wiemken, A., and Sanders, I.R., 1998. Mycorrhizal fungal diversity determines plant biodiversity, ecosystem variability and productivity. *Nature*, 396(6706), 69–72.

van der Heijden, M.G.A., Martin, F.M., Selosse, M.-A., and Sanders, I.R., 2015. Mycorrhizal ecology and evolution: The past, the present, and the future. *New Phytologist*, 205(4), 1406–1423.

Helander, M., Saloniemi, I., Omacini, M., Druille, M., Salminen, J.-P., and Saikkonen, K., 2018. Glyphosate decreases mycorrhizal colonization and affects plant-soil feedback. *Science of the Total Environment*, 642, 285–291.

Heuck, M.K., Powell, J.R., Kath, J., Birnbaum, C., and Frew, A., 2024. Evaluating the usefulness of the C-S-R framework for understanding AM fungal responses to climate change in agroecosystems. *Global Change Biology*, 30(11), e17566.

Hewitt, J.E., Thrush, S.F., and Lundquist, C., 2017. Scale-dependence in ecological systems. *In*: *Encyclopedia of Life Sciences (eLS)*. John Wiley & Sons, Ltd (Chichester), 1–7.

Hoang, D.T.T., Rashtbari, M., Anh, L.T., Wang, S., Tu, D.T., Hiep, N.V., and Razavi, B.S., 2022. Mutualistic interaction between arbuscular mycorrhiza fungi and soybean roots enhances drought resistant through regulating glucose exudation and rhizosphere expansion. *Soil Biology and Biochemistry*, 171, 108728.

Hong, P., Schmid, B., De Laender, F., Eisenhauer, N., Zhang, X., Chen, H., Craven, D., De Boeck, H.J., Hautier, Y., Petchey, O.L., Reich, P.B., Steudel, B., Striebel, M., Thakur, M.P., and Wang, S., 2022. Biodiversity promotes ecosystem functioning despite environmental change. *Ecology Letters*, 25(2), 555–569.

Hu, H., He, L., Ma, H., Wang, J., Li, Y., Wang, J., Guo, Y., Ren, C., Bai, H., and Zhao, F., 2022. Responses of AM fungal abundance to the drivers of global climate change: A meta-analysis. *Science of the Total Environment*, 805, 150362.

Hyde, K.D., Noorabadi, M.T., Thiyagaraja, V., He, M.Q., Johnston, P.R., Wijesinghe, S.N., Armand, A., Biketova, A.Y., Chethana, K.W.T., Erdoğdu, M., Ge, Z.W., Groenewald, J.Z., Hongsanan, S., Kušan, I., Leontyev, D.V., Li, D.W., Lin, C.G., Liu, N.G., Maharachchikumbura, S.S.N., Matočec, N., May, T.W., McKenzie, E.H.C., Mešić, A., Perera, R.H., Phukhamsakda, C., Piątek, M., Samarakoon, M.C., Selcuk, F., Senanayake, I.C., Tanney, J.B., Tian, Q., Vizzini, A., Wanasinghe, D.N., Wannasawang, N., Wijayawardene, N.N., Zhao, R.L., Abdel-Wahab, M.A., Abdollahzadeh, J., Abeywickrama, P.D., Absalan, S., Acharya, K., Afshari, N., Afshan, N.S., Afzalinia, S., Ahmadpour, S.A., Akulov, O., Alizadeh, A., Alizadeh, M., Al-Sadi, A.M., Alves, A., Alves, V.C.S., Alves-Silva, G., Antonín, V., Aouali, S., Aptroot, A., Apurillo, C.C.S., Arias, R.M., Asgari, B., Asghari, R., Assis, D.M.A., Assyov, B., Atienza, V., Aumentado, H.D.R., Avasthi, S., Azevedo, E., Bakhshi, M., Bao, D.F., Baral, H.O., Barata, M., Barbosa, K.D., Barbosa, R.N., Barbosa, F.R., Baroncelli, R., Barreto, G.G., Baschien, C., Bennett, R.M., Bera, I., Bezerra, J.D.P., Bhunjun, C.S., Bianchinotti, M.V., Błaszkowski, J., Boekhout, T., Bonito, G.M., Boonmee, S., Boonyuen, N., Bortnikov, F.M., Bregant, C., Bundhun, D., Burgaud, G., Buyck, B., Caeiro, M.F., Cabarroi-Hernández, M., Feng, C.M., Cai, L., Calabon, M.S., Calaça, F.J.S., Callalli, M., Câmara, M.P.S., Cano-Lira, J., Cao, B., Carlavilla, J.R., Carvalho, A., Carvalho, T.G., Castañeda-Ruiz, R.F., Catania, M.D.V., Cazabonne, J., Cedeño-Sanchez, M., Chaharmiri-Dokhaharani, S., Chaiwan, N., Chakraborty, N., Cheewankoon, R., Chen, C., Chen, J., Chen, Q., Chen, Y.P., Chinaglia,

S., Coelho-Nascimento, C.C., Coleine, C., Costa-Rezende, D.H., Cortés-Pérez, A., Crouch, J.A., Crous, P.W., Cruz, R.H.S.F., Czachura, P., Damm, U., Darmostuk, V., Daroodi, Z., Das, K., Das, K., Davoodian, N., Davydov, E.A., daSilva, G.A., daSilva, I.R., daSilva, R.M.F., daSilvaSantos, A.C., Dai, D.Q., Dai, Y.C., deGrootMichiel, D., DeKesel, A., DeLange, R., deMedeiros, E.V., deSouza, C.F.A., deSouza, F.A., delaCruz, T.E.E., Decock, C., Delgado, G., Denchev, C.M., Denchev, T.T., Deng, Y.L., Dentinger, B.T.M., Devadatha, B., Dianese, J.C., Dima, B., Doilom, M., Dissanayake, A.J., Dissanayake, L.S., Diniz, A.G., Dolatabadi, S., Dong, J.H., Dong, W., Dong, Z.Y., Drechsler-Santos, E.R., Druzhinina, I.S., Du, T.Y., Dubey, M.K., Dutta, A.K., Elliott, T.F., Elshahed, M.S., Egidi, E., Eisvand, P., Fan, L., Fan, X., Fan, X.L., Fedosova, A.G., Ferro, L.O., Fiuza, P.O., Flakus, A., Fonseca, E.O., Fryar, S.C., Gabaldón, T., Gajanayake, A.J., Gannibal, P.B., Gao, F., García-Sánchez, D., García-Sandoval, R., Garrido-Benavent, I., Garzoli, L., Gasca-Pineda, J., Gautam, A.K., Gené, J., Ghobad-Nejhad, M., Ghosh, A., Giachini, A.J., Gibertoni, T.B., Gentekaki, E., Gmoshinskiy, V.I., Góes-Neto, A., Gomdola, D., Gorjón, S.P., Goto, B.T., Granados-Montero, M.M., Griffith, G.W., Groenewald, M., Grossart, H.P., Gu, Z.R., Gueidan, C., Gunarathne, A., Gunaseelan, S., Guo, S.L., Gusmão, L.F.P., Gutierrez, A.C., Guzmán-Dávalos, L., Haelewaters, D., Haituk, H., Halling, R.E., He, S.C., Heredia, G., Hernández-Restrepo, M., Hosoya, T., Hoog, S.D., Horak, E., Hou, C.L., Houbraken, J., Htet, Z.H., Huang, S.K., Huang, W.J., Hurdeal, V.G., Hustad, V.P., Inácio, C.A., Janik, P., Jayalal, R.G.U., Jayasiri, S.C., Jayawardena, R.S., Jeewon, R., Jerônimo, G.H., Jin, J., Jones, E.B.G., Joshi, Y., Jurjević, Justo, A., Kakishima, M., Kaliyaperumal, M., Kang, G.P., Kang, J.C., Karimi, O., Karunarathna, S.C., Karpov, S.A., Kezo, K., Khalid, A.N., Khan, M.K., Khuna, S., Khyaju, S., Kirchmair, M., Klawonn, I., Kraisitudomsook, N., Kukwa, M., Kularathnage, N.D., Kumar, S., Lachance, M.A., Lado, C., Latha, K.P.D., Lee, H.B., Leonardi, M., Lestari, A.S., Li, C., Li, H., Li, J., Li, Q., Li, Y., Li, Y.C., Li, Y.X., Liao, C.F., Lima, J.L.R., Lima, J.M.S., Lima, N.B., Lin, L., Linaldeddu, B.T., Linn, M.M., Liu, F., Liu, J.K., Liu, J.W., Liu, S., Liu, S.L., Liu, X.F., Liu, X.Y., Longcore, J.E., Luangharn, T., Luangsa-ard, J.J., Lu, L., Lu, Y.Z., Lumbsch, H.T., Luo, L., Luo, M., Luo, Z.L., Ma, J., Madagammana, A.D., Madhushan, A., Madrid, H., Magurno, F., Magyar, D., Mahadevakumar, S., Malosso, E., Malysh, J.M., Mamarabadi, M., Manawasinghe, I.S., Manfrino, R.G., Manimohan, P., Mao, N., Mapook, A., Marchese, P., Marasinghe, D.S., Mardones, M., Marin-Felix, Y., Masigol, H., Mehrabi, M., Mehrabi-Koushki, M., de, M.O.A., Melo, R.F.R., Mendes-Alvarenga, R.L., Mendieta, S., Meng, Q.F., Menkis, A., Menolli, N., Mikšík, M., Miller, S.L., Moncada, B., Moncalvo, J.M., Monteiro, J.S., Monteiro, M., Mora-Montes, H.M., Moroz, E.L., Moura, J.C., Muhammad, U., Mukhopadhyay, S., Nagy, G.L., NajamulSehar, A., Najafiniya, M., Nanayakkara, C.M., Naseer, A., Nascimento, E.C.R., Nascimento, S.S., Neuhauser, S., Neves, M.A., Niazi, A.R., Yong, N., Nilsson, R.H., Nogueira, P.T.S., Novozhilov, Y.K., Noordeloos, M., Norphanphoun, C., NuñezOtaño, N., O'Donnell, R.P., Oehl, F., Oliveira, J.A., Oliveira, I., Oliveira, N.V.L., Oliveira, P.H.F., Orihara, T., Oset, M., Pang, K.L., Papp, V., Pathirana, L.S., Peintner, U., Pem, D., Pereira, O.L., Pérez-Moreno, J., Pérez-Ortega, S., Péter, G., Pires-Zottarelli, C.L.A., Phonemany, M., Phongeun, S., Pošta, A., Prazeres, J.F.S.A., Quan, Y., Quandt, C.A., Queiroz, M.B., Radek, R., Rahnama, K., Raj, K.N.A., Rajeshkumar, K.C., Soumyadeep, R., Ralaiveloarisoa, A.B., Rämä, T., Ramírez-Cruz, V., Rambold, G., Rathnayaka, A.R., Raza, M., Ren, G.C., Rinaldi, A.C., Rivas-Ferreiro, M., Robledo, G.L., Ronikier, A., Rossi, W., Rusevska, K., Ryberg, M., Safi, A., Salimi, F., Salvador-Montoya, C.A., Samant, B., Samaradiwakara, N.P., Sánchez-Castro, I., Sandoval-Denis, M., Santiago, A.L.C.M.A., Santos, A.C.D.S., dos Santos, L.A., Sarma, V.V., Sarwar, S., Savchenko, A., Savchenko, K., Saxena, R.K., Schoutteten, N., Selbmann, L., Ševčíková, H., Sharma, A., Shen, H.W., Shen, Y.M., Shu, Y.X., Silva, H.F., Silva-Filho, A.G.S., Silva, V.S.H., Simmons, D.R., Singh, R., Sir, E.B., Sohrabi, M., Souza, F.A., Souza-Motta, C.M., Sri-indrasutdhi, V., Sruthi, O.P., Stadler, M., Stemler, J., Stephenson, S.L., Stoyneva-Gaertner, M.P., Strassert, J.F.H., Stryjak-Bogacka, M., Su, H., Sun, Y.R., Svantesson, S., Sysouphanthong, P., Takamatsu, S., Tan, T.H., Tanaka, K., Tang, C., Tang, X., Taylor, J.E., Taylor, P.W.J., Tennakoon, D.S., Thakshila, S.A.D., Thambugala, K.M., Thamodini, G.K., Thilanga, D., Thines, M., Tiago, P.V., Tian, X.G., Tian, W.H., Tibpromma, S., Tkalčec, Z., Tokarev, Y.S., Tomšovský, M., Torruella, G., Tsurykau, A., Udayanga, D., Ulukapı, M., Untereiner, W.A., Usman, M., Uzunov, B.A., Vadthanarat, S., Valenzuela, R., VandenWyngaert, S., VanVooren, N., Velez, P., Verma, R.K., Vieira, L.C., Vieira, W.A.S., Vinzelj, J.M., Tang, A.M.C., Walker, A., Walker, A.K., Wang, Q.M., Wang, Y., Wang, X.Y., Wang, Z.Y., Wannathes, N., Wartchow, F., Weerakoon, G., Wei, D.P., Wei, X., White, J.F., Wijesundara, D.S.A., Wisitrassameewong, K., Worobiec, G., Wu, H.X., Wu, N., Xiong, Y.R., Xu, B., Xu, J.P., Xu, R., Xu, R.F., Xu, R.J., Yadav, S., Yakovchenko, L.S., Yang, H.D., Yang, X., Yang, Y.H., Yang, Y., Yang, Y.Y., Yoshioka, R., YoussefNoha, H., Yu, F.M., Yu, Z.F., Yuan, L.L., Yuan, Q., Zabin, D.A., Zamora, J.C., Zapata, C.V., Zare, R., Zeng, M., Zeng, X.Y., Zhang, J.F., Zhang, J.Y., Zhang, S.,

Zhang, X.C., Zhao, C.L., Zhao, H., Zhao, H.J., Zhao, Q., Zhou, H.M., Zhu, X.Y., Zmitrovich, I.V., Zucconi, L., and Zvyagina, E., 2024. The 2024 outline of *Fungi* and fungus-like taxa. *Mycosphere*, 15(1), 5146–6239.

Ijdo, M., Cranenbrouck, S., and Declerck, S., 2011. Methods for large-scale production of AM fungi: Past, present, and future. *Mycorrhiza*, 21(1), 1–16.

Janoušková, M. and Jansa, J., 2025. More for less: How mycorrhizal fungal diversity benefits plant communities. *New Phytologist*, 248(2), 438–440.

Jansa, J., Smith, F.A., and Smith, S.E., 2008. Are there benefits of simultaneous root colonization by different arbuscular mycorrhizal fungi? *New Phytologist*, 177(3), 779–789.

Jin, L., Wang, X., Zheng, J., Shi, G., and Jiang, Y., 2025. Deciphering the mycorrhizal nexus: Root economic trade-offs shape arbuscular mycorrhizal fungal assembly dynamics to enhance crop productivity. *Journal of Ecology*, 113, 3225–3237.

Jin, Z., Jiang, F., Wang, L., Declerck, S., Feng, G., and Zhang, L., 2024. Arbuscular mycorrhizal fungi and *Streptomyces*: Brothers in arms to shape the structure and function of the hyphosphere microbiome in the early stage of interaction. *Microbiome*, 12(1), 83.

Johnson, N.C., Graham, J.-H., and Smith, F.A., 1997. Functioning of mycorrhizal associations along the mutualism–parasitism continuum. *New Phytologist*, 135(4), 575–585.

Jones, M.D. and Smith, S.E., 2004. Exploring functional definitions of mycorrhizas: Are mycorrhizas always mutualisms? *Canadian Journal of Botany*, 82(8), 1089–1109.

Kapulnik, Y. and Koltai, H., 2014. Strigolactone involvement in root development, response to abiotic stress, and interactions with the biotic soil environment. *Plant Physiology*, 166(2), 560–569.

Khade, S.W. and Adholeya, A., 2009. Arbuscular mycorrhizal association in plants growing on metal-contaminated and noncontaminated soils adjoining Kanpur Tanneries, Uttar Pradesh, India. *Water, Air, and Soil Pollution*, 202(1), 45–56.

Khan, A.G., 2006. Mycorrhizoremediation: An enhanced form of phytoremediation. *Journal of Zhejiang University Science B*, 7(7), 503–514.

Kier, G., Kreft, H., Lee, T.M., Jetz, W., Ibisch, P.L., Nowicki, C., Mutke, J., and Barthlott, W., 2009. A global assessment of endemism and species richness across island and mainland regions. *Proceedings of the National Academy of Sciences of the United States of America*, 106(23), 9322–9327.

Kiers, E.T., Duhamel, M., Beesetty, Y., Mensah, J.A., Franken, O., Verbruggen, E., Fellbaum, C.R., Kowalchuk, G.A., Hart, M.M., Bago, A., Palmer, T.M., West, S.A., Vandenkoornhuyse, P., Jansa, J., and Bücking, H., 2011. Reciprocal rewards stabilize cooperation in the mycorrhizal symbiosis. *Science*, 333(6044), 880–882.

Kivlin, S.N., Hawkes, C.V., and Treseder, K.K., 2011. Global diversity and distribution of arbuscular mycorrhizal fungi. *Soil Biology and Biochemistry*, 43(11), 2294–2303.

Kjøller, R. and Rosendahl, S., 2000. Effects of fungicides on arbuscular mycorrhizal fungi: Differential responses in alkaline phosphatase activity of external and internal hyphae. *Biology and Fertility of Soils*, 31(5), 361–365.

Klironomos, J.N., 2003. Variation in plant response to native and exotic arbuscular mycorrhizal fungi. *Ecology*, 84(9), 2292–2301.

Klironomos, J.N., McCune, J., Hart, M., and Neville, J., 2000. The influence of arbuscular mycorrhizae on the relationship between plant diversity and productivity. *Ecology Letters*, 3(2), 137–141.

Kohler, A., Kuo, A., Nagy, L.G., Morin, E., Barry, K.W., Buscot, F., Canbäck, B., Choi, C., Cichocki, N., Clum, A., Colpaert, J., Copeland, A., Costa, M.D., Doré, J., Floudas, D., Gay, G., Girlanda, M., Henrissat, B., Herrmann, S., Hess, J., Högberg, N., Johansson, T., Khouja, H.-R., LaButti, K., Lahrmann, U., Levasseur, A., Lindquist, E.A., Lipzen, A., Marmeisse, R., Martino, E., Murat, C., Ngan, C.Y., Nehls, U., Plett, J.M., Pringle, A., Ohm, R.A., Perotto, S., Peter, M., Riley, R., Rineau, F., Ruytinx, J., Salamov, A., Shah, F., Sun, H., Tarkka, M., Tritt, A., Veneault-Fourrey, C., Zuccaro, A., Tunlid, A., Grigoriev, I.V., Hibbett, D.S., and Martin, F., 2015. Convergent losses of decay mechanisms and rapid turnover of symbiosis genes in mycorrhizal mutualists. *Nature Genetics*, 47(4), 410–415.

Koyama, A., Maherali, H., and Antunes, P.M., 2019. Plant geographic origin and phylogeny as potential drivers of community structure in root-inhabiting fungi. *Journal of Ecology*, 107(4), 1720–1736.

Koziol, L., McKenna, T.P., and Bever, J.D., 2025. Meta-analysis reveals globally sourced commercial mycorrhizal inoculants fall short. *New Phytologist*, 246(3), 821–827.

Lalaymia, I., Cranenbrouck, S., and Declerck, S., 2014. Maintenance and preservation of ectomycorrhizal and arbuscular mycorrhizal fungi. *Mycorrhiza*, 24(5), 323–337.

Lanfranco, L., Fiorilli, V., Venice, F., and Bonfante, P., 2018. Strigolactones cross the kingdoms: Plants, fungi, and bacteria in the arbuscular mycorrhizal symbiosis. *Journal of Experimental Botany*, 69(9), 2175–2188.

Lang, A.K., LaRue, E.A., Kivlin, S.N., Edwards, J.D., Phillips, R.P., Gallion, J., Kong, N., Parker, J.D., McCormick, M.K., Domke, G., and Fei, S., 2023. Forest structural diversity is linked to soil microbial diversity. *Ecosphere*, 14(11), e4702.

Latham, J., Cumani, R., Rosati, I., and Bloise, M., 2014. *Global Land Cover (GLC-SHARE) Beta-Release 1.0 Database*. Food and Agriculture Organization of the United Nations, Land and Water Division: Rome, Italy.

Leake, J.R., Donnelly, D.P., and Boddy, L., 2003. Interactions between ecto-mycorrhizal and saprotrophic ungi. *In*: M.G.A. van der Heijden and I.R. Sanders, eds. *Mycorrhizal Ecology*. Berlin, Heidelberg: Springer, 345–372.

Li, A., Meidl, P., Wang, S., Tang, B., Rillig, M.C., Yu, G., Chen, J., Liu, R., Lie, Z., Wu, A., Rong, L., Peng, C., Liu, Z., Zhang, W., Lu, X., Liu, J., Ye, Q., Mo, J., and Zheng, M., 2025. Atmospheric nitrogen deposition has minor impacts on the abundance and diversity of arbuscular mycorrhizal fungi and their contribution to soil carbon stock in tropical forests. *Soil Biology and Biochemistry*, 204, 109746.

Li, P., Yang, G., Zhou, X., Zhou, L., Shao, A., and Chen, M., 2015. Effect of five fungicides on growth of *Glycyrrhiza uralensis* and efficiency of mycorrhizal symbiosis. *China Journal of Chinese Materia Medica*, 40(23), 4591–4596.

Lutz, S., Bodenhausen, N., Hess, J., Valzano-Held, A., Waelchli, J., Deslandes-Hérold, G., Schlaeppi, K., and van der Heijden, M.G.A., 2023. Soil microbiome indicators can predict crop growth response to large-scale inoculation with arbuscular mycorrhizal fungi. *Nature Microbiology*, 8(12), 2277–2289.

Lutz, S., Mikryukov, V., Labouyrie, M., Bahram, M., Jones, A., Panagos, P., Delgado-Baquerizo, M., Maestre, F.T., Orgiazzi, A., Tedersoo, L., and van der Heijden, M.G.A., 2025. Global richness of arbuscular mycorrhizal fungi. *Fungal Ecology*, 74, 101407.

Ma, X., Geng, Q., Zhang, H., Bian, C., Chen, H.Y.H., Jiang, D., and Xu, X., 2021. Global negative effects of nutrient enrichment on arbuscular mycorrhizal fungi, plant diversity and ecosystem multifunctionality. *New Phytologist*, 229(5), 2957–2969.

Ma, X., Xu, X., Geng, Q., Luo, Y., Ju, C., Li, Q., and Zhou, Y., 2023. Global arbuscular mycorrhizal fungal diversity and abundance decreases with soil available phosphorus. *Global Ecology and Biogeography*, 32(8), 1423–1434.

Ma, X., Zhu, B., Nie, Y., Liu, Y., and Kuzyakov, Y., 2021. Root and mycorrhizal strategies for nutrient acquisition in forests under nitrogen deposition: A meta-analysis. *Soil Biology and Biochemistry*, 163, 108418.

Maciá-Vicente, J.G. and Popa, F., 2021. Local endemism and ecological generalism in the assembly of root-colonizing fungi. *Ecological Monographs*, 92(1), e01489.

Maherali, H. and Klironomos, J.N., 2007. Influence of phylogeny on fungal community assembly and ecosystem functioning. *Science*, 316(5832), 1746–1748.

Manoharan, L., Rosenstock, N.P., Williams, A., and Hedlund, K., 2017. Agricultural management practices influence AMF diversity and community composition with cascading effects on plant productivity. *Applied Soil Ecology*, 115, 53–59.

Mao, H., Whalen, J.K., Zhang, Z., Sheng, X., Hu, G., Chen, B., and Ma, M., 2024. Arbuscular mycorrhizal fungi associated with alpine meadow multifunctionality in a warmer climate with variable precipitation. *Soil Biology and Biochemistry*, 198, 109555.

Martin, F.M. and van der Heijden, M.G.A., 2024. The mycorrhizal symbiosis: Research frontiers in genomics, ecology, and agricultural application. *New Phytologist*, 242(4), 1486–1506.

Martin, F.M., Uroz, S., and Barker, D.G., 2017. Ancestral alliances: Plant mutualistic symbioses with fungi and bacteria. *Science*, 356(6340), eaad4501.

Mason, R.E., Craine, J.M., Lany, N.K., Jonard, M., Ollinger, S.V., Groffman, P.M., Fulweiler, R.W., Angerer, J., Read, Q.D., Reich, P.B., Templer, P.H., and Elmore, A.J., 2022. Evidence, causes, and consequences of declining nitrogen availability in terrestrial ecosystems. *Science*, 376(6590), eabh3767.

Monroe, A.P., Heinrichs, J.A., Whipple, A.L., O'Donnell, M.S., Edmunds, D.R., and Aldridge, C.L., 2022. Spatial scale selection for informing species conservation in a changing landscape. *Ecosphere*, 13(12), e4320.

Morrison, E., Lagos, L., Al-Agely, A., Glaab, H., Johnson, W., Jorquera, M.A., and Ogram, A., 2017. Mycorrhizal inoculation increases genes associated with nitrification and improved nutrient retention in soil. *Biology and Fertility of Soils*, 53(3), 275–279.

Munkvold, L., Kjøller, R., Vestberg, M., Rosendahl, S., and Jakobsen, I., 2004. High functional diversity within species of arbuscular mycorrhizal fungi. *New Phytologist*, 164(2), 357–364.

Nayak, S.K., Baliyarsingh, B., Singh, A., Mannazzu, I., and Mishra, B.B., eds., 2022. *Advances in Agricultural and Industrial Microbiology: Volume-2: Applications of Microbes for Sustainable Agriculture and In-Silico Strategies*. Singapore: Springer.

Oberhuber, W., Swidrak, I., Pirkebner, D., and Gruber, A., 2011. Temporal dynamics of nonstructural carbohydrates and xylem growth in Pinus sylvestris exposed to drought. *Canadian Journal of Forest Research*, 41(8), 1590–1597.

Oehl, F., Laczko, E., Bogenrieder, A., Stahr, K., Bösch, R., van der Heijden, M., and Sieverding, E., 2010. Soil type and land use intensity determine the composition of arbuscular mycorrhizal fungal communities. *Soil Biology and Biochemistry*, 42(5), 724–738.

Olsson, P.A., Jakobsen, I., and Wallander, H., 2003. Foraging and resource allocation strategies of mycorrhizal fungi in a patchy environment. *In*: M.G.A. van der Heijden and I.R. Sanders, eds. *Mycorrhizal Ecology*. Berlin, Heidelberg: Springer, 93–115.

Öpik, M., Davison, J., Moora, M., Pärtel, M., and Zobel, M., 2016. Response to comment on "Global assessment of arbuscular mycorrhizal fungus diversity reveals very low endemism". *Science*, 351(6275), 826–826.

Öpik, M., Metsis, M., Daniell, T.J., Zobel, M., and Moora, M., 2009. Large-scale parallel 454 sequencing reveals host ecological group specificity of arbuscular mycorrhizal fungi in a boreonemoral forest. *New Phytologist*, 184(2), 424–437.

Ortaş, I., Rafique, M., and Ahmed, İ.A.M., 2017. Application of arbuscular mycorrhizal fungi into agriculture. *In*: Q.-S. Wu, ed. *Arbuscular Mycorrhizas and Stress Tolerance of Plants*. Singapore: Springer, 305–327.

Pandit, A., Gunsola, D., Apaarna, Kumar, R., Panneerselvam, P., and Mitra, D., 2025. Advancing innovative techniques in arbuscular mycorrhizal fungi propagation: A key to sustainable agriculture and ecosystem management. *Journal of Microbiological Methods*, 236, 107200.

Peršoh, D., 2015. Plant-associated fungal communities in the light of meta'omics. *Fungal Diversity*, 75(1), 1–25.

Poiani, K.A., Richter, B.D., Anderson, M.G., and Richter, H.E., 2000. Biodiversity conservation at multiple scales: Functional sites, landscapes, and networks. *BioScience*, 50(2), 133–146.

Powell, J.R. and Rillig, M.C., 2018. Biodiversity of arbuscular mycorrhizal fungi and ecosystem function. *New Phytologist*, 220(4), 1059–1075.

Prescott, C.E., Grayston, S.J., Helmisaari, H.-S., Kaštovská, E., Körner, C., Lambers, H., Meier, I.C., Millard, P., and Ostonen, I., 2020. Surplus carbon drives allocation and plant–soil interactions. *Trends in Ecology & Evolution*, 35(12), 1110–1118.

Querejeta, J.I., Allen, M.F., Caravaca, F., and Roldán, A., 2006. Differential modulation of host plant $\delta^{13}C$ and $\delta^{18}O$ by native and nonnative arbuscular mycorrhizal fungi in a semiarid environment. *New Phytologist*, 169(2), 379–387.

Quiroga, G., Castagneyrol, B., Abdala-Roberts, L., and Moreira, X., 2024. A meta-analysis of the effects of climate change-related abiotic factors on aboveground and belowground plant-associated microbes. *Oikos*, 2024(7), e10411.

Ramana, J.V., Tylianakis, J.M., Allen, W.J., Ridgway, H.J., Waller, L.P., Orwin, K.H., and Dickie, I.A., 2025. Invasive plants decrease arbuscular mycorrhizal fungal diversity and promote generalist fungal partners. *New Phytologist*, 247(5), 2381–2389.

Ramana, J.V., Tylianakis, J.M., Ridgway, H.J., and Dickie, I.A., 2023. Root diameter, host specificity and arbuscular mycorrhizal fungal community composition among native and exotic plant species. *New Phytologist*, 239(1), 301–310.

Remy, W., Taylor, T.N., Hass, H., and Kerp, H., 1994. Four hundred-million-year-old vesicular arbuscular mycorrhizae. *Proceedings of the National Academy of Sciences of the United States of America*, 91(25), 11841–11843.

Riaz, M., Kamran, M., Fang, Y., Wang, Q., Cao, H., Yang, G., Deng, L., Wang, Y., Zhou, Y., Anastopoulos, I., and Wang, X., 2021. Arbuscular mycorrhizal fungi-induced mitigation of heavy metal phytotoxicity in metal contaminated soils: A critical review. *Journal of Hazardous Materials*, 402, 123919.

Rillig, M.C. and Mummey, D.L., 2006. Mycorrhizas and soil structure. *New Phytologist*, 171(1), 41–53.

Rillig, M.C., Wright, S.F., Kimball, B.A., Pinter, P.J., Wall, G.W., Ottman, M.J., and Leavitt, S.W., 2001. Elevated carbon dioxide and irrigation effects on water stable aggregates in a Sorghum field: A possible role for arbuscular mycorrhizal fungi. *Global Change Biology*, 7(3), 333–337.

Rosendahl, S., Mcgee, P., and Morton, J.B., 2009. Lack of global population genetic differentiation in the arbuscular mycorrhizal fungus *Glomus mosseae* suggests a recent range expansion which may have coincided with the spread of agriculture. *Molecular Ecology*, 18(20), 4316–4329.

Rowe, A.R. and Pringle, A., 2005. Morphological and molecular evidence of arbuscular mycorrhizal fungal associations in Costa Rican epiphytic bromeliads. *Biotropica*, 37(2), 245–250.

Runnel, K., Tedersoo, L., Krah, F.-S., Piepenbring, M., Scheepens, J.F., Hollert, H., Johann, S., Meyer, N., and Bässler, C., 2025. Toward harnessing biodiversity–ecosystem function relationships in fungi. *Trends in Ecology & Evolution*, 40(2), 180–190.

Ruotsalainen, A.L., Markkola, A., and Kozlov, M.V., 2007. Root fungal colonisation in *Deschampsia flexuosa*: Effects of pollution and neighbouring trees. *Environmental Pollution*, 147(3), 723–728.

Salomon, M.J., Demarmels, R., Watts-Williams, S.J., McLaughlin, M.J., Kafle, A., Ketelsen, C., Soupir, A., Bücking, H., Cavagnaro, T.R., and van der Heijden, M.G.A., 2022. Global evaluation of commercial arbuscular mycorrhizal inoculants under greenhouse and field conditions. *Applied Soil Ecology*, 169, 104225.

Salomon, M.J., Watts-Williams, S.J., McLaughlin, M.J., Bücking, H., Singh, B.K., Hutter, I., Schneider, C., Martin, F.M., Vosatka, M., Guo, L., Ezawa, T., Saito, M., Declerck, S., Zhu, Y.-G., Bowles, T., Abbott, L.K., Smith, F.A., Cavagnaro, T.R., and van der Heijden, M.G.A., 2022. Establishing a quality management framework for commercial inoculants containing arbuscular mycorrhizal fungi. *iScience*, 25(7), 104636.

Sanders, I.R., Streitwolf-Engel, R., van der Heijden, M.G.A., Boller, T., and Wiemken, A., 1998. Increased allocation to external hyphae of arbuscular mycorrhizal fungi under CO_2 enrichment. *Oecologia*, 117(4), 496–503.

Savin, M.C., Purcell, L.C., Daigh, A., and Manfredini, A., 2009. Response of mycorrhizal infection to glyphosate applications and P fertilization in glyphosate-tolerant soybean, maize, and cotton. *Journal of Plant Nutrition*, 32(10), 1702–1717.

Schwartz, M.W. and Hoeksema, J.D., 1998. Specialization and resource trade: Biological markets as a model of mutualisms. *Ecology*, 79(3), 1029–1038.

Schwartz, M.W., Hoeksema, J.D., Gehring, C.A., Johnson, N.C., Klironomos, J.N., Abbott, L.K., and Pringle, A., 2006. The promise and the potential consequences of the global transport of mycorrhizal fungal inoculum. *Ecology Letters*, 9(5), 501–515.

Sielaff, A.C., Polley, H.W., Fuentes-Ramirez, A., Hofmockel, K., and Wilsey, B.J., 2019. Mycorrhizal colonization and its relationship with plant performance differs between exotic and native grassland plant species. *Biological Invasions*, 21(6), 1981–1991.

Sikes, B.A., Hawkes, C.V., and Fukami, T., 2016. Plant and root endophyte assembly history: Interactive effects on native and exotic plants. *Ecology*, 97(2), 484–493.

da Silva Maia, R., Vasconcelos, S.S., Viana-Junior, A.B., Castellani, D.C., and Kato, O.R., 2021. Oil palm (*Elaeis guineensis*) shows higher mycorrhizal colonization when planted in agroforestry than in monoculture. *Agroforestry Systems*, 95(4), 731–740.

Smith, S.E. and Read, D.J., 2008. *Mycorrhizal Symbiosis*. 3rd edn. London: Academic Press.

Sokornova, S., Malygin, D., Terentev, A., and Dolzhenko, V., 2022. Arbuscular mycorrhiza symbiosis as a factor of *Asteraceae* species invasion. *Agronomy*, 12(12), 3214.

Soliveres, S., van der Plas, F., Manning, P., Prati, D., Gossner, M.M., Renner, S.C., Alt, F., Arndt, H., Baumgartner, V., Binkenstein, J., Birkhofer, K., Blaser, S., Blüthgen, N., Boch, S., Böhm, S., Börschig, C., Buscot, F., Diekötter, T., Heinze, J., Hölzel, N., Jung, K., Klaus, V.H., Kleinebecker, T., Klemmer, S., Krauss, J., Lange, M., Morris, E.K., Müller, J., Oelmann, Y., Overmann, J., Pašalić, E., Rillig, M.C., Schaefer, H.M., Schloter, M., Schmitt, B., Schöning, I., Schrumpf, M., Sikorski, J., Socher, S.A., Solly, E.F., Sonnemann, I., Sorkau, E., Steckel, J., Steffan-Dewenter, I., Stempfhuber, B., Tschapka, M., Türke, M., Venter, P.C., Weiner, C.N., Weisser, W.W., Werner, M., Westphal, C., Wilcke, W., Wolters, V., Wubet, T., Wurst, S., Fischer, M., and Allan, E., 2016. Biodiversity at multiple trophic levels is needed for ecosystem multifunctionality. *Nature*, 536(7617), 456–459.

Staddon, P.L., 2005. Mycorrhizal fungi and environmental change: The need for a mycocentric approach. *New Phytologist*, 167(3), 635–637.

Staddon, P.L., Thompson, K., Jakobsen, I., Grime, J.P., Askew, A.P., and Fitter, A.H., 2003. Mycorrhizal fungal abundance is affected by long-term climatic manipulations in the field. *Global Change Biology*, 9(2), 186–194.

Stewart, J.D., Corrales, A., Canteiro, C., Qin, C., Gupta, M.M., Otgonsuren, B., Peña-Venegas, C.P., Van Nuland, M.E., Kohout, P., Větrovský, T., Kokkoris, V., and Manley, B.F., 2025. Advancing knowledge on the biogeography of arbuscular mycorrhizal fungi to support sustainable development goal 15: Life on land. *FEMS Microbiology Letters*, 372, fnaf055.

Tao, J. and Liu, X., 2024. Does arbuscular mycorrhizal fungi inoculation influence soil carbon sequestration? *Biology and Fertility of Soils*, 60(2), 213–225.

Tedersoo, L., Bahram, M., Põlme, S., Kõljalg, U., Yorou, N.S., Wijesundera, R., Ruiz, L.V., Vasco-Palacios, A.M., Thu, P.Q., Suija, A., Smith, M.E., Sharp, C., Saluveer, E., Saitta, A., Rosas, M., Riit, T., Ratkowsky, D., Pritsch, K., Põldmaa, K., Piepenbring, M., Phosri, C., Peterson, M., Parts, K., Pärtel, K., Otsing, E., Nouhra, E., Njouonkou, A.L., Nilsson, R.H., Morgado, L.N., Mayor, J., May, T.W., Majuakim, L., Lodge, D.J., Lee, S.S., Larsson, K.-H., Kohout, P., Hosaka, K., Hiiesalu, I., Henkel, T.W., Harend, H.,

Guo, L., Greslebin, A., Grelet, G., Geml, J., Gates, G., Dunstan, W., Dunk, C., Drenkhan, R., Dearnaley, J., De Kesel, A., Dang, T., Chen, X., Buegger, F., Brearley, F.Q., Bonito, G., Anslan, S., Abell, S., and Abarenkov, K., 2014. Global diversity and geography of soil fungi. *Science*, 346(6213), 1256688.

Tedersoo, L., Magurno, F., Alkahtani, S., and Mikryukov, V., 2024. Phylogenetic classification of arbuscular mycorrhizal fungi: New species and higher-ranking taxa in Glomeromycota and Mucoromycota (class Endogonomycetes). *MycoKeys*, 107, 273–325.

Tedersoo, L., Mikryukov, V., Zizka, A., Bahram, M., Hagh-Doust, N., Anslan, S., Prylutskyi, O., Delgado-Baquerizo, M., Maestre, F.T., Pärn, J., Öpik, M., Moora, M., Zobel, M., Espenberg, M., Mander, Ü., Khalid, A.N., Corrales, A., Agan, A., Vasco-Palacios, A.-M., Saitta, A., Rinaldi, A.C., Verbeken, A., Sulistyo, B.P., Tamgnoue, B., Furneaux, B., Ritter, C.D., Nyamukondiwa, C., Sharp, C., Marín, C., Gohar, D., Klavina, D., Sharmah, D., Dai, D.Q., Nouhra, E., Biersma, E.M., Rähn, E., Cameron, E.K., De Crop, E., Otsing, E., Davydov, E.A., Albornoz, F.E., Brearley, F.Q., Buegger, F., Zahn, G., Bonito, G., Hiiesalu, I., Barrio, I.C., Heilmann-Clausen, J., Ankuda, J., Kupagme, J.Y., Maciá-Vicente, J.G., Fovo, J.D., Geml, J., Alatalo, J.M., Alvarez-Manjarrez, J., Põldmaa, K., Runnel, K., Adamson, K., Bråthen, K.A., Pritsch, K., Tchan, K.I., Armolaitis, K., Hyde, K.D., Newsham, K.K., Panksep, K., Lateef, A.A., Tiirmann, L., Hansson, L., Lamit, L.J., Saba, M., Tuomi, M., Gryzenhout, M., Bauters, M., Piepenbring, M., Wijayawardene, N., Yorou, N.S., Kurina, O., Mortimer, P.E., Meidl, P., Kohout, P., Nilsson, R.H., Puusepp, R., Drenkhan, R., Garibay-Orijel, R., Godoy, R., Alkahtani, S., Rahimlou, S., Dudov, S.V., Põlme, S., Ghosh, S., Mundra, S., Ahmed, T., Netherway, T., Henkel, T.W., Roslin, T., Nteziryayo, V., Fedosov, V.E., Onipchenko, V.G., Yasanthika, W.A.E., Lim, Y.W., Soudzilovskaia, N.A., Antonelli, A., Kõljalg, U., and Abarenkov, K., 2022. Global patterns in endemicity and vulnerability of soil fungi. *Global Change Biology*, 28(22), 6696–6710.

Tilman, D., Isbell, F., and Cowles, J.M., 2014. Biodiversity and ecosystem functioning. *Annual Review of Ecology, Evolution, and Systematics*, 45, 471–493.

Tisserant, E., Malbreil, M., Kuo, A., Kohler, A., Symeonidi, A., Balestrini, R., Charron, P., Duensing, N., dit Frey, N.F., Gianinazzi-Pearson, V., Gilbert, L.B., Handa, Y., Herr, J.R., Hijri, M., Koul, R., Kawaguchi, M., Krajinski, F., Lammers, P.J., Masclaux, F.G., Murat, C., Morin, E., Ndikumana, S., Pagni, M., Petitpierre, D., Requena, N., Rosikiewicz, P., Riley, R., Saito, K., Clemente, H.S., Shapiro, H., Tuinen, D. van, Bécard, G., Bonfante, P., Paszkowski, U., Shachar-Hill, Y.Y., Tuskan, G.A., Young, J.P.W., Sanders, I.R., Henrissat, B., Rensing, S.A., Grigoriev, I.V., Corradi, N., Roux, C., and Martin, F., 2013. Genome of an arbuscular mycorrhizal fungus provides insight into the oldest plant symbiosis. *Proceedings of the National Academy of Sciences of the United States of America*, 110(50), 20117–20122.

Toju, H., Peay, K.G., Yamamichi, M., Narisawa, K., Hiruma, K., Naito, K., Fukuda, S., Ushio, M., Nakaoka, S., Onoda, Y., Yoshida, K., Schlaeppi, K., Bai, Y., Sugiura, R., Ichihashi, Y., Minamisawa, K., and Kiers, E.T., 2018. Core microbiomes for sustainable agroecosystems. *Nature Plants*, 4(5), 247–257.

Toussaint, A., Bueno, G., Davison, J., Moora, M., Tedersoo, L., Zobel, M., Öpik, M., and Pärtel, M., 2020. Asymmetric patterns of global diversity among plants and mycorrhizal fungi. *Journal of Vegetation Science*, 31(2), 355–366.

Treseder, K.K. and Allen, M.F., 2000. Mycorrhizal fungi have a potential role in soil carbon storage under elevated CO_2 and nitrogen deposition. *New Phytologist*, 147(1), 189–200.

Treseder, K.K. and Cross, A., 2006. Global distributions of arbuscular mycorrhizal fungi. *Ecosystems*, 9(2), 305–316.

Turrini, A., Avio, L., Bedini, S., and Giovannetti, M., 2008. In situ collection of endangered arbuscular mychorrhizal fungi in a Mediterranean UNESCO Biosphere Reserve. *Biodiversity and Conservation*, 17(3), 643–657.

Turrini, A. and Giovannetti, M., 2012. Arbuscular mycorrhizal fungi in national parks, nature reserves and protected areas worldwide: A strategic perspective for their in situ conservation. *Mycorrhiza*, 22(2), 81–97.

Tylianakis, J.M., Didham, R.K., Bascompte, J., and Wardle, D.A., 2008. Global change and species interactions in terrestrial ecosystems. *Ecology Letters*, 11(12), 1351–1363.

Vályi, K., Mardhiah, U., Rillig, M.C., and Hempel, S., 2016. Community assembly and coexistence in communities of arbuscular mycorrhizal fungi. *The ISME Journal*, 10, 2341–2351.

Van Der Heijden, M.G.A. and Scheublin, T.R., 2007. Functional traits in mycorrhizal ecology: their use for predicting the impact of arbuscular mycorrhizal fungal communities on plant growth and ecosystem functioning. *New Phytologist*, 174(2), 244–250.

Van Nuland, M.E., Averill, C., Stewart, J.D., Prylutskyi, O., Corrales, A., van Galen, L.G., Manley, B.F., Qin, C., Lauber, T., Mikryukov, V., Dulia, O., Furci, G., Marín, C., Sheldrake, M., Weedon, J.T., Peay, K.G., Cornwallis, C.K., Větrovský, T., Kohout, P., Baldrian, P., Tedersoo, L., West, S.A., Crowther, T.W., Kiers, E.T., and van den Hoogen, J., 2025. Global hotspots of mycorrhizal fungal richness are poorly protected. *Nature*, 645, 1–9.

Vellend, M., 2016. *The Theory of Ecological Communities*. Princeton, NJ; Oxford, GB: Princeton University Press.

Větrovský, T., Kolaříková, Z., Lepinay, C., Awokunle Hollá, S., Davison, J., Fleyberková, A., Gromyko, A., Jelínková, B., Kolařík, M., Krüger, M., Lejsková, R., Michalčíková, L., Michalová, T., Moora, M., Moravcová, A., Moulíková, Š., Odriozola, I., Öpik, M., Pappová, M., Piché-Choquette, S., Skřivánek, J., Vlk, L., Zobel, M., Baldrian, P., and Kohout, P., 2023. GlobalAMFungi: A global database of arbuscular mycorrhizal fungal occurrences from high-throughput sequencing metabarcoding studies. *New Phytologist*, 240(5), 2151–2163.

Vogelsang, K.M., Reynolds, H.L., and Bever, J.D., 2006. Mycorrhizal fungal identity and richness determine the diversity and productivity of a tallgrass prairie system. *New Phytologist*, 172(3), 554–562.

Vorholt, J.A., 2012. Microbial life in the phyllosphere. *Nature Reviews Microbiology*, 10(12), 828–840.

Wagg, C., Jansa, J., Schmid, B., and van der Heijden, M.G.A., 2011. Belowground biodiversity effects of plant symbionts support aboveground productivity. *Ecology Letters*, 14(10), 1001–1009.

Waller, L.P., Allen, W.J., Barratt, B.I.P., Condron, L.M., França, F.M., Hunt, J.E., Koele, N., Orwin, K.H., Steel, G.S., Tylianakis, J.M., Wakelin, S.A., and Dickie, I.A., 2020. Biotic interactions drive ecosystem responses to exotic plant invaders. *Science*, 368(6494), 967–972.

Wang, L., Zhang, L., George, T.S., and Feng, G., 2022. A core microbiome in the hyphosphere of arbuscular mycorrhizal fungi has functional significance in organic phosphorus mineralization. *New Phytologist*, 238(2), 859–873.

Wang, Y., Bao, X., and Li, S., 2021. Effects of arbuscular mycorrhizal fungi on rice growth under different flooding and shading regimes. *Frontiers in Microbiology*, 12, 756752.

Weber, S.E., Bascompte, J., Kahmen, A., and Niklaus, P.A., 2025. AMF diversity promotes plant community phosphorus acquisition and reduces carbon costs per unit of phosphorus. *New Phytologist*, 248(2), 886–896.

Wilson, G.W.T., Rice, C.W., Rillig, M.C., Springer, A., and Hartnett, D.C., 2009. Soil aggregation and carbon sequestration are tightly correlated with the abundance of arbuscular mycorrhizal fungi: results from long-term field experiments. *Ecology Letters*, 12(5), 452–461.

Wipf, D., Krajinski, F., Tuinen, D. van, Recorbet, G., and Courty, P.-E., 2019. Trading on the arbuscular mycorrhiza market: From arbuscules to common mycorrhizal networks. *New Phytologist*, 223(3), 1127–1142.

Wolfe, B.E., Parrent, J.L., Koch, A.M., Sikes, B.A., Gardes, M., and Klironomos, J.N., 2009. Spatial Heterogeneity in mycorrhizal populations and communities: Scales and mechanisms. *In*: C. Azcón-Aguilar, J.M. Barea, S. Gianinazzi, and V. Gianinazzi-Pearson, eds. *Mycorrhizas - Functional Processes and Ecological Impact*. Berlin, Heidelberg: Springer, 167–185.

Wu, S., Zhang, X., Sun, Y., Wu, Z., Li, T., Hu, Y., Su, D., Lv, J., Li, G., Zhang, Z., Zheng, L., Zhang, J., and Chen, B., 2015. Transformation and immobilization of chromium by arbuscular mycorrhizal fungi as revealed by SEM–EDS, TEM–EDS, and XAFS. *Environmental Science & Technology*, 49(24), 14036–14047.

Wu, Y., Chen, C., and Wang, G., 2024. Inoculation with arbuscular mycorrhizal fungi improves plant biomass and nitrogen and phosphorus nutrients: A meta-analysis. *BMC Plant Biology*, 24(1), 960.

Wu, Z., McGrouther, K., Huang, J., Wu, P., Wu, W., and Wang, H., 2014. Decomposition and the contribution of glomalin-related soil protein (GRSP) in heavy metal sequestration: Field experiment. *Soil Biology and Biochemistry*, 68, 283–290.

Wyatt, G.A.K., Kiers, E.T., Gardner, A., and West, S.A., 2014. A biological market analysis of the plant-mycorrhizal symbiosis. *Evolution*, 68(9), 2603–2618.

Xiao, Y. and Chen, L., 2022. Arbuscular mycorrhizal fungi reduce potassium, cadmium and ammonium losses but increases nitrate loss under high intensity leaching events. *BMC Plant Biology*, 22(1), 365.

Xie, X., Yoneyama, K., and Yoneyama, K., 2010. The strigolactone story. *Annual Review of Phytopathology*, 48, 93–117.

Yang, C., Ellouze, W., Navarro-Borrell, A., Taheri, A.E., Klabi, R., Dai, M., Kabir, Z., and Hamel, C., 2014. Management of the arbuscular mycorrhizal symbiosis in sustainable crop production. *In*: Z.M. Solaiman, L.K. Abbott, and A. Varma, eds. *Mycorrhizal Fungi: Use in Sustainable Agriculture and Land Restoration*. Berlin, Heidelberg: Springer, 89–118.

Yang, S.K., Kim, M.K., Seo, Y.W., Choi, K.J., Lee, S.T., Kwak, Y.-S., and Lee, Y.H., 2012. Soil microbial community analysis of between no-till and tillage in a controlled horticultural field. *World Journal of Microbiology and Biotechnology*, 28(4), 1797–1801.

Zhou, L., Zhou, X., He, Y., Fu, Y., Du, Z., Lu, M., Sun, X., Li, C., Lu, C., Liu, R., Zhou, G., Bai, S.H., and Thakur, M.P., 2022. Global systematic review with meta-analysis shows that warming effects on terrestrial plant biomass allocation are influenced by precipitation and mycorrhizal association. *Nature Communications*, 13(1), 4914.

Zhou, Z., Wang, C., and Luo, Y., 2020. Meta-analysis of the impacts of global change factors on soil microbial diversity and functionality. *Nature Communications*, 11(1), 3072.

Zhu, Q., Riley, W.J., Tang, J., and Bouskill, N.J., 2024. Plant responses to elevated CO_2 under competing hypotheses of nitrogen and phosphorus limitations. *Ecological Applications*, 34(3), e2967.

Ziane, H., Meddad-Hamza, A., Beddiar, A., and Gianinazzi, S., 2017. Effects of arbuscular mycorrhizal fungi and fertilization levels on industrial tomato growth and production. *International Journal of Agriculture and Biology*, 19(02), 341–347.

Zurell, D., Bocedi, G., Velazco, S.J.E., Gonzalez, A., Purvis, A., Wintle, B., Merow, C., Lundquist, C., Guillera-Arroita, G., Settele, J., Serra-Diaz, J.M., Cabral, J.S., Travis, J.M.J., Schifferle, K., Buckley, L., Briscoe, N.J., Isaac, N.J.B., Peres-Neto, P.R., Keuth, R., Gascoigne, S.J.L., Ferrier, S., and Urban, M.C., 2025. Predicting the way forward for the global biodiversity framework. *Proceedings of the National Academy of Sciences of the United States of America*, 122(41), e2501695122.

Part V

Biodiversity in Urban Environments

18 Conserving Wildlife Biodiversity in Urban Ecosystems

Ruijia Hu and Susanna T.Y. Tong

18.1 INTRODUCTION

18.1.1 Urban Ecosystems in the Anthropocene: From Ecological Degradation to Conservation Challenges and Opportunities

During the present Anthropocene era, the Earth's ecosystem has been transformed at an unprecedented rate through extensive land-use change, overexploitation of natural resources, alteration of habitat and species composition, disruption of hydrological processes, and modification of energy flow and nutrient cycles (Alberti et al., 2003). Among all these drivers, urbanisation is the most prominent one, contributing significantly to declines in species diversity and ecosystem integrity (Pickett et al., 2011). Suppose the current trends in population growth and urban expansion persist, in that case, the extent of urban land globally is projected to increase by approximately 1.2 million km^2 by 2030, nearly tripling the urban area in 2000 (Seto et al., 2012), with much of this growth expected in biodiversity hotspots previously undisturbed by development at the start of the century (G. Li et al., 2022).

The definitions of "urban" vary but generally involve high population density, extensive built infrastructure, and altered biophysical processes (Pickett et al., 2011). Urban ecosystems extend beyond dense urban cores to include suburban and peri-urban areas that are functionally connected through commuting networks, utilities, or shared services (Pickett et al., 2001). Unlike rural ecosystems, which tend to exhibit more contiguous natural landscapes, urban regions have a fragmented mosaic of natural areas, managed parks and green spaces, residential and commercial zones, and industrial land (Kowarik, 2011). These landscapes are shaped by complex interactions between socioeconomic factors and biophysical processes, and are constantly exposed to deteriorating abiotic and biotic conditions, such as higher temperatures and drier soils, elevated levels of artificial light, greater habitat fragmentation, and increased air and soil pollution as shown in the red arrow in Figure 18.1 (Francis & Chadwick, 2013). Combining these factors creates a new environment that reshapes species assemblages and influences natural selection (Figure 18.1, red arrow) (Partecke, 2013). Rapid environmental change due to human impact can accelerate evolutionary change in species' traits. According to Cadotte et al. (2011), traits can be defined as any measurable features of an individual that potentially affect performance or fitness and can be physical, biochemical, behavioural, temporal, or phenological. At the individual level, traits can be classified as functional or performance traits. Functional traits (e.g., leaf area, root depth, or body size) indirectly affect fitness through their impact on growth, reproduction, and survival across environments. In contrast, performance traits (e.g., vegetative biomass, reproductive output, and plant survival) directly relate to an individual's survival, growth, or reproductive output under specific conditions (Violle et al., 2007). When species interact with the changing environments, traits can also be regarded as response traits or effect traits. Response traits determine how species respond to environmental changes, whereas effect

DOI: 10.1201/9781003640899-23

traits reflect how species influence environmental conditions, community structure, or ecosystem properties (Violle et al., 2007). These categories are not mutually exclusive. Under various ecological contexts, a given trait may serve both response and effect roles (Violle et al., 2007). Regardless, traits can shape species' environmental tolerances and habitat requirements, which can influence how species interact with one another and their environments, ultimately influencing population dynamics, community structure, and ecosystem functioning (Cadotte et al., 2011). These reciprocal interactions between ecological and evolutionary processes are called eco-evolutionary dynamics (as denoted by the grey arrows in Figure 18.1) (Alberti, 2015). While these transformations can threaten native biodiversity and ecosystem integrity, they can also create space for innovative, adaptive, and place-based conservation strategies (as denoted by the blue arrow in Figure 18.1) (Alberti 2015).

The recent paradigmatic shift in urban ecology, from viewing cities as ecological voids in contrast to nature ("cities versus nature") to recognising cities as embedded within nature ("cities with nature"), has given rise to research and planning frameworks that treat urban environments as integral components of ecological systems (Alberti et al., 2003). This shift challenges traditional ecological models by calling for the full integration of human activity into ecosystem research and fostering green infrastructure (GI) implementation and nature-based solutions (NbS) within the design and governance of human-dominated landscapes (Alberti, 2015). Central to this paradigmatic shift is the conservation of urban biodiversity, not only for its intrinsic value but also for its vital role in supporting ecosystem functioning and long-term human well-being (as denoted by the green arrow in Figure 18.1).

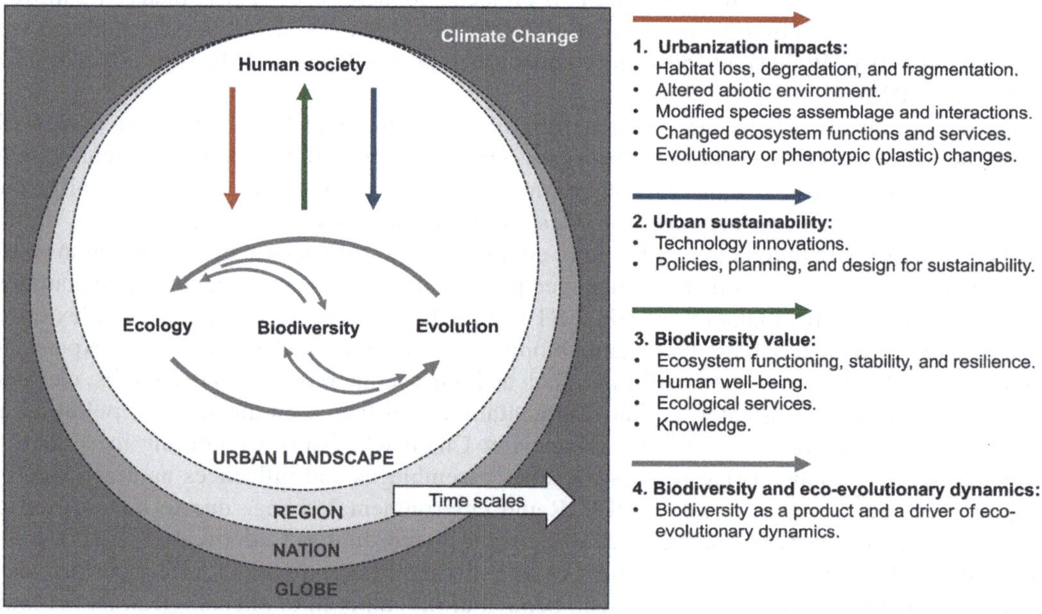

FIGURE 18.1 A conceptual diagram illustrating the role of urban biodiversity within socio-eco-evolutionary dynamics. Biodiversity patterns both shape and are reshaped by eco-evolutionary dynamics driven by human social processes. Achieving sustainability requires integrating biodiversity conservation with planning, innovations, and knowledge generation across scales, aligned with societal governance structures and priorities. (This figure is adapted from Des Roches et al., 2021 and used under CC BY 4.0 license. It has been modified by including additional information abstracted from Alberti, 2015; Alberti et al., 2020; Wu, 2014.)

18.1.2 The Value of Urban Biodiversity

Biodiversity refers to the variability among living organisms in terrestrial, marine and other aquatic ecosystems, as well as the ecological complexes in which they are a part of (Convention on Biological Diversity, 2006). Its value to ecological functioning and human well-being has been studied through biodiversity–ecosystem functioning (BEF) and biodiversity–ecosystem services (BES) frameworks. Higher biodiversity enhances ecosystem functioning through several key mechanisms. The selection effect suggests that communities with more species are more likely to include particularly productive species; however, dominant species may sometimes have low functional performance, diminishing overall system efficiency (Duffy, 2009). The complementarity effect arises when species with different traits interact positively, such as through niche partitioning or facilitation, where one species modifies the environment to benefit the other (Cardinale et al., 2012). Response diversity, the diversity of species' responses to environmental stress, provides a buffer against species loss (Folke et al., 2004). When one species is adversely impacted, and if other species can maintain similar ecosystem functions, the stability of the ecosystem will be ensured (Folke et al., 2004).

Understanding these interactions is crucial for assessing how species loss may influence urban ecosystem functioning and resilience (Larondelle & Haase, 2013). Although definitions of urban ecosystem resilience differ (see Meerow et al., 2016), a commonly accepted interpretation is the degree of changes an ecosystem can tolerate before reorganising into a new set of structures and processes (Alberti et al., 2003). In urban settings, this resilience emerges from the dynamic interactions between human activities and ecological processes (Alberti et al., 2003). It ensures the consistent provision of ecosystem services, including air purification, climate regulation, and opportunities for recreation and mental restoration (Ring et al., 2010).

Urban settlements have historically emerged in ecologically rich locations, particularly along river corridors and at the intersection of distinct ecological zones, due to their strategic locations and easy access to water, fertile soils, and transportation routes (Dearborn & Kark, 2010). As cities expand, their roles as habitats for urban wildlife and in conserving regional or even global biodiversity are becoming increasingly important (Dearborn & Kark, 2010). Small habitat patches within cities can act as stepping stones or corridors, linking fragmented habitats (Hu & Tong, 2025). Given growing evidence that urbanisation drives species adaptation at a global scale (e.g., Santangelo et al., 2022), preserving natural areas within urban matrices not only facilitates species adjustment to urbanisation, but also provides critical opportunities for observing and evaluating species' responses to varied management strategies in situ (Dearborn & Kark, 2010).

Urban biodiversity also plays a vital role in human–nature connectedness, which fosters pro-environmental awareness and pro-environmental behaviour (Richardson, 2025). Urban biodiversity conservation thus offers the dual benefit of protecting imperilled species and wildlife habitats while providing opportunities for urban residents to engage with nature (Ives et al., 2016). As such, it contributes to a broad array of Sustainable Development Goals (SDGs) as outlined in the 2030 Agenda of the United Nations (United Nations, 2015), including SDG 14 (Life Below Water) and SDG 15 (Life on Land). It supports co-benefits across public health (SDG 3), education and awareness (SDG 4), food security (SDG 2), and equitable access to clean energy and GI (SDG 7).

However, cities present a unique socio-ecological context that challenges traditional understandings of how biodiversity influences ecosystem functioning, resilience, and service provision (Schwarz et al., 2017). Compared to rural areas, urban ecosystem services face disproportionately higher demand due to the density of human populations. Moreover, these services function as ecological outputs and public goods, with their distribution and accessibility deeply tied to issues of social equity and justice (Costanza, 2020). In cities, the conservation value of wildlife depends not only on biodiversity but also on human use (e.g., recreation, emotional attachment, and perceived uniqueness) (Wang et al., 2019). Trade-offs, limited resources, and competing interests further complicate urban biodiversity planning. In highly human-dominated systems, the conservation of biodiversity is often contested. One common debate concerns the inclusion of exotic, or alien, species (Dearborn & Kark, 2010). While such species may enhance species richness and increase wildlife exposure to residents, they may not support long-term ecological function or sustainability (Kowarik, 2011).

Given the complexity of urban systems, critical questions arise: Do biodiversity conservation strategies—such as those targeting species, genetic, or functional diversity—deliver measurable co-benefits for human well-being and urban sustainability? Besides, do ecosystem service-based approaches in biodiversity conservation, like GI and NbS, incorporate clear metrics to assess their effectiveness in conserving biodiversity, including its genetic and functional dimensions?

To address these questions, this chapter begins by introducing key biodiversity concepts and metrics and outlining the evolution of the field, from urban ecology to urban evolutionary ecology, as a foundation for conservation. It then reviews strategies and tools for safeguarding urban biodiversity, including technological, policy approaches, and planning frameworks. Finally, it highlights key gaps and opportunities to advance future research, policy, and practice in urban wildlife biodiversity conservation.

18.2 URBAN ECOLOGY: EVOLVING PARADIGMS AND TOOLS FOR BIODIVERSITY RESEARCH

18.2.1 Measuring Biodiversity in Urban Ecosystems

The term "biodiversity" means different biological characteristics. Species diversity refers to the variety and abundance of various species within a community. Commonly used metrics for species diversity are species richness (the number of different species within a specific area), the Shannon Index, and Simpson's index (Weaver, 1949; Simpson, 1949). While both indices account for species evenness, the Shannon Index emphasises richness and rare species, whereas the Simpson Index gives more weight to dominant species (Peet, 1974). These metrics are widely used in ecological research in assessing the impacts of habitat loss and fragmentation (Simamora et al., 2021) or in evaluating the effectiveness of biodiversity conservation policies, including those associated with the International Union for Conservation of Nature (IUCN) Red List (Guadilla-Saez et al., 2019). Their popularity stems from their being relatively easy to measure, effective in communicating with policymakers, and readily incorporated into legal and conservation frameworks (Duffy, 2009). Although species diversity is vital for ecosystem functioning, growing evidence suggests that functional diversity (the range of traits among species; see below) may be a more reliable predictor of ecosystem processes than species count alone (Booy et al., 2000). Additionally, relying solely on species richness or diversity indices can yield misleading conclusions. As an example, while local species counts may appear stable or even increase, they can mask the loss of evolutionarily distinct native species and the proliferation of closely related generalists like grasses and ornamental plants (Duffy, 2009).

Functional diversity, or trait diversity, measures the value, range, and distribution of functional traits within communities and is key to link species assemblages with ecosystem processes (Cadotte et al., 2011). Depending on which traits (e.g., response or effect traits) are considered, the resulting functional diversity may differ (Hordley et al., 2021). Functional group diversity refers to the diversity of groups of organisms (such as pollinators, nitrogen-fixers, or primary producers) that share similar functions or ecological roles (Folke et al., 2004). Functional response diversity, or response diversity, enhances ecosystem resilience by ensuring that not all species within a group respond identically to disturbances (Folke et al., 2004). In practice, ecosystems that harbour a wide range of traits tend to maintain greater functional stability over time, especially in managed or restored ecosystems (Cadotte et al., 2011). Common metrics, like functional richness, functional evenness, and functional divergence, are typically derived using trait-value-based, dendrogram-based, or distance-based approaches (Chao et al., 2014).

Genetic diversity describes the variation in a DNA sequence among individuals or populations of a species and plays a fundamental role in shaping population's ability to persist under environmental change (Ellegren & Galtier, 2016). This variation supports adaptive potential by providing new raw material for natural selection. Populations with reduced genetic diversity often face diminished capacity to adapt and may become more vulnerable to inbreeding depression, which can lower fitness and increase extinction risk (Booy et al., 2000). This diversity can be quantified using several key metrics. Allelic richness, the number of different alleles at a genetic locus in a population, reflects the overall genetic variation and is sensitive to the effects of genetic drift (random changes in allele frequencies due to stochastic processes) in small or isolated populations (Greenbaum et al., 2014). Heterozygosity, often interpreted in relation to the Harvy-Weinberg equilibrium (expected frequencies of heterozygotes under a model with no inbreeding, selection, or other evolutionary forces), measures the proportion of individuals carrying different alleles at a given locus and is commonly used to infer levels of inbreeding (Gregorius, 1978). Another widely used measure is the fixation index. An example is the F_ST score, which quantifies genetic differentiation among populations by estimating the reduction in heterozygosity due to population structure or inbreeding (Nagylaki, 1998). Together, these metrics help assess populations' current genetic health and their potential for long-term evolutionary resilience (Allendorf, 1986).

Phylogenetic diversity evaluates the evolutionary relationships among species in a community. One of the indices, Faith's Phylogenetic Diversity, calculates the sum of branch lengths of a phylogenetic tree encompassing all species in a community, reflecting the total evolutionary history (Faith, 1992). Closely related species often share similar functional traits due to common ancestry (Huang et al., 2020). In communities dominated by such species, their overlapping functions may reduce the uniqueness of individual contributions, potentially narrowing the range of ecosystem services. However, this overlap can also provide functional redundancy that enhances resilience to loss in function following environmental change (Biggs et al., 2020). In some cases, strong interspecific competition can drive divergence on key traits among closely related species, resulting in high functional performance even with low phylogenetic diversity in a community (Grab et al., 2019). As such, it remains an open question whether the loss of closely related species (with similar traits) or distantly related species (with unique traits) poses a greater threat to ecosystem functioning (Grab et al., 2019). Nevertheless, phylogenetic diversity offers a useful lens for understanding how environmental pressures, such as urbanisation, reshape community structures and affect the delivery of ecosystem functions and services (Grab et al., 2019).

18.2.2 Integrating Biodiversity Metrics in Urban Research

Over the past decade, research on urban diversity has taken more integrated approaches by combining taxonomic, phylogenetic, functional, and genetic metrics. This shift has enriched our understanding of how urbanisation shapes species assemblages, drives evolutionary processes, and influences ecosystem functioning and resilience. By analysing both species and phylogenetic richness, Knapp et al. (2017) have revealed whether species turnover occurs randomly or follows phylogenetically clustering. This distinction offers deeper insights into biodiversity loss patterns and informs more effective conservation strategies in urban landscapes. Similarly, Biswas and Mallik (2011) suggested that understanding how species diversity is distributed across functional traits can provide a better understanding of ecosystem stability across different levels of disturbance.

Note that taxonomic, phylogenetic, and functional diversity often provide complementary but not necessarily correlated insights into ecosystem functioning. These different dimensions of biodiversity may reflect ecological processes through distinct or even opposing ways. For instance, Le Bagousse-Pinguet et al. (2019) found that subordinate (less dominant) species, especially those with early-diverging evolutionary lineages and functionally redundant traits, were crucial for maintaining multiple ecosystem functions. Their findings suggest that restoration and conservation strategies should extend beyond dominant species to explicitly account for the contributions of functionally and phylogenetically diverse subordinate species to enhance both resilience and ecosystem performance.

Moreover, life-history traits (mating and reproductive strategies, lifespan, and dispersal ability) influence the response of species to environmental change and genetic structure within and between populations (De Kort et al., 2021). It is therefore important to move beyond species-level assessments to conservations framework that incorporates population-scale genetic diversity and life-history variation. Together, these studies highlight the value of a multidimensional approach that integrates different biodiversity attributes. Such holistic frameworks are becoming more and more important for preserving the evolutionary potential and ecological functionality of urban ecosystems, particularly in fragmented or rapidly changing environments.

18.2.3 From Urban Ecology to Urban Evolutionary Ecology

Cities evolve through complex interactions between human decisions (made by households, businesses, developers, and governments) and biophysical factors regarding geomorphology, climate, and natural disturbance regimes (Alberti et al., 2003). These interactions give rise to diverse urban development patterns, land use, and infrastructure density, each influencing ecosystem processes locally and at a distance through land conversion, resource consumption, and waste production. These environmental changes, in turn, impact human health and well-being (Alberti et al., 2003).

Urban ecology is an integrative subdiscipline of ecology that examines the distribution and abundance of organisms in and around cities, along with the biogeochemical processes operating in urban environments (Pickett et al., 2001). By bridging natural and social sciences, it views cities as socio-ecological systems, with the aim of informing urban planning practices that enhance environmental quality and reduce ecological footprints of cities (Pickett et al., 2001). Two core concepts are included in this framework: (1) social differentiation, where accesses to resources and environmental benefits are shaped by wealth, power, and knowledge, and (2) human-landscape interactions, which emphasise the reciprocal relationships between spatial heterogeneity and sociocultural and biophysical processes (Pickett et al., 2001).

Urban ecology has then developed along several complementary trajectories (Wu, 2014). The "ecology in cities" perspective focuses on the distribution and dynamics of non-human organisms within urban areas; "ecology of cities" conceptualises cities as ecosystems with flows of energy, materials, and species; and "cities as social–ecological systems" emphasises the integrated interactions between humans and nature in urban contexts (Wu et al., 2013). The maturation of these approaches coincides with the rise of landscape ecology in the 1990s, propelled by advances in remote sensing, geographic information systems (GIS), and spatial analysis techniques (Wu et al., 2013). Urban landscape ecology investigates how landscape patterns influence ecological processes, with the broader goal of informing more sustainable urban development (Wu et al., 2013). Foundational frameworks, such as the patch–corridor–matrix model (Forman, 1995) and the hierarchical patch dynamics paradigm (Wu & Loucks, 1995), conceptualise urban areas as spatially heterogeneous human–environment systems. These models are complemented by graph-based approaches that analyse habitat connectivity and gradient-based approaches that assess spatial heterogeneity across continuous surfaces (Frazier & Wang, 2013). These tools have been used widely to study the influence of landscape composition and configuration on urban sprawl (Ji et al., 2006), biodiversity (Schindler et al., 2013), and the provision of ecosystem functioning and services (Syrbe & Walz, 2012). More recently, researchers have applied these tools to assess trade-offs and synergies among multiple ecosystem services (e.g., Karimi et al., 2021), contributing to the development of evidence-based strategies that support resilient and biodiversity-rich urban environments.

Over the last two decades, there has been growing recognition that urbanisation contributes to rapid evolutionary changes (Verrelli et al., 2022). Traditionally, studies of local adaptation have focused on spatial scales larger than both the spatial scale of environmental variation and the typical dispersal distances of the focal organism (Richardson et al., 2014). This focus assumed that high gene flow overwhelms local selective pressure and prevents divergence (Richardson et al., 2014). However, recent studies have demonstrated that "microgeographic divergence" (fine-scale adaptation despite gene flow) can emerge when strong selection pressures are coupled with ecological barriers caused by differences in habitat type, breeding time, or other factors (e.g., Richter-Boix et al., 2013). Urban ecology has increasingly moved toward an eco-evolutionary perspective, acknowledging that evolved traits can alter competition, predation, mutualism, or other species interactions (Alberti et al., 2020). Ignoring such subtle evolutionary processes risks underestimating species' adaptive capacities and misjudging the mechanisms driving ecological outcomes.

However, detecting evolutionary change is not straightforward. Phenotypic traits like physiology, appearance, or behaviours may vary in response to urban conditions due to phenotypic plasticity, adaptive evolution, or maternal effects (Partecke, 2013). Phenotypic plasticity, the reversible, non-genetic responses to environmental conditions, can help organisms survive urban challenges like noise or pollution, and in some cases, may itself evolve (Partecke, 2013; Vinton et al., 2022). Distinguishing plasticity from genetic changes can be an arduous task, as it requires the identification of ecologically relevant traits from their underlying genetic makeup (Partecke, 2013). But since urbanisation impacts species traits and population dynamics (e.g., population densities and growth rates), it is paramount to understand how plasticity and adaptation interact to influence species persistence in cities (Vinton et al., 2022).

Urban evolutionary ecology provides a conceptual and methodological framework to explore how urbanisation alters ecological and evolutionary processes across spatial and temporal scales (Figure 18.2) (Verrelli et al., 2022). Drawing from multiple disciplines and leveraging advances in

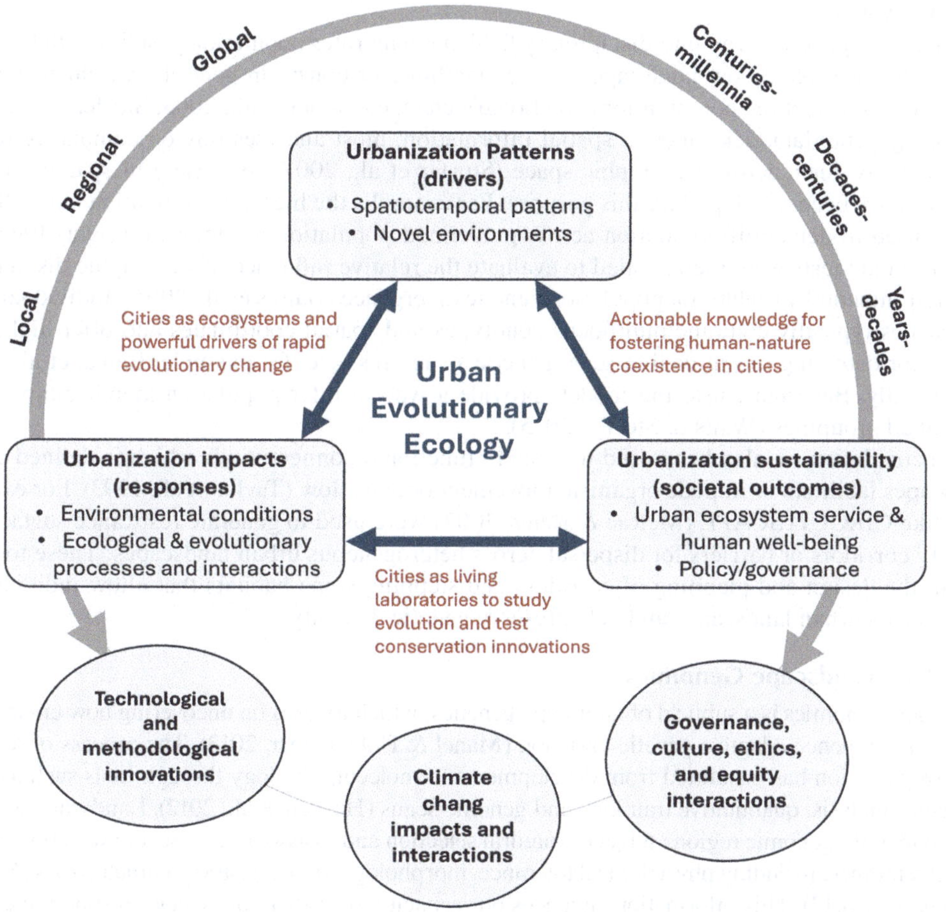

FIGURE 18.2 A triadic conceptualisation of urban evolutionary ecology showing how spatiotemporal patterns, environmental and socioeconomic impacts, and sustainability challenges of urbanisation interact. (This figure is adapted from Wu, 2014, with permission from Elsevier. It is modified to include relabelling and additional information on technological innovations, sociopolitical considerations, and climate change interactions abstracted from Rivkin et al., 2019; Verrelli et al., 2022.)

genomics, remote sensing, and spatial modelling, it aims to build a shared research agenda for understanding how these dynamics shape biodiversity, ecosystem functioning, and human well-being (Figure 18.2) (Verrelli et al., 2022). By acknowledging cities as ecosystems and powerful drivers of rapid evolutionary change, urban evolutionary ecology offers opportunities to generate actionable knowledge to support urban sustainability (Figure 18.2).

18.2.3.1 Landscape Genetics

The long-term survival of wildlife in fragmented urban landscapes depends on the availability of habitats as well as a combination of demographic, environmental, and genetic factors (Booy et al., 2000). Urban development often leads to habitat reduction and population isolation, limiting gene flow and increasing inbreeding and genetic drift, processes that ultimately erode genetic diversity and adaptive potential (Miles et al., 2019; Rivkin et al., 2019). While habitat loss and fragmentation can undermine connectivity and genetic diversity, it may also promote local adaptation when environmental gradients across urban matrices impose divergent selective pressures (Richter-Boix et al., 2013). These contradictory outcomes raise the following questions (Rivkin et al., 2019): Which landscape features are the most effective in influencing population size and gene flow? How do they shape both neutral genetic processes (e.g., drift and inbreeding) and adaptive processes (e.g., selection and local adaptation)? Under what conditions does urbanisation lead to nonadaptive vs. adaptive evolution?

Landscape genetics is an interdisciplinary field that integrates population genetics with landscape ecology to investigate how landscape features facilitate or constrain gene flow, influence genetic structure, and affect the potential for evolutionary change (Manel et al., 2003; Storfer et al., 2007). Because genetic data lack inherent spatial information, most analyses rely on comparing populations or individuals across geographic space (Storfer et al., 2007). A variety of spatially explicit methods have been developed for this purpose. For example, the hierarchical fixation index "F_CT" can be used to detect differentiation across predefined population groupings (Wright, 1965). The partial Mantel tests were then applied to evaluate the relative influence of geographic distance versus environmental variables on population genetic divergence (Banks et al., 2005). Individual-based approaches typically combine multilocus genotypes and spatial coordinates and offer higher sensitivity for detecting recent or fine-scale genetic structures (see, for example, Banks et al., 2005). Additionally, Bayesian clustering models provide a way to infer population membership without predefined groupings (Waits & Storfer, 2015).

Genetic data have also been used to model "functional connectivity", which is defined as how landscapes facilitate or impede organism movement or gene flow (Taylor et al., 1993). For example, tools like CIRCUITSCAPE (McRae & Beier, 2007) were used to generate resistance surfaces and identify corridors or barriers for dispersal across heterogeneous urban landscapes. These tools can inform the design and planning of corridors and stepping stone habitats that allow individuals to move across urban landscapes and help preserve genetic diversity.

18.2.3.2 Landscape Genomics

Landscape genomics is a subfield of landscape genetics, which focuses on uncovering how environmental factors influence adaptive genetic variation (Manel & Holderegger, 2013). The process of detecting adaptive variation has benefitted from developments in molecular biology through tools such as candidate gene analysis, quantitative trait loci, and genome scans (Hansen et al., 2012). Landscape genomics aims to identify genomic regions subject to natural selection and to associate these loci with traits of ecological relevance, including physiological tolerance, morphology, or life-history characteristics (Manel & Holderegger, 2013). This information enhances our capacity to predict how species respond to environmental gradients and stressors in fragmented or rapidly changing habitats.

In recent years, combining genomic data with spatial modelling frameworks has become a key approach for anticipating how biodiversity may respond to climate change (Manel & Holderegger, 2013). One notable example is the integration of landscape genomics with species distribution models, which helps identify populations both genetically suitable and genetically equipped to persist under future climates (Waldvogel et al., 2020). Similarly, Chen et al. (2022) introduced a framework that combines ecological genomics and ecological niche modelling to estimate genomic offset, the mismatch between current genotype-environment associations and future climate projections. This approach helps identify high-conservation-priority populations, such as those characterised by both low genomic offset and minimal habitat loss.

Despite this potential, urban environments remain underexplored because of their unique ecological and evolutionary challenges (Manel & Holderegger, 2013). Realising the full potential of landscape genomics in conservation will require coordinated, expansive data collection across diverse taxa (primarily keystone species) and the development of standardised modelling frameworks that merge genomic, ecological, and environmental data to account for eco-evolutionary processes in conservation planning and forecasting (Waldvogel et al., 2020).

18.3 REMOTE SENSING AS A TOOL IN BIODIVERSITY CONSERVATION

18.3.1 Measuring, Monitoring, and Simulating Biodiversity

Advances in earth observation technologies have made satellite-based monitoring essential for environmental and biodiversity research. However, direct detection of individual species is still plagued with many problems, especially for mobile or cryptic fauna. Nevertheless, remote sensing provides essential indirect measures, or proxies, for biodiversity analysis, especially for plants (Kuenzer et al., 2014). These proxies, derived from multispectral, hyperspectral, or Light Detection and Ranging (LiDAR) data, represent geophysical (e.g., surface temperature), index-based (e.g., Normalized Difference Vegetation Index [NDVI], Enhanced Vegetation Index), thematic (e.g., land cover classification), topographic (e.g., elevation and slope), and texture (e.g., spatial heterogeneity) variables (Kuenzer et al., 2014). They are widely applied to map and model land cover change, monitor vegetation health and phenology, and characterise habitat conditions through both direct indicators (e.g., canopy structure and vegetation composition) and indirect drivers (e.g., temperature and precipitation) (Kuenzer et al., 2014). Applied across diverse terrestrial and aquatic ecosystems, these technologies have proven valuable for a wide range of plant species, greatly expanding their utility in ecological research and biodiversity conservation planning (Lausch et al., 2016).

A range of remote sensing imagery has been used in urban biodiversity research and conservation. Airborne LiDAR offers high-resolution, 3D vegetation structural information (e.g., canopy height, leaf area density, and leaf area per unit of volume), essential for habitat quality assessment and fragmentation modelling, especially for arboreal and forest-dependent species (e.g., Hu & Tong, 2022, 2025). Passive satellite remote sensing technologies, such as the Landsat imagery acquired by the National Aeronautics and Space Administration (NASA) and US Geological Survey (USGS), Sentinel-2 multispectral imagery from the Copernicus Programme, and NASA's Airborne Visible/Infrared Imaging Spectrometer (AVIRIS) hyperspectral imagery, enable long-term, landscape-scale monitoring of land cover dynamics (Rogan & Chen, 2004), urban heat island effects (Mackey et al., 2012), vegetation health, and phenological shifts (Dronova & Taddeo, 2022). These tools also facilitate invasive species mapping (Diao & Wang, 2014) and, when integrated with Global Positioning System tracking, can even support analysis of animal movement patterns (e.g., Robb et al., 2019). Unmanned aerial systems (UASs) equipped with multispectral (e.g., red, green, blue, and near-infrared), hyperspectral, thermal, and LiDAR sensors offer flexible, cost-effective, near-real-time data collection for fine-scale applications, including rooftop-nesting bird surveys (Blight et al., 2019), avian distribution and abundance modelling (Shokirov et al., 2023), and vegetation and crop patterns mapping (Wagner & Egerer, 2022).

Despite advances in sensor and processing techniques, a trade-off in spatial, spectral, and temporal resolution remains a constraint (Neyns & Canters, 2022). As such, researchers have been adopting multisource data fusion, combining complementary datasets from high and low-spatial resolution images, multispectral and hyperspectral data, or satellite imagery and LiDAR, to enhance spatial, spectral, and temporal coverage (Belgiu & Stein, 2019).

In data processing, fusion techniques can be categorised into three levels (Ghassemian, 2016):

- Low-level (pixel-level) fusion combines geometrically registered raw images of the same scene, as in fusing high spatial resolution data with high spectral resolution imagery, to produce a composite image rich in both detail and spectral information.
- Feature level (intermediate) fusion extracts informative features from each dataset (e.g., vegetation indices like NDVI or canopy structural metrics from LiDAR) and integrates them to improve classification performance.
- Decision-level (high-level) fusion aggregates the outputs of multiple classifiers trained on different data sources or feature sets using ensemble methods, such as majority voting or weighted averaging, to generate a final prediction.

These fusion methods have been effectively applied in various ecological applications, enabling 2-D and 3-D individual tree segmentation and tree species classification at local to regional scales (e.g., Hartling et al., 2021; Kukunda et al., 2018). In urban landscapes, where environmental changes are rapid, patchy, and complex, combining spatiotemporal fusion with multi-sensor data holds significant promise (Meng et al., 2022). As an illustration, fusing multi-temporal Landsat and MODIS imagery has been used to track changes in land surface temperature, surface biophysical characteristics, and built-up area, offering critical insights into the urban heat island effect and its ecological consequences (Ezimand et al., 2021).

Incorporating these advanced fusion techniques into Cloud-based platforms, like Google Earth Engine, facilitates near-real-time data collection (Belgiu & Stein, 2019) and has been used for urban tree species classification and mapping (e.g., Vizzari et al., 2025). However, scaling species-level mapping to landscape or continental levels encounters various barriers in terms of the availability of very high-resolution imagery, insufficient comprehensive ground-truth data, and signal noise from topography, shadows, and sensor inconsistencies (Belgiu & Stein, 2019).

To address these challenges, future research must improve fusion techniques that reduce noise while preserving fine-scale features and develop robust algorithms compatible with emerging airborne and satellite sensors (Li et al., 2017). Both supervised and unsupervised machine learning approaches can enhance classification accuracy but demand high computational resources. Inconsistencies in sensor types, processing methods, and sampling strategies further hinder the generalisability of models across studies (Wang et al., 2020). J. Li et al. (2022) suggested that developing a framework for applying deep learning to multisource datasets (e.g., from satellites and UASs) will be crucial to harness the full potential of multimodal remote sensing for biodiversity monitoring.

18.3.2 Remote Sensing for Trait-based Ecological Studies

Plant functional traits, such as chlorophyll content, leaf chemistry, canopy structure, phenology, and the like, regulate key plant processes, including light capture, photosynthesis, and resource use. These processes, in turn, drive ecosystem functions such as primary productivity, nutrient cycling, carbon sequestration, water use efficiency, and climate regulation (Rao et al., 2024). Earth observation technologies can detect spectral reflectance patterns linked to biochemical, structural, and physiological properties of vegetation (Lausch et al., 2016). As a result, remote sensing offers a powerful means of inferring plant functional traits, which reflect both the internal states of plants and their responses to abiotic and biotic pressures (Lausch et al., 2016). These traits serve as valuable indicators for forecasting ecosystem function under shifting abiotic and biotic conditions (Pettorelli et al., 2018).

The spectral variation hypothesis assumes that greater spectral heterogeneity in vegetation reflects higher biodiversity, as variations in pigment concentration, water content, and canopy architecture influence spectral signatures (Sun et al., 2024). Consequently, earth observation supports the characterisation of taxonomic, functional, and structural plant diversities across landscapes

(Lausch et al., 2016). As plant traits are shaped by taxonomic, phylogenetic, and environmental drivers and closely linked to species' ecological strategies, fitness, and adaptive responses (Lausch et al., 2016), monitoring trait variation within and across species via earth observation reveals signals of trait divergence, local adaptation, and selection pressures. These insights are essential for understanding eco-evolutionary dynamics and informing biodiversity conservation under global change.

Advancements in sensor technology have significantly enhanced the capabilities of remote sensing to map plant traits at increasingly fine spatial and spectral resolutions. The combination of UAS-derived hyperspectral data with LiDAR, for instance, has enabled high-resolution (0.2 m) 3D trait mapping and ecological assessments in local wetland environments (Fu et al., 2024). Likewise, imaging spectroscopy (e.g., hyperspectral imaging) was successfully used to identify and scale leaf traits to landscape-level maps in eastern North America (Wang et al., 2020).

Remote sensing is now widely applied to estimate species richness (Rossi et al., 2022), functional diversity, and polygenetic diversity of plants (Schneider et al., 2017). However, the relationship between spectral diversity and plant diversity is context-dependent, varying across ecosystems with different structural and compositional properties (Sun et al., 2024). Forests, wetlands, and grasslands differ in canopy complexity and species assemblages, which influence how plant traits are expressed in spectral signals (Schweiger & Laliberté, 2022). As a result, generalising spectral diversity–biodiversity relationships across regions continues to pose difficulties (Sun et al., 2024). Nevertheless, integrating cross-regional field data with remote sensing observations can be instrumental in developing robust, scalable biodiversity monitoring frameworks (Wang et al., 2020). Data fusion techniques also can offer promising solutions to improve spatial and spectral resolution and enhance biodiversity assessments across scales (Rossi et al., 2022).

While early remote sensing applications have focused on identifying species richness "hotspots", there is a growing emphasis on mapping the variation in species composition between locations to capture spatial patterns of biodiversity turnover (β-diversity), ecological transitions, and habitat connectivity (Rocchini et al., 2018) to improve the understanding of community dynamics. Advances in spectral diversity metrics are making β-diversity assessments more feasible across a range of ecosystems (Schweiger & Laliberté, 2022), further expanding the utility of remote sensing in examining how urbanisation drives species turnover and community assembly and in evaluating the effectiveness of GI in improving functional connectivity within urban landscapes.

18.4 PLANNING AND POLICY APPROACHES

Because of their competitive goals, there are always tensions between urban development and habitat and biodiversity conservation. Addressing these challenges requires viewing cities as coupled human–natural systems, where ecological processes and human decisions are interdependent (Pickett et al., 2017). Urban ecosystems are shaped by decisions at various governance levels—from individuals and communities to municipal, regional, and national governments. In turn, ecological outcomes can affect policy and planning, influencing governance structure and public behaviour (Figure 18.2) (Pickett et al., 2017). Effective biodiversity planning and management must therefore consider these intersecting governance scales, considering administrative boundaries, land tenure systems, legislation, and international agreements. This alignment is crucial for understanding how species interact with the urban matrix and how these ecological patterns and dynamics correspond to planning, policy, and management decisions.

While the intrinsic value of biodiversity is gradually acknowledged, urban biodiversity conservation efforts still emphasise its instrumental (e.g., ecosystem services) and relational (e.g., cultural identity, emotional attachment) values. Biodiversity contributes to human well-being across scales: locally, it enhances air quality, regulates temperature, supports physical and mental health by providing access to nature, and contributes to cultural and spiritual well-being; regionally, it underpins ecosystem resilience and services, such as water regulation, soil stabilisation, and pollination; globally, it safeguards genetic resources essential for long-term ecological stability and adaptive capacity in the face of environmental change.

In essence, biodiversity conservation strategies in urban areas must be guided by planning and policy approaches that recognise the complex interplay of ecological and social systems, operate across governance levels, and leverage direct protection of species and habitats with broader landscape planning. Such strategies are essential to sustain ecological integrity and human well-being in rapidly urbanising environments.

18.4.1 Conservation Policies and Strategies in the United States of America

18.4.1.1 National Strategies

Urban biodiversity conservation is often shaped by national legislations and policies aimed at protecting species, habitats, and ecosystem functions. In the U.S., several foundational environmental laws and domestic initiatives contribute to biodiversity outcomes in urban areas.

The US Endangered Species Act (ESA) of 1973 mandates the protection of threatened and endangered species and their critical habitats, some of which are located within or near urban environments. The ESA has supported San Diego's Multiple Species Conservation Plan, which safeguards 85 species unique to the region, including the California gnatcatcher and the western burrowing owl (City of San Diego, 2025). Despite the successes of ESA, many species are now experiencing more threats than ever due to accelerating urbanisation and climate change (Gerber et al., 2023). Moreover, funding and protection efforts tend to focus on listed species. Species that are not listed or hardly studied with deficient data receive little attention, even though many are declining in their distribution and abundance (Gerber et al., 2023).

The US Clean Water Act (CWA), originally enacted in 1948 and amended in 1972, regulates both point source pollution (e.g., industrial waste) and non-point source pollution (e.g., agricultural and urban runoff) to protect freshwater ecosystems, many of which intersect urban landscapes. The CWA mandates the US Environmental Protection Agency to "promote the use of GI in permitting enforcement, planning, research, technical assistance, and funding programs" (US Environmental Protection Agency, 2025). By integrating GI with traditional engineered systems, the US Environmental Protection Agency has supported cities and municipalities in reducing water pollution, particularly from combined sewer overflows. A notable initiative is the "Five Star and Urban Waters Restoration Program", a public–private partnership between the US Environmental Protection Agency, the National Fish and Wildlife Foundation non-profit organisation, and the Wildlife Habitat Council. Drawing on federal, private, and corporate funding, the programme supports state and local level projects that restore wetland, forest, riparian, and coastal habitat. It promotes community stewardship through initiatives like K–12 environmental education and incentivises locally driven solutions by engaging nonprofits and municipal agencies.

The US National Environmental Policy Act (NEPA) of 1969 mandates federal agencies to assess the environmental impacts of proposed projects and engage the public in the decision-making process. NEPA has become a key tool for minimising harm to biodiversity and ecosystem services, especially in rapidly growing urban areas, but with varying degrees of success. Projects such as the "Los Angeles River Revitalization Plan" aim to prevent biodiversity loss and consider ecological parameters, including seral stages, habitat fragmentation, riparian condition, species composition, and snags presence (Council on Environmental Quality, 1993). However, they rarely integrate biodiversity metrics, such as genetic or functional diversity, to guide more effective biodiversity outcomes.

In the U.S., legislations are enacted at the national level; nevertheless, biodiversity conservation programmes operate within a highly decentralised governance structure, where planning and development decisions are made at the federal, state, regional, and local levels. The federal government owns substantial land, particularly in the western U.S., but it has limited authority to enforce land-use reforms at the local level. Federal agencies like the US Environmental Protection Agency and Department of Transportation are often peripheral to day-to-day planning decisions (Schilling & Logan, 2008). This tiered and fragmented administrative system frequently causes piecemeal implementation and limited alignment of conservation priorities across jurisdictions (Mell, 2014).

18.4.1.2 State, Regional, and Local Planning Tools

At the state level, the cornerstone of the conservation programme is the "State Wildlife Grants Program", established in 2002 following the Wildlife Conservation and Reinvestment Act of 2001. The programme provides annual federal funding to support the conservation of species of greatest conservation need. Each state must develop a state wildlife action plan approved by the US Fish and Wildlife Service and revised every 10 years, detailing priority species, habitat conditions, threats, conservation actions, and monitoring strategies (Ohio Department of Natural Resources, 2015). The planning process involves coordination with landowners, stakeholders, resource agencies, and the public to ensure comprehensive planning.

At the local level, park systems and natural resource management strategies often support biodiversity conservation. Many U.S. counties and cities develop Natural Resource or Urban Forest Management Plans to guide the management of street trees, urban forests, and other publicly owned lands. These plans commonly align with the Ten-Year Urban Forestry Action Plan developed by the US Forest Service and aim to improve biodiversity, prioritise native tree species, increase the number of large trees, and expand overall tree canopy cover across urban neighbourhoods (Grant et al., 2022).

Most of these plans also focus on enhancing urban tree ecosystem services, primarily stormwater mitigation, carbon sequestration, and air pollution abatement, using strategies like tree replacement, reducing maintenance cycles, and careful species selection (Grant et al., 2022). Analytical tools, such as i-Tree (https://landscape.itreetools.org/) developed by the US Forest Service, support communities in quantifying and managing urban forest benefits (Van Sambeek, 2017).

BOX 18.1 A CASE STUDY OF LOCAL CONSERVATION PLAN

The city of Cincinnati in Ohio, USA, has adopted the 2015 Ohio Wildlife Action Plan. Developed in collaboration with universities and conservation partners, the plan uses scoring metrics to evaluate species based on population trends and conservation urgency, whereas action scores identify knowledge gaps and priority interventions (Ohio Department of Natural Resources, 2015). GIS is used to track changes in habitat quantity, and efforts are underway to incorporate habitat quality indicators. In Ohio, all conservation actions at the regional and local levels must align with Ohio Division of Wildlife guidelines and the broader strategic direction of the state wildlife action plan. However, building a robust spatial and temporal database is a challenging endeavour.

The state funds the action plan. The $400 million Clean Ohio Fund (Ohio Public Works Commission, 2025), administered by the Ohio Public Works Commission, supports land acquisition, open space preservation, riparian corridor protection, and brownfield remediation, enhancing both ecological integrity and public access to nature (Ohio Public Works Commission, 2025).

The city's street tree plan prioritises native and disease-resistant species and incorporates climate change projections from US Forest Service models to guide management (City of Cincinnati, 2020). It covers forest regeneration, invasive species removal, and protection of ecologically sensitive habitats. Outside city boundaries, the Great Parks of Hamilton County oversees forests, wetlands, brushlands, and prairies, explicitly incorporating habitat and wildlife conservation. Together with the city park system, they manage roughly one-third of the undeveloped areas in Hamilton County. Species presence and abundance data are collected with the help of researchers, volunteers, and citizen science initiatives, notably iNaturalist (n.d.), along with state and local agencies and universities. However, biodiversity assessments beyond tree species remain sparse, and the plan does not emphasise genetic diversity.

Despite all these conservation efforts across national, regional, and local levels, obstacles continue to persist. The U.S. lacks a unified national biodiversity strategy and has not conducted a comprehensive national biodiversity assessment since 1999 (Gerber et al., 2023). Biodiversity monitoring is uneven, often limited to regionally coordinated efforts, such as Ohio's wetland breeding bird surveys (Ohio Department of Natural Resources, 2015). Funding disparities further hinder conservation: game species receive robust financial support through excise taxes on hunting and fishing gear (e.g., Pittman-Robertson and Dingell-Johnson programmes), while non-game and non-ESA-listed species—such as songbirds, amphibians, and rare plants—rely on voluntary donations and special conservation licence plates, often insufficient to meet conservation needs (Ohio Department of Natural Resources, 2015). As a result, basic population and distribution data of many of these species are still lacking.

The absence of a coherent framework to balance biodiversity with economic development and other societal priorities is equally concerning. Overlapping and sometimes contradictory laws and policies impede effective coordination, which results in gaps between stated biodiversity goals and measurable outcomes (Hu & Tong, 2025). These shortcomings underscore the urgent need for a comprehensive national biodiversity strategy that provides clear directions, aligns cross-sectoral efforts, secures sustainable funding for non-game species, and ensures accountability for conservation outcomes (Gerber et al., 2023).

Urban green space networks serve as important biodiversity refuges but often fall short of their potential due to fragmented governance. Although funding opportunities (such as the "Ohio Open Space Program", "Green Infrastructure Program", and the "Brownfield Program" from the US Environmental Protection Agency) can support the creation of green space and the restoration of wildlife habitats, municipal agencies typically manage green spaces independently. This results in missed opportunities for building ecological networks that improve functional connectivity. Addressing these governance gaps calls for integrated planning frameworks that promote transparent data sharing and support collaborative neighbourhood-scale planning (Schilling & Logan, 2008). Such frameworks must also address persistent challenges, including limited financing, resident displacement, and ambiguous legal authority. In addition, land banks should be developed to prioritise biodiversity outcomes while aligning with municipal budget constraints (Schilling & Logan, 2008).

18.4.2 International Conservation Efforts

Besides the U.S., many other nations also have biodiversity conservation programmes. They often align their conservation efforts with global initiatives, such as the Convention on Biological Diversity (CBD) and the Economics of Ecosystems and Biodiversity (TEEB) initiatives. These international instruments provide a framework to integrate biodiversity and ecosystem services into policy and planning processes.

18.4.2.1 Convention on Biological Diversity (CBD)

The CBD, established as a legally binding global agreement in 1993, aims to advance three broad goals: "the conservation of biological diversity", "the sustainable use of its components", and "the fair and equitable sharing of benefits arising from genetic resources" (Convention on Biological Diversity, 2020). With 196 participating parties, the CBD has provided the institutional foundation for a series of global biodiversity agendas, beginning with the 2010 Biodiversity Targets, followed by the 2020 Aichi Targets, and most recently, the Post-2020 Global Biodiversity Framework (Convention on Biological Diversity, 2020). These plans set long-term goals on halting biodiversity loss, enhancing ecosystem resilience, and securing nature's contributions to people (Convention on Biological Diversity, 2020). Countries that are Parties to the CBD are required to develop National Biodiversity Strategies and Action Plans (NBSAPs) as a means to contribute to achieving global biodiversity targets. However, despite progress under earlier NBSAPs, most national plans failed to reach the 2020 Aichi Targets (Convention on Biological Diversity, 2020).

BOX 18.2 GAPS IN THE EARLY NBSAPS AND RECOMMENDATIONS FOR THE POST-2020 GLOBAL BIODIVERSITY FRAMEWORK

Key implementation gaps in the NBSAPs were identified to circumvent the earlier failures. Major gaps in the NBSAPs include:

- **Weak legal integration**: Many biodiversity objectives were not legally binding, as they were not ratified through national legislation. This lack of legal grounding limited both enforceability and long-term accountability (Xu et al., 2021).
- **Fragmented governance and competing sectoral priorities**: Biodiversity conservation continues to be undermined by sectoral competition, inadequate cross-ministerial coordination, and the prioritisation of short-term economic growth over ecological sustainability (Sarkki et al., 2016).
- **Limited dedicated funding**: A lack of funding at both national and international levels hindered programme implementation (Sarkki et al., 2016; Xu et al., 2021).
- **Insufficient stakeholder engagement**: In many cases, targets were developed without broader input from diverse stakeholders and review mechanisms, leading to limited iterative planning and poor tracking of implementation progress (Xu et al., 2021).
- **Inadequate genetic indicators**: Existing indicators, like the Red List Index, do not adequately capture patterns of within- and among-population genetic diversity (Hoban et al., 2021, 2023).
- **Low inclusion of non-agricultural species**: Genetic monitoring largely focuses on agricultural species, ignoring broader biodiversity concerns (Hoban et al., 2021).

Based on the experience from earlier efforts, recommendations were made for the post-2020 Global Biodiversity Framework.

RECOMMENDATIONS FOR STRENGTHENING POST-2020 IMPLEMENTATION

- Enhance integration of genetic diversity into the post-2020 Global Biodiversity Framework by adopting more scientifically grounded and rigorous targets, including the concept of maintaining "adaptive potential" and commitment to "maintain at least 99% of within population genetic diversity and all distinct populations" (Hoban et al., 2023).
- Incorporate practical and measurable indicators into NBSAPs, especially where DNA-based monitoring programmes are limited. Key indicators include:
  - **Effective population size ≥500**: the threshold to retain adaptive potential and prevent genetic drift (Hoban et al., 2023).
  - **Population maintenance**: the ratio of the currently existing populations to the original number of populations (Mastretta-Yanes et al., 2024).
- Use indirect estimation methods in data-poor contexts. When genetic data are unavailable, Effective Population Size can be approximated using census data, range-based estimates, or semi-qualitative inputs derived from expert knowledge, citizen science, or local ecological observations (Mastretta-Yanes et al., 2024).
- Track progress in genetic monitoring by reporting the number of species or populations assessed using DNA-based methods. This indicator helps evaluate national capacity, identify gaps, and prioritise species for future genetic assessment and management (Mastretta-Yanes et al., 2024).

- Strengthen legal mandates by embedding biodiversity goals into legislation, as seen in Finland's legally binding NBSAP, which creates ministerial accountability across election cycles (Sarkki et al., 2016).
- Foster stakeholder engagement and iterative negotiation processes to align biodiversity targets with broader development priorities and shift the sole responsibility from environmental monitoring agencies to all policy sectors (Sarkki et al., 2016; Xu et al., 2021).
- Targeted funding opportunities and incentive mechanisms can help address gaps in biodiversity valuation and decision-making. The CBD can be paired with funding sources such as the Global Environment Facility to support developing countries in conserving biodiversity, as well as result-based financing strategies that provide financial incentives for meeting biodiversity targets (Xu et al., 2021).

18.4.2.2 Economics of Ecosystems and Biodiversity (TEEB)

The TEEB initiative was launched in 2007 by the German government and the European Commission, following a proposal from the G8+5 Environment Ministers. It emphasises the economic value of biodiversity and highlights the rising costs associated with its degradation and loss (Ring et al., 2010). By challenging conventional economic models that prioritise short-term consumption, private wealth, and the exploitation of finite natural resources, TEEB supports broader policy frameworks, such as the CBD and the EU Biodiversity Strategy (Hedden-Dunkhorst et al., 2015; Ring et al., 2010).

The core aim of TEEB is to integrate the value of biodiversity and ecosystem services into national planning and accounting systems. Through a variety of case studies and policy applications, the initiative also demonstrates the co-benefits of biodiversity conservation and human well-being (Ring et al., 2010). An example of the TEEB-led effort is the Mapping and Assessment of Ecosystems and Their Services, which addresses challenges like trade-offs between ecosystem services, spatial and temporal scale mismatches, and the use of scenario analysis for forward-looking policy (Hedden-Dunkhorst et al., 2015). As a result, many nations have since adopted similar approaches in evaluating and incorporating biodiversity values into their national assessments and planning frameworks (Hedden-Dunkhorst et al., 2015).

18.4.2.3 The New Urban Agenda

Another key global policy framework, the New Urban Agenda (United Nations, 2016), indirectly supports urban biodiversity by promoting goals for greener, more inclusive, and resilient cities. Organised around five key pillars (national urban policies, legislation and regulations, urban planning and design, local economy and finance, and local implementation), it establishes standards and principles for the planning, construction, development, management, and improvement of cities (United Nations, 2016). While biodiversity is not its central focus, the agenda's emphasis on integrated and sustainable urban development creates entry points for incorporating biodiversity considerations into urban policy and planning across multiple governance levels and sectors.

Despite growing technological capabilities and data availability in urban monitoring technologies, challenges remain in translating these resources into actionable insights at the local level. As Robin et al. (2019) noted, persistent gaps still exist in how urban biodiversity data are shared with local actors and whether such information aligns with decision-making needs. This disconnect raises concerns about the usability of national or global biodiversity targets and indicators when developing locally relevant conservation strategies (Robin et al., 2019).

18.4.2.4 Global South

The global biodiversity crisis manifests unevenly across regions. Megadiverse countries that harbour the majority of Earth's species (Mittermeier et al., 1997) are geographically diverse and vary widely in their capacity to implement effective conservation strategies. In many parts of the Global South, the lack of local scientific systems hampers the application of existing knowledge, the generation of new research tailored to national contexts, and the development of context-based conservation solutions (Zhang et al., 2023). These gaps limit domestic conservation outcomes and hinder global efforts to apply and expand the biodiversity knowledge base. Compared to developed countries where urban growth has slowed in many places, cities in the Global South continue to face difficult trade-offs between biodiversity conservation and urban development (Leveau et al., 2025). Realising synergies between ecological protection and human well-being will be crucial to advancing sustainability in such dynamic urban landscapes (de Camargo et al., 2022).

BOX 18.3 SÃO PAULO, BRAZIL: CO-BENEFITS AND GAPS IN URBAN ECOSYSTEM GOVERNANCE

As one of the 17 megadiverse countries, Brazil is disproportionately responsible for global biodiversity conservation (Zhang et al., 2023). However, its representation in global biodiversity governance remains limited; for instance, it is underrepresented in the IUCN leadership (Zhang et al., 2023). This governance mismatch underscores the need for enhanced national engagement and international support.

From 2002 to 2011, São Paulo experienced a 57% increase in urban land areas, including the ecologically sensitive southern zones and Environmental Protection Areas (Moreno-Peñaranda, 2021). These developments have led to habitat loss and fragmentation, as well as biodiversity decline, highlighting the tension between urban development and ecological sustainability, a challenge shared by many fast-growing cities in the Global South (Leveau et al., 2025).

Despite some promising efforts, urban ecosystem governance in São Paulo has not fully delivered on its potential for human well-being and sustainability co-benefits (Moreno-Peñaranda, 2021). According to Moreno-Peñaranda (2021), key urban ecosystem governance initiatives and limitations include:

1. **Greenhouse gas emissions inventory**: São Paulo's pioneered greenhouse gas emissions inventory includes terrestrial ecosystems in emissions accounting. However, its narrow focus on carbon mitigation overlooks the biodiversity and multifunctional ecosystem services provided by urban green spaces, limiting the case for NbS.
2. **Green belt biosphere reserve**: The Green Belt Biosphere Reserve, designated under the Global Network of Biosphere Reserves framework, was established to conserve biodiversity from urbanisation. While it includes locally managed youth education and eco-training programmes, its heavy reliance on supra-local actors can impede locally led actions. Even though São Paulo lies at the heart of the Green Belt Biosphere Reserve, the reserve still lacks sufficient capacity, resources, and governance mechanisms to manage urban growth and promote social inclusion effectively.
3. **Environmental protection areas**: These areas support biodiversity and local livelihoods, particularly for vulnerable groups. However, they face significant resource constraints, weak cross-sector planning, and limited scalability, which reduce their potential for systematic biodiversity protection.

4. **Clean agriculture program**: The programme promotes agroecological practices and urban–rural linkages to reduce pollution and the distance between food production and consumption. However, it lacks a systemic approach and faces challenges, such as aging farmers, invasive species pressures, and low productivity.

São Paulo's experience reflects many challenges shared by CBD signatories in developed countries, including fragmented governance and siloed implementation, inconsistent biodiversity monitoring and data availability, and weak integration of biodiversity, social equity, and multifunctionality in urban planning (de Camargo et al., 2022; Moreno-Peñaranda, 2021). At the same time, São Paulo confronts challenges typical of rapidly urbanising Global South regions, such as informal urbanisation occurring outside formal planning and regulatory frameworks, marginalisation of slum communities, and persistent land-use conflicts (Moreno-Peñaranda, 2021). While international collaboration is essential to strengthen conservation outcomes, equally important is the development of localised environmental policies and governance mechanisms that reflect cultural and behavioural contexts (Moreno-Peñaranda, 2021). Improving local management with attention to these contextual factors can better support both biodiversity conservation and social well-being. As an example, inclusive governance that fosters public participation through decision-making processes and citizen science initiatives not only improves policy effectiveness and fosters community ownership but also contributes valuable knowledge to biodiversity research and conservation (de Camargo et al., 2022).

18.4.3 Green Infrastructure Planning and Nature-based Solutions in Biodiversity Conservation

The European Commission defines GI as "a strategically planned network of natural and semi-natural areas with other environmental features, designed and managed to deliver a wide range of ecosystem services, while also enhancing biodiversity" (European Commission, n.d.). This definition explicitly incorporates both green spaces (parks, forests, and green roofs) and blue infrastructure (rivers, wetlands, and coastal systems), positioning GI as a tool to support biodiversity, ecosystem resilience, and human well-being. In 2013, the European Commission adopted a strategy to employ GI to strengthen investment in Europe's natural capital and deliver biodiversity objectives by 2020. Although member states are not mandated to implement the GI strategy, it has stimulated numerous landscape- and local-scale projects aimed at improving ecological resilience and halting biodiversity loss (Lähde & Di Marino, 2019).

NbS builds on GI by explicitly using natural processes to address societal challenges while advancing sustainability, resilience, and equity (European Commission, n.d.). NbS aligns with global frameworks, such as the SDGs, and often links biodiversity restoration with urban innovation. In Turin, Italy, the ProGIreg Farfalle in ToUr project promoted butterfly habitat restoration while engaging socially underrepresented communities. It addressed both ecological and social inclusion objectives (proGIreg, n.d.).

BOX 18.4 COMPARATIVE APPROACHES TO GREEN INFRASTRUCTURE (GI) PLANNING: ITALY, THE UK, AND THE USA

Despite growing interest in GI as a means to enhance urban sustainability, significant gaps remain in incorporating biodiversity as a central objective in GI planning. In many cases, ecosystem services, such as stormwater control and heat mitigation, were prioritised over goals related to ecological connectivity and native species conservation. A comparative look at GI approaches in Italy, the UK, and the USA reveals important differences in national priorities, institutional structures, and levels of biodiversity integration.

ITALY: BIODIVERSITY-DRIVEN GREEN INFRASTRUCTURE NETWORKS

Italy has pursued a biodiversity-driven model of GI, evidenced by national legislation, such as the Law on the Development of Green Urban Areas (European Environment Agency, 2020). This strategy prioritises the identification and protection of key biodiversity areas that lie outside existing conservation zones. These areas are integrated into broader ecological networks through the use of habitat corridors, linear green spaces, and other landscape elements that enhance connectivity (European Environment Agency, 2020).

In Rome, GI planning efforts have explicitly prioritised biodiversity conservation alongside ecosystem service delivery, with the aim of reinforcing ecological resilience in the urban landscape (Capotorti et al., 2016). Several pilot GI projects have demonstrated how enhancements to urban biodiversity can also generate positive health outcomes, thereby illustrating the potential of GI to support multiple policy objectives (Capotorti et al., 2019). However, many GI initiatives still focus primarily on ecosystem service efficiency (e.g., rapid carbon sequestration or temperature regulation) over biodiversity, often favouring fast-growing, service-oriented species at the expense of slower-establishing native species. Addressing these trade-offs requires balanced species selection, greater use of local ecological knowledge, and long-term planning (Capotorti et al., 2016).

UNITED KINGDOM: STRATEGIC AND MULTIFUNCTIONAL GI APPROACHES

The UK adopts a strategic, multifunctional approach to GI planning, embedded in local development plans and supported by non-statutory strategic frameworks (Mell, 2014).

The Urban Fringe Action Plan emphasises landscape connectivity, habitat protection, and ecosystem service delivery through a hierarchical green space network (Kambites & Owen, 2006). However, the decentralisation of planning authority, from regional to local scale, has weakened the evidence base for multifunctionality (Mell, 2014). In the absence of robust regional datasets and frameworks, decision-makers often emphasise short-term economic returns over long-term ecological benefits. As a result, current GI implementation focuses on using evidence-based arguments and advisory guidance to influence local policy and investment decisions (Mell, 2014). In England, the responsibilities of GI implementation are distributed across several government departments, each overseeing aspects such as environmental policy, urban green space planning, national development policy, and funding for local authorities (Green Infrastructure Partnership, 2025).

The Green Infrastructure Framework provides practical guidance for local planning authorities, developers, and communities to enhance existing green spaces and develop new, high-quality natural areas (Green Infrastructure Partnership, 2025). GI planning is closely linked to Biodiversity Net Gain and the Nature Recovery Network, which require all development projects to achieve a 10% net gain in biodiversity and create connected habitats.

A notable biodiversity-led GI initiative in London redeveloped brownfield sites through collaboration among developers, local authorities, ecologists, landscape architects, and national environmental agencies (Connop et al., 2016). The pilot project combined conventional urban landscaping (e.g., ornamental beds and hedging) with brownfield habitat features, demonstrating that biodiversity objectives could coexist with urban aesthetics and low-maintenance design to support ecologically rich, multifunctional urban redevelopment.

UNITED STATES: GI FOR STORMWATER, LESS FOR BIODIVERSITY

In contrast, GI in the U.S. emerged primarily as a stormwater management solution, emphasising filtration, infiltration, and evapotranspiration to manage stormwater at its source (US Environmental Protection Agency, 2025).

While GI initiatives are beginning to address co-benefits in improving air quality, mitigating urban heat, and supporting community well-being, their implementation is still largely regulatory and engineering-driven (Mell, 2014).

In Cincinnati, Ohio, the targets of its Metropolitan Sewer District GI programme are mainly driven by combined sewer overflow compliance under the Clean Water Act. Although effective in managing aquatic systems and watersheds (Shifflett et al., 2019), these initiatives have not evolved to integrate biodiversity strategies and often overlook habitat fragmentation, terrestrial species, and ecological connectivity (Buffam et al., 2022; Grabowski et al., 2023).

18.5 GAPS AND OPPORTUNITIES

18.5.1 Do Ecosystem Service-based Approaches Like GI and NbS Integrate Metrics for Assessing Biodiversity Outcomes?

While the ecosystem services concept is useful and can help justify conservation, it cannot replace biodiversity conservation as a goal (Salomaa et al., 2016). Relying solely on ecosystem services risks prioritising short-term human benefits over long-term ecological integrity, potentially neglecting species or ecosystems that don't have obvious or immediate economic value (Salomaa et al., 2016). Notwithstanding the progress in implementing GI and NbS in environmental management, many projects, as in the case in Cincinnati, are still focusing on ecosystem services in terms of stormwater management, temperature regulation, or recreation, and have yet to embrace explicit biodiversity objectives (Xie & Bulkeley, 2020). When biodiversity is considered, some projects overlook the genetic dimension, such as genetic metrics or proxies like Effective Population Size. This results in limited contributions to the CBD's genetic diversity goals and missed opportunities to enhance functional connectivity through well-designed GI networks. Even if biodiversity metrics are applied, they are commonly limited to taxonomic abundance or diversity, which are weaker predictors of ecosystem processes than functional trait-based metrics (see Sections 18.2.1 and 18.2.2).

This gap stems partly from cost, the complexity in collecting genetic data, and other scientific impediments. Although the link between biodiversity, ecosystem functioning, and ecosystem resilience is well established, societal benefits of biodiversity—beyond provisioning and regulating services—remain under debate (Cardinale et al., 2012). Urban contexts amplify this uncertainty: only 24% of the reviewed studies in Schwarz et al. (2017) tested biodiversity-ecosystem service relationships either empirically or statistically despite the assumption that "more biodiversity always equals better outcomes" often fails under urban conditions (Ziter, 2016).

18.5.2 Do Biodiversity Conservation Strategies Deliver Measurable Co-benefits for Human Well-Being and Urban Sustainability?

Urban planning often overlooks cultural and relational values associated with nature, even when biodiversity is prioritised. Conservation efforts typically emphasise biophysical indicators (e.g., species richness, canopy cover) while neglecting a sense of place, local ecological knowledge, cultural identity, and other social factors (Xie & Bulkeley, 2020). This limits the ability of urban biodiversity strategies to fully capture their contribution to equity, social cohesion, and human well-being.

Urban ecosystem services are public goods, deeply embedded within social structures of accessibility and equity (Costanza, 2020). The conservation value of urban natural environment depends not only on biodiversity but also on human use, through activities, like jogging or dog walking, and through emotional attachment and a sense of uniqueness (Wang et al., 2019). Aesthetic preferences, for instance, may favour non-native species in street tree planting. Yet, these species can also generate ecological disservices in suppressing native biodiversity or disrupting ecological interactions, potentially leading to unintended social or economic costs (Schwarz et al., 2017).

Optimising urban BES outcomes requires balancing ecological and social considerations (Kowarik, 2011). However, empirical research on how people perceive ecosystem services and disservices remains scanty, particularly in culturally diverse urban settings. Emerging tools, such as structural equation modelling, provide a way forward by analysing complex, non-linear interactions between biodiversity, human perceptions, and human well-being outcomes (Chen et al., 2025; Leong et al., 2020; Wang et al., 2019). Therefore, sustainable urban planning requires strategies that integrate ecological, social, and cultural dimensions, ensuring that biodiversity conservation delivers ecological resilience alongside equitable, inclusive, and culturally relevant benefits to urban populations.

18.5.3 A Trait-based Framework to Link Biodiversity and Ecosystem Services

Bridging science and policy requires tools that explicitly link biodiversity to the benefits ecosystems provide to people. Developing such tools depends on predictive models that are grounded in empirically quantified ecological mechanisms, capable of forecasting changes in ecosystem services at policy-relevant scales and connected to social, economic, and political systems (Cardinale et al., 2012). A key direction is to bridge BEF and BES by extending their traditional scopes, linking BEF research's local, mechanistic focus with the regional patterns emphasised in BES (Cardinale et al., 2012).

Functional trait-based approaches, which explicitly link response traits (species' sensitivity to environmental change) with effect traits (contributions to ecosystem services), offer deeper insights into how community composition changes influence service delivery (Schwarz et al., 2017). Swartz et al. (2023) applied a response-and-effect trait framework to evaluate how urban green-space characteristics shape bird communities and associated ecosystem services. By modelling species abundances as a function of response traits (e.g., body size, nest location, and foraging behaviour) and linking these to effect traits relevant to cultural (e.g., plumage coloration and song complexity) and regulating (e.g., diet-related traits) services, they identified habitat features that promote high-service bird assemblages. Such trait-based approaches can also inform planning decisions by revealing synergies and trade-offs among multiple services, helping design multifunctional urban ecosystems that support biodiversity and human needs. As illustrated in Figure 18.3, such a framework involves several steps: Step 1: Identifying response traits using species-trait-environment models (1a), effect traits through BEF experiments (1b), and traits influencing cultural services using statistical models (1c) (Echeverri et al., 2020; Tribot et al., 2016). Step 2: Linking effect and response traits to detect environmentally sensitive effect traits (Swartz et al., 2023). Step 3: Aggregating traits at community and landscape scales (Swartz et al., 2023). Step 4: Modelling trait–environment relationship, accounting for spatiotemporal variability and phylogenetic information to improve predictions (Luck et al., 2012; Swartz et al., 2023). Step 5: Integrating these relationships into spatially explicit decision-support models (e.g., InVEST) (Natural Capital Project, n.d.) to simulate ecosystem service trade-offs (Cardinale et al., 2012). Step 6: Using model feedback to refine trait selection, validate trait-function assumptions, and improve the representation of spatial dynamics and service delivery across scales (Cardinale et al., 2012; Luck et al., 2012).

Following similar procedures, Tran et al. (2020) proposed a GI planning tool for maximising the multifunctionality of GI by optimising both site selection and plant traits. Their approach allows a better understanding of synergies and trade-offs among multiple services in urban contexts.

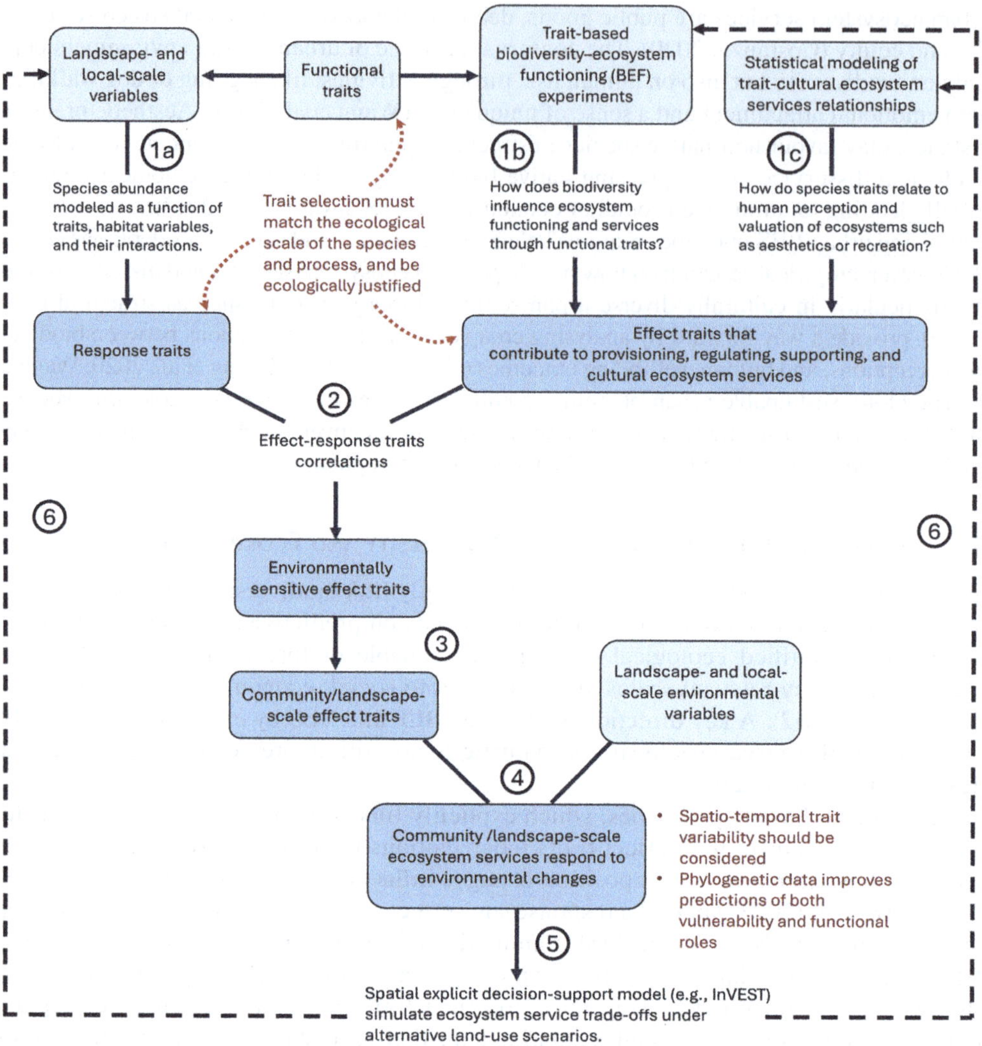

FIGURE 18.3 A trait-based framework evaluating how ecosystem services respond to environmental changes and the resulting trade-offs. (This figure is adapted and modified based on the methodologies described in Cardinale et al., 2012; Luck et al., 2012; Swartz et al., 2023.)

18.5.4 Future Directions

Predicting how ecosystems will respond to future biodiversity loss at a global scale is important, but it requires a deeper understanding of how complementarity operates across different ecological contexts. Combining functional and phylogenetic diversities provides useful perspectives for this task. For example, Huang et al. (2020) found that phylogenetic diversity positively correlates with the complementarity effect, suggesting that evolutionary distinct plant species are more functionally divergent and thus better complement each other to enhance biomass production. In contrast, greater variation in plant height, a key functional trait, was linked to reduced selection effects, potentially limiting the dominance of a single species. These findings underscore the importance of integrating multiple biodiversity dimensions to maintain ecosystem functioning under anthropogenic pressures, such as urbanisation and climate change.

Data-driven tools also offer promise. For instance, using a trait-based machine learning approach, Chua et al. (2019) analysed data on nine functional traits, including body size, diet, and mouth position, for 893 freshwater fish species to simulate extinction scenarios across freshwater ecoregions. Their results reveal that biodiversity loss reduces species richness and erodes functional redundancy and resilience, ultimately threatening ecosystem functioning. Remote sensing technologies, notably through advances in machine learning and data fusion techniques, now enable the quantification of key ecosystem functions (e.g., biomass) and the mapping of taxonomic and functional diversity across spatial scales (see Section 18.3.2; Fu et al., 2024; Wang et al., 2020). These advances create new opportunities to uncover consistent patterns in BEF relationships at regional to global scales. Future opportunities lie in integrating satellite remote sensing with deep learning approaches to improve classification accuracy of critical urban resources for wildlife, enable fine-scale monitoring of wildlife populations, and leverage time-series data to detect ecological trends over time (Pettorelli et al., 2025).

18.6 CONCLUSION

This chapter examines the conservation of wildlife biodiversity in urban ecosystems, emphasising how these environments differ fundamentally from natural and rural systems. Urban areas represent tightly coupled human–natural systems, where human activities interact with eco-evolutionary dynamics and shape biodiversity patterns while being influenced by them. BES and BEF research continue to demonstrate that biodiversity underpins the stability of populations, communities, and ecosystems, ensuring their capacity to contribute to human well-being. These fields also provide scientific evidence of how human activities affect biodiversity, identify thresholds beyond which ecosystems may lose resilience, and inform strategies to maintain or restore ecosystem stability.

Urban ecology has advanced significantly through technological and methodological innovations. Remote sensing and other geospatial tools can enable fine-scale biodiversity monitoring. At the same time, landscape genetics and genomics help disentangle the mechanisms driving spatial genetic patterns, assess adaptive potential, and inform strategies to preserve genetic diversity in fragmented urban landscapes. Likewise, trait-based functional diversity research has grown rapidly over the last two decades, providing a powerful framework for linking species' ecological roles to ecosystem processes and their responses to environmental stressors. Machine learning plays a critical role in processing global-scale datasets and the integration of diverse remote sensing inputs for fine-scale biodiversity and habitat monitoring. These advancements are essential for understanding how biodiversity contributes to urban sustainability and for identifying leverage points for conservation action.

While advances in scientific understanding are essential, they are not sufficient on their own to ensure meaningful biodiversity outcomes in urban settings. Robust governance structures, rooted in legislative frameworks and planning instruments at international, national, state, regional, and municipal levels, are pivotal in embedding biodiversity within the urban agenda. Moving forward, strengthening analytical capacity, improving ecological data collection, and fostering cross-departmental coordination are all critical steps towards integrating biodiversity into mainstream urban planning processes. To achieve these aims, targeted funding mechanisms and clearly articulated policy instruments must be established to elevate biodiversity alongside other urban development goals. Such support would empower municipalities to design and implement greening strategies that address both ecosystem service provision and the conservation of species and habitats.

Bridging current gaps in governance and implementation is vital to shaping urban landscapes that are ecologically resilient, socially equitable, and capable of sustaining both biodiversity and human well-being in the face of accelerating urbanisation. Doing so requires integrating science, policy, and practice to ensure that biodiversity is not an afterthought but a core element of sustainable urban futures.

DECLARATION OF GENERATIVE AI USE

During the preparation of this manuscript, the authors used ChatGPT 4.0 to assist literature searching, improve clarity and organisation of writing, and check grammar. The authors subsequently reviewed and edited the content and take full responsibility for the final version of the published chapter.

REFERENCES

Alberti, M. (2015). Eco-evolutionary dynamics in an urbanizing planet. *Trends in Ecology and Evolution*, *30*(2), 114–126. https://doi.org/10.1016/j.tree.2014.11.007

Alberti, M., Marzluff, J. M., Shulenberger, E., Bradley, G., Ryan, C., & Zumbrunnen, C. (2003). Integrating humans into ecology: Opportunities and challenges for studying urban ecosystems. *BioScience*, *53*(12), 1169–1179. https://doi.org/10.1641/0006-3568(2003)053[1169:IHIEOA]2.0.CO;2

Alberti, M., Palkovacs, E. P., Roches, S. D., Meester, L. D., Brans, K. I., Govaert, L., Grimm, N. B., Harris, N. C., Hendry, A. P., Schell, C. J., & Szulkin, M. (2020). The complexity of urban eco-evolutionary dynamics. *BioScience*, *70*(9), 772–793. https://doi.org/10.1093/biosci/biaa079

Allendorf, F. W. (1986). Genetic drift and the loss of alleles versus heterozygosity. *Zoo Biology*, *5*(2), 181–190. https://doi.org/10.1002/zoo.1430050212

Banks, S. C., Lindenmayer, D. B., Ward, S. J., & Taylor, A. C. (2005). The effects of habitat fragmentation via forestry plantation establishment on spatial genotypic structure in the small marsupial carnivore, Antechinus agilis. *Molecular Ecology*, *14*(6), 1667–1680. https://doi.org/10.1111/j.1365-294X.2005.02525.x

Belgiu, M., & Stein, A. (2019). Spatiotemporal image fusion in remote sensing. *Remote Sensing*, *11*(7), Article 818. https://doi.org/10.3390/rs11070818

Biggs, C. R., Yeager, L. A., Bolser, D. G., Bonsell, C., Dichiera, A. M., Hou, Z., Keyser, S. R., Khursigara, A. J., Lu, K., Muth, A. F., Negrete Jr., B. & Erisman, B. E. (2020). Does functional redundancy affect ecological stability and resilience? A review and meta-analysis. *Ecosphere*, *11*(7), Article e03184. https://doi.org/10.1002/ecs2.3184

Biswas, S. R., & Mallik, A. U. (2011). Species diversity and functional diversity relationship varies with disturbance intensity. *Ecosphere*, *2*(4), 1–10. https://doi.org/10.1890/ES10-00206.1

Blight, L. K., Bertram, D. F., & Kroc, E. (2019). Evaluating UAV-based techniques to census an urban-nesting gull population on Canada's Pacific coast. *Journal of Unmanned Vehicle Systems*, *7*(4), 312–324. https://doi.org/10.1139/juvs-2019-0005

Booy, G., Hendriks, R. J. J., Smulders, M. J. M., Van Groenendael, J. M., & Vosman, B. (2000). Genetic diversity and the survival of populations. *Plant Biology*, *2*(4), 379–395. https://doi.org/10.1055/s-2000-5958

Buffam, I., Hagemann, F. A., Emilsson, T., Gamstetter, D., Palsdottir, A. M., Randrup, T. B., Yeshitela, K., & Ode Sang, Å. (2022). Priorities and barriers for urban ecosystem service provision: A comparison of stakeholder perspectives from three cities. *Frontiers in Sustainable Cities*, *4*, Article 838971. https://doi.org/10.3389/frsc.2022.838971

Cadotte, M. W., Carscadden, K., & Mirotchnick, N. (2011). Beyond species: Functional diversity and the maintenance of ecological processes and services. *Journal of Applied Ecology*, *48*(5), 1079–1087. https://doi.org/10.1111/j.1365-2664.2011.02048.x

Capotorti, G., Del Vico, E., Anzellotti, I., & Celesti-Grapow, L. (2016). Combining the conservation of biodiversity with the provision of ecosystem services in urban green infrastructure planning: Critical features arising from a case study in the metropolitan area of Rome. *Sustainability*, *9*(1), Article 10. https://doi.org/10.3390/su9010010

Capotorti, G., Ortí, M. M. A., Copiz, R., Fusaro, L., Mollo, B., Salvatori, E., & Zavattero, L. (2019). Biodiversity and ecosystem services in urban green infrastructure planning: A case study from the metropolitan area of Rome (Italy). *Urban Forestry and Urban Greening*, *37*, 87–96. https://doi.org/10.1016/j.ufug.2017.12.014

Cardinale, B. J., Duffy, J. E., Gonzalez, A., Hooper, D. U., Perrings, C., Venail, P., Narwani, A., Mace, G. M., Tilman, D., Wardle, D. A., & Kinzig, A. P. (2012). Biodiversity loss and its impact on humanity. *Nature*, *486*, 59–67. https://doi.org/10.1038/nature11148

Chao, A., Chiu, C. H., & Jost, L. (2014). Unifying species diversity, phylogenetic diversity, functional diversity, and related similarity and differentiation measures through Hill numbers. *Annual Review of Ecology, Evolution, and Systematics*, *45*(1), 297–324. https://doi.org/10.1146/annurev-ecolsys-120213-091540

Chen, S., Wang, H., & Xu, W. (2025). Bird richness as a mediator between greenspace and mental health relationships. *Landscape and Urban Planning, 259*, Article 105360. https://doi.org/10.1016/j.landurbplan.2025.105360

Chen, Y., Jiang, Z., Fan, P., Ericson, P. G., Song, G., Luo, X., Lei, F., & Qu, Y. (2022). The combination of genomic offset and niche modelling provides insights into climate change-driven vulnerability. *Nature Communications, 13*, Article 4821. https://doi.org/10.1038/s41467-022-32546-z

Chua, K. W., Tan, H. H., & Yeo, D. C. (2019). Loss of endemic fish species drives impacts on functional richness, redundancy and vulnerability in freshwater ecoregions of Sundaland. *Biological Conservation, 234*, 72–81. https://doi.org/10.1016/j.biocon.2019.03.019

City of Cincinnati. (2020). *Natural resources management plan*. https://www.cincinnati-oh.gov/cincyparks/natural-resources/

City of San Diego. (2025). *Biodiversity*. https://www.sandiego.gov/planning/work/biodiversity

Connop, S., Vandergert, P., Eisenberg, B., Collier, M. J., Nash, C., Clough, J., & Newport, D. (2016). Renaturing cities using a regionally-focused biodiversity-led multifunctional benefits approach to urban green infrastructure. *Environmental Science and Policy, 62*, 99–111. https://doi.org/10.1016/j.envsci.2016.01.013

Convention on Biological Diversity. (2006). *Article 2. Use of terms*. https://www.cbd.int/convention/articles?a=cbd-02

Convention on Biological Diversity. (2020). *Global biodiversity outlook 5*. Secretariat of the Convention on Biological Diversity. https://www.cbd.int/gbo5

Costanza, R. (2020). Valuing natural capital and ecosystem services toward the goals of efficiency, fairness, and sustainability. *Ecosystem Services, 43*, Article 101096. https://doi.org/10.1016/j.ecoser.2020.101096

Council on Environmental Quality (US). (1993). *Incorporating biodiversity considerations into environmental impact analysis under the National Environmental Policy Act*. Executive Office of the President. https://ceq.doe.gov/docs/ceq-publications/Incorporating_Biodiversity_1993.pdf

Dearborn, D. C., & Kark, S. (2010). Motivations for conserving urban biodiversity. *Conservation Biology, 24*(2), 432–440. https://doi.org/10.1111/j.1523-1739.2009.01328.x

de Camargo, J. F., da Silva, F. L. & Smith, W. S. (2022). City biodiversity index and the cities-biodiversity relationship: A case study for Sorocaba, SP, Brazil. *Urban Ecosystems, 25*, 673–689. https://doi.org/10.1007/s11252-021-01178-4

De Kort, H., Prunier, J. G., Ducatez, S., Honnay, O., Baguette, M., Stevens, V. M., & Blanchet, S. (2021). Life history, climate and biogeography interactively affect worldwide genetic diversity of plant and animal populations. *Nature Communications, 12*(1), Article 516. https://doi.org/10.1038/s41467-021-20958-2

Des Roches, S., Brans, K. I., Lambert, M. R., Rivkin, L. R., Savage, A. M., Schell, C. J., Correa, C., Meester, L. D., Diamond, S. E., Grimm, N. B., Harris, N. C., Govaert, L., Hendry, A. P., Johnson, M. T. J., Munshi-South, J., Palkovacs, E. P., Szulkin, M., Urban, M. C., Verrelli, B. C., … & Alberti, M. (2021). Socio-eco-evolutionary dynamics in cities. *Evolutionary Applications, 14*(1), 248–267. https://doi.org/10.1111/eva.13065

Diao, C., & Wang, L. (2014). Development of an invasive species distribution model with fine-resolution remote sensing. *International Journal of Applied Earth Observation and Geoinformation, 30*, 65–75. https://doi.org/10.1016/j.jag.2014.01.015

Dronova, I., & Taddeo, S. (2022). Remote sensing of phenology: Towards the comprehensive indicators of plant community dynamics from species to regional scales. *Journal of Ecology, 110*(7), 1460–1484. https://doi.org/10.1111/1365-2745.13897

Duffy, J. E. (2009). Why biodiversity is important to the functioning of real-world ecosystems. *Frontiers in Ecology and the Environment, 7*(8), 437–444. https://doi.org/10.1890/070195

Echeverri, A., Karp, D. S., Naidoo, R., Tobias, J. A., Zhao, J., & Chan, K. M. A. (2020). Can avian functional traits predict cultural ecosystem services? *People and Nature, 2*(1), 138–151. https://doi.org/10.1002/pan3.10058

Ellegren, H., & Galtier, N. (2016). Determinants of genetic diversity. *Nature Reviews Genetics, 17*(7), 422–433. https://doi.org/10.1038/nrg.2016.58

European Commission. (n.d.). *Environmental topics*. https://environment.ec.europa.eu/topics_en

European Environment Agency. (2020). *Green infrastructure in Italy*. Biodiversity Information System for Europe. https://biodiversity.europa.eu/countries/italy/green-infrastructure

Ezimand, K., Chahardoli, M., Azadbakht, M., & Matkan, A. A. (2021). Spatiotemporal analysis of land surface temperature using multi-temporal and multi-sensor image fusion techniques. *Sustainable Cities and Society, 64*, Article 102508. https://doi.org/10.1016/j.scs.2020.102508

Faith, D. P. (1992). Conservation evaluation and phylogenetic diversity. *Biological Conservation, 61*(1), 1–10. https://doi.org/10.1016/0006-3207(92)91201-3

Folke, C., Carpenter, S., Walker, B., Scheffer, M., Elmqvist, T., Gunderson, L., & Holling, C. S. (2004). Regime shifts, resilience, and biodiversity in ecosystem management. *Annual Review of Ecology, Evolution and Systematics*, *35*(1), 557–581. https://doi.org/10.1146/annurev.ecolsys.35.021103.105711

Forman, R. T. (1995). Some general principles of landscape and regional ecology. *Landscape Ecology*, *10*(3), 133–142. https://doi.org/10.1007/BF00133027

Francis, R. A., & Chadwick, M. A. (2013). *Urban Ecosystems: Understanding the Human Environment*. Routledge. https://doi.org/10.4324/9780203133644

Frazier, A. E., & Wang, L. (2013). Modeling landscape structure response across a gradient of land cover intensity. *Landscape Ecology*, *28*, 233–246. https://doi.org/10.1007/s10980-012-9839-8

Fu, B., Deng, L., Sun, W., He, H., Li, H., Wang, Y., & Wang, Y. (2024). Quantifying vegetation species functional traits along hydrologic gradients in karst wetland based on 3D mapping with UAV hyperspectral point cloud. *Remote Sensing of Environment*, *307*, Article 114160. https://doi.org/10.1016/j.rse.2024.114160

Gerber, L. R., Schwartz, M. W., & Dreiss, L. M. (2023). The US needs a national biodiversity strategy. *Conservation Science and Practice*, *5*(11), Article e13028. https://doi.org/10.1111/csp2.13028

Ghassemian, H. (2016). A review of remote sensing image fusion methods. *Information Fusion*, *32*, 75–89. https://doi.org/10.1016/j.inffus.2016.03.003

Grab, H., Branstetter, M. G., Amon, N., Urban-Mead, K. R., Park, M. G., Gibbs, J., Blitzer, E. J., Poveda, K., Loeb, G., & Danforth, B. N. (2019). Agriculturally dominated landscapes reduce bee phylogenetic diversity and pollination services. *Science*, *363*(6424), 282–284. https://doi.org/10.1126/science.aat6016

Grabowski, Z. J., McPhearson, T., & Pickett, S. T. (2023). Transforming US urban green infrastructure planning to address equity. *Landscape and Urban Planning*, *229*, Article 104591. https://doi.org/10.1016/j.landurbplan.2022.104591

Grant, A., Millward, A. A., Edge, S., Roman, L. A., & Teelucksingh, C. (2022). Where is environmental justice? A review of US urban forest management plans. *Urban Forestry and Urban Greening*, *77*, Article 127737. https://doi.org/10.1016/j.ufug.2022.127737

Green Infrastructure Partnership. (2025). *GI policy in the UK*. https://gipartnership.org.uk/gi-policy-in-the-uk/

Greenbaum, G., Templeton, A. R., Zarmi, Y., & Bar-David, S. (2014). Allelic richness following population founding events–a stochastic modeling framework incorporating gene flow and genetic drift. *PloS One*, *9*(12), Article e115203. https://doi.org/10.1371/journal.pone.0115203

Gregorius, H. R. (1978). The concept of genetic diversity and its formal relationship to heterozygosity and genetic distance. *Mathematical Biosciences*, *41*(3–4), 253–271. https://doi.org/10.1016/0025-5564(78)90040-8

Guadilla-Saez, S., Pardo-de-Santayana, M., Reyes-García, V., & Svenning, J. C. (2019). Biodiversity conservation effectiveness provided by a protection status in temperate forest commons of north Spain. *Forest Ecology and Management*, *433*, 656–666. https://doi.org/10.1016/j.foreco.2018.11.040

Hansen, M. M., Olivieri, I., Waller, D. M., Nielsen, E. E., & GeM Working Group. (2012). Monitoring adaptive genetic responses to environmental change. *Molecular Ecology*, *21*(6), 1311–1329. https://doi.org/10.1111/j.1365-294X.2011.05463.x

Hartling, S., Sagan, V., & Maimaitijiang, M. (2021). Urban tree species classification using UAV-based multi-sensor data fusion and machine learning. *GIScience and Remote Sensing*, *58*(8), 1250–1275. https://doi.org/10.1080/15481603.2021.1974275

Hedden-Dunkhorst, B., Braat, L., & Wittmer, H. (2015). TEEB emerging at the country level: Challenges and opportunities. *Ecosystem Services*, *14*, 37–44. https://doi.org/10.1016/j.ecoser.2015.03.007

Hoban, S., Bruford, M. W., da Silva, J. M., Funk, W. C., Frankham, R., Gill, M. J., Grueber, C. E., Heuertz, M., Hunter, M. E., Kershaw, F., & Lacy, R. C. (2023). Genetic diversity goals and targets have improved, but remain insufficient for clear implementation of the post-2020 global biodiversity framework. *Conservation Genetics, 24*, 181–191. https://doi.org/10.1007/s10592-022-01492-0

Hoban, S., Campbell, C. D., da Silva, J. M., Ekblom, R., Funk, W. C., Garner, B. A., Godoy, J. A., Kershaw, F., MacDonald, A. J., Mergeay, J., & Minter, M. (2021). Genetic diversity is considered important but interpreted narrowly in country reports to the Convention on Biological Diversity: Current actions and indicators are insufficient. *Biological Conservation*, *261*, Article 109233. https://doi.org/10.1016/j.biocon.2021.109233

Hordley, L. A., Gillings, S., Petchey, O. L., Tobias, J. A., & Oliver, T. H. (2021). Diversity of response and effect traits provides complementary information about avian community dynamics linked to ecological function. *Functional Ecology*, *35*(9), 1938–1950. https://doi.org/10.1111/1365-2435.13865

Hu, R., & Tong, S. T. Y. (2022). The use of remotely sensed data to model habitat selections of pileated woodpeckers (*Dryocopus pileatus*) in fragmented landscapes. *Forest Ecology and Management*, *521*, Article 120433. https://doi.org/10.1016/j.foreco.2022.120433

Hu, R., & Tong, S. T. Y. (2025). Habitat connectivity modeling for urban conservation planning: A case study of pileated woodpecker (*Dryocopus pileatus*) in Hamilton County, Ohio, USA. *Ecological Indicators*, *176*, Article 113656. https://doi.org/10.1016/j.ecolind.2025.113656

Huang, M., Liu, X., Cadotte, M. W., & Zhou, S. (2020). Functional and phylogenetic diversity explain different components of diversity effects on biomass production. *Oikos*, *129*(8), 1185–1195. https://doi.org/10.1111/oik.07032

iNaturalist (n.d.). *iNaturalist*. https://www.inaturalist.org

Ives, C. D., Lentini, P. E., Threlfall, C. G., Ikin, K., Shanahan, D. F., Garrard, G. E., Bekessy, S. A., Fuller, R. A., Mumaw, L., Rayner, L., & Rowe, R. (2016). Cities are hotspots for threatened species. *Global Ecology and Biogeography*, *25*(1), 117–126. https://doi.org/10.1111/geb.12404

Ji, W., Ma, J., Twibell, R. W., & Underhill, K. (2006). Characterizing urban sprawl using multi-stage remote sensing images and landscape metrics. *Computers, Environment and Urban Systems*, *30*(6), 861–879. https://doi.org/10.1016/j.compenvurbsys.2005.09.002

Kambites, C., & Owen, S. (2006). Renewed prospects for green infrastructure planning in the UK. *Planning, Practice and Research*, *21*(4), 483–496. https://doi.org/10.1080/02697450601173413

Karimi, J. D., Corstanje, R., & Harris, J. A. (2021). Bundling ecosystem services at a high resolution in the UK: Trade-offs and synergies in urban landscapes. *Landscape Ecology*, *36*(6), 1817–1835. https://doi.org/10.1007/s10980-021-01252-4

Knapp, S., Winter, M., & Klotz, S. (2017). Increasing species richness but decreasing phylogenetic richness and divergence over a 320-year period of urbanization. *Journal of Applied Ecology*, *54*(4), 1152–1160. https://doi.org/10.1111/1365-2664.12826

Kowarik, I. (2011). Novel urban ecosystems, biodiversity, and conservation. *Environmental Pollution*, *159*(8–9), 1974–1983. https://doi.org/10.1016/j.envpol.2011.02.022

Kuenzer, C., Ottinger, M., Wegmann, M., Guo, H., Wang, C., Zhang, J., Dech, S., & Wikelski, M. (2014). Earth observation satellite sensors for biodiversity monitoring: Potentials and bottlenecks. *International Journal of Remote Sensing*, *35*(18), 6599–6647. https://doi.org/10.1080/01431161.2014.964349

Kukunda, C. B., Duque-Lazo, J., González-Ferreiro, E., Thaden, H., & Kleinn, C. (2018). Ensemble classification of individual Pinus crowns from multispectral satellite imagery and airborne LiDAR. *International Journal of Applied Earth Observation and Geoinformation*, *65*, 12–23. https://doi.org/10.1016/j.jag.2017.09.016

Lähde, E., & Di Marino, M. (2019). Multidisciplinary collaboration and understanding of green infrastructure results from the cities of Tampere, Vantaa and Jyväskylä (Finland). *Urban Forestry and Urban Greening*, *40*, 63–72. https://doi.org/10.1016/j.ufug.2018.03.012

Larondelle, N., & Haase, D. (2013). Urban ecosystem services assessment along a rural–urban gradient: A cross-analysis of European cities. *Ecological Indicators*, *29*, 179–190. https://doi.org/10.1016/j.ecolind.2012.12.022

Lausch, A., Bannehr, L., Beckmann, M., Boehm, C., Feilhauer, H., Hacker, J. M., Heurich, M., Jung, A., Klenke, R., Neumann, C., & Pause, M. (2016). Linking earth observation and taxonomic, structural and functional biodiversity: Local to ecosystem perspectives. *Ecological Indicators*, *70*, 317–339. https://doi.org/10.1016/j.ecolind.2016.06.022

Le Bagousse-Pinguet, Y., Soliveres, S., Gross, N., Torices, R., Berdugo, M., & Maestre, F. T. (2019). Phylogenetic, functional, and taxonomic richness have both positive and negative effects on ecosystem multifunctionality. *Proceedings of the National Academy of Sciences*, *116*(17), Article 8419–8424.

Leong, R. A., Fung, T. K., Sachidhanandam, U., Drillet, Z., Edwards, P. J., & Richards, D. R. (2020). Use of structural equation modeling to explore influences on perceptions of ecosystem services and disservices attributed to birds in Singapore. *Ecosystem Services*, *46*, Article 101211. https://doi.org/10.1016/j.ecoser.2020.101211

Leveau, L. M., Pereyra, L. C., & Ortega-Álvarez, R. (2025). Urban biodiversity in the Global South. *Frontiers in Ecology and Evolution, 13*, Article 1591037. https://doi.org/10.3389/fevo.2025.1591037

Li, G., Fang, C., Li, Y., Wang, Z., Sun, S., He, S., Qi, W., Bao, C., Ma, H., Fan, Y., & Feng, Y. (2022). Global impacts of future urban expansion on terrestrial vertebrate diversity. *Nature Communications*, *13*(1), Article 1628. https://doi.org/10.1038/s41467-022-29324-2

Li, J., Hong, D., Gao, L., Yao, J., Zheng, K., Zhang, B., & Chanussot, J. (2022). Deep learning in multimodal remote sensing data fusion: A comprehensive review. *International Journal of Applied Earth Observation and Geoinformation, 112*, Article 102926. https://doi.org/10.1016/j.jag.2022.102926

Li, S., Kang, X., Fang, L., Hu, J., & Yin, H. (2017). Pixel-level image fusion: A survey of the state of the art. *Information Fusion, 33*, 100–112. https://doi.org/10.1016/j.inffus.2016.05.004

Luck, G. W., Lavorel, S., McIntyre, S., & Lumb, K. (2012). Improving the application of vertebrate trait-based frameworks to the study of ecosystem services. *Journal of Animal Ecology, 81*(5), 1065–1076. https://doi.org/10.1111/j.1365-2656.2012.01974.x

Mackey, C. W., Lee, X., & Smith, R. B. (2012). Remotely sensing the cooling effects of city scale efforts to reduce urban heat island. *Building and Environment, 49*, 348–358. https://doi.org/10.1016/j.buildenv.2011.08.004

Manel, S., & Holderegger, R. (2013). Ten years of landscape genetics. *Trends in Ecology and Evolution, 28*(10), 614–621. https://doi.org/10.1016/j.tree.2013.05.012

Manel, S., Schwartz, M. K., Luikart, G., & Taberlet, P. (2003). Landscape genetics: Combining landscape ecology and population genetics. *Trends in Ecology and Evolution, 18*(4), 189–197. https://doi.org/10.1016/S0169-5347(03)00008-9

Mastretta-Yanes, A., da Silva, J. M., Grueber, C. E., Castillo-Reina, L., Köppä, V., Forester, B. R., Funk, W. C., Heuertz, M., Ishihama, F., Jordan, R., Mergeay, J., Paz-Vinas, I., Rincon-Parra, V. J., Rodriguez-Morales, M. A., Arredondo-Amezcua, L., Brahy, G., DeSaix, M., Durkee, L., Hamilton, A., … & Hoban, S. (2024). Multinational evaluation of genetic diversity indicators for the Kunming-Montreal Global Biodiversity Framework. *Ecology Letters, 27*(7), Article e14461.

McRae, B. H., & Beier, P. (2007). Circuit theory predicts gene flow in plant and animal populations. *Proceedings of the National Academy of Sciences, 104*(50), 19885–19890. https://doi.org/10.1073/pnas.0706568104

Meerow, S., Newell, J. P., & Stults, M. (2016). Defining urban resilience: A review. *Landscape and Urban Planning, 147*, 38–49. https://doi.org/10.1016/j.landurbplan.2015.11.011

Mell, I. C. (2014). Aligning fragmented planning structures through a green infrastructure approach to urban development in the UK and USA. *Urban Forestry and Urban Greening, 13*(4), 612–620. https://doi.org/10.1016/j.ufug.2014.07.007

Meng, X., Liu, Q., Shao, F., & Li, S. (2022). Spatio–temporal–spectral collaborative learning for spatio–temporal fusion with land cover changes. *IEEE Transactions on Geoscience and Remote Sensing, 60*, 1–16. https://doi.org/10.1109/TGRS.2022.3185459

Miles, L. S., Rivkin, L. R., Johnson, M. T., Munshi-South, J., & Verrelli, B. C. (2019). Gene flow and genetic drift in urban environments. *Molecular Ecology, 28*(18), 4138–4151. https://doi.org/10.1111/mec.15221

Mittermeier, R. A., Mittermeier, C. G., & Gil P. R. (1997). *Megadiversity: Earth's Biologically Wealthiest Nations*. CEMEX.

Moreno-Peñaranda, R. (2021). Synergies in urban environmental policy: Ecosystem services and biodiversity co-benefits in São Paulo city, Brazil. In K. Ito (Ed.), *Urban Biodiversity and Ecological Design for Sustainable Cities*. Springer. https://doi.org/10.1007/978-4-431-56856-8_11

Nagylaki, T. (1998). Fixation indices in subdivided populations. *Genetics, 148*(3), 1325–1332. https://doi.org/10.1093/genetics/148.3.1325

Natural Capital Project. (n.d.). *InVEST*. https://naturalcapitalproject.stanford.edu/software/invest

Neyns, R., & Canters, F. (2022). Mapping of urban vegetation with high-resolution remote sensing: A review. *Remote Sensing, 14*(4), 1031. https://doi.org/10.3390/rs14041031

Ohio Department of Natural Resources. (2015). *Ohio's state wildlife action plan 2015*. Division of Wildlife. https://dam.assets.ohio.gov/image/upload/ohiodnr.gov/documents/wildlife/wildlife-management/OH_SWAP_2015.pdf

Ohio Public Works Commission. (2025). *District 2: Clean Ohio Program*. https://publicworks.ohio.gov/districts/districts-1-19/district-2

Partecke, J. (2013). Mechanisms of phenotypic responses following colonization of urban areas: from plastic to genetic adaptation. In D. Gil & H. Brumm (Eds.), *Avian Urban Ecology: Behavioural and Physiological Adaptations* (pp. 131–142). Oxford University Books.

Peet, R. K. (1974). The measurement of species diversity. *Annual Review of Ecology and Systematics, 5*, 285–307. https://www.jstor.org/stable/2096890

Pettorelli, N., Schulte to Bühne, H., Tulloch, A., Dubois, G., Macinnis-Ng, C., Queirós, A. M., Keith, D. A., Wegmann, M., Schrodt, F., Stellmes, M., & Sonnenschein, R. (2018). Satellite remote sensing of ecosystem functions: Opportunities, challenges and way forward. *Remote Sensing in Ecology and Conservation, 4*(2), 71–93. https://doi.org/10.1002/rse2.59

Pettorelli, N., Williams, J., Schulte to Bühne, H., & Crowson, M. (2025). Deep learning and satellite remote sensing for biodiversity monitoring and conservation. *Remote Sensing in Ecology and Conservation, 11*(2), 123–132. https://doi.org/10.1002/rse2.415

Pickett, S. T. A., Cadenasso, M. L., Grove, J. M., Boone, C. G., Groffman, P. M., Irwin, E., Kaushal, S. S., Marshall, V., McGrath, B. P., Nilon, C. H., & Pouyat, R. V. (2011). Urban ecological systems: Scientific foundations and a decade of progress. *Journal of Environmental Management, 92*(3), 331–362. https://doi.org/10.1016/j.jenvman.2010.08.022

Pickett, S. T. A., Cadenasso, M. L., Grove, J. M., Nilon, C. H., Pouyat, R. V., Zipperer, W. C., & Costanza, R. (2001). Urban ecological systems: Linking terrestrial ecological, physical, and socioeconomic components of metropolitan areas. *Annual Review of Ecology and Systematics*, *32*(1), 127–157. https://www.jstor.org/stable/2678637

Pickett, S. T. A., Cadenasso, M. L., Rosi-Marshall, E. J., Belt, K. T., Groffman, P. M., Grove, J. M., Irwin, E. G., Kaushal, S. S., LaDeau, S. L., Nilon, C. H., & Swan, C. M. (2017). Dynamic heterogeneity: A framework to promote ecological integration and hypothesis generation in urban systems. *Urban Ecosystems*, *20*(1), 1–14. https://doi.org/10.1007/s11252-016-0574-9

proGIreg. (n.d.). *Co-creation of nature-based solutions: Living labs for nature and people*. https://progireg.eu/news/?c=search&uid=vVla9KBs

Rao, A. S., Chhawri, R., Chauhan, A., Yadav, S. S., Meena, K. C., & Bansal, P. (2024). Plant functional traits: A key framework for understanding and managing ecosystem responses to global environmental challenges. In N. Kumar & H. Singh (Eds.), *Plant Functional Traits for Improving Productivity* (pp. 287–299). Springer.

Richardson, J. L., Urban, M. C., Bolnick, D. I., & Skelly, D. K. (2014). Microgeographic adaptation and the spatial scale of evolution. *Trends in Ecology and Evolution*, *29*(3), 165–176. https://doi.org/10.1016/j.tree.2014.01.002

Richardson, M. (2025). Modelling nature connectedness within environmental systems: Human-nature relationships from 1800 to 2020 and beyond. *Earth, 6*(3), 82. https://doi.org/10.3390/earth6030082

Richter-Boix, A., Quintela, M., Kierczak, M., Franch, M., & Laurila, A. (2013). Fine-grained adaptive divergence in an amphibian: Genetic basis of phenotypic divergence and the role of nonrandom gene flow in restricting effective migration among wetlands. *Molecular Ecology*, *22*(5), 1322–1340. https://doi.org/10.1111/mec.12181

Ring, I., Hansjürgens, B., Elmqvist, T., Wittmer, H., & Sukhdev, P. (2010). Challenges in framing the economics of ecosystems and biodiversity: The TEEB initiative. *Current Opinion in Environmental Sustainability*, *2*(1–2), 15–26. https://doi.org/10.1016/j.cosust.2010.03.005

Rivkin, L. R., Santangelo, J. S., Alberti, M., Aronson, M. F. J., de Keyzer, C. W., Diamond, S. E., Fortin, M. J., Frazee, L. J., Gordon, A. J., Hendry, A. P., Liu, Y., Losos, J. B., MacIvor, J. S., Martin, R. A., McDonnell, M. J., Miles, L. S., Munshi-South, J., Ness, R. W., Newman, A. E. M., … & Johnson, M. T. J. (2019). A roadmap for urban evolutionary ecology. *Evolutionary Applications*, *12*(3), 384–398. https://doi.org/10.1111/eva.12734

Robb, B., Huang, Q., Sexton, J. O., Stoner, D., & Leimgruber, P. (2019). Environmental differences between migratory and resident ungulates: Predicting movement strategies in rocky mountain mule deer (*Odocoileus hemionus*) with remotely sensed plant phenology, snow, and land cover. *Remote Sensing*, *11*(17), 1980. https://doi.org/10.3390/rs11171980

Robin, E., Steenmans, K., & Acuto, M. (2019). Harnessing inclusive urban knowledge for the implementation of the New Urban Agenda. *Urban Research and Practice*, *12*(2), 137–155. https://doi.org/10.1080/17535069.2017.1414870

Rocchini, D., Luque, S., Pettorelli, N., Bastin, L., Doktor, D., Faedi, N., Feilhauer, H., Féret, J. B., Foody, G. M., Gavish, Y., & Godinho, S. (2018). Measuring β-diversity by remote sensing: A challenge for biodiversity monitoring. *Methods in Ecology and Evolution, 9*(8), 1787–1798. https://doi.org/10.1111/2041-210X.12941

Rogan, J., & Chen, D. (2004). Remote sensing technology for mapping and monitoring land-cover and land-use change. *Progress in Planning*, *61*(4), 301–325. https://doi.org/10.1016/S0305-9006(03)00066-7

Rossi, C., Kneubühler, M., Schütz, M., Schaepman, M. E., Haller, R. M., & Risch, A. C. (2022). Spatial resolution, spectral metrics and biomass are key aspects in estimating plant species richness from spectral diversity in species-rich grasslands. *Remote Sensing in Ecology and Conservation, 8*(3), 297–314. https://doi.org/10.1002/rse2.244

Salomaa, A., Paloniemi, R., Kotiaho, J. S., Kettunen, M., Apostolopoulou, E., & Cent, J. (2016). Can green infrastructure help to conserve biodiversity? *Environment and Planning C: Politics and Space*, *35*(2), 265–288. https://doi.org/10.1177/0263774X16649363

Santangelo, J. S., Ness, R. W., Cohan, B., Fitzpatrick, C. R., Innes, S. G., Koch, S., Miles, L. S., Munim, S., Peres-Neto, P. R., Prashad, C., Tong, A. T., Aguirre, W. E., Akinwole, P. O., Alberti, M., Álvarez, J., Anderson, J. T., Anderson, J. J., Ando, Y., Andrew, N. R., … & Lampei, C. (2022). Global urban environmental change drives adaptation in white clover. *Science,* 375(6586), 1275–1281. https://doi.org/10.1126/science.abk0989

Sarkki, S., Niemelä, J., Tinch, R., Jäppinen, J. P., Nummelin, M., Toivonen, H., & Von Weissenberg, M. (2016). Are national biodiversity strategies and action plans appropriate for building responsibilities for mainstreaming biodiversity across policy sectors? The case of Finland. *Journal of Environmental Planning and Management*, *59*(8), 1377–1396. https://doi.org/10.1080/09640568.2015.1076384

Schilling, J., & Logan, J. (2008). Greening the rust belt: A green infrastructure model for right sizing America's shrinking cities. *Journal of the American Planning Association, 74*(4), 451–466. https://doi.org/10.1080/01944360802354956

Schindler, S., Von Wehrden, H., Poirazidis, K., Wrbka, T., & Kati, V. (2013). Multiscale performance of landscape metrics as indicators of species richness of plants, insects and vertebrates. *Ecological Indicators, 31*, 41–48. https://doi.org/10.1016/j.ecolind.2012.04.012

Schneider, F. D., Morsdorf, F., Schmid, B., Petchey, O. L., Hueni, A., Schimel, D. S., & Schaepman, M. E. (2017). Mapping functional diversity from remotely sensed morphological and physiological forest traits. *Nature Communications, 8*(1), 1441. https://doi.org/10.1038/s41467-017-01530-3

Schwarz, N., Moretti, M., Bugalho, M. N., Davies, Z. G., Haase, D., Hack, J., Hof, A., Melero, Y., Pett, T. J., & Knapp, S. (2017). Understanding biodiversity-ecosystem service relationships in urban areas: A comprehensive literature review. *Ecosystem Services, 27*, 161–171. https://doi.org/10.1016/j.ecoser.2017.08.014

Schweiger, A. K., & Laliberté, E. (2022). Plant beta-diversity across biomes captured by imaging spectroscopy. *Nature Communications, 13*(1), 2767. https://doi.org/10.1038/s41467-022-30369-6

Seto, K. C., Güneralp, B., & Hutyra, L. R. (2012). Global forecasts of urban expansion to 2030 and direct impacts on biodiversity and carbon pools. *Proceedings of the National Academy of Sciences, 109*(40), 16083–16088. https://doi.org/10.1073/pnas.1211658109

Shifflett, S. D., Newcomer-Johnson, T., Yess, T., & Jacobs, S. (2019). Interdisciplinary collaboration on green infrastructure for urban watershed management: An Ohio case study. *Water, 11*(4), 738. https://doi.org/10.3390/w11040738

Shokirov, S., Jucker, T., Levick, S. R., Manning, A. D., Bonnet, T., Yebra, M., & Youngentob, K. N. (2023). Habitat highs and lows: Using terrestrial and UAV LiDAR for modelling avian species richness and abundance in a restored woodland. *Remote Sensing of Environment, 285*, Article 113326. https://doi.org/10.1016/j.rse.2022.113326

Simamora, T. I., Purbowo, S. D., & Laumonier, Y. (2021). Looking for indicator bird species in the context of forest fragmentation and isolation in West Kalimantan, Indonesia. *Global Ecology and Conservation, 27*, Article e01610. https://doi.org/10.1016/j.gecco.2021.e01610

Simpson, E. H. (1949). Measurement of diversity. *Nature, 163*(4148), 688–688. https://doi.org/10.1038/163688a0

Storfer, A., Murphy, M. A., Evans, J. S., Goldberg, C. S., Robinson, S., Spear, S. F., Dezzani, R., Delmelle, E., Vierling, L., & Waits, L. P. (2007). Putting the 'landscape' in landscape genetics. *Heredity, 98*(3), 128–142. https://doi.org/10.1038/sj.hdy.6800917

Sun, W., Chen, D., Li, Z., Li, S., Cheng, S., Niu, X., Cai, Y., Shi, Z., Wu, C., Yang, G. & Yang, X. (2024). Monitoring wetland plant diversity from space: Progress and perspective. *International Journal of Applied Earth Observation and Geoinformation, 130*, Article 103943. https://doi.org/10.1016/j.jag.2024.103943

Swartz, T. M., Gleditsch, J. M., & Behm, J. E. (2023). A functional trait approach reveals the effects of landscape context on ecosystem services provided by urban birds. *Landscape and Urban Planning, 234*, Article 104724. https://doi.org/10.1016/j.landurbplan.2023.104724

Syrbe, R. U., & Walz, U. (2012). Spatial indicators for the assessment of ecosystem services: Providing, benefiting and connecting areas and landscape metrics. *Ecological Indicators, 21*, 80–88. https://doi.org/10.1016/j.ecolind.2012.02.013

Taylor, P. D., Fahrig, L., Henein, K., & Merriam, G. (1993). Connectivity is a vital element of landscape structure. *Oikos*, 571–573. https://doi.org/10.2307/3544927

Tran, T. J., Helmus, M. R., & Behm, J. E. (2020). Green infrastructure space and traits (GIST) model: Integrating green infrastructure spatial placement and plant traits to maximize multifunctionality. *Urban Forestry and Urban Greening, 49*, Article 126635. https://doi.org/10.1016/j.ufug.2020.126635

Tribot, A. S., Mouquet, N., Villéger, S., Raymond, M., Hoff, F., Boissery, P., Holon, F., & Deter, J. (2016). Taxonomic and functional diversity increase the aesthetic value of coralligenous reefs. *Scientific Reports, 6*(1), Article 34229. https://doi.org/10.1038/srep34229

United Nations. (2015). *Transforming our world: The 2030 agenda for sustainable development*. https://sdgs.un.org/2030agenda

United Nations. (2016). *The new urban agenda*. https://habitat3.org/the-new-urban-agenda

US Environmental Protection Agency. (2025). *Green infrastructure program*. U.S. Environmental Protection Agency. https://www.epa.gov/climate-change-water-sector/green-infrastructure-program

Van Sambeek, J. (2017). Ten-year urban forestry action plan. *Green Horizons. 21*(1), 4. https://www.fs.usda.gov/nrs/pubs/jrnl/2017/nrs_2017_vansambeek_001.pdf

Verrelli, B. C., Alberti, M., Des Roches, S., Harris, N. C., Hendry, A. P., Johnson, M. T., Savage, A. M., Charmantier, A., Gotanda, K. M., Govaert, L., & Miles, L. S. (2022). A global horizon scan for urban evolutionary ecology. *Trends in Ecology and Evolution, 37*(11), 1006–1019. https://doi.org/10.1016/j.tree.2022.07.012

Vinton, A. C., Gascoigne, S. J., Sepil, I., & Salguero-Gómez, R. (2022). Plasticity's role in adaptive evolution depends on environmental change components. *Trends in Ecology and Evolution, 37*(12), 1067–1078. https://doi.org/10.1016/j.tree.2022.08.008

Violle, C., Navas, M. L., Vile, D., Kazakou, E., Fortunel, C., Hummel, I., & Garnier, E. (2007). Let the concept of trait be functional! *Oikos, 116*(5), 882–892. https://doi.org/10.1111/j.0030-1299.2007.15559.x

Vizzari, M., Antonielli, F., Bonciarelli, L., Grohmann, D., & Menconi, M. E. (2025). Urban greenery mapping using object-based classification and multi-sensor data fusion in Google Earth Engine. *Urban Forestry and Urban Greening*, Article 128697. https://doi.org/10.1016/j.ufug.2025.128697

Wagner, B., & Egerer, M. (2022). Application of UAV remote sensing and machine learning to model and map land use in urban gardens. *Journal of Urban Ecology, 8*(1), Article juac008. https://doi.org/10.1093/jue/juac008

Waits, L. P., & Storfer, A. (2015). Basics of population genetics: quantifying neutral and adaptive genetic variation for landscape genetic studies. In N. Balkenhol, S. Cushman, A. Storfer, & L. P. Waits (Eds.), *Landscape Genetics: Concepts, Methods, Applications* (pp. 35–57). Wiley. https://doi.org/10.1002/9781118525258.ch03

Waldvogel, A. M., Feldmeyer, B., Rolshausen, G., Exposito-Alonso, M., Rellstab, C., Kofler, R., Mock, T., Schmid, K., Schmitt, I., Bataillon, T., & Savolainen, O. (2020). Evolutionary genomics can improve prediction of species' responses to climate change. *Evolution Letters, 4*(1), 4–18. https://doi.org/10.1002/evl3.154

Wang, Y., Kotze, D. J., Vierikko, K., & Niemelä, J. (2019). What makes urban greenspace unique–Relationships between citizens' perceptions on unique urban nature, biodiversity and environmental factors. *Urban Forestry and Urban Greening, 42*, 1–9. https://doi.org/10.1016/j.ufug.2019.04.005

Wang, Z., Chlus, A., Geygan, R., Ye, Z., Zheng, T., Singh, A., Couture, J. J., Cavender-Bares, J., Kruger, E. L., & Townsend, P. A. (2020). Foliar functional traits from imaging spectroscopy across biomes in eastern North America. *New Phytologist, 228*(2), 494–511. https://doi.org/10.1111/nph.16711

Weaver, W. (1949). Recent contributions to the mathematical theory of communication. In C. E. Shannon & W. Weaver (Eds.), *A mathematical theory of communication* (pp. 11–20). Urbana, IL: University of Illinois Press.

Wright, S. (1965). The interpretation of population structure by F-statistics with special regard to systems of mating. *Evolution, 19*(3), 395–420. https://doi.org/10.1111/j.1558-5646.1965.tb01731.x

Wu, J. (2014). Urban ecology and sustainability: The state-of-the-science and future directions. *Landscape and Urban Planning, 125*, 209–221. https://doi.org/10.1016/j.landurbplan.2014.01.018

Wu, J., He, C., Huang, G., & Yu, D. (2013). Urban landscape ecology: Past, present, and future. In B. Fu & K. B. Jones (Eds.), *Landscape Ecology for Sustainable Environment and Culture* (pp. 37–53). Springer. https://doi.org/10.1007/978-94-007-6530-6_3

Wu, J., & Loucks, O. L. (1995). From balance of nature to hierarchical patch dynamics: A paradigm shift in ecology. *The Quarterly Review of Biology, 70*(4), 439–466. https://doi.org/10.1086/419172

Xie, L., & Bulkeley, H. (2020). Nature-based solutions for urban biodiversity governance. *Environmental Science and Policy, 110*, 77–87. https://doi.org/10.1016/j.envsci.2020.04.002

Xu, H., Cao, Y., Yu, D., Cao, M., He, Y., Gill, M., & Pereira, H. M. (2021). Ensuring effective implementation of the post-2020 global biodiversity targets. *Nature Ecology and Evolution, 5*(4), 411–418. https://doi.org/10.1038/s41559-020-01375-y

Zhang, L., Yang, L., Chapman, C. A., Peres, C. A., Lee, T. M., & Fan, P. F. (2023). Growing disparity in global conservation research capacity and its impact on biodiversity conservation. *One Earth, 6*(2), 147–157. https://doi.org/10.1016/j.oneear.2023.01.003

Ziter, C. (2016). The biodiversity–ecosystem service relationship in urban areas: A quantitative review. *Oikos, 125*(6), 761–768. https://doi.org/10.1111/oik.02883

19 Evaluation of the Ecological Functions of Urban Rivers and Strategies for Their Restoration

A Case Study of the Dasha River in Shenzhen

Zhipeng Guo, Guiping Fu, and Xiaohan Yi

19.1 INTRODUCTION

Rivers are the lifeblood of cities and essential landscape elements, serving functions such as water storage, flood control, drainage, and transportation, while also acting as vital ecological corridors within urban areas (Chen 2017). Urban rivers play a crucial role in the development of sponge cities. Integrating natural and artificial measures to maximize rainwater storage, infiltration, and purification within urban environments—while ensuring effective drainage and flood safety—has garnered widespread attention. However, rapid urbanization poses significant threats to river ecosystem health, resulting in water quality degradation, loss of hydrological regulation, and sharp declines in biodiversity (Pan et al. 2015). Although urban areas cover only about 2% of the Earth's surface, they contain some of the most severely degraded ecosystems worldwide (Wen, Schoups and Van de Giesen 2017). Therefore, effectively assessing the ecological functions of rivers in urban built-up areas and proposing targeted restoration measures to promote sustainable biodiversity management have become important issues in the management of urban river ecosystems and sponge city planning today.

Many scholars have extensively studied river ecological functions in the context of urbanization (Han et al. 2016; Hoyer and Chang 2014; Jiang 2019; Wang et al. 2018). Common evaluation methods include modeling, value assessment, and indicator-based evaluation, such as Spain's Riparian Forest Quality Index (Munné et al. 2003), the UK's River Invertebrate Prediction and Classification System` (Wright 1995), Australia's River Assessment Scheme (Sims et al. 2013), Denmark's 5-S model (Verdonschot 2000), and Sweden's Riparian, Channel, and Environmental Index (Petersen 1992). Among these, indicator-based evaluation is the most widely used approach for assessing river ecological functions. Chinese researchers have applied this method to the Yangtze River basin (Cao, Zhou and Wu 2019; Chen 2019), Huai River basin (Li et al. 2016), Xin'an River basin (Yang et al. 2018), and Taizi River basin (Wan et al. 2013). Key factors considered by most researchers include water quality, riparian zone conditions, hydrology, and river channel morphology. Water quality reflects a river's ability to provide a suitable habitat for aquatic organisms and is primarily represented by indicators such as dissolved oxygen (DO), nutrient levels, and the comprehensive pollution index. Riparian zone conditions describe the vegetation cover, stability, and width of the riverbanks, emphasizing their buffering and supportive roles in sustaining river habitats. Hydrological conditions indicate fluctuations in water volume and the extent to which ecological water requirements are met, serving as the foundation for maintaining healthy river ecosystems. River channel morphology is characterized by indicators such as sinuosity, longitudinal connectivity, and the degree

DOI: 10.1201/9781003640899-24

of channel modification, illustrating the natural state of the river channel and its influence on habitat diversity. Indicator-based evaluation typically aligns with the ecological management needs of the study area, making it highly applicable for management purposes. However, it also has limitations: it often focuses on natural river functions while neglecting social functions (Wan et al. 2013); it tends to evaluate rivers at large spatial scales, overlooking small-scale rivers (Wan et al. 2013); and it usually selects rivers with continuous flow, ignoring artificially controlled rivers with interrupted flow due to dams and sluices (Cao and Li 2018). In addition, evaluation index systems for river ecological functions that emphasize aquatic organisms and aim to sustain biodiversity remain scarce. Consequently, existing index system methods are challenging to apply when assessing river ecological functions in urban built-up areas.

Shenzhen has over 310 rivers of various sizes, most of which are short, rain-fed, and have small runoff; only 13 have basins larger than 100 km^2 (Li 2019). Among these, the Dasha River in Nanshan District has undergone pollution interception, water replenishment, and greening projects since 1989, leading to continuous habitat changes and varying degrees of biological stress. However, the ecological status of the restored river channel remains unclear. This study builds upon and enhances existing indicator evaluation systems by incorporating the specific characteristics of the river. It employs a multi-factor comprehensive evaluation method, primarily based on aquatic organisms and supplemented by the river's natural functions. The study assesses the ecological functional status of the Dasha River at the river segment scale and proposes targeted restoration strategies. These findings provide valuable references for ecological function evaluation, biodiversity conservation, and ecosystem management of rivers in urban built-up areas.

To establish an integrated river ecological management approach, this study proposes a systematic analytical framework grounded in existing evaluation systems and restoration strategies. The framework emphasizes four key dimensions: (1) identifying the current ecological status and core issues of urban rivers; (2) developing an ecological function evaluation system at the river segment scale; (3) recommending targeted ecological restoration measures based on evaluation results; and (4) integrating innovative practices—such as the resource utilization of invasive riparian plants—to create novel governance and management support mechanisms. This framework directly addresses the primary challenges faced by urban river ecosystems, reinforcing the logical progression from problem diagnosis to system maintenance. While enhancing the representativeness of the Dasha River case, it also offers theoretical and practical guidance for the ecological management of similar urban rivers.

19.2 OVERVIEW OF THE STUDY AREA

The Dasha River originates from Yangtai Mountain (elevation 587.3 m) in western Shenzhen. It has a basin area of 92.99 km^2 and a main channel length of 13.7 km (Cai 2020; Bian et al. 2023). The tidal-influenced section extends 3.6 km, beginning at the outlet of the flood discharge energy dissipation pool of Changlingpi Reservoir. The river flows east to west through Tanglang and Pingshan villages, intersects with the Xili Reservoir flood discharge channel, then turns south and discharges into Shenzhen Bay at the Binhai Dasha River Interchange (Figure 19.1). The Dasha River traverses the entire Nanshan District, connecting Yangtai Mountain, Tanglang Mountain, and the coastal leisure belt. It serves as an important green ecological and innovation corridor in Shenzhen. University Town, High-Tech Park, and Nanshan Central Park are located along both banks. The basin is highly developed with a dense population, representing a typical urban built-up river area. The basin experiences a subtropical marine climate, with an average annual temperature of 22.4°C and average annual precipitation of 1,981.0 mm, which is unevenly distributed; 85.3% of the rainfall occurs from April to September (the flood season).

Currently, the upper riverbanks are primarily protected by vegetation and gabions, featuring a moderately meandering channel with a gravel substrate. The upper-middle reaches have vegetated concrete banks, a large water volume, a wide water surface, and a sedimentary mud substrate. The middle reach banks are dominated by grass and shrubs, with low water volume and a

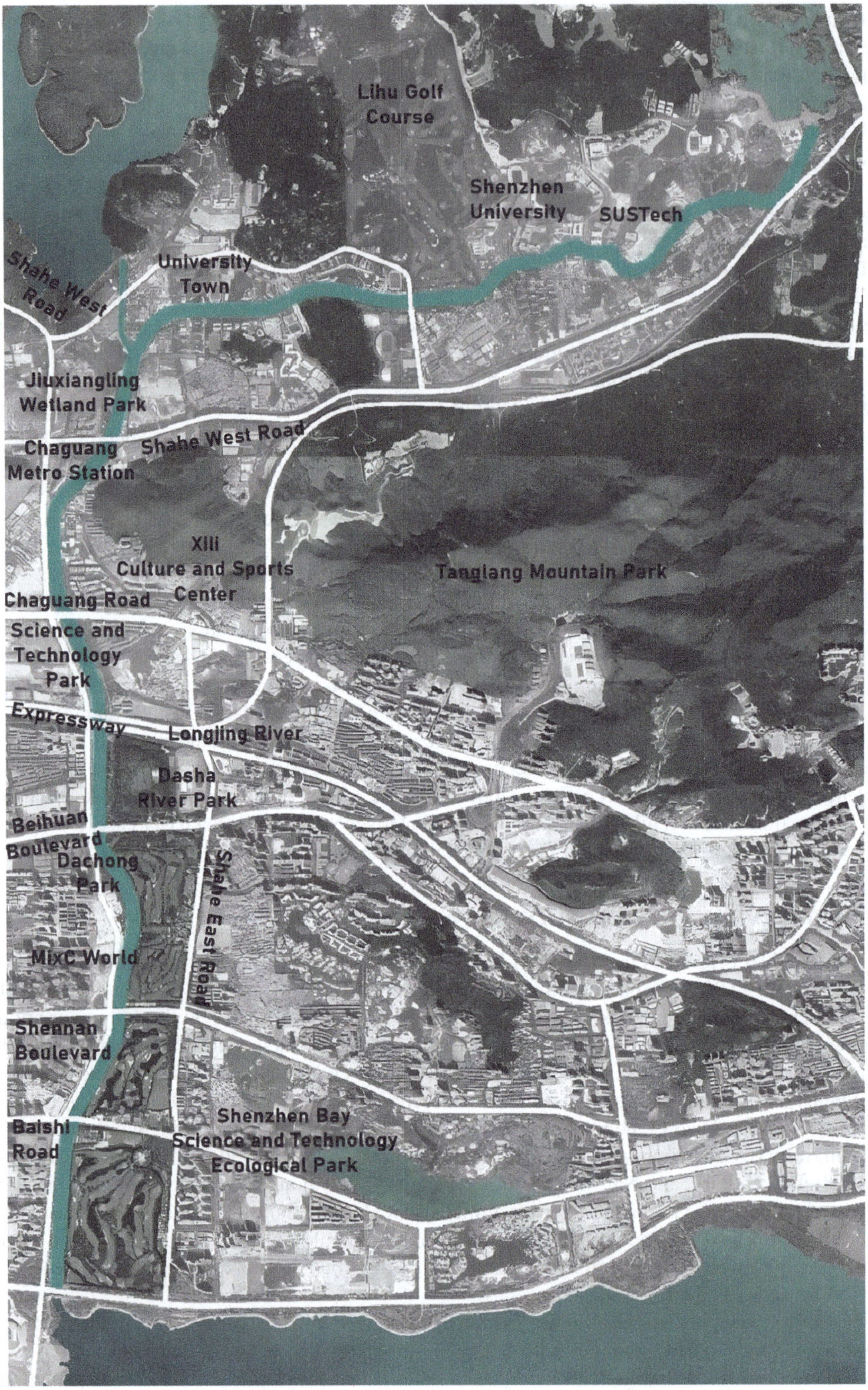

FIGURE 19.1 Map of the Dasha River Basin. (Base map/imagery: Google Maps (Imagery), © Google.)

muddy substrate. The lower reach connects to Shenzhen Bay Park and is characterized by a straight channel, heavily hardened banks, and poor water flow. Sewage interception box culverts have been installed on both banks. Two sewage treatment facilities are located in the upper reaches—within the Southern University of Science and Technology and University Town—treating 10,000 and 4,000 m^3/day, respectively, and discharging treated water into the river. Additionally, 50,000 m^3/day of reclaimed water from the Xili Reclaimed Water Plant is introduced at the Xili Reservoir flood discharge channel. Four sluices in the middle and lower reaches create three scenic water bodies in the Southern University of Science and Technology section, University Town section, and estuary section, with a total water volume of approximately 640,000 m^3 (Cai 2020; Li 2019; Bian et al. 2023; field observations during 2018–2019).

19.3 RESEARCH METHODS

Current evaluations of river ecological functions are often conducted at large spatial scales, such as basin or regional levels, producing broad results that are challenging to apply in management (Li, Wang and Lou 2017; Peng et al. 2014). To overcome this limitation and better reflect the specific conditions of the Dasha River, four spatial zones were delineated to enable a more precise assessment of ecological functions:

Upper reach: from Changlingpi Reservoir to the upstream area of the University Town scenic water body. This section includes two sewage treatment plants, is heavily channelized, has a narrow water surface, and exhibits fast flow.

Upper-middle reach: University Town scenic water body, featuring sluices or flap gates at both ends, with deep water, a wide surface, and slow flow.

Middle reach: from downstream of University Town to Dasha River Park (a relatively natural channel with low water volume and some flow).

Lower reach: tidal estuary section characterized by deep water, a wide surface area, and a steel sluice at the estuary; tidal effects are not considered in this study.

Eight sampling points were established from upstream to downstream (Figure 19.2, Table 19.1). Field surveys, sampling, and analyses of basin background, river geomorphology, water quality, plankton, benthic fauna, fish, and vegetation were conducted four times between January 2018 and January 2019.

19.4 ECOLOGICAL FUNCTION ASSESSMENT OF THE DASHA RIVER

19.4.1 Development of the Evaluation Index System

Based on existing studies and field investigations, an indicator evaluation system was developed, focusing on hydrological support, water environment support, habitat maintenance, and aquatic biodiversity maintenance functions. Considering representativeness, data availability, and operability, ten evaluation indicators and 14 evaluation indices closely related to aquatic organisms were selected to construct the ecological function evaluation system. To highlight differences and the relative importance among indicators and indices, the Analytic Hierarchy Process (AHP) was employed to determine their weights (Zhen and Luo 2019; Lu 2018). Weighted sums of the evaluation indices produced indicator-level scores, which were then further weighted to obtain function-level scores. AHP is fundamentally a "decomposition-judgment-synthesis" decision-making process that ensures rational weighting.

The main steps of the AHP are as follows:

1. Establish a hierarchical structure by classifying and grouping evaluation indicators into a three-tier hierarchy: goal layer, indicator layer, and index layer.
2. Construct pairwise comparison judgment matrices by comparing the importance of factors within the indicator layer to assign values and build the matrices; apply the same process for the index layers.

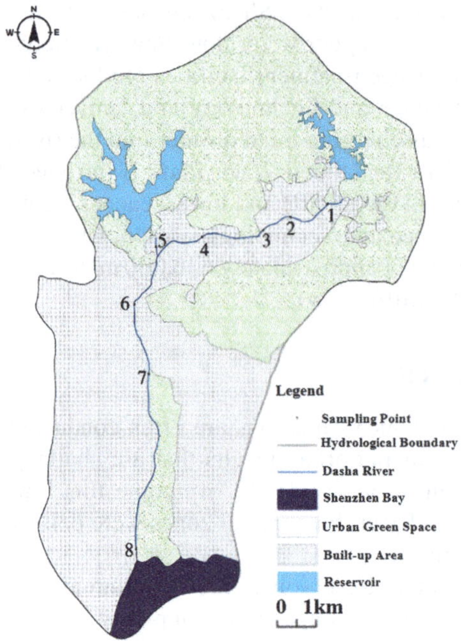

FIGURE 19.2 Overview of the Dasha River Basin and sampling locations.

TABLE 19.1
Sampling Point for the Ecological Investigation of the Dasha River in Shenzhen

Sampling Areas	Sampling Point	Location
Upstream	1	Outflow of Southern University of Science and Technology Wastewater Treatment Facility
	2	Shenzhen University Xili Campus
	3	Xili Golf Course
Mid-Upper Stream	4	Graduate School of Tsinghua University
Midstream	5	Outflow of Xili Reclaimed Water Plant
	6	Nanshan Wenli Experimental School
	7	Dasha River Park
Downstream	8	Dasha River Shenzhen Bay Estuary

3. Perform single-level ranking by calculating the maximum eigenvalues and corresponding eigenvectors of the judgment matrices to determine the weights of indicators relative to the goal, as well as the weights of indices relative to the indicators.
4. Perform an overall ranking by integrating single-level rankings to derive the relative importance weights of the indices with respect to the target layer.

The composition of the ecological function evaluation index system for the Dasha River in Shenzhen is shown in Table 19.2.

TABLE 19.2
Evaluation Indices of River Ecological Function of the Dasha River in Shenzhen

Ecological Functions	Evaluation Indicator	Indicator Weight	Indicator Description	Assessment Index	Index Weight	Data Source
Aquatic organisms	Aquatic organisms	0.30	Reflects the tolerance indicators of benthic aquatic organisms in the river	Shannon Diversity Index (H)	1	Field Survey
Diversity	Diversity	0.30	Reflects the tolerance indicators of planktonic aquatic organisms in the river	Shannon Diversity Index (H′)	Average of both	Field Survey
Maintenance function	Maintenance function	0.40	Reflects the richness of rare fish species in the river section	Rarity Index (C)	1	Field Survey
Habitat maintenance function	Habitat naturalness	0.20	Indicates the degree of human impact on river habitats	Description of River Channel and Riparian Zone Morphology	0.5	Field Survey
				Description of River Connectivity	0.5	Field Survey
	Habitat diversity	0.15	Refers to the complexity of river morphology and riparian landscape diversity	River Meandering Degree	Both	Field Survey and Historical Data
				Landscape Diversity Index (P)	Maximum	
	Habitat importance	0.50	Reflecting important characteristics of regional aquatic ecology	Description of Important Biological Habitats	0.8	Field Survey
				Riparian Vegetation Coverage	0.2	Field Survey
	Riparian zone stability	0.15	Provide a buffering function for habitats and waterways	Description of Bank Stability	0.5	Field Survey
				Description of Bank Protection Methods	0.5	Field Survey
Water environment support function	Water quality level	1.00	Indicates the river's ability to provide a living environment for aquatic organisms	Water quality parameter concentration scores	Calculate the average of each item	Field Survey
Water environment support function	Water quantity status	0.50	Basic Conditions for Ensuring the Existence of Aquatic Ecosystems	Water quantity status score	1	Field Survey

Shannon Diversity Index: H or $H = -\sum^{i}(P_i \ln P_i)$, where P_i represents the relative abundance of species i. The Fish Rarity Index (C) is defined as $c = \sum_{i=1}^{n} N_i / \sum_{j=1}^{m} N_j$ where N_i and N_j denote the counts of rare species in the analysis unit and the protected area, respectively. The Landscape Diversity Index (P) is given by $P = 22 - \sum_i (P_i' \text{In } P_i')$ where P_i' is the proportion of landscape type i within 500 m on both sides of the river channel.

19.4.2 Scoring Criteria for Evaluation Indices

Multi-indicator evaluation requires establishing rating levels and grading standards for each indicator, typically using odd-numbered scales such as 3 or 5 levels, along with corresponding scoring methods (Chen et al. 2019). For the Dasha River evaluation system, a 5-level scale was adopted, incorporating both quantitative and qualitative analyses, where 4 points represent the highest level and 0 the lowest. The water quality parameters include Chemical Oxygen Demand (COD_{Cr}), 5-day Biochemical Oxygen Demand (BOD_5), ammonia nitrogen concentration (NH_4^+-N), DO concentration, and total phosphorus concentration (TP). Detailed scoring criteria are presented in Table 19.3.

TABLE 19.3
Scoring Criteria for Evaluation Indicators of River Ecological Function in the Dasha River

Evaluation Index	Scoring Criteria				
	4	3	2	1	0
Benthic animal diversity index	>3.0	≤3.0	≤2.0	≤1.5	≤1.0
Planktonic plant and animal diversity index	>4.0	≤4.0	≤3.0	≤2.0	≤1.0
Fish rarity index	>0.4	≤0.4	≤0.3	≤0.2	≤0.1
Riverbank morphology	No channelization or siltation, the river maintains its natural form	There are minor occurrences of channel widening and deepening, with no significant channelization	Partial channelization exists, with embankments built on both banks	Severe channelization, embankments built on both banks, riverbed not channelized	Severely channelized; riverbanks and riverbed have been channelized, greatly altering the habitat within the river channel
River connectivity	No dams or weirs, biological migration is unobstructed	There are a few small weirs built, which somewhat hinder the migration of small aquatic organisms	A certain number of small to medium-sized weirs and dams have been constructed, causing some obstruction to biological migration	Numerous weirs or large dams and reservoirs constructed, but with fish passage systems. Biological migration is greatly obstructed	Large dams and reservoirs constructed, no fish passages; biological corridors are completely blocked
Degree of river meandering	Natural meandering is preserved, no straightening or cutting of bends	After straightening the bend, a significant degree of recovery was achieved (≥60%)	After straightening the river bends, a certain degree of restoration has been carried out (≥30%)	After straightening the river, a small degree of restoration was carried out (≥10%)	After straightening and cutting bends, the river is very straight

(Continued)

TABLE 19.3 (*Continued*)
Scoring Criteria for Evaluation Indicators of River Ecological Function in the Dasha River

Evaluation Index	Scoring Criteria 4	3	2	1	0
Diversity of riparian zone landscapes	>1.5	≤1.5	≤1.3	≤1.2	≤1.1
Riverbank protection types	Natural soil banks covered with vegetation	Near-natural sloped ecological embankments are present	Waterfront platforms with bank protection or bare soil slopes without vegetation	Step-like artificial revetments or masonry block stone revetments	Vertical reinforced concrete revetments
Riverbank stability	Riverbanks are stable, with no significant erosion	The riverbanks are stable, with minor erosion in some areas (<20%)	The riverbanks are relatively unstable, with moderate erosion (20%~50%)	Riverbanks are unstable, severely eroded, with flood risks (50%~80%)	Riverbanks are highly unstable, with erosion in the vast majority of areas (80%~100%)
Important habitat value	Habitat for internationally and nationally protected first-class species, serving as refuge, conservation area, feeding ground, and spawning site	There are national secondary protected species refuges, conservation areas, feeding grounds, and spawning sites	Refuges, conservation areas, feeding grounds, and spawning sites for common species	Important habitat for aquatic organisms	Not an important activity area for aquatic organisms
Proportion of vegetated riverbank	95%~100%	80%~94%	65%~79%	40%~64%	0~39%
Water quality parameters/(mg/L)	$\rho\,(COD_{Cr}) \leq 15.0$; $\rho\,(BOD_5) \leq 3.0$; $\rho(NH_4^+\text{-}N) \leq 1.0$; $\rho\,(DO) \geq 7.5$; $\rho\,(TP) \leq 0.05$	$\rho\,(COD_{Cr}) \leq 20.0$; $\rho\,(BOD_5) \leq 4.0$; $\rho(NH_4^+\text{-}N) \leq 1.5$; $\rho\,(DO) \geq 6.0$; $\rho\,(TP) \leq 0.1$	$\rho\,(COD_{Cr}) \leq 30.0$; $\rho\,(BOD_5) \leq 4.0$; $\rho(NH_4^+\text{-}N) \leq 2.0$; $\rho\,(DO) \geq 5.0$; $\rho\,(TP) \leq 0.2$	$\rho\,(COD_{Cr}) \leq 40.0$; $\rho\,(BOD_5) \leq 10.0$; $\rho(NH_4^+\text{-}N) \leq 8.0$; $\rho\,(DO) \geq 3.0$; $\rho\,(TP) \leq 0.5$	$\rho\,(COD_{Cr}) > 40.0$; $\rho\,(BOD_5) > 10.0$; $\rho(NH_4^+\text{-}N) > 8.0$; $\rho\,(DO) \leq 2.0$; $\rho\,(TP) > 0.5$
Flow velocity conditions	Both ends open, with a certain flow velocity, uneven flow velocity across cross-sections	Both ends open, flow velocity relatively uniform, slight variations in flow velocity across cross-sections	Both ends open, flow velocity uniform, no variation in flow velocity across cross-sections	One end of the river open, water flow obstructed	No water flow, isolated from other rivers
Water quantity conditions	Water level reaches both banks; only a small amount of substrate exposed	Water coverage >75%; substrate exposed<25%	Water coverage 50%~75%; substrate exposed<50%	Water coverage<25%, most of the shallow area exposed	Very little water, substrate almost completely exposed

19.4.3 Evaluation Criteria

Based on the health status and developmental patterns of urban built-up area river ecosystems, and considering local conditions and literature references, a tiered interval classification method (Cao and Li 2018) was employed to categorize the ecological functions of different sections of the Dasha River into four levels, ranked from high to low: sound, relatively sound, partially threatened, and basic ecological function only (Table 19.4).

TABLE 19.4
The Standard of the Ecological Function of the Different Reaches of the Dashe River

Level	Score Range	Ecological Characteristics	Functional Features
I	12~16	Aquatic ecosystems are unaffected or minimally affected by human activities	Ecological function: Intact
II	10~12	Aquatic ecosystems are affected by a small degree of human activities	Ecological function: Relatively intact
III	8~10	Aquatic ecosystems are affected by a moderate degree of human activities	Some ecological functions are threatened
IV	≤8	Aquatic ecosystems are affected by a high degree of human activities	Only the most basic ecological functions are maintained

19.5 RESULTS AND ANALYSIS

19.5.1 Ecological Survey Results of the Dasha River

Water quality monitoring indicated that DO ranged from 5.22 to 8.12 mg/L, chemical oxygen demand (COD_{Cr}) from 15.75 to 45.33 mg/L, biochemical oxygen demand over 5 days (BOD_5) from 2.77 to 5.73 mg/L, ammonium nitrogen (NH_4^+-N) from 0.66 to 7.66 mg/L, and total phosphorus (TP) from 0.43 to 0.93 mg/L, with significant variation observed among different reaches.

Four surveys conducted in January, March, June, and September 2018 recorded 12 species of large benthic animals: six gastropods, two bivalves, two annelids, and two crustacean shrimps. The middle reach exhibited relatively greater benthic fauna richness and higher total abundance compared to the upper-middle reach. The Shannon-Wiener diversity index ranged from 1.024 to 1.457, indicating low benthic diversity and community richness across all reaches.

Planktonic algae identified encompassed seven phyla and 50 species (genera), dominated by Bacillariophyta (18 species), Chlorophyta (14 species), and Cyanophyta (ten species). Dominant species included diatoms such as Navicula, Cyclotella, and Fragilaria; cyanobacteria such as Oscillatoria and Microcystis; and naked algae. The Shannon-Wiener diversity index ranged from 0.229 to 2.860, indicating conditions characteristic of moderately to heavily polluted waters.

Planktonic animals included representatives from seven phyla and 17 species (genera). Dominant species in the upper reach were copepods, cladocerans, and chironomid larvae, while the middle reach featured indicator species of clean water, such as water fleas and hydras, suggesting better water quality. The lower tidal reach was dominated by crustacean larvae associated with barnacles in Shenzhen Bay.

Fish surveys identified 16 species across six orders, nine families, and 14 genera. The order Cypriniformes represented 50% of the species, comprising three families, seven genera, and eight species. Notably, four species were invasive alien species: Nile tilapia, Mozambique tilapia, mosquitofish, and channel catfish, with Nile and Mozambique tilapia being the dominant species. Invasive species accounted for 25% of the total species caught, a proportion significantly higher than that observed in other domestic rivers (Yang and Lu 2016).

19.5.2 Results of the Ecological Function Evaluation

Using the evaluation method described in Section 19.3, the comprehensive ecological function levels of the Dasha River basin were as follows: no Class I reaches; the upper and upper-middle reaches classified as Class III due to significant human impact; the middle reach classified as Class II,

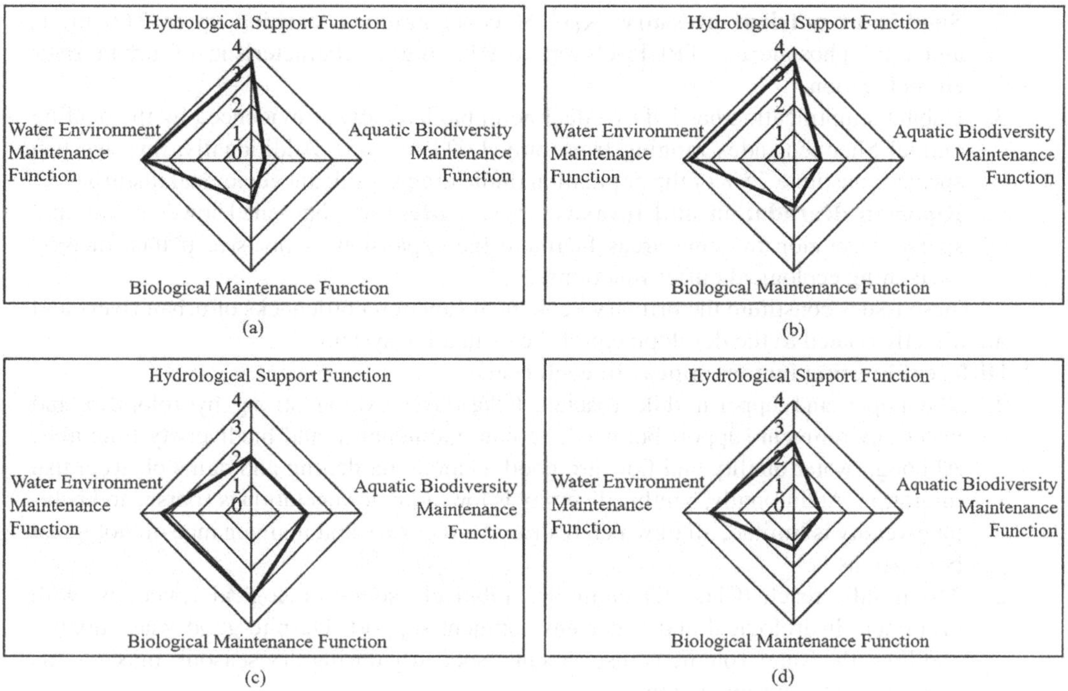

FIGURE 19.3 Evaluation results of ecological function in each reach of the Dasha River.

indicating relatively sound ecological function; and the lower reach classified as Class IV, possessing only basic ecological functions.

Detailed analysis by reach: The upper and upper-middle reaches exhibit relatively high water surface coverage, moderate flow, and good water quality, which contribute to strong hydrological and water environment support functions. However, channelization and the presence of small to medium dams impede biological migration, resulting in reduced habitat maintenance and diminished aquatic biodiversity functions (Figure 19.3a and b).

The middle reach retains several natural characteristics, including minor channel straightening and widening, vegetated natural soil banks with over 80% vegetation coverage, diverse plant species, the absence of large dams, unobstructed biological migration, good water quality, and a relatively abundant aquatic species population. These features collectively contribute to effective habitat maintenance and support of the aquatic environment (Figure 19.3c).

The lower reach is a tidal estuary with a large water volume but heavily channelized banks constructed from masonry or concrete, resulting in a very low habitat maintenance function. Water flow is obstructed by estuary sluices; therefore, despite high hydrological support, habitat maintenance and biodiversity functions remain low (Figure 19.3d).

19.5.3 Ecological Restoration Strategies for the Dasha River

1. Based on the ecological assessment, urbanization has intensified structural ecological conflicts in rivers within built-up areas, such as the Dasha River, including:
    1. **The coexistence of weakened hydrodynamics and reliance on water replenishment**: multiple reaches are controlled by sluices, which increase water stagnation and disrupt natural hydrological rhythms.

2. **Significant nutrient pressure**: NH_4^+-N concentrations reaching up to 7.66 mg/L and total phosphorus (TP) levels up to 0.93 mg/L, characteristic of urban river eutrophication.
3. Habitat simplification has led to a decline in biodiversity, as evidenced by the benthic animal Shannon index ranging from only 1.024 to 1.457. Additionally, invasive fish species constitute 25% of the population, indicating significant ecological disturbance.
4. **Riparian degradation and invasive species pressure**: hardened lower banks and sparse vegetation in some areas facilitate the expansion of invasive plants, thereby weakening ecological buffer functions.

These issues constitute the primary ecological function bottlenecks of urban rivers and are directly related to the development of the evaluation system.

2. **Different limiting factors appear in each reach**:
   1. The upper and upper-middle reaches (Class III) exhibit strong hydrological and water environment support but weak habitat maintenance and biodiversity functions. Although water quality and flow are good, channel hardening and dams obstruct fish migration. Additionally, benthic diversity is low, aquatic vegetation is sparse, and habitat diversity is limited, all of which restrict the recovery and maintenance of biological populations.
   2. The middle reach (Class II) maintains relatively sound ecological functions, with prominent hydrological and water environment support. Despite good water quality and flow, the water volume is insufficient, especially during dry seasons, making this reach prone to eutrophication.
   3. The lower reach (Class IV) experiences severe ecological degradation, primarily due to hardened channels and dams that cause habitat loss. Water flow is reduced, habitats are limited, and biodiversity declines sharply.

Therefore, restoration strategies strictly adhere to evaluation results to achieve targeted, function shortfall-oriented restoration.

19.5.3.1 Key Ecological Issues and Restoration Objectives

Based on the ecological assessment results, the fundamental ecological conflict in the Dasha River is essentially an imbalance in the system structure caused by urbanization disturbances. This imbalance is manifested not only by the deterioration of individual indicators but also by a disruption in the connectivity among the four core functional components: hydrology, water quality, habitat, and biology. The sluice gates weaken hydrodynamics and alter hydrological rhythms, leading to a decline in the water body's self-purification capacity and exacerbating the accumulation of nutrients such as nitrogen and phosphorus (NH_4^+-Nreaching 7.66 mg/L, TP reaching 0.93 mg/L). The simplification and hardening of aquatic and riparian habitats directly result in low benthic animal diversity indices (H index ranging from 1.024 to 1.457) and create conditions favorable for the expansion of invasive alien plants like* Bidens pilosa*, further eroding the already fragile ecological buffer zone function and perpetuating a vicious cycle of "degradation-invasion."

Therefore, restoration strategies must strictly adhere to the principle of "function shortcoming orientation," transforming structural conflicts into clearly prioritized restoration goals based on the evaluation results for each river section (Figure 19.2):

Upstream and upper-middle reaches (Class III): The primary challenge is the weakest habitat maintenance and biodiversity support functions. Although water quality is relatively good and water flow is moderate in these sections, river channel hardening and partial damming restrict fish migration, reduce benthic community diversity, and cause a lack of aquatic vegetation and diverse habitat types. These factors limit the recovery and maintenance of biological populations. The goal is to completely eliminate the physical barriers caused by channel hardening and damming while preserving existing hydrological advantages and reconstructing a diversified habitat structure.

Middle reaches (Class II): The core conflict is insufficient water volume and underutilized habitat potential within a relatively intact and functional environment. The goal is to focus on both water quality assurance and habitat expansion to fully utilize the ecosystem's carrying capacity and restoration potential.

Downstream reaches (Class IV): The primary issues are disrupted connectivity and severe habitat degradation. Water flow is poor, biological habitats are limited, and biodiversity has sharply declined. The objective is to prioritize restoring physical connectivity, systematically enhance habitat heterogeneity, and initiate biodiversity recovery processes.

19.5.3.2 Graded and Classified Systematic Restoration Strategies

To achieve the objectives outlined above, fragmented measures must be abandoned in favor of developing a systematic restoration toolkit specifically tailored to different sections of the river.

1. **Hydrological connectivity restoration strategy**: Addressing water stagnation and ecological barriers caused by sluice gates through targeted scheduling and structural modifications.

   Key measures include constructing smart sluice gates at the downstream estuary to enable ecological scheduling based on tidal rhythms and fish migration cycles, thereby restoring river-sea ecological connectivity. Additionally, modify fish passages or replace ecological dams associated with sluice gates or weirs in the upper-middle reaches that obstruct fish migration.
2. **Water quality improvement and nutrient management strategy**: Develop a three-dimensional purification system that addresses eutrophication by targeting point sources, tailwaters, river channels, and riparian zones.

   Key measures include enhancing deep denitrification treatment through constructed wetlands for sewage plant effluent upstream. In the upper-middle and middle reaches, fully utilize green spaces on both banks to establish distributed purification wetland systems and intra-section circulation systems. These should be combined with in-situ restoration using submerged and floating-leaf plants to continuously reduce total nitrogen (TN) and total phosphorus (TP) levels.
3. **Habitat diversity reconstruction strategy**: Actively develop complex habitats using engineering and biological methods to counteract trends of habitat simplification.

   Key measures include installing large-scale ecological floating islands and underwater fish nests in the downstream and some middle reaches to provide refuge and spawning grounds for aquatic organisms and waterfowl. Outside the floodplain areas of the middle reaches, construct a "wetland-pond system" that serves as a biological germplasm resource bank, offering functions such as rainstorm regulation, water purification, and sensitive species conservation.
4. **Riparian ecological function restoration and integrated invasive plant management strategy**: Addressing the intertwined core issues of riparian degradation and invasive plant expansion, this strategy proposes treating the riparian zone as a critical interface between aquatic and terrestrial ecosystems. It adopts a systematic governance approach of "removal–blockage–improvement–reconstruction" to thoroughly break the vicious cycle of "habitat degradation → invasion intensification → functional loss." This study innovatively integrates invasive plant and riverbed sediment co-composting resource utilization technology as a core component, aiming to safely dispose of ecologically risky waste and recycle it, thereby providing the material basis and technical support for riparian ecological restoration.
   1. **Technical pathway**: Develop a closed-loop governance system encompassing: Thorough manual removal: Focus on manually removing major invasive plants such as *Bidens pilosa*, emphasizing the elimination of their reproductive stolons and root segments.

Ensure centralized management of the removed biomass to prevent secondary invasion caused by residual fragments.

Coordinated composting resource utilization: Co-compost the removed invasive plants with dredged riverbed sediment at a mass ratio of 10:1. This method employs a sustained high-temperature process of 50°C–60°C over several days (Figure 19.4), effectively eliminating invasive plant propagules and pathogens, thereby achieving biological harmlessness. Experiments confirm that adding sediment significantly accelerates compost maturation and enhances the fertilizer efficacy of the final product (Figure 19.5).

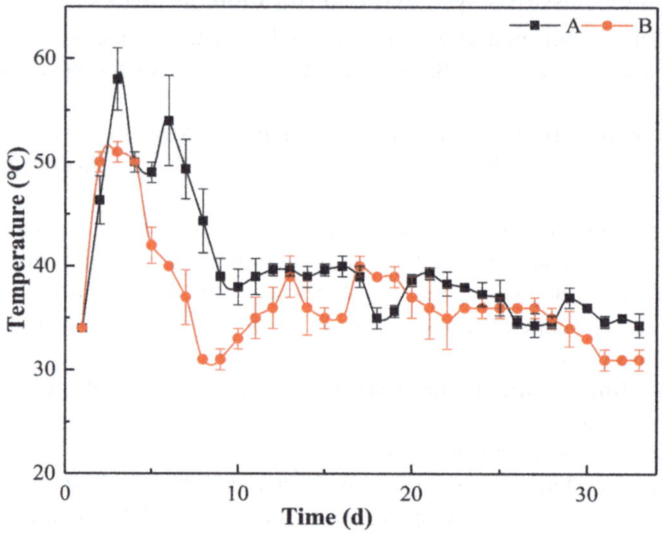

FIGURE 19.4 Changes of temperature. (a) Without adding sediment, (b) adding sediment. (*S. trilobata*: sediment = 10:1, weight ratio). (Error bars represent mean ± SD (n = 3)).

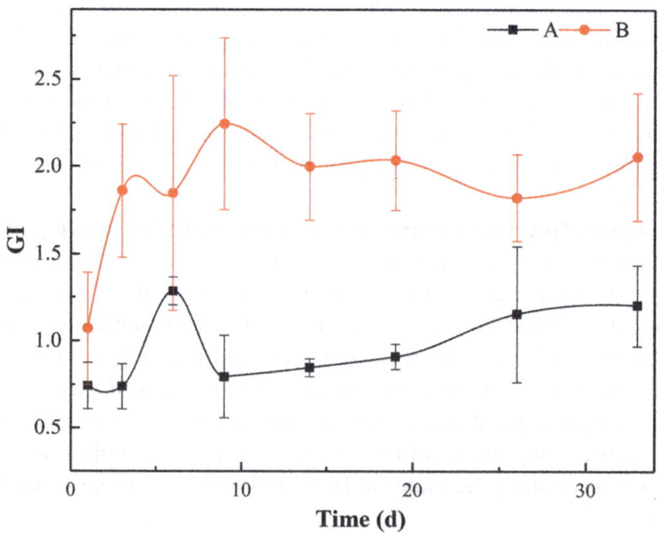

FIGURE 19.5 Changes of GI during composting. (a) Without adding sediment, (b) adding sediment. (*S. trilobata*: sediment = 10:1, weight ratio). (Error bars represent mean ± SD (n = 3)).

Riparian soil improvement: Apply fully decomposed compost to nutrient-poor riparian soil at an approximate rate of 5% by volume. Studies indicate that this proportion optimally enhances soil structure, increases organic matter and nutrient content, and promotes subsequent plant growth—resulting in a 15% increase in germination rate and a 20% increase in plant height (Figure 19.6). These improvements create favorable conditions for the restoration of native vegetation.

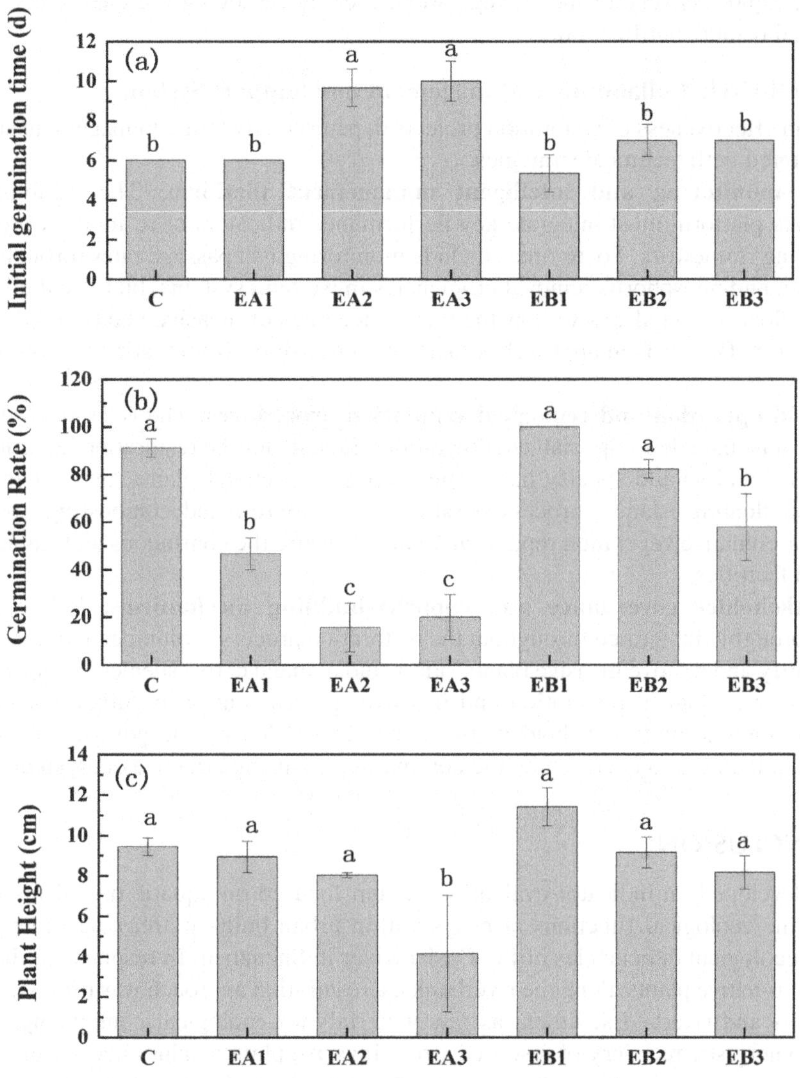

FIGURE 19.6 Seed initial germination time (a), germination rate and (b) plant height (c) with different amounts of compost products. Control group (c): soil, EA1: 95% soil+5% compost product A, EA2: 90% soil+10% compost product A, EA3: 80% soil+20% compost product A, EB1: 95% soil+5% compost product B, EB2: 90% soil+10% compost product B, EB3: 80% soil+20% compost product B. Product A: without adding sediment. Product B: adding sediment (*S. trilobata*: sediment=10:1, weight ratio). (Values are mean±SD (n=3). Data were analyzed using one-way ANOVA followed by Duncan's multiple range test (P<0.05) in SPSS (v26.0). Different letters indicate significant differences among treatments.)

Reconstruction of native plant communities: On improved soil, densely plant selected native pioneer species with strong competitive abilities and slope-stabilizing properties (such as bermudagrass and reed-like plants) to rapidly establish a closed canopy community. This approach suppresses the re-colonization of invasive plants by occupying physical space and competing for resources, thereby creating a stable and disturbance-resistant riparian vegetation zone.

The composting results demonstrate that co-composting invasive plants with river sediment effectively prevents the risk of spread during removal while simultaneously recycling sediment and green waste. This process provides a continuous source of material for riparian soil improvement. This approach supports habitat restoration, suppresses the expansion of invasive plants, and promotes the rapid recovery of native vegetation, thereby achieving the dual goals of ecological restoration and resource utilization.

19.5.3.3 Full-Cycle Collaborative Management and Support System

The long-term effectiveness of restoration projects depends on dynamic management support that is closely integrated with technical strategies.

Adaptive monitoring and intelligent management platform: The established aquatic ecological data platform must integrate key performance indicators of restoration projects into its core monitoring framework. Focus areas include monitoring fish passage rates through smart sluice gates, nitrogen and phosphorus removal efficiencies in wetland systems, biological utilization rates of ecological floating islands, as well as the recurrence rates of invasive plants and the coverage of native vegetation. Data-driven approaches facilitate a transition from static restoration to adaptive management.

Specialized operation and ecological regulation procedures: The river ecosystem management system must include a "Specialized Operation Manual" for the restoration facilities mentioned above. This manual should specify harvesting cycles for wetland plants, maintenance standards for ecological floating islands, process parameters for coordinated composting operations, and phenology-based native vegetation replanting plans to ensure the continuous and stable functioning of ecological facilities.

Multi-stakeholder governance and capacity-building mechanisms: Public participation should be thoroughly integrated throughout the restoration process. Volunteer activities can be organized to identify and remove invasive plants, adopt and manage native species, and offer specialized courses on river ecological restoration and invasive species control in "urban classrooms." This approach promotes a multi-stakeholder governance model led by the government, implemented by professional teams, and overseen by the community, ensuring effective ecosystem stewardship.

19.6 CONCLUSION

This study developed an indicator evaluation system focused on aquatic organisms to systematically assess the ecological functions of rivers within urban built-up areas, accurately identifying the primary ecological deficiencies of the Dasha River in Shenzhen. In response to the widespread invasion of non-native plants along the riverbanks, a restoration approach was proposed that utilizes invasive plants and riverbed sediment as raw materials for ecological composting. This method enables the synergistic recovery of resources from invasive plant residues and sediment, which are then used to improve riparian soil and promote the restoration of native vegetation. This approach not only effectively controls the spread of invasive species and gradually restores riparian habitats but also facilitates the resource utilization of green waste and sediment, providing a sustainable local material source for soil enhancement along the riverbanks. The study embodies the ecological restoration concept of "treating waste with waste and recycling," offering a demonstrative technical reference for improving the stability and sustainability of urban river ecosystems.

Compared with widely used river indices that focus on a single component or are typically implemented at basin/regional scales, our framework is organized at the river-segment scale that matches practical management units. By integrating hydrology, water quality, habitat structure, and biota-based metrics, it improves the diagnosis of biodiversity-related deficits in short, regulated urban rivers. Moreover, the grading results are directly translated into reach-specific restoration toolkits (including circular invasive-plant–sediment composting), strengthening the link from evaluation to actionable management. For application in other cities, the indicator set and AHP weights should be recalibrated with local data and validated through long-term monitoring.

DECLARATION OF GENERATIVE AI USE

During the preparation of this chapter, the authors used ChatGPT to refine the language, improve readability, and expand/clarify parts of the text based on the authors' original research and previously published work from the research group. The authors subsequently reviewed, edited, and verified all AI-assisted content and take full responsibility for the accuracy, originality, and integrity of the final version of the published chapter. Generative AI was not used to generate or alter any data, results, or figures.

REFERENCES

Bian, W., Huang, D., Ma, X., et al., 2023. Impacts of environmental color landscape in urban waterfront areas on public mental health: a case study of Shenzhen Dasha River Ecological Corridor. Journal of Chinese Urban Forestry, 21(5): 65–73+113. (in Chinese).

Cai, S., 2020. Practice and reflection on ecological landscape construction of Shenzhen Dasha River. Guangdong Landscape Architecture, 42(4): 46–51. (in Chinese).

Cao, C., Li, X., 2018. Health assessment of river ecosystem at district and county scales: a case study of Fangshan District in Beijing. Acta Ecologica Sinica, 38(12): 4296–4306. (in Chinese).

Cao, L., Zhou, F., Wu, M., 2019. Study on the ecological compensation mechanism of a watershed based on an urban agglomeration using the Yangtze River basin as an example. Acta Ecologica Sinica, 39(1): 85–96. (in Chinese).

Chen, W.Y., 2017. Environmental externalities of urban river pollution and restoration: a hedonic analysis in Guangzhou (China). Landscape and Urban Planning, 157(1): 170–179.

Chen, X., Jin, T., Su, H., et al., 2019. Construction and application of health assessment index system for Lhasa River. Acta Ecologica Sinica, 39(3): 799–809. (in Chinese).

Chen, Y., 2019. Restoring aquatic ecosystem health of Yangtze River basin under multiple human disturbances. Yangtze River, 50(2): 19–23. (in Chinese).

Han, L., Xu, Y., Lei, C., et al., 2016. Degrading river network due to urbanization in Yangtze River Delta. Journal of Geographical Sciences, 26(6): 694–706.

Hoyer, R., Chang, H.J., 2014. Assessment of freshwater ecosystem services in the Tualatin and Yamhill basins under climate change and urbanization. Applied Geography, 53(9): 402–416.

Jiang, H., 2019. Design strategy of urban river comprehensive regulation project and ecological restoration. China Resources Comprehensive Utilization, 37(8): 155–157. (in Chinese).

Li, J., 2019. Water environment and pollution control measures of main rivers in Shenzhen City. Guangdong Chemical Industry, 46(11): 154–155. (in Chinese).

Li, Y., Wang, R., Lou, G., 2017. Overall framework of water ecological protection and restoration in Huangshui River basin. Journal of Hydroecology, 38(6): 11–18. (in Chinese).

Li, Y., Yu, L., Lu, X., et al., 2016. Health evaluation and repairing mode on river ecosystem of Huaihe River basin (Henan section). Environmental Science & Technology, 39(7): 185–192. (in Chinese).

Lu, H., 2018. Weight analysis of river health evaluation index. Water Resources Development and Management, 9(9): 14–18. (in Chinese).

Munné, A., Prat, N., Solà, C., et al., 2003. A simple field method for assessing the ecological quality of riparian habitat in rivers and streams: QBR index. Aquatic Conservation: Marine and Freshwater Ecosystems, 13(2): 147–163.

Pan, G., Xu, Y., Yu, Z., et al., 2015. Analysis of river health variation under the background of urbanization based on entropy weight and matter-element model: a case study in Huzhou City in the Yangtze River Delta, China. Environmental Research, 139(5): 31–35.

Peng, B., Wang, H., Wang, R., et al., 2014. Development of a river health assessment system for the lower Yellow River. Journal of Hydroecology, 35(6): 81–87. (in Chinese).

Petersen, R.C., 1992. The RCE: a riparian, channel, and environmental inventory for small streams in the agricultural landscape. Freshwater Biology, 27(2): 295–306.
Sims, A., Zhang, Y., Gajaraj, S., et al., 2013. Toward the development of microbial indicators for wetland assessment. Water Research, 47(5): 1711–1725.
Verdonschot, P.F.M., 2000. Integrated ecological assessment methods as a basis for sustainable catchment management. Hydrobiologia, 422/423: 389–412.
Wan, J., Liu, H., Zhang, Y., et al., 2013. Ecological function evaluation and related management strategies of river ecosystem in Taizi River basin, North China. Chinese Journal of Applied Ecology, 24(10): 2933–2940. (in Chinese).
Wang, S., Mao, Y., Liang, Z., et al., 2018. Study on the evaluation indicator system of river ecological restoration in Chongqing. Environmental Impact Assessment, 40(4): 39–42. (in Chinese).
Wen, Y., Schoups, G., Van de Giesen, N., 2017. Organic pollution of rivers: combined threats of urbanization, livestock farming and global climate change. Scientific Reports, 7(1): 43289.
Wright, J.F., 1995. Development and use of a system for predicting the macroinvertebrate fauna in flowing waters. Austral Ecology, 20(1): 181–197.
Yang, W., Zhao, Y., Zhao, K., et al., 2018. Evaluation on the ecosystem services value of the Xin'anjiang River in Huangshan. Environmental Management of China, 10(4): 100–106. (in Chinese).
Yang, Z., Lu, Q., 2016. Exotic fish potential hazards in Dongguan reaches of Dongjiang River and its coping strategies. Journal of Green Science and Technology, 2016(10): 149–151.
Zhen, B., Luo, W., 2019. On the health evaluation index system and weight of river ecosystem. Hydropower and New Energy, 33(8): 60–65. (in Chinese).

20 The Trade of Six Songbird Species in Hong Kong

Evidence from Direct Count Records in a Local Market

Louis S. H. Lee and Grace Y. L. Chan

20.1 ASIAN SONGBIRD CRISIS

Unsustainable trapping has endangered many songbird species. Large-scale trapping of songbirds has been recorded in the past (McClure & Chaiyaphun, 1971; Nash, 1993). Population decline, shrinkage in distribution range, reduction in population density, and even local extinction have been observed for resident and migratory songbirds in Southeast Asia (Yong et al., 2015). Frugivore and nectarivore birds, especially those with larger body size, were most heavily associated with desirable market traits (Almeida, Mazza, & Lockwood, 2024). Unsustainable harvest has impacted the wild populations of Asian songbirds and deprived their original habitats of the ecosystem services related to wild birds. These devastating consequences are directly a result of ongoing trapping activities.

In response to the crisis, conservation organisations and stakeholders gathered in Singapore in 2015 for the first Asian Songbird Trade Crisis Summit. An action plan composed of four key themes of work was published afterwards (Lee, Chng, & Eaton, 2016). Alongside the overarching action plan, species-specific action plans were formulated for 28 focus species, 12 of which were classified as first-tier species requiring urgent conservation priority. As a result, the monitoring of songbird markets in Southeast Asia has become important for the identification of heavily traded bird species and the analysis of the temporal trends in their trading volume.

Bird-keeping is a traditional hobby in many Asian cultures. Songbirds are particularly favoured by bird-keepers for their beautiful plumage (Senior, Oliveira, Dale, & Scheffers, 2022) and melodious songs (Burivalova et al., 2017). The reasons for keeping songbirds varied geographically. In Hong Kong, life satisfaction (Leung & Yung, 2021) and psychological well-being (Leung, 2022) were mentioned as the main motivating factors. In Guiyang, China, 46% of correspondents in a questionnaire survey cited companionship for keeping birds (Dai, Chen, & Zheng, 2021). In Singapore, a survey found that 73% of correspondents cited influences from family and friends as the highest motivation for keeping birds (Chiok, Lee, Lee, & Jain, 2022). Large numbers of birds were trapped in Southeast Asia or imported from other countries to support bird-keeping. The nature of data sources may affect how bird trading activities are portrayed. 86% of the 93,986 CITES-listed birds imported into China during 2010–2019 were parrots (Psittaciformes) (Wang, Shi, Shen, & Zhao, 2021). However, trade figures may not fully reflect the genuine extent of bird-keeping. Domestic trapping and wildlife trafficking, which are absent in official importation records, may be revealed through bird market observation. For example, 20% of the 95,912 birds in Huadiwan Animal Market in Guangzhou, China were munias (*Lonchura spp.*) (Fiennes, Zhang, Sun, & Lee, 2021). Observations can be made from the consumer side. Market survey is a viable approach to estimating the impacts of songbird trade.

DOI: 10.1201/9781003640899-25

The direct count method has been used in songbird market surveys. A plethora of market survey studies have focused on Indonesia due to the prevalence and history of bird-keeping. General pastimes (Jepson & Ladle, 2009), bounties from bird song contests (Angguni, Mulyani, & Mardiastuti, 2021), and appreciation of bird songs (Burivalova et al., 2017) were common reasons for bird-keeping. An estimated 65,831,000–82,811,000 birds were kept across the six provinces in Java in 2018 (Marshall et al., 2020). A market survey found 25,298 birds from 151 species in Kalimantan in 2015–2016 (Rentschlar et al., 2018). Case studies were conducted on critically endangered, endemic species, such as Javan Pied Myna (*Gracupica jalla*) (Nijman et al., 2021), Javan White-eye (*Zoesterops falvus*) (van Balen, Saryanthi, & Marsden, 2023), Javan Green Magpie (*Cissa thalassina*) (Nijman, Sari, Siriwat, Sigaud, & Nekaris, 2017), and Bali Myna (*Leucopsar rothschildi*) (Harris et al., 2015). Species of conservation interests also received research attention, such as Sunda laughingthrush (*Garrulax palliatus*), Chestnut-capped laughingthrush (*Garrulax mitratus*) (Shepherd, Eaton, & Chng, 2016), Greater Green Leafbird (*Chloropsis sonnerati*) (Chng & Eaton, 2016), and Black-winged Myna (*Acridotheres melanopterus*) (Nijman, Langgeng, Birot, Imron, & Nekaris, 2018). In recent years, market surveys have extended to other Southeast Asian locations, including Cambodia (Gilbert, Sokha, Joyner, Thomson, & Poole, 2012), China (Z. Liang et al., 2024), Singapore (Eaton, Leupen, & Krishnasamy, 2017), South Africa (Shivambu, Shivambu, & Downs, 2022), Taiwan (Su, Cassey, Vall-llosera, & Blackburn, 2015), Thailand (Siriwat & Nijman, 2020), and Vietnam (Thang et al., 2024). A comprehensive geographical coverage of market surveys can assist in refining conservation actions by tracking songbird trade.

Hong Kong is an overlooked market for songbirds in the existing literature. As an international logistics hub with a developed economy, the bird-keeping community in Hong Kong can afford songbirds imported from various biogeographic regions. Only two studies on songbirds for religious release were conducted (Chan, 2006; Kadoorie Farm and Botanic Garden, 2011). Although songbird trading volume had plunged (Nash, 1993), songbird species with conservation priorities can be found in Hong Kong's songbird market. Among them are first-tier focus species identified by the Asian Songbird Working Group, such as Common Hill Myna (*Gracula religiosa*) and Silver-eared Mesia (*Leiothrix argentauris*) (Lee et al., 2016). However, second-tier focus species have been traded in large numbers, particularly White-eye (*Zosterops spp.*) and Oriental Magpie-robin (*Copsychus saularis*). Additionally, songbird species of research interest were traded, including Chinese Hwamei (*Garrulax canorus*) (Nelson & Shepherd, 2023; Shepherd, Leupen, Siriwat, & Nijman, 2020) and Black-throated Laughingthrush (*Pterorhinus chinensis*) (Shepherd et al., 2023). Regular market surveys in Hong Kong can fill a geographic niche in songbird trade research. This chapter is based on monthly counts of six consistently traded species in Hong Kong from 2017 to 2025. Conservation implications are drawn from the trends of their numbers in the market.

20.2 METHODS OF INVESTIGATION

20.2.1 History and Current Conditions of Bird Street

This investigation was based in Yuen Po Street Bird Garden in Hong Kong (22°19′34.1″N 114°10′26.2″E). The Garden is located in the Yau Tsim Mong District at the heart of urban Hong Kong. Locally, the Garden is figuratively referred to as *Bird Street* due to its unique historical roots. Since the 1960s, bird keepers used to meet in Chinese restaurants on what is now a demolished section of Hong Lok Street (22°19′05.4″N 114°10′07.0″E). Hawkers were soon seen with live birds, cages, decorations, accessories, and bird food under balconies and patios. The characteristic street section earned itself the informal but famous name, Bird Street. In the early 1990s, the self-organised agglomeration was disbanded because Bird Street was gazetted for urban renewal. For the purpose of resettlement, the Yuen Po Street Bird Garden was established at 950 m northeast of the former Bird Street. When the construction was completed in 1997, 70 stores were reserved for the shopkeepers moving into the garden. Despite the location shift, the nickname Bird Street has been retained. For concise communication, the term *Bird Street* is used in this chapter to refer to the Yuen Po Street Bird Garden.

Currently, there are 70 stores on Bird Street. In Hong Kong, the sale of live birds involves a wide range of duties, including importation, health certification, public hygiene, and business registration. As a result, the stores are managed by the Leisure and Cultural Services Department, Agriculture, Fisheries and Conservation Department, and Food and Environmental Hygiene Department. A shopkeeper may rent adjacent stores and combine them into one large shopfront. Only a portion of shops specialise in live bird trade. Shopkeepers may buy or sell their bird trade business, resulting in a dynamic number of shops in business in Bird Street. In this study, a shop is strictly defined as a discrete business unit that offers live birds and has a distinct shopkeeper. Under this definition, a shop may consist of one or more stores as long as the shopkeeper in charge of the shop can be distinguished from those of adjacent shops. Therefore, despite the physical presence of 70 stores, only 17 shops can be recognised as of this writing.

20.2.2 Long-Term Monthly Market Surveys

Bird Street was visited monthly from February 2017 to April 2025. Surveys were carried out on weekends near the middle of the month, i.e., the 15th of each month. Vendors would cover bird cages in fabrics or even close early in the rain. Therefore, visits were postponed by a week in case of rainy weather. Data were not collected in April 2018 due to a temporary market closure following the detection of avian influenza (HKSARG, 2018). Data were missing from June 2019 to November 2020 due to local social movements (Holbig, 2020) and the COVID-19 pandemic (Wong et al., 2022). Consequently, a total of 80 visits were conducted.

To maximise the detectability of birds on Bird Street, visits were conducted on the weekend closest to the middle of the month. Such an arrangement aligns with the local socio-economic situation. In Hong Kong, Saturday and Sunday are the days off for most people. Caged birds are typically displayed prominently to attract weekend customers. Also, a monthly payroll cycle is common among most of Hong Kong's working population. Salary payments are typically made on the last working day of the month. Bills are usually settled at the beginning or end of each month. As a result, bird vendors tend to benefit from a peak propensity to spend in the middle of the month and showcase their stock accordingly. Therefore, the detectability issue was addressed through the careful scheduling of the visits.

Using a direct counting method, the number of individuals of each detected species was recorded in each shop. The direct counting method proved to be simple yet reliable in capturing the market situation at a given moment (Fiennes et al., 2021). This method has been described as providing a "snapshot" of a market (Chng & Eaton, 2016). Two surveyors visited each of the 17 shops on Bird Street. One surveyor verbally reported the bird species and their quantities to the other surveyor, who marked the information in a notebook discreetly. Birds sold for merit release were kept in large groups in bamboo cages. In cases of great difficulty in counting, it was assumed that each cage held 30 birds. To minimise suspicion, the two surveyors dressed in plain clothes, avoided interactions with shopkeepers and customers, and randomised the order of their visits to different shops on Bird Street.

Birds were identified to the species level whenever possible. Before the first visit to Bird Street, the bird species reported in previously published market surveys in Hong Kong, along with their biogeographic realms, were consolidated (Chan, 2006; Kadoorie Farm and Botanic Garden, 2011). The two surveyors then studied field guides of bird species from the Afrotropical (Barlow & Wacher, 1997; van Perlo, 1999), Neotropical (Schulenberg, 2007), Australasian (Day & Simpson, 1996), Palaearctic (Arlott, 2007; Viney, Phillipps, & Lam, 2005), and Indomalayan (Kazmierczak, 2000; Robson, 2018) realms were studied before heading out to Bird Street. When uncertain, unidentified birds were photographed and examined later.

Additional information was obtained opportunistically. In some shops, information tags were affixed to bird cages, with descriptions that sometimes included prices, species, age, or origin. Despite their questionable authenticity, the descriptions were recorded for examination. To realistically depict the market, photographs were taken, but only sporadically to avoid suspicion. Although the surveyors avoided interactions with anyone on Bird Street, observations were made passively by overhearing conversations and paying attention to market behaviours related to songbird trade.

20.2.3 Analysis of Market Observations

Species- and genus-accumulation curves were plotted with dates on the *x*-axis. Discontinuities in the data collection period were represented by dashed lines. Both curves were visually examined to assess the adequacy of the monthly market surveys in representing the bird species on Bird Street and to evaluate the impact of the missing data periods.

Six songbird species were classified as *consistently encountered species* (CES) because they were recorded in every monthly survey. The six CES were: warbling white-eye (*Zosterops japonicus*); yellow-fronted canary (*Crithagra mozambica*); oriental magpie-robin (*Copsychus saularis*); Siberian rubythroat (*Calliope calliope*); pied bush-chat (*Saxicola caprata*); and Chinese hwamei (*Garrulax canorus*).

The temporal evolution of frequency, defined as the number of individuals recorded in a survey throughout the data collection period, was plotted and examined for each CES. Seasonal variation in bird counts was studied by dividing the calendar year into four seasons: spring (March–May), summer (June–August), autumn (September–November), and winter (December–February). Also, the data collection period was segmented into three periods with respect to the COVID-19 pandemic, namely before (February 2017 to May 2019), during (December 2020 to February 2023), and after COVID-19 (March 2023 to April 2025).

Additionally, the vernacular names of the CES were provided. Vernacular names help enhance the understanding of the market chain system (Iskandar, Iskandar, Mulyanto, Alfian, & Partasasmita, 2020). In Hong Kong, Cantonese is the predominant spoken language. Thus, vernacular names were supplemented with the Jyutping Cantonese Romanization Scheme (Linguistic Society of Hong Kong, 2022). The information related to vernacular names can be found in Appendix 20.1. The conservation implications related to the six CES were discussed.

General market observations were provided. Cages used to house traded birds on Bird Street were illustrated. The immediate living conditions and animal welfare issues affecting the traded birds were explored. Species other than the six CES were cited as supplementary examples. Observations regarding the behaviours of bird sellers and customers were also described. By generalising from market observations, conservation considerations were discussed in local and regional contexts.

Linear models were constructed to examine the effects of seasons and COVID-19 on the monthly counts. Mixed-effects models were first constructed with the month of the year as the random effect. However, the random effect was found to be negligible. Thus, linear models with only fixed and interaction effects were constructed. By the principle of parsimony, insignificant effects were discarded. Each CES was modelled individually, followed by post-hoc comparisons in order to interpret any significant factors.

Based on the market observations and conservation-related incidents during the data collection period, a SWOT analysis (strengths, weaknesses, opportunities, threats) was performed. In particular, the role of legal enforcement in combatting unsustainable songbird trade was emphasised. The SWOT analysis consisted mainly of qualitative information. Therefore, the contents were more related to the Discussion section.

20.3 MARKET SITUATION ON BIRD STREET

20.3.1 Species- and Genus-Accumulation Curves

85 species were discovered in the initial market survey conducted in February 2017. Substantial species accumulation of 11–21 species was recorded in the following 3 months (Figure 20.1a). Subsequently, stable monthly increments by ≤10 species were observed. A major increase from 234

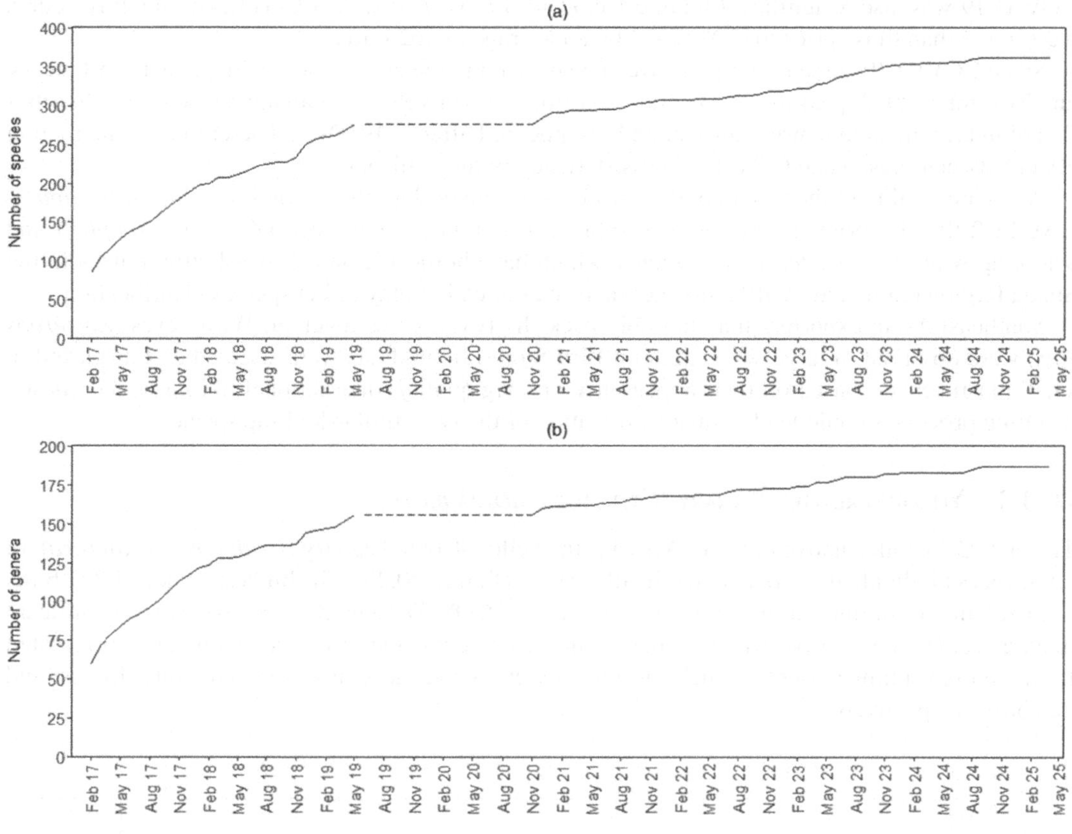

FIGURE 20.1 Accumulation curves of (a) species and (b) genera of birds encountered in the monthly surveys conducted in Yuen Po Street Bird Garden, Hong Kong, from February 2017 to April 2025. Missing data in April 2018 and from June 2019 to November 2020 were indicated by a dashed line.

species to 251 species was noticed in December 2018. Before the data gap, the species-accumulation curve exhibited a logarithmic growth. The resumption of data collection in December 2020 was followed by a period of slower species accumulation rates. A transient increase in the species discovery by ≤7 species per month occurred from April to November 2023. However, a plateau at 362 species persisted beginning in September 2024.

The overall evolution of the genus-accumulation curve was similar to that of species-accumulation curve. 60 genera were recorded in the initial survey (Figure 20.1b). Logarithmic growth was maintained until March 2018. A marked increase of seven genera was observed in December 2018. After the data gap, the accumulation of genera slowed down and progressed almost linearly, with ≤3 genera discovered per month and no addition in 33 out of 52 months. Except for August and September 2024, no new genera were found starting from February 2024. By the end of the data collection period, 187 genera were recorded.

20.3.2 Warbling White-Eye (*Zosterops japonicus*)

All monthly counts added up to a total of 28,454 warbling white-eyes (*Zosterops japonicus*), marking the highest aggregate count among all CES (Figure 20.2a). Annual maximum counts occurred mostly in summer or autumn, reaching 764 birds in July 2017 and 828 birds in September 2018.

Season of the year was a significant main effect ($F=4.96$, $p<0.01$). Spring had significantly lower counts than summer, autumn, and winter, with respective mean differences of 134, 137, and 123 birds. Annual minima, which ranged from 60 to 311 birds, occurred exclusively in spring. Meanwhile,

COVID-19 was also a significant main effect ($F=9.19$, $p<0.001$). Post-COVID-19 monthly counts were lower than those prior to COVID-19 by an average of 162 birds.

Since COVID-19, annual maxima have been delayed to the cool season, with plateaus in January and November 2022, peaking at 733 and 525 birds, respectively. Such an altered seasonal distribution of maximum counts was transient and disappeared after COVID-19. Therefore, the interaction effect between season and COVID-19 was statistically insignificant.

At the beginning of the research, this species was named Japanese white-eye (*Zosterops japonicus*). In 2018, the original species was split into Swinhoe's white-eye (*Zosterops simplex*) and Warbling white-eye (*Zosterops japonicus*), which have bordering native distribution ranges in the Sunda Region (Lim et al., 2019). Taxonomic re-classification may affect species identification.

Southeast-Asian exporters may possibly pack the two species together. White-eyes (*Zosterops spp.*) were observed in large wire cages on Bird Street. Individuals were selected and displayed in small plastic cages, based on their appearance or song quality. International trading and domestic handling processes could lead to misidentification of the two similar-looking species.

20.3.3 YELLOW-FRONTED CANARY (*CRITHAGRA MOZAMBICA*)

The only CES with a native range in Africa is the yellow-fronted canary (*Crithagra mozambica*). In total, 16,984 individuals were counted in all surveys (Figure 20.2b). The highest peak of 1,377 birds occurred in the summer immediately after the end of COVID-19 in 2023. Season was a significant main effect ($F=6.94$, $p<0.001$). Post-hoc comparisons confirmed that the significance originated from the high summer counts, which significantly exceeded those in spring and winter by 255 and 226 birds, respectively.

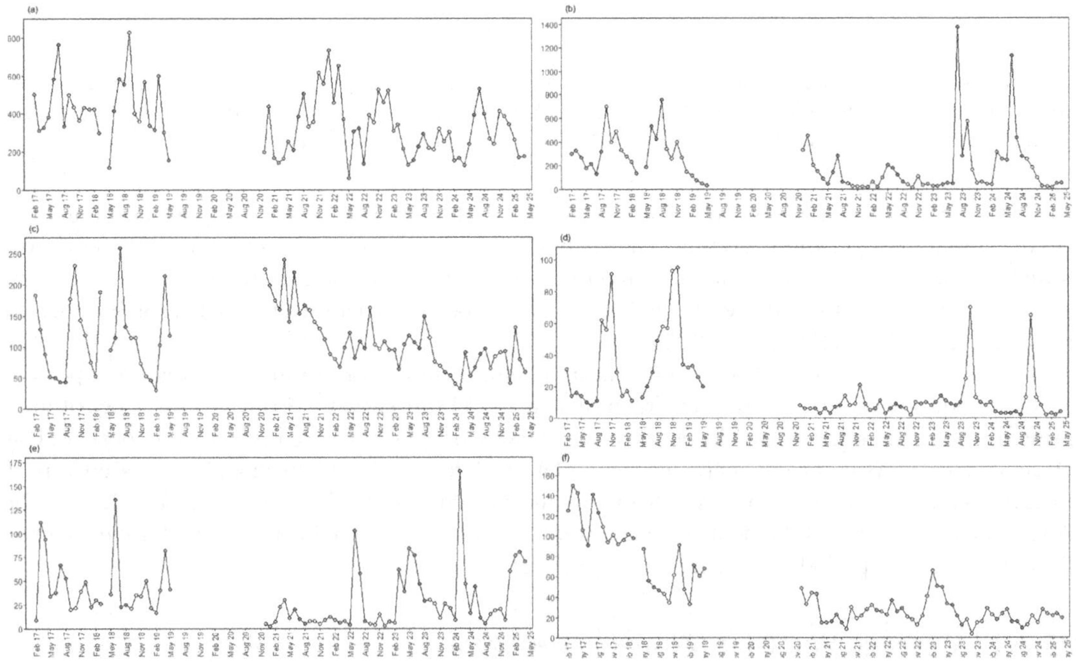

FIGURE 20.2 Monthly counts of six songbird species during the data collection period from February 2017 to April 2025. (also see Table 20.1). These species were termed consistently encountered species (CES): (a) warbling white-eye (*Zosterops japonicus*), (b) yellow-fronted canary (*Crithagra mozambica*), (c) oriental magpie-robin (*Copsychus saularis*), (d) Siberian rubythroat (*Copsychus saularis*), (e) pied bush-chat (*Saxicola caprata*), and (f) Chinese hwamei (*Garrulax canorus*)

TABLE 20.1
Summary of Maximum Monthly Counts of Six Consistently Encountered Species (CES) from Monthly Market Surveys on Bird Street during Three Periods: Pre-COVID-19 (February 2017 to May 2019), COVID-19 (December 2020 to February 2023), and Post-COVID-19 (March 2023 to April 2025)

	Pre-COVID-19	COVID-19	Post-COVID-19
• Warbling white-eye (*Zosterops japonicus*)	• 828 Autumn, 2018	• 733 Winter, 2022	• 529 Summer, 2024
• Yellow-fronted canary (*Crithagra mozambica*)	• 758 Summer, 2018	• 454 Winter, 2021	• 1,377 Summer, 2023
• Oriental magpie-robin (*Copsychus saularis*)	• 258 Summer, 2018	• 240 Spring, 2021	• 149 Summer, 2023
• Siberian rubythroat (*Calliope calliope*)	• 95 Winter, 2018	• 21 Winter, 2021	• 70 Autumn, 2023
• Pied bush-chat (*Saxicola caprata*)	• 136 Summer, 2018	• 103 Summer, 2022	• 166 Spring, 2024
• Chinese hwamei (*Garrulax canorus*)	• 150 Spring, 2017	• 66 Spring, 2023	• 51 Spring, 2023

The numbers displayed in this table occurred in single months. The months in which the maximum counts occurred were classified into various seasons of the year: spring (March to May), summer (June to August), autumn (September to November), and winter (December to February).

During COVID-19, the peaks became rather inconspicuous, causing a significant main effect of COVID-19 ($F=6.71$, $p<0.01$). During COVID-19, the monthly counts were 186 birds fewer than those prior to COVID-19. The suppressed peaks during COVID-19 were reflected in a significant interaction term ($F=3.33$, $p<0.01$). Post-hoc comparisons revealed that the significance was attributed to significantly lower autumn counts by 332 birds during COVID-19.

The annual peaks in yellow-fronted canaries (*Crithagra mozambica*) were primarily attributed to the observation of large numbers of birds in large wire cages. It was uncertain whether the birds were transported to Hong Kong at the high density of birds as observed in the cages on Bird Street. The packaging of small-bodied songbirds in international shipping is regulated (International Air Transport Association, 2019). However, the cage conditions on Bird Street were far from optimal in terms of animal welfare. Examples of cages can be found in Appendix 20.2.

It appeared that the shops lacked the space to display all birds in small cages. Customers holding spray bottles were observed wetting the desired individuals inside large cages so that shopkeepers could retrieve the marked birds for further examination afterwards. The exact source of this species on Bird Street could not be determined. Yellow-fronted canary (*Crithagra mozambica*) has been known as the *Green Singer*, which can be captive-bred in aviculture. Meanwhile, information tags claimed that some displayed birds were from South Africa.

20.3.4 Oriental Magpie-Robin (*Copsychus saularis*)

Oriental magpie-robin (*Copsychus saularis*) recorded the third-highest total count at 8,785 birds across all monthly surveys (Figure 20.2c). The monthly count of this species fluctuated each year. The magnitudes of annual fluctuation were 187, 205, and 183 birds in 2017, 2018, and 2019, respectively. However, season was not a significant main effect. After the data collection gap, the magnitude progressively decreased from 128 birds in 2021 to 65 birds in 2024.

A long-term, downward trend was discerned and corroborated with COVID-19 being a highly significant main effect ($F=7.89$, $p<0.001$). The number of birds recorded before and during COVID-19 showed insignificant differences. But when COVID-19 ended, the monthly counts significantly decreased by 53 birds on average when compared to the counts made during COVID-19.

All oriental magpie-robins (*Copsychus saularis*) were displayed individually in small cages or traditional Chinese-style cages. Most were male, presumably due to their better song quality and more handsome appearance. A male black-bellied magpie-robin was observed in 2017. This individual may belong to a black-bellied race, such as *amoenus* or *adamsi*, or it may be a completely different species, namely Seychelles magpie-robin (*Copsychus sechellarum*) (Cavill, Hernández-Alonso, Ogden, & Gilbert, 2023). Given the rarity of the latter, the black-bellied individual was believed to be an oriental magpie-robin (*Copsychus saularis*).

20.3.5 Siberian Rubythroat (*Calliope calliope*)

Siberian rubythroat (*Calliope calliope*) had the lowest total count among the six CES, with an aggregate of 1,476 birds throughout the data collection period (Figure 20.2d). Its full migrant status was evidenced by peak counts in the autumn, causing a highly significant main effect of season ($F=16.2$, $p<0.001$). In particular, monthly counts made in autumn significantly exceeded those from other seasons ($p<0.001$).

During COVID-19, the peaks diminished, with only up to 21 birds in December 2021. COVID-19 was a significant main effect ($F=33.62$, $p<0.001$). Using post-hoc comparisons, differences related to COVID-19 were significant only in the case of autumn and winter months. The largest contrast was a decrease by 61 birds in the autumn months during COVID-19 times than those before. The COVID-related effects with respect to seasons created a significant interaction effect ($F=4.83$, $p<0.001$).

All Siberian rubythroats (*Calliope calliope*) recorded on Bird Street were male. Birds were transported to the market in large wire cages but were quickly transferred to individual small cages. During some surveys, it was observed that playback tapes were looped through speakers, probably training or stimulating the birds to sing. The price of Siberian rubythroats (*Calliope calliope*) likely depends primarily on song quality. In June 2018, the asking prices for birds without obvious feather damage ranged from HK$ 380 to HK$ 1,000. In September 2021, prices of up to HK$ 1,500 per bird were observed.

20.3.6 Pied Bush-Chat (*Saxicola caprata*)

By summing all monthly counts, a total of 2,668 individuals of pied bush-chat (*Saxicola caprata*) were recorded (Figure 20.2e). Season was a significant main effect ($F=6.88$, $p<0.001$). Annual maxima and minima occurred mostly in spring and autumn, respectively. The largest seasonal differences between these two seasons were detected after COVID-19, reaching a mean difference of 50 birds.

Furthermore, COVID-19 acted as a significant main effect on the monthly count ($F=8.41$, $p<0.001$). Much-reduced numbers were recorded during the pandemic. The contrast during and after COVID-19 was the most noticeable, reaching 57 birds. Yet, no interaction effects were found.

The dramatic rise and fall of the monthly counts of Pied bush-chat (*Saxicola caprata*) suggested that the strong demand was satisfied by an ample supply. Non-peak counts were relatively higher between 2017 and 2019, reaching a mean level of 35 birds/month. Since COVID-19, monthly counts seldom exceeded 20 birds outside the peak supply season of each year.

Thanks to sexual dimorphism, all adult Pied bush-chats (*Saxicola caprata*) sold on Bird Street were identified as male. Fledglings were found during peak season but could not be sexed. Large shipments of birds arrived in large wire cages. Birds were hand-fed for a period and subsequently selected for individual display. Price levels may vary between individuals, depending on song attractiveness, fullness of feathers, and distinctiveness of the white wing bars.

20.3.7 Chinese Hwamei (*Garrulax canorus*)

Among the six CES, Chinese hwamei (*Garrulax canorus*) had the largest body size. Altogether, 3,652 individuals were recorded when all monthly counts were summed (Figure 20.2f). Records were concentrated during the early data collection period. Approximately two-thirds (63.6%, n=2,323) of the records were made in the first 27 months (2017–2019), which accounted for one-third of the research period.

A downward trend was observed. COVID-19 acted as a significant main effect ($F=70.74$, $p<0.001$). The maximum count of 150 birds, attained in March 2017, dropped steadily to as low as 33 birds in February 2019. The difference in monthly counts between before and during the pandemic was 59 birds. The difference rose to 64 birds when the difference before and after the pandemic was considered.

Chinese hwamei (*Garrulax canorus*) were mainly kept in rectangular bamboo cages. The birds frequently peeked their heads through the narrow, abrasive cage bars with fraying fibres. As a possible result, wounds with exposed flesh were observed on the crowns, lore, and foreheads of some birds. Shopkeepers may group seriously wounded birds in larger wire cages. However, their subspecies and sexes were impossible to tell because of their highly similar appearances.

20.4 CONSERVATION IMPLICATIONS

20.4.1 Were Songbirds in Hong Kong Trapped?

Small-scale local trapping of songbirds may occur in Hong Kong. With the exception of yellow-fronted canary (*Crithagra mozambica*) and the pied bush-chat (*Saxicola caprata*), the CES occur naturally in urban green spaces and the countryside of Hong Kong. All wild birds in Hong Kong are protected under the Wild Animals Protection Ordinance (Cap. 170). The legal consequences are far-reaching (Table 20.2a). Possessing and/or selling any live wild birds or their nests and eggs taken in Hong Kong could result in a maximum fine of HK$ 100,000 and 1-year imprisonment.

However, offenders must be caught in the act for subsequent prosecution. Evidence such as photographic records taken by third parties is useful only if the record can reliably show that the possessed or sold birds are wild and taken from Hong Kong. Another potential loophole is that locally

TABLE 20.2
SWOT Analysis of the Role of Hong Kong in Songbird Conservation

(a) Strengths	(b) Weaknesses
• Aggregation of songbird trading activities on Bird Street for consistent monitoring • Established legal protection for the possession and sale of wild-caught birds	• Unchartered local songbird trapping activities within the territory of Hong Kong • Laundering of locally caught songbirds into legally imported songbirds offered on Bird Street
(c) Opportunities	**(d) Threats**
• Potential inter-agency collaboration between HKSAR and mainland China • Introduction of captive-bred bird species into the market for the reduction of the exploitation of wild songbird populations	• Unchartered local trapping and potential influx of trapping tools • Competition caused by captive escapees of non-native songbirds

The table is divided into (a) strengths, (b) weaknesses, (c) opportunities, and (d) threats.

poached birds could be laundered into legally imported birds. Local poachers and shop owners may exploit these judicial hindrances.

Local poachers are likely to weigh the potential revenues of poaching against the difficulties and legal consequences. On Bird Street, visitors may offer songbirds to shopkeepers for a price. Several incidents of reverse selling were noted in the monthly surveys during the local bird breeding season. The visitors cannot be definitively identified as poachers. Nonetheless, the offered birds are likely wild-caught as they were claimed to be "from local areas". Nestlings with begging calls were also offered, sometimes even in an intact nest. Such nestlings would struggle to survive cross-border travel without intensive care. These observations provided counter-evidence to the proposition of the absence of local bird trapping in a study conducted in the late 2000s (Kadoorie Farm and Botanic Garden, 2011). Still, the magnitude, distribution, and affected species of local poaching require further research.

Small-scale local poaching in Hong Kong is currently unorganised, undetected, and unprosecuted (Table 20.2b). Fortunately, live bird trapping in Hong Kong is still far from prevalent, unlike neighbouring Chinese regions such as Hainan (W. Liang, Cai, & Yang, 2013) and Guiyang (Dai & Zhang, 2017).

Still, trapping tools are inexpensive and available online, as they are in other regions along the East Asian-Australasian Flyway (Yong et al., 2015). Confiscating trapping tools at the border is impractical because these tools are small, unobtrusive items hidden among common goods (Table 20.2d). Trappers and vendors are potential targets for outreach interventions in the songbird trade (Mirin & Klinck, 2021). Law enforcement officials in Hong Kong have yet to press serious charges against local songbird poachers. Strict law enforcement serves as a means of intervention, creating a strong deterrent effect on the trapping and trading of live songbirds in Hong Kong.

20.4.2 Where Did the Birds Come From?

Market dynamics provided insights into the trade of non-native songbirds. Among the six CES, pied bush-chat (*Saxicola caprata*) and yellow-fronted canary (*Crithagra mozambica*) are exotic to Hong Kong. Their peak counts occurred somewhat later than the breeding season in their native range. Thus, a hypothetical scenario is formulated. Poachers could easily and abundantly collect nestlings during the breeding season. Large numbers of hand-raised individuals would be shipped afterwards via rapid transportation due to the belief that younger songbirds have a greater potential to develop better song quality. This hypothesis is supported by observations of some birds showing juvenile plumage and yellowish gape flanges during peak counts. However, the substantially reduced counts after the peaks may not necessarily indicate a popular demand because unsold birds can be released to make room for new arrivals. The sexing of juvenile birds is difficult. When adult plumage emerges, female birds may be among those released. Although escaped individuals of pied bush-chat (*Saxicola caprata*) and yellow-fronted canary (*Crithagra mozambica*) have been sighted in Hong Kong, ecological impacts such as invasiveness and competition have yet to be examined.

The time gap between the breeding season and the peak counts of non-native CES may indicate the potential source-to-market distance. Pied bush-chat (*Saxicola caprata*) numbers peaked mainly in early summer, showing a time gap of 1–3 months after the breeding season (Figure 20.2e). Fledglings were possibly imported from the Indochinese Peninsula, i.e., Cambodia, Laos, Malaysia, Myanmar, Thailand, and Vietnam. The other non-native species, yellow-fronted canary (*Crithagra mozambica*), showed a peak later in mid-summer (Figure 20.2b). The relatively longer time gap suggests a longer transportation time and a more distant origin, possibly from Sub-Saharan African countries in the Northern Hemisphere. Yellow-fronted canaries (*Crithagra mozambica*) may be imported from Southern Africa, where the breeding season starts in August. The second wave of imports may explain the secondary peaks observed in November 2017 and 2018. Nonetheless, Southern African suppliers played a secondary role due to the lower magnitude of the corresponding peak. This identification strategy can help map out wildlife trafficking routes in the absence of official records.

The seasonal variation of trapped migratory birds deserves special attention. As the only true migrant in CES, Siberian rubythroat (*Calliope calliope*) showed distinct peaks in autumn (Figure 20.2d). In the wild, this species migrates along the coastline of China, arriving in Hong Kong in good numbers by mid-October. By contrast, the numbers on Bird Street surged much earlier, exceeding the extreme dates of their wild counterparts. Thus, trapping likely took place at migration stopovers north of Hong Kong. With rapid transportation, the trapped birds outpaced their wild counterparts and appeared on Bird Street, but not in their natural habitats. By comparing with the natural migration schedule, the sources of migratory songbirds can be approximated.

For Siberian rubythroat (*Calliope calliope*), the surges before the annual peak suggest that shopkeepers continually requested birds from suppliers in separate lots. The magnitude of peak counts of birds depends on the availability of harvested birds and the quantity purchased by each shopkeeper. Trappers may only be able to supply a limited number of birds to multiple markets before bird migration is in full swing. Shipping smaller batches can test trafficking routes. On the other hand, shopkeepers may sell an initial small batch to test market response, check the quality of birds, and encourage customers to review before sourcing more songbirds in bulk. Therefore, the number of migratory species on Bird Street resulted from the interplay between suppliers and shopkeepers.

Spatial-temporal estimation is difficult for species with both resident and migratory populations. For instance, the warbling white-eye (*Zosterops japonicus*) may be poached at multiple locations. The passage of migratory birds may result in a boosted catch in trapping locations. However, Bird Street serves as a terminal in the songbird trade network. Monthly counts can at best represent the aggregate impact over a period. An additional complicating factor is the entry of locally trapped birds into the domestic market (Table 20.2b). The domestic trapping and transport of wild birds can be difficult to trace (Shepherd et al., 2023; Wang et al., 2021). Trapping sites function as nodes in the songbird trafficking ring. The trapping levels at individual sites can only be revealed through other means, such as confiscation records. Frequent and stringent border checks, police raids, and local patrols are required to produce adequate data for analysis.

20.4.3 Can Listing as China's State Key Protected Species Bird?

Recognising the conservation importance of heavily traded species within the legal framework is fundamental to law enforcement against bird trapping. In this research, Siberian rubythroat (*Calliope calliope*) and Chinese hwamei (*Garrulax canorus*) were credibly sourced from mainland China. In fact, these two species have been added to the revised List of State Key Protected Wild Animals as second-tier protected species in 2021, aiming to curb the poaching, transportation, and trading of wildlife (National Forestry and Grassland Administration of PRC & Ministry of Agriculture and Rural Affairs of PRC, 2021). Without a permit from the provincial government, any trapping activity will lead to prosecution. China is also a range country for three other CES, namely warbling white-eye (*Zosterops japonicus*), oriental magpie-robin (*Copsychus saularis*), and pied bush-chat (*Saxicola caprata*). Their continuous presence in the bird trade warrants consideration of their inclusion in the next revised List.

The spillover effect of the revision has yet to be observed in Hong Kong. The low numbers of Siberian rubythroat (*Calliope calliope*) in 2021 and 2022 may be due to COVID-19, but not solely due to the deterrent effect of the revision (Figure 20.2d). The peaks in the autumn of 2023 and 2024 suggested that trappers and traffickers took risks and continued trading. The revised list responded to the conservation need of Chinese hwamei (*Garrulax canorus*) based on previous global trade records (Angguni et al., 2021; Leupen, 2022; Shepherd et al., 2020). However, the downward trend for this species began three years earlier (Figure 20.2f). Continuous market surveys will be required to assess the increased conservation status.

In mainland China, the possession of even a small number of bird species in the List of State Key Protected Wild Animals can lead to swift prosecution. Many keepers of listed species, such as Chinese hwamei (*Garrulax canorus*) (Public Security Bureau of Jinhua, 2022) and Siberian rubythroat (Public Security Bureau of Shanwei, 2024), have been successfully prosecuted. A large number of successful convictions resulted from frequent patrols and raids of hotspots for bird trapping, songbird contests, and pet bird trade. Cross-border, inter-agency cooperation is required to effectively curb songbird smuggling (Table 20.2c). The difficulty of law enforcement is greater in the case of songbird trafficking.

A large volume of goods and commodities from China arrives in Hong Kong every day. A few cages of densely packed songbirds can re-stock the shops on Bird Street (HKSARG Agriculture, Fisheries and Conservation Department, 2011; HKSARG Customs and Excise Department, 2018). More careful border checks may be the only effective solution at the moment.

Stronger customs control at the border, coupled with harsher sentences upon convictions, is essential in the combat against cross-border trafficking, missing either of which would fail to deliver a deterring message to the public. For example, in April 2019, a Hong Kong citizen was intercepted at the Sha Tau Kok Control Point, one of the land border crossing points between Hong Kong and Shenzhen, China. The arrest involved the smuggling of six African grey parrots (*Psittacus erithacus*) and another 115 live birds of unknown species from mainland China to the Hong Kong Special Administrative Region. In particular, African grey parrots can be found on Bird Street in Hong Kong. This species is listed as endangered on the IUCN Red List and is listed under CITES Appendix 20.1. In Hong Kong, the trade of species under CITES Appendix 20.1 is governed under the Protection of Endangered Species of Animals and Plants Ordinance (Cap. 586), which requires specific pre-approval and a license from the Agriculture, Fisheries and Conservation Department.

The smuggling case was brought to the District Court in May 2020. The smuggler was charged with breaches of the Protection of Endangered Species of Animals and Plants Ordinance (Cap. 586) and the Public Health (Animals and Birds) Ordinance (Cap. 139), the latter of which prohibits the import of any birds into Hong Kong without a relevant permit. At the hearing, the defendant admitted to bringing the birds from China and an intention to keep the African grey parrots as pets and merit-release the other 115 live birds. Involvement in any commercial bird trade was denied. Upon conviction, the court sentenced the smuggler to 8 months of imprisonment, suspended for 18 months, along with a fine of HK$ 55,000. As for the birds, half of the African grey parrots died afterwards, and all the other 115 live birds were euthanized due to the lack of relevant health certificates and permits.

While the Protection of Endangered Species of Animals and Plants Ordinance (Cap. 586) was amended in 2018 to increase penalties for violation to a fine of up to HK$10 million and imprisonment for up to 10 years, the Hong Kong judiciary, following common law principles, has still been careful in passing on sentences considering precedent cases of a similar nature. Therefore, the maximum sentences have yet to be applied to any cases.

20.4.4 Which Species Can Benefit from Captive Breeding?

The only captive breeding success recorded for a CES is yellow-fronted canary (*Crithagra mozambica*), known as the *green singer* in aviculture. Congeneric species, such as lemon-breasted canary (*Crithagra citrinipectus*), can also be bred in captivity (Jolliffe, 2000). However, Hong Kong is outside the native range of these species. Despite this, heavy trapping and trade exert pressure on the wild population. For example, yellow-fronted canaries (*Crithagra mozambica*) were exported from Uganda to South Africa in large quantities (Rossi, 2018). A small-scale, ex-situ captive breeding may help mitigate the impact of the bird trade on this species. Given that birdkeepers may release some of their birds with disappointing song quality, the first captive generation can be established through a collection programme. The collection of birds which would otherwise be released can lessen the ecological impact caused by non-native songbirds. Birdkeepers may re-sell their birds at very low prices, but caution must be exercised to ensure that these prices must be substantially lower than the import price per individual bird. Otherwise, the programme may be exploited by shopkeepers with unsold birds.

Ample space is required for the captive breeding of yellow-fronted canary (*Crithagra mozambica*). Currently, empty outdoor aviaries are available in Hong Kong Park and the Zoological and Botanic Garden. It remains questionable whether birdkeepers can distinguish between captive-bred and wild-caught birds (Burivalova et al., 2017; Marshall et al., 2025). Thus, captive-bred canaries can be officially sold at competitive prices on Bird Street (Table 20.2c). Online songbird trade has thrived in many other regions (Canlas, Sy, & Chng, 2017; Grimwood, Menner, & Marsden, 2024; Indraswari et al., 2020; Yin et al., 2020). Nevertheless, an online songbird trading community has yet to emerge in Hong Kong. Therefore, captive-bred birds from any captive breeding programmes should be channelled into brick-and-mortar markets in Hong Kong.

Bird-keeping is a part of the traditional Chinese culture. The rectification of this cultural practice is a long-term operation. Thanks to global logistics, birdkeepers in Hong Kong can obtain exotic birds from different parts of the world, contributing to the unsustainability of the bird-keeping culture. If captive breeding is viable in Hong Kong, the ecological impact of songbird trade on wild songbird populations can be lessened but not completely uprooted because of a potential preference for wild-caught birds.

Conservation breeding for the other CES is unlikely to take place in Hong Kong in the near future. The lack of captive breeding records implies that suitable breeding conditions cannot be met. The availability of source birds is very limited. In Hong Kong, most confiscated songbirds were euthanized due to a lack of health certification (HKSARG Agriculture Fisheries and Conservation Department, 2011; HKSARG Customs and Excise Department, 2018). Even if confiscated songbirds are used as the source in captive breeding, the preservation of cryptic diversity may be hindered by the mixing of birds from geographically distinct populations during trafficking.

The stress faced by migratory species, such as the warbling white-eye (*Zosterops japonicus*) and the Siberian rubythroat (*Calliope calliope*), in stationary aviaries can impede breeding. Stress is also a problem for territorial species, such as the oriental magpie-robin (*Copsychus saularis*) and Chinese hwamei (*Garrulax canorus*). For songbirds with limited captive breeding possibilities, strict border control and public education may be necessary measures to deter suppliers from smuggling and help potential customers to realise the adverse impacts of purchasing wild-caught birds.

20.5 CONCLUSION

The Asian Songbird Crisis is closely related to the unsustainable songbird trade. The local scale of songbird trade can be tapped through songbird market observations. Six species of songbirds were recorded throughout the market surveys conducted on Bird Street, which is an active market of songbird trade in Hong Kong. Based on the market monitoring results and songbird trade in the region, the following proposed action may benefit the conservation of songbirds:

- Strengthening cross-border cooperation between the HKSAR and mainland China in detecting and raiding the smuggling of songbirds and trapping tools at border checkpoints;
- Introducing captive-bred songbirds to the songbird trade under strict genetic and welfare protocols for reducing the exploitation of wild songbird populations; and
- Launching public education campaigns targeting bird-keeping communities for balancing sustainability and traditional culture.

Small-scale local trapping of songbirds is arguably present. However, the bulk of the songbirds found on Bird Street are internationally shipped into Hong Kong. The mapping of logistic routes may help in regulating and reducing the volume of songbird trade. National conservation policies and captive breeding programmes may help curb unsustainable songbird trade if properly implemented. There is still much work to be done in the conservation of frequently traded songbirds.

APPENDIX 20.1 VERNACULAR NAMES OF CONSISTENTLY ENCOUNTERED SPECIES (CES)

Species	Vernacular Name	Meaning
• Warbling white-eye (*Zosterops japonicus*)	• 相思 (soeng1 si1)	• The vernacular name is an old Chinese vocab which poetically expresses the feeling of *being lovesick*.
• Yellow-fronted canary (Crithagra mozambica)	• 石燕 (sek6 jin3)	• The literal meaning of the vernacular name is *rock swallow*, which is taxonomically inaccurate.
• Oriental magpie-robin • (*Copsychus saularis*)	• 吱喳 (zi1 zaa6) • 豬屎喳 (zyu1 si2 zaa6)	• Multiple vernacular names are used, all describing the intonation of the vocalisation of this species.
• Siberian rubythroat (*Calliope calliope*)	• 紅波 (hung4 bo1)	• The meaning is *red ball*, which describes the red throat on male birds.
• Pied bush-chat (*Saxicola caprata*)	• 黑白仔 (hak1 baak6 zai2)	• The Chinese vernacular name, *black-and-white boy*, implies that most traded individuals are male.
• Chinese hwamei (*Garrulax canorus*)	• 畫眉 (waa6 mei4)	• The English common name is a homophonous adoption of the Chinese name, meaning *decorated eyebrow*.

APPENDIX 20.2 DESCRIPTION OF CAGES FOUND ON BIRD STREET

Birds were displayed in various types of cages. Cube-shaped plastic cages were the most common. Each side of these small cages usually measured ≤250 mm (Figure 20.S1a). Cages were stacked along the walls of most shops. Thin linen or cotton sheets separated these cages. Small-bodied, low-price songbirds, such as pied bush-chat (*Saxicola caprata*) and Hainan blue flycatcher (*Cyornis hainanus*), were kept in these small cages. From May to July, multiple nestlings and fledglings of species such as warbling white-eye (*Zosterops japonicus*) and Java sparrow (*Lonchura oryzivora*) were seen in one cage. Sometimes, two to three adult birds may be placed in one cage to save storage space upon the arrival of large shipments.

Traditional-style cages were hung on the doorpost of most shops. These cages had a round shape with a dome-shaped top and came in non-standard sizes (Figure 20.S1b). The material was either bamboo or stainless steel. Larger-bodied birds, such as Chinese hwamei (*Garrulax canorus*) and white-rumped shama (*Copsychus malabaricus*), were displayed individually in large Chinese-style cages. Rare birds and expensive species, such as greater green leafbird (*Chloropsis sonnerati*) and orange-headed thrush (*Geokichla citrina*), were kept in expensive hand-crafted cages.

Acrylic containers were predominantly used for parrots. These containers were available in various shapes and sizes, accommodating parrots from small species such as Pacific parrotlet (*Forpus coelestis*) to large species such as yellow-crowned amazon (*Amazona ochrocephala*) can be accommodated. The four transparent sides made acrylic containers ideal display boxes for colourful parrots (Figure 20.S1c). Shopkeepers may remove the tops of the containers to allow potential customers to interact with parrots. Macaws and cockatoos were also seen on metal display stands (Figure 20.S1d). On most occasions, shopkeepers allowed petting of parrots.

FIGURE 20.S1 Types of cages observed on Bird Street: (a) plastic cage; (b) traditional style; (c) open-top acrylic container; (d) metal display stand; (e) wire cages; and (f) bamboo cages.

Large cages served multiple purposes. Songbirds such as common hill myna (*Gracula religiosa*) were isolated in wire cages to prevent them from pecking each other. Larger wire cages were used to house an assortment of true and estrildid finches, such as Atlantic canary (*Serinus canaria*), red-cheeked cordon-bleu (*Uraeginthus bengalus*), Sunda zebra finch (*Taeniopygia guttata*), or other small-bodied species (Figure 20.S1e). Sometimes, CES were observed in groups in large wire cases. Birds for merit release were kept in large rectangular bamboo cages (Figure 20.S1f). Bamboo cages were often stacked. White-eyes (*Zosterops spp.*) and munias (*Lonchura spp.*) were used as prayer birds. However, larger birds, such as greater necklaced laughingthrush (*Pterorhinus pectoralis*), were recorded as prayer birds.

Some customers expressed an intention to bring purchased birds back to mainland China. Shopkeepers would offer a specialised transport service for a surcharge. If customers insisted on crossing the border with the birds themselves, the shopkeepers would provide an oblong black bag made of glossy velvet or non-woven fabric to conceal the birds at border checkpoints. Eurasian tree sparrow (*Passer montanus*) and oriental magpie-robin (*Copsychus saularus*) were involved in such transactions. A similar procedure was observed for grasshoppers sold as bird food. However, no conversations regarding domestic delivery services within Hong Kong were overheard.

It could not be confidently determined whether such customers were local Hong Kong Special Administrative Region or mainland China residents, as Cantonese is spoken in both the Hong Kong Special Administrative Region and Guangzhou. It was uncertain whether services related to certification and quarantine would be provided. Otherwise, the shopkeepers may hint that the birds would be smuggled into mainland China. Still, such observations evidenced that Bird Street satisfied part of the demand for songbirds from across the border.

REFERENCES

Almeida, R. J., Mazza, A. N., & Lockwood, J. L. (2024). Does fortune follow function? Exploring how consumer preferences drive the functional trait composition of the global songbird trade. *People and Nature*, *6*(3), 1366–1377. https://doi.org/10.1002/pan3.10652

Angguni, T., Mulyani, Y. A., & Mardiastuti, A. (2021). Bird species contested at songbird competition in Jabodetabek Region, Indonesia. *IOP Conference Series: Earth and Environmental Science*, *762*(1), 012014. https://doi.org/10.1088/1755-1315/762/1/012014

Arlott, N. (2007). *Birds of Europe, Russia, China, and Japan: Passerines: Tyrant Flycatchers to Buntings*. Princeton: Princeton University Press.

Barlow, C., & Wacher, T. (1997). *A Field Guide to Birds of the Gambia and Senegal*. New Haven: Yale University Press.

Burivalova, Z., Lee, T. M., Hua, F., Lee, J. S. H., Prawiradilaga, D. M., & Wilcove, D. S. (2017). Understanding consumer preferences and demography in order to reduce the domestic trade in wild-caught birds. *Biological Conservation*, *209*, 423–431. https://doi.org/10.1016/j.biocon.2017.03.005

Canlas, C. P., Sy, E. Y., & Chng, S. (2017). A rapid survey of online trade in live birds and reptiles in the Philippines. *TRAFFIC Bulletin*, *29*(2), 58–63.

Cavill, E. L., Hernández-Alonso, G., Ogden, R., & Gilbert, M. Thomas. P. (2023, April 15). *Never judge a bird by its feathers: genetics unveils the true target of trafficking in morphologically similar* Copsychus *species*. https://doi.org/10.1101/2023.04.14.536889

Chan, S. W. (2006). *Religious release of birds in Hong Kong*. University of Hong Kong, Hong Kong.

Chiok, W. X., Lee, R. Y. Y., Lee, J. G. H., & Jain, A. (2022). The dynamics of songbird ownership and community interconnectedness in Singapore. *Bird Conservation International*, *32*(4), 573–589. https://doi.org/10.1017/S0959270921000393

Chng, S. C. L., & Eaton, J. A. (2016). Snapshot of an on-going trade: an inventory of birds for sale in Chatuchak weekend market, Bangkok, Thailand. *BirdingASIA*, *25*, 24–29.

Dai, C., Chen, Y., & Zheng, Q. (2021). The social and cultural importance of keeping wild birds as pets in an ethnic community in Guiyang City, China. *Human Ecology*, *49*(1), 69–79. https://doi.org/10.1007/s10745-020-00208-7

Dai, C., & Zhang, C. (2017). The local bird trade and its conservation impacts in the city of Guiyang, Southwest China. *Regional Environmental Change*, *17*(6), 1763–1773. https://doi.org/10.1007/s10113-017-1141-5

Day, N., & Simpson, K. (1996). *Field Guide to the Birds of Australia* (5th ed.). Ringwood: Viking.

Eaton, J. A., Leupen, B. T. C., & Krishnasamy, K. (2017). Songsters of Singapore: an overview of the bird species in Singapore pet shops. *TRAFFIC Report*.

Fiennes, S., Zhang, M., Sun, F., & Lee, T. M. (2021). Understanding retail dynamics of a regionally important domestic bird market in Guangzhou, China. *Conservation Science and Practice*, *3*(10). https://doi.org/10.1111/csp2.487

Gilbert, M., Sokha, C., Joyner, P. H., Thomson, R. L., & Poole, C. (2012). Characterizing the trade of wild birds for merit release in Phnom Penh, Cambodia and associated risks to health and ecology. *Biological Conservation*, *153*, 10–16. https://doi.org/10.1016/j.biocon.2012.04.024

Grimwood, E. I., Menner, J. K., & Marsden, S. J. (2024). Comparisons of songbirds on sale across online and physical markets in Indonesia. *Biodiversity and Conservation*, *33*(5), 1591–1604. https://doi.org/10.1007/s10531-024-02825-w

Harris, J. B. C., Green, J. M. H., Prawiradilaga, D. M., Giam, X., Giyanto, Hikmatullah, D., … & Wilcove, D. S. (2015). Using market data and expert opinion to identify overexploited species in the wild bird trade. *Biological Conservation*, *187*, 51–60. https://doi.org/10.1016/j.biocon.2015.04.009

HKSARG. (2018, April 13). *H5 virus detected on swab sample from Yuen Po Street Bird Garden*. Hong Kong: HKSARG. Retrieved from https://www.info.gov.hk/gia/general/201804/13/P2018041300374.htm

HKSARG Agriculture Fisheries and Conservation Department. (2011, October 7). Illegal import of live birds seized at Man Kam To. *Press Release*. Retrieved from https://www.afcd.gov.hk/english/publications/publications_press/pr1691.html

HKSARG Customs and Excise Department. (2018, December 18). Hong Kong Customs seizes suspected illegally imported live birds. *Press Release*. Retrieved from https://www.customs.gov.hk/en/customs-announcement/press-release/index_id_3218.html

Holbig, H. (2020). Be water, my friend: Hong Kong's 2019 anti-extradition protests. *International Journal of Sociology*, *50*(4), 325–337. https://doi.org/10.1080/00207659.2020.1802556

Indraswari, K., Friedman, R. S., Noske, R., Shepherd, C. R., Biggs, D., Susilawati, C., & Wilson, C. (2020). It's in the news: characterising Indonesia's wild bird trade network from media-reported seizure incidents. *Biological Conservation*, *243*, 108431. https://doi.org/10.1016/j.biocon.2020.108431

International Air Transport Association. (2019). *Live Animals Regulations (IATA Resolution 620, Attachment 'A') Effective 1 January-31 December 2020* (46th ed.). Montreal, International Air Transport Association.

Iskandar, J., Iskandar, B. S., Mulyanto, D., Alfian, R. L., & Partasasmita, R. (2020). Traditional ecological knowledge of the bird traders on bird species bird naming, and bird market chain: a case study in bird market Pasty Yogyakarta, Indonesia. *Biodiversitas Journal of Biological Diversity*, *21*(6). https://doi.org/10.13057/biodiv/d210631

Jepson, P., & Ladle, R. J. (2009). Governing bird-keeping in Java and Bali: evidence from a household survey. *Oryx*, *43*(03), 364. https://doi.org/10.1017/S0030605309990251

Jolliffe, T. (2000). Breeding the lemon-breasted canary Serinus Citrinipectus and notes on some other Serinus species. *The Avicultural Magazine*, *106*(4), 178–182.

Kadoorie Farm and Botanic Garden. (2011). *A survey of the prayer bird trade at the Yuen Po Bird Market, Hong Kong August 2008-August 2009*. Hong Kong. Retrieved from https://www.kfbg.org/images/download/BirdMarketReport_20111027.pdf

Kazmierczak, K. (2000). *A Field Guide to the Birds of the Indian Subcontinent*. New Haven: Yale University Press.

Lee, J. G. H., Chng, S. C. L., & Eaton, J. A. (2016). *Conservation Strategy for Southeast Asian Songbirds in Trade. Recommendations from the First Asian Songbird Trade Crisis Summit 2015 Held in Jurong Bird Park,* Singapore. Singapore: Wildlife Reserves Singapore and TRAFFIC.

Leung, H. H. (2022). Urban space and leisure in Hong Kong: the health and social life of elderly bird-keepers. *Journal of Urban Affairs*, *44*(2), 243–258. https://doi.org/10.1080/07352166.2020.1722029

Leung, H. H., & Yung, E. H. K. (2021). Leisure, life satisfaction, and urban space. In L. A. Reese (Ed.), *Animals in the City* (1st ed., pp. 179–201). London: Routledge.

Leupen, B. T. C. (2022). A brief overview of the online bird trade in Vietnam. *Asian Journal of Conservation Biology*, *11*(2), 176–188. https://doi.org/10.53562/ajcb.71823

Liang, W., Cai, Y., & Yang, C.-C. (2013). Extreme levels of hunting of birds in a remote village of Hainan Island, China. *Bird Conservation International*, *23*(1), 45–52. https://doi.org/10.1017/S0959270911000499

Liang, Z., Hu, S., Zhong, J., Wei, Q., Ruan, X., Zhang, L., … & Liu, Y. (2024). Nationwide law enforcement impact on the pet bird trade in China. *Proceedings of the National Academy of Sciences*, 121(25). https://doi.org/10.1073/pnas.2321479121

Lim, B. T. M., Sadanandan, K. R., Dingle, C., Leung, Y. Y., Prawiradilaga, D. M., Irham, M., … & Rheindt, F. E. (2019). Molecular evidence suggests radical revision of species limits in the great speciator white-eye genus Zosterops. *Journal of Ornithology*, *160*(1), 1–16. https://doi.org/10.1007/s10336-018-1583-7

Linguistic Society of Hong Kong. (2022). Jyutping Cantonese Romanization Scheme. Retrieved from https://lshk.org/jyutping-scheme/

Marshall, H., Collar, N. J., Lees, A. C., Moss, A., Yuda, P., & Marsden, S. J. (2020). Spatio-temporal dynamics of consumer demand driving the Asian Songbird Crisis. *Biological Conservation*, *241*, 108237. https://doi.org/10.1016/j.biocon.2019.108237

Marshall, H., Collar, N. J., Lees, A. C., Moss, A., Yuda, P., & Marsden, S. J. (2025). Messaging with appeal to intrinsic or relational values shows potential to shift demand for wildlife as pets. *People and Nature*, *7*(1), 4–14. https://doi.org/10.1002/pan3.10756

McClure, H. E., & Chaiyaphun, S. (1971). The sale of birds at the Bangkok "Sunday Market" Thailand. *Natural History Bulletin of the Siam Society*, *24*, 41–78.

Mirin, B. H., & Klinck, H. (2021). Bird singing contests: looking back on thirty years of research on a global conservation concern. *Global Ecology and Conservation*, *30*, e01812. https://doi.org/10.1016/j.gecco.2021.e01812

Nash, S. V. (1993). *Sold for a Song: The Trade in Southeast Asian Non-CITES Birds*. Cambridge: TRAFFIC International.

National Forestry and Grassland Administration of PRC, & Ministry of Agriculture and Rural Affairs of PRC. (2021, February). *List of state key protected wild animals*. Beijing. Retrieved from https://www.gov.cn/xinwen/2021-02/09/5586227/files/e007df5cdb364bcdbcb89d169047d6c5.pdf

Nelson, S. S., & Shepherd, C. R. (2023). Assessing the trade of Chinese Hwamei *Garrulax canorus* in the USA. *Bird Conservation International, 33*, e49. https://doi.org/10.1017/S0959270923000096

Nijman, V., Campera, M., Ardiansyah, A., Balestri, M., El Bizri, H. R., Budiadi, B., ... & Nekaris, K. A. I. (2021). Large-scale trade in a songbird that is extinct in the wild. *Diversity, 13*(6), 238. https://doi.org/10.3390/d13060238

Nijman, V., Langgeng, A., Birot, H., Imron, M. A., & Nekaris, K. A. I. (2018). Wildlife trade, captive breeding and the imminent extinction of a songbird. *Global Ecology and Conservation, 15*, e00425. https://doi.org/10.1016/j.gecco.2018.e00425

Nijman, V., Sari, S. L., Siriwat, P., Sigaud, M., & Nekaris, K. A.-I. (2017). Records of four critically endangered songbirds in the markets of Java suggest domestic trade is a major impediment to their conservation. *BirdingASIA, 27*, 20–25.

Public Security Bureau of Jinhua. (2022, February 25). *Breaking the law: keeping a national second-class protected animal as a pet*. Retrieved from https://gaj.jinhua.gov.cn/art/2022/2/25/art_1229180965_59007710.html

Public Security Bureau of Shanwei. (2024, November 6). *Four typical cases of the destruction of forest resource: illegal occupation of forests, deforestation, and poaching*. https://www.shanwei.gov.cn/swsgaj/gkmlpt/content/1/1094/post_1094669.html

Rentschlar, K. A., Miller, A. E., Lauck, K. S., Rodiansyah, M., Bobby, Muflihati, & Kartikawati. (2018). A silent morning: the songbird trade in Kalimantan, Indonesia. *Tropical Conservation Science, 11*. https://doi.org/10.1177/1940082917753909

Robson, C. (2018). *Field Guide to the Birds of South-East Asia*. London: Bloomsbury Publishing.

Rossi, A. (2018). *Uganda Wildlife Trafficking Assessment*. Cambridge: TRAFFIC International.

Schulenberg, T. S. (2007). *Birds of Peru*. Princeton: Princeton University Press.

Senior, R. A., Oliveira, B. F., Dale, J., & Scheffers, B. R. (2022). Wildlife trade targets colorful birds and threatens the aesthetic value of nature. *Current Biology, 32*(19), 4299–4305.e4. https://doi.org/10.1016/j.cub.2022.07.066

Shepherd, C. R., Eaton, J. A., & Chng, S. C. L. (2016). Nothing to laugh about: the ongoing illegal trade in laughingthrushes (*Garrulax* species) in the bird markets of Java, Indonesia. *Bird Conservation International, 26*(4), 524–530. https://doi.org/10.1017/S0959270916000320

Shepherd, C. R., Leupen, B. T. C., Nelson, S. S., Gomez, L., Allcock, J. A., Bruslund, S., ... & Nijman, V. (2023). Global commercialisation, under-reported trade, and the need for increased international regulation of a non-CITES listed songbird. *Journal of Asian Ornithology, 39*, 71–77.

Shepherd, C. R., Leupen, B. T. C., Siriwat, P., & Nijman, V. (2020). International wildlife trade, avian influenza, organised crime and the effectiveness of CITES: the Chinese hwamei as a case study. *Global Ecology and Conservation, 23*, e01185. https://doi.org/10.1016/j.gecco.2020.e01185

Shivambu, T. C., Shivambu, N., & Downs, C. T. (2022). An assessment of avian species sold in the South African pet trade. *African Journal of Ecology, 60*(4), 980–995. https://doi.org/10.1111/aje.13029

Siriwat, P., & Nijman, V. (2020). Wildlife trade shifts from brick-and-mortar markets to virtual marketplaces: a case study of birds of prey trade in Thailand. *Journal of Asia-Pacific Biodiversity, 13*(3), 454–461. https://doi.org/10.1016/j.japb.2020.03.012

Su, S., Cassey, P., Vall-llosera, M., & Blackburn, T. M. (2015). Going cheap: determinants of bird price in the Taiwanese pet market. *PLoS One, 10*(5), e0127482. https://doi.org/10.1371/journal.pone.0127482

Thang, H., Trung, L. N. T., Hung, L. M., Hoa, N. N., Thach, N. D., & Giang, H. T. C. (2024). Wild birds traded in the central coastal provinces, Vietnam. *Hue University Journal of Science: Agriculture and Rural Development, 133*(3D). https://doi.org/10.26459/hueunijard.v133i3D.7693

van Balen, S. (Bas), Saryanthi, R., & Marsden, S. (2023). Evidence of steep declines in the heavily traded Javan White-eye *Zosterops flavus* from repeated standardised surveys. *Bird Conservation International, 33*, e19. https://doi.org/10.1017/S0959270922000144

van Perlo, B. (1999). *Birds of Southern Africa*. Princeton: Princeton University Press.

Viney, C., Phillipps, K., & Lam, C. Y. (2005). *The Birds of Hong Kong and South China* (8th ed.). Hong Kong: Information Services Department.

Wang, Q., Shi, J., Shen, X., & Zhao, T. (2021). Characteristics and patterns of international trade in CITES-listed live birds in China from 2010 to 2019. *Global Ecology and Conservation*, *30*, e01786. https://doi.org/10.1016/j.gecco.2021.e01786

Wong, S. C., Au, A. K. W., Lo, J. Y. C., Ho, P. L., Hung, I. F. N., To, K. K. W., … & Cheng, V. C. C. (2022). Evolution and control of COVID-19 epidemic in Hong Kong. *Viruses*, *14*(11), 2519. https://doi.org/10.3390/v14112519

Yin, R. Y., Ye, Y. C., Newman, C., Buesching, C. D., Macdonald, D. W., Luo, Y., & Zhou, Z.-M. (2020). China's online parrot trade: generation length and body mass determine sales volume via price. *Global Ecology and Conservation*, *23*, e01047. https://doi.org/10.1016/j.gecco.2020.e01047

Yong, D. L., Liu, Y., Low, B. W., Espanola, C. P., Choi, C.-Y., & Kawakami, K. (2015). Migratory songbirds in the East Asian-Australasian Flyway: a review from a conservation perspective. *Bird Conservation International*, *25*(1), 1–37. https://doi.org/10.1017/S0959270914000276

Index

G

H

I

J

T

U

V

W

X

Y

Z

For Product Safety Concerns and Information please contact our EU representative GPSR@taylorandfrancis.com Taylor & Francis Verlag GmbH, Kaufingerstraße 24, 80331 München, Germany

Batch number: 10733693

Printed by Printforce, the Netherlands